Heating, Ventilation, and Air Conditioning

A Residential and Light Commercial Text and Lab Book

Cecil Johnson

THOMSON

DELMAR LEARNING

Australia   Canada   Mexico   Singapore   Spain   United Kingdom   United States

THOMSON

DELMAR LEARNING

Heating, Ventilation, and Air Conditioning:
A Residential and Light Commercial Text and Lab Book
Cecil Johnson

Vice President, Technology and Trades SBU:
Alar Elken

Editorial Director:
Sandy Clark

Senior Acquisitions Editor:
James Devoe

Senior Development Editor:
John Fisher

Marketing Director:
Dave Garza

Channel Manager:
Dennis Williams

Marketing Coordinator:
Stacey Wiktorek

Production Director:
Mary Ellen Black

Production Manager:
Andrew Crouth

Senior Production Editor:
Stacy Masucci

Technology Project Manager:
Kevin Smith

Technology Project Specialist:
Linda Verde

Editorial Assistant:
Tom Best

Library of Congress Cataloging-in-Publication Data:

Heating, ventilation, and air conditioning : a residential and light commercial text and lab book / Cecil Johnson.
   p. cm.
ISBN 1-4018-8472-5 (pbk. : alk. paper)
1. Dwellings—Heating and ventilation—Textbooks. 2. Dwellings—Air conditioning—Textbooks. I. Title.
TH7684.D9 J64 2006
697—dc22   2005018330

NOTICE TO THE READER

Contents

Preface

Introducing a completely current and innovative way to teach the basics of HVAC-R! Featuring more than 150 practical competencies, this "how to" guide has been carefully designed and thoroughly modernized to provide a complete learning system for the fundamentals and applications of core HVAC concepts. It combines straightforward theory lessons with useful "hands-on" opportunities for learning about the industry's hottest topics, including electricity and electrical controls, refrigeration fundamentals, heat pumps, oil and gas heat, safety, and more. Enhancements to this edition include an updated tool identification chart, new and improved graphics, expanded information on calculator usage, and a pressure temperature chart for use by technicians in the field.

Acknowledgments

This book is dedicated to:

    My wife: Amy Johnson

    My daughters and sons-in-law: Heather and Kelly Andres, Courtney and Adam Shamenek

    My son: Justin Johnson

    My parents: Paul and Barb Rogers

    My sisters and brothers-in-law: Dr. Skip and Mary Lynn Kingston, Teresa and Dan Daum, Chris and Doug Jerew, Charlene and Mike Tennar, Connie and Bob Blevins, Rhonda and Terry Hero, Betty, Diane

    My brother: Clarence Fisher and family

    Beloved friends: Jim and Jacquelyn Evock; Ernie and Marlena Singer; Jim and Sue Albright; John, Sandy, Crystal, and Sara Ganoe; Larry and Sue Jones; Stan and Arlene Machamer; Tom and Deb Ferguson

Special thanks to Bridie Eckenrode, Tai Nguyen, Mike Morris, and my Dauphin County Technical School HVAC-R students.

And thanks to Cadmus Professional Communications for production services.

Safety Precautions

PRESSURE VESSELS AND PIPING

Move the cylinder *only while a protective cap is on it, if it is designed for one.*

Pressurized cylinders *should be chained to and removed safely on an approved cart. The protective cap must be secured.*

Refrigerant cylinders *should be stored and transported in the upright position to keep pressure relief valve in contact with the vapor space, not the liquid inside the cylinder.*

Refrigerant cylinder *should never have direct heat applied. To keep cylinder pressure up, set refrigerant cylinder in a container of warm water, with a temperature of no higher than 90 degrees (F).*

Refrigerant leaks – *If a leak develops with refrigerant escaping, do not try and stop the leak with your hands. Look for a shut-off valve near leak area. If none can be found, EPA mandates the area should be ventilated and vacated.*

Refrigerant transportation – *All refrigerant cylinders must be Department of Transportation (DOT) approved cylinders.*

Refrigerant transportation – *It is against the law to transport refrigerant in nonapproved cylinders.*

Refrigerant recovery cylinders – *DOT-approved recovery cylinder color code is gray bodies with yellow tops.*

Pressurized oxygen – *Never use pressurized oxygen to pressurize lines for leak testing. Oil or oil residue in a system will lead to an explosion.*

Safety glasses
- *Should be worn when working with refrigerants or any pressurized sealed system or container.*

Safety gloves
- *Should be worn when working with refrigerants and sealed system service valves and hoses.*

ELECTRICAL POWER HAZARDS

Electrical Power
- *Should always be shut-off at the distribution or entrance panel and locked out in an approved manner when installing equipment.*

Electrical Testing
- *Extreme care should be taken when making electrical tests.*
- *Ensure that your hands touch only the meter probes.*
- *Ensure that your arms and the rest of your body stay clear of all electrical terminals and connections.*
- *Make sure your meter is set to the proper function and proper range.*
- *Use proper procedures and discharge all capacitors before removing from a system.*
- *Replace capacitors which are becoming disfigured (even if they are working in a system).*

Electrical Shock
- *To prevent electrical shock, do not become a conductor between two live wires or a live (hot) wire and ground.*
- *Electrical shock occurs when you become part of the circuit.*

- *The higher voltage you work with, the greater the potential to become injured.*
- *Always know the power supply you are working with.*
- *Use only properly grounded power tools connected to properly grounded circuits.*
- *When using the **three-to two-prong adapters**, the third wire must be connected to a ground for the circuit to give protection. Fastening the wire under the wall plate screw does not provide protection unless the electrical boxes are grounded. In most cases the electrical box is nailed to a wooden stud – this does not protect you! Make sure that the third or ground wire is properly connected to a good ground.*
- *When using extension cords for power tools, plug extension cord into a GFCI receptacle.*

Electrical Burns

- *Do not wear jewelry (rings and watches) while working on live electric circuits.*
- *Never use a screwdriver or other tools in an electrical panel when power is applied.*
- *Nonconducting ladders should be used on all jobs.*

LADDER SAFETY

- *Ladders should extend at least 3 feet above the landing surface when working on a roof.*
- *The horizontal distance from the top support to the foot of the ladder should be approximately one-fourth the working length of the ladder.*
- *Ensure that ladders are placed on a stable, level surface.*
- *Do not use ladders on slippery surfaces.*
- *Ladders should be provided with slip-resistant feet.*
- *Ladder should be secured at the feet.*
- *Do not overload the ladder with more weight than it was designed to handle.*
- *Ladders should be free of oil, grease, and any other slippery hazards.*
- *When angles are too steep, safety belt should be used and tied off in such a manner that a fall would be stopped.*
- *Technicians should always work in pairs when doing rooftop work or working on high ladders.*
- *Ladder should be secured at the top.*
- *Never use the cross bracing on the far side of stepladders for climbing.*

HEAT

- *When soldering using concentrated heat, you should have a fire extinguisher close by, and you should know exactly where it is and how to use it.*
- *Never solder tubing lines that are sealed.*
- *Service valves or Schrader ports should be open before soldering is attempted.*
- *Hot refrigerant lines, hot heat exchangers, and hot motors can burn your skin and leave a permanent scar.*
- *Working in very hot weather or in hot attics can be very hard on the body.*
- *Technicians should be aware of how their body and the bodies of those around them are reacting to the working conditions.*
- *Watch for signs of overheating, such as someone's face turning very red or someone who has stopped sweating. Get them out of the heat and cool them off.*
- *Overheating can be life threatening; call for emergency assistance.*

COLD

- *Liquid refrigerant can freeze the skin or eyes instantaneously.*
- *Long exposure to cold is harmful and can cause frostbite.*
- *Wear proper clothing and waterproof boots.*

- *Be aware of the wind chill when working outside.*
- *Cold weather gear should be worn when working inside of freezers.*

MECHANICAL EQUIPMENT

- *Loose clothing should never be worn around rotating machinery.*
- *When starting an open motor, stand well to the side of the motor drive mechanism.*
- *Make sure that all nuts are tight on couplings and other components.*
- *When a large motor is coasting to a stop, do not try to stop it.*
- *Never wear jewelry while working on a job that requires much movement.*
- *When using a grinder to sharpen tools, remove burrs, or for other reasons, wear safety glasses and a face shield.*
- *Use the correct grinding stone for the metal you are grinding.*
- *The tool rest should be adjusted to approximately 1/16 inch from the grinding stone.*
- *A grinding stone must not be used on a grinder that turns faster than the stone's rated maximum revolutions per minute (rpm).*

MOVING HEAVY OBJECTS

- *Plan out the best way to move heavy objects.*
- *Do not try to lift heavy equipment by yourself.*
- *When you lift, use your legs, not your back, and wear an approved back brace belt.*
- *Keep your back straight when lifting.*

REFRIGERANTS IN YOUR BREATHING SPACE

- *Fresh refrigerant vapors and many other gases are heavier than air and can displace the oxygen in a closed in space.*
- *Low-level ventilation should be used in closed-in spaces.*
- *When working with refrigerants or other gases you should be aware of the following symptoms: dizzy feeling, lips become numb. **Should you feel this way, move quickly to a place with fresh air.***
- *Proper ventilation should be set up in advance of starting a job that has the potential of gas vapors escaping.*
- *Use fans to push and pull fresh air in to the confined areas.*
- *Cross ventilation can help prevent a buildup of fumes.*
- *Special leak detectors with alarms are required in some installations.*
- *When alarm sounds, take proper precautions, which may require special breathing gear.*
- *Open flames around refrigerant vapors can create chloroflorocarbons and hydrochloroflorocarbons, or phosgene gas. **All are VERY TOXIC and DANGEROUS.***
- *If you are soldering in a close place, keep your head below the rising fumes and have plenty of ventilation.*

USING CHEMICALS

- *Wear safety glasses and protective clothing when working with chemicals.*
- *Handle chemicals according to manufacturer's directions.*
- *If you spill chemicals on your skin **or splash them in your eyes,** follow the manufacturer's direction and **seek medical attention.***
- *Refrigerant and oil form a motor burnout can be harmful, take proper precautions when working with these.*
- *Keep your distance if a line is subject to rupture or any amount of refrigerant is allowed to escape from a sealed system.*

Tools and Equipment

T–01 (A) Phillips tip. (B) Straight or slot blade. (C) Offset. *(Photos by Bill Johnson)*

FASTENER TYPES AND DRIVER TIPS.

KEYSTONE	CABINET	PHILLIPS	TORX®

CLUTCH HEAD	HEX HEAD	REED & PRINCE (FREARSON)	SQUARE RECESS

T-02 Standard screwdriver bit types. *(Courtesy Klein Tools)*

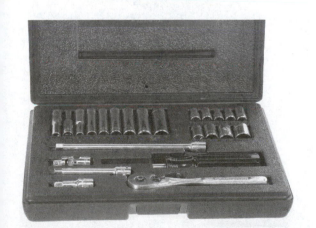

T-03 Socket with ratchet handle. *(Photo by Bill Johnson)*

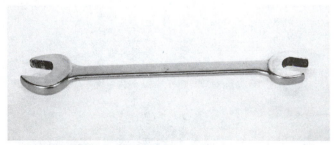

T-04 Open end wrench. *(Photo by Bill Johnson)*

T-05   Box end wrench. *(Photo by Bill Johnson)*

T-06   Combination wrench. *(Photo by Bill Johnson)*

T-07   Adjustable open-end wrench. *(Photo by Bill Johnson)*

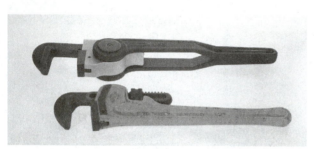

T-08   Ratchet box wrench. *(Photo by Bill Johnson)*

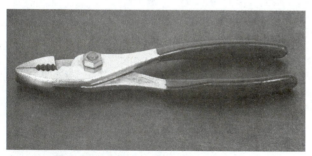

T-09   Pipe wrench. *(Photo by Bill Johnson)*

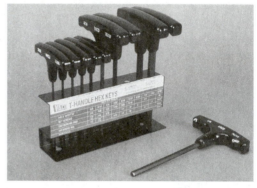

T-10   T-Handle hex keys. *(Courtesy Klein Tools)*

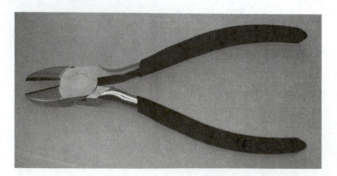

T-11   General purpose pliers. *(Photo by Bill Johnson)*

T-12   Needle-nose pliers. *(Photo by Bill Johnson)*

T-13   Side-cutting pliers. *(Photo by Bill Johnson)*

T-14   Slip joint pliers. *(Photo by Bill Johnson)*

T-15   Locking pliers. *(Photo by Bill Johnson)*

T-16   Ball peen hammer. *(Photo by Bill Johnson)*

T-17   Soft head hammer. *(Photo by Bill Johnson)*

T-18   Carpenter's claw hammer. *(Photo by Bill Johnson)*

T-19   Cold chisel. *(Photo by Bill Johnson)*

T-20   File. *(Photo by Bill Johnson)*

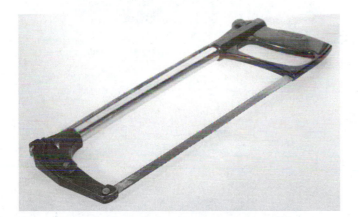

T-21   Hacksaw. *(Photo by Bill Johnson)*

T-22   Drill bits. *(Photo by Bill Johnson)*

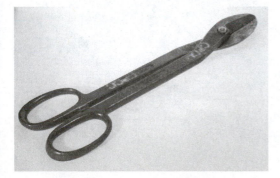

T-23   Straight metal snips. *(Photo by Bill Johnson)*

T-24   Aviation metal snips. *(Photo by Bill Johnson)*

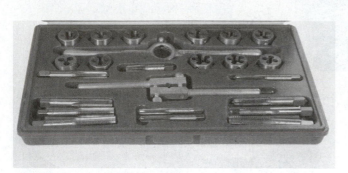

T-25 Tap and die set. (*Photo by Bill Johnson*)

T-26 Pipe-threading die. (*Photo by Bill Johnson*)

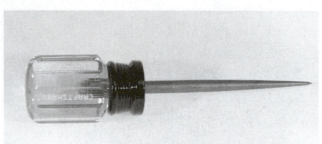

T-27 Awl. (*Photo by Bill Johnson*)

T-28 Tape rule. (*Photo by Bill Johnson*)

T-29 Flashlight. (*Photo by Bill Johnson*)

T-30 Extension cord with lights. (*Courtesy Klein Tools*)

T-31 Portable electric drill, cord type. *(Photo by Bill Johnson)*

T-32 Portable electric drill, cordless. *(Photo by Bill Johnson)*

T-33 Hole saw. *(Photo by Bill Johnson)*

T-34 Square. *(Photo by Bill Johnson)*

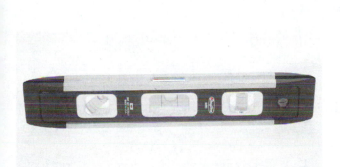

T-35 Level. *(Photo by Bill Johnson)*

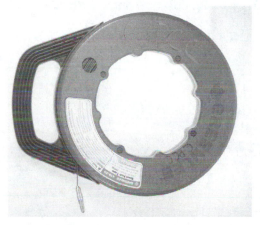

T-36 Fish tape. *(Photo by Bill Johnson)*

T-37 Utility knife. *(Photo by Bill Johnson)*

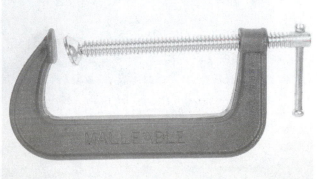

T-38 C-clamp. *(Photo by Bill Johnson)*

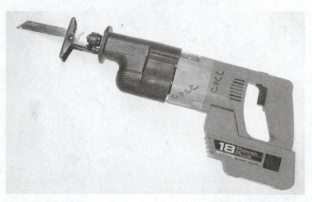

T-39   Reciprocating saw. *(Photo by Bill Johnson)*

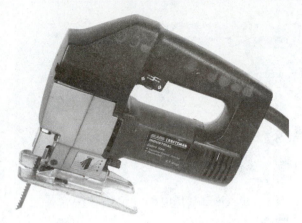

T-40   Jigsaw. *(Photo by Bill Johnson)*

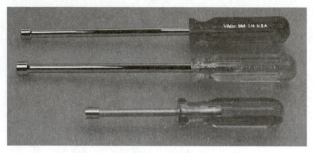

T-41   Assorted nut drivers. *(Photo by Bill Johnson)*

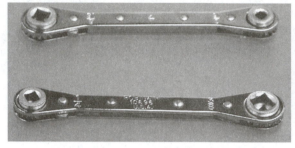

T-42   Air-conditioning and refrigeration reversible ratchet box wrenches. *(Photo by Bill Johnson)*

T-43   Combination crimping and stripping tool. *(Photo by Bill Johnson)*

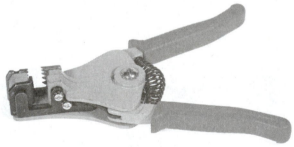

T-44   Automatic wire stripper. *(Courtesy Klein Tools)*

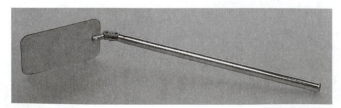

T-45   Inspection mirror. *(Photo by Bill Johnson)*

T-46   Stapling tacker. *(Photo by Bill Johnson)*

T-47   Tube cutter. *(Photo by Bill Johnson)*

T-48   Small tube cutter. *(Photo by Bill Johnson)*

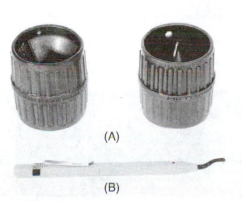

(A)

(B)

T-49   (A) Inner-outer reamers. (B) Deburring tool. *(Photo by Bill Johnson)*

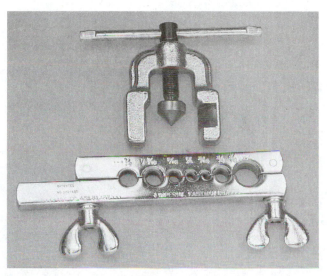

T-50   Flaring tool – flaring bar – yoke – feed screw with a flaring cone. *(Photo by Bill Johnson)*

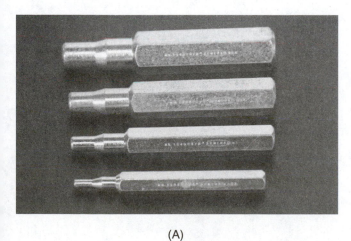

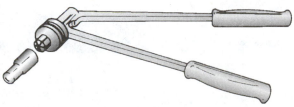

(A)   (B)

T-51   (A) Swaging tool. (B) Lever type. *(Photo by Bill Johnson)*

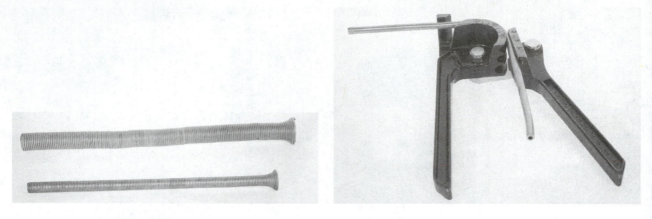

(A)   (B)

T-52   (A) Spring tube benders. (B) Lever type. *(Photos by Bill Johnson)*

T-53   Tube brushes.
(Courtesy Shaefer Brushes)

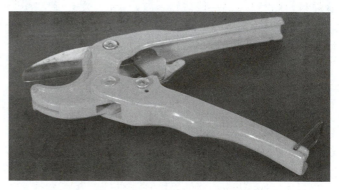

T-54   Plastic tubing shear. *(Photo by Bill Johnson)*

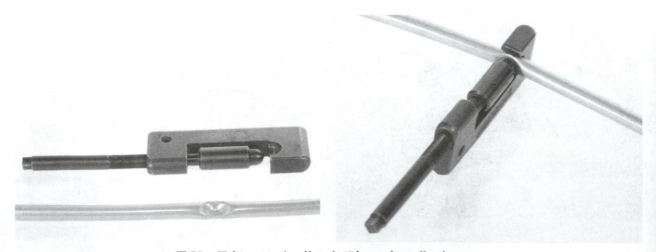

T-55   Tubing pinch-off tool. *(Photos by Bill Johnson)*

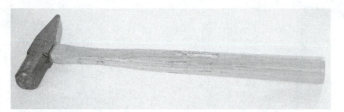

T-56   Metalworker's hammer. *(Photo by Bill Johnson)*

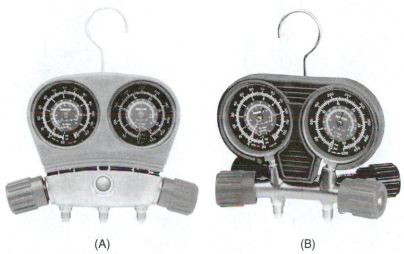

(A)   (B)

T-57   Gage manifolds. (A) Two valve gage manifold with hoses.
(B) Four-valve gage manifold with four hoses. *(Courtesy Robinair SPX Corporation)*

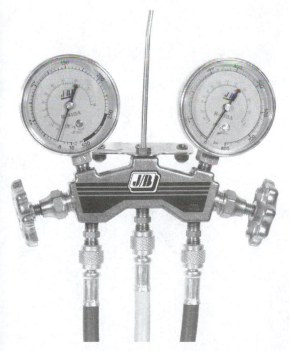

T-58   R-410A gage manifold. *(Photo by Bill Johnson)*

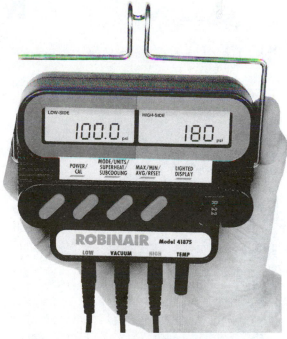

T-59   Electronic gage manifold. *(Courtesy Robinair SPX Corporation)*

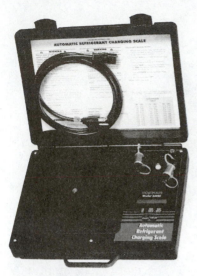

T-60 Programmable charging scale. *(Courtesy Robinair SPX Corporation)*

T-61 U-Tube mercury manometer. *(Photo by Bill Johnson)*

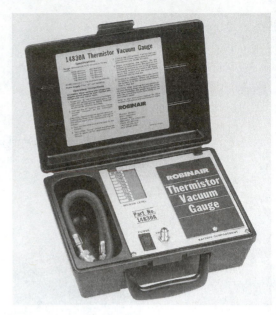

T-62 Electronic thermistor vacuum gage. *(Courtesy Robinair SPX Corporation)*

T-63 Typical vacuum pump. *(Courtesy Robinair SPX Corporation)*

T-64 Refrigerant recovery recycling station. *(Courtesy Robinair SPX Corporation)*

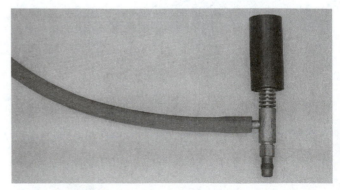

T-65 Halide leak detector used with acetylene. *(Photo by Bill Johnson)*

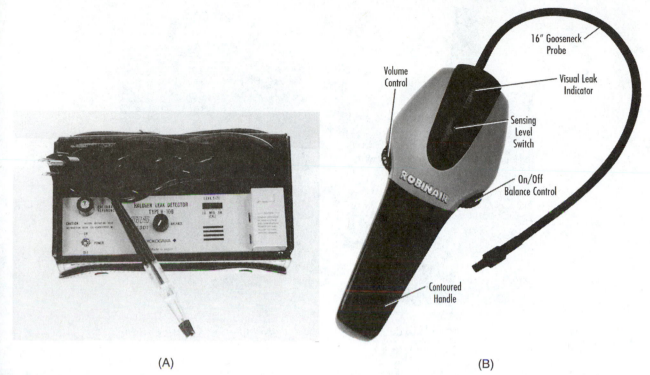

(A)   (B)

T-66   (A) AC-powered electronic leak detector. (B) Heated diode type. *(A, Photo by Bill Johnson; B, Courtesy Robinair SPX Corporation)*

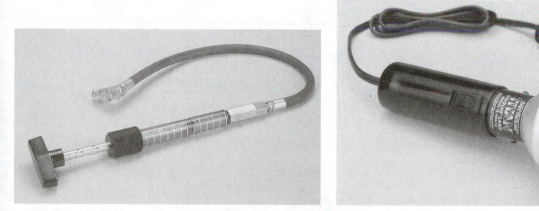

T-67   Fluorescent refrigerant leak detection system using an additive with a high-intensity ultraviolet lamp. *(Courtesy Spectronics Corporation)*

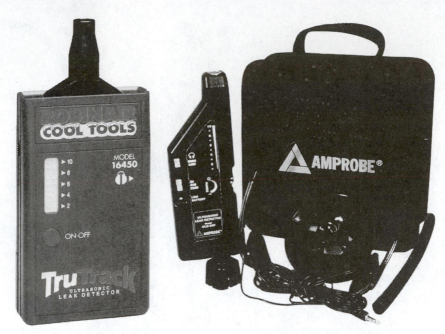

T-68   Ultrasound leak detectors. *(A, Courtesy Robinair SPX Corporation; B, Courtesy Amprobe)*

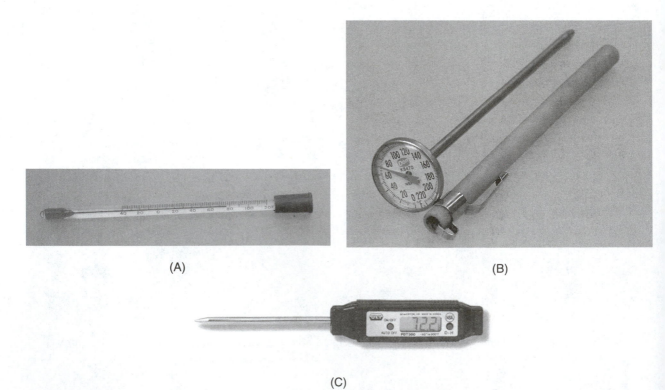

(A)

(B)

(C)

T-69   (A) Glass stem thermometers. (B) Pocket dial indicator. (C) Pocket digital indicators. *(Photos A and B by Bill Johnson, Photo C Courtesy UEi)*

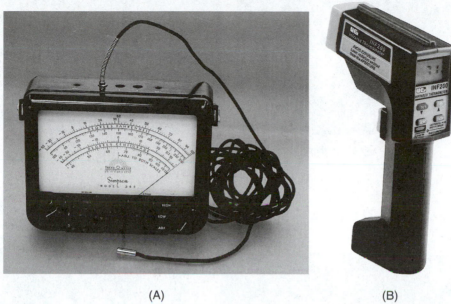

(A)   (B)

T-70   (A) Electronic thermometer. (B) Infrared thermometer. *(A, Photo by Bill Johnson; B, Courtesy UEi)*

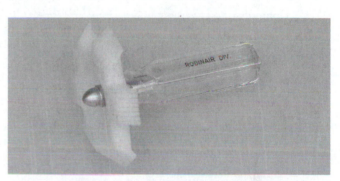

T-72   Condenser or evaporator coil fin straightener. *(Photo by Bill Johnson)*

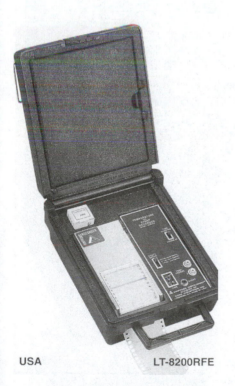

T-71   Recording thermometer. *(Courtesy Amprobe Instruments)*

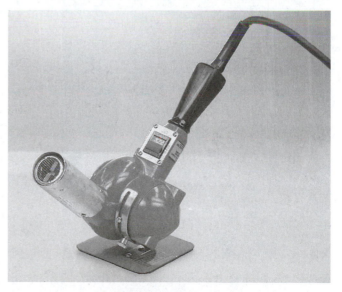

T-73   Heat gun. *(Photo by Bill Johnson)*

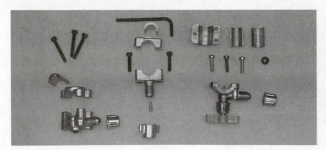 

T-74   Tubing piercing valve. *(Photos by Bill Johnson)*

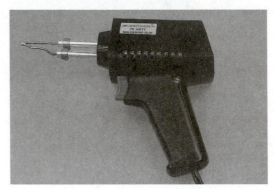

T-75   Compressor oil charging pump. *(Photo by Bill Johnson)*

T-76   Soldering gun. *(Photo by Bill Johnson)*

T-77   Propane gas torch with a disposable propane gas tank. *(Photo by Bill Johnson)*

T-78   Air acetylene unit. *(Photo by Bill Johnson)*

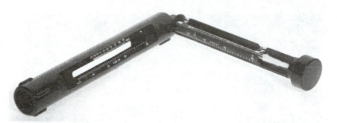

T-80   Sling psychrometer. *(Photo by Bill Johnson)*

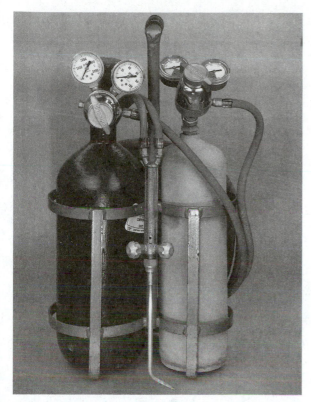

T-79   Oxyacetylene welding unit. *(Photo by Bill Johnson)*

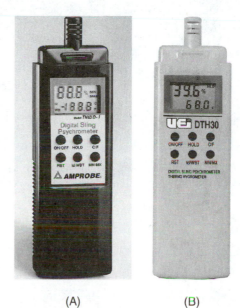

(A)   (B)

T-81   Two styles of digital sling psychrometers. *(A, Courtesy Amprobe; B, Courtesy UEi)*

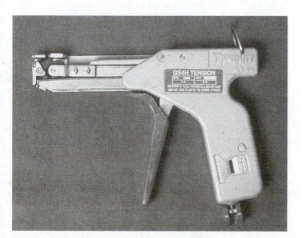

T-82   Tool used to install nylon strap clamps around flexible duct. *(Photo by Bill Johnson)*

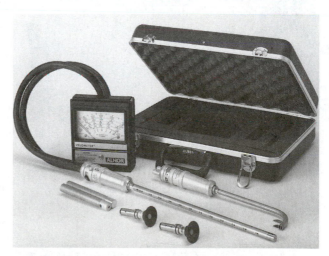

T-83   Air velocity measuring kit. *(Courtesy Alnor Instrument Company)*

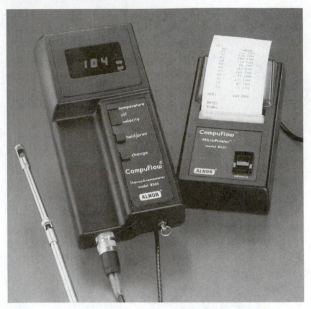

T-84   Air velocity measuring instrument with microprocessor. *(Courtesy Alnor Instrument Company)*

T-85   Air balancing meter. *(Courtesy Alnor Instrument Company)*

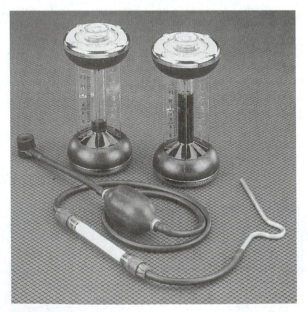

T-86   Carbon dioxide and oxygen indicators. *(Courtesy Bacharach, Inc.)*

T-87   Carbon monoxide indicator. *(Courtesy Bacharach, Inc.)*

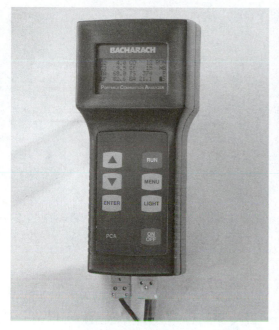

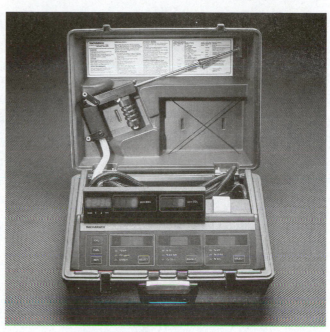

T-88   Combustion analyzers. *(Courtesy Bacharach, Inc.)*

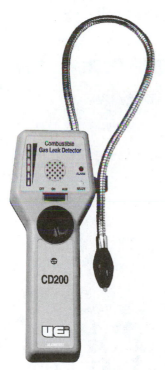

T-89   Combustion gas leak
detector. *(Courtesy UEi)*

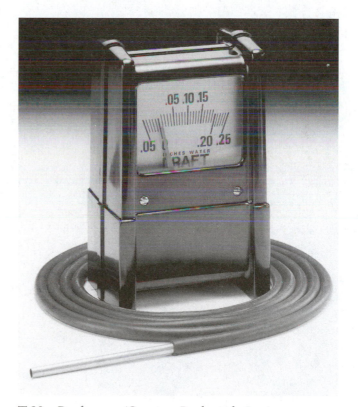

T-90   Draft gage. *(Courtesy Bacharach, Inc.)*

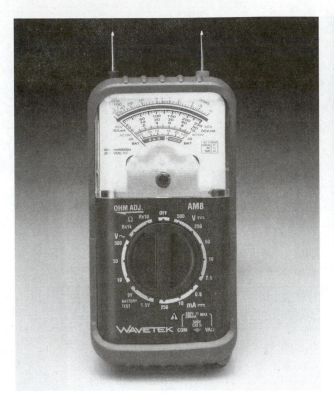

(A)

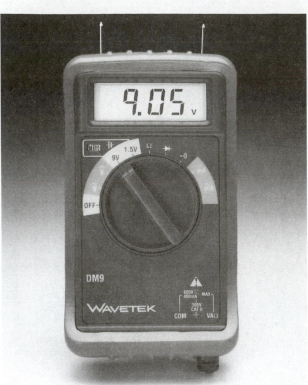

(B)

T-91 Volt-ohm–milliameter (VOM). (A) Analog. (B) Digital. *(Courtesy Wavetek)*

(A)   (B)

T-92 AC clamp-on ammeters. (A) Analog. (B) Digital. *(Courtesy Amprobe)*

T-93 This megohmmeter will measure very high resistances. *(Reproduced with permission of Fluke Corporation)*

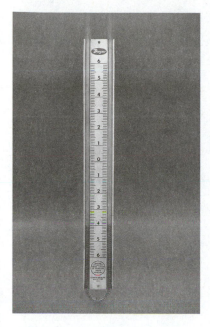

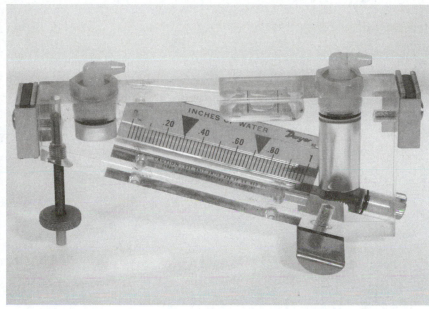

(A)   (B)

T-94   Manometers. (A) Water U-tube. (B) Inclined water. *(Photos by Bill Johnson)*

T-95   Digital electronic
manometer. *(Courtesy UEi)*

1 ELECTRICITY

Theory Lesson: Electricity

SUGGESTED MATERIALS

Textbook
Refrigeration & Air Conditioning Technology, 5th Edition, Thomson Delmar Learning
Unit 12—Basic Electricity and Magnetism

Review Topics
Movement of Electrons; Direct Current; Alternating Current; Electrical Units of Measurements; The Electrical Circuit; Ohm's Law; Characteristics of Series Circuits; Characteristics of Parallel Circuits

Key Terms
alternating current • amps • complete circuit • conductor • direct current • electrical current • electricity • electron • hertz • like charges • load series circuit • negative charged atoms • neutral charged atoms • neutron • ohms • Ohm's law formulas • parallel circuit • positive charged atoms • proton • unlike charges • volts

OVERVIEW

Comprehending the theory of electricity starts with understanding that everything in our universe is made up of a combination of atoms. An atom consists of three major types of particles: **electrons, protons,** and **neutrons** (**Figure 1–1**).

The electron is the smallest part of the atom and the only particle in the atom that can be easily moved from atom to atom. In fact, it would take 28 billion, billion, billion electrons to total 1 ounce. All electrons are alike and therefore can be moved to and from both like and unlike atoms, but not all electrons are easily moved. The movement of an electron from one atom to another atom is called **electric current** or **electricity.**

The fact that not all electrons are easily moved from atom to atom makes certain atoms poor conductors of electricity. Atoms that contain three or more electrons are not considered good electrical conductors because too much energy is required to force the electrons to move out of the outer shell of these atoms (**Figure 1–2**).

Atoms that contain eight or more electrons are considered **good electrical insulators.**

Because electricity is defined as the movement of electrons from atom to atom, **good electrical conductors** are those atoms that have one or two electrons in their outer shell. Gold, silver, and copper atoms contain one electron in their outer shells, whereas aluminum contains two. Because copper and aluminum are the least expensive of the elements listed, they are the most widely used electrical conductors. **Figure 1–3** shows a copper atom that contains twenty-nine protons and twenty-nine electrons. Notice that there is a free-floating electron in the outer shell, which can be easily removed by another electron striking it (**Figure 1–4**).

There are many ways of causing electrons to move in copper and aluminum conductors. Heat, chemical reaction, and magnetism are a few methods of generating enough force to cause the electrons in these two conductors to move out of their outer shells and into the outer shells of other atoms in the same conductor. Magnetism is the most popular of these methods of moving electrons from one atom to another.

ELECTRICAL CHARGES

The charge of the proton of an atom is (+) positive and the charge of the electron of an atom is (−) negative. An atom has a **neutral charge** when it contains the same number of protons and electrons. Because the electron can be moved from atom to atom in the conductor, the charge of an atom can be changed.

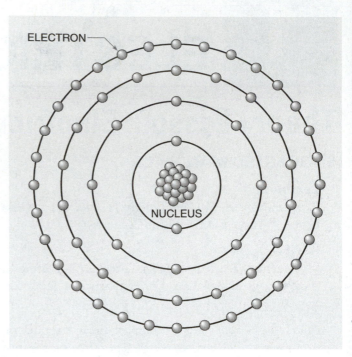

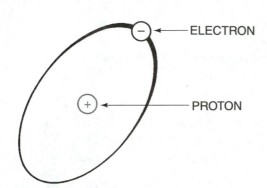

FIGURE 1–1 A hydrogen atom, with one electron and one proton.

FIGURE 1–2 Electron orbits.

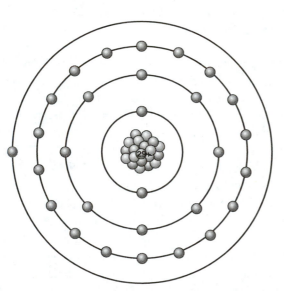

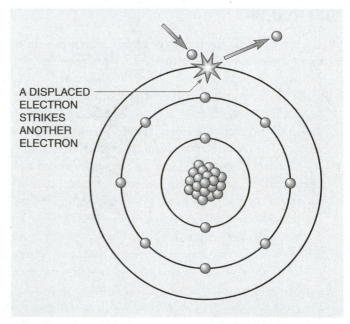

FIGURE 1–3 A copper atom, with twenty-nine protons and twenty-nine electrons.

FIGURE 1–4 A loose electron can strike and knock another atom out of orbit.

An atom that has more protons (*positive particles*) than electrons (*negative particles*) has a **positive charge**. An atom that contains more electrons (*negative particles*) than protons (*positive particles*) is said to have a **negative charge**. A basic law of electrical charges is that *like charges* repel (**Figure 1–5**) and *unlike charges* attract. (**Figure 1–6**).

Because electricity is the movement of electrons from atom to atom, in a good conductor the charge of the atoms is constantly changing. This creates a *repelling* and *attracting* action. In fact, in a 60-hertz AC (*alternating current*) this repelling and attracting action is happening 60 times a second. This means that every 1/60 of a second the charge of just one atom in a conductor is changing from **positive** to **negative** or vice versa, creating a repelling and attracting action at the same time.

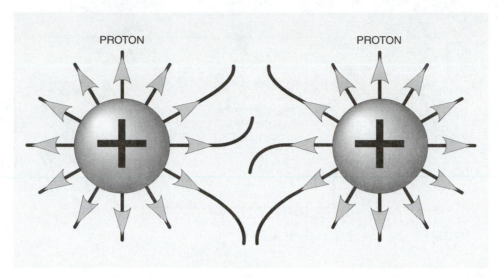

FIGURE 1–5   Like charges repel each other.

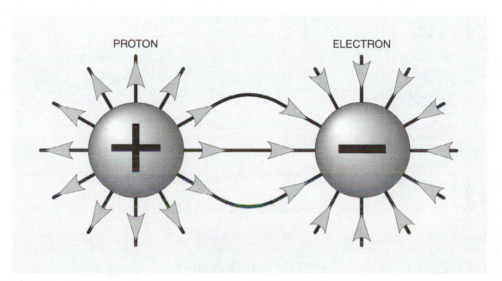

FIGURE 1–6   Opposite charges attract each other.

The repelling and attracting action of the atoms in a conductor is taking place at a speed of 186,000 miles per second. At this speed, the energy that is produced from the repelling and attracting action in a conductor **creates a source of heat energy that can be turned into consumable energy.**

Electricity is generated in two forms: **direct current (DC)** and **alternating current (AC)**. **Direct current (DC)** *is electron flow in only one direction.* **Figure 1–7** shows the Electron Flow Theory, which states that current always flows from the **negative terminal** to the **positive terminal** of the (DC) power supply.

Alternating current (AC) *is produced in most cases from an electrical generator.* **Figure 1–8** shows one sine wave cycle of an AC generator.

As the rotor coil is turned it causes the electrons in a conductor to move in one direction from point **A** to **B** and then to point **C**. A fraction of a second later, as the rotor continues to turn in the generator, the electrons move in the other direction from point **C** to **D** and back to **A**. The change in direction of electron movement through a conductor in alternating current is measured in **hertz**, which *means cycles per second.* In simple terms, in alternating current, electrons move through the conductor in one direction and 1/60 of a second later the electrons move in the other direction. Thus, the electron flow alternates.

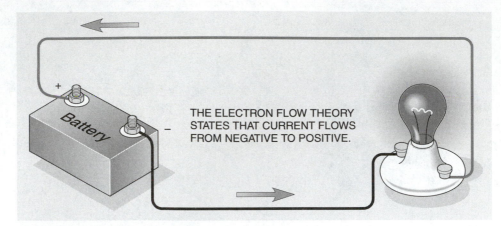

FIGURE 1–7 Direction of electron flow.

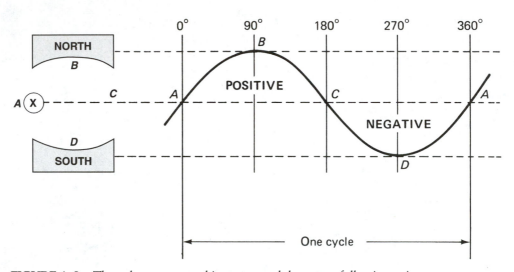

FIGURE 1–8 The voltage generated increases and decreases, following a sine wave pattern.

UNITS OF MEASURE OF ELECTRICAL ENERGY

To put electrical energy into consumable use, a way is needed to measure the value of this energy and to control it so that it can be used safely. The values of electrical energy are measured in voltage, amperage, and ohms.

Voltage

Voltage *is used* to *indicate the electrical pressure or force* needed to move quantities of electrons through a good conductor. Voltage is also referred to as **electromotive force (EMF)**. *EMF is the moving force behind an electric current.*

Amperage

Amperage *is used to measure the quantity of electrons flowing in a good conductor past a given point, in 1 second.* The quantity of 6,250,000,000,000,000,000 of electrons moving across a given point in a conductor in 1 second is equal to *1 amp.*

Ohms (Resistance)

Resistance *is defined as the opposition to current flow in a circuit, or could be referred to as the **brakes**, which slow down or control the flow of the quantity of electrons in a circuit.* The ease with which electrons move in a material determines the material's resistance. A **good conductor** of electricity, such as copper, aluminum, or silver, has electrons that move freely. Low voltage can move a large quantity of electrons in these conductors. In **good insulators** with high resistance such as glass, or plastic, even high voltage will move only a few electrons.

The unit of measurement for resistance is the **ohm,** established by George S. Ohm in the 1800s. Mr. Ohm found that a relationship existed between resistance, voltage, and amperage, and this became known as **Ohm's Law.** Simply stated, *one volt of pressure will push 1 amp of electrons through 1 ohm of resistance in 1 second.* In other words, it takes 1 volt of pressure to push 1 ampere of electrons through 1 ohm of resistance in 1 second.

Resistance in a circuit is a way of controlling how fast and how many electrons can flow through the circuit without destroying any conductors or electrical loads in the circuit. When a conductor or electrical load burns out in a circuit, it is because the resistance of the conductor or load became so low that a larger quantity of electrons were able to flow, creating so much more additional heat than the conductor could handle. The additional heat created as a result of the larger electron movement is what causes the conductor or electrical load to burn out.

Electrical Circuits

A circuit is used to take the electrons to where they belong or are needed. Electrons need a complete circuit or a path to flow in a conductor. **Figure 1–9** shows an electrical circuit.

A proper circuit requires a **power source of voltage, a conducting path** to and from the power source, and **a resistance, usually called the load.** An **electrical load** *is any device that takes electrical energy and turns it into consumable energy.* This consumable energy could be in the form of light, heat, mechanical motion, magnetism, and so forth.

The power source (voltage) could be direct current (DC) or alternating current (AC). The **wires** are used to create a conducting path from the power source to the electrical load and back to the power source. In **a complete circuit** all three elements of **voltage, current,** and **resistance** are present.

There are *three types of electrical circuits:* **series circuit, parallel circuit,** and **series-parallel circuit.** Electrical equipment may contain one or many combinations of all of these circuits. Each type of electrical circuit is discussed in detail on the Student Practical Competency assignments.

OHM'S LAW

As mentioned previously, **George Ohm** *discovered the relationship between voltage, current, and resistance in a closed electrical circuit.* **Figure 1–10** shows the **Ohm's law triangle or Ohm's pie.**

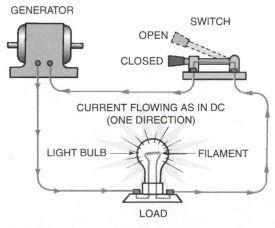

FIGURE 1–9   An electric circuit.

FIGURE 1–10   Graphic representation of Ohm's Law.

Ohm's Law Formulas

To find **voltage**:	$E = I \times R$
To find **amperage**:	$I = E \div R$
To find **resistance**:	$R = E \div I$

Ohm's law gives us the ability to determine or evaluate what effect each one of the individual electrical values (voltage, current, and resistance) will have on the others in an electrical circuit before it is even turned on. A simple explanation of this theory is the following: An **increase in voltage** causes an **increase in current flow** through a circuit, and a **decrease in voltage** causes a **decrease in current flow** through a circuit. So if the voltage on a circuit is increased, the current flow through the circuit will also increase. On the other hand, if the voltage on a circuit is decreased, the current flow will also decrease.

With a set voltage for a circuit, the amount of resistance in the circuit also determines the amount of current that will flow through the circuit. The **lower the resistance** in the circuit, the **higher the current flow** will be in the circuit. The **higher the resistance** in the circuit, the **lower the current flow** will be in the circuit.

Theory Lesson: Series Circuits

SUGGESTED MATERIALS

Textbook
Refrigeration & Air Conditioning Technology, 5th Edition, Thomson Delmar Learning
Unit 12—Basic Electricity and Magnetism

Review Topics
The Electrical Circuit; Characteristics of a Series Circuit; Ohm's Law

Key Terms
conductors • current in series circuit • electrical power • lead • open circuit • protective device • resistance in a series circuit • voltage drop in a series circuit

OVERVIEW

The path through which the electrical current flows is called a circuit. It has the following components: a source of electrical power, conductors (wires), the load(s), and some means of controlling the flow of electricity.

A series circuit is the simplest of all electrical circuits. A series circuit with two identical light bulbs connected one after the other in the circuit is shown in **Figure 1–11**.

The current in the series circuit has only one conducting path (*one wire*) to flow through. The current starts through one leg of the conducting path at the power source and will flow through each electrical component of the circuit and return to the power source on the other side of the conducting path.

Series circuits are commonly **used in system control circuits** and **normally contain protective devices and control switches**. A series circuit can be identified in two ways: There is always only one conductor (*wire*) connected to one terminal, and there is one path for the current to flow through from the power source through the load(s) and back to the power source. If there is an open circuit in any part of a series circuit, the current flow will cease to flow. This means that the current flow has stopped and no electrical component will work. A good example of this is a string of Christmas lights wired in series. When one bulb burns out, none of the remaining bulbs will light.

OHM'S LAW CHARACTERISTICS OF A SERIES CIRCUIT

1. **Figure 1–12** shows resistance in a series circuit. In a series circuit all of the loads are connected into the circuit, one after the other, so the total resistance of the circuit is the sum of the resistances of all the individual loads in the circuit.

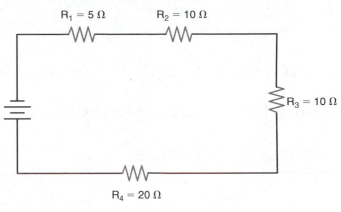

$R_T = 5\ \Omega + 10\ \Omega + 10\ \Omega + 20\ \Omega = 45\ \Omega$

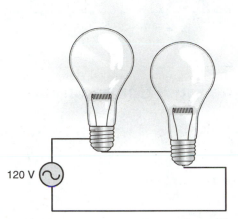

FIGURE 1–11   Series circuit with two identical light bulbs.

FIGURE 1–12   Calculating resistance in a series circuit.

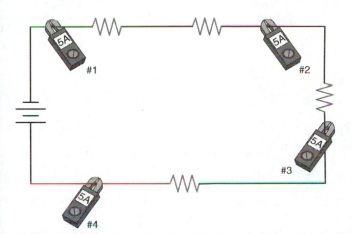

FIGURE 1–13   The current is the same at all points in a series circuit.

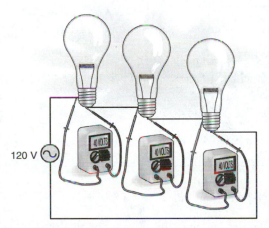

FIGURE 1–14   Voltage is divided equally among the three light bulbs.

2. **Figure 1–13** shows current in a series circuit. The current is the same throughout the complete circuit because there is only one path for it to follow. This means that a current (*amps*) reading will be the same no matter where the reading is taken in the circuit.
3. **Figure 1–14** shows voltage drop in a series circuit. As the current flows through a series circuit, the voltage drops at each load component in the circuit. The total of these voltage drops at the circuit loads will always be equal to the voltage applied to the circuit.

Practical Competency 1

Wiring Three Loads (Light Bulbs) and a Switch in a Series Circuit

SUGGESTED MATERIALS

Textbook
Refrigeration & Air Conditioning Technology, 5th Edition, Thomson Delmar Learning
Unit 12—Basic Electricity and Magnetism

Review Topics
The Electrical Circuit; Characteristics of Series Circuits; Ohm's Law

COMPETENCY OBJECTIVE

The student will be able to wire loads into a series circuit and learn how a single-pole, single-throw switch is wired into the same circuit to control the circuit. (Refer to **Figure 1–15**.)

FIGURE 1–15   Series circuit with a single-pole, single-throw switch.

OVERVIEW

A series circuit is the simplest of all electrical circuits. It consists of electrical components connected one after the other in the circuit. The current in the series circuit has only one conducting path (*one wire*) to flow through. The current starts through one leg of the conducting path at the power source and will flow through each electrical component of the circuit and return to the power source on the other side of the conducting path.

Series circuits are commonly used in system control circuits and normally contain protective devices and control switches. A series circuit can be identified in two ways: there is always only one conductor (*wire*) connected to one terminal, and there is one path for the current to flow through from the power source through the load(s) and back to the power source. If there is an open circuit in any part of a series circuit, the current flow will cease to flow. This means that the current flow has stopped and no electrical component will work.

OHM'S LAW CHARACTERISTICS OF A SERIES CIRCUIT

1. Resistance in a series circuit. In a series circuit all of the loads are connected into the circuit, one after the other, so the total resistance of the circuit is the resistance of each individual load in the circuit added together. (Refer to **Figure 1–12**.)
2. Current in a series circuit. The current is the same throughout the complete circuit because there is only one path for it to follow. This means that a current (*amps*) reading will be the same no matter where the reading is taken in the circuit. (Refer to **Figure 1–13**.)
3. Voltage drop in a series circuit. As the current flows through a series circuit, the voltage drops at each load component in the circuit. The total of these voltage drops at the circuit loads will always be equal to the voltage applied to the circuit. (Refer to **Figure 1–14**.)

EQUIPMENT REQUIRED

A power source
3 Light bulbs
3 Light bulb receptacles
1 Jumper cord with alligator clips at the ends
4 Jumper wires with alligator clips at each end
1 Single-pole, single-throw switch

SAFETY PRACTICES

Use the proper power source, size of conductors (*wires*), and loads for this exercise. Make sure that all conductors are properly insulated. Unless directed otherwise, do not plug circuit into the power source until circuit is completely wired and checked by your instructor.

> NOTE: *Switches are not load devices and are always wired in series to the power source in any circuit. All kinds of switches are used in electricity. Their main function is to break the conducting path of current flow in the total circuit or break the conducting path of individual electrical components in a circuit.*

COMPETENCY PROCEDURES   Checklist

1. Put the three light bulbs into the light bulb receptacles. ❑
2. Take one leg of the jumper cord and attach it to one leg or terminal of one of the light bulb receptacles. ❑
3. From the opposite side of this first light bulb receptacle, attach one side of a jumper wire. ❑
4. Attach the opposite end of the jumper wire from the first light bulb receptacle to one leg or terminal of a second light bulb receptacle. ❑
5. From the opposite side of the second light bulb receptacle attach another leg of another jumper wire. ❑
6. Attach the opposite end of the jumper wire from the second light bulb receptacle to one leg or terminal of the third light bulb receptacle. ❑
7. Attach one leg of a jumper wire to the other leg or terminal of the third light bulb receptacle. ❑
8. Attach the opposite end of the jumper wire from the third light bulb terminal or leg to one leg or terminal of the single-pole, single-throw switch. ❑
9. Attach the other leg of the jumper cord to the other leg or terminal of the single-pole, single-throw switch. ❑
10. Have the circuit checked by your instructor. ❑
11. With the approval of the instructor, plug the jumper cord into the proper power source. ❑
12. Turn the single-pole, single-throw switch on. ❑
13. All three light bulbs should light. ❑
14. Carefully unscrew one of the light bulbs (*the bulbs will be hot*). *All the lights should go out.* ❑
15. *Carefully* screw the bulb back into the receptacle. *All the lights should come back on.* ❑
16. Turn the single-pole, single-throw switch off and then back on. *The lights should go off and on.* ❑
17. With the approval of your instructor, unplug your series circuit and disconnect all jumper wires, jumper cord, lights, and switch and put materials back into their proper location. ❑

RESEARCH QUESTIONS

1. Electrical power is measured in what?

_____

_____

2. What is the formula for measuring electrical power?

_____

_____

3. How can resistances or loads be wired into circuits?

_____

_____

4. What are materials that allow electrons to move from one atom to another called?

_____

_____

5. What are materials that make it hard for electrons to move from one electron to another called?

_____

_____

Passed Competency _____ **Failed Competency** _____

Instructor Signature _____ **Grade** _____

Practical Competency 2

Wiring Two Loads (Light Bulbs) in Series Using a Switch to Turn Them Off and On, Keeping a Third Light in the Circuit On Continuously

SUGGESTED MATERIALS

Textbook
Refrigeration & Air Conditioning Technology, 5th Edition, Thomson Delmar Learning
Unit 12—Basic Electricity and Magnetism

Review Topics
The Electrical Circuit; Characteristics of Series Circuits; Ohm's Law

COMPETENCY OBJECTIVE

The student will be able to wire loads into a series circuit and learn how a single-pole, single-throw switch is wired into the same circuit to control the circuit, with one light being energized all the time (**Figure 1–16**).

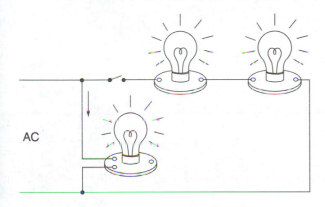

FIGURE 1–16   Series circuit with a single-pole, single-throw switch.

OVERVIEW

A series circuit is the simplest of all electrical circuits. It consists of electrical components connected one after the other in the circuit. The current in the series circuit has only one conducting path (*one wire*) to flow through. The current starts through one leg of the conducting path at the power source and will flow through each electrical component of the circuit and return to the power source on the other side of the conducting path.

Series circuits are commonly used in system control circuits and normally contain protective devices and control switches. A series circuit can be identified in two ways: there is always only one conductor (*wire*) connected to one terminal, and there is one path for the current to flow through from the power source through the load(s) and back to the power source. If there is an open circuit in any part of a series circuit, the current flow will cease to flow. This means that the current flow has stopped and no electrical component will work.

OHM'S LAW CHARACTERISTICS OF A SERIES CIRCUIT

1. Resistance in a series circuit. In a series circuit all of the loads are connected into the circuit, one after the other, so the total resistance of the circuit is the sum of the resistances of all individual loads in the circuit. (Refer to **Figure 1–12**.)

2. Current in a series circuit. The current is the same throughout the complete circuit because there is only one path for it to follow. This means that a current (*amps*) reading will be the same no matter where the reading is taken in the circuit. (Refer to **Figure 1–13**.)

3. Voltage drops in a series circuit. As the current flows through a series circuit, the voltage drops at each load component in the circuit. The total of these voltage drops at the circuit loads will always be equal to the voltage applied to the circuit. (Refer to **Figure 1–14**.)

EQUIPMENT REQUIRED

A power source
3 Light bulbs
3 Light bulb receptacles
1 Jumper cord with alligator clips at the ends
4 Jumper wires with alligator clips at each end
1 Single-pole, single-throw switch

SAFETY PRACTICES

Use the proper power source, size of conductors (*wires*), and loads for this exercise. Make sure that all conductors are properly insulated. Unless directed otherwise, do not plug circuit into the power source until the circuit is completely wired and checked by your instructor.

> NOTE: *Switches are not load devices* and are always wired in series to the power source in any circuit. There are all kinds of switches used in electricity. Their main function is to break the conducting path of current flow in the total circuit or break the conducting path of individual electrical components in a circuit.

COMPETENCY PROCEDURES

Checklist

1. Put the three light bulbs into the light bulb receptacles. ❏
2. Take one leg of the jumper cord and attach it to one leg or terminal of one of the light bulb receptacles. ❏
3. Attach the other leg of the jumper cord to the other leg or terminal of the same light bulb receptacle. ❏
4. Attach one end of a jumper wire to one leg of the jumper cord at the attached light bulb receptacle. ❏
5. Attach the opposite end of the jumper wire from the attached receptacle to one leg or terminal of the single-pole, single-throw switch. ❏
6. Attach one end of a jumper wire to the other leg or terminal of the single-pole, single-throw switch. ❏
7. Attach the opposite end of the jumper wire from the single-pole, single-throw switch to one leg or terminal of one of the other light bulb receptacle. ❏
8. Attach one end of another jumper wire to the opposite leg or terminal of a second light bulb receptacle. ❏
9. Attach the opposite end of the jumper wire from the second light bulb receptacle to one leg or terminal of the third light bulb receptacle. ❏
10. Attach one end of a jumper wire to the leg or terminal of the third light bulb receptacle. ❏
11. Attach the other end of the jumper wire from the third light bulb receptacle to the opposite side of the jumper cord at the first light bulb receptacle. ❏
12. With the approval of the instructor, and the single-pole, single-throw switch turned off; plug the jumper cord into the proper power source. ❏
13. One light should come on and the other two light bulbs should be off. ❏

14. Turn on the single-pole, single-throw switch. *The other two light bulbs should come on and the first light bulb should stay on.* ❏
15. (*The bulbs will be hot.*) Unscrew the first light bulb that is wired to the direct power source. *It should go out and the other two light bulbs should stay on.* ❏ ❏
16. *Carefully* screw the bulb back into the receptacle. *All the lights should come back on.* ❏
17. Turn the single-pole, single-throw switch off and then back on. *The two lights wired in series with the switch should go off and on.* ❏
18. With the approval of your instructor, unplug your series circuit and disconnect all jumper wires, jumper cord, lights, and switch. ❏

RESEARCH QUESTIONS

1. What are a couple of ways that electricity can be produced?

   _____

   _____

2. How is electricity produced from magnetism?

   _____

   _____

3. What is the electrical unit of measurement of current flow?

   _____

   _____

4. What is the electrical unit of measurement for the force that causes electron flow?

   _____

   _____

5. What is resistance in an electrical circuit?

   _____

   _____

Passed Competency _____ Failed Competency _____

Instructor Signature _____ Grade _____

Theory Lesson: Parallel Circuits

SUGGESTED MATERIALS

Textbook
Refrigeration & Air Conditioning Technology, 5th Edition, Thomson Delmar Learning
Unit 12—Basic Electricity and Magnetism

Review Topics
The Electrical Circuit; Characteristics of a Parallel Circuit; Ohm's Law

Keys Terms
circuits • conductors • current • electrical power • load • parallel circuit • resistance • voltage drop

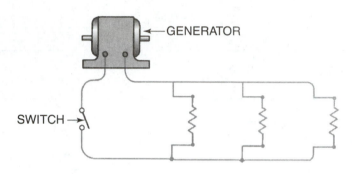

FIGURE 1–17   Multiple resistances shown in parallel.

OVERVIEW

The path through which the electrical current flows is called the circuit. It has the following components: a source of electrical power, conductors (*wires*), the load or loads, and some means of controlling the flow of electricity.

A parallel circuit has two or more paths for current to flow. The current flow will start at the source of the power supply and will divide to follow different paths through the circuit loads before returning to the source of power supply.

Visualize a parallel circuit as a ladder. The ladder has two main supports—one on each side holding rungs together in the center so you can climb the ladder. If one of the rungs in the ladder breaks, you could still climb the ladder if you bypass the broken rung.

This is also how a parallel circuit works. The two main beams on the ladder represent the two legs from the power supply. The rungs in the center of the ladder represent individual electrical circuits. These circuits are actually series circuits with their own individual loads and controls stacked one after the other, receiving their power supply from the same power source.

Like the rungs in a ladder, the individual circuits of a parallel circuit can be bypassed when they are broken. If something in one of the individual circuits burned out, or the circuit became open, everything in that particular circuit would stop working because no current could move from one leg of the power supply through the loads in the circuit and back to the other leg of the power supply. But because all of the other circuits are receiving their own source of power from the same power supply, they would continue to work as long as the circuit was complete. Each time a new component is added to a parallel circuit, a new path for current flow is created.

Parallel circuits are most commonly used in **load circuits** to **supply full-line voltage** to equipment such as motors, accessory devices, and household appliances.

OHM'S LAW CHARACTERISTICS OF A PARALLEL CIRCUIT

1. Resistance in a parallel circuit. The total circuit resistance will always be smaller than the smallest resistance in the circuit because the total circuit current is always greater than the current through any individual resistance in a parallel circuit and cannot be added to yield the total resistance. Ohm's law has two different formulas for finding the total resistance in a parallel circuit (**Figure 1–18**).
2. The total current flowing in the circuit is the total of each of the individual branch circuit currents (**Figure 1–19**).
3. Voltage drops in a parallel circuit. The voltage across each circuit that makes up a parallel circuit will be the same as the supply voltage because each circuit gets its voltage from the same power supply (**Figure 1–20**).

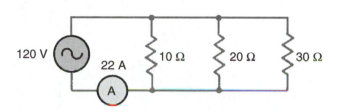

For fewer than two resistances:

$$R_T = \frac{R_1 \times R_2}{R_1 + R_2}$$

For two or more resistances:

$$R_T = \frac{1}{\frac{1}{R_1} + \frac{1}{R_2} + \frac{1}{R_3} + \dots}$$

FIGURE 1–18   Parallel circuit and two formulas for calculating total resistance.

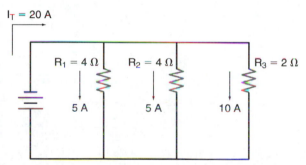

FIGURE 1–19   Calculating branch currents in a parallel circuit.

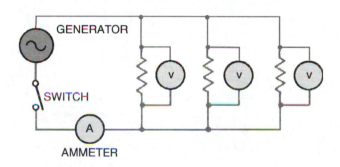

FIGURE 1–20   Voltage readings across electrical loads wired in parallel.

Practical Competency 3

Wiring Three Loads (Light Bulbs) in a Parallel Circuit and Using a Single-Pole, Single-Throw Switch to Control the Circuit

SUGGESTED MATERIALS

Textbook
Refrigeration & Air Conditioning Technology, 5th Edition, Thomson Delmar Learning
Unit 12—Basic Electricity and Magnetism

Review Topics
The Electrical Circuit; Characteristics of a Parallel Circuit; Ohm's Law

COMPETENCY OBJECTIVE

The student will be able to wire a single-pole, single-throw switch into a parallel circuit to control the operation of three load devices. (Refer to **Figure 1–21**.)

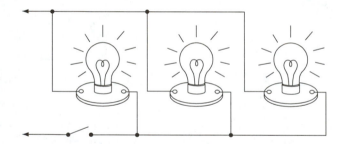

FIGURE 1–21   Parallel circuit with three light bulbs controlled by a single-pole, single-throw switch.

OVERVIEW

The path through which the electrical current flows is the circuit. It has the following components: a source of electrical power, conductors (*wires*), the load or loads, and some means of controlling the flow of electricity.

A parallel circuit has two or more paths for current to flow. The current flow will start at the source of the power supply and will divide to follow different paths through the circuit loads before returning to the source of power supply.

Visualize a parallel circuit as a ladder. The ladder has two main supports—one on each side holding rungs together in the center so you can climb the ladder. If one of the rungs in the ladder breaks, you could still climb the ladder if you bypass the broken rung.

This is also how a parallel circuit works. The two main beams on the ladder would represent the two legs from the power supply. The rungs in the center of the ladder represent individual electrical circuits. These circuits are actually series circuits with their own individual loads and controls stacked one after the other, receiving their power supply from the same power source.

Like the rungs in a ladder, the individual circuits of a parallel circuit can be bypassed when they are broken. If something in one of the individual circuits burned out, or the circuit became open, everything in that circuit would stop working because no current could move from one leg of the power supply through the loads in the circuit and back to the other leg of the power supply.

But because all of the other circuits are receiving their own source of power from the same power supply, they would continue to work as long as the circuit was complete. Each time a new component is added to a parallel circuit, a new path for current flow is created. Parallel circuits are most commonly used in load circuits to supply full-line voltage to equipment such as motors, accessory devices, and household appliances.

OHM'S LAW CHARACTERISTICS OF A PARALLEL CIRCUIT

1. Resistance in a parallel circuit. The total circuit resistance will always be smaller than the smallest resistance in the circuit because the total circuit current is always greater than the current through any individual resistance in a parallel circuit and cannot be added to yield the total resistance. Ohm's law has two different formulas for finding the total resistance in a parallel circuit. (Refer to **Figure 1–18.**)
2. Current in a parallel circuit. The total current flowing in the circuit is the total of each of the individual branch circuit currents. (Refer to **Figure 1–19.**)
3. Voltage drops in a parallel circuit. The voltage across each circuit that makes up a parallel circuit will be the same as the supply voltage because each circuit gets its voltage from the same power supply. (Refer to **Figure 1–20.**)

EQUIPMENT NEEDED

Power supply
3 Light bulbs
3 Light bulb receptacles
1 Jumper cord with alligator clips at each end
6 Jumper wires with alligator clips at each end
1 Single-pole, single-throw switch

SAFETY PRACTICES

Use the proper power source, size of conductors (*wires*), and loads for this exercise. Make sure that all conductors are properly insulated. Unless directed otherwise, do not plug circuit into the power source until the circuit is completely wired and checked by your instructor.

> NOTE: *Switches are not load devices and should always be wired in series with a load(s) in any circuit. All kinds of switches are used in electricity. Their main functions are to break the conducting path of current flow in the total circuit or to break the conducting path of individual electrical components in a circuit.*

COMPETENCY PROCEDURES
Checklist

1. Put the light bulbs in the light bulb receptacles and align the receptacles in a straight line. ❏
2. Take one leg of the jumper cord and attach it to one leg or terminal of the single-pole, single-throw switch. ❏
3. Take one end of a jumper wire and attach it to the other leg or terminal of the single-pole, single-throw switch. ❏
4. Attach the other end of the jumper wire from the single-pole, single-throw switch and attach it to one leg or terminal of the first light bulb receptacle. ❏
5. Take the other leg of the jumper cord and attach it to the other leg or terminal of the first light bulb receptacle. ❏
6. Take a jumper wire and attach one end of it to the leg or terminal of the first light bulb receptacle where one side of the cord leg is attached. ❏
7. Take another jumper wire and attach one end of it to the leg or terminal of the first light bulb receptacle at the side where the single-pole, single-throw switch is attached. ❏

8. Take the opposite end of one of the jumper wires from the first light bulb receptacle and attach it to one leg or terminal of the second light bulb receptacle. ❑
9. Take the other jumper wire end from the first light bulb receptacle and attach it to the other side of the second light bulb receptacle. ❑
10. Take another jumper wire and attach it to one leg or terminal of the second light bulb receptacle. ❑

> **NOTE:** *You will have two jumper wire ends attached at both sides of the second light bulb receptacle.*

11. Take another jumper wire and attach one end of it to the other leg or terminal of the second light bulb receptacle. ❑
12. Take the opposite end of one of the jumper wires from the second light bulb receptacle and attach it to the leg or terminal of the third light bulb receptacle. ❑
13. Take the opposite jumper wire end from the second light bulb receptacle and attach it to the other leg or terminal of the third light bulb receptacle. ❑
14. Have your instructor check your parallel circuit, and with the instructor's approval plug the jumper cord into the power source and turn on the single-pole, single-throw switch. *All three bulbs should come on.* ❑
15. Turn the single-pole, single-throw switch off. *All the lights should go out.* ❑

RESEARCH QUESTIONS

1. How many watts are equal to a kilowatt?

   _____

   _____

2. How can the magnetic field created around a conductor wire of electricity be increased?

   _____

   _____

3. What is a solenoid?

   _____

   _____

4. What does EMF stand for in electricity?

   _____

   _____

5. What is inductive reactance?

   _____

   _____

> **NOTE:** *From this exercise you could see where and how the single-pole, single-throw switch was wired into the parallel circuit to control the circuit's operation.*

Passed Competency _____   Failed Competency _____

Instructor Signature _____   Grade _____

Practical Competency 4

Wiring Three Loads (Light Bulbs) in Parallel Using a Single-Pole, Single-Throw Switch to Turn Two Light Bulbs Off and Leave One Light Bulb On All the Time

SUGGESTED MATERIALS

Textbook
Refrigeration & Air Conditioning Technology, 5th Edition, Thomson Delmar Learning
Unit 12—Basic Electricity and Magnetism

Review Topics
The Electrical Circuit; Characteristics of a Parallel Circuit; Ohm's Law

COMPETENCY OBJECTIVE

The student will be able to wire a single-pole, single-throw switch at the correct location in a parallel circuit to be able to control certain load circuits without stopping current flow in the complete circuit when the switch is turned off or on (**Figure 1–22**).

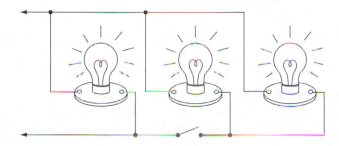

FIGURE 1–22 Parallel circuit with two bulbs operated by a single-pole, single-throw switch, and one bulb on all the time.

OVERVIEW

The path through which the electrical current flows is called the circuit. It has the following components: a source of electrical power, conductors (*wires*), the load or loads, and some means of controlling the flow of electricity.

A parallel circuit has two or more paths for current to flow. The current flow will start at the source of the power supply and will divide to follow different paths through the circuit loads before returning to the source of power supply.

Visualize a parallel circuit as a ladder. The ladder has two main supports—one on each side holding rungs together in the center so you can climb the ladder. If one of the rungs in the ladder breaks, you could still climb the ladder if you bypass the broken rung. This is also how a parallel circuit works.

The two main beams on the ladder represent the two legs from the power supply. The rungs in the center of the ladder represent individual electrical circuits. These circuits are actually series circuits with their own individual loads and controls stacked one after the other, receiving their power supply from the same power source.

Like the rungs in a ladder, the individual circuits of a parallel circuit can be bypassed when they are broken. If something in one of the individual circuits burned out, or the circuit became open, everything in that circuit would stop working because no current could move from one leg of the power supply through the loads in the circuit and back to the other leg of the power supply. But because all of the other circuits are receiving their own source of power from the same power supply they would continue to work as long as the circuit was complete. Each time a new component is added to a parallel circuit, a new path for current flow is created.

Parallel circuits are most commonly used in load circuits to supply full-line voltage to equipment such as motors, accessory devices, and household appliances.

OHM'S LAW CHARACTERISTICS OF A PARALLEL CIRCUIT

1. Resistance in a parallel circuit. The total circuit resistance will always be smaller than the smallest resistance in the circuit because the total circuit current is always greater than the current through any individual resistances in a parallel circuit and cannot be added to yield the total resistance. Ohm's law has two different formulas for finding the total resistance in a parallel circuit. (Refer to **Figure 1–18**.)
2. Current in a parallel circuit. The total current flowing in the circuit is the total of each of the individual branch circuit currents. (Refer to **Figure 1–19**.)
3. Voltage drops in a parallel circuit. The voltage across each circuit that makes up a parallel circuit will be the same as the supply voltage because each circuit gets its voltage from the same power supply. (Refer to **Figure 1–20**.)

EQUIPMENT REQUIRED

Power supply
3 Light bulbs
3 Light bulb receptacles
1 Jumper cord with alligator clips at each end
6 Jumper wires with alligator clips at each end
1 Single-pole, single-throw switch

SAFETY PRACTICES

Use the proper power source, size of conductors (*wires*), and loads for this exercise. Make sure that all conductors are properly insulated. Unless directed otherwise, do not plug circuit into the power source until circuit is completely wired and checked by your instructor.

---

*NOTE: **Switches are not load devices** and should always be wired in series with a load(s) in any circuit. Many kinds of switches are used in electricity. Their main functions are to break the conducting path of current flow in the total circuit or to break the conducting path of individual electrical components in a circuit.*

COMPETENCY PROCEDURES
Checklist

1. Put the light bulbs in the light bulb receptacles and line up the receptacles in a straight line. ❏
2. Take one leg of the jumper cord and attach it to one leg or terminal of the first light bulb receptacle. ❏
3. Take the opposite leg of the jumper cord and attach it to the other leg or terminal of the first light bulb receptacle. ❏
4. Take a jumper wire and attach one end of it to the leg or terminal of the first light bulb receptacle where one side of the cord leg is attached. ❏
5. Take another jumper wire and attach one end of it to the leg or terminal of the first light bulb receptacle where the other side of the cord leg is attached. ❏
6. Attach one of the ends of the jumper wires from the first light bulb receptacle to one leg or terminal of the single-pole, single-throw switch. ❏
7. Attach the opposite end of the jumper wire from the first light bulb receptacle to the leg or terminal of the second light bulb receptacle. ❏
8. Attach one end of another jumper wire to the other leg or terminal of the single-pole, single-throw switch. ❏

9. Attach the opposite end of the jumper wire from the single-pole, single-throw switch to the leg or terminal of the second light bulb receptacle. ❏
10. Take another jumper wire and attach it to one leg or terminal of the second light bulb receptacle. ❏

> **NOTE:** *You will have two jumper wire ends attached at both sides of the second light bulb receptacle.*

11. Take another jumper wire and attach one end of it to the other leg or terminal of the second light bulb receptacle. ❏
12. Take the opposite end of one of the jumper wires from the second light bulb receptacle and attach it to the leg or terminal of the third light bulb receptacle. ❏
13. Take the other jumper wire end from the second light bulb receptacle and attach it to the other leg or terminal of the third light bulb receptacle. ❏
14. Have your instructor check your parallel circuit. With the instructor's approval, making sure the single-pole, single-throw switch is turned off, plug the jumper cord into the power supply. *At this point one light should be on and two of the lights turned off.* ❏
15. Turn on the single-pole, single-throw switch. *The other two lights should come on. The other light should still be on.* ❏
16. (*Lights will be hot.*) Unscrew the light bulb that stays on all the time. *It should go out, and the other two lights should still be on with the switch turned on.* ❏
17. Leave the bulb unscrewed and turn the single-pole, single-throw switch off and on. *All light bulbs in the circuit should be off when the switch is turned off. When the switch is turned on, only two lights should come on.* ❏
18. Unplug the jumper cord from the power supply and remove all light bulbs, jumper cord, jumper wires, and switch from the circuit. ❏

RESEARCH QUESTIONS

1. What is a transformer?

_____

_____

2. The following formula: volts ∞ amps, would be used to find what factor?

_____

_____

3. What would be the electrical formula for finding the amount of kilowatts used?

_____

_____

4. What Ohm's law formula would be used to find the total resistance of a parallel circuit?

_____

_____

5. The following formula: I ∞ R, would be used to find what factor?

_____

_____

Theory Lesson: Series-Parallel Circuits

SUGGESTED MATERIALS

Textbook
Refrigeration & Air Conditioning Technology, 5th Edition, Thomson Delmar Learning
Unit 12—Basic Electricity and Magnetism

Review Topics
The Electrical Circuit; Characteristics of Series Circuits; Characteristics of Parallel Circuits; Series Circuits; Parallel Circuits

Key Terms
current • parallel circuit • resistance • series circuit • voltage drop

OVERVIEW

Series-parallel circuits are a combination of both series circuits and parallel circuits. These circuits can be simple or quite complicated depending on the purpose of the circuit. When making any Ohm's law calculation in a series-parallel circuit, it must first be determined what part of the series-parallel circuit is wired in series, and what part of the circuit is wired in parallel.

The part of the circuit that is wired in series will be calculated with series circuit formulas and those that are wired in parallel will be calculated with the parallel circuit formulas.

OHM'S LAW CHARACTERISTICS OF A SERIES CIRCUIT

1. Resistance in a series circuit. In a series circuit all of the loads are connected into the circuit, one after the other, so the total resistance of the circuit is the resistance of each individual load in the circuit added together. (Refer to **Figure 1–12.**)
2. Current in a series circuit. The current is the same throughout the complete circuit because there is only one path for it to follow. This means that a current (*amps*) reading will be the same no matter where the reading is taken in the circuit. (Refer to **Figure 1–13.**)
3. Voltage drop in a series circuit. As the current flows through a series circuit, the voltage drops at each load component in the circuit. The total of these voltage drops at the circuit loads will always be equal to the voltage applied to the circuit. (Refer to **Figure 1–14.**)

OHM'S LAW CHARACTERISTICS OF A PARALLEL CIRCUIT

1. Resistance in a parallel circuit. The total circuit resistance will always be smaller than the smallest resistance in the circuit because the total circuit current is always greater than the current through any individual resistance in a parallel circuit and cannot be added to yield the total resistance. Ohm's law has two different formulas for finding the total resistance in a parallel circuit. (Refer to **Figure 1–18.**)
2. Current in a parallel circuit. The total current flowing in the circuit is the total of each of the individual branch circuit currents. (Refer to **Figure 1–19.**)
3. Voltage drops in a parallel circuit. The voltage across each circuit that makes up a parallel circuit will be the same as the supply voltage because each circuit gets its voltage from the same power supply. (Refer to **Figure 1–20.**)

Practical Competency 5

Wiring a Series-Parallel Circuit with a Single-Pole, Single-Throw Switch Controlling the Total Circuit, Where Three Loads (Light Bulbs) Are Wired in Parallel and Three Loads (Light Bulbs) Are Wired in Series

SUGGESTED MATERIALS

Textbook
Refrigeration & Air Conditioning Technology, 5th Edition, Thomson Delmar Learning
Unit 12—Basic Electricity and Magnetism

Review Topics
The Electrical Circuit; Characteristics of Series Circuits; Characteristics of Parallel Circuits; Series Circuits; Parallel Circuits

COMPETENCY OBJECTIVE

The student will be able to wire loads into a series-parallel circuit by practicing the procedures of wiring from the power supply to the individual loads to create an individual series circuit and an individual parallel circuit that uses the same power supply. The student will also learn how to add a single-pole, single-throw switch in a location in the circuit to turn the total circuit off and on (**Figure 1–23**).

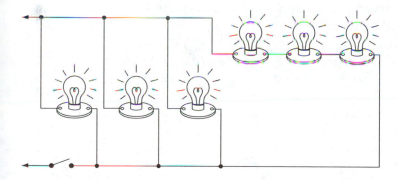

FIGURE 1–23 Series-parallel circuit with a single-pole, single-throw switch controlling the total circuit.

OVERVIEW

Series-parallel circuits are a combination of both series circuits and parallel circuits. These circuits can be simple or quite complicated depending on their purpose. When making any Ohm's law calculation in a series parallel circuit, it must first be determined what part of the series-parallel circuit is wired in series, and what part of the circuit is wired in parallel. The part of the circuit that is wired in series will be calculated with series circuit formulas and those that are wired in parallel will be calculated with the parallel circuit formulas.

> **NOTE:** *Switches are not load devices and should always be wired in series with a load(s) in any circuit. There are all kinds of switches used in electricity. Their main function is to break the conducting path of current flow in the total circuit or to break the conducting path of individual electrical components in a circuit.*

EQUIPMENT REQUIRED

Power supply
6 Light bulbs
6 Light bulb receptacles
1 Jumper cord with alligator clips at each end
10 Jumper wires with alligator clips at each end
1 Single-pole, single-throw switch

SAFETY PRACTICES

Use the proper power source, size of conductors (*wires*), and loads for this exercise. Make sure that all conductors are properly insulated. Unless directed otherwise, do not plug circuit into the power source until the circuit is completely wired and checked by your instructor.

COMPETENCY PROCEDURES

Checklist

1. Put the six light bulbs in the six light bulb receptacles and align the receptacles in a straight line. ❏
2. Take one leg of the jumper cord and attach it to one leg or terminal of the single-pole, single-throw switch. ❏
3. Attach one end of a jumper wire to the other leg or terminal of the single-pole, single-throw switch. ❏
4. Attach the opposite end of the jumper wire from the single-pole, single-throw switch to the leg or terminal of the first light bulb receptacle. ❏
5. Take the other leg of the jumper cord and attach it to the other leg or terminal of the first light bulb receptacle. ❏
6. Take a jumper wire and attach one end of it to the leg or terminal of the first light bulb receptacle, where one side of the cord leg is attached. ❏
7. Take another jumper wire and attach one end of it to the leg or terminal of the first light bulb receptacle at the other side, where the jumper cord leg is attached. ❏
8. Take the opposite end of one of the jumper wires from the first light bulb receptacle and attach it to one leg or terminal of the second light bulb receptacle. ❏
9. Take the other jumper wire end from the first light bulb receptacle and attach it to the other side of the second light bulb receptacle. ❏
10. Take another jumper wire and attach it to one leg or terminal of the second light bulb receptacle. ❏

> **NOTE:** *You will have two jumper wire ends attached at both sides of the second light bulb receptacle.*

11. Take another jumper wire and attach one end of it to the other terminal of the second light bulb receptacle. ❏
12. Take the opposite end of one of the jumper wires from the second light bulb receptacle and attach it to the leg or terminal of the third light bulb receptacle. ❏
13. Take the other jumper wire end from the second light bulb receptacle and attach it to the other leg or terminal of the third light bulb receptacle. ❏

14. Take another jumper wire and attach one end of it to one leg or terminal of the third light bulb receptacle. *You will have two jumper wire ends attached to the leg or terminals of the third light bulb receptacle when the circuit is completed.* ❑

15. Attach the opposite end of the jumper wire from the third light bulb receptacle to one leg or terminal of the fourth light bulb receptacle. ❑

16. From the opposite side of the fourth light bulb receptacle, attach one end of another jumper wire. ❑

17. Attach the other end of the jumper wire from the fourth light bulb receptacle to one leg or terminal of the fifth light bulb receptacle. ❑

18. From the opposite leg or terminal of the fifth light bulb receptacle attach one end of another jumper wire. ❑

19. Attach the end of the jumper wire from the fifth light bulb receptacle to one leg or terminal of the sixth light bulb receptacle. ❑

20. Attach one end of another jumper wire from the other leg or terminal of the sixth light bulb receptacle. ❑

21. **(Important step)** Attach the end of the last jumper wire from the sixth light bulb receptacle back to the opposite side of the third light bulb receptacle. ❑

22. Have the instructor check your series-parallel circuit. With the instructor's permission and the single-pole, single-throw switch off, plug the jumper cord into the power supply. *All six lights should be off.* ❑

23. Turn the single-pole single-throw switch on. *All the lights should come on.* ❑

24. Turn the switch off. *All the lights should go out.* ❑

> *NOTE: What you should notice from this competency is the location at which the single-pole, single-throw switch was wired into this circuit. Because the switch was supposed to control the total series-parallel circuit, it had to be put in the circuit in a location where it could stop total current flow through the whole circuit. By wiring it into one of the main power supply legs, current would have to pass through the switch first before being able to pass through any circuit or load device to make a completed path for current to flow through the total circuit. The switch could have been wired on either leg of the jumper cord. The important thing for you to understand is that if you want to control a total circuit with a switch, you must break one of the power source legs with the switch before any load devices or circuits.*

25. Unplug jumper cord from the power supply and remove all light bulbs, jumper cord, switch, and jumper wires. ❑

RESEARCH QUESTIONS

1. What happens to the resistance of a conductor wire as the temperature of the wire increases?

_____

_____

2. What happens to the amperage in a circuit if the resistance decreases?

_____

_____

3. If the voltage increases in a circuit, what happens to the amperage?

_____

_____

4. If the resistance increases in a series circuit, what happens to the amperage?

_____

_____

5. What causes a conductor or load device in a circuit to burn out?

_____

_____

Passed Competency _____ **Failed Competency** _____

Instructor Signature _____ **Grade** _____

Practical Competency 6

Wiring a Series-Parallel Circuit with a Single-Pole, Single-Throw Switch Controlling the Series Circuit, Where Three Loads (Light Bulbs) Are Wired in Parallel

SUGGESTED MATERIALS

Textbook
Refrigeration & Air Conditioning Technology, 5th Edition, Thomson Delmar Learning
Unit 12—Basic Electricity and Magnetism

Review Topics
The Electrical Circuit; Characteristics of Series Circuits; Characteristics of Parallel Circuits; Series Circuits; Parallel Circuits

COMPETENCY OBJECTIVE

The student will be able to wire loads into a series-parallel circuit by practicing the procedures of wiring from the power supply to the individual loads to create an individual series circuit and an individual parallel circuit that uses the same power supply. The student will also learn where to wire a single-pole, single-throw switch to control the series portion of the circuit. (Refer to **Figure 1–24**.)

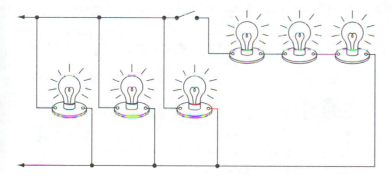

FIGURE 1–24 Series-parallel circuit with a single-pole, single-throw switch controlling the series circuit.

OVERVIEW

Series-parallel circuits are a combination of both series circuits and parallel circuits. These circuits can be simple or quite complicated, depending on the purpose of the circuit. When making any Ohm's law calculation in a series parallel circuit, it must first be determined what part of the series-parallel circuit is wired in series, and what part of the circuit is wired in parallel. The part of the circuit that is wired in series will be calculated with series circuit formulas and those that are wired in parallel will be calculated with the parallel circuit formulas.

> NOTE: *Switches are not load devices and should always be wired in series with a load(s) in any circuit. Many kinds of switches are used in electricity. Their main functions are to break the conducting path of current flow in the total circuit or to break the conducting path of individual electrical components in a circuit.*

EQUIPMENT REQUIRED

Power supply
6 Light bulbs
6 Light bulb receptacles
1 Jumper cord with alligator clips at each end
10 Jumper wires with alligator clips at each end
1 Single-pole single-throw switch

SAFETY PRACTICES

Use the proper power source, size of conductors (*wires*), and loads for this exercise. Make sure that all conductors are properly insulated. Unless directed otherwise, do not plug circuit into the power source until circuit is completely wired and checked by your instructor.

COMPETENCY PROCEDURES

Checklist

1. Put the six light bulbs in the six light bulb receptacles and align the receptacles in a straight line. ❏
2. Take one leg of the jumper cord and attach it to one leg or terminal of the first light bulb receptacle. ❏
3. Take the opposite leg of the jumper cord and attach it to the other leg or terminal of the first light bulb receptacle. ❏
4. Take a jumper wire and attach one end of it to the leg or terminal of the first light bulb receptacle where one side of the cord leg is attached. ❏
5. Take another jumper wire and attach one end of it to the leg or terminal of the first light bulb receptacle at the opposite side where the jumper cord leg is attached. ❏
6. Take the other end of one of the jumper wires from the first light bulb receptacle and attach it to one leg or terminal of the second light bulb receptacle. ❏
7. Take the opposite jumper wire end from the first light bulb receptacle and attach it to the other side of the second light bulb receptacle. ❏
8. Take another jumper wire and attach it to one leg or terminal of the second light bulb receptacle. ❏

> **NOTE:** *You will have two jumper wire ends attached at both sides of the second light bulb receptacle.*

---

9. Take another jumper wire and attach one end of it to the other leg or terminal of the second light bulb receptacle. ❏
10. Take the opposite end of one of the jumper wires from the second light bulb receptacle and attach it to the leg or terminal of the third light bulb receptacle. ❏
11. Take the other jumper wire end from the second light bulb receptacle and attach it to the other leg or terminal of the third light bulb receptacle. ❏
12. Take another jumper wire and attach one end of it to one leg or terminal of the third light bulb receptacle. *You will have two jumper wire ends attached to the leg or terminals of the third light bulb receptacle when the circuit is completed.* ❏
13. Attach the opposite end of the jumper wire from the third light bulb receptacle to one leg or terminal of the single-pole, single-throw switch. ❏
14. Take another jumper wire and attach one end of it to the other leg or terminal of the single-pole, single-throw switch. ❏
15. Attach the opposite end of the jumper wire from the switch to one leg or terminal of the fourth light bulb receptacle. ❏
16. From the opposite side of the fourth light bulb receptacle attach one end of another jumper wire. ❏
17. Attach the other end of the jumper wire from the fourth light bulb receptacle to one leg or terminal of the fifth light bulb receptacle. ❏

18. From the opposite leg or terminal of the fifth light bulb receptacle attach one end of another jumper wire. ❏
19. Attach the end of the jumper wire from the fifth light bulb receptacle to one leg or terminal of the sixth light bulb receptacle. ❏
20. Attach one end of another jumper wire from the other leg or terminal of the sixth light bulb receptacle. ❏
21. (**Important step**) Attach the end of the last jumper wire from the sixth light bulb receptacle back to the opposite side of the third light bulb receptacle. ❏
22. Have the instructor check your series-parallel circuit. With the instructor's permission and the switch turned off, plug the jumper cord into the power supply. *The first three lights should come on and the other lights should be off.* ❏
23. Turn on the single-pole, single-throw switch. *The three lights in the series circuit should come on.* ❏

> NOTE: *From this exercise you should have learned how and where to wire the switch in this circuit to control the series circuit. Notice again that the switch was wired in series with one leg of the power supply before the loads.*

24. Unplug jumper cord from the power supply and remove all light bulbs, jumper cord, and jumper wires. ❏

RESEARCH QUESTIONS

1. What are the three types of electrical loads?

   _____

   _____

2. Which voltage can do more work, alternating current (AC) or direct current (DC)?

   _____

   _____

3. What is impedance?

   _____

   _____

4. In what type of circuit does the voltage lead the current?

   _____

   _____

5. In what type of circuit are the voltage and the current in phase with each other?

   _____

   _____

Passed Competency _____   Failed Competency _____

Instructor Signature _____   Grade _____

Theory Lesson: Electrical Meters and Usage

SUGGESTED MATERIALS

Textbook
Refrigeration & Air Conditioning Technology, 5th Edition, Thomson Delmar Learning
Unit 12—Basic Electricity and Magnetism

Review Topics
Electrical Units of Measurement; Electrical Measuring Instruments

Key Terms
ammeter • analog meter • clamp-on ammeter • digital meter • multimeter • ohmmeter • voltmeter

OVERVIEW

Electricity is something that you cannot see, hear, taste, or smell. This is why electrical testing meters have been developed. Once you have detected the presence of electricity, you need to know how much of it is present. Another reason for using electrical testing meters is so that you can detect the presence of electricity without having to feel it, as feeling electricity can be deadly. It takes a meter to measure electricity safely.

Electrical meters help determine the three properties of a circuit: voltage, current, and resistance. Because electrical meters can aid in determining these circuit properties, they give technicians the ability of troubleshooting electricity and electrical equipment.

Electrical meters come in two types of face reading capabilities, the **analog meter** (**Figure 1–25**) and the **digital meter** (**Figure 1–26**).

The analog meter uses a needle movement on the meter face to indicate value readings, and the digital meter displays the value readings in digits on a LCD face.

Many types of electrical meters are available on the market today that can be used to measure the electrical values of voltage, amperage, and resistance. Each type of meter comes with its own operating instructions; however, there are general rules that should be understood about using meters no matter what type it is or who made it.

First, all meters are very delicate instruments and should be handled with care. Before using any meter, you should become familiar with the operation instructions of that meter or have someone who knows how to properly use the meter show you. Always set the meter on the highest value range if you are not sure of the value range of the element of electricity you are dealing with.

For ohms readings, never take an ohms reading on a control or electrical equipment with the power source applied to it. Before using the meter, make sure that the functions switches are set properly.

VOLTMETERS

The voltmeter measures electrical pressure, or volts. The voltmeter leads are connected directly (*parallel*) to the power source or parallel to the load in the circuit. (Refer to **Figure 1–27**.)

AMMETER (AMP METER)

The ammeter is designed to measure the amount of current flow in a circuit. The current reading can be obtained two ways: with an **in-line ammeter** or a **clamp-on ammeter**. The in-line method can be done only with a meter that has the selection and range scale to read in-line amperage in the circuit. With an in-line ammeter, the meter actually gets wired in series to the load in the circuit or in line with one of the main line power source legs. (Refer to **Figure 1–28**.)

The **clamp-on ammeter** is used in the field by most technicians today. To use this meter, the jaw of the meter is clamped around one of the conductors (*wires*) supplying the power to the load or loads. (Refer to **Figure 1–29**.)

FIGURE 1–25   Analog meter. (*Courtesy of Wavetek*)

FIGURE 1–26   Digital meter. (*Courtesy of Wavetek*)

FIGURE 1–27   Meter set to the 200-volt scale to read voltage at a wall outlet.

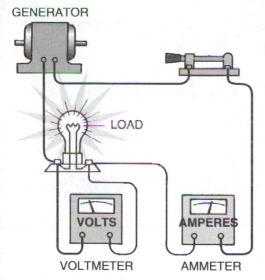

FIGURE 1–28   Voltage is measured across the load; amperage is measured in series.

FIGURE 1–29   Taking a current reading at the compressor.

> **NOTE:** *If the meter is clamped around more than one conductor (wires), the magnetic fields of the wires cancel each other and the meter will indicate an infinity reading.*

OHMMETER

The ohmmeter is used to measure resistance in a circuit or an individual load device. (Refer to **Figure 1–30**.)

The ohmmeter provides its own power to measure resistance by a battery in the meter. For accurate resistance readings, the analog ohmmeter must first be zeroed in. This is done with the meter's ohms adjustment control. Connect the meter leads together and adjust the ohms adjustment knob until the meter indicates (0)

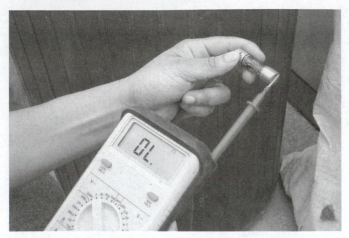

FIGURE 1–30   The OL indicates that the meter is out of range or that there is no continuity through the component being checked. (*Photo by Eugene Silberstein*)

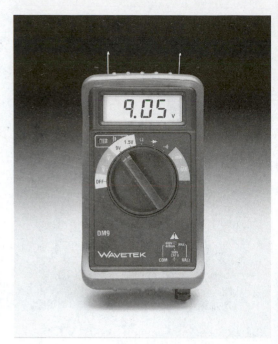

FIGURE 1–31   VOM with digital readout. (*Courtesy of Wavetek*)

on the meter scale. When the meter leads are separated, the meter will return to an infinity resistance. Digital ohmmeters do not require adjustment for zeroing.

> **NOTE:** *Never connect the ohmmeter to the circuit or load when there is power present—the meter could be damage or destroyed.*

MULTIMETER

This type of a meter has the capabilities of being used to measure all of or a couple of the electrical elements of voltage, amperage, and resistance. It is important for you to become familiar with the operating instructions of this type of meter. They are manufactured in digital or analog style. (Refer to **Figure 1–31.**)

Practical Competency 7

Using a Voltmeter to Measure Voltage and Record Voltage Drop in a Series Circuit

SUGGESTED MATERIALS

Textbook
Refrigeration & Air Conditioning Technology, 5th Edition, Thomson Delmar Learning
Unit 12—Basic Electricity and Magnetism

Review Topics
The Electrical Circuit; Making Electrical Measurements; Characteristics of Series Circuit

COMPETENCY OBJECTIVE

Students will be able to use a voltmeter to measure a series circuit's voltage. They will also use Ohm's law and calculate the voltage drop of each load in a series circuit. (Refer to **Figure 1–32**.)

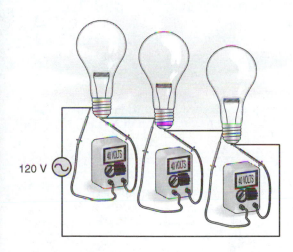

FIGURE 1–32   The voltage is divided equally among the three bulbs.

OVERVIEW

The voltmeter measures electrical pressure, or volts. The voltmeter leads get connected (parallel) to the power source or parallel to the load in the circuit.

EQUIPMENT REQUIRED

Power supply
1 Electrical meter capable of reading voltage
3 Light bulbs
3 Light bulb receptacles
1 Jumper cord with alligator clips at each end
4 Jumper wires with alligator clips at each end

SAFETY PRACTICES

Use the proper power source, size of conductors (*wires*), and loads for this exercise. Make sure that all conductors are properly insulated. Make sure that you are familiar with the meter you are using and that all settings for the meter are set properly before taking voltage readings on circuit. Unless directed otherwise, do not plug circuit into the power source until circuit is completely wired and checked by your instructor.

> NOTE: *The student should have completed and passed the Practical Competency 1, Wiring three loads (light bulbs) in a series circuit.*

COMPETENCY PROCEDURES

Checklist

MEASURING VOLTAGE VALUE AT EACH LIGHT

1. Using the equipment required for this competency, put together a series circuit. ❑
2. Set meter to proper function and value range for the power source you are working with. ❑
3. Make sure that meter leads are in the proper meter jacks. ❑
4. Plug jumper cord into power source. *All lights should come on.* ❑

> NOTE: *The voltmeter leads get connected (parallel) to the power source or parallel to the load in the circuit.*

5. Take meter leads and place or attach them at the two ends of the jumper cord of the circuit. ❑
6. *Record your voltage reading.* _____ ❑
7. Take the meter leads and place or attach them to the terminals or legs of the first light bulb in the series circuit. ❑
8. *Record your voltage reading at the first light bulb.* _____ ❑
9. Take the meter leads and place or attach them to the terminals or legs of the second light bulbin the series circuit. ❑
10. *Record you voltage reading at the second light bulb.* _____ ❑
11. Take the meter leads and place or attach them to the terminals or legs of the third light bulb. ❑
12. *Record your voltage reading at the third light bulb.* _____ ❑

> NOTE: *Ohm's law on voltage drop in a series circuit states: As the current flows through a series circuit, the voltage drops at each load component in the circuit. The total of these voltage drops at the circuit loads will always be equal to the voltage applied to the circuit.*

DETERMINE THE VOLTAGE DROP ACROSS EACH LIGHT.

13. What was the voltage drop at the first light bulb? _____ (voltage drop = power source voltage – voltage drop across the load) ❑
14. What was the voltage drop at the second light bulb? _____ (voltage drop = power source voltage – voltage drop across the load) ❑
15. What was the voltage drop at the third light bulb? _____ (voltage drop = power source voltage – voltage drop across the load) ❑
16. What was the total voltage drop of the circuit? _____ (Total voltage drop is the total of the voltage drop readings from light bulb one, two, and three.) ❑
17. Was the total voltage drops of the circuit equal to the applied source voltage? _____ ❑
18. Have your instructor review your voltage readings. ❑
19. Unplug power source and disassemble the series circuit and return materials to proper location. ❑

RESEARCH QUESTIONS

1. What happens to the voltage in a series circuit?

   _____

   _____

2. What happens to the voltage in a parallel circuit?

   _____

   _____

3. What happens to the current in a series circuit?

   _____

   _____

4. A watt is a measurement of what?

   _____

   _____

5. Ohms are used to represent what value?

   _____

   _____

Passed Competency _____   Failed Competency _____

Instructor Signature _____   Grade _____

Practical Competency 8

Using a Voltmeter to Measure Voltage and Record Voltage Drop in a Parallel Circuit

SUGGESTED MATERIALS

Textbook
Refrigeration & Air Conditioning Technology, 5th Edition, Thomson Delmar Learning
Unit 12—Basic Electricity and Magnetism

Review Topics
The Electrical Circuit; Making Electrical Measurements; Characteristics of a Parallel Circuit

COMPETENCY OBJECTIVE

Students will be able to use a voltmeter to measure a parallel circuit's voltage. They will also use Ohm's law and calculate the voltage drop of each load in a parallel circuit. (Refer to **Figure 1–33.**)

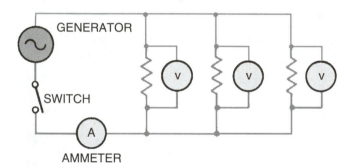

FIGURE 1–33   Voltage readings are taken across the resistances in the circuit.

OVERVIEW

The voltmeter measures electrical pressure, or volts. The voltmeter leads get connected (**parallel**) to the power source or parallel to the load in the circuit.

EQUIPMENT REQUIRED

Power supply
An electrical meter capable of reading voltage
3 Light bulbs
3 Light bulb receptacles
1 Jumper cord with alligator clips at each end
4 Jumper wires with alligator clips at each end

SAFETY PRACTICES

Use the proper power source, size of conductors (*wires*), and loads for this exercise. Make sure that all conductors are properly insulated. Make sure that you are familiar with the meter you are using and that all settings for the meter are set properly before taking voltage readings on circuit. Unless directed otherwise, do not plug circuit into the power source until circuit is completely wired and checked by your instructor.

> **NOTE:** *The student should have completed and passed the Practical Competency 3, Wiring three loads (light bulbs) in a parallel circuit.*

COMPETENCY PROCEDURES

MEASURING VOLTAGE VALUE AT EACH LIGHT
1. Using the equipment required for this competency put together a parallel circuit. ❏
2. Set meter to proper function and value range for the power source you are working with. ❏

> **NOTE:** *The voltmeter leads get connected (parallel) to the power source or parallel to the load in the circuit. (Refer to Figure 1–10 and Figure 1–11.)*

3. Make sure that meter leads are in the proper meter jacks. ❏
4. Plug jumper cord into power source. *All lights should come on.* ❏
5. Take meter leads and place or attach them at the two ends of the jumper cord of the circuit. ❏
6. *Record your voltage reading.* _____ ❏
7. Take the meter leads and place or attach them to the terminals or legs of the first light bulb in the parallel circuit. ❏
8. *Record your voltage reading at the first light bulb.* _____ ❏
9. Take the meter leads and place or attach them to the terminals or legs of the second light bulb in the parallel circuit. ❏
10. *Record you voltage reading at the second light bulb.* _____ ❏
11. Take the meter leads and place or attach them to the terminals or legs of the third light bulb. ❏
12. *Record your voltage reading at the third light bulb.* _____ ❏

> **NOTE:** *Ohm's law for voltage drops in parallel circuits states: The voltage across each circuit that makes up a parallel circuit will be the same as the supply voltage because each circuit gets its voltage from the same power supply.*

DETERMINE THE VOLTAGE DROP ACROSS EACH LIGHT.
13. What was the voltage drop at the number one light bulb? _____
    (voltage drop = power source voltage – voltage drop across the load) ❏
14. What was the voltage drop at the number two light bulb? _____
    (voltage drop = power source voltage – voltage drop across the load) ❏
15. What was the voltage drop at the number three light bulb? _____
    (voltage drop = power source voltage – voltage drop across the load) ❏
16. What was the total voltage drop of the circuit? _____ (Total voltage drop is the total of the voltage drop readings from light bulb one, two, and three.) ❏
17. Was the total voltage drop of the circuit equal to the applied source voltage? _____ ❏
18. Have your instructor review your circuit and voltage readings. ❏
19. Unplug power source and disassemble the series circuit and return materials to proper location. ❏

RESEARCH QUESTIONS

1. What are three ways electricity can be produced?

   _____

   _____

2. How fast does the repelling and attracting action of the atoms of a good conductor take place?

   _____

   _____

3. What is required in a circuit to have a capacitance load?

   _____

   _____

4. What is capacitance?

   _____

   _____

5. What is the movement of electrons from atom to atom is called?

   _____

   _____

Passed Competency _____   Failed Competency _____

Instructor Signature _____   Grade _____

Practical Competency 9

Using a Clamp-On Ammeter to Measure Current in a Series Circuit

SUGGESTED MATERIALS

Textbook
Refrigeration & Air Conditioning Technology, 5th Edition, Thomson Delmar Learning
Unit 12—Basic Electricity and Magnetism

Review Topics
The Electrical Circuit; Making Electrical Measurements; Characteristics of a Series Circuit

COMPETENCY OBJECTIVE

Students will be able to use a clamp-on ammeter to measure current flow in a series circuit (**Figure 1–34**).

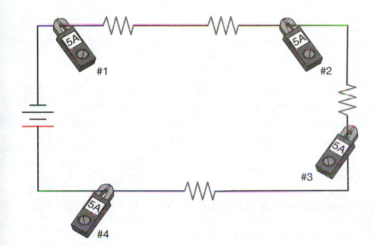

FIGURE 1–34   The current is the same at all points in a series circuit.

OVERVIEW

The ammeter is designed to measure the amount of current flow in a circuit. The current reading can be obtained two ways: with an in-line ammeter, or a clamp-on ammeter. The in-line method can only be done with a meter that has the selection and range scale to read in-line amperage in the circuit. With an in-line ammeter, the meter actually gets wired in series to the load in the circuit or in line with one of the main line power source legs. (Refer to **Figure 1–28**.)

The clamp-on ammeter is used in the field by most technicians today. To use this meter, the jaw of the meter is clamped around one of the conductors (*wires*) supplying the power to the load or loads. (Refer to **Figure 1–29**.)

> **NOTE:** *If the meter is clamped around more than one conductor (wires), the magnetic fields of the wires cancel each other and the meter will indicate a zero reading.*

EQUIPMENT REQUIRED

Power supply
1 Clamp-on ammeter
3 Light bulbs
3 Light bulb receptacles
1 Jumper cord with alligator clips at each end
4 Jumper wires with alligator clips at each end

> **NOTE:** *If using an analog clamp-on ammeter, amperage readings maybe hard to interrupt. In this case, use a 10-wrap multiplier.*

This requires that a piece of electrical wire be wrapped around the jaw of the ammeter 10 times. Jumper wires with alligator clips can be placed at each end of the coiled wire so that amp readings can be obtained. The jumper wires of the coil of wire would then be placed in series with the electrical circuit or electrical load. (Refer to **Figure 1–35**.)

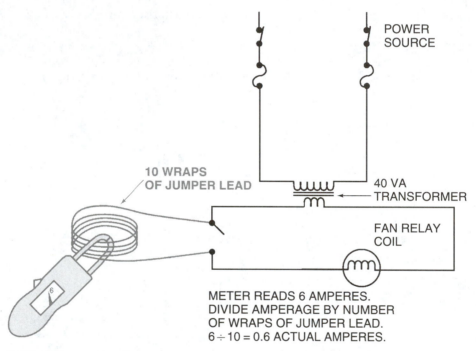

FIGURE 1–35   An illustration using 10-wrap multiplier with the ammeter.

Once an amperage reading is recorded, divide the amperage reading by the number of wraps of the jumper wire. In this case 10 would be used to divide into the amperage reading. This would equal the actual amperage of the circuit or electrical load.

SAFETY PRACTICES

Use the proper power source, size of conductors (*wires*), and loads for this exercise. Make sure that all conductors are properly insulated. Make sure that you are familiar with the meter you are using and that all settings for the meter are set properly before taking amperage readings on circuit. Unless directed otherwise, do not plug circuit into the power source until circuit is completely wired and checked by your instructor.

> **NOTE:** *Student should have completed and passed the Practical Competency 1, Wiring three loads (light bulbs) in a series circuit.*

COMPETENCY PROCEDURES

MEASURING AMPERAGE IN A SERIES CIRCUIT

1. Using the equipment required for this competency put together a series circuit. ❑
2. Set meter to proper function and amperage value range. *If you are not sure of the amperage range of this circuit, set meter to the highest value range.* ❑
3. Plug jumper cord into the power source. *All lights should come on.* ❑

NOTE: *The jaw of the meter is clamped around one of the conductors (wires) supplying the power to the load or loads.*

---

4. Clamp the jaws of the ammeter around one leg of the power jumper cord. ❑
5. *Record your amperage reading. _____* ❑
6. Clamp the jaws of the ammeter around the conductor wire between the first and second light bulb receptacle. ❑
7. *Record your amperage reading. _____* ❑
8. Clamp the jaws of the ammeter around the wire conductor between the second and third light bulb receptacle. ❑
9. *Record your amperage reading. _____* ❑
10. Clamp the jaws of the ammeter around the other leg of the jumper Cord. ❑
11. *Record you amperage reading. _____* ❑

NOTE: *Ohm's law on current in a series circuit states: The current is the same throughout the complete circuit because there is only one path for it to follow. This means that a current (amp) reading will be the same no matter where the reading is taken in the circuit.*

---

12. Were the amperage readings at all test points the same? _____ ❑
13. Have your instructor review your voltage readings. ❑
14. Unplug power source and disassemble the series circuit and return materials to proper location. ❑

RESEARCH QUESTIONS

1. What is a coulomb?

   _____

   _____

2. The charge of one coulomb of electrons is how many electrons?

   _____

   _____

3. What is electromotive force (EMF)?

   _____

   _____

4. What is a volt?

   _____

   _____

5. What is the controlled movement of electrons through a substance is called?

_____

_____

Practical Competency 10

Using a Clamp-On Ammeter to Measure Amperage in a Parallel Circuit

SUGGESTED MATERIALS

Textbook
Refrigeration & Air Conditioning Technology, 5th Edition, Thomson Delmar Learning
Unit 12—Basic Electricity and Magnetism

Review Topics
The Electrical Circuit; Making Electrical Measurements; Characteristics of a Parallel Circuit

COMPETENCY OBJECTIVE

Students will be able use a clamp-on ammeter to measure amperage in a parallel circuit.

OVERVIEW

The clamp-on ammeter is used in the field by most technicians today. To use this meter, the jaw of the meter is clamped around one of the conductors (*wires*) supplying the power to the load or loads. (Refer to **Figure 1–29**.)

> *NOTE: If the meter is clamped around more than one conductor (wires), the magnetic fields of the wires cancel each other and the meter will indicate a zero reading.*

EQUIPMENT REQUIRED

Power supply
1 Clamp-on ammeter
3 Light bulbs
3 Light bulb receptacles
1 Jumper cord with alligator clips at each end
4 Jumper wires with alligator clips at each end

> *NOTE: If using an analog clamp-on ammeter, amperage readings maybe hard to interrupt. In this case, use a 10-wrap multiplier.*

This requires that a piece of electrical wire be wrapped around the jaw of the ammeter 10 times. Jumper wires with alligator clips can be placed at each end of the coiled wire so that amp readings can be obtained. The jumper wires of the coil of wire would then be placed in series with the electrical circuit or electrical load.

Once an amperage reading is recorded, divide the amperage reading by the number of wraps of the jumper wire. In this case 10 would be used to divide into the amperage reading. This would equal the actual amperage of the circuit or electrical load.

SAFETY PRACTICES

Use the proper power source, size of conductors (*wires*), and loads for this exercise. Make sure that all conductors are properly insulated. Make sure that you are familiar with the meter you are using and that all settings for the meter are set properly before taking amperage readings on circuit. Unless directed otherwise, do not plug the circuit into the power source until circuit is completely wired and checked by your instructor.

> **NOTE:** *Students should have completed and passed Practical Competency 3, Wiring three loads (light bulbs) into a parallel circuit.*

COMPETENCY PROCEDURES

Checklist

DETERMINE AMPERAGE AT EACH CIRCUIT.
1. Using the equipment required for this competency put together a parallel circuit. ❏
2. Set the meter to proper function and value range for the amperage you are working with. ❏

> **NOTE:** *The jaw of the meter is clamped around one of the conductors (wires) supplying the power to the load or loads.*

3. Clamp the jaws of the meter around one leg of the power jumper cord. ❏
4. *Record your amperage reading.* _____ ❏
5. Clamp the jaws of the ammeter around either one of the conductor wires between the first and second light bulb receptacle. ❏
6. *Record your amperage reading.* _____ ❏
7. Clamp the jaws of the ammeter around either one of the conductor wires between the second and third light bulb receptacle. ❏
8. *Record your amperage reading.* _____ ❏
9. Clamp the jaws of the meter around the other leg of the jumper cord. ❏
10. *Record you amperage reading.* _____ ❏

> **NOTE:** *Ohm's law on a current in a parallel circuit states: The total current flowing in the circuit is the total of each of the individual branch circuit currents.*

DETERMINE TOTAL CIRCUIT AMPERAGE.
11. What was the amperage reading at the first cord jumper leg? _____ ❏
12. What was the amperage reading at the conductor wire between the first and second light bulb receptacle? _____ ❏
13. What was the amperage reading at the conductor wire between the second and third light bulb receptacle? _____ ❏
14. What was the amperage reading at the second jumper cord leg? _____ ❏
15. What was the total amperage of the circuit? _____ ❏
16. Have your instructor check your circuit and amperage readings. ❏
17. Unplug the power source and disassemble the parallel circuit and return materials to proper location. ❏

RESEARCH QUESTIONS

1. One ampere is equal to how many electrons?

_____

_____

2. Define ohms.

_____

_____

3. What three elements are present in any electrical circuit?

_____

_____

4. Electrons can only move when there is what?

_____

_____

5. Are all electrons alike?

_____

_____

Passed Competency _____ **Failed Competency** _____

Instructor Signature _____ **Grade** _____

Practical Competency 11

Taking an In-Line Amperage Reading with a VOM

SUGGESTED MATERIAL

Textbook

Refrigeration & Air Conditioning Technology, 5th Edition, Thomson Delmar Learning
Unit 12—Basic Electricity and Magnetism

Review Topics

The Electrical Circuit; Making Electrical Measurements; Characteristics of a Parallel Circuit

COMPETENCY OBJECTIVE

Students will be able to use a VOM ammeter to take an in-line amperage reading in a series circuit.

OVERVIEW

The ammeter is designed to measure the amount of current flow in a circuit. The current reading can be obtained two ways: with an in-line ammeter or a clamp-on ammeter. The in-line method can only be done with a meter that has the selection and range scale to read in-line amperage in the circuit. With an in-line ammeter, the meter actually gets wired in series to the load in the circuit or in line with one of the main line power source legs. (Refer to **Figure 1–36**.)

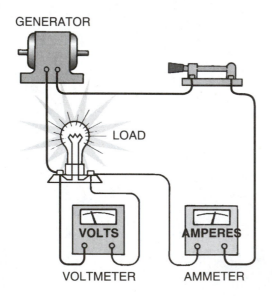

FIGURE 1–36   Voltage is measured across the load; amperage is measured in series.

EQUIPMENT REQUIRED

Power supply
1 VOM (volt-ohm-milliammeter)
3 Light bulbs
3 Light bulb receptacles

1 Jumper cord with alligator clips at each end
4 Jumper wires with alligator clips at each end

SAFETY PRACTICES

Use the proper power source, size of conductors (*wires*), and loads for this exercise. Make sure that all conductors are properly insulated. Make sure that you are familiar with the meter you are using and that all settings for the meter are set properly before taking amperage readings on circuit. Unless directed otherwise, do not plug circuit into the power source until the circuit is completely wired and checked by your instructor.

> *NOTE: Students should have completed and passed the Practical Competency 1, Wiring three loads (light bulbs) in a series circuit.*

> *NOTE: To take an in-line amperage reading on a circuit or load, the VOM ammeter is wired in series to the circuit or load (Figure 1–36).*

COMPETENCY PROCEDURES

Checklist

DETERMINE AMPERAGE AT EACH CIRCUIT.
1. Using the equipment required for this competency put together a series circuit. ❑
2. Set meter to proper function and amperage value range. *If you are not sure of the amperage range of this circuit, set meter to the highest value range.* ❑
3. Attach one leg of the jumper cord to one of the meter leads of the VOM ammeter. ❑
4. Attach one end of a jumper wire to the other VOM ammeter lead. ❑
5. Attach the other end of the jumper wire hooked to the meter lead to the light bulb receptacle that completes the circuit. ❑
6. Plug jumper cord into power source. *All lights should come on.* ❑
7. *Record your in-line amperage reading.* _____ ❑
8. Unplug the jumper cord from the power source. ❑
9. Now attach the VOM ammeter in series between the second and third light bulb. ❑
10. Plug the jumper cord into the power source. *All three lights should come on.* ❑
11. *Record your amperage reading.* _____ ❑

> *NOTE: Ohm's law on a current in a series circuit states: The current is the same throughout the complete circuit because there is only one path for it to follow. This means that a current (amps) reading will be the same no matter where the reading is taken in the circuit.*

DETERMINE TOTAL CIRCUIT AMPERAGE.
12. Were the amperage readings the same at both locations? _____ ❑
13. Have your instructor review your circuit and amperage readings. ❑
14. Unplug power source and disassemble the series circuit and return materials to proper location. ❑

RESEARCH QUESTIONS

1. What happens to like charges?

_____

_____

2. What is another name for potential difference?

_____

_____

3. What happens to unlike charges?

_____

_____

4. What does hertz stand for?

_____

_____

5. What is a volt defined as?

_____

_____

Passed Competency _____ **Failed Competency** _____

Instructor Signature _____ **Grade** _____

Practical Competency 12

Checking the Resistance of a Series Circuit with an Ohmmeter

SUGGESTED MATERIALS

Textbook
Refrigeration & Air Conditioning Technology, 5th Edition, Thomson Delmar Learning
Unit 12—Basic Electricity and Magnetism

Review Topics
Electrical Values; Meters; Ohmmeters

COMPETENCY OBJECTIVE

Students will be able to use an ohmmeter to check the resistance of a series circuit. (Refer to **Figure 1–37**.)

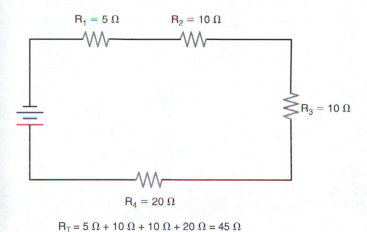

$R_T = 5\ \Omega + 10\ \Omega + 10\ \Omega + 20\ \Omega = 45\ \Omega$

FIGURE 1–37   Calculating resistance in a
series circuit.

OVERVIEW

The ohmmeter is used to measure resistance in a circuit or of individual load devices. The ohmmeter provides its own power to measure resistance by a battery in the meter. For accurate resistance readings with an analog meter, the ohmmeter must first be zeroed in. This is done with the meter's ohms adjustment control. Connect the meter leads together and adjust the ohms adjustment knob until the meter indicates (0) on the meter scale. When the meter leads are separated, the meter will return to an infinity resistance.

> **NOTE:** *Never connect the ohmmeter to the circuit or load when there is power present because the meter could be damaged or destroyed.*

EQUIPMENT REQUIRED

Power supply
1 VOM or ohmmeter
3 Light bulbs
3 Light bulb receptacles
1 Jumper cord with alligator clips at each end
4 Jumper wires with alligator clips at each end

SAFETY PRACTICES

Use the proper power source, size of conductors (*wires*), and loads for this exercise. Make sure that all conductors are properly insulated. Make sure that you are familiar with the meter you are using and that all settings for the meter are set properly before taking amperage readings on circuit. Do not plug the circuit into the power source until circuit is completely wired and checked by your instructor.

> **NOTE:** *Student should have completed and passed Practical Competency 1, Wiring three loads (light bulbs) in a series circuit.*

OHMMETER

Become familiar with the function and operation of the meter you are working with. If you are using a digital meter you will not have to zero the meter in before taking a resistance reading. All analog meters (*ones that use a needle to indicate value readings*) should be zeroed in before checking resistance on a circuit or load devices.

> **NOTE:** *Never check resistance on a circuit or load with live voltage applied.*

COMPETENCY PROCEDURES

Checklist

1. Using the equipment required for this competency, put together a series circuit. ☐
2. Set the meter on the ohms scale and a resistance range of $R \infty 1$. ☐ ☐
3. To obtain the circuit's total resistance, attach a 1-meter lead to one leg of the jumper cord end. ☐
4. Attach the other meter lead to the other leg of the jumper cord end. ☐
5. *Record the resistance value recorded by the meter.* _____ ☐
6. Attach the meter leads across the terminals or legs of the first light bulb receptacle. ☐
7. *Record the resistance value of the first light bulb.* _____ ☐ ☐
8. Attach the meter leads across the terminals or legs of the second light bulb receptacle. ☐
9. *Record the resistance value of the second light bulb.* _____ ☐
10. Attach the meter leads across the terminals or legs of the third light bulb receptacle. ☐
11. *Record the resistance reading of the third light bulb.* _____ ☐

> **NOTE:** *Ohm's law on resistance in a series circuit states: In a series circuit all of the loads are connected into the circuit, one after the other, so the total resistance of the circuit is the resistance of each individual load in the circuit added together.*

12. Total the resistances of the first, second, and third light bulbs and record it. _____ ❏
13. What was the resistance value when you took a resistance reading across the jumper cord leads? _____ ❏
14. The resistance reading recorded for procedure #13 should be the same as the resistance recorded for procedure #12. Was it? _____ ❏
15. Have your instructor review your circuit and resistance readings. ❏
16. Unplug power source and disassemble the series circuit and return materials to proper location. ❏

RESEARCH QUESTIONS

1. What does the term conductivity mean?

   _____

   _____

2 What is George Ohm known for?

   _____

   _____

3. What two factors will determine the amount of current flow in a circuit?

   _____

   _____

4. How many watts equal one horsepower?

   _____

   _____

5. What does a kilowatt-hour mean?

   _____

   _____

Passed Competency _____ **Failed Competency** _____

Instructor Signature _____ **Grade** _____

Practical Competency 13

Using a VOM or OHM Meter to Measure Resistance and Calculate Resistance in a Parallel Circuit

SUGGESTED MATERIALS

Textbook
Refrigeration & Air Conditioning Technology, 5th Edition, Thomson Delmar Learning
Unit 12—Basic Electricity and Magnetism

Review Topics
Making Electrical Measurements; Characteristics of a Parallel Circuit

COMPETENCY OBJECTIVE

Students will be able to use a VOM or OHM Meter to measure resistance and calculate resistance in a parallel according to Ohm's law. (Refer to **Figure 1–38.**)

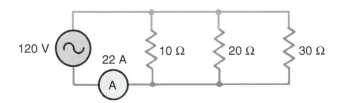

FIGURE 1–38   Total resistance calculations.

OVERVIEW

The ohmmeter is used to measure resistance in a circuit or of individual load devices. The ohmmeter provides its own power to measure resistance by a battery in the meter. For accurate resistance readings with an analog meter, the ohmmeter must first be zeroed in. This is done with the meter's ohms adjustment control. Connect the meter leads together and adjust the ohms adjustment knob until the meter indicates (0) on the meter scale. When the meter leads are separated, the meter will return to an infinity resistance.

> **NOTE:** *Never connect the ohmmeter to the circuit or load when there is power present because the meter could be damaged or destroyed.*

EQUIPMENT REQUIRED

1 VOM meter or Ohmmeter
3 Light bulbs
3 Light bulb receptacles
1 Jumper cord with alligator clips at each end
4 Jumper wires with alligator clips at each end

SAFETY PRACTICES

Use the proper power source, size of conductors (*wires*), and loads for this exercise. Make sure that all conductors are properly insulated. Make sure that you are familiar with the meter you are using and that all settings for the meter are set properly before taking voltage readings on circuit. Unless directed otherwise, do not plug circuit into the power source until circuit is completely wired and checked by your instructor.

NOTE: *Students should have completed and passed the Practical Competency 3, Wiring three loads (light bulbs) in a parallel circuit.*

COMPETENCY PROCEDURES

Checklist

MEASURING RESISTANCE OF EACH LOAD (LIGHT BULBS)
1. Using the equipment required for this competency put together a parallel circuit. ❑
2. Set VOM or Ohmmeter to proper function and value range for measuring resistance ❑

NOTE: *The meter leads get connected (parallel) to each load in the circuit.*

---

3. Make sure that meter leads are in the proper meter jacks. ❑
4. **DO NOT** plug jumper cord into power source. *All lights should be off.* ❑
5. Place meter leads across the terminals or legs of the first light bulb in the parallel circuit. ❑
6. *Record the resistance value reading at the first light bulb.* _____ ❑
7. Place meter leads across the terminals or legs of the second light bulb in the parallel circuit. ❑
8. *Record the resistance value at the second light bulb.* _____ ❑
9. Place meter leads across the terminals or legs of the third light bulb. ❑
10. *Record the resistance value at the third light bulb.* _____ ❑

NOTE: *Ohm's law on resistance in a parallel circuit states: The total circuit resistance will always be smaller than the smallest resistance in the circuit because the total circuit current is always greater than the current through any individual resistance in a parallel circuit and cannot be added to yield the total resistance. Ohm's law has two different formulas for finding the total resistance in a parallel circuit.*

DETERMINE THE TOTAL RESISTANCE.
11. Divide the resistance value of the first light bulb into (**1**) **one** and record the value. _____ ❑
12. Divide the resistance value of the second light bulb into (**1**) **one** and record the value. _____ ❑
13. Divide the resistance value of the third light bulb into (**1**) **one** and record the value. _____ ❑
14. Add the recorded values together:
    ***Bulb one** _____ **(plus) Bulb two** _____ **(plus) Bulb three** _____* ❑
15. *Record added values.* _____ ❑
16. Divide added value into (**1**) **one.** ❑
17. *Record total circuit resistance.* _____ ❑
18. Have your instructor review your circuit and voltage readings. ❑
19. Disassemble the series circuit and return materials to proper location. ❑

RESEARCH QUESTIONS

1. The _____ is that part of the atom that moves from one atom to another.
   - A. Electron
   - B. Proton
   - C. Neutron

2. If the resistance in a 120-volt circuit were 40 ohms, what would the current be in amperes?
   - A. 4800
   - B. 48
   - C. 4
   - D. 3

3. Electrical power is measured in:
   - A. Amperes divided by the resistance
   - B. Amperes divided by the voltage
   - C. Watts
   - D. Voltage divided by the amperes

4. The unit of measurement for the charge a capacitor can store is the:
   - A. Microfarad
   - B. Ohm
   - C. Joule
   - D. Inductive reactance

5. What is the formula for determining electrical power:
   - A. Resistance times amperage
   - B. Voltage divided by amperage
   - C. Voltage times amperage
   - D. Voltage divided by resistance

Passed Competency _____   **Failed Competency** _____

Instructor Signature _____   **Grade** _____

2

ELECTRICAL CONTROLS

Theory Lesson: Transformers

SUGGESTED MATERIALS

Textbook
Refrigeration & Air Conditioning Technology, 5th Edition, Thomson Delmar Learning
Unit 12—Basic Electricity & Magnetism

Review Topics
Inductance; Transformers; Magnetism

Key Terms
electricity • electromagnet • induced voltage • induction – turns of wire • isolation transformer • primary coil • secondary coil • step-down transformer • step-up transformer

OVERVIEW

To understand how a transformer works you need to understand three important facts:

1. It has long been understood that anytime electrical current passes through a conductor wire, a magnetic field is produced around the wire (**Figure 2–1**). If this wire is coiled and electrical current passes through it, the magnetic field is made even stronger, especially through the center of the coil (**Figure 2–2**). If this coil of wire is wrapped around an iron core and current is allowed to pass through it, the magnetic field through the center of the coil will magnetize the iron core (**Figure 2–3**).

   The apparatus just described is called an **electromagnet**. As long as current passes through the coil, the iron coil is magnetized. So coiling wire around an iron core and passing electrical current through it can create an electromagnet.

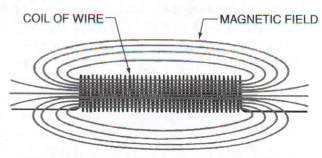

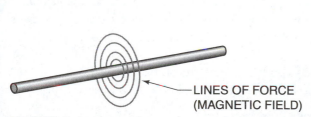

FIGURE 2–1   A straight conductor with a magnetic field.

FIGURE 2–2   There is a stronger magnetic field surrounding wire formed into a coil.

2. The strength of the electromagnet can be increased two ways: first, by *increasing the number of turns of wire in the coil* or second, *increasing the amount of supply voltage.*
3. It is also known that if a piece of conductor wire is passed through the center of a magnetic field, the conductor wire will cut the lines of force of the magnet, which causes electron flow in the wire (**Figure 2–4**). *Remember that electricity is the flow of electrons in a good conductor.*

Taking the conductor wire and cutting the lines of force of the magnet creates current flow. This type of current flow is called **induced voltage** because it is being induced in the conductor wire by the wire cutting the lines of force of the magnetic field.

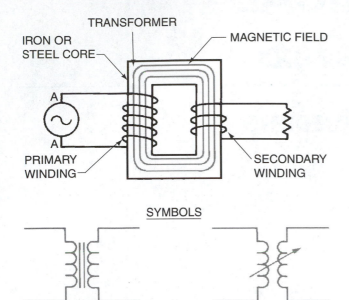

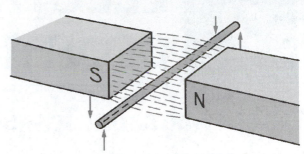

FIGURE 2–4 The movement of wire up and down cuts the lines of force, causing electric current flow in the wire.

FIGURE 2–3 Voltage applied across terminals produces a magnetic field around an iron or steel core.

These three principles are used to form a transformer. Sending electricity into a coil of wire wrapped around an iron core creates a magnetic field referred to as an electromagnet, and then this electromagnet is used to create another source of electricity by coiling another wire around the electromagnet on the other side; this is to induce electron flow in the coil.

The first coil of wire around the iron core creates the electromagnet when a power source is applied and is referred to as the **primary winding**. Wrapping another coil of wire around the other side of the iron core cuts the lines of force of the electromagnet, inducing current flow into the coil, thus producing a secondary voltage. This coil is referred to as the **secondary winding**.

Most of the transformers used in the air conditioning, heating, and refrigeration industry are known as **isolation transformers**. This means that the primary and secondary windings are magnetically coupled (*wrapped around the same iron core*), but electrically isolated from each other. The **primary winding** is the one that **is connected to the power source** and brings power to the transformer. The **secondary winding is used to supply power to the loads**. A transformer can be either a **step-up or a step-down transformer**. This is based on the number of turns of wire in the primary winding compared to the number of turns in the secondary winding of the transformer.

A **step-up transformer** has more turns of wire in the secondary than in the primary (**Figure 2–5**).

A **step-down transformer** has more turns of wire in the primary than in the secondary (**Figure 2–6**).

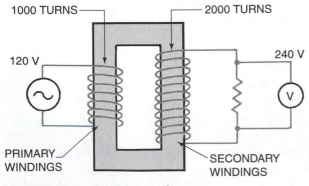

FIGURE 2–5 Step-up transformer.

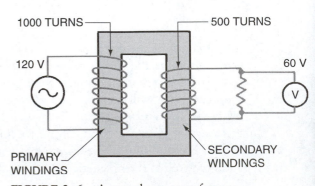

FIGURE 2–6 A step-down transformer.

Most air conditioning and refrigeration transformers are step-down, normally from 120 to 240 volts as the supply voltage to the primary, and 24 volts output on the secondary. These transformers are normally used as **control transformers** for control circuits. Transformers can be purchased with **fixed primary** and secondary voltages, or with **multitap primary** input voltages and fixed, or **multitap secondary voltages (Figure 2–7)**.

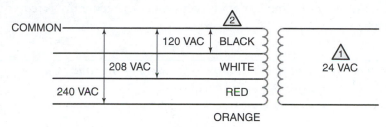

△1 SECONDARY CONNECTIONS ARE SCREW TERMINALS.

△2 BLACK IS COMMON WITH RESPECT TO THE TRANSFORMER WINDING ONLY AND NOT THE EXTERNAL CIRCUIT.

FIGURE 2–7   Multitap transformer schematic. (*Courtesy of Honeywell, Inc.*)

Multitap voltage transformers will have multiple colored leads for correct connection for proper input and output voltages. Usually the correct colored lead combinations for proper voltages are shown on the transformer or transformer instructions.

The power of a transformer or the work that can be done by it is rated in **volt-amperes (VA)**. The VA Rating refers to the transformer's secondary winding and is used to determine its output capability. If the control circuit's resistance wired to the transformers secondary becomes too low, current flow in the transformer's secondary will increase. If this current exceeds the transformer's low-voltage circuit current, the transformer will be damaged.

To find the maximum allowable current draw that can be drawn on the transformer's secondary, **divide the transformer's rated secondary output voltage into the transformer's rated VA:**

$$\textbf{Output in amperes} = \frac{\text{VA Rating}}{\text{Voltage}}$$

This rating will tell you the maximum amperage the transformer can be expected to carry effectively in a low-voltage circuit (**Figure 2–8**).

There are a couple of ways to determine the actual amperage being drawn on the secondary coil of the transformer in a low-voltage circuit. An in-line amperage reading will give you the exact secondary amperage draw.

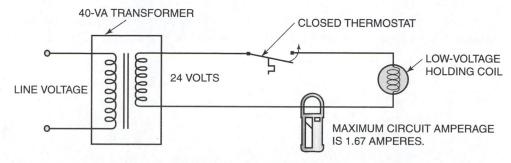

24 VOLTS X 1.67 AMPERES = 40 VA

FIGURE 2–8   Control circuit amperage multiplied by the voltage must not exceed the VA rating of the transformer.

Another way is to use a clamp-on ammeter and a coil of wire. Use a jumper wire and coil it 10 times around one of the jaws of the meter. Place the jumper leads in series with the low-voltage circuit. Measure your amperage reading and divide by 10.

This will equal the amount of current draw on the transformer's secondary while it is used in a low-voltage circuit. If the measured secondary amp draw is greater than the maximum allowable amp draw, a transformer with a higher rated VA would have to be installed to handle the controls and loads of the low-voltage circuit of the equipment. Air conditioning system transformers are generally rated at 40 VA, with heating system transformers rated at 20 VA.

> NOTE: *It is important to check the low voltage circuit current output when adding Air conditioning to an existing heating system. A higher VA transformer may be required.*

TESTING TRANSFORMERS

Ohmmeter Method

Normally, you can use an ohmmeter to check the coils of the transformer. Check for continuity (*resistance*) through each winding. If the windings are good, you should get a resistance value reading on each coil.

> NOTE: *The resistance values will not be the same because of the different number of turns of wire in the coils, which is dependent on whether the transformer is a step-up or a step-down. You could determine if a transformer is a step-up or step-down by the different resistant value readings, as long as you know which winding of the transformer is the primary.*

Remember that a step-up transformer has more turns of wire in the secondary than in the primary. This means that on a step-up transformer, you should have a higher resistance value reading on the secondary than on the primary and vice versa for a step-down transformer.

Voltmeter Method

Hooking an AC voltmeter to the primary and secondary winding leads can test the input and output voltage of the transformer. If the input voltage is correct and the output voltage is not close to the rated voltage, it is probably defective.

> NOTE: *If the test is being done without a load connected to the secondary, the secondary voltage may be a little higher than the rated voltage.*

Grounded Winding Test

Check with an ohmmeter to be sure there is no continuity between the windings and the case or iron core of the transformer. If you get a resistance reading on this test the transformer is grounded and should be replaced.

Practical Competency 14

Connect a Step-Down Transformer to a Two-Pole Relay Contactor

SUGGESTED MATERIALS

Textbook
Refrigeration & Air Conditioning Technology, 5th Edition, Thomson Delmar Learning
Unit 12—Basic Electricity and Magnetism
Unit 19—Motor Controls

Review Topics
Transformers; Inductance; The Contactor

COMPETENCY OBJECTIVE

The student will be able to wire a step-down transformer into a circuit to provide low-voltage supply to a relay contactor. (Refer to **Figure 2–9**).

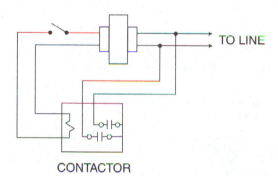

FIGURE 2–9 Wiring a step-down transformer to control a relay contactor.

OVERVIEW

A transformer has two coils. The input side where the power source is applied is called the primary coil. The output side of the transformer where the voltage is induced from the electromagnet field is called the secondary coil.

Most of the transformers used in the air conditioning, heating, and refrigeration industry are known as isolation transformers. This means that the primary and secondary windings are magnetically coupled (*wrapped around the same iron core*), but electrically isolated from each other. The primary winding is the one that is connected to the power source and brings power to the transformer. The secondary winding is used to supply power to the low-voltage loads.

A transformer can be either a step-up or a step-down transformer based on the number of turns of wire in the primary winding compared to the number of turns in the secondary winding.

A step-up transformer has more turns of wire in the secondary than in the primary. (*Refer to Figure 2–5, Theory Lesson: Transformers.*)

A step-down transformer has more turns of wire in the primary than in the secondary. (*Refer to Figure 2–6, Theory Lesson: Transformers.*)

Most air conditioning and refrigeration transformers are step-down, normally from 120 to 240 volts as the supply voltage to the primary and 24 volts output on the secondary. These transformers are normally used as control transformers for control circuits. Transformers can be purchased with fixed primary and secondary voltages, or with multitap primary input voltages and a fixed, or multitap secondary voltages. (*Refer to Figure 2–7, Theory Lesson: Transformers.*)

EQUIPMENT REQUIRED

120-Volt power supply
1 Step-down transformer (*120 volts to 24 volts*)
1 Relay contactor with a 24-volt coil
1 Single-pole, single-throw switch
1 Jumper cord with insulated alligator clip ends
3 Jumper wire with insulated alligator clips at ends

SAFETY PRACTICES

Use the proper power source, size of conductors (*wires*), and loads for this exercise. Make sure that all conductors are properly insulated. Unless directed otherwise, do not plug the circuit into the power source until the circuit is completely wired and checked by your instructor.

COMPETENCY PROCEDURES

Checklist

1. Take one leg or terminal of the jumper cord and attach it to one leg or terminal of the primary side of the transformer. ❑
2. Take the other leg or terminal of the jumper cord and attach it to the other leg or terminal of the primary side of the transformer. ❑
3. Take one end of a jumper wire and attach it to the leg or terminal of the secondary side of the transformer. ❑
4. Take the other end of the jumper wire and attach it to one side of the single-pole, single-throw switch. ❑
5. Use another jumper wire and attach one end of it to the other side of the single-pole, single throw switch. ❑
6. Take the other end of the switch jumper wire and attach it to one of the low-voltage terminals on the contactor relay coil. ❑
7. Take another jumper wire and attach one end of it to the other leg or terminal of the secondary winding of the transformer. ❑
8. Take the other end of this jumper wire and attach it to the other low-voltage terminal of the contactor relay coil. ❑
9. With your instructor's permission, make sure the switch is in the OFF position and plug the jumper cord into the proper power source. ❑
10. Turn the switch to the ON position. You should hear and see the contactor relay coil energize and close the main contacts on the contactor. ❑
11. Turn the switch OFF and ON a few times and you will see the 24 volts from the transformer controlling the relay coil. ❑
12. Turn the switch to the OFF position. ❑
13. Unplug the power cord and remove all jumper wires from the transformer, switch, and relay contactor. ❑

NOTE: *In this circuit, the step-down transformer is taking a supply voltage of 120 volts and stepping it down to 24 volts on the secondary side of the transformer, and supplying the 24 volts to the contactor relay coil once the switch is turned ON. The coil being energized by the 24 volts creates a magnetic field around the armature of the contactor, causing it to be drawn into the center of the coil. The armature stem has contact plates attached to it and therefore closes the circuit to the stationary contacts on the relay.*

14. Practice this circuit until you can do it without instructions or observing the competency drawing. ❏
15. Have your instructor check your practical skills with this competency once you feel confident that you can perform it without any type of assistance. ❏
16. On completion of the competency, return all materials to their proper location. ❏

RESEARCH QUESTIONS

1. What is a lodestone?

_____

_____

2. What are the two poles of a magnet?

_____

_____

3. What is magnetic induction?

_____

_____

4. What is magnetic flux?

_____

_____

5. Like poles of magnets do what?

_____

_____

Passed Competency _____   Failed Competency _____

Instructor Signature _____   Grade _____

Practical Competency 15

Evaluating the Condition of a Transformer with an Ohmmeter

SUGGESTED MATERIALS

Textbook
Refrigeration & Air Conditioning Technology, 5th Edition, Thomson Delmar Learning
Unit 12—Basic Electricity and Magnetism

Review Topics
Transformers; Making Electrical Measurements

COMPETENCY OBJECTIVE

The student will be able to use an ohmmeter and evaluate the condition of a transformer.

OVERVIEW

Normally, you can use an ohmmeter to check the coils of the transformer. Check for continuity (*resistance*) through each winding. If the windings are good, you should get a resistance value reading on each one.

> *NOTE: The resistance values will not be the same due to the different number of turns of wire in each coil. This is also dependent on whether the transformer is a step-up or a step-down. You could determine if a transformer is a step-up or step-down by the different resistant value readings between the two coils, as long as you know which winding of the transformer is the primary.*

Remember that a step-up transformer has more turns of wire in the secondary than the primary. This means that on a step-up transformer, you should have a higher resistance value reading on the secondary than on the primary and vice versa for a step-down transformer.

EQUIPMENT REQUIRED

1 Step-down transformer or step-up transformer
1 Ohmmeter

SAFETY PRACTICES

Be sure that all electrical power has been turned off or that the transformer has been removed from the power source.

COMPETENCY PROCEDURES   **Checklist**

1. Set the ohmmeter on the ohms scale or 'Beeper' scale.   ❑
2. Attach the meter leads to the two legs or terminals of the primary side of the transformer.   ❑

> **NOTE:** *You should get an ohms value reading if the primary coil is good.* NO *ohms value reading indicates an open coil on the primary. This also means that the secondary coil will not work, so the transformer has to be replaced.*

3. *Record your ohms value reading from the primary side of the transformer. Ohms reading* ❑
   _____
4. Attach the meter leads to the legs or terminals of the secondary coil of the transformer.
   A.   You should get an ohms value reading if the secondary coil is good. ❑
   B.   NO ohms value reading indicates an open coil on the secondary winding. There would be no voltage on the secondary even if the primary side of the transformer were OK. (*The transformer would have to be replaced.*) ❑
5. *Record your ohms value reading from the secondary side of the transformer ohms. Record ohms value from transformer's secondary.* _____ ❑
6. *From the ohms readings you recorded, were both coils of the transformer good?* _____ ❑
7. *From the ohms readings you recorded, is this transformer a step-up or step-down transformer?* _____ ❑
8. Have your instructor check your work. ❑
9. Return components to their proper location. ❑

REVIEW QUESTIONS

1. When current flows through a conductor, what is produced around the conductor?
   _____
   _____

2. What determines the strength of an electromagnet?
   _____
   _____

3. What determines the magnetic polarity of an electromagnet?
   _____
   _____

4. What happens to the magnetic field of a coil of wire if an iron core is placed in the middle of the coil?
   _____
   _____

5. What is a solenoid?
   _____
   _____

Practical Competency 16

Evaluating a Transformer's Primary and Secondary Voltages

SUGGESTED MATERIALS

Textbook
Refrigeration & Air Conditioning Technology, 5th Edition, Thomson Delmar Learning
Unit 12—Basic Electricity and Magnetism

Review Topics
Transformers; Making Electrical Measurements

COMPETENCY OBJECTIVE

The student will be able to measure a transformer's primary and secondary voltages

OVERVIEW

Hooking an AC voltmeter to the primary and secondary winding leads can test the input and output voltage of the transformer. If the input voltage is correct and the output voltage is not close to the rated voltage, it is probably defective.

> NOTE: *If the test is being done without a load connected to the secondary, the secondary voltage may be a little higher than the rated voltage.*

EQUIPMENT REQUIRED

Voltmeter Method
120-Volt power supply
1 VOM meter
1 Step-down transformer (*120 volts to 24 volts*)
1 Relay contactor with a 24-volt coil
1 Single-pole, single-throw switch
1 Jumper cord with insulated alligator clip ends
3 Jumper wire with insulated alligator clips at ends

SAFETY PRACTICES

Use the proper power source, size of conductors (*wires*), and loads for this exercise. Make sure that all conductors are properly insulated. Unless directed otherwise, do not plug circuit into the power source until the circuit is completely wired and checked by your instructor.

> NOTE: *If you are checking the voltage of a step-up transformer, be sure that the meter you are using has a voltage range that is high enough to measure the transformer's output voltage.*

COMPETENCY PROCEDURES

1. Build the circuit as explained in Practical Competency 14 (*Connect a Step-Down Transformer to a Two-Pole Relay Contactor*). ❑

> **NOTE:** *If Practical Competency 14 has not been completed, proceed to that competency before completing this competency.*

---

2. Turn the switch to the OFF position. ❑
3. Make sure that your voltmeter is set on the proper function and voltage range. ❑
4. With your instructor's permission, plug the jumper cord into the proper power source. ❑
5. Touch the voltmeter leads to the primary leads of the transformer. ❑
6. *What were the transformers rated primary voltage?* _____ *(This information should be marked on the transformer.)* ❑
7. *Record the primary input voltage.* _____ ❑
8. *Touch the meter leads to the low-voltage terminals of the relay contactor.* ❑
9. *Record the voltage recorded at the low-voltage terminals of the relay contactor.* _____ ❑

> **NOTE:** *In this test "Zero Voltage" should have been observed because the switch is in the OFF position, which means that the primary winding of the transformer is not being supplied with input voltage. Without voltage to the primary, no voltage is being induced into the transformer's secondary coil, providing output voltage.*
>
> *Another point to be aware of is the primary voltage, which was recorded in Step 7. Notice that although there was voltage present at the transformer's primary leads, the primary coil was not being energized because the switch was in the OFF position. This same situation could occur if the primary coil of the transformer was defective. Point is, a recorded voltage at the transformer's primary leads does not mean that the primary coil of the transformer is good.*

---

10. *Record the transformer's rated secondary voltage.* _____ *(This information should be marked on the transformer.)* ❑
11. Turn the circuit switch to the ON position. ❑
12. Place meter's leads at the low-voltage terminals of the relay contactor coil. ❑
13. *Record the measured voltage of the secondary winding terminal leads.* _____ ❑
14. *Was the output voltage close to the rated voltage?* _____ ❑

> **NOTE:** *If the output voltage is not close to the rated voltage, it is probably defective. If the test is being done without a load connected to the secondary, the secondary voltage may be a little higher than the rated voltage.*

---

15. *Would you say this transformer is good or bad according to your voltage readings?* _____ ❑
16. Turn the circuit switch OFF. ❑
17. Have your instructor check your procedure and recorded voltage. ❑
18. Unplug the jumper cord and remove all wires and components and return them to their proper location. ❑

RESEARCH QUESTIONS

1. What do NO Contacts and NC Contacts represent?

   _____

   _____

2. What does VA stand for?

   _____

   _____

3. A transformer that has 500 turns of wire in the primary and 1000 turns of wire in the secondary would be what type of transformer?

   _____

   _____

4. What would be the voltage output of a transformer with 1000 turns of wire in the primary and a 120-volt input with a secondary winding with 500 turns of wire?

   _____

   _____

5. What is a variable transformer?

   _____

   _____

Passed Competency _____   Failed Competency _____

Instructor Signature _____   Grade _____

Student Name _____ Grade _____ Date _____

Practical Competency 17

Measuring the Volt-Amperes of a Control Transformer

SUGGESTED MATERIALS

Textbook
Refrigeration & Air Conditioning Technology, 5th Edition, Thomson Delmar Learning
Unit 12—Basic Electricity and Magnetism

Review Topics
Induction; Transformer; Volt-Amperes

COMPETENCY OBJECTIVE

Student will be able to use a clamp-on ammeter to determine the volt-amperes of a control transformer.

OVERVIEW

The power of a transformer or the work that can be done by it is rated in **volt-amperes (VA)**. The VA Rating refers to the transformer's secondary winding and is used to determine its output capability. If the control circuit's resistance wired to the transformer's secondary becomes too low, current flow in the transformer's secondary will increase. If this current exceeds the transformer's low-voltage circuit current, the transformer will be damaged.

To find the maximum allowable current draw that can be drawn on the transformer's secondary, **divide the transformer's rated secondary output voltage into the transformer's rated VA:**

$$\text{Output in amperes} = \frac{\text{VA Rating}}{\text{Voltage}}$$

This rating will tell you the maximum amperage the transformer can be expected to carry effectively in a low-voltage circuit.

There are a couple of ways to determine the actual amperage being drawn on the secondary coil of the transformer in a low-voltage circuit. An in-line amperage reading will give you the exact secondary amperage draw.

Another way is to use a clamp-on ammeter and a coil of wire. Use a jumper wire and coil it 10 times around one of the jaws of the meter. Place the jumper leads in series with the low-voltage circuit. Measure your amperage reading and divide by 10.

This will equal the amount of current draw on the transformer's secondary while it is used in a low-voltage circuit. If the measured secondary amp draw is greater than the maximum allowable amp draw, a transformer with a higher rated VA would have to be installed to handle the controls and loads of the low-voltage circuit of the equipment. Air conditioning system transformers are generally rated at 40 VA, with heating system transformers rated at 20 VA.

> **NOTE:** *It is important to check the low voltage circuit current output when adding air conditioning to an existing heating system. A higher VA transformer may be required.*

EQUIPMENT REQUIRED

1 Operational A/C unit or heat pump
1 Clamp-on ammeter
18" of light gage wire

NOTE: *This competency could also be performed on any other low-voltage circuit.*

SAFETY PRACTICES

The student should follow all electrical safety rules when working with live electricity. Make sure all electrical meters are set to the proper function and value ranges.

COMPETENCY PROCEDURES   **Checklist**

NOTE: *This test could be performed at the control thermostat of the unit or at the transformer's secondary lead terminals.*

---

1. **Turn OFF power** to the unit.   ❏
2. Gain access to the thermostat terminals.   ❏
3. Set the ammeter to at least a 10-amp scale.   ❏
4. Take the length of wire and coil it around one of the jaws of the clamp-on ammeter.   ❏
5. *Record the transformer's volt-ampere (VA) rating.* _____   ❏

PROCEDURES FOR CHECKING VOLT-AMPERE AT THERMOSTAT

6. Attach one end of the coil of wire from around the ammeter jaw to the (R) terminal of the thermostat.   ❏
7. Attach the opposite end of the coil of wire around the ammeter jaw to the (Y) terminal of the thermostat.   ❏
8. **Turn power ON** to the unit.   ❏
9. Set the thermostat to air conditioning mode (*Cooling*).   ❏
10. Make sure that the unit has gone into the cooling mode before measuring low-voltage circuit amperage.   ❏

NOTE: *Some thermostats have a built-in "Delay" to allow the compressor pressures to equalize.*

---

11. *Record your ammeter reading.* _____   ❏
12. *Divide your ammeter reading by 10.*   ❏
13. *Record your actual amperage value.* _____   ❏

NOTE: *This is the actual amp draw on the secondary low-voltage control circuit.*

DETERMINING THE MAXIMUM AMPERAGE THE TRANSFORMER CAN BE EXPECTED TO CARRY EFFECTIVELY IN A LOW-VOLTAGE CIRCUIT

14. *Record the VA rating of the transformer being used for this competency.* _____   ❏
15. *Record the transformer's rated secondary voltage.* _____   ❏

16. **Divide** the transformer's **VA rating** _____ by the transformer's rated **secondary voltage.** _____   ❑
17. Record the transformer's **maximum amperage.** _____   ❑

NOTE: *This amperage value is the maximum amperage that can be pulled off of the transformer's secondary.*

---

18. Was the recorded *actual amperage* value in Step 13 *above or below* the transformer's *actual amperage value?* _____   ❑
19. Have your instructor check your work.   ❑
20. Disconnect equipment and return all tools and test equipment to their proper location.   ❑

RESEARCH QUESTIONS

1. What effect does an amp draw on a transformer over the rated VA have on a low-voltage circuit of a unit?

   _____

   _____

2. The coil of a relay contactor is good and you are getting a 24-volt reading at the relay coil terminals, yet the relay coil cannot pull the armature closed. *What is the possible problem?*

   _____

   _____

3. Could higher than maximum amp draw on the secondary winding of a transformer cause the secondary winding to burn out?

   _____

   _____

4. To which terminal on a thermostat does the hot leg of the secondary winding of a transformer get connected?

   _____

   _____

5. Which terminal on the thermostat must be energized to start the unit heating cycle?

   _____

   _____

Passed Competency _____   **Failed Competency** _____

Instructor Signature _____   **Grade** _____

Theory Lesson: Relay Contactor and General-Purpose Relay

SUGGESTED MATERIALS

Textbook

Refrigeration & Air Conditioning Technology, 5th Edition, Thomson Delmar Learning
Unit 12—Basic Electricity and Magnetism
Unit 19—Motor Controls

Review Topics

The Relay; The Contactor

Key Terms

general-purpose relay • normally closed switch (NC) • normally open switch (NO) • relay contactor—solenoid

OVERVIEW

The term *relay* is often used to describe any type of magnetically operated switch. There are many types of relays used in the HVAC industry. They each perform a special function, but all mostly work the same way. It is important to understand the principle of operation of all relays. Most relays work on the solenoid principle. A **solenoid** is an electrical device that converts electrical energy into mechanical motion.

All relays have some type of a coil of wire, except for solid-state relays. This coil of wire is wound around an iron core referred to as the armature or plunger (**Figure 2–10**).

When current flows through the coil, a magnetic field is developed in the center of the coil (**Figure 2–11**).

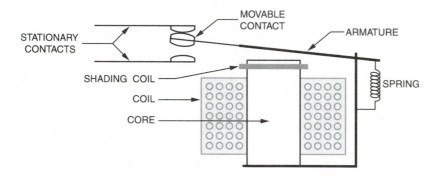

FIGURE 2–10   Simple relay in the de-energized position.

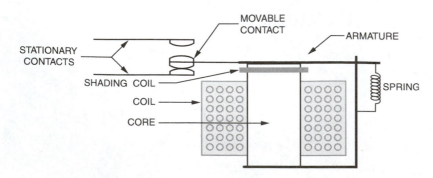

FIGURE 2–11   The coil of this relay is energized and the contacts have changed position.

This magnetic field pulls the armature or plunger of the relay into the center of the coil. Attached to the armature or plunger are moveable contacts that make contact with stationary contacts attached to the relay body. A relay may have multiple sets of contacts attached to the armature or plunger. Contacts can be *normally closed* (**NC**) or *normally open* (**NO**). (Refer to **Figure 2–12** and **Figure 2–13**.)

FIGURE 2–12   This electrical symbol represents a set of **normally open contacts.**

FIGURE 2–13   This electrical symbol represents a set of **normally closed contacts.**

This means that a set of normally open contacts (**NO**) will not allow current to flow through the set of contacts until they are closed. When they are closed, current will flow through the contacts to control or operate any circuit or load(s) that are wired through them. A set of normally closed (**NC**) contacts will automatically allow current flow through the contacts to the circuit or loads that are wired through them when power is applied to the circuit.

When power is applied to the relay's low-voltage coil, the armature or plunger of the relay becomes magnetized, and any contact attached to the relay will change position and stay that way as long as the relay's coil is energized. In other words, a set of *normally open* (NO) *contacts will close* sending current through them to control any circuit or loads that are wired to them. A set of *normally closed* (NC) *contacts will open*, stopping current flow to any circuit or loads that are wired to them.

A relay may contain one set of contacts or multiple sets of contacts that could be normally open (NO) or normally closed (NC) or have both. Keep in mind that when the relay coil is energized, all the contacts on the relay change position and stay in that position as long as the coil is energized.

Figure 2–14 shows **general-purpose relays** that contain contacts that are enclosed inside of the relay housing.

These are contacts that you cannot see, so on the housing of the relay the manufacturers have identified the contacts by numbers and by electrical symbols for the type of contacts used in the relay. The manufacturer will also identify which set of contact terminals is normally open and which set of contact terminals are normally closed. (Refer to **Figure 2–15**.)

The following electrical symbol represents the relay's low-voltage coil (**Figure 2–16**).

FIGURE 2–14   General-purpose relay. (*Photo by Bill Johnson*)

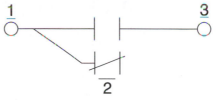

FIGURE 2–15   Contact switch terminals 1 and 3 are NO; switch terminals 1 and 2 are NC.

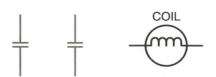

FIGURE 2–16   Electrical symbols for relay coils and contacts.

A **relay contactor** is a type of relay, except that the contactor contains large-load contacts designed to control large amounts of current (**Figure 2–17**).

FIGURE 2–17   Two-pole contactors.

There are different types of contactors, although all contactors have common features. These features are: the contacts, both moveable and stationary; the springs that hold the contacts; and the holding coil.

Some contactors can be rebuilt; however, most used in residential and light commercial equipment require replacement if found to be defective.

The contacts make the electrical circuit when the energized relay coil closes them. The coil of the contactor normally is energized by a low-voltage power supply.

In general, a 24-volt step-down transformer is used for contactor coil operation in residential and light commercial HVAC equipment. The load-carrying contacts can be NO, NC, or both; in addition there can be multiple sets of contacts on the contactor. Most contactors are used for *load-carrying devices* that operate with a current draw of 15 amps or higher.

Practical Competency 18

Wiring a Relay Contactor into a Circuit

SUGGESTED MATERIALS

Textbook
Refrigeration & Air Conditioning Technology, 5th Edition, Thomson Delmar Learning
Unit 12—Basic Electricity and Magnetism
Unit 19—Motor Controls

Review Topics
The Relay; The Contactor

COMPETENCY OBJECTIVE

The student will be able to make an electrical circuit utilizing a relay contactor (**Figure 2–18**).

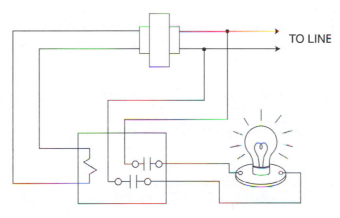

FIGURE 2–18   Wiring a relay contactor into a
circuit.

OVERVIEW

A relay contactor is a type of relay, except that the contactor contains large-load contacts designed to control large amounts of current. There are different types of contactors, although all contactors have common features. These features are: the contacts, both moveable and stationary; the springs that hold the contacts; and the holding coil.

Some contactors can be rebuilt; however, most used in residential and light commercial equipment require replacement if found to be defective.

The contacts make the electrical circuit when the energized relay coil closes them. The coil of the contactor normally is energized by a low-voltage power supply. In general a 24-volt step-down transformer is used for contactor coil operation in residential and light commercial HVAC equipment. The load-carrying contacts can be NO, NC, or both; in addition there can be multiple sets of contacts on the contactor. Most contactors are used for *load-carrying devices* that operate with a current draw of 15 amps or higher.

EQUIPMENT REQUIRED

125-Volt proper power supply
1 Single-pole, single-throw switch
1 Step-down transformer (*24 volts secondary*)
1 Relay contactor
1 Load device (*light bulb*)
1 Light bulb receptacle
1 Jumper cord with insulated alligator clip ends
7 Jumper wires with insulated alligator clips

SAFETY PRACTICES

Use the proper power source, size of conductors (*wires*), and loads for this exercise. Make sure that all conductors are properly insulated. Unless directed otherwise, do not plug circuit into power source until the circuit is completely wired and checked by your instructor.

COMPETENCY PROCEDURES   Checklist

1. Put the light bulb into the light bulb receptacle.   ❏
2. Attach the jumper cord ends to the transformer's primary coil terminals or wire leads.   ❏
3. Attach one end of a jumper wire to one leg or terminal of the secondary coil of the
    step-down transformer.   ❏
4. Attach the opposite end of this jumper wire to one of the terminal screws or leads of the
    single-pole, single-throw switch.   ❏
5. Attach another jumper wire to the opposite terminal screw or lead wire of the single-pole,
    single-throw switch.   ❏
6. Attach the opposite end of the jumper wire to one of the low-voltage terminals of the relay
    contactor coil.   ❏
7. Attach another jumper wire to the opposite leg or terminal of the secondary coil of the
    step-down transformer.   ❏
8. Attach the opposite end of this jumper wire to the other low-voltage terminal of the relay
    contactor coil.   ❏
9. Take another jumper wire and attach one end of it to one leg of the **main jumper cord wire**
    attached at the primary coil side of the transformer.   ❏
10. Attach the opposite end of the jumper wire to one of the line terminals of the relay contactor.   ❏

> **NOTE:** *Normally the line side terminals of the contactors are marked L-1 and L-2; these are the terminals for the legs of the line voltage. If the contactor has a single set of contacts, make sure to make this connection at the contactor terminal where the contacts will break and close the line voltage circuit.*

11. Take another jumper wire and attach one end of it to the other side of the jumper cord wire
    attached to the opposite leg of the primary coil of the transformer.   ❏
12. Attach the opposite end of this jumper wire to the other line terminal of the
    relay contactor (*either L-1 or L-2*).   ❏

> **NOTE:** *The other sides of the contactor contact terminals are normally marked T-1 and T-2.*

13. Attach another jumper wire to the opposite side of the relay load contact terminals
    (T-1 and T-2).   ❏

14. Attach the opposite end of this jumper wire to one leg or terminal of the light bulb receptacle. ❏
15. Attach another jumper wire to the other load terminal of the relay contactor (T-1 or T-2). ❏
16. Attach the opposite end of the jumper wire to the other leg or terminal of the light bulb receptacle. ❏
17. Have your instructor check your circuit. ❏
18. Turn the switch to the OFF position ❏
19. With the instructor's approval, plug the power cord into the proper power supply ❏
20. Turn the switch to the ON position ❏

> NOTE: *You should hear the relay contactor coil snap the load contacts closed and see the light bulb come on.*

21. Remove one of the jumper wire leads at the contractor's low-voltage coil terminals. ❏

> NOTE: *You should see the spring of the load contacts push the contacts open and shut the light bulb off.*

22. Remove the light bulb from the light bulb receptacle. ❏
23. Reattach the low-voltage wire to the low-voltage coil terminal of the relay contactor. ❏

> NOTE: *You should see and hear the contactor coil close the load contacts on the contactor. This is to prove that the step-down transformer's 24 volt controls the contactor coil.*

24. Turn the switch to the OFF position and have your instructor approve your procedures and circuit. ❏
25. Unplug the power cord from the power supply and remove all wires and jumpers, transformer, contactor, light bulb, and receptacle. ❏
26. Replace supplies to their proper location. ❏

RESEARCH QUESTIONS

1. What is a relay?

_____

_____

2. What principle do most relays operate on?

_____

_____

3. A solenoid is a device that does what?

_____

_____

4. The iron core of a relay is called what?

_____

_____

5. NO stands for what type of electrical contacts?

_____

_____

Passed Competency _____   Failed Competency _____

Instructor Signature _____   Grade _____

Practical Competency 19

Checking a Relay Contactor with a Voltmeter

SUGGESTED MATERIALS

Textbook

Refrigeration & Air Conditioning Technology, 5th Edition, Thomson Delmar Learning
Unit 12—*Basic Electricity and Magnetism*
Unit 15—*Troubleshooting Basic Controls*
Unit 19—*Motor Controls*

Review Topics

The Relay; The Contactor; Making Electrical Measurements

COMPETENCY OBJECTIVE

The student will be able to perform the proper procedure of using a voltmeter to check the functions of a relay contactor in a circuit.

> *NOTE: Student should know how to build a circuit with a step-down transformer and a relay contactor. (Refer to Practical Competency 18.)*

OVERVIEW

A relay contactor is a type of relay, except that the contactor contains large-load contacts designed to control large amounts of current. There are different types of contactors, although all contactors have common features. These features are: the contacts, both moveable and stationary; the springs that hold the contacts; and the holding coil.

Some contactors can be rebuilt; however, most used in residential and light commercial equipment require replacement if found to be defective.

The contacts make the electrical circuit when the energized relay coil closes them. The coil of the contactor normally is energized by a low-voltage power supply. In general a 24-volt step-down transformer is used for contactor coil operation in residential and light commercial HVAC equipment. The load-carrying contacts can be NO, NC, or both; in addition there can be multiple sets of contacts on the contactor. Most contactors are used for *load-carrying devices* that operate with a current draw of 15 amps or higher.

EQUIPMENT REQUIRED

125-Volt proper power supply
1 Single-pole, single-throw switch
1 Voltmeter
1 Step-down transformer (*24 volts secondary*)
1 Relay contactor
1 Load device (*light bulb or other load-carrying device*)
1 Jumper cord with insulated alligator clip ends
7 Jumper wires with insulated alligator clips

SAFETY PRACTICES

Use the proper power source, size of conductors (*wires*), and loads for this exercise. Make sure that all conductors are properly insulated. Unless directed otherwise, do not plug circuit into power source until the circuit is completely wired and checked by your instructor.

COMPETENCY PROCEDURES

Checklist

1. Put together an electrical circuit using a relay contactor or use electrical circuit as described in *Practical Competency 18*. ❏
2. Have your instructor check your circuit before plugging the jumper cord into the power supply. ❏
3. Plug your jumper cord into the proper power supply and make sure the circuit is operating. ❏
4. Set your voltmeter on the proper function and value ranges in order to check voltage readings on the relay contactor. ❏

> **NOTE:** *If you are not sure of your voltage value range, always start at the highest value.*

5. Take your meter leads and touch one of them to one of the low-voltage terminals of the relay contactor coil, and the other meter lead to the other relay contactor coil terminal. You should get a low-voltage reading. *Record your voltage reading.* _____ ❏

> **NOTE:** *What this reading tells you is that the secondary side of the transformer is supplying the voltage for the relay contactor coil. This reading does not tell you that the coil is necessarily using the voltage to magnetize the armature to close or open contacts on the relay contactor.*

6. Remove one of the low-voltage coil terminal wires. *Did the contacts open up or close when the coil was de-energized?* _____ (*The coil should de-energize causing all contacts on the relay contactor to return to their normal position of NO or NC.*) ❏

> **NOTE:** *This shows that the relay coil is using low voltage from the secondary of the transformer to create magnetism to pull the contactor armature in and close the main contacts.*

7. Leave one of the low-voltage wires disconnected from the low-voltage contactor coil and place the meter leads to the main power contactor line terminals (L-1 and L-2). (You should have gotten a supply voltage reading.) *Record your voltage reading.* _____ ❏

> **NOTE:** *To obtain a voltage reading where main power is wired through one set of the contactor contacts, place one meter lead to the power lead at the contactor terminal and the other meter lead at the opposite line power lead.*

> **NOTE:** *This test tells you that there is line voltage supplied to the relay contactor line contact terminals. It does not tell you that the load contacts of the relay are either opened or closed.*

8. With a low-voltage wire disconnected from the contactor coil, touch your meter leads to the relay contractor's load terminals (**terminal T-1 or T-2**). *Record your voltage reading.*

_____ ❏

NOTE: *If the contacts on the relay are NO contacts, you should have gotten an (infinity) voltage reading in this test, telling you that the load contacts are open and no current can pass through them to operate any load devices. If the contacts on the relay are NC contacts, you should have gotten a line voltage reading in this test, telling you that the load contacts are closed and current can pass to the load devices.*

9. *Were the contacts on the relay you tested NO or NC?* _____ ❑
10. Reattach the wire to the low-voltage coil of the contactor and make sure that the switch is in the ON position. ❑
11. Again, touch your meter leads to the relay load contacts (**terminals T-1** or **T-2**). *Record your voltage reading.* _____ ❑

NOTE: *If the contacts on the relay were NO contacts, they should have closed and you should have gotten a line voltage reading. If the contacts on the relay were NC contacts, they should have opened and you should have gotten a (0) voltage reading.*

12. Have your instructor check your circuit and readings. ❑
13. Turn circuit switch to the OFF position and unplug the jumper cord from the power supply. ❑
14. Remove all wire and components of the circuit and put them in their proper location. ❑

RESEARCH QUESTIONS

1. A NO set of contacts means that current (*will or will not*) flow to the circuit or load device?

   _____

   _____

2. A NC set of contacts means that current (*will or will not*) flow to the circuit or load device?

   _____

   _____

3. What voltage operates the coil of contactors normally used in residential and light commercial HVAC equipment?

   _____

   _____

4. Contactors are used in circuits that have load devices that have an operating amperage of _____ or higher.

5. An ohms value reading on contactor coils indicates that the coil is *Good or Defective*.

   _____

Passed Competency _____   Failed Competency _____

Instructor Signature _____   Grade _____

Practical Competency 20

Checking a Relay Contactor with an Ohmmeter

SUGGESTED MATERIALS

Textbook

Refrigeration & Air Conditioning Technology, 5th Edition, Thomson Delmar Learning
Unit 12—Basic Electricity and Magnetism
Unit 15—Troubleshooting Basic Controls
Unit 19—Motor Controls

Review Topics

Relays; Relay Contactors – Making Electrical Measurements

COMPETENCY OBJECTIVE

The student will be able to check the function of a relay contactor by using an ohmmeter.

OVERVIEW

A relay contactor is a type of relay, except that the contactor contains large-load contacts designed to control large amounts of current. There are different types of contactors, although all contactors have common features. These features are: the contacts, both moveable and stationary; the springs that hold the contacts; and the holding coil.

Some contactors can be rebuilt; however, most used in residential and light commercial equipment require replacement if found to be defective.

The contacts make the electrical circuit when the energized relay coil closes them. The coil of the contactor normally is energized by a low-voltage power supply. In general a 24-volt step-down transformer is used for contactor coil operation in residential and light commercial HVAC equipment. The load-carrying contacts can be NO, NC, or both; in addition there can be multiple sets of contacts on the contactor. Most contactors are used for *load-carrying devices* that operate with a current draw of 15 amps or higher.

Ohmmeter Test

The ohmmeter is used to measure resistance in a circuit or individual load device. The ohmmeter provides its own power to measure resistance by a battery in the meter. For accurate resistance readings, the analog ohmmeter must first be zeroed in. This is done with the meter's ohms adjustment control. Connect the meter leads together and adjust the ohms adjustment knob until the meter indicates (0) on the meter scale. When the meter leads are separated, the meter will return to an infinity resistance.

> **NOTE:** *Never connect the ohmmeter to the circuit or load when there is power present: You will damage or destroy the meter.*

EQUIPMENT REQUIRED

1 Ohmmeter
1 Relay contactor

SAFETY PRACTICES

Make sure that voltage has been disconnected from the contactor.

COMPETENCY PROCEDURES

Checklist

1. Set your meter to the ohms scale. ☐
2. Touch your meter leads to the relay contactor low-voltage coil terminals. ☐
3. *Record your ohms value reading.* _____ ☐

> **NOTE:** *You should have gotten an ohms value with this test. This means the relay coil is good and if current were passed through the coil, it would magnetize the armature and open or close the load contacts of the relay.*
>
> *If you did not get an ohms value reading on this test, it means that the relay contactor coil is open and no current can pass through the coil, which means that the contactor will not work.*

4. Determine how many sets of contacts are attached to the armature of the relay contactor you are checking. ☐
5. *How many sets of contacts are on the relay contactor you are checking?* _____ ☐
6. With the relay contactor in its normal position, touch your meter leads across a set of contacts. *Record your ohms value reading.* _____ ☐
   Set 1 ohms value _____
   Set 2 ohms value _____
   Set 3 ohms value _____

> **NOTE:** *If there is more than one set of contacts on the relay, check all of them and record the ohms value for each set of contacts on the contactor.*

> **NOTE:** *A (0) ohms value reading on any set of contacts of the contactor means that they are closed and good. A high ohms value or infinity value reading on a set of contacts of the contactor means that they are open.*

> **NOTE:** *When checking across a NO set of contacts, you should get a high ohms or infinity value reading.*
>
> *When checking across a good NC set of contacts, you should get a (0) ohms value reading. If the resistance is greater than 2 or 3 ohms, the contacts or contactor should be replaced.*

7. Manually push in the relay contactor armature switch and take an ohms reading on each set of contacts again. *Record your ohms reading for all sets of contacts on the contactor.* ☐
   Set 1 ohms value _____
   Set 2 ohms value _____
   Set 3 ohms value _____
8. *On the contactor you are checking, indicate if the sets of contacts are NO or NC contacts.* ☐
   Set 1 _____
   Set 2 _____
   Set 3 _____
9. *Are the contacts on the relay functioning properly?* _____ ☐

10. *Is the relay contactor coil good or bad?* _____ ❑
11. Have your instructor check your readings. ❑
12. Return meter and contactor to their proper locations. ❑

RESEARCH QUESTIONS

1. The part of the contactor that gets magnetized by the relay coil is called what?

   _____

   _____

2. Does a voltage reading at the relay coil terminals tell you that the coil is good or not?

   _____

   _____

3. Would a secondary voltage reading indicate that the transformer is good?

   _____

   _____

4. Draw the electrical symbol for a set of NO relay contacts.

   _____

   _____

5. Draw the electrical symbol for a set of NC relay contacts.

   _____

   _____

Passed Competency _____   Failed Competency _____

Instructor Signature _____   Grade _____

Theory Lesson: General-Purpose Relays

SUGGESTED MATERIALS

Textbook
Refrigeration & Air Conditioning Technology, 5th Edition, Thomson Delmar Learning
Unit 12—Basic Electricity and Magnetism
Unit 19—Motor Controls

Review Topics
The Relay

Key Terms
light duty relays • main load terminals • normally closed contacts • normally open contacts • relay coil

OVERVIEW

General-purpose relays are considered to be **light-duty relays.**

The contacts of the relay are capable of handling an amperage load up to 10 amps at 120 volts and are normally enclosed by a molded case with the terminal and contact identification printed on top of the relay. The relay coil could be designed to operate at line voltage or low voltage.

These relays normally can contain multiple sets of switch contacts that can be NO, NC, or both (**Figure 2–19**).

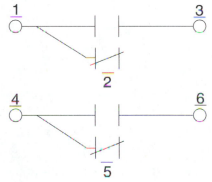

FIGURE 2–19   Switch terminals and contact sets.

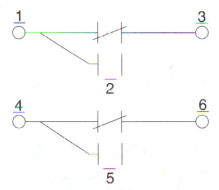

FIGURE 2–20   Switch contact position with the relay coil energized.

These contacts will change position when the relay coil is energized.

It is important to understand the operation of the switch contacts contained in these relays. Understanding this can help you determine which type and set of contacts you should use for a given circuit or load function.

Look at the following switch relay contacts (**Figure 2–21**).

In the example, there are *four sets of contacts*: **two contact sets are NO (normally open)** and **two sets of contacts are NC (normally closed)**. Look at the contact terminal layout. The *NO contacts are terminals 1 and 3*, and *terminals 4 and 6*. These contacts would not permit current flow through them until the relay coil is energized.

Once the relay **coil is energized,** the *NO contacts 1 and 3, and 4 and 6* **will close** and the contact sets will look like **Figure 2–22.**

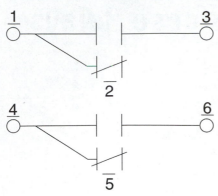

FIGURE 2–21 NC (normally closed contact terminals).

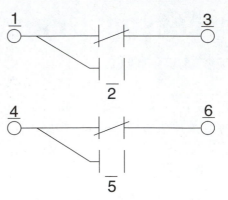

FIGURE 2–22 Contacts switched position once relay coil is energized.

At this point, current could pass through the contacts and provide line voltage to a circuit and load. Look at the contact terminal layout in **Figure 2–23**.

The **NC contacts** are *terminals 1 and 2*, and *terminals 4 and 5*. These contacts would permit current to flow to a circuit and load device until the relay coil is energized.

Once the **relay coil is energized,** the *NC contacts 1 and 2, and 4 and 5* **will open** and the contact sets will look like **Figure 2–24**.

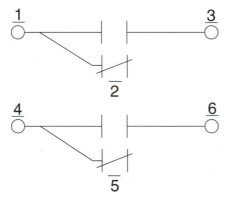

FIGURE 2–23 NC contacts coil de-energized.

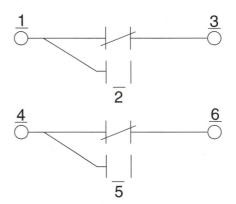

FIGURE 2–24 NC contacts opened when relay coil is energized.

At this point, *no current can pass through them*. This set of contacts can be used to turn off a circuit and load *when the relay coil is energized*.

Also notice that *terminal numbers 1 and 4 are used in both NO and NC contact sets*, these are the **main load terminals** for the switch contact sets of the relay. One leg of the supply voltage will always have to be attached to one or both of these terminals to be able to feed current through the relay switch contact sets.

> *NOTE: The switch terminals 1, 2, and 3 are a separate bank of switches from terminals 4, 5, and 6. Depending on the operation that the relay is being used for, there may only be one bank of switch contacts being used on the relay, or both banks of contacts.*

Troubleshooting

NOTE: *When using an ohmmeter to check the relay, all power and conductor wires must be removed from the relay coil terminals and contact terminals.*

Troubleshooting general-purpose relays is simple; you can use an ohmmeter to check the relay coil and also check the contacts position. Remember, you should get an ohms value reading on a good relay coil—an Infinity ohms reading on a coil means the coil is open and will not allow current to pass through it. In most cases, when the relay coil is bad, the whole relay must be replaced even if the switch contacts are OK.

When checking contacts on the relay, remember that you should get a (0) ohms value reading on a set of good NC contacts. If you don't or the resistance is greater than 2 or 3 ohms, the contacts are pitted and should be replaced.

If you get a good ohms reading on a set of NC contacts, you will also want to see if they open when the relay coil is energized.

When checking a set of NO contacts, you should get a high or infinity ohms value reading on your meter. If you get a low ohms value reading, it means the contacts are welded shut and will not open. You will also want to check to see if the NO contacts close when the relay coil is energized.

Practical Competency 21

Using a General-Purpose Relay to Control Two Loads in a Circuit

SUGGESTED MATERIALS

Textbook
Refrigeration & Air Conditioning Technology, 5th Edition, Thomson Delmar Learning
Unit 12—Basic Electricity and Magnetism
Unit 15—Troubleshooting Basic Controls
Unit 19—Motor Controls

Review Topics
The Relay

COMPETENCY OBJECTIVE

The student will be able to operate a circuit with two loads (*light bulbs*) using a general-purpose relay.

OVERVIEW

General-purpose relays are considered to be **light-duty relays.**

The contacts of the relay are capable of handling an amperage load up to 10 amps at 120 volts and are normally enclosed by a molded case with the terminal and contact identification printed on top of the relay. The relay coil could be designed to operate at line voltage or low voltage.

These relays normally can contain multiple sets of switch contacts that can be NO, NC, or both.

These contacts will change position when the relay coil is energized.

It is important to understand the operation of the switch contacts contained in these relays. Understanding this can help you determine which type and set of contacts you should use for a given circuit or load function. (Refer to *Theory Lesson: General-Purpose Relays.*)

EQUIPMENT REQUIRED

125-Volt proper power supply
1 Single-pole, single-throw switch
1 Step-down transformer (*24 volts secondary*)
1 General-purpose relay
2 Load devices (*light bulb*)
2 Light bulb receptacles
1 Jumper cord with insulated alligator clip end
6 Jumper wires with insulated alligator clips

SAFETY PRACTICES

Use the proper power source, size of conductors (*wires*), and loads for this exercise. Make sure that all conductors are properly insulated. Unless directed otherwise, do not plug circuit into power source until the circuit is completely wired and checked by your instructor.

COMPETENCY PROCEDURES

CONTROL VOLTAGE CIRCUIT CONNECTION TO THE GENERAL-PURPOSE RELAY

1. Put the light bulbs into the light bulb receptacles. ❑
2. Attach one leg of the *jumper cord* to one leg or terminal of the single-pole, single-throw switch. ❑
3. Attach another jumper wire to the other side of the single-pole, single-throw switch. ❑
4. Attach the opposite end of this jumper wire to one of the terminals or legs of the primary side of the transformer. ❑
5. Attach the opposite *jumper cord* end to the other transformer's primary side terminal. ❑
6. Attach one end of a jumper wire to one leg or terminal of the secondary side of the step-down transformer. ❑
7. Attach the opposite end of this jumper wire to one of the low-voltage terminals of the general-purpose relay coil. ❑
8. Attach another jumper wire and attach it to the other leg or terminal of the secondary side of the step-down transformer. ❑
9. Attach the opposite end of this jumper wire to the other low-voltage terminal of the general-purpose relay coil (**Figure 2–25**). ❑

NORMALLY CLOSED CIRCUIT CONNECTION

10. Attach a jumper wire to the jumper cord connection at the switch. ❑
11. Attach the opposite end of this jumper wire to load terminal **1** of the general-purpose relay. (Refer to **Figure 2–26**.) ❑

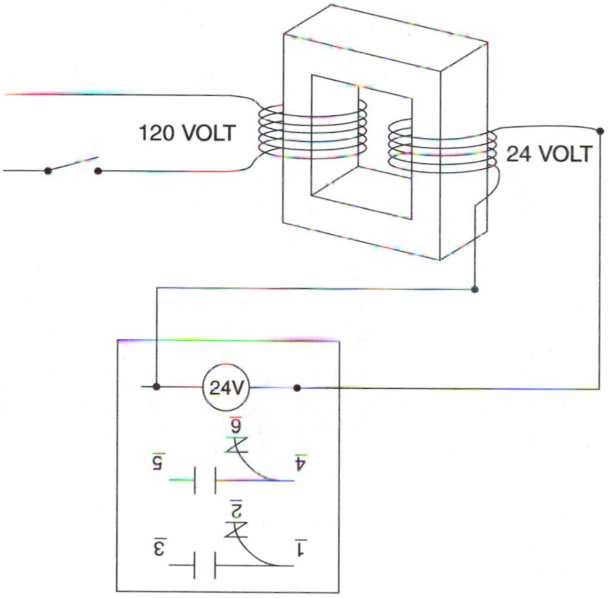

FIGURE 2–25   Transformer's secondary voltage attached to the general-purpose relay coil terminals.

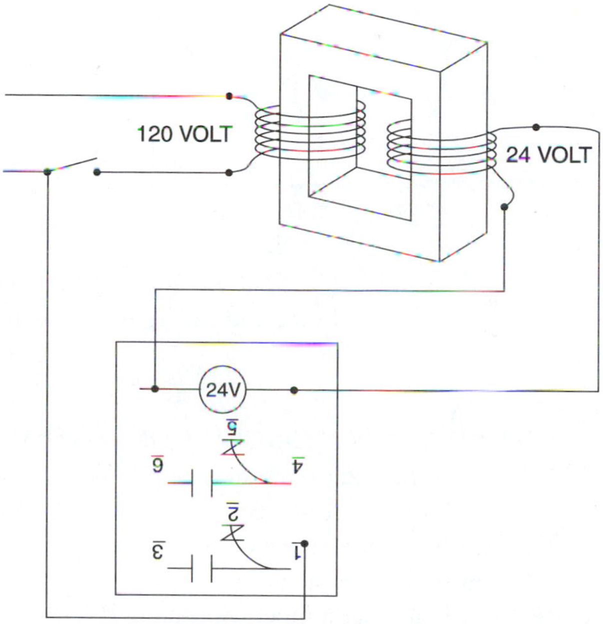

FIGURE 2–26   Jumper wire from power cord at switch to the general-purpose relay, relay terminal 1.

12. Attach another jumper wire to terminal **2** of the general-purpose relay. (*This would be the NC contact of the general-purpose relay.*) ❑
13. Attach the opposite end of this jumper wire to one leg or terminal of one of the light bulb receptacles. ❑
14. Take another jumper wire and attach one end of it to the other leg or terminal of the light bulb receptacle. ❑
15. Attach the other end of this jumper wire from the first light bulb receptacle to the opposite leg of the jumper cord. ❑

> **NOTE:** *This completed circuit will allow this light to be ON when the relay coil is de-energized because it is wired through the NC terminals* **1** *and* **2** *(**Figure 2–27**).*

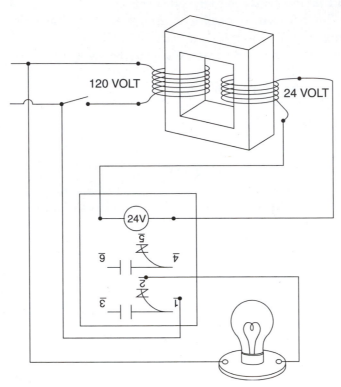

FIGURE 2–27   Jumper wire connection at relay terminal 2 and one leg or terminal of the first light bulb.

NORMALLY OPEN CIRCUIT CONNECTION

16. Attach the end of another jumper wire to terminal 3 of the general-purpose relay. (*This is the NO set of contacts between terminals* **1** *and* **3.**)   ❑
17. Attach the opposite end of the jumper wire to one of the legs or terminals of the second light bulb receptacle.   ❑
18. Attach another jumper wire end to the other leg or terminal of the second light bulb receptacle.   ❑
19. Attach the other end of this jumper wire from the second light bulb receptacle to the opposite leg of the jumper cord.   ❑

> **NOTE:** *This is the jumper cord leg that is not attached to the single-pole, single-throw switch.*

> **NOTE:** *This completed circuit will allow the second light to be OFF while the relay coil is de-energized because it is wired through the NC terminals* **1** *and* **3** *(**Figure 2–28**).*

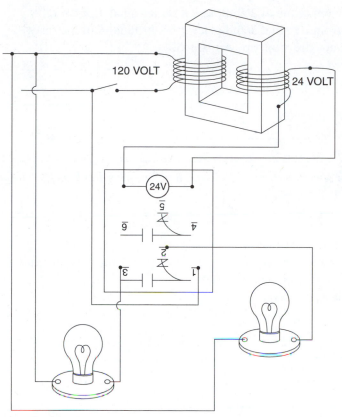

FIGURE 2–28   Contact terminals 1 and 3 are normally open.

20. Have your instructor check your circuit. ❑
21. Make sure the switch is in the OFF position. With the instructor's approval, plug the power cord into the proper power supply voltage. ❑
     Explain what happened. _____
22. Carefully remove the jumper wire from terminal 2 of the general-purpose relay. ❑
     Explain what happened. _____
23. Remove jumper wire at terminal **3** and place jumper wire from terminal **2** there. ❑
     Explain what happened. _____
24. Carefully place jumper wire from terminal **3** to terminal **2**. ❑

    CAUTION: You may get a small arc when jumper terminal is placed on terminal **2**.

    Explain what happened. _____
25. Turn the switch to the ON position. ❑
     Explain what happened. _____
26. With switch in the ON position, remove one of the low-voltage jumper wires from the general-purpose relay coil. ❑
     Explain what happened. _____

NOTE: *The second light should turn ON and the first light should come OFF. This shows that there are 24 volts to the relay coil and that switch contacts are changing position when the coil is energized and de-energized.*

---

27. Turn the switch to the OFF position and unplug the jumper cord. ❑
28. Remove jumper wire from terminal *1* and place it on terminal *4* of the relay. ❑

29. Remove jumper wire from terminal *2* and place it on terminal *5* of the relay. ❑
30. Remove jumper wire from terminal *3* and place it on terminal *6* of the relay. ❑
31. Plug the jumper cord into the proper power source. ❑
    Explain what happened. _____
32. Turn the switch to the ON position. ❑
    Explain what happened. _____
33. Turn switch to the OFF position. ❑
    Explain what happened. _____

NOTE: *This shows that switch terminals 4, 5, and 6 are another bank of switches that operate the same way as terminals 1, 2, and 3.*

---

34. Have your instructor check your work. ❑
35. Unplug the power cord and remove all wires and jumpers from transformer, light bulbs, general-purpose relay, and receptacle. ❑
36. Replace supplies to their proper location. ❑

RESEARCH QUESTIONS

1. General-purpose relay contacts are used on a load device with amperage rating of up to how many amps?

    _____

    _____

2. What is the voltage of the general-purpose relay coil?

    _____

    _____

3. What would a NO set of contacts on a general-purpose relay do when the relay coil is energized?

    _____

    _____

4. What would a NC set of contacts on a general-purpose relay do when the relay coil is energized?

    _____

    _____

5. What type of contacts would you wire an indicator light through, to indicate that a motor was de-energized (**NO or NC**)?

    _____

    _____

Passed Competency _____   **Failed Competency** _____

Instructor Signature _____   **Grade** _____

Theory Lesson: Current Starting Relays (Figure 2–29)

FIGURE 2–29   Current relay.

SUGGESTED MATERIALS

Textbook
Refrigeration & Air Conditioning Technology, 5th Edition, Thomson Delmar Learning
Unit 17—Types of Electrical Motors

Review Topics
The Current Relay

Key Terms
coil • current magnetic relay • current relay • infinite • magnetic field • normally open contacts • phase shift • run winding • start winding

OVERVIEW

The **current relay** is also known as a **current magnetic relay** and operates on the principle of a magnetic field. It is normally used to remove the start winding and start capacitor from the motor once the motor has reached 75% of its rated speed. The current relay contains a **coil** of large wire and a set of normally open contacts (**Figure 2–30**).

The coil of the relay is connected in series with the **run winding** of a motor. The contacts of the relay are connected in series with the **start winding** of the motor.

When power is applied to the motor, the starting relay's contacts are in the open position, which means that no power is being applied to the starting winding of the motor. This prevents the motor from starting. This causes a current of about three times the normal load amperage to flow through the run winding. This high current flows through the current relay coil, producing a strong magnetic field in the center of the coil. The **magnetic field** is strong enough to attract the armature (*which contains the normally open starting contacts*) to the center of the coil. (Refer to **Figure 2–31**.)

When the starting contacts close, power is applied to the start winding and the motor begins to turn because of what is called a **phase shift**. As the motor continues to run, its speed picks up. As the motor

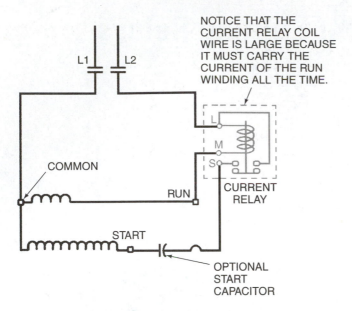

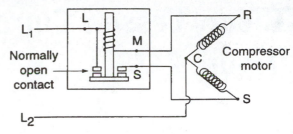

FIGURE 2–31 Current relay NO contacts.

FIGURE 2–30 The wiring diagram of a current relay.

runs faster, the current flow through the run winding decreases rapidly, also causing the current flow through the relay coil to decrease. This causes the strength of the magnetic field in the coil to become weaker. When the motor reaches about 75% of its rated speed, the magnetic field is weak enough to permit the armature to drop from the center of the coil, opening the starting contacts of the relay. When this happens, it disconnects the starting winding from the motor and the motor continues to operate on its run winding.

The current relay is used to disconnect the start winding of the motor only and does not provide any overload protection for the motor. Current relays are normally used on fractional horsepower motors of up to 1/2 horsepower (**Figure 2–32**).

FIGURE 2–32 Compressor with a current relay that plugs into the compressor motor terminals.

These relays are normally position sensitive and should be mounted in the correct position. If the relay was installed upside down, the starting contacts would be closed all the time, keeping the start winding energized in the motor circuit and causing it to overheat and burn out. Current relays come in different sizes, which are matched to the horsepower and amperage draw of the motor.

TESTING A CURRENT RELAY

In most cases, only an ohmmeter is needed to check the current relay. If you hold the relay in the upright position and put the ohmmeter test leads into the relay contact terminals, the ohmmeter should give an infinite reading, indicating an open circuit. An infinite reading on an ohmmeter means that there is no reading indicated on the meter display if it is a digital meter, and no needle movement if it is an analog meter.

CHECKING THE CONTACTS OF A CURRENT RELAY

Hold the relay in the **upright position** and place the meter leads at the **M** and **S** terminals of the relay (**Figure 2–33**).

> NOTE: *In this test there should be an Infinity value ready indicating the contacts are open. If there is a measurable resistance in this position, this is an indication that the contacts are closed, and would energize the motor start winding all the time. This is would require a replacement of the relay.*

With the meter leads placed in the M and S terminals of the relay, turn the relay **upside down** (**Figure 2–34**).

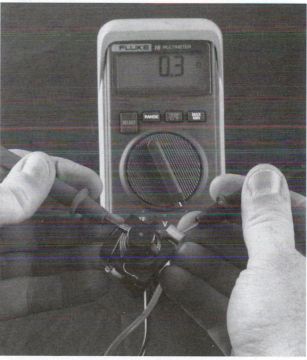

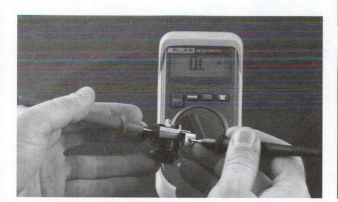

FIGURE 2–33   Continuity test of contacts with relay in the upright position.

FIGURE 2–34   Continuity test of contacts with relay in the upside down position.

> NOTE: *The contacts should close and continuity should be indicated. This shows that the contacts are good. If there is an infinity reading in this position, the contacts are stuck open, which would prevent the motor start winding being energized during motor start-up. Replacement of the relay would be required.*

CHECKING THE RELAY COIL

The coil of the relay can also be checked with an ohmmeter. A visual inspection will indicate a burnt relay coil. An ohms test on the relay coil will also indicate if the coil is open or not. To check a coil with an ohmmeter, attach the meter leads to the relay coil terminals. Some are marked L and M (**Figure 2–35**).

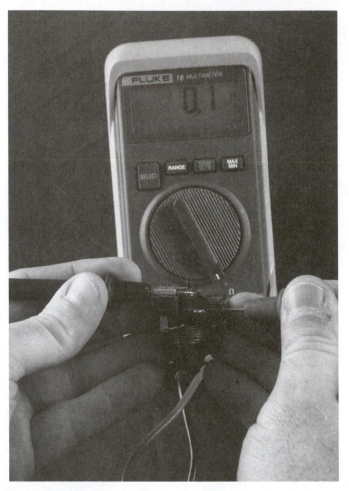

FIGURE 2–35   Testing the coil across terminals L and M.

If the coil is good, the meter should indicate an ohms value reading. If there is an infinity reading between the coil terminals, it means the coil is defective and the relay should be replaced.

Ammeter Test

The current relay can also be checked with a clamp-on ammeter or an in-line ammeter. With a clamp-on ammeter, clamp the jaws around the power supply line attached to terminal L of the relay. Start the motor and watch the amperage reading on the meter.

If the relay is *functioning properly, you should see the amperage reading rise high for a split second at start-up and then see the amperage drop back to the rated run load amperage of the motor.*

NOTE: *If the amperage reading stays high for more than 5 seconds, it means that the start relay contacts are not opening and taking out the start winding of the motor. Eventually the start winding will burn out if the action continues.*

Practical Competency 22

Wiring a Current Relay into a Fractional Horsepower Hermetic Compressor Circuit

SUGGESTED MATERIALS

Textbook
Refrigeration & Air Conditioning Technology, 5th Edition, Thomson Delmar Learning
Unit 17—Types of Electrical Motors
Unit 19—Motor Controls

Review Topics
The Current Relay

COMPETENCY OBJECTIVE

Student will be able to wire a current relay into a motor circuit.

OVERVIEW

The current relay is also known as a **current magnetic relay** and operates on the principle of a magnetic field. It is normally used to remove the start winding and start capacitor from the motor once the motor has reached 75% of its rated speed. The current relay contains a coil of large wire and a set of normally open contacts.

The coil of the relay is connected in series with the **run winding** of a motor. The contacts of the relay are connected in series with the **start winding** of the motor.

When power is applied to the motor, the starting relay's contacts are in the open position, which means that no power is being applied to the starting winding of the motor. This prevents the motor from starting. This causes a current of about three times the normal load amperage to flow through the run winding. This high current flows through the current relay coil, producing a strong magnetic field in the center of the coil. The magnetic field is strong enough to attract the armature (*which contains the normally open starting contacts*) to the center of the coil.

When the starting contacts close, power is applied to the start winding and the motor begins to turn because of what is called a phase shift. As the motor continues to run, its speed picks up. As the motor runs faster, the current flow through the run winding decreases rapidly, also causing the current flow through the relay coil to decrease. This causes the strength of the magnetic field in the coil to become weaker. When the motor reaches about 75% of its rated speed, the magnetic field is weak enough to permit the armature to drop from the center of the coil, opening the starting contacts of the relay. When this happens, it disconnects the starting winding from the motor and the motor continues to operate on its run winding.

The current relay is used to disconnect the start winding of the motor only and does not provide any overload protection for the motor. Current relays are normally used on fractional horsepower motors of up to 1/2 horsepower.

These relays are normally position sensitive and should be mounted in the correct position. If the relay was installed upside down, the starting contacts would be closed all the time, keeping the start winding energized in the motor circuit and causing it to overheat and burn out. Current relays come in different sizes, which are matched to the horsepower and amperage draw of the motor.

EQUIPMENT REQUIRED

Proper power supply
1 Fractional horsepower hermetic compressor
1 Push-on or hard-wire current relay sized for the hermetic compressor
1 Jumper cord with insulated alligator clip ends
1 Current overload protector (*sized for the current draw of the compressor being used*)

SAFETY PRACTICES

Use a jumper cord with insulated alligator clips. Make all connections to circuit with the power off. Unless directed otherwise, have your instructor check your electrical connections to the circuit before proceeding to supply line voltage to the circuit.

> *NOTE: Student should be competent in determining which windings of a compressor are the run, start, and common windings.*

COMPETENCY PROCEDURES

Checklist

1. Determine which windings of the compressor are run, start, and common. ❏
2. Make sure that the current relay being used is the right size for the hermetic compressor being used. ❏
3. Identify the S and M terminal of the current relay. ❏

> *NOTE: If this is a push-on current relay, make sure that the compressor terminal layout is the same as the S and M terminal layout of the current relay.*

4. If this is a push-on current relay, attach the relay to the hermetic compressor's start and run terminals. ❏

> *NOTE: If this is a hard wire current relay, attach a wire with electrical terminals to terminal S of the relay and to the start winding terminal of the compressor. Then attach a wire with electrical terminals to terminal R of the relay and to the run terminal of the compressor.*

5. Attach one leg of the jumper cord to terminal L of the current relay. ❏
6. Attach the opposite leg of the jumper cord to one leg of the overload protector. ❏
7. Attach the opposite leg of the overload protector to the common terminal of the compressor. ❏
8. Unless directed otherwise, have your instructor check your circuit. ❏
9. Make sure that the compressor is braced in some manner and plug the jumper cord into the proper power supply. (*The compressor should start.*) ❏
10. With your instructor's approval, unplug the jumper cord from the power supply and disconnect the circuit hookup and return materials to proper location. ❏

RESEARCH QUESTIONS

1. Current starting relays operate on the principle of what?

_____

_____

2. Are the switch contacts of the current relay NO or NC?

_____

_____

3. Current relays are used in motors of up to how many horsepower?

_____

_____

4. The coil of the relay is connected in series with which winding of the motor?

_____

_____

5. The contacts of the current relay are connected in series with which winding of the motor?

_____

_____

Passed Competency _____   Failed Competency _____

Instructor Signature _____   Grade _____

Practical Competency 23

Checking the Performance of a Current Relay in a Motor Circuit

SUGGESTED MATERIALS

Textbook

Refrigeration & Air Conditioning Technology, 5th Edition, Thomson Delmar Learning
Unit 17—Types of Electrical Motors
Unit 19—Motor Controls

Review Topics

Current Relays

COMPETENCY OBJECTIVE

Student will learn the proper procedures for wiring a current relay into a motor circuit and how to check the performance of the relay using an ammeter and an ohmmeter.

OVERVIEW

The current relay is also known as a **current magnetic relay** and operates on the principle of a magnetic field. It is normally used to remove the start winding and start capacitor from the motor once the motor has reached 75% of its rated speed. The current relay contains a coil of large wire and a set of normally open contacts.

The coil of the relay is connected in series with the **run winding** of a motor. The contacts of the relay are connected in series with the **start winding** of the motor.

When power is applied to the motor, the starting relay's contacts are in the open position, which means that no power is being applied to the starting winding of the motor. This prevents the motor from starting. This causes a current of about three times the normal load amperage to flow through the run winding. This high current flows through the current relay coil, producing a strong magnetic field in the center of the coil. The magnetic field is strong enough to attract the armature (*which contains the normally open starting contacts*) to the center of the coil.

When the starting contacts close, power is applied to the start winding and the motor begins to turn because of what is called a phase shift. As the motor continues to run, its speed picks up. As the motor runs faster, the current flow through the run winding decreases rapidly, also causing the current flow through the relay coil to decrease. This causes the strength of the magnetic field in the coil to become weaker. When the motor reaches about 75% of its rated speed, the magnetic field is weak enough to permit the armature to drop from the center of the coil, opening the starting contacts of the relay. When this happens, it disconnects the starting winding from the motor and the motor continues to operate on its run winding.

The current relay is used to disconnect the start winding of the motor only and does not provide any overload protection for the motor. Current relays are normally used on fractional horsepower motors of up to 1/2 horsepower.

These relays are normally position sensitive and should be mounted in the correct position. If the relay was installed upside down, the starting contacts would be closed all the time, keeping the start winding energized in the motor circuit and causing it to overheat and burn out. Current relays come in different sizes, which are matched to the horsepower and amperage draw of the motor.

Ammeter Test

The current relay can also be checked with a clamp-on ammeter or an in-line ammeter. With a clamp-on ammeter, clamp the jaws around the power supply line attached to terminal L of the relay. Start the motor and watch the amperage reading on the meter.

If the relay is *functioning properly, you should see the amperage reading rise high for a split-second at start up and then see the amperage drop back to the rated run load amperage of the motor.*

> NOTE: *If the amperage reading stays high for more than 5 seconds, it means that the start relay contacts are not opening and taking out the start winding of the motor. Eventually the start winding will burn out if the action continues.*

EQUIPMENT REQUIRED

Proper power supply
1 Single-pole, single-throw switch
1 Clamp-on ammeter
1 Ohmmeter
1 Fractional horsepower hermetic compressor
1 Current relay sized for the hermetic compressor being used in this competency
1 Jumper cord with insulated alligator clip ends
1 Current overload protector (sized for the current draw of the compressor being used)

SAFETY PRACTICES

Use a jumper cord with insulated alligator clips. Make all connections to circuit with the power OFF. Unless directed otherwise, have your instructor check your electrical connections to the circuit before proceeding to supply line voltage to the circuit. Make sure to set meters to proper settings and functions. You will be making amperage checks on the circuit with power supplied.

> NOTE: *Student should be competent in determining which windings of a compressor are the run, start, and common windings. Student should be competent in the use of an ammeter and ohmmeter.*

COMPETENCY PROCEDURES   Checklist

PUSH-ON CURRENT RELAY

1. Determine which windings of the compressor are run, start, and common.   ❑
2. Make sure that the current relay being used is the right size for the hermetic compressor being used.   ❑
3. Identify the S and the M terminals of the current relay.   ❑

> NOTE: *If this is a push on current relay, make sure that the compressor terminal layout is the same as the S and M terminal layout of the current relay.*

4. Attach relay to the hermetic compressors start and run terminals.   ❑
5. Attach one leg of the **jumper cord** to terminal or lead wire of the single-pole, single-throw switch.   ❑
6. Attach a jumper wire from the other terminal or lead wire of the single-pole, single-throw switch.   ❑
7. Attach the other end of the jumper wire to the L or power lead terminal on the current relay.   ❑

8. Attach the opposite leg of the **jumper cord** to one leg of the overload protector. ❑
9. Attach the opposite leg of the overload protector to the common terminal of the compressor. ❑
10. Clamp the ammeter around the **jumper cord** lead wire attached to the compressor overload protector. ❑

HARD-WIRE RELAY CONNECTIONS

11. If this is a hard wire current relay, attach a wire with electrical terminals to **terminal S** of the relay and to the **start winding terminal of the compressor.** Then attach a wire with electrical terminals to **terminal R** of the relay and to the **run terminal of the compressor.** ❑
12. Have your instructor check your circuit. ❑
13. Make sure that the compressor is braced and the switch is in the OFF position and plug the jumper cord into the proper power supply. ❑
14. Turn the switch to the ON position (*The compressor should start.*) ❑
15. *Record the compressor running amperage.* _____ ❑

> **NOTE:** *This is the compressor running amperage.*

16. Turn the switch to the OFF position. (*The compressor will stop*) ❑
17. With the ammeter still around one jumper cord, set the meter in a position where you will be able to read the meter scale as soon as the switch is turned ON. (*This will be the compressor's start-up amperage.*) ❑
18. *Turn the switch to the* **ON** *position and record the amperage reading right at start-up.* _____ ❑
19. *Was the start-up amperage higher than the compressor's running amperage?* _____ ❑
20. *Did the motor return to the compressor running amperage as recorded in Step 15?* _____ ❑

> **NOTE:** *If the amperage was high for a split second and dropped to the compressor running amperage, this means that the current relay is working. It is removing the start winding of the compressor from the motor circuit when the compressor reaches 75% of its rated speed.*

21. Turn the switch to the OFF position and unplug the jumper cord from the power supply. ❑
22. Remove the current relay from the compressor circuit and hold it in the upright position. ❑
23. Set the ohmmeter to measure resistance and take the meter leads and plug them into terminals S and M of the current relay. ❑
24. *Record your resistance reading.* _____ ❑

> **NOTE:** *You should get no resistance value reading with the test, showing that the contacts of the relay are open like they should be. If you get a resistance value reading with this test, the contacts of the relay are closed for some reason and the relay will have to be replaced.*

25. Leave the meter leads in terminals S and M of the relay and turn it over. ❑
26. *Record the resistance value reading.* _____ ❑

> **NOTE:** *You should get a resistance value reading with this test, showing that the starting contacts of the relay are closed and will work in a motor circuit. If you did not get a resistance reading in this test, it means that the starting contacts of the relay are not closing and will not work in a motor circuit; therefore, it would have to be replaced.*

CHECKING THE RELAY COIL

27. Attach the ohmmeter leads to terminals L and M of the relay. ❑

28. *Record your resistance value reading.* _____ ❑

> **NOTE:** *This test is checking the current relay's coil. With this test you should get a resistance value reading, showing that the relay coil is good. If you did not get a resistance value reading in this test, it means the coil is open and the relay will not work in a motor circuit; therefore it will have to be replaced.*

29. Have your instructor check your meter test and replace all equipment and supplies to their proper location. ❑

RESEARCH QUESTIONS

1. The current relay is used in the motor to do what?

   _____

   _____

2. When checking a current relay's starting contacts in the proper position with an ohmmeter, should you or should you not get an ohms value reading to indicate that the contacts are good and in the proper position?

   _____

   _____

3. When turning the relay over and checking its starting contacts with an ohmmeter, should you or should you not get an ohms value reading?

   _____

   _____

4. When using a clamp-on ammeter to check the starting contacts of the current relay, what should you see happen on the ammeter to indicate to that the relay contacts are working?

   _____

   _____

5. To check the relay coil with an ohmmeter, which terminals should you check across to see if the coil is good or not?

   _____

   _____

Passed Competency _____ Failed Competency _____

Instructor Signature _____ Grade _____

Theory Lesson: Positive Temperature Coefficient Start Devices

SUGGESTED MATERIALS

Textbook
Refrigeration & Air Conditioning Technology, 5th Edition, Thomson Delmar Learning
Unit 12—Basic Electricity and Magnetism
Unit 17—Types of Electric Motors
Unit 19—Motor Controls

Review Topics
Start devices; Positive Temperature Coefficient (PTC)

Key Terms
overload protector • permanent-split capacitor motor • positive temperature coefficient (PTC) • potential relay • run capacitor • run winding • start winding • thermistor

OVERVIEW

Positive temperature coefficient (PTC) start devices are electronic devices used in many applications (**Figure 2–36**). One area you will see these devices in is a **permanent-split capacitor (PSC) motor** circuit. These motors are designed to operate with a **run capacitor** in the motor circuit to give them better running efficiency.

FIGURE 2–36 Positive temperature coefficient device (PTC). (*Photo by Bill Johnson*)

Sometimes the PSC motor needs help in getting started and will require the addition of a **start capacitor**. To install the starting capacitor in the motor circuit, a **potential relay** must be added. This combination will give the motor greater starting torque during start-up, and remove the start capacitor once the motor has reached 75% of its rated speed.

A PTC device may also be used with a run capacitor to accomplish the same thing, although it will not provide the same amount of starting torque as a starting capacitor. Using a PTC device instead of a potential relay does have some advantages because it does not have any moving parts and does not require a starting capacitor. *These PTC devices are wired in parallel with the run capacitor, which is wired in parallel across the start and run winding of the motor.* Because the PTC device is wired parallel to the run capacitor, it acts like a *short across the capacitor during starting*. This allows full-line voltage to the start winding of the motor during start-up. This gives the motor a stronger **phase shift** to get the motor going.

The PTC device is referred to as a thermistor and works on the principle of resistance and heat. The **thermistor is a solid-state** device that allows fewer electrons to flow through it as its temperature increases. The resistance changes about 3% for each 1 degree Fahrenheit of temperature change. At ambient temperature, the **resistance on the thermistor is very low** (Figure 2–37).

Because the thermistor is installed to a motor's windings, when the motor is started the current draw through the windings of the motor causes an increase in temperature of the PTC device, which causes its resistance to rise to about **10,000 to 20,000 ohms (Figure 2–38)**.

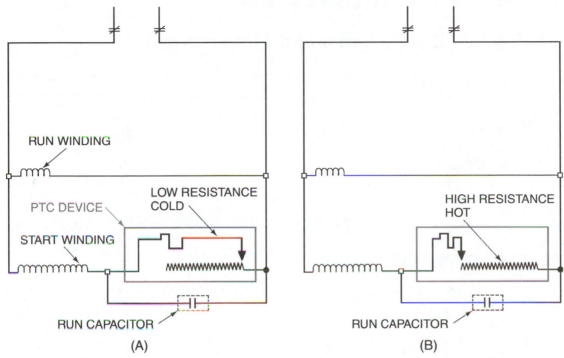

FIGURE 2–37   PTC device at ambient temperature (low resistance).

FIGURE 2–38   PTC device at ambient temperature (high resistance).

With the thermistor, the start winding is not removed from the motor circuit once the motor gets started. The start winding actually stays in the motor circuit all the time as long as the motor is running without being damaged. This is accomplished by the PTC device inside the thermistor, which is wired in parallel to the run capacitor and both the start and run winding of the motor.

During start-up of the motor, both run and starting winding are used to get the motor going. This creates a high current draw through the windings of the motor, which generates an increase of heat on the PTC device, causing its **resistance to rise to about 10,000 to 20,000 ohms**. This amount of high resistance on the PTC device limits the amount of current flow through **the starting winding to about 0.03 to 0.05 amps** while the motor continues to run.

Once the motor circuit is shut down, it takes about **2 to 5 minutes for the PTC device inside the thermistor to cool down** to a point that its resistance reaches a point where the motor could be started again. If the motor tries to start before the PTC device has cooled enough, the motor's **overload protector** will trip the motor circuit, giving more time for the PTC device to cool down enough to allow the motor to start.

Testing a PTC Device with an Ohmmeter
You can test a thermistor with an ohmmeter or an ammeter. With an ohmmeter set at a low ohms scale, allow the thermistor to cool to ambient temperature. Connect the meter leads across the side terminals of the thermistor after removing the capacitor. You should read very low resistance.

Testing a PTC Device with a Clamp-on Ammeter
To test with an ammeter you must install a thermistor and run capacitor into the motor circuit. With the motor running, clamp the jaws of the clamp-on ammeter around the start winding lead of the relay and read the amperage draw. You should see amperage of about (0.03–0.05 Amps). This shows you that the PTC device is limiting the current flow through the start winding of the motor.

Practical Competency 24

Installing a PTC Device into a PSC Motor Circuit

SUGGESTED MATERIALS

Textbook
Refrigeration & Air Conditioning Technology, 5th Edition, Thomson Delmar Learning
Unit 12—Basic Electricity and Magnetism
Unit 17—Types of Electric Motors
Unit 19—Motor Controls

Review Topics
PTC Devices

COMPETENCY OBJECTIVE

The student will be able to wire a PTC device and run capacitor into a hermetic motor circuit (**Figure 2–39**).

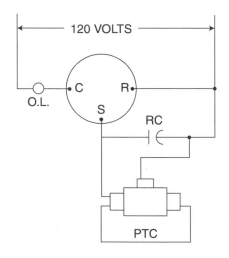

FIGURE 2–39   Motor circuit with a
PTC device.

OVERVIEW

Positive temperature coefficient (PTC) start devices are electronic devices used in many applications. One area in which you will see these devices is a permanent-split capacitor (PSC) motor circuit. These motors are designed to operate with a run capacitor in the motor circuit to give them better running efficiency.

Sometimes the PSC motor needs help in getting started and will require the addition of a start capacitor. To install the starting capacitor in the motor circuit, a potential relay must be added. This combination will give the motor greater starting torque during start-up, and remove the start capacitor once the motor has reached 75% of its rated speed.

A PTC device may also be used with a run capacitor to accomplish the same thing, although it will not provide the same amount of starting torque as a starting capacitor. Using a PTC device instead of a potential relay does have some advantages because it does not have any moving parts and does not require a starting capacitor. *These PTC devices are wired in parallel with the run capacitor, which is wired in parallel across the start and run winding of the motor*. Because the PTC device is wired parallel to the run capacitor, it acts like a *short across the capacitor during starting*. This allows full-line voltage to the start winding of the motor during start up. This gives the motor a stronger phase shift to get the motor going.

The PTC device is referred to as a thermistor and works on the principle of resistance and heat. The thermistor is a solid-state device that allows fewer electrons to flow through it as its temperature increases. The resistance changes about 3% for each 1 degree Fahrenheit of temperature change. At ambient temperature, the resistance on the thermistor is very low. Because the thermistor is installed to a motor's windings, when the motor is started the current draw through the windings of the motor causes an increase in temperature of the PTC device, which causes its resistance to rise to about 10,000 to 20,000 ohms.

With the thermistor, the start winding is not removed from the motor circuit once the motor gets started. The start winding actually stays in the motor circuit all the time as long as the motor is running without being damaged. This is accomplished by the PTC device inside the thermistor, which is wired in parallel to the run capacitor and both the start and run winding of the motor.

During start-up of the motor, both run and starting winding are used to get the motor going. This creates a high current draw through the windings of the motor, which generates an increase of heat on the PTC device, causing its resistance to rise to about 10,000 to 20,000 ohms. This amount of high resistance on the PTC device limits the amount of current flow through the starting winding to about 0.03 to 0.05 amps while the motor continues to run.

Once the motor circuit is shut down, it takes about 2 to 5 minutes for the PTC device inside the thermistor to cool down to a point that its resistance reaches a point where the motor could be started again. If the motor tries to start before the PTC device has cooled enough, the motor's overload protector will trip the motor circuit, giving more time for the PTC device to cool down enough to allow the motor to start.

EQUIPMENT REQUIRED

1 Jumper cord with insulated alligator clip ends
1 Single-pole, single-throw switch
1 Hermetic motor
1 PTC (thermistor) (sized for the hermetic motor used in this competency)
1 Run capacitor (sized for hermetic motor used in competency)
6 Jumper wires with alligator clips at each end
1 Overload protector (sized for the hermetic motor used for competency)

SAFETY PRACTICES

Use the proper power source, size of conductors (*wires*), and loads for this exercise. Make sure that all conductors are properly insulated. Unless directed otherwise, do not plug circuit into power source until the circuit is completely wired and checked by your instructor.

COMPETENCY PROCEDURES   Checklist

1. Attach one lead of the **jumper cord** to one of the terminals or lead wire of the single-pole, single-throw switch. ❏
2. Attach a jumper wire to the other terminal or lead wire of the switch. ❏
3. Attach the other end of the jumper wire to the power lead of the overload protector. ❏
4. Take a jumper wire and attach one end of it to the other terminal or leg of the overload protector. ❏
5. Attach the opposite end of the jumper wire from the overload to the common terminal of the motor. ❏
6. Attach the opposite leg of the **jumper cord** to one of the terminals of the run capacitor. ❏
7. Attach one end of jumper wire to the terminal on the run capacitor where the leg of the **jumper cord** is attached. ❏
8. Attach the opposite end of this jumper wire to the run terminal of the compressor. ❏
9. Attach one end of another jumper wire to the same terminal on the run capacitor where the cord leg is attached. ❏

10. Attach the opposite end of this jumper wire to the center terminal on the PTC device (**Figure 2–40**). ❏
11. Attach one end of another jumper wire to one of the side terminals on the PTC device. ❏
12. Attach the opposite end of this jumper wire to the other side terminal of the PTC device. (*This will look like a jumper from one end of the PTC device to the other end [**Figure 2–41**].*) ❏

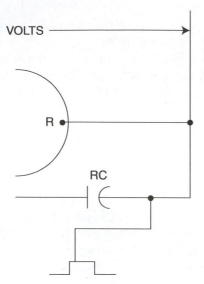

FIGURE 2–40   Leg of cord wire and two jumper wires attached to one terminal of the run capacitor; one jumper wire is attached to the run terminal of the compressor and the other end of the jumper wire to the center terminal of the PTC device.

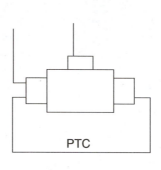

FIGURE 2–41   Jumper wire across PTC device terminals.

13. Take another jumper wire and attach one end of it to the other terminal of the run capacitor. ❏
14. Attach the opposite end of this jumper wire to the start terminal of the compressor. ❏
15. Take another jumper wire and attach one end of it to the same terminal of the run capacitor that the start winding of the compressor is attached. ❏
16. Take the opposite end of this jumper wire and attach it to one of the side terminals of the PTC device (**Figure 2–42**). ❏
17. Have your instructor check your circuit (**Figure 2–43**). ❏

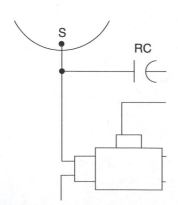

FIGURE 2–42   The other side of the run capacitor hookup to the start terminal of the compressor, and one side terminal of the PTC device.

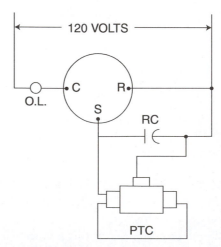

FIGURE 2–43   Completed motor circuit with a PTC device and run capacitor.

18. Turn the switch to the OFF position and plug the jumper cord into the power supply. ❑

19. Turn the switch to the ON position. (*The compressor should start.*) ❑

20. Turn the switch OFF and right back ON. (*The compressor should not start and should trip out on overload.*) ❑

21. Turn the switch OFF and wait 2 to 5 minutes for the PTC to cool down and turn the switch back on again. (*The compressor should start.*) ❑

22. Turn the switch OFF and have the circuit checked by your instructor. ❑

23. Disconnect the circuit and place all materials in their proper location. ❑

RESEARCH QUESTIONS

1. The thermistor is wired in _____ to the run capacitor used in the motor circuit.

2. The thermistor limits the _____ flow through the start winding of the motor.

3. A PTC device is a device that operates on what principle?

   _____

   _____

4. The resistance on a PTC device should be around _____ ohms when it is at ambient temperature.

5. Due to the heat generated by the windings of a motor, the resistance of the PTC device will rise to about _____ to _____ ohms.

Passed Competency _____   Failed Competency _____

Instructor Signature _____   Grade _____

Practical Competency 25

Testing a PTC Device In and Out of a Motor Circuit

SUGGESTED MATERIALS

Textbook
Refrigeration & Air Conditioning Technology, 5th Edition, Thomson Delmar Learning
Unit 12—*Basic Electricity and Magnetism*
Unit 17—*Types of Electric Motors*
Unit 19—*Motor Controls*

Review Topics
Positive Temperature Coefficient Device

COMPETENCY OBJECTIVE

The student will be able to test a PTC device in and out of the motor circuit using an ohmmeter and clamp-on ammeter.

OVERVIEW

Positive temperature coefficient (PTC) start devices are electronic devices used in many applications. One area you will see these devices in is a permanent-split capacitor (PSC) motor circuit. These motors are designed to operate with a run capacitor in the motor circuit to give them better running efficiency.

Sometimes the PSC motor needs help in getting started and will require the addition of a start capacitor. To install the starting capacitor in the motor circuit, a potential relay must be added. This combination will give the motor greater starting torque during start-up, and remove the start capacitor once the motor has reached 75% of its rated speed.

A PTC device may also be used with a run capacitor to accomplish the same thing, although it will not provide the same amount of starting torque as a starting capacitor. Using a PTC device instead of a potential relay does have some advantages because it does not have any moving parts and does not require a starting capacitor. *These PTC devices are wired in parallel with the run capacitor, which is wired in parallel across the start and run winding of the motor.* Because the PTC device is wired parallel to the run capacitor, it acts like a *short across the capacitor during starting.* This allows full-line voltage to the start winding of the motor during start up. This gives the motor a stronger phase shift to get the motor going.

The PTC device is referred to as a thermistor and works on the principle of resistance and heat. The thermistor is a solid-state device that allows fewer electrons to flow through it as its temperature increases. The resistance changes about 3% for each 1 degree Fahrenheit of temperature change. At ambient temperature, the resistance on the thermistor is very low. Because the thermistor is installed to a motor's windings, when the motor is started the current draw through the windings of the motor causes an increase in temperature of the PTC device, which causes its resistance to rise to about 10,000 to 20,000 ohms.

With the thermistor, the start winding is not removed from the motor circuit once the motor gets started. The start winding actually stays in the motor circuit all the time as long as the motor is running without being damaged. This is accomplished by the PTC device inside the thermistor, which is wired in parallel to the run capacitor and both the start and run winding of the motor.

During start-up of the motor, both run and starting winding are used to get the motor going. This creates a high current draw through the windings of the motor, which generates an increase of heat on the PTC device, causing its resistance to rise to about 10,000 to 20,000 ohms. This amount of high resistance on the PTC device limits the amount of current flow through the starting winding to about 0.03 to 0.05 amps while the motor continues to run.

Once the motor circuit is shut down, it takes about 2 to 5 minutes for the PTC device inside the thermistor to cool down to a point that its resistance reaches a point where the motor could be started again. If the motor tries to start before the PTC device has cooled enough, the motor's overload protector will trip the motor circuit, giving more time for the PTC device to cool down enough to allow the motor to start.

Testing a PTC Device with an Ohmmeter

You can test a thermistor with an ohmmeter or an ammeter. With an ohmmeter set at a low ohms scale; allow the thermistor to cool to ambient temperature. Connect the meter leads across the side terminals of the thermistor after removing the capacitor. You should read very low resistance.

Testing a PTC Device with a Clamp-on Ammeter

To test with an ammeter you must install a thermistor and run capacitor into the motor circuit. With the motor running, clamp the jaws of the clamp-on ammeter around the start winding lead of the relay and read the amperage draw. You should see amperage of about 0 amps. This shows you that the PTC device is limiting the current flow through the start winding of the motor.

EQUIPMENT REQUIRED

1 Jumper cord with insulated alligator clip ends
1 Single-pole, single-throw switch
1 Hermetic motor
1 PTC (thermistor) (sized for the hermetic motor used in competency)
1 Run capacitor sized for hermetic motor used in competency
6 Jumper wires with alligator clips at each end
1 Overload protector (sized for the hermetic motor used for competency)
1 Clamp-on ammeter
1 Ohmmeter

SAFETY PRACTICES

Use the proper power source, size of conductors (*wires*), and loads for this exercise. Make sure that all conductors are properly insulated. Unless directed otherwise, do not plug circuit into power source until the circuit is completely wired and checked by your instructor.

COMPETENCY PROCEDURES   Checklist

CLAMP-ON AMMETER TEST
1. Put together a motor circuit containing a PTC device (thermistor) for this competency. Procedures for this are covered in *Practical Competency 24: Installing a PTC Device into a PSC Motor Circuit.*   ❏
2. Once the motor circuit is complete, clamp the ammeter jaw around the motor's start terminal. ❏
3. Turn the switch to the ON position and observe the motor's **start-up amperage**.   ❏
4. *Record the motor's start-up amperage.* _____   ❏
5. Leave the clamp-on ammeter around the start motor lead and let the motor run for a few seconds. ❏
6. *Record motor's start winding running amperage.* _____   ❏

> NOTE: *This is how you check a solid-state relay with an ammeter. What you are looking for in this test is to see a high amp draw for the start-up amperage and then see the amperage for the running motor drop down to almost (0) amperage. This tells you that the thermistor is working OK. If the start-up amperage did not drop to almost (0) after the motor ran for a few seconds, the thermistor is defective and should be replaced.*

7. Turn the switch OFF and remove the thermistor from the motor circuit. ❑

> NOTE: *Care should be taken when removing the PTC device because of heat.*

OHMMETER TEST

8. Take the ohmmeter and set it on 10K ohms scale and put the leads of the meters across the two side terminals of the PTC. ❑

> NOTE: *You are testing on the 10K scale because the PTC was just removed from the motor circuit and will have a high resistance until it cools down to ambient temperature. As the resistance drops, make adjustment on the ohms scale to obtain the best ohms value reading.*

9. Once your resistance reading stops going down, record the ohms value reading on the thermistor. ❑
10. *Record the ohms value once the PTC has cooled.* _____ ❑

> NOTE: *What you are looking for with this test is to see that the resistance value is high at first on the PTC. As the thermistor cools, the resistance value will drop as the ambient temperature drops. The ohms value reading should be low once the PTC cools. This tells you the thermistor is OK. If the resistance value on the thermistor does not drop below 100 ohms after cooling, a defective PTC device should be suspected and should be replaced.*

11. Have your instructor check your circuit and test readings. ❑
12. Disconnect the circuit and place all materials in their proper location. ❑

RESEARCH QUESTIONS

1. PTC devices are referred to as _____.
2. PTC devices are used on what type of motors?

   _____

   _____
3. The PTC device limits the amperage flow through the start winding to about _____ amps.
4. It takes about _____ minutes for the PTC device to cool before it will allow the motor to start again.
5. When checking the thermistor with the ammeter, you would read the amperage of what lead of the motor while it is operating?

   _____

   _____

Passed Competency _____   Failed Competency _____

Instructor Signature _____   Grade _____

Theory Lesson: Overloads

SUGGESTED MATERIALS

Textbook
Refrigeration & Air Conditioning Technology, 5th Edition, Thomson Delmar Learning
Unit 12—*Basic Electricity and Magnetism*
Unit 19—*Motor Controls*

Review Topics
External Motor Protection; Magnetic Overload Device; Circuit Protection Device; Inherent Overload Protection

Key Terms
external line-breaking type • external overload • internal overload • line-breaking type • manual reset • thermostatic type

OVERVIEW

Overloads are designed to protect a motor from high current draw or high temperatures. They can be installed internally or externally, in series to the motor circuit. The contacts of the overload are normally closed and open at a predesigned amperage or temperature determined by the manufacturer.

Internal Overloads

Internal overloads come in two types. There is the **thermostatic type,** which is usually wired into the motor's control circuit. When this overload opens, it disconnects the control voltage from the motor starting device. The second type is the **line-break type,** which is wired into the common winding of the motor circuit and will disconnect supply voltage to the windings if an overload situation exists. They are inserted directly at a precise location in the motor windings, allowing them to sense the winding temperature (**Figure 2–44**).

When either of these overloads has opened and stopped the motor, they must be allowed to cool before the motor will start again. It is important to understand that depending on how hot the motor is, the cooling of the internal overload may take a great amount of time. It may take hours to give the motor a chance to cool down enough to allow the overload(s) to reset.

Technicians must not misdiagnose this problem. It has been reported by motor manufacturers that 50% of the motors returned by technicians for open winding failures were OK. The problem was that the technicians did not allow the motor to cool enough to allow the internal overload(s) to reset. The standard rule is that if the motor is at ambient temperature and the overload(s) has not reset, the motor circuit is probably bad. Technicians may also use water to speed the cooling process by allowing it to run over the top of the compressor. But, care should be taken to avoid wetting electrical controls.

Three phase motors will have line-break type **internal overloads** on each winding of the motor. This makes it very important for the technician to be sure to diagnose an open winding on these motors correctly (**Figure 2–45**).

External Overloads

These overloads are also designed to protect the motor from overheating due to temperature, or an overload of current draw on the windings, or both. They can be either the line-break type or the thermostatic type. They are installed externally at a location designed by the manufacturer. One advantage of these external overloads is that they can be replaced if found to be defective.

The **external line-breaking type** of overload consists of a bimetal disk with a set of contacts that are normally closed (**Figure 2–46**).

In series with the stationary contacts of the overload is a resistive heater that responds in temperature to the motor current draw while starting and running.

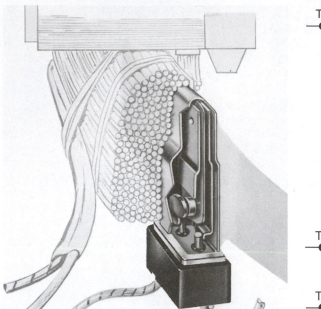

FIGURE 2–44   An internal compressor overload protection device that breaks the line circuit.

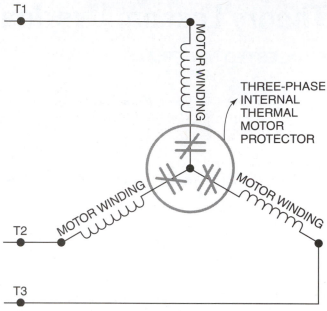

FIGURE 2–45   A three-phase WYE wound motor with internal overload protection.

FIGURE 2–46   Bimetal overload protector. (*Photo by Bill Johnson*)

If the current draw becomes too high for whatever reason, the heater becomes too hot, causing the bimetal disk to heat up and warp open, which breaks the contacts to the motor circuit.

These contacts are wired to the common winding of the motor, so that when the contacts open, line voltage is removed from both the run and start winding of the motor. The motor will not be able to restart until the bimetal disk has had a chance to cool and close the contacts to the motor circuit again.

The **thermostatic type** of external overload reacts the same way as the line-breaking overload, except that the contacts of the overload are wired in series to the motor's control circuit. When the contacts open, it removes the control voltage from the motor's control circuit device. These relays must also be allowed to cool before they will reset to start the motor again.

If the overloads do not reset with the motor at ambient temperature, it is defective and should be replaced with the exact size required by the manufacturer. Some external overloads are **manual reset.**

Even with the manual reset type of external overload, once tripped, time must be allowed for the bimetal to cool before it can be reset.

TESTING EXTERNAL OVERLOADS

External overloads are easy to check. With the overload at ambient temperature, check across the overload leads with an ohmmeter. The contacts of the overload are normally closed, so an ohms value reading should be seen in this test if the overload is good. If there is an infinity value reading across the contacts, the overload is defective and should be replaced.

Temperature Test

Another test is to see if the overload contacts will open on a temperature rise. Using a lighter or match can test this. Heat the bimetal disk and check to see if the disk of the overload warps and opens the contacts. Then allow the disk to cool and see if it warps closed again. If the disk performs in this manner, the overload is OK. If it does not, the overload is defective and should be replaced.

Clamp-on Ammeter Test

A clamp-on ammeter can also be used to check the overloads performance by clamping the ammeter jaw around the motor lead wire and intentionally causing the motor to trip by cycling the motor on and off. In this test, you would see the ammeter indicate high amp draw on start up and then return to zero, indicating that the overload has tripped open. Once the overload cools and resets, the motor would start again and operate normally.

Practical Competency 26

Checking an External Line-Breaking Overload Protector

SUGGESTED MATERIALS

Textbook
Refrigeration & Air Conditioning Technology, 5th Edition, Thomson Delmar Learning
Unit 12—Basic Electricity and Magnetism
Unit 19—Motor Controls

Review Topics
External Motor Protection; Magnetic Overload Device; Circuit Protection Device; Inherent Overload Protection

COMPETENCY OBJECTIVE

Student will be able to check an external line-breaking overload protector with an ohmmeter and clamp-on ammeter.

OVERVIEW

The external line-breaking type of overload consists of a bimetal disk with a set of contacts that are normally closed. In series with the stationary contacts of the overload is a resistive heater that responds in temperature to the motor current draw while starting and running. If the current draw becomes too high for whatever reason, the heater becomes too hot causing the bimetal disk to heat up and warp open, which breaks the contacts to the motor circuit. These contacts are wired to the common winding of the motor, so that when the contacts open, line voltage is removed from both the run and start winding of the motor. The motor will not be able to restart until the bimetal disk has had a chance to cool and close the contacts to the motor circuit again.

TESTING EXTERNAL OVERLOADS

External overloads are easy to check. With the overload at ambient temperature, check across the overload leads with an ohmmeter. The contacts of the overload are normally closed, so an ohms value reading should be seen in this test if the overload is good. If there is an infinity value reading across the contacts, the overload is defective and should be replaced.

Temperature Test
Another test is to see if the overload contacts will open on a temperature rise. Using a lighter or match can test this. Heat the bimetal disk and check to see if the disk of the overload warps and opens the contacts. Then allow the disk to cool and see if it warps closed again. If the disk performs in this manner, the overload is OK. If it does not, the overload is defective and should be replaced.

Clamp-on Ammeter Test
A clamp-on ammeter can also be used to check the overloads performance by clamping the ammeter jaw around the motor lead wire and intentionally causing the motor to trip by cycling the motor on and off. In this test, you would see the ammeter indicate high amp draw on start up and then return to zero, indicating that the overload has tripped open. Once the overload cools and resets, the motor would start again and operate normally.

EQUIPMENT REQUIRED

Proper power supply for competency
1 Jumper cord with insulated alligator clamps
3 Jumper wires with insulated alligator clips
1 Single-pole, single-throw switch
1 Hermetic compressor
1 External line-breaking overload protector
1 Clamp-on ammeter
1 VOM or ohmmeter

SAFETY PRACTICES

Make sure that the overload protector is removed from the motor circuit and that the VOM or ohmmeter is set to the ohms value scale. Unless directed otherwise, have instructor check any electrical controls before plugging into power supply.

COMPETENCY PROCEDURES   Checklist

1. Set the ohmmeter to the ohms scale   ❑
2. Attach the ohmmeter leads to the overload protector terminals.   ❑

> NOTE: *The contacts of the overload are normally closed and you should get an ohms value reading with this test to show the overload contacts are closed. If you did not get an ohms value reading in this test, the contacts are open and the overload should be replaced.*

AMMETER TEST

3. Attach one leg of the **jumper cord** to leg or terminal lead of the single-pole, single-throw switch.   ❑
4. Use a jumper wire and attach one end of it to the other terminal or lead wire of the single-pole, single-throw switch.   ❑
5. Attach the other end to one leg or terminal of the overload protector.   ❑
6. Attach one end of a jumper wire to the other side of the **overload protector**. ❑
7. Attach the opposite end of the jumper wire to the **common terminal** of the compressor. ❑
8. Attach the opposite end of the **jumper cord** to the **run terminal** of the hermetic compressor. ❑
9. Attach the clamp-on ammeter around the wire attached to the **common terminal** of the compressor.   ❑
10. Have your instructor check your circuit.   ❑
11. Turn the switch to the OFF position and plug the jumper cord into the power supply. ❑
12. Turn the switch to the ON and OFF and ON again, and watch the ammeter.   ❑

> NOTE: *You may have to turn the switch ON and OFF a couple of times to get the overload to trip open.*

> NOTE: *You should see the amperage go up and return to zero. This shows that the overload protector contacts opened up due to an increase in current draw on the contacts. This causes the bimetal disk to heat up and warps open the overload contacts, which remove the line voltage from the motor circuit.*
>
> *If the amperage reading did not return to zero on this test, it means the overload contacts are not opening up and the motor windings could be damaged. The overload protector should be replaced.*

13. Once the overload trips turn the switch to the OFF position and allow the overload protector to cool. When the disk cools, it allows the contacts to reset so that line voltage can flow through the motor circuit once the switch is turned back ON. ❏
14. Unplug the jumper cord from the power supply and return all supplies to their proper location. ❏

RESEARCH QUESTIONS

1. Internal overloads are located in the _____ of the motor.
2. Overloads must be allowed to _____ before the motor will start again.
3. What two methods can be used to allow the motor to cool enough to allow the overload to reset?

_____

_____

4. Three-phase motors may have an internal overload in each _____ of the motor.
5. External overloads are located where on the motor?

_____

_____

Passed Competency _____   Failed Competency _____

Instructor Signature _____   Grade _____

Practical Competency 27

Testing the Operation of a Two-Wire Compressor Overload Protector

SUGGESTED MATERIALS

Textbook
Refrigeration & Air Conditioning Technology, 5th Edition, Thomson Delmar Learning
Unit 12—Basic Electricity and Magnetism
Unit 19—Motor Controls

Review Topics
Compressors; Motor Overloads; Testing of Motors

COMPETENCY OBJECTIVE

Student will be able to check the operation of a two-wire motor overload protector.

OVERVIEW

The external line-breaking type of overload consists of a bimetal disk with a set of contacts that are normally closed. In series with the stationary contacts of the overload is a resistive heater that responds in temperature to the motor current draw while starting and running. If the current draw becomes too high for whatever reason, the heater becomes too hot, causing the bimetal disk to heat up and warp open, which breaks the contacts to the motor circuit. These contacts are wired to the common winding of the motor, so that when the contacts open, line voltage is removed from both the run and start winding of the motor. The motor will not be able to restart until the bimetal disk has had a chance to cool and close the contacts to the motor circuit again.

TESTING EXTERNAL OVERLOADS

External overloads are easy to check. With the overload at ambient temperature, check across the overload leads with an ohmmeter. The contacts of the overload are normally closed, so an ohms value reading should be seen in this test if the overload is good. If there is an infinity value reading across the contacts, the overload is defective and should be replaced.

Temperature Test
Another test is to see if the overload contacts will open on a temperature rise. Using a lighter or match can test this. Heat the bimetal disk and check to see if the disk of the overload warps and opens the contacts. Then allow the disk to cool and see if it warps closed again. If the disk performs in this manner, the overload is OK. If it does not, the overload is defective and should be replaced.

Clamp-on Ammeter Test
A clamp-on ammeter can also be used to check the overloads performance by clamping the ammeter jaw around the motor lead wire and intentionally causing the motor to trip by cycling the motor on and off. In this test, you would see the ammeter indicate high amp draw on start up and then return to zero, indicating that the overload has tripped open. Once the overload cools and resets, the motor would start again and operate normally.

EQUIPMENT REQUIRED

1 Two-wire overload protector in a motor circuit
1 VOM
1 Clamp-on ammeter

SAFETY PRACTICES

Make sure meters are set to the proper functions and ranges. You will be working with a live electrical motor circuit; be sure to follow all electrical safety rules.

COMPETENCY PROCEDURES

AMMETER TESTING METHOD

1. Gain access to the compressor two-wire overload protector. ❑
2. Place an ammeter around the common electric line to the compressor. ❑
3. Start the compressor and observe the ammeter. ❑
4. *Record the compressors amp draw on start-up.* _____ ❑

> NOTE: *The ammeter should show a momentary current flow of 3 to 5 times the running amperage of the compressor, and then drop back to the rated compressor's running amperage or below. This should happen within a couple of seconds.*
>
> *If the overload cycles the motor after the motor starts and runs at the correct amperages, the overload protector is defective and needs to be replaced.*
>
> *If the amperage remains in the start-up amperage range without dropping back to the running amperage, the overload is OK, and the problem is elsewhere in the motor circuit.*

5. *Record the compressor's running amperage once motor is running.* _____ ❑
6. *Record the manufacturer's required running amperage for the compressor you tested.* _____ ❑
7. *Was the compressor running within the rated amperage range?* (**Yes or No**) ❑
8. *From your test, did the overload perform properly?* (**Yes or No**) ❑

VOLTMETER TESTING METHOD

9. Shut the compressor motor circuit off. ❑
10. Set voltmeter to AC function and a value scale range equal to or above the operating voltage of the motor circuit you are working with. ❑

> NOTE: *If you are not sure of the motor's operating voltage range, set meter voltage scale to the highest scale.*

11. Take the voltmeter leads and place them across the overload terminals. ❑
12. *Record the voltage reading across the overload terminals.* _____ ❑

> NOTE: *The overload is a normally closed switch and should not indicate a voltage reading in the above test. This tells you that the overload is closed and would allow the motor to start when electrical power is applied to the circuit.*

13. Leave voltmeter leads on the overload terminals and start the compressor. Shut it off and try to restart the compressor immediately. This should cause the overload to trip open the motor circuit and stop the compressor. ❑

> *NOTE: You may have to do this a couple of times to get the overload to trip the circuit.*

14. *Once the overload trips and opens the motor circuit, record the voltage reading across the overload terminals.* _____ ❑
15. *Once the overload trips, leave the motor circuit plugged into the power source and observe the ammeter. Once the ammeter resets, record the voltage reading across the overload terminals.* _____ ❑

> *NOTE: In the above test, you should have gotten a voltage reading once the overload tripped and opened the electrical circuit to the motor to provide protection. Once the overload cooled down, it should have reset, closing the electrical contacts so that the motor would start again. This test tells you that the overload is operating properly.*

16. Disconnect the power source to the motor circuit. ❑
17. Have your instructor check your work. ❑
18. Return all equipment to its original condition and return all meters and supplies to their proper location. ❑

RESEARCH QUESTIONS

1. What is the purpose of a motor centrifugal switch?

   _____

   _____

2. What is the purpose of a motor overload protector?

   _____

   _____

3. Who sets the standard for all motor overload protection?

   _____

   _____

4. What two factors will cause the overload to remove the motor from the circuit?

   _____

   _____

5. What type of overload is not affected by ambient temperature?

   _____

   _____

Passed Competency _____ Failed Competency _____

Instructor Signature _____ Grade _____

Theory Lesson: High- and Low-Pressure Switches

SUGGESTED MATERIALS

Textbook
Refrigeration & Air Conditioning Technology, 5th Edition, Thomson Delmar Learning
Unit 14—Automatic Control Components and Applications
Unit 25—Special Refrigeration System Components

Review Topics
Pressure Sensing Devices; High-Pressure Controls; Low-Pressure Controls

Key Terms
close on drop • close on rise • cut-in • cut-out • differential low event • dual-pressure switch • high event • high-pressure switch • low-pressure switch • oil pressure controls • open on drop • open on rise

OVERVIEW

The pressure switch works off of pressure, which could be air, water, or refrigerant pressure, and so forth. Depending on the design, the pressure used to activate the switching action could come from a force created by low or high pressure. The designed switch position could be **normally closed** or **normally open,** which can be **opened or closed on an increase or decrease in pressure.** This action can be used to control electrical circuits, electrical loads, and the flow of fluids. For the most part, pressure switches used in the refrigeration and air conditioning industry are used as safety control switches.

Low-pressure switches can also be used to control temperature inside of a refrigerated box.

Pressure switches with the purpose of breaking electrical current flow to a circuit or load device consist of electrical terminals for connection to the electrical circuits or loads and a set of switch contacts that can be opened or closed by pressure, with the electrical diagram indicating the switching action of the switch. The **electrical symbol** for the pressure switch indicates if the switch action of the pressure switch opens or closes due to a drop or rise in pressure.

Listed are electrical symbols that represent the role of the pressure switch in HVAC-R equipment. Technicians should become familiar with these electrical symbols and the electrical interpretation action of the switch so they can determine their purpose in the equipment. The electrical symbols shown are used to identify a pressure switch (**Figure 2–47**).

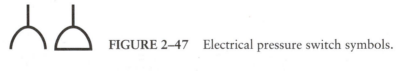

FIGURE 2–47 Electrical pressure switch symbols.

These switch symbols with the addition of the switch arm help the technician determine the function of the pressure switch and what electrical circuit(s) or loads are controlled by it.

Electrical symbols for normally open pressure switches are shown in **Figure 2–48**.

Electrical symbols for normally closed pressure switches are shown in **Figure 2–49**.

Pressure switches are not load devices and are wired in series with an electrical circuit or load device. The majority of pressure switches are adjustable (**Figure 2–50**).

A few come as nonadjustable (**Figure 2–51**).

Nonadjustable pressure switches are designed to function at a preset pressure by the manufacturer and cannot be adjusted.

COULD BE APPLIED AS A
CONDENSER FAN-CYCLE CONTROL

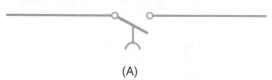

(A)

COULD BE APPLIED AS A SWITCH
TO SHOW A DROP IN PRESSURE

(B)

FIGURE 2–48   (A) NO switch that will close on a rise in pressure. (B) NO switch that will close on a drop in pressure.

COULD BE APPLIED AS A
LOW-PRESSURE CUT-OUT CONTROL

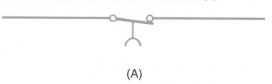

(A)

COULD BE APPLIED AS A
HIGH-PRESSURE CUT-OUT CONTROL

(B)

FIGURE 2–49   (A) NC switch that will open on a drop in pressure. (B) NC that which will open on a rise in pressure.

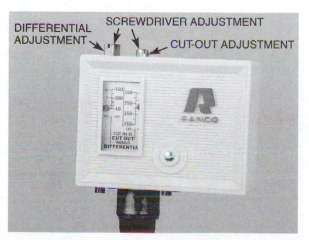

FIGURE 2–50   High-pressure switch adjustment controls.

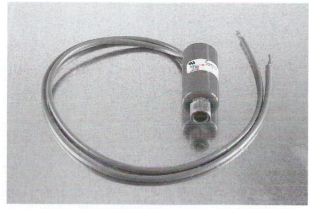

FIGURE 2–51   Nonadjustable pressure switch.

Pressure switches come as **low-pressure, high-pressure,** and **dual-pressure switches** (**Figure 2–52**).

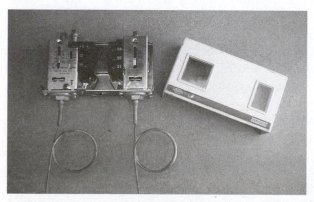

FIGURE 2–52   Dual-pressure switch.

Depending on the design of the pressure switch, the presence of high or low pressures will open or close the switch. Some used in the refrigeration and air conditioning industry are designed to operate at pressure ranges for specific refrigerants used in the industry and can be manual reset switches or automatic reset switches.

The **manual reset pressure switch** requires someone to reset the switch contacts once it has tripped. The **automatic reset switches** will reset themselves once the pressure has lowered or risen past the pressure point that caused it to trip.

On **adjustable pressure switches** of either low- or high-pressure function, it is important for the technician to understand how to set the proper pressures for the pressure switch to perform the function for which it is to operate within the equipment. This requires the technician to adjust **cut-in** and **cut-out** pressures so that the switch can function properly in the system.

Adjustable pressure switches have adjustment screws so that the desired cut-in and cut-out pressures can be set. A low-pressure switch scale plate will look similar to the one in the example (**Figure 2–53**). At the top of the scale plate is the indication of how the switch activates. This switch "**closes on a pressure rise.**"

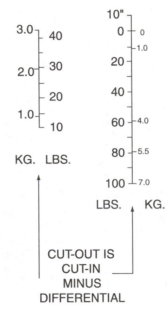

FIGURE 2–53   Closes on a rise.

Notice at the bottom of the switch scale plate where it states: "**Cutout is cut in minus differential**" (**Figure 2–54**).

The **cut-in pressure** is the point at which the contacts of the pressure switch **will close** and allow electrical current flow to the circuit or load(s). The **cut-out pressure** is the point at which the pressure switch contacts **will open** and stop the electrical current flow through the circuit or load(s). The **differential** is the difference between the cut-in and cut-out pressure settings.

High-pressure switch scale plates will look similar to the scale plate shown below. Notice the top of the scale plate where it indicates that the switch: "**opens on a pressure rise**" (Refer to *Figure 2–54*.)

At the bottom of the switch scale plate it states: "**Cut-in is Cut-out minus differential.**" Proper interpretation is required so technicians can establish correct settings that will allow equipment to operate in design range.

Other pressure switches will have the following listed on the pressure scale plate: "**Switch low event is high event minus the differential.**" Depending on the type of pressure switch, *low event is the pressure at which the* **switch contacts will open or close,** *stopping or allowing electrical current flow through the*

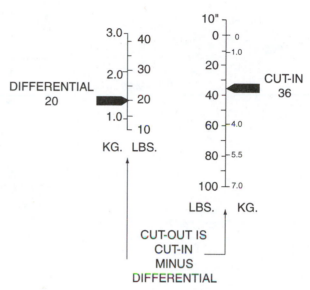

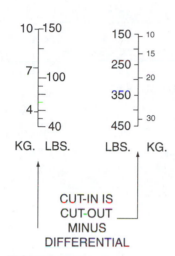

FIGURE 2–54   Opens on a rise in pressure.

FIGURE 2–55   Example of a low-pressure switch scale plate. (*Courtesy of Ranco Controls*)

circuit or load(s). Switch **high event** is the **pressure at which the switch contacts will open or close,** *stopping or allowing electrical current flow to the circuit or load(s). The differential is the difference between the low-event and high-event pressure settings.*

Example of Setting Cut-in and Cut-out Pressures for High- and Low-Pressure Switches

In the example of the low-pressure switch scale plate, it indicates that the **cut-in pressure is 36 psi** and the **differential is 20 psi** (**Figure 2–55**).The *cut-out is found by subtracting the differential pressure from the cut-in pressure.*

Cut-in – differential = cut-out

36 psi – 20 psi = 16 psi

In the example of the high-pressure switch scale plate, it indicates that the **cut-out pressure is 360 psi** and the **differential is 120 psi**. The cut-in is found by subtracting the differential pressure from the cut-out pressure (**Figure 2–56** and **Figure 2–57**).

Cut-out – differential = cut-in

360 psi – 120 psi = 240

Example: Low-Pressure Cut-in and Cut-out and Differential Points Along with Temperature Switch with cut-in and Cut-out and Differential Points

Only qualified technicians should make differential adjustments. A *differential setting that is too low* can cause short cycling of the compressor and any other load devices in the circuit. This can cause overheating of the compressor motor windings and can cause pitting of electrical contacts. A *differential that is set too high* may never allow the motor to shut off, raising electrical consumption.

The cut-out setting will be changed automatically anytime the cut-in setting is changed (**Figure 2–58**).

All control settings should be set at the manufacturer specifications. If this information is not available, the technician must consider what function the pressure switch is to serve in the sealed system in which it is to be used. Once this decision is made, the proper pressure switch can be selected. Technicians can determine the settings on a particular pressure switch as long as there is a firm understanding of the operation of

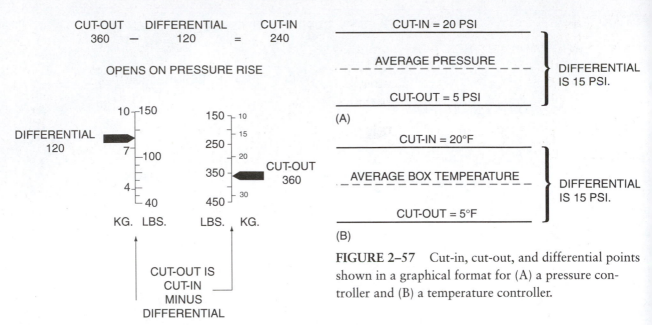

FIGURE 2–57   Cut-in, cut-out, and differential points shown in a graphical format for (A) a pressure controller and (B) a temperature controller.

FIGURE 2–56   Example of a high-pressure switch scale plate. (*Courtesy of Ranco Controls*)

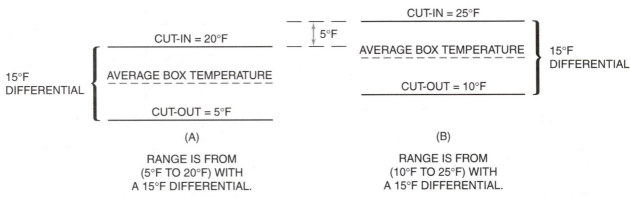

FIGURE 2–58   A change in the cut-in adjustment by 5 degrees (F) in (B) will automatically change the cut-out setting by 5 degrees (F). This changes the range. Notice that the differential is the same in both examples (A) and (B).

refrigeration or air conditioning sealed system. In addition there must be an understanding of the temperature–pressure relationship of the refrigerant being used in the system and the design operating temperature ranges.

Checking Low-Pressure Switches

Low-pressure switches can be checked with or without the switch installed in a refrigeration system. If in a sealed system, turn the system OFF and place a *jumper wire across the pressure switch leads* (**Figure 2–59**).

Turn the unit back on. Bypassing the low-pressure switch will allow the unit to continue to operate. After that, use an ohmmeter and connect the meter leads across the pressure switch terminals. Connect a manifold gage to the suction line service valve. Close the liquid line valve and allow the system to pump down.

Observe the ohmmeter scale to determine the low-pressure switch's opening or closing pressure. Open the liquid valve and allow the pressure to rise; observe the ohmmeter scale to determine the opening or closing pressure of the pressure switch. Make adjustments on the pressure switch settings as needed. If pressures cannot be set to operate the switching action in the required pressure ranges desired, the pressure switch should be replaced.

FIGURE 2–59   Bypassing the low-pressure control. (*Photo by Bill Johnson*)

Another method is by *blocking the load crossing the evaporator*. This can be done in some cases by *disconnecting the evaporator or blower fan* and checking the switch cut-in or cut-out pressures. Nitrogen can also be used to pressurize the switch so the switch action and pressure settings can be tested with an ohmmeter in the same manner.

Checking High-Pressure Switching

High-pressure switches can be checked by attaching a gage manifold set to the refrigeration system and *disconnecting the condenser fan or by blocking the airflow across the condenser coil.* Then observe the cut-out pressure. On *water cooled condensers, reduce or shut off the water flow* and observe the cut-out pressure. If the pressure required cannot be obtained, the pressure switch should be replaced. Nitrogen can also be used to pressurize the switch so the switch action and pressure settings can be tested with an ohmmeter in the same manner.

Practical Competency 28

Checking Calibration and Switch Activation of an Adjustable Low-Pressure Control

SUGGESTED MATERIALS

Textbook
Refrigeration & Air Conditioning Technology, 5th Edition, Thomson Delmar Learning
Unit 14—Automatic Control Components and Applications
Unit 25—Special Refrigeration System Components

Review Topics
Pressure Sensing Devices; High-Pressure Controls; Low-Pressure Controls

COMPETENCY OBJECTIVE

The student will be able to check the calibration and switch action of a low-pressure switch.

OVERVIEW

The pressure switch works off of pressure, which could be air, water, or refrigerant pressure, and so forth. Depending on the design, the pressure used to activate the switching action could come from a force created by low or high pressure. The designed switch position could be **normally closed** or **normally open** which can be **opened or closed on an increase or decrease in pressure**. This action can be used to control electrical circuits, electrical loads, and the flow of fluids. For the most part, pressure switches used in the refrigeration and air conditioning industry are used as safety control switches. **Low-pressure switches** can also be used to control temperature inside of a refrigerated box. Pressure switches with the purpose of breaking electrical current flow to a circuit or load device consist of electrical terminals for connection to the electrical circuits or loads and a set of switch contacts that can be opened or closed by pressure, with the electrical diagram indicating the switching action of the switch. Pressure switches are not load devices and are wired in series with an electrical circuit or load device. The majority of pressure switches are adjustable. (Refer to *Figure 2–50, Theory Lesson: High- and Low-Pressure Switches*.) Nonadjustable pressure switches are designed to function at a preset pressure by the manufacturer and cannot be adjusted. Depending on the design of the pressure switch, the presence of high or low pressures will open or close the switch. Some used in the refrigeration and air conditioning industry are designed to operate at pressure ranges for specific refrigerants used in the industry and can be manual reset switches or automatic reset switches. On **adjustable pressure switches** of either low- or high-pressure function, it is important for the technician to understand how to set the proper pressures for the pressure switch to perform the function for which it is to operate within the equipment. This requires the technician to adjust **cut-in** and **cut-out** pressures so that the switch can function properly in the system. Adjustable pressure switches have adjustment screws so that the desired cut-in and cut-out pressures can be set. The **cut-in pressure** is the point at which the contacts of the pressure switch **will close** and allow electrical current flow to the circuit or load(s). The **cut-out pressure** is the point at which the pressure switch contacts **will open** and stop the electrical current flow through the circuit or load(s). The **differential** is the difference between the cut-in and cut-out pressure settings. Other pressure switches will have the following listed on the pressure scale plate: "**Switch low event is high event minus the differential.**" Depending on the type of pressure switch, *low event is the pressure at which the* **switch contacts will open** or close, *stopping or allowing electrical current flow through the circuit or load(s).* Switch **high event** is the **pressure at which the switch contacts will open or close,** *stopping or allowing electrical current flow to the circuit or load(s).* The *differential is the difference between the low-event and high-event pressure settings.*

Cut-in – differential = cut-out
36 psi – 20 psi = 16 psi

Only qualified technicians should make differential adjustments. A *differential setting that is too low* can cause short cycling of the compressor and any other load devices in the circuit. This can cause overheating of the compressor motor windings and can causing pitting of electrical contacts. A *differential that is set too high* may never allow the motor to shut off, raising electrical consumption. The cut-out setting will be changed automatically anytime the cut-in setting is changed. (Refer to *Figure 2–58, Theory Lesson: High- and Low-Pressure Switches.*) All control settings should be set at the manufacturer specifications. If this information is not available, the technician must consider what function the pressure switch is to serve in the sealed system in which it is to be used. Once this decision is made, the proper pressure switch can be selected. Technicians can determine the settings on a particular pressure switch as long as there is a firm understanding of the operation of refrigeration or air conditioning sealed system. In addition there must be an understanding of the temperature–pressure relationship of the refrigerant being used in the system and the design operating temperature ranges.

Checking Low-Pressure Switches

Low-pressure switches can be checked with or without the switch installed in a refrigeration system. If in a sealed system, turn the system OFF and place a jumper wire across the pressure switch leads. (Refer to *Figure 2–59, Theory Lesson: High- and Low-Pressure Switches.*) Turn the unit back on. Bypassing the low-pressure switch will allow the unit to continue to operate. Next, use an ohmmeter and connect the meter leads across the pressure switch terminals. Connect a manifold gage to the suction line service valve. Close the liquid line service valve and allow the system to pump down. Observe the ohmmeter scale to determine the low-pressure switch's opening or closing pressure. Open the liquid valve and allow the pressure to rise; observe the ohmmeter scale to determine the opening or closing pressure of the pressure switch. Make adjustments on the pressure switch settings as needed. If pressures cannot be set to operate the switching action in the required pressure ranges desired, the pressure switch should be replaced. Another method is by *blocking the load crossing the evaporator.* This can be done in some cases by *disconnecting the evaporator or blower fan* and checking the switch cut-in or cut-out pressures. Nitrogen can also be used to pressurize the switch so the switch action and pressure settings can be tested with an ohmmeter in the same manner.

EQUIPMENT REQUIRED

1 Ohmmeter
1 Adjustable low-pressure switch
1 Refrigeration manifold gage set
1 Bottle of refrigerant or nitrogen bottle with regulator
Safety glasses

SAFETY PRACTICES

Be sure to wear safety glasses and make sure ohmmeter is set to proper function and value scale.

NOTE: *If pressure switch is in an existing sealed system, disconnect power from the switch or remove switch from system for testing.*

COMPETENCY PROCEDURES

Checklist

1. Remove pressure switch cover to gain access to the switch's electrical terminals. ❑

2. With the ohmmeter set to read resistance. Attach ohmmeter leads to the switch's electrical terminals. ❑
   Record if the switch contacts are in the open or closed position. _____

3. Remove the low-side hose from the manifold gage and attach the low-pressure control to the manifold low-side gage port (**Figure 2–60**). ❏

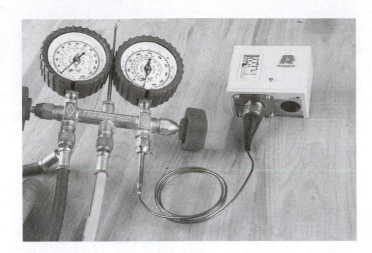

FIGURE 2–60   Connecting a pressure control to refrigerant gage manifold.

4. Attach the center hose from the manifold gage set to the refrigerant bottle or nitrogen bottle regulator (**Figure 2–61**). ❏
5. Review the pressure switch's scale plate and determine the set cut-in and cut-out pressure settings. ❏
   Record: cut-in pressure _____ cut-out pressure _____
6. Determine if the switch opens on a rise in pressure or on a drop in pressure. ❏
   Record: Did the switch open or close on a rise in pressure? _____
7. Open the valve on the refrigeration cylinder or nitrogen bottle. ❏

8. Slowly open the low-side manifold valve and observe the low-side gage pressure and ohmmeter. ❏
9. *Record the pressure at which the ohmmeter indicated that the switch opened or closed.*
   _____ ❏
10. *During the increase of pressure, did the switch open or close? Record your answer. _____* ❏
11. *At what pressure did the switch open or close? _____*
12. *Did this pressure match the cut-in pressure as recorded in* **step 5**? _____ ❏
13. Close the valve on the refrigerant or nitrogen cylinder. ❏
14. Slowly unscrew the center hose attached to the gage manifold and observe the ohmmeter scale and low-side gage. ❏
15. *At what pressure did the ohmmeter indicate that the pressure switch opened or closed? _____* ❏

FIGURE 2–61 Connecting the gage manifold to a nitrogen tank regulator.

16. *Did this pressure match the cut-out pressure as recorded in* **step 5**? _____ ❑
17. Did the switch activate at the preset cut-in and cut-out pressures? _____ ❑

NOTE: *If the switch did not operate in the preset cut-in and cut-out range, make adjustments as needed and retest as described in procedures 8–15.*

---

18. Have your instructor check your work. ❑
19. Turn off the valve to the refrigerant or nitrogen cylinder. ❑

NOTE: *If using nitrogen, back out the pressure regulator valve.*

---

20. Slowly unscrew the hose connected to the refrigerant or nitrogen cylinder and allow the pressure to bleed. ❑
21. Disconnect pressure switch from the gage manifold.
22. If pressure switch being tested was removed from a sealed system, replace it if working properly. If the switch could not be adjusted to the desired cut-in and cut-out range, see your instructor for a replacement. ❑
23. Return all materials to their proper location. ❑

RESEARCH QUESTIONS

1. Pressure switches work off of some form of _____
   _____
   _____.

2. In the refrigeration and air conditioning industry, pressure switches are normally used as what type?

_____

_____

3. The switch contacts of the pressure switch could be _____ or _____.
4. Cut-in pressure is the point at which the switch will do what?

_____

_____

5. Switch low event is the point at which the switch will do what?

_____

_____

Passed Competency _____   Failed Competency _____

Instructor Signature _____   Grade _____

Practical Competency 29

Checking Calibration and Switch Activation of an Adjustable High-Pressure Control

SUGGESTED MATERIALS

Textbook

Refrigeration & Air Conditioning Technology, 5th Edition, Thomson Delmar Learning
Unit 14—Automatic Control Components and Applications
Unit 25—Special Refrigeration System Components

Review Topics

Pressure Sensing Devices; High-Pressure Controls; Low-Pressure Controls

COMPETENCY OBJECTIVE

The student will be able to check the calibration and switch action of a high-pressure switch.

OVERVIEW

The pressure switch works off of pressure, which could be air, water, or refrigerant pressure, and so forth. Depending on the design, the pressure used to activate the switching action could come from a force created by low or high pressure. The designed switch position could be **normally closed** or **normally open,** which can be **opened or closed on an increase or decrease in pressure.** This action can be used to control electrical circuits, electrical loads, and the flow of fluids. For the most part, pressure switches used in the refrigeration and air conditioning industry are used as safety control switches. **Low-pressure switches** can also be used to control temperature inside of a refrigerated box. Pressure switches with the purpose of breaking electrical current flow to a circuit or load device consist of electrical terminals for connection to the electrical circuits or loads and a set of switch contacts that can be opened or closed by pressure, with the electrical diagram indicating the switching action of the switch. Pressure switches are not load devices and are wired in series with an electrical circuit or load device. The majority of pressure switches are adjustable. (Refer to *Figure 2–50, Theory Lesson: High- and Low-Pressure Switches.*) Nonadjustable pressure switches are designed to function at a preset pressure by the manufacturer and cannot be adjusted. Depending on the design of the pressure switch, the presence of high or low pressures will open or close the switch. Some used in the refrigeration and air conditioning industry are designed to operate at pressure ranges for specific refrigerants used in the industry and can be manual reset switches or automatic reset switches. On **adjustable pressure switches** of either low- or high-pressure function, it is important for the technician to understand how to set the proper pressures for the pressure switch to perform the function for which it is to operate within the equipment. This requires the technician to adjust **cut-in** and **cut-out** pressures so that the switch can function properly in the system. Adjustable pressure switches have adjustment screws so that the desired cut-in and cut-out pressures can be set. The **cut-in pressure** is the point at which the contacts of the pressure switch **will close** and allow electrical current flow to the circuit or load(s). The **cut-out pressure** is the point at which the pressure switch contacts **will open** and stop the electrical current flow through the circuit or load(s). The **differential** is the difference between the cut-in and cut-out pressure settings. Other pressure switches will have the following listed on the pressure scale plate: **"Switch low event is high event minus the differential."** Depending on the type of pressure switch, *low event is the pressure at which the* **switch contacts will open or close***, stopping or allowing electrical*

131

current flow through the circuit or load(s). Switch **high event** is the **pressure at which the switch contacts will open or close,** *stopping or allowing electrical current flow to the circuit or load(s).* The *differential is the difference between the low-event and high-event pressure settings.*

$$\text{Cut-out} - \text{differential} = \text{cut-in}$$
$$360 \text{ psi} - 120 \text{ psi} = 240$$

Only qualified technicians should make differential adjustments. A *differential setting that is too low* can cause short cycling of the compressor and any other load devices in the circuit. This can cause overheating of the compressor motor windings and can causing pitting of electrical contacts. A *differential that is set too high* may never allow the motor to shut off, raising electrical consumption. The cut-out setting will be changed automatically anytime the cut-in setting is changed. (*Refer to Figure 2–58, Theory Lesson: High- and Low-Pressure Switches.*) All control settings should be set at the manufacturer specifications. If this information is not available, the technician must consider what function the pressure switch is to serve in the sealed system in which it is to be used. Once this decision is made, the proper pressure switch can be selected. Technicians can determine the settings on a particular pressure switch as long as there is a firm understanding of the operation of refrigeration or air conditioning sealed system. In addition there must be an understanding of the temperature–pressure relationship of the refrigerant being used in the system and the design operating temperature ranges.

Checking High-Pressure Switching

High-pressure switches can be checked by attaching a gage manifold set to the refrigeration system and *disconnecting the condenser fan or by blocking the airflow across the condenser coil.* Then observe the cut-out and cut-in pressures. On *water cooled condensers, reduce or shut off the water flow* and observe the cut-out and cut-in pressures. If the pressure required cannot be obtained, the pressure switch should be replaced. Nitrogen can also be used to pressurize the switch so the switch action and pressure settings can be tested with an ohmmeter in the same manner.

EQUIPMENT REQUIRED

1 Ohmmeter
1 Adjustable high-pressure switch
1 Refrigeration manifold gage set
1 Bottle of refrigerant or nitrogen bottle with regulator
Safety glasses

SAFETY PRACTICES

Be sure to wear safety glasses and make sure ohmmeter is set to proper function and value scale.

> **NOTE:** *If pressure switch is in an existing sealed system, disconnect power from the switch or remove switch from system for testing.*

COMPETENCY PROCEDURES

Checklist

1. Remove pressure switch cover to gain access to the switch's electrical terminals. ❏
2. With the ohmmeter set to read resistance, attach ohmmeter leads to the switch's electrical terminals. ❏
   Verify that the switch contacts are in the closed position. Are they? _____

> **NOTE:** *If the switch contacts are closed, there should be continuity. If the contacts are open there should be no continuity or infinity.*

3. Remove the high-side hose from the manifold gage and attach the high-pressure control to the manifold high-side gage port. ❏

NOTE: *Make sure manifold gage valve is shut.*

4. Attach the center hose from the manifold gage set to the refrigerant bottle or nitrogen bottle regulator. ❏
5. Review the pressure switch's scale plate and determine the set cut-out and cut-in pressure settings. ❏
   Record: cut-out pressure _____ cut-in pressure _____
6. Open the valve on the refrigeration cylinder or nitrogen cylinder. ❏

NOTE: *If using a nitrogen cylinder to perform the test, set the regular pressure higher than the desired cut-out pressure.*

7. Slowly open the high-side gage manifold valve and observe the high-side gage and ohmmeter. ❏
8. *Record the pressure at which the ohmmeter indicated that the switch opened.* _____ ❏
9. *During the increase of pressure, did the switch open? Record your answer.* _____ ❏
10. *Record the pressure at which the switch opened?* _____ ❏
11. *Did this pressure match the cut-out pressure as recorded in step 5?* _____ ❏
12. Close the refrigerant cylinder or nitrogen cylinder valve. ❏
13. Slowly unscrew the center hose attached to the gage manifold and observe the ohmmeter and high-side gage. ❏

NOTE: *Most high-pressure controls are factory set with a 100 psig differential.*

NOTE: *If the high-pressure control is a manual reset switch, continually push the reset switch as the pressure is being bled. Watch and record the pressure at which the reset switch holds.*

14. *At what pressure did the ohmmeter indicate that the pressure switch closed?* _____ ❏
15. *Did this pressure match the cut-in pressure as recorded in step 5?* _____ ❏
16. *Did the switch activate at the preset cut-out and cut-in pressures?* _____ ❏

NOTE: *If the switch did not operate in the preset cut-out and cut-in ranges, make adjustments as needed and retest as describe in procedures 7–14.*

17. Have your instructor check your work. ❏
18. Turn off the valve to the refrigerant or nitrogen cylinder. ❏

NOTE: *If using nitrogen, back out the pressure regulator valve.*

19. Slowly unscrew the hose connected to the refrigerant or nitrogen cylinder and allow the pressure to bleed out. ❑
20. Disconnect the pressure switch from the manifold. ❑
21. If the pressure switch being tested was removed from a sealed system, replace it if working properly. If the switch could not be adjusted for proper operation at the desired cut-out and cut-in range, see your instructor for a replacement. ❑
22. Return all materials to their proper location. ❑

RESEARCH QUESTIONS

1. Electrically how are pressure switches wired into an electrical circuit?
   _____
   _____

2. Are nonadjustable pressure switches adjustable?
   _____
   _____

3. On adjustable pressure switches, technicians must set the _____ and _____ pressures for the switch to operate in the system properly.

4. High event is the pressure at which the switch contacts of the pressure switch will do what?
   _____
   _____

5. The differential setting of a pressure switch is the difference between what and what?
   _____
   _____

Passed Competency _____   Failed Competency _____

Instructor Signature _____   Grade _____

Practical Competency 30

Calculating and Setting Cut-in, Differential, and Cut-out Pressures for Low-Pressure Control Used for Low-Charge Protection

SUGGESTED MATERIALS

Textbook

Refrigeration & Air Conditioning Technology, 5th Edition, Thomson Delmar Learning
Unit 14—Automatic Control Components and Applications
Unit 25—Special Refrigeration System Components
Unit 29—Troubleshooting and Typical Operating Conditions for Commercial Refrigeration

Review Topics

Pressure Switches; Typical Operating Conditions

COMPETENCY OBJECTIVE

The student will be able to calculate and set cut-in, cut-out, and differential pressures for a low-pressure control being used for low-charge protection.

OVERVIEW

For the most part, pressure switches used in the refrigeration and air conditioning industry are used as a *safety control* to protect the sealed system from low-charge or low-pressure conditions. When used in this manner, the cut-out pressure must be well below the system's normal operating pressure, yet not so low that it would allow the system to pull below atmospheric pressure. As a low-charge protection device, the cut-in setting should be the pressure that corresponds to the highest low-side pressure range the system's low-side pressure would normally operate under design conditions. An example would be an R-22 central air conditioning unit, which normally has a low-side operating pressure range of 60 to 70 psig under standard conditions. This means that the refrigerant boiling range in the evaporator would be approximately 35 to 40 degrees (F). At approximately 57 psig, the refrigerant in the evaporator would be boiling at 32 degrees (F) causing moisture in the air to start to freeze on the evaporator. The cut-out pressure would be set just above this 57 psig to prevent icing of the evaporator coil. The cut-in setting would be set in this range. Low-pressure switches used in this manner are normally manual reset switches, although some are automatic reset. The manual reset type prevents the compressor from short cycling and possibly pulling into a vacuum, contaminating the system by sucking air and moisture into the sealed system. An automatic reset switch would cut OFF and ON the compressor (**short cycling**). This would continue as long as there was enough refrigerant left in the system that during the "OFF cycle," system pressures would rise and stabilize between low-side and high-side, causing the *cut-in* pressure to be reached, bring the compressor back ON. At this point the compressor would quickly suck the low-side pressure down to the *cut-out* pressure, shutting the system OFF again. The short cycling would continue until the system was serviced. An owner's complaint may be that the system is turning ON and OFF, but no refrigerant effect is taking place.

The majority of pressure switches are adjustable. On **adjustable pressure switches** of either low- or high-pressure function, it is important for the technician to understand how to set the proper operating pressures for the switch to function and maintain designed equipment operation. This requires that the cut-in and cut-out pressures be adjusted so that the switch can function properly in the system. Adjustable pressure switches have adjustment screws so that the desired cut-in and cut-out pressures can be set.

The **cut-in pressure** is the point at which the contacts of the pressure switch **will close** and allow electrical current flow to the circuit or load(s). The **cut-out pressure** is the point at which the pressure switch contacts **will open** and stop electrical current flow through the circuit or load(s). The **differential** is the difference between the *cut-in* and *cut-out* pressure settings. The cut-out pressure on most pressure switch is determined by the following formulas:

$$\text{Cut-in} - \text{the Differential} = \text{Cut-out}$$
$$\text{OR}$$
$$\text{Cut-in} - \text{Cut-out} = \text{Differential}$$

DETERMINING DIFFERENTIAL SETTING OR CUT-OUT PSIG

In the example of the R-22 central air conditioning unit referred to earlier which normally has low-side operating pressure range of 60 to 70 psig under standard conditions. The cut-in setting would be set in this range; lets say it's set at **70 psig**. It was also stated that at 57 psig, the evaporator coil would reach 32 degrees (F) and moisture from the air would start icing the evaporator coil, so *the cut-out pressure should be set above this point.* Between **58** and **59 psig**, the refrigerant in the evaporator would be boiling between 33 and 34 degrees (F), preventing the evaporator from icing. Using the formulas listed above:

Finding the Differential

$$\text{Cut-in} - \text{Cut-out} = \text{Differential}$$
70 PSIG – 59 PSIG = 11

OR

Finding the Cut-out psig

$$\text{Cut-in} - \text{Differential} = \text{Cut-out}$$
70 PSIG – 11 = 59 PSIG

NOTE: In this example, cut-in would be 70 psig with a differential of 11 psig and would allow the pressure switch to stop the compressor once the system's pressure reached 58 psig, also preventing icing of the evaporator coil.

Some pressure switches will have the following listed on the pressure scale plate: **"Switch low event is high event minus the differential."** Depending on the switch design, *low event is the pressure at which the* switch contacts **will open or close**, *stopping or allowing electrical current flow through the circuit or load(s).* Switch **high event** is the **pressure at which the switch contacts will open or close**, *stopping or allowing electrical current flow to the circuit or load(s).* The *differential is the difference between the low-event and high-event pressure settings.* Pressure settings of this type of switch would be determined in the same manner as listed above.

High Event – differential = Low Event
70 psig – 11 psig = 59 psig

Only qualified technicians should make differential adjustments. A *differential setting that is too low* can cause short cycling of the compressor and any other load devices in the circuit. This can cause overheating of the compressor motor windings and can causing pitting of electrical contacts. A *differential that is set too high* may never allow the motor to shut off, causing higher electrical consumption. The cut-out setting will be changed automatically anytime the cut-in setting is changed. (*Refer to Figure 2–58, Theory Lesson: High- and Low-Pressure Switches.*) All control settings should be set at the manufacturer specifications. If this information is not available, the technician must consider what function the pressure switch is to serve in the sealed system in which it is to be used. Once this decision is made, the proper pressure switch can be selected. Technicians can determine the settings on a particular pressure switch as long as they have a firm understanding of the operation of refrigeration or air conditioning sealed system. In addition they must have an understanding of the temperature–pressure relationship of the refrigerant being used in the system and the **design operating temperature ranges**.

EQUIPMENT REQUIRED

1 Ohmmeter
1 Adjustable low-pressure switch
1 Operational system with low-pressure control being used as low-charge protection
1 Refrigeration manifold gage set
1 Bottle of refrigerant or nitrogen bottle with regulator
Safety glasses
1 Standard blade screwdriver

SAFETY PRACTICES

Be sure to wear safety glasses when working with refrigerants or any pressure equipment

COMPETENCY PROCEDURES   Checklist

> NOTE: *This competency can be performed on an operational system with a low-pressure control or on a low-pressure control removed from a sealed system.*

COMPETENCY PROCEDURES USING A LOW-PRESSURE CONTROL REMOVED FROM THE SEALED SYSTEM

SYSTEM FACTS: This is a medium-temperature cooler being controlled by a temperature control set to maintain a box temperature range of 45 degrees (F) to 34 degree (F). At the cut-out temperature of 34 degrees (F), the evaporator coil will be approximately 15 degrees colder than the box air temperature of 34 degrees (F). The system is charged with R-134a. The lowest pressure the system would operate to maintain a 34-degree (F) box temperature would be 18 psig. This is determined by the fact that at 34 degrees (F) box temperature minus the 15 degree (F) evaporator temperature differential, the evaporator coil is approximately 19 degrees (F) which converts to 18 psig. The cut-out control should be set well below this pressure, but not low enough to allow the compressor to pull into a vacuum. For this exercise we will use 5 psig as the cut-out pressure.

The cut-in pressure should be set at approximately 15 degrees below the highest expected box temperature of 45 degrees.

EXERCISE: Using the information stated above, set the proper cut-in, differential, and cut-out pressure on the low pressure control to be used in this cooler.

> NOTE: *Low-Pressure Control states the following:*
> **Cut-out is cut-in − the differential**

DETERMINE CUT-IN PRESSURE
1. *The temperature control will turn the cooler ON when the box temperature reaches what temperature?* _____   ❑
2. *What would be the system's low-side pressure at the box temperature recorded in Step 1?* _____   ❑
3. *The pressure control's cut-in pressure should be set for how many degrees below the pressure recorded in Step 2?* _____   ❑

4. Based on information determined in Steps 2 and 3, what should be the **cut-in pressure** for the control? _____ ❑

DETERMINE CUT-OUT PRESSURE

5. The temperature control will turn the compressor OFF to the cooler at what temperature? _____ ❑
6. What would be the low-side pressure at the point that the temperature control turns the compressor OFF? _____ ❑
7. What pressure was it determined that the pressure control would be set to shut the compressor OFF to prevent pulling into a vacuum in case of a low-charge situation? _____ ❑

DETERMINING THE DIFFERENTIAL SETTING

8. What is the established **cut-in** pressure? _____ ❑
9. What is the established **cut-out** pressure? _____ ❑
10. Record the differential. _____ ❑
11. Set the cut-in and differential settings on the low-pressure control. ❑
12. Have your instructor check your calculations and pressure control settings. ❑

NOTE: *The instructor may have you check calibration and settings following procedures listed in Practical Competency 28: Checking Calibration and Switch Activation of an Adjustable Low-Pressure Control.*

---

13. On completion of Competency, return all materials to their proper location. ❑

RESEARCH QUESTIONS

1. The differential setting on a pressure control is determined how?

   _____

   _____

2. If the cut-in pressure is 150 psi and the cut-out is 300 psi, what would be the differential setting?

   _____

   _____

3. If the cut-in is 45 psi and the differential is 20 psi, what is the cut-out pressure setting?

   _____

   _____

4. If you were adding a low-pressure switch on a sealed system, what side of the sealed system would you access pressure for the switch to operate?

   _____

   _____

5. A high-side pressure switch would be used in a sealed system to protect the system from what?

   _____

   _____

Passed Competency _____   Failed Competency _____

Instructor Signature _____   Grade _____

Practical Competency 31

Calculating and Setting Cut-in, Differential, and Cut-out Pressures for a High-Pressure Control

SUGGESTED MATERIALS

Textbook
Refrigeration & Air Conditioning Technology, 5th Edition, Thomson Delmar Learning
Unit 14—Automatic Control Components and Applications
Unit 22—Condensers
Unit 25—Special Refrigeration System Components
Unit 29—Troubleshooting and Typical Operating Conditions for Commercial Refrigeration

Review Topics
High Pressure Control; Standard Air-Cooled Condenser; High-Efficiency Air-Cooled condenser; Water-Cooled condenser; Cut-out; Cut-in; Differential; Normally closed; Manual reset; Automatic reset

COMPETENCY OBJECTIVE

The student will be able to calculate and set cut-out, cut-in, and differential settings on a high-pressure control.

OVERVIEW

High-pressure controls are used to keep the compressor from operating with a high head pressure. They are *normally closed switches,* designed to open on a raise in pressure and may be either *manual reset* or *automatic reset* switches. High-pressure switches can be found on some air-cooled condensers and are necessary on all water-cooled condensers. Establishing the **cut-out, cut-in,** and **differential** settings on these controls is not as complicated as calculations required to set low-pressure controls.

Before calculating and setting the cut-in and cut-out pressures on these controls, it is important to recognize the design condensing temperature for air-cooled and water-cooled condensers currently used on refrigerated and air conditioning equipment. There are two types of air-cooled condensers: **standard efficiency** and **high efficiency.** Based on **standard conditions** (*inside air temperature of 95 degrees Fahrenheit with 50% humidity and an outside air temperature of 95 degrees*), *standard air-cooled condensers* are designed to *condense the refrigerant vapor at approximately 30 degrees above this temperature.* This means that condensing of the refrigerant vapor will take place at approximately 125 degrees (F) (**Figure 2–62**).

High-efficiency air-cooled condensers are designed to condense the refrigerant vapor at approximately 15 to 20 degrees (F) above standard conditions. This means that condensing of the refrigerant vapor will take place at approximately 110 to 115 degrees (F) (**Figure 2–63**).

Water-cooled condensers are designed to condense the refrigerant vapor at approximately 10 degrees (F) above the leaving water temperature. This means that condensing of the refrigerant vapor will take place at approximately 105 degrees (F) based on a leaving water temperature of 95 degrees (F) (**Figure 2–64**).

Understanding the designed condensing temperatures of the type of condenser being used on the equipment will aid in establishing the cut-out pressure. Most importantly, the cut-out pressure should be set above the design saturation temperature of the condenser being used. **For example,** standard air-cooled condensers condense at approximately 30 degrees above the design condition of 95 degrees (F). This means the design saturation temperature is 125 degrees (F). If this were an R-22 system the design head operating head pressure would

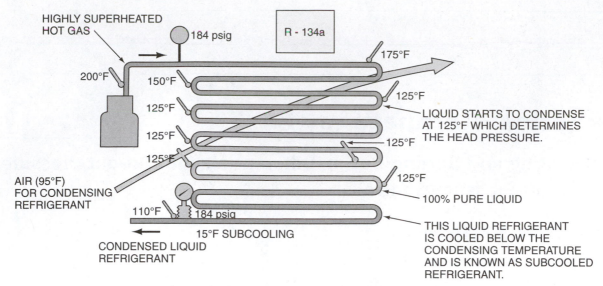

FIGURE 2–62 R-134a unit with standard air-cooled condenser performing at standard condition of 95 degree outside air temperature and saturation temperature of 125 degrees (F), with a head pressure of 184 psig.

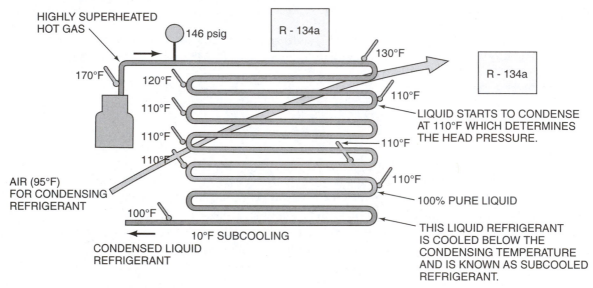

FIGURE 2–63 R-134a unit with high-efficiency air-cooled condenser performing at standard condition of 95 degrees outside air temperature and saturation temperature of 110 degrees (F), with a head pressure of 146 psig.

be 277 psig. This means that the high-pressure control's cut-out pressure would have to be set well above 277 psig, so that it doesn't stop the refrigeration cycle during normal system operation, yet is low enough to protect the equipment. The **standard rule** *is that the cut-out pressure should not be any higher than 20 % over the maximum expected operating pressure for the system. Consideration of the maximum operating temperature range where the condenser is located would have determined for cut-out pressure setting based on this rule.*

Most high-pressure switches come factory set with a differential of 50 psig or 100 psig. If the switch is set to cut-out at 350 psig and has a factory set differential of 100 psig, the cut-in would happen once the pressure dropped 100 psig below the cut-in pressure, which would equal a cut-in pressure of 250 psig. If the high-pressure control does not have a fixed differential setting, the rule of thumb for setting the differential is 65 psig.

Checking High-Pressure Switching

High-pressure switches can be checked by attaching a gage manifold set to the refrigeration system and *disconnecting the condenser fan or by blocking the airflow across the condenser coil*. Then observe the cut-out and cut-in pressures. On *water-cooled condensers, reduce or shut off the water flow* and observe

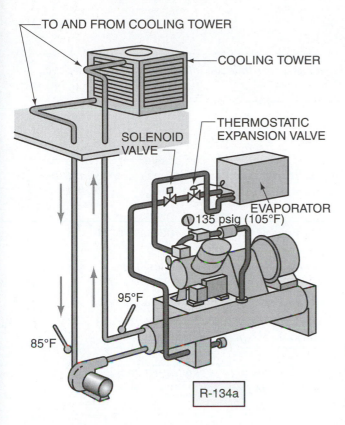

FIGURE 2–64 R-134a unit with a recirculated water-cooled condenser with a leaving water temperature of 95 degrees and saturation temperature at 105 degrees (F), with head pressure of 135 psig.

the cut-out and cut-in pressures. If the pressure required cannot be obtained, the pressure switch should be replaced. Nitrogen can also be used to pressurize the switch so the switch action and pressure settings can be tested with an ohmmeter in the same manner.

EQUIPMENT REQUIRED

1 Ohmmeter
1 Adjustable high-pressure switch
1 Refrigeration manifold gage set
1 Bottle of refrigerant or nitrogen bottle with regulator
Safety glasses
1 Standard blade screwdriver

SAFETY PRACTICES

Be sure to wear safety glasses.

COMPETENCY PROCEDURES

Checklist

> **NOTE:** *Competency can be performed using a high-pressure switch that is in an operational system or one that is removed from the system.*

COMPETENCY PROCEDURES FOR HIGH-PRESSURE CONTROL REMOVED FROM THE SYSTEM

> **NOTE:** *If performing competency on a system with a high-pressure control, turn power to unit OFF before testing calibration.*

1. *The sealed system uses what type of condenser?* _____ ❏

> **NOTE:** *If control is removed from a system, have instructor indicate what type of condenser the control would be used in.*

2. *What type of refrigerant is used in the sealed system?* _____ ❏

> **NOTE:** *If control is removed from a system, have instructor indicate what type of refrigerant the control would be used with.*

3. *What is the design condensing temperature of the condenser being used in this exercise based on standard conditions?* _____ ❏
4. *Under standard conditions, what would be the system's design operating high-side pressure?* _____ ❏
5. *Using the rule of 20% over the system design operating high-side pressure, calculate the high-pressure control's cut-out pressure.* _____ ❏
6. *If the pressure control has a **preset differential**, calculate the pressure control's cut-in pressure.* _____ ❏
7. *If the pressure control **does not have a preset differential**, using the rule stated above for setting the differential, calculate the pressure switch's cut-in pressure.* _____ ❏
8. Set the calculated cut-out pressure on the high-pressure control. ❏
9. If required, set the pressure control differential. ❏

> **NOTE:** *To check cut-out and cut-in pressures refer to procedures listed in Practical Competency 29: Checking Calibration and Switch Activation of an Adjustable High-Pressure Control, or checked by attaching a gage manifold set to the refrigeration system and disconnecting the condenser fan or by blocking the airflow across the condenser coil. Then observe the cut-out and cut-in pressures.*

10. Have your instructor check your readings. ❏
11. If the pressure switch was removed from a sealed system, replace in the system if performance check was OK. If the switch was defective or out of calibration, check with your instructor for a replacement. ❏
12. Return all materials to their proper location. ❏

RESEARCH QUESTIONS

1. Cut-out pressure is the pressure at which the switch contacts of a high-pressure switch will do what?

_____

_____

2. If a pressure switch opens on a rise in pressure, is the switch a NO or NC switch?

_____

_____

3. If a pressure switch closes on a drop in pressure, is it a NO or NC switch?

_____

_____

4. How will a manual reset pressure switch reset?

_____

_____

5. How will an automatic reset switch reset?

_____

_____

Passed Competency _____ Failed Competency _____

Instructor Signature _____ Grade _____

Practical Competency 32

Water-Cooled Condenser, High-Pressure Control Calculation and Settings

SUGGESTED MATERIALS

Textbook
Refrigeration & Air Conditioning Technology, 5th Edition, Thomson Delmar Learning
Unit 14—Automatic Control Components and Applications
Unit 22—Condensers
Unit 25—Special Refrigeration System Components
Unit 29—Troubleshooting and Typical Operating Conditions for Commercial Refrigeration

Review Topics
High-Pressure Control; Water-Cooled Condenser; Cut-out; Cut-in; Differential; Normally closed; Manual reset; Automatic reset

COMPETENCY OBJECTIVE

The student will be able to calculate and set cut-out, cut-in, and differential settings on a high-pressure control used in a water-cooled condenser.

OVERVIEW

High-pressure controls are used to keep the compressor from operating with a high head pressure. They are *normally closed switches,* designed to open on a raise in pressure and may be either *manual reset* or *automatic reset* switches. High-pressure switches can be found on some air-cooled condensers and are necessary on all water-cooled condensers. Establishing the **cut-out, cut-in,** and **differential** settings on these controls is not as complicated as calculations required to set low-pressure controls.

Before calculating and setting the cut-in and cut-out pressures on these controls, it is important to recognize the design condensing temperature for air-cooled and water-cooled condensers currently used on refrigerated and air conditioning equipment.

Water-cooled condensers are designed to condense the refrigerant vapor at approximately 10 degrees (F) above the leaving water temperature. This means that condensing of the refrigerant vapor will take place at approximately 105 degrees (F) based on a leaving water temperature of 95 degrees (F).

Understanding the designed condensing temperatures of the type of condenser being used on the equipment will aid in establishing the cut-out pressure. Most importantly, the cut-out pressure should be set above the design saturation temperature of the condenser being used. **For example,** standard air-cooled condensers condense at approximately 30 degrees above the design condition of 95 degrees (F). This means the design saturation temperature is 125 degrees (F). If this were an R-22 system the design head operating head pressure would be 277 psig. This means that the high-pressure control's cut-out pressure would have to be set well above 277 psig, so that it doesn't stop the refrigeration cycle during normal system operation, yet is low enough to protect the equipment. **Standard rule** *is that the cut-out pressure should not be any higher than 20% over the maximum expected operating pressure for the system. Consideration of the maximum operating temperature range where the condenser is located would have determined for cut-out pressure setting based on this rule.*

Most high-pressure switches come factory set with a differential of 50 psig or 100 psig. If the switch is set to cut-out at 350 psig and has a factory set differential of 100 psig, the cut-in would happen once the pressure dropped 100 psig below the cut-in pressure, which would equal a cut-in pressure of 250 psig. If the high-pressure control does not have a fixed differential setting, the rule of thumb for setting the differential is 65 psig.

Checking High-Pressure Switching

On *water-cooled condensers, reduce or shut off the water flow* and observe the cut-out pressure. If the pressure required cannot be obtained, the pressure switch should be replaced. Nitrogen can also be used to pressurize the switch so the switch action and pressure settings can be tested with an ohmmeter in the same manner.

EQUIPMENT REQUIRED

1 Ohmmeter
1 Adjustable high-pressure switch
1 Temperature tester
1 Refrigeration manifold gage set
1 Bottle of refrigerant or nitrogen bottle with regulator
Safety glasses
1 Standard blade screwdriver

SAFETY PRACTICES

Be sure to wear safety glasses.

COMPETENCY PROCEDURES   Checklist

> NOTE: *Competency can be performed using a high-pressure switch that is in an operational system or one that is removed from the system.*

COMPETENCY PROCEDURES FOR HIGH-PRESSURE CONTROL REMOVED FROM THE SYSTEM

> NOTE: *If performing competency on a water-cooled condenser, unit must be operational for proper temperature measurements and calibrations.*

1. *What type of refrigerant is used in the sealed system?* _____   ❑
   _____
   _____

> NOTE: *If control is removed from a system, have instructor indicate what type of refrigerant the control would be used with.*

2. Attach high-side manifold gage to the system's high-side service valve.   ❑
3. Mid-seat the high-side service valve (*if required*).   ❑
4. Turn water supply to system ON (*if required*).   ❑
5. Attach a temperature probe to the leaving water line of the condenser.   ❑
6. Turn system ON and let operate for 15 minutes.   ❑
7. *Record the leaving water temperature after unit operates for 15 minutes.* _____   ❑

8. *Record the system's high-side operating pressure.* _____ ❑
9. *Using the rule of 20% over the system design operating high-side pressure, calculate the high-pressure control cut-out pressure.* _____ ❑
10. *If the pressure control has a **preset differential**, calculate the pressure control's cut-in pressure.* _____ ❑
11. *If the pressure control **does not have a preset differential**, using the rule stated above for setting the differential, calculate the pressure switch's cut-in pressure.* _____ ❑
12. Set the calculated cut-out pressure on the high-pressure control. ❑
13. If required, set the pressure control differential. ❑

NOTE: *To check cut-out and cut-in pressures refer to procedures listed in Practical Competency 29: Checking Calibration and Switch Activation of an Adjustable High-Pressure Control, or on water-cooled condensers, reduce or shut off the water flow and observe the cut-out pressure and the cut-in pressure.*

---

14. Have your instructor check your readings. ❑
15. If the pressure switch was removed from a sealed system, replace in the system if the performance check was OK. If the switch was defective or out of calibration, check with your instructor for a replacement. ❑
16. Return all materials to their proper location. ❑

RESEARCH QUESTIONS

1. The type of condenser that has the lower operating head pressure is the _____ cooled condenser.

2. What are the two places that the water from water-cooled condensers can be deposited?

   _____

   _____

3. The condensing temperature for standard air-cooled condensers take place how many degrees above ambient temperature?

   _____

   _____

4. The condensing temperature for water-cooled condensers takes place how many degrees above the leaving water temperature?

   _____

   _____

5. **True or False:** Water-cooled condensers are more efficient than high-efficient air-cooled condensers.

   _____

   _____

Passed Competency _____   Failed Competency _____

Instructor Signature _____   Grade _____

Theory Lesson: Oil Pressure Safety Switch (Figure 2–65)

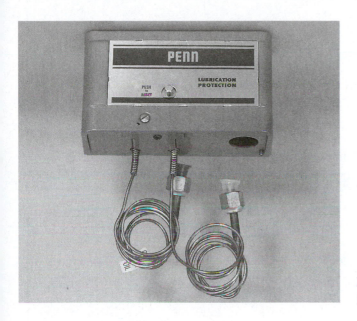

FIGURE 2–65   Oil safety control. (*Photo by Bill Johnson*)

SUGGESTED MATERIALS

Textbook
Refrigeration & Air Conditioning Technology, 5th Edition, Thomson Delmar Learning
Unit 14—Automatic Control Components and Applications
Unit 25–Special Refrigeration System Components

Review Topics
Basic Electricity and Magnetism; Motor Starting; Electrical Controls

Key Terms
control contacts • differential switch • net oil pressure • oil pressure • time delay heater • timer

OVERVIEW

The oil pressure control is used as a safety switch on large air conditioning and refrigeration compressors to ensure that the compressor has **oil pressure** when operating. Numerous large compressors, normally over 5 hp, used in the refrigeration and air-conditioning industry employ a forced oil system instead of the splash system used in smaller compressors. The larger compressor's lubrication system is contained in the compressor crankcase. An oil pump is located externally on these compressors along with an oil pressure safety switch. The oil safety switch has two capillary tubes with flare nuts attached at the ends so they can be attached to fittings on the compressor's suction line and oil pump.

In these compressors it is important that the proper oil pressure be established by the time the compressor comes up to speed, and maintained for as long as the compressor is operating. If a loss of oil pressure occurs, the compressor can encounter extreme damage. To protect the compressor from damage due to loss of oil pressure, oil pressure safety switches are installed to monitor the compressor's oil pressure.

If the proper oil pressure is not established in a predetermined amount of time after the compressor starts, or if there is a loss of oil pressure during the running cycle, the oil pressure switch will open the circuit to the **motor starting relay coil** and remove the compressor from the circuit (**Figure 2–66**).

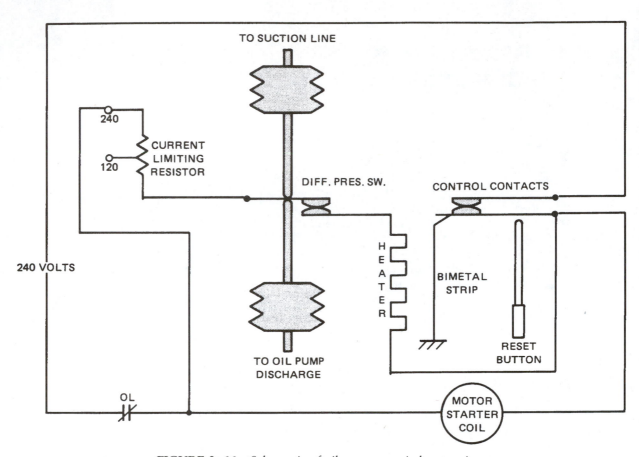

FIGURE 2–66   Schematic of oil pressure switch operation.

Once the oil pressure safety switch opens the circuit to the motor starter relay and removes the compressor from the circuit, it will have to be **manually reset** before the compressor can be restarted again.

The oil pressure safety switch contains several control functions that are incorporated in the same switch. There is a **differential switch**, a **timer**, and a set of **control contacts**. The oil safety pressure switch must be able to determine the compressor's **net oil pressure**. This is the pressure difference of the system's suction pressure and the discharge pressure of the oil pump. This is measured by the oil safety switch's **differential switch**.

Here is an example of a system's net oil pressure:
Total pressure reading at the oil discharge port = 55 psig
The system's suction pressure = 35 psig
Net oil pressure = discharge oil pressure − suction pressure
Net oil pressure = 20 psig
Most compressors need a net oil pressure of 30 psig to 45 psig to ensure proper oil lubrication.

Because it takes time for oil pressure to build in a compressor on start up, it must be given time to bring the oil pressure up to operating pressures. This allowable time is accomplished by the oil pressure switch's *time delay system*. The manufacturer normally preset the amount of the time for the delay, and in most cases it is preset for around 90 seconds (**Figure 2–67**).

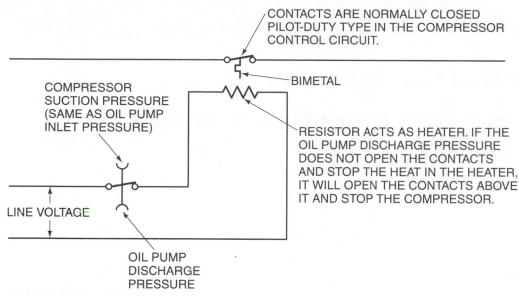

CONTACTS ARE NORMALLY CLOSED
PILOT-DUTY TYPE IN THE COMPRESSOR
CONTROL CIRCUIT.

BIMETAL

COMPRESSOR
SUCTION PRESSURE
(SAME AS OIL PUMP
INLET PRESSURE)

RESISTOR ACTS AS HEATER. IF THE
OIL PUMP DISCHARGE PRESSURE
DOES NOT OPEN THE CONTACTS
AND STOP THE HEAT IN THE HEATER,
IT WILL OPEN THE CONTACTS ABOVE
IT AND STOP THE COMPRESSOR.

LINE VOLTAGE

OIL PUMP
DISCHARGE
PRESSURE

FIGURE 2–67   Oil pressure control time-delay circuit.

If the oil pressure does not build up and overcome the opposing force of the system's suction pressure plus the differential spring setting within the time limit, the differential switch contacts will remain closed and the time delay heater will remain energized. This will heat the bimetal switch to a point that it warps, opening the time switch contacts and stopping the compressor operation. If the oil pressure builds to the designed oil pressure level in the allowable time limit, it will force the differential switch contacts open, removing the **time delay heater** from the circuit and the compressor will continue to operate.

During operation, if the oil level falls below the designed level, *the suction pressure and the spring differential pressure will overcome the oil pressure, closing the **differential switch contacts**.* The contacts will re-energize the **time delay heater**, causing it to heat the bimetal switch to a point that it warps, opening the time switch contacts and stopping the compressor operation.

Troubleshooting the Oil Pressure Safety Control

Leave the oil pressure safety switch in the refrigeration or air conditioning system. Attach manifold low-side gage to the suction service valve and connect the high-side gage to the oil pump discharge port. Connect leads of a voltmeter across the differential switch contacts. These are the *switch terminals 1 and 2 on the oil safety switch. The voltmeter should have no indication of voltage present at this point.* Start the system and observe the manifold gages and voltmeter. In a short time of system operation, you should see a difference in the two pressures of at least 10 psi. This is the minimum required pressure difference for the differential switch contacts to open and remove the time delay circuit of the oil pressure safety switch. When the difference of 10 psi or more is reached between the suction pressure and oil pump discharge pressure, the contacts between terminals 1 and 2 should open and the voltmeter should indicate a voltage reading. *This tells you the switch is working OK.*

If the differential contacts do not open once the minimum pressure or higher is reached, the time delay heater will continue to heat the bimetal switch to a point that it warps and opens the time switch contacts, stopping supply voltage to the compressor circuit. *This means the oil safety switch is not operating properly and should be replaced.*

Practical Competency 33

Oil Pressure Safety Control Performance Check in an Operational System

SUGGESTED MATERIALS

Textbook

Refrigeration & Air Conditioning Technology, 5th Edition, Thomson Delmar Learning
Unit 14—Automatic Control Components and Applications
Unit 25—Special Refrigeration System Components

Review Topics
Oil Pressure Safety Control

COMPETENCY OBJECTIVE

The student will be able to check the oil pressure safety switches operation in a refrigeration or air conditioning system.

> **NOTE:** *This competency is for checking an oil pressure safety switch that is already operational in a refrigeration or air conditioning system.*

OVERVIEW

The oil pressure control is used as a safety switch on large air conditioning and refrigeration compressors to ensure that the compressor has **oil pressure** when operating. Numerous large compressors, normally over 5 hp, used in the refrigeration and air-conditioning industry employ a forced oil system instead of the splash system used in smaller compressors. The larger compressor's lubrication system is contained in the compressor crankcase. An oil pump is located externally on these compressors along with an oil pressure safety switch. The oil safety switch has two capillary tubes with flare nuts attached at the ends so they can be attached to fittings on the compressors suction line and oil pump.

In these compressors it is important that the proper oil pressure be established by the time the compressor comes up to speed and maintained for as long as the compressor is operating. If a loss of oil pressure occurs, the compressor could encounter extreme damage. To protect the compressor from damage due to loss of oil pressure, oil pressure safety switches are installed to monitor the compressors oil pressure.

If the proper oil pressure is not established in a predetermined amount of time after the compressor starts or if there is a loss of oil pressure during the running cycle, the oil pressure switch will open the circuit to the motor starting relay coil and remove the compressor from the circuit. Once the oil pressure safety switch opens the circuit to the motor starter relay and removes the compressor from the circuit, it will have to be manually reset before the compressor can be re-started again.

Most compressors need a net oil pressure of 30 psig to 45 psig to ensure proper oil lubrication.

Because it takes time for oil pressure to build in a compressor on start up, it must be given time to bring the oil pressure up to operating pressures. This allowable time is accomplished by the oil pressure switch's time delay system. The amount of the time for the delay system of the switch is normally preset by the manufacturer, and in most cases allows around 60 to 120 seconds for compressor operation for the oil pressure to build up before the compressor operation is terminated.

If the oil pressure does not build up and overcome the opposing force of the system's suction pressure plus the differential spring setting within the time limit, the differential switch contacts will remain closed and the time delay heater will remain energized. This will heat the bimetal switch to a point that it warps, opening the time switch contacts and stopping the compressor operation. If the oil pressure builds to the designed oil pressure level in the allowable time limit, it will force the differential switch contacts open, removing the time delay heater from the circuit and the compressor will continue to operate.

Troubleshooting the Oil Pressure Safety Control

Leave the oil pressure safety switch in the refrigeration or air conditioning system. Take a set of manifold gages and attach the low-side gage to the suction service valve and connect the high-side gage to the oil pump discharge port. Connect the leads of a voltmeter across the differential switch contacts. These are the switch terminals 1 and 2 on the oil safety switch.

Start the system and observe the manifold gages and voltmeter.

In a short time of system operation, you should see a difference in the two pressures of at least 10 psi. This is the minimum required pressure difference for the differential switch contacts to open and remove the time delay circuit of the oil pressure safety switch. When the difference of 10 psi or more is reached between the suction pressure and oil pump discharge pressure, the contacts between terminals 1 and 2 should open and the voltmeter should indicate a voltage reading. This tells you the switch is working OK.

If the differential contacts do not open once the minimum pressure or higher is reached, the time delay heater will continue to heat the bimetal switch to a point that it warps and opens the time switch contacts, stopping supply voltage to the compressor circuit. This means the oil safety switch is not operating properly and should be replaced.

EQUIPMENT REQUIRED

1 Refrigeration or air conditioning system with an operational oil pressure safety switch in the system
1 Manifold gage set
1 Refrigeration wrench (*if needed*)
1 VOM or voltmeter
Safety glasses
Clock or watch

SAFETY PRACTICES

Be sure to wear safety glasses and make sure the meter is set to proper function and hand valves are closed on the manifold gage set. Be aware of high pressures and live voltage.

COMPETENCY PROCEDURES   Checklist

1. Turn unit OFF.   ❏
2. Attach the low-side hose of the manifold gage set to the suction port of the refrigeration
   or air conditioning system.   ❏
3. Attach the high-side hose of the manifold gage set to the oil pump discharge port.   ❏
4. Set the voltmeter to the proper voltage and function range.   ❏
5. Attach the voltmeter leads to the oil pressure safety control *switch terminals 1 and 2*.   ❏
6. Have your instructor check your hookup.   ❏
7. With your instructor's permission, **start the refrigeration or air conditioning unit** and
   observe the gages and voltmeter scale plate.   ❏
8. *Once there is a pressure difference of at least 10 psig, record the voltage reading of the*
   voltmeter. _____   ❏
9. *Did the differential switch contacts open?* (**Yes or No**) _____   ❏

10. *If differential switch contacts opened would you say the oil safety switch is working or not? Record your response and explain.* _____ ❑
11. *Record the system's operating suction pressure.* _____ ❑
12. *Record the oil pump's discharge pressure.* _____ ❑
13. *What is the system's net oil pressure?* _____ ❑
14. Turn the system OFF. ❑
15. Disconnect the oil pressure safety switch's capillary tube hookup from the oil pump discharge port and put a cap on the oil pump discharge port. ❑
16. Have your instructor check your setup, and **with the instructor's permission, restart the system.** ❑
17. Observe the gages, the voltmeter scale, and timer or watch. ❑
18. *Record the voltmeter reading once the differential pressure between the suction and oil discharge is at least 10 psig. Record voltmeter reading.* _____ ❑
19. *Did the differential switch contacts open once the differential pressure was 10 psig or greater? Record your answer.* _____ ❑
20. If the differential switch contacts did not open, let the system continue to operate and record the amount of time it takes for the system to shut down automatically. ❑
21. Record the amount of time it took for the oil safety switch to shut the system down automatically. _____ ❑
22. *Explain what shut the system down.* ❑

NOTE: *You should have gotten a voltage reading in this test. If you did not, you were not set up properly with the oil safety pressure switch's capillary tube disconnected from the oil pump discharge port. Recheck your hookup.*

---

23. Have your instructor check your performance checks in this section. ❑
24. With your instructor's permission, disconnect your voltmeter and gages from the system and put the system back into operational condition. ❑
25. Return all equipment and test materials to their proper location. ❑

RESEARCH QUESTIONS

1. Oil pressure switches are used as what type switch on large compressors?

   _____

   _____

2. Large compressors use a _____ oil system instead of a splash system used in smaller compressors.
3. Oil pumps on large compressors are located _____ on the compressor.
4. Oil safety pressure switches work off of what two pressures from the compressor in a large refrigeration or air conditioning system? _____ _____
5. Why is it important to ensure that the proper oil pressure is maintained while the compressor is operating?

   _____

   _____

Passed Competency _____   Failed Competency _____

Instructor Signature _____   Grade _____

Practical Competency 34

Checking the Differential Function of an Oil Pressure Safety Switch

SUGGESTED MATERIALS

Textbook

Refrigeration & Air Conditioning Technology, 5th Edition, Thomson Delmar Learning
Unit 14—*Automatic Control Components and Applications*
Unit 25—*Special Refrigeration System Components*

Review Topics

Oil Pressure Safety Switches

COMPETENCY OBJECTIVE

The student will be able to check the oil pressure controls differential function outside of an operational system.

> NOTE: *This competency is for checking an oil pressure safety switch that is outside an operational system.*

OVERVIEW

Oil pressure switches act as safety switches on large air conditioning and refrigeration compressors. These compressors employ a forced oil system instead of the splash system used in smaller compressors. The larger compressor's lubrication system is contained in the compressor crankcase. An oil pump is located externally on these compressors along with an oil pressure safety switch. The oil safety switch has two capillary tubes with flare nuts attached at the ends so they can be attached to fittings on the compressor's suction line and oil pump.

If the oil pressure does not build up and overcome the opposing force of the system's suction pressure plus the differential spring setting within the time limit, the differential switch contacts will remain closed and the time delay heater will remain energized. This will heat the bimetal switch to a point that it warps, opening the time switch contacts and stopping the compressor operation.

If the oil pressure builds to the designed oil pressure level in the allowable time limit, it will force the differential switch contacts open, removing the **time delay heater** from the circuit and the compressor will continue to operate.

During operation, if the oil level falls below the designed level, *the suction pressure and the spring differential pressure will overcome the oil pressure, closing the **differential switch contacts***. The contacts will re-energize the **time delay heater**, causing it to heat the bimetal switch to a point that it warps, opening the time switch contacts and stopping the compressor operation.

Troubleshooting the Oil Pressure Safety Control

Leave the oil pressure safety switch in the refrigeration or air conditioning system. Take a set of manifold gages and attach the low-side gage to the suction service valve and connect the high-side gage to the oil pump discharge port. Connect the leads of a voltmeter across the differential switch contacts. These are the switch terminals 1 and 2 on the oil safety switch. Start the system and observe the manifold gages and voltmeter.

In a short time of system operation, you should see a difference in the two pressures of at least 10 psi. This is the minimum required pressure difference for the differential switch contacts to open and remove the time delay circuit of the oil safety pressure switch. When the difference of 10 psi or more is reached between the suction pressure and oil pump discharge pressure, the contacts between terminals 1 and 2 should open and the voltmeter should indicate a voltage reading. This tells you the switch is working OK.

If the differential contacts do not open once the minimum pressure or higher is reached, the time delay heater will continue to heat the bimetal switch to a point that it warps and opens the time switch contacts, stopping supply voltage to the compressor circuit. This means the oil safety switch is not operating properly and should be replaced.

EQUIPMENT REQUIRED

1 Oil pressure safety switch rated to operate on 120 volts
120 Voltage supply
1 120-Volt light bulb and light bulb receptacle

> *NOTE: Your instructor may allow you to use any load device that operates on 120 volts.*

1 Jumper cord with insulated alligator ends
2 Jumper wires with insulated alligator clips
1 Light bulb receptacle
1 Light (*120 Volt*)
1 Bottle of freon 22 or nitrogen bottle with pressure regulator
2 1/4' Flared union to fit the manifold gage hose and oil pressure safety switch flare nuts
1 Manifold gage set
1 VOM or voltmeter
Safety glasses
Clock or watch

SAFETY PRACTICES

Be sure to wear safety glasses. Make sure meter is set to the proper function and hand valves are closed on manifold gage set. Be aware of high pressures and live voltage.

COMPETENCY PROCEDURES

Checklist

1. Put the light bulb in the light bulb receptacle. ❑
2. Connect one leg of the **jumper cord** to terminal 1 of the oil pressure safety switch. ❑
3. Connect one leg of a **jumper wire** to terminal 2 of the oil pressure safety switch. ❑
4. Connect the opposite end of the **jumper wire** from the switch terminal 2 to one of the terminals or legs of the light bulb receptacle. ❑
5. Connect one end of the other **jumper wire** to the opposite terminal or leg of the light bulb receptacle. ❑
6. Attach the opposite end of the second **jumper wire** to the other leg of the **jumper cord**. ❑
7. Connect a 1/4' union to the suction pressure port and the oil discharge port of the oil pressure safety control. ❑
8. Connect the low-side hose to the suction port of the oil pressure control flared fitting. ❑
9. Connect the high-side hose from the manifold gage set to the oil discharge flared fitting of the oil pressure safety control. ❑
10. Make sure the valves on the manifold gage set are closed and connect the center hose of the manifold gage set to a refrigerant cylinder or nitrogen regulator. ❑
11. Set the voltmeter to the proper voltage and function range. ❑
12. Attach the voltmeter leads to the oil pressure safety control switch terminals 1 and 2. ❑

13. Have your instructor check your hookup. ❑
14. With your instructor's permission, plug the jumper cord into the proper power supply. ❑
15. *Did the light bulb come on? (**Yes or No**)* _____ ❑

> **NOTE:** *The light should have come on, if it did not, check the electrical circuit and oil pressure safety switch.*

16. Turn the refrigerant or nitrogen bottle on. ❑

> **NOTE:** *If using nitrogen cylinder for this competency, Adjust the regulator to a working pressure of 70 psig.*

17. Open the low-side manifold gage and allow the pressure to build to 10 psig. ❑
18. *Record the voltmeter voltage reading.* _____ ❑

> **NOTE:** *You should have gotten a (0) voltage reading at this point.*

19. Open the high-side gage manifold valve and allow the pressure to build to 50 psig. ❑
20. *Did the light bulb or load device go out or stop? (**Yes or No**)* _____ ❑

> **NOTE:** *The light bulb or whatever load device is being used in this competency should have gone out or stopped.*

21. *Record the voltmeter voltage reading at this point.* _____ ❑

> **NOTE:** *You should have gotten a supply voltage reading at this point.*

> **NOTE:** *This test was to see if the differential switch contacts opened or not. If they did open, the differential switch contacts are good and working. If the light bulb or load device did not go out or stop in this test, the switch is bad and should be replaced because the differential switch contacts did not open.*

22. Have your instructor check your procedures. ❑
23. Shut the refrigeration of nitrogen bottle OFF. ❑
24. Slowly loosen the manifold gage hoses and bleed the pressure from each. ❑
25. Unplug the jumper cord from the power supply. ❑
26. Disconnect all materials and return them to their proper location. ❑

RESEARCH QUESTIONS

1. How much time is normally allowed for the compressor to build operating pressure before the safety switch shuts the compressor off?

_____

_____

2. Which set of switch contacts on the oil safety switch opens the compressor starting circuit if the required oil pressure is not reached in the allowable time?

_____

_____

3. Which pressure on the compressor must overcome the suction pressure and spring differential pressure to disconnect the time delay heater to allow the compressor to continue to operate?

_____

_____

4. The differential switch contacts are used to energize and de-energize what function of the oil safety switch?

_____

_____

5. If the differential switch does not open once the differential switch pressure is reached and the time delay opens the compressor circuit, what should be done with the oil safety switch?

_____

_____

Theory Lesson: Heating and Cooling Thermostats (Figure 2–68)

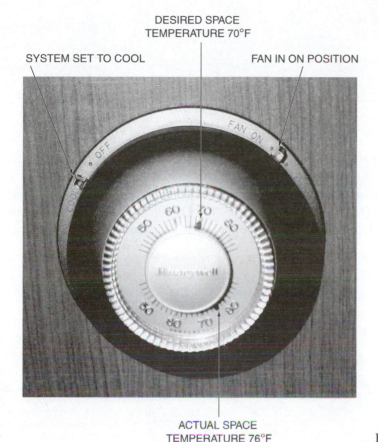

SYSTEM SET TO COOL

DESIRED SPACE
TEMPERATURE 70°F

FAN IN ON POSITION

ACTUAL SPACE
TEMPERATURE 76°F

FIGURE 2–68   Low-voltage thermostat.

SUGGESTED MATERIALS

Textbook
Refrigeration & Air Conditioning Technology, 5th Edition, Thomson Delmar Learning
Unit 14—Automatic Control Components and Applications

Key Terms
bimetallic thermostat • control contacts • cooling anticipator • dissimilar metals • heat anticipator • mercury bulb thermostat • overshoot • 24-volt control voltage

OVERVIEW

Heating and cooling thermostats play a very important part in heating and cooling equipment because they control the temperature of the conditioned space as well as being the command center for the operation of the total heating or cooling equipment. Most residential heating and cooling thermostats operate off of a **24-volt control circuit** power source. The **bimetallic thermostat** is the most common thermostat used in residential and commercial heating and cooling applications.

The bimetallic thermostat uses two **dissimilar metals** that are bonded together (**Figure 2–69**).

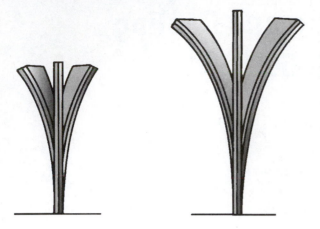

FIGURE 2–69 Two dissimilar metals reaction to temperature changes.

The metals are designed to expand in one direction or the other in response to temperature changes. One of the metals is designed to have very little expansion in low temperature, but greater expansion in higher temperatures. The other metal is designed to have a greater expansion in lower temperature, but little expansion in higher temperatures. This type of reaction to temperatures can be used to open or close the electrical **switch contacts** that are attached to the bimetal strip of the thermostat (**Figure 2–70**).

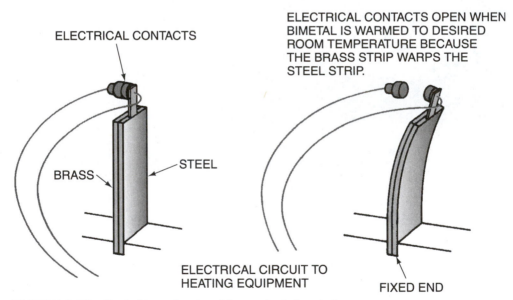

FIGURE 2–70 Basic bimetal strip with attached electrical contacts.

The **control contacts** of both types of thermostats are designed to close rapidly to avoid burnt or welded contacts. This is accomplished by using a magnetic snap action system, a spring action operation, or springs and toggles to produce a snap action of opening or closing of the contacts.

The **mercury bulb thermostat** is the most common low-voltage thermostat used for heating and cooling applications (**Figure 2–71**).

The **mercury bulb** of the thermostat is used to make and break the electrical circuit's contacts. The mercury bulb is fastened to a wound spiral bimetal coil. The **bimetal coil** will move in response to temperature change. This causes the mercury in the bulb to move back and forth over the electrical probes inserted in the bulb. As the mercury settles over the particular probes in the bulb, a 24-volt electrical circuit is

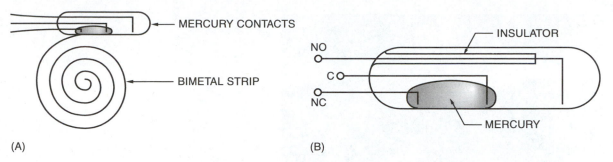

FIGURE 2–71   (A) Bimetal spiral. (B) Mercury bulb.

completed across the probes. This action completes the circuit for a particular function of the heating or cooling equipment.

The mercury bulb thermostats are designed with a **heat anticipator** that is designed to shut the heating equipment off a couple of degrees, normally 2 degrees (F), before the thermostat is satisfied at the desired temperature. This is to allow the additional heat in the heat exchanger to be used to bring the temperature of the conditioned space up to the thermostat set point. If this were not incorporated in the thermostat, there would be an overshoot of the conditioned space temperature.

The **heat anticipator** is a small coil of wire that is wired in series with the heating contacts of the thermostat (**Figure 2–72**).

This allows a small amount of current to pass through the heat anticipator coil, which heats the thermostat bimetal a couple of degrees higher than the space temperature. The current flow through the heat anticipator coil must be set at the proper setting for the heating equipment being used (**Figure 2–73**).

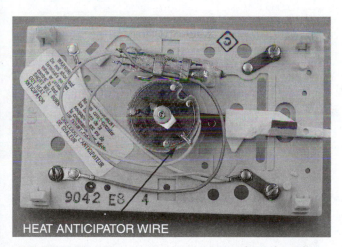

HEAT ANTICIPATOR WIRE

FIGURE 2–72   Heat anticipator wire. (*Photo by Bill Johnson*)

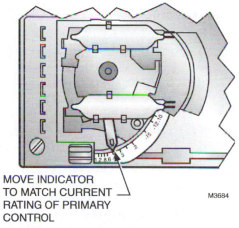

MOVE INDICATOR
TO MATCH CURRENT
RATING OF PRIMARY
CONTROL

M3684

FIGURE 2–73   Adjustable heat anticipator. The indicator must match the current rating of the heating control circuit. (*Courtesy of Honeywell, Inc.*)

The **cooling anticipator** is designed to prevent the conditioned space temperature from rising higher than the desired temperature setting before turning on the air conditioning equipment. The cooling anticipator coil or resistor is wired parallel with the cooling thermostat contacts (**Figure 2–74**).

When the thermostat is in the OFF position, a small amount of current passes through the resistor or anticipator coil, heating the bimetal coil a couple of degrees higher than the room temperature. This creates a false temperature reading by the thermostat and brings the air conditioning equipment on a couple of degrees lower than the thermostat setting.

FIGURE 2–74   Cold anticipator fixed resistor. (*Photo by Bill Johnson*)

> *NOTE: Additional information on heat anticipator and cooling anticipator settings will be covered in other Theory Lessons and Practical Competencies.*

Low-voltage thermostats can be purchased for *cooling only equipment, heating only equipment, heating and cooling equipment*, and *multistage equipment*. Regardless of the type of thermostat required for a particular installation, it is important to understand the universal low-voltage terminal functions of these thermostats.

Universally, low-voltage thermostat terminal functions are widely accepted by manufacturers of heating and cooling equipment and thermostat manufacturers (**Figure 2–75**).

Basic low-voltage terminal functions are as follows:
　R Terminal—*24-Volt power source*
　G Terminal—*Indoor fan/blower circuit*
　W Terminal—*Heating circuit*
　Y Terminal—*Cooling circuit*

> *NOTE: Programmable thermostats will have the addition of a C terminal. This is for the neutral lead of the transformer's 24-volt power supply. This is used to complete the 24-volt circuit for the thermostat's LED display.*

Thermostats with **RC and RH terminals** allow for two step-down transformer control circuit power supplies. They can also be used with a single transformer control circuit power supply, but require that a *jumper be placed between the RC and RH terminals on the thermostat.*

It's true that most technicians observe a practice of matching color of thermostat conductor wire to thermostat terminal lettering.

　R Terminal—*Red conductor wire*
　G Terminal—*Green conductor wire*
　W Terminal—*White conductor wire*
　Y Terminal—*Yellow conductor wire*

> *NOTE: Not all technicians follow this practice, so it's more important to recognize the terminal circuit functions of the thermostat as stated above.*

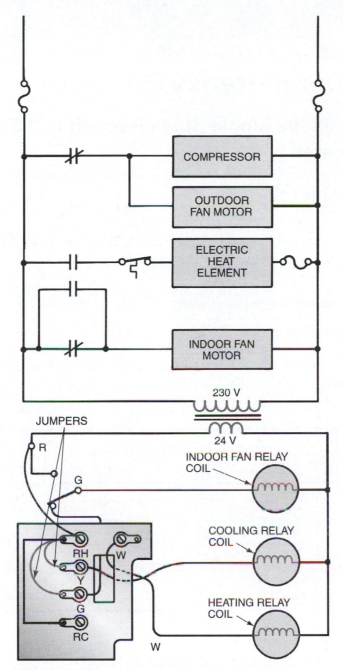

FIGURE 2–75   Thermostat terminal circuits.

Multistage thermostats provide multiple terminals for heating and cooling stages, and with thermostat terminals shown as:

W1—Heating—Stage 1
W2—Heating—Stage 2
Y1—Cooling—Stage 1
Y2—Cooling—Stage 2

These types of thermostats are *normally used on heat-pump systems.*

Besides the standard thermostat terminals discussed, the thermostat being used may have additional terminals. To determine the additional terminal circuit functions, refer to the thermostat instructions.

Student Name _____ Grade _____ Date _____

Practical Competency 35

Wiring a Low-Voltage Single-Stage Heating and Single-Stage Cooling Thermostat

SUGGESTED MATERIALS

Textbook
Refrigeration & Air Conditioning Technology, 5th Edition, Thomson Delmar Learning
Unit 14—Automatic Control Components and Applications

Review Topics
Space Temperature Controls; Low Voltage

COMPETENCY OBJECTIVE

The student will be able to wire a low-voltage single-stage heating and single-stage cooling thermostat for proper operation (**Figure 2–76**).

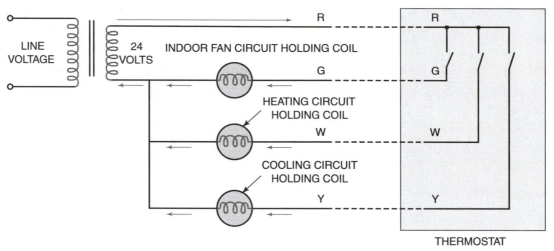

FIGURE 2–76   Simplified low-voltage thermostat control circuit.

OVERVIEW

The heating and cooling thermostat plays a very important part in heating and cooling equipment because it controls the temperature of the conditioned space as well as being the command center for the operation of the total heating or cooling equipment. Most residential heating and cooling thermostats operate off of a 24-volt control circuit power source.

Low-voltage thermostats can be purchased for *cooling only equipment, heating only equipment, heating and cooling equipment*, and *multistage equipment*. Regardless of the type of thermostat required for a particular installation, it is important to understand the universal low-voltage terminal functions of these thermostats. Universally, low-voltage thermostat terminal functions are widely accepted by manufacturers of heating and cooling equipment and thermostat manufacturers.

Basic low voltage terminal functions are as follows:

    R Terminal—*24-Volt power source*
    G Terminal—*Indoor fan/blower circuit*
    W Terminal—*Heating circuit*
    Y Terminal—*Cooling circuit*

NOTE: *Programmable thermostats will have the addition of a C terminal. This is for the neutral lead of the transformer's 24-volt power supply. This is used to complete the 24-volt circuit for the thermostat's LED display.*

Thermostats with **RC and RH terminals** allow for two step-down transformer control circuit power supplies. They can also be used with a single transformer control circuit power supply, but require that a *jumper be placed between the RC and RH terminals on the thermostat.*

It's true that most technicians observe a practice of matching color of thermostat conductor wire to thermostat terminal lettering.

    R Terminal—*Red conductor wire*
    G Terminal—*Green conductor wire*
    W Terminal—*White conductor wire*
    Y Terminal—*Yellow conductor wire*

NOTE: *Not all technicians follow this practice, so it's more important to recognize the terminal circuit functions of the thermostat as stated above.*

EQUIPMENT REQUIRED

1 Single-stage heating and cooling thermostat
1 120/24 Volt step-down transformer
3 Contactors with 24-volt coil
1 Jumper cord with insulated alligator clips
7 Jumper wires with alligator clips

SAFETY PRACTICES

Use the proper power source, size of conductors (wires), and loads for this exercise. Make sure that all conductors are properly insulated. Unless otherwise directed by your instructor, do not plug the circuit into the power source until the circuit is completely wired and checked by your instructor.

COMPETENCY PROCEDURES

Checklist

1. Attach one of the **jumper cord leads** to one of the transformer's primary coil wires. ❑
2. Attach the opposite **jumper cord lead** to the other transformer primary coil wire. ❑
3. Attach one end of a jumper wires to the transformer's secondary coil. ❑
4. Attach the opposite end of this wire to **terminal (R)** of the thermostat. ❑
5. Attach one end of a jumper wire to the opposite wire of the transformer's secondary coil. ❑
6. Attach the opposite end of this jumper wire to one of the contactor's 24-volt coil terminals. ❑
7. Attach one end of another jumper wire to the same contactor's 24-volt coil terminal used in procedure 6. ❑
8. Attach the opposite end of this jumper wire to one of the 24-volt coil terminals of the second contactor. ❑

9. Take another jumper wire and attach it to the same coil terminal used on the second contactor's 24-volt coil terminal used in procedure 8. ❑
10. Attach the other end of this jumper wire to one of the 24-volt coil terminals of the third contactor. ❑

> **NOTE:** *Determine which contactor you want to use for the A/C blower operation of the thermostat, heating operation of the thermostat, and the cooling operation of the thermostat. Once you have determined this, proceed.*

11. Take another jumper wire and attach one end of it to the thermostat's **terminal G.** ❑
12. Attach the opposite end of this jumper wire to the open coil terminal on the contactor that you decided to use for the thermostat's A/C blower operation. ❑
13. Take another jumper wire and attach one end of it to the thermostat's **terminal W.** ❑
14. Attach the opposite end of this jumper wire to the open coil terminal on the contactor that you decided to use for the thermostat's heating operation. ❑
15. Take another jumper wire and attach one end of it to the thermostat's **terminal Y.** ❑
16. Attach the opposite end of this jumper wire to the open coil terminal on the contactor that you decided to use for the thermostat's cooling operation. ❑

> **NOTE:** You have completed the *low-voltage control circuit wiring for the thermostat.*

CHECKING THE THERMOSTAT'S FAN ON AND AUTO OPERATION

17. Lay the thermostat on a flat surface and turn the fan switch to **Auto** and the selector switch to OFF. ❑
18. Unless directed otherwise, have your instructor check your wiring, and with approval, plug the jumper cord of the circuit into a 120-volt power source. ❑

> **NOTE:** *None of the contactor's should have been energized at this point.*

19. Turn the fan switch to the ON position. ❑
20. *Explain what happened when the fan switch was turned to ON.* _____ ❑
21. Turn the fan switch to **Auto** again.
22. *Explain what happened when the fan switch was turned back to Auto.* _____ ❑

> **NOTE:** *You should have realized in this test that the fan ON switch just energizes the blower contactor to turn just the blower on. This means that the customer could run the blower by itself to move air through the house or building without being in the heating or cooling mode.*

CHECKING THE THERMOSTAT'S COOLING OPERATION

23. With the fan switch in the **Auto** setting, move the thermostat's temperature control to its highest setting. ❑
24. Turn the thermostat's selector switch to the **Cooling** cycle. ❑
25. *Explain what happened when you moved the selector switch to the cooling mode.* _____ ❑
26. Now move the thermostat's **temperature control** to its **lowest setting.** ❑
27. *Explain what happened when the temperature control was moved to its lowest setting.* _____ ❑
28. Now move the **temperature control** back to the thermostat's **highest setting.** ❑

29. *Explain what happened when the temperature switch was moved back to its highest setting.* _____ ❏

> **NOTE:** *In this test you should have learned that just because the thermostat is set on the cooling mode, it does not mean that the A/C will come on unless the temperature control is set below the conditioned space temperature. You also should have learned that when the temperature control is moved below the conditioned space temperature, two contactors are energized at the same time: the A/C blower contactor and the A/C unit contactor. This is what the fan Auto function is for on the thermostat. Any time the thermostat is set to call for cooling, the A/C unit and indoor blower come on together and shut off together.*

CHECKING THE THERMOSTAT'S HEATING OPERATION
30. Set the thermostat's **temperature control** to the **lowest setting.** ❏
31. Move the thermostat's selector switch to the **Heating** cycle. ❏
32. Explain what happened when you moved the selector switch to the heating mode. _____ ❏ ❏
33. Move the **temperature control** switch to the thermostat **highest temperature** setting. ❏
34. *Explain what happened when you moved the temperature control switch to the thermostat's highest setting.* _____ ❏
35. Now move the thermostat's **temperature control** switch to the **lowest setting** a gain. ❏
36. *Explain what happened when you moved the thermostat's temperature control to the lowest setting.* _____ ❏

> **NOTE:** *From this test you should have learned that just because the thermostat is set to the selected heating mode does not mean that the heating cycle will come on. The temperature control must be set above the ambient temperature of the conditioned area.*

> **NOTE:** *In this test you should have noticed that the blower or fan relay was not energized as it was in the cooling cycle. The thermostat does not energize the fan/blower of the furnace when in the heating mode. In a forced hot air furnace the heating cycle would come on first and allow the heat exchanger to heat up before the fan/blower would be turned on. Another means would be used to bring the fan/blower on when the heat exchanger reached a set temperature. A fan-limit switch, stack switch, or a time relay accomplishes this. Once it was hot enough, the blower would be turned on by one of these methods.*

37. Turn the thermostat to the OFF position and have your instructor check your work.
38. With your instructor's permission, unplug the circuit and return all controls and supplies to their proper location.

RESEARCH QUESTIONS

1. In the heating mode of a forced hot air furnace, why would the fan/blower be delayed in coming on once the furnace began to heat?

   _____

   _____

2. In an oil or gas forced air furnace, what methods are used to bring the fan/blower on?

   _____

   _____

3. Why are low-voltage thermostats used for heating and cooling equipment instead of line-voltage thermostats?

_____

_____

4. A heating thermostat does what on a rise in temperature?

_____

_____

5. A cooling thermostat does what on a rise in temperature?

_____

_____

Practical Competency 36

Checking the Control Operation of a Single-Stage Heating and Cooling Thermostat

SUGGESTED MATERIALS

Textbook
Refrigeration & Air Conditioning Technology, 5th Edition, Thomson Delmar Learning
Unit 14—Automatic Control Components and Applications

Review Topics
Space Temperature Controls; Low Voltage

COMPETENCY OBJECTIVE

The student will be able to identify the thermostat terminal functions; wire the low-voltage power source for the heating and cooling thermostat; identify switching sequence of operation; and observe how the thermostat's low-voltage control circuits energize the main loads.

OVERVIEW

The heating and cooling thermostat plays a very important part in heating and cooling equipment because it controls the temperature of the conditioned space as well as being the command center for the operation of the total heating or cooling equipment. Most residential heating and cooling thermostats operate off of a 24-volt control circuit power source.

Low-voltage thermostats can be purchased for *cooling only equipment, heating only equipment, heating and cooling equipment*, and *multistage equipment*. Regardless of the type of thermostat required for a particular installation, it is important to understand the universal low-voltage terminal functions of these thermostats. Universally, low-voltage thermostat terminal functions are widely accepted by manufacturers of heating and cooling equipment and thermostat manufacturers.

Basic low voltage terminal functions are as follows:
   R Terminal—*24-Volt power source*
   G Terminal—*Indoor fan/blower circuit*
   W Terminal—*Heating circuit*
   Y Terminal—*Cooling circuit*

> **NOTE:** *Programmable thermostats will have the addition of a C terminal. This is for the neutral lead of the transformer's 24-volt power supply. This is used to complete the 24-volt circuit for the thermostat's LED display.*

Thermostats with **RC and RH terminals** allow for two step-down transformer control circuit power supplies. They can also be used with a single transformer control circuit power supply, but require that a *jumper be placed between the RC and RH terminals on the thermostat.*

It's true that most technicians observe a practice of matching color of thermostat conductor wire to thermostat terminal lettering.

R Terminal—*Red conductor wire*
G Terminal—*Green conductor wire*
W Terminal—*White conductor wire*
Y Terminal—*Yellow conductor wire*

> *NOTE: Not all technicians follow this practice, so it's more important to recognize the terminal circuit functions of the thermostat as stated above.*

EQUIPMENT REQUIRED

1 Single-stage heating and cooling thermostat
1 120/24-Volt step-down transformer
3 Contactors with 24-volt coil
1 120-Volt refrigerator evaporator fan motor or equivalent
1 Defrost heater or equivalent
1 Light bulb and light socket
1 Jumper cord with insulated alligator clips
1 Piece of masking tape (4")
18 Jumper wires with alligator clips

> *NOTE: This Competency can also be completed with three light bulbs and sockets or any other 120-volt load devices.*

SAFETY PRACTICES

Use the proper power source, size of conductors (*wires*), and loads for this exercise. Make sure that all conductors are properly insulated. Do not plug circuit into power source until circuit is completely wired and checked by your instructor.

COMPETENCY PROCEDURES   Checklist

1. Take the piece of masking tape and tear in half. With an ink pen, letter one piece of tape L-1 and other piece of tape L-2.   ❑
2. Attach the **L-1 jumper cord** lead to one of the transformer's primary coil wires.   ❑
3. Attach the **L-2 jumper cord** lead to the other transformer's primary coil wire.   ❑
4. Attach one end of a jumper to one of the wires of the transformer secondary coil.   ❑
5. Attach the opposite end of this wire to **terminal R** of the thermostat.   ❑
6. Attach one end of a **jumper wire** to the opposite wire of the transformer's secondary coil. ❑
7. Attach the opposite end of this jumper wire to one of the contactor's coil terminals.   ❑
8. Take another jumper wire and attach one end of it to the same contactor coil terminal used in procedure 6.   ❑
9. Attach the opposite end of this jumper wire to one coil terminal of the **second contactor**. ❑
10. Attach another jumper wire end to the same coil terminal on the **second contactor** coil used in procedure 8.   ❑
11. Attach the opposite end of this jumper wire to one of the coil terminals of the **third contactor**. ❑

12. Take another jumper wire and attach one end of it to the thermostat **terminal G.** ❏
13. Attach the opposite end of this jumper wire to the open coil terminal on the contactor that you decided to use for the thermostat's A/C blower operation. ❏
14. Take another jumper wire and attach one end of it to the thermostat **terminal W.** ❏
15. Attach the opposite end of this jumper wire to the open coil terminal on the contactor that you decided to use for the thermostat heating operation. ❏
16. Take another jumper wire and attach one end of it to the thermostat **terminal Y.** ❏
17. Attach the opposite end of this jumper wire to the open coil terminal on the contactor that you decided to use for the thermostat cooling operation. ❏

18. Take another jumper wire and attach one end of it to the **L-1 leg** of the jumper cord. ❏
19. At the contactor **terminal L-1** being used for the A/C blower operation, attach the opposite end of this jumper wire. ❏
20. Take another jumper wire and attach one end of it to **terminal T-1** on the contactor you used for the A/C blower. ❏
21. Attach the opposite end of this jumper wire to one of the fan motor power leads. ❏
22. Using the same fan motor, attach another jumper wire to the opposite fan motor power lead. ❏
23. Attach the other end of this jumper wire to the **L-2 leg of the jumper cord.** ❏
24. Take another jumper wire and attach one end of it to the **L-1 leg of the jumper cord.** ❏
25. Attach the opposite end to the **L-1 terminal** of the contactor being used for heating. ❏
26. Take another jumper wire and attach one end of it to the **T-1 contactor terminal** used for heating. ❏
27. Attach the opposite end of this jumper wire to one of the power leads of the **electric heater.** ❏
28. Take another jumper wire and attach it to the opposite power lead of the **electric heater.** ❏
29. Attach the opposite end of this jumper to the **L-2 lead of the jumper cord.** ❏
30. Take another jumper wire and attach one end of it to the **L-1 leg of the jumper cord.** ❏
31. Attach the opposite end of this jumper to the **L-1 terminal** of the contactor being used for the A/C condensing unit operation. ❏
32. Take another jumper wire and attach one end of it to the **T-1 contactor terminal** being used for the A/C condensing unit operation. ❏
33. Attach the opposite end of this jumper wire to one of the **power legs of the lamp socket** being used as the condensing unit. ❏
34. Take another jumper wire and attach one end of it to the opposite power leg of the lamp socket. ❏
35. Attach the other end of this jumper to the **L-2 lead of the jumper cord.** ❏
36. Lay the thermostat on a flat surface and turn the fan switch to **Auto** and the selector switch to OFF. ❏
37. Unless directed otherwise, have your instructor check your wiring, and with approval, plug the jumper cord of the circuit into a 120-volt power source. ❏

38. Turn the fan switch to the ON position. ❏

39. *Explain what happened when the fan switch was turned to ON.* ❏

_____

40. Turn the fan switch to **Auto** again.

41. *Explain what happened when the fan switch was turned back to Auto.* ❏

_____

> **NOTE:** *In this test you should have realized that the fan ON switch energizes the blower contactor to turn just the blower ON. This means that the customer could run the blower by itself as long as he wanted to move air through the house or building without being in the heating or cooling mode.*

CHECKING THE THERMOSTAT'S COOLING OPERATION

42. With the fan switch in the **Auto** setting, move the thermostat's temperature control to its **highest setting.** ❏

43. Turn the thermostat's selector switch to the **Cooling** cycle. ❏

44. *Explain what happened when you moved the selector switch to the cooling mode.* ❏

_____ ❏

45. Now move the thermostat's temperature control to its lowest setting.

46. *Explain what happened when the temperature control was moved to its lowest setting.* ❏

_____

47. Now move the **temperature control** back to the thermostat's **highest setting.** ❏

48. *Explain what happened when the temperature switch was moved back to its highest setting.* ❏

_____

> **NOTE:** *In this test you should have learned that just because the thermostat is set on the cooling mode, it does not mean that the A/C will come on unless the temperature control is set below the conditioned space present temperature. You also should have learned that when the temperature control is moved below the conditioned area's present temperature, two contactors are energized at the same time—the A/C blower contactor and the A/C unit contactor. This is what the fan Auto function is for on the thermostat. Any time the thermostat is set to call for cooling, the A/C unit and indoor blower come on together and shut off together.*

CHECKING THE THERMOSTAT'S HEATING OPERATION

49. Set the thermostat's **temperature control** to the **lowest setting.** ❏

50. Move the thermostat selector switch to the **HEATING** mode. ❏

51. *Explain what happened when you moved the selector switch to the heating mode.* ❏

_____

52. Move the **temperature control** switch to the thermostat's **highest temperature** setting. ❏

53. *Explain what happened when you moved the temperature control switch to the thermostat's highest setting.* _____ ❏

54. Now move the thermostat's **temperature control** switch to the **lowest setting** again. ❏

55. *Explain what happened when you moved the thermostat's temperature control to the lowest setting.* _____ ❏

> **NOTE:** *From this test you should have learned that although the thermostat is set to the selected heating mode, it does not mean that the heating cycle will come on. The temperature control must be set above the ambient temperature of the conditioned area.*

NOTE: *In this test you should have noticed that the blower or fan relay was not energized as it was in the cooling cycle. The thermostat does not energize the fan/blower of the furnace when in the heating mode. In a forced hot air furnace the heating cycle would come on first and allow the heat exchanger to heat up before the fan/blower would be turned on. Another means would be used to bring the fan/blower on when the heat exchanger reached a set temperature. A fan-limit switch, stack switch, or a time relay accomplishes this. Once it was hot enough, the blower would be turned on by one of these methods.*

56. Turn the thermostat to the OFF position and have your instructor check your work. ❑

57. With your instructor's permission, unplug the circuit and return all controls and supplies to their proper location. ❑

RESEARCH QUESTIONS

1. What is the purpose of heating and cooling thermostats?

   _____

   _____

2. What is the operating voltage of mercury bulb thermostats?

   _____

   _____

3. What are the two most common thermostats used for heating and cooling applications?

   _____

   _____

4. What is the purpose of the thermostat heat anticipator?

   _____

   _____

5. Why is the fan/blower energized at the same time the A/C condenser is energized on a call for cooling?

   _____

   _____

Passed Competency _____   Failed Competency _____

Instructor Signature _____   Grade _____

Practical Competency 37

Checking Internal Switching Functions of a Honeywell Heating and Cooling Thermostat

SUGGESTED MATERIALS

Textbook
Refrigeration & Air Conditioning Technology, 5th Edition, Thomson Delmar Learning
Unit 14—Automatic Control Components and Applications

Review Topics
Space Temperature Controls, Low Voltage

COMPETENCY OBJECTIVE

The student will be able to use an ohmmeter to check the switching operation of a single-stage heating and cooling thermostat.

OVERVIEW

The heating and cooling thermostat plays a very important part in heating and cooling equipment because it controls the temperature of the conditioned space as well as being the command center for the operation of the total heating or cooling equipment. Most residential heating and cooling thermostats operate off of a 24-volt control circuit power source.

Low-voltage thermostats can be purchased for *cooling only equipment, heating only equipment, heating and cooling equipment*, and *multistage equipment*. Regardless of the type of thermostat required for a particular installation, it is important to understand the universal low-voltage terminal functions of these thermostats. Universally, low-voltage thermostat terminal functions are widely accepted by manufacturers of heating and cooling equipment and thermostat manufacturers.

Basic low-voltage terminal functions are as follows:
  R Terminal—*24-Volt power source*
  G Terminal—*Indoor fan/blower circuit*
  W Terminal—*Heating circuit*
  Y Terminal—*Cooling circuit*

> **NOTE:** *Programmable thermostats will have the addition of a C terminal. This is for the neutral lead of the transformer's 24-volt power supply. This is used to complete the 24-volt circuit for the thermostat's LED display.*

Thermostats with **RC and RH terminals** allow for two step-down transformer control circuit power supplies. They can also be used with a single transformer control circuit power supply, but require that a *jumper be placed between the RC and RH terminals on the thermostat.*

It's true that most technicians observe a practice of matching color of thermostat conductor wire to thermostat terminal lettering.

R Terminal—*Red conductor wire*
G Terminal—*Green conductor wire*
W Terminal—*White conductor wire*
Y Terminal—*Yellow conductor wire*

> *NOTE: Not all technicians follow this practice, so it's more important to recognize the terminal circuit functions of the thermostat as stated above.*

EQUIPMENT REQUIRED

1 Single-stage heating and cooling thermostat
1 Ohmmeter or VOM

SAFETY PRACTICES

Make sure meter is set to the proper function and value.

COMPETENCY PROCEDURES

Checklist

1. Set the ohmmeter or VOM meter to measure resistance. ☐
2. Gain access to thermostat's terminals. ☐
3. Set the fan function switch to the **Auto** setting. ☐
4. Set the selector switch to the OFF position. ☐
5. Turn the **temperature control** setting to the **lowest setting.** ☐
6. Take one of the meter leads and attach it to **terminal R** on the thermostat. ☐

> *NOTE: This meter lead will stay on **terminal R** through the testing procedure to test the fan switch function.*

7. Take the other meter lead and attach it to **terminal G** on the thermostat. ☐
8. Turn the **fan** selector switch to the **ON** position and observe the ohmmeter scale. *Record what action you observed on the ohmmeter scale in this operation.*

_____ ☐

> *NOTE: You should have gotten a (0) ohms value reading in this test to show that the fan ON switch function is working.*

9. Turn the **fan** selector switch to the **Auto** position and notice the ohmmeter scale. *Record what action you observed on the ohmmeter scale in this operation.*

_____ ☐

> *NOTE: You should not have gotten an ohms value reading as long as the function selector switch is in the OFF mode and the fan switch is in the Auto setting.*

10. Leave the **fan switch** in the **Auto** and the temperature setting at the **lowest setting**. Move the thermostat's selector switch to the **Cooling function** and notice the ohmmeter scale. *Record what action you observed on the ohmmeter scale in this operation.*

_____ ☐

11. Move the temperature selector switch to the highest temperature setting and observe the ohmmeter scale. *Record what action you observed on the ohmmeter scale in this operation.*

_____ ❏

> **NOTE:** *In this test you should have noticed that when the selector switch is moved to the cooling function, you should have gotten an ohms value reading on your meter. When you moved the temperature selector switch to the highest setting, you should have noticed the ohmmeter scale return to an infinity value reading. This test tells you that the **fan Auto switch** function is working OK. On this thermostat, you can see that the fan switch function is controlled between **terminals R and G.***

CHECKING THE COOLING FUNCTION
12. Move the meter lead from terminal G on the thermostat to **terminal Y.** ❏
13. Make sure the thermostat selector switch is set on the **Cooling** function. ❏
14. Move the temperature selector switch to the **lowest setting** and notice the ohmmeter scale. *Record what action you observed on the ohmmeter scale in this operation.*

_____ ❏

15. Leave the thermostat settings where they are and move your ohmmeter lead to **terminal G** and observe the ohmmeter scale. *Record what action you observed on the ohmmeter scale in this operation.* _____ ❏

> **NOTE:** *You should have noticed an ohmmeter value of (0) on **terminal G** in this setting. As you can see, in the cooling mode function, the thermostat also closes the switch from terminal R to **terminals Y and G** at the same time, so that the fan comes on right away when the cooling function is called for.*

16. Move the temperature selector switch to the **highest setting** and take the ohmmeter test lead and check at **terminals G and Y** of the thermostat again. *Record the action you observed on the ohmmeter scale in this operation.*

_____ ❏

> **NOTE:** *In this test you should have noticed that with the thermostat selector switch set on the **cooling function**, and the temperature selector switch set at the highest temperature setting, the ohmmeter showed an **infinity value** reading at both **terminals G and Y.** This shows you that when the temperature of the thermostat is satisfied, it shuts off both the blower from **terminals R and G** and the A/C unit from **terminals R and Y** at the same time.*

This test shows you that the cooling function of the thermostat is working OK. It does not tell you that the thermostat is maintaining the correct or desired temperature. This test will be in another competency.

CHECKING THE HEATING FUNCTION
17. Set the temperature setting to the **lowest setting**, the **fan switch to Auto**, and the selector switch to the **Heating** function. ❏
18. Place your ohmmeter leads across **terminals R and W** and notice the ohmmeter scale. *Record what action you observed on the ohmmeter scale in this operation.*

_____ ❏

19. Move the temperature selector switch to the **highest setting** and check across **terminals R and W** again. *Record the action you observed on the ohmmeter scale in this operation.* ❏

_____

> NOTE: *In this test, you should have noticed an* **infinity ohms** *value with the temperature setting at the* **lowest setting**, *and a (0) ohms value reading when the temperature setting was moved to the* **highest setting**. *When the temperature selector was set at the* **highest setting**, *the thermostat closed the switch between terminals R and W to turn on the heating equipment.*
>
> *You should have also noticed that although the thermostat is set at the heating function, it does not mean that the heating equipment will come on unless the temperature control is set above the ambient temperature in the conditioned space.*

20. Leave the thermostat set at the heating function, the temperature switch set at the highest setting and check across **terminals R and G** of the thermostat. *Record what action you observed on the ohmmeter scale in this operation.* ❏

_____

> NOTE: *In this test, you should have seen an infinity ohms value reading, indicating that the blower or fan does not come on during the heating mode by the command of the thermostat. This is true for all forced hot air furnaces, except for some electric heating furnaces and heat pumps.*
>
> *Once the heat exchanger reaches a certain temperature it is then brought on by another means. The same is true for the fan operation at the end of the heating cycle. Once the thermostat is satisfied, it will shut off the heating equipment, but not the blower. Whatever means in the furnace that was used to delay the blower coming on at the beginning of the heating cycle will also delay the blower going off at the end of the heating cycle.*

21. Have your instructor check your work. ❏
22. Return tools and equipment to their proper location. ❏

RESEARCH QUESTIONS

1. During the thermostat's call for cooling, why does the blower come on at the same time as the air conditioning equipment?

_____

_____

2. During the thermostat's call for heating, why is the blower delayed coming on?

_____

_____

3. During the thermostat's call for heating, why is the blower delayed in going off at the end of the heating cycle?

_____

_____

4. What two methods are used to delay the blower coming on or going off in the heating cycle of a forced hot air furnace?

_____

_____

5. Which contact terminals of the thermostat are closed during the cooling cycle?

_____

_____

Passed Competency _____ Failed Competency _____

Instructor Signature _____ Grade _____

Theory Lesson: Capacitors (Figure 2–77)

FIGURE 2–77   Starting and running capacitor.
(*Photo by Bill Johnson*)

SUGGESTED MATERIALS

Textbook
Refrigeration & Air Conditioning Technology, 5th Edition, Thomson Delmar Learning
Unit 17—Types of Electrical Motors
Unit 20—Troubleshooting Electric Motors

Review Topics
Capacitor Start Motor; Capacitor Start; Capacitor Run Motors; Checking Capacitors; Identifications of Capacitors

Key Terms
bleed resistor • capacitance • capacitor • CS motor • CSCR motor • MFD (microfarads, μF) • PSC motor • run capacitor • start capacitor

OVERVIEW

A capacitor is a device that stores excess electrons on its plates (**Figure 2–78**).

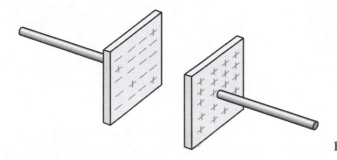

FIGURE 2–78   A charged capacitor.

The excess electrons are propelled ahead of the applied voltage of a circuit once the motor circuit is turned on. Capacitance is the term used to measure the amount of electrical charge of the capacitor. The **capacitance** of the capacitor is marked on the side of the capacitor's case. The capacitance rating of the capacitor will be shown as MFD (**microfarads**) or μF, which also stands for microfarad. The Greek letter μ is the symbol for micro. Capacitors have the ability to store this energy over a long period of time.

Because of this, care should be taken to make sure that the capacitor is discharged once it is removed from a motor circuit. Some capacitors will have a **bleed resistor** soldered across the capacitor terminals. This resistor is used to discharge the capacitor plates each time the circuit of the motor is shut down. Even with a bleed resistor on a capacitor, it is still recommended that the plates be manually discharged.

The applied voltage of the capacitor will also be marked on the capacitor case, normally as VAC. No voltage higher than the listed voltage on the capacitor should be applied. A higher voltage will destroy the capacitor plates and could also cause the capacitor to explode.

There are two types of capacitors used in the HVAC-R equipment—the **run capacitor** and the **start capacitor**. The **run capacitor is referred to as the tin can** because its case is made of tin. Run capacitors are used to **increase a motor's operating efficiency**. The **start capacitor comes in a plastic case** and **is installed in a motor circuit to give a motor a greater starting torque.** There are times when both the run and start capacitors may be used in the same motor circuit. Both capacitors operate in the same manner. The only difference between them is the amount of capacitive energy the capacitor can store.

Run capacitors *come in sizes from 1.5 microfarads to 60 microfarads and are wired parallel into the motor circuit with the run and start winding of the motor.* Run capacitors are left in the motor circuit during the motor's complete operation. **Start capacitors** *come in sizes from 75 microfarads to 600 microfarads. The start capacitor is wired in series with the start winding of the motor and must be removed from the motor circuit once the motor reaches 75% of its rated speed.* This is accomplished by a centrifugal switch, or motor starting relays (**Figure 2–79**).

Motors that are equipped with capacitors are classified as **PSC** (*permanent-split capacitor*) *motors* (*Figure 2–80*), **CS** (*capacitor-start*) *motors,* and **CSCR** (*capacitor-start capacitor-run*) *motors* (*Figure 2–81*).

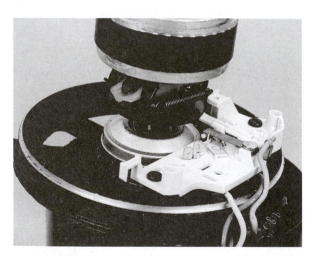

FIGURE 2–79 The centrifugal switch. (*Photo by Bill Johnson*)

FIGURE 2–80 A permanent split-capacitor motor. (*Courtesy of Universal Electric Company*)

A *PSC motor* is equipped with a run capacitor to give it better running efficiency. A *CS motor* is equipped with a start capacitor to give it better starting torque. A *CSCR motor* is equipped with both a run and start capacitor to give it better running efficiency and greater starting torque.

In any of the motors, if the capacitor fails, the capacitor would have to be replaced with one that had the same MFD rating. The voltage rating should be the same, but a higher voltage rated capacitor could be used in the motor circuit as long as the capacitance of the capacitor was the same. The following formula

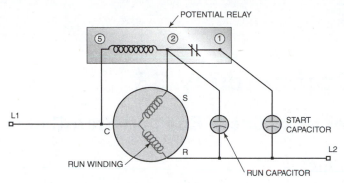

FIGURE 2–81 Wiring diagram for a CSCR motor with a potential relay to remove the start capacitor.

can be used to determine the capacitor capacitance range if the information was not available on the capacitor being replaced:

$$\text{Capacitance (uf)} = \frac{2452 \times \text{Amps}}{\text{Rated (Applied) Volts}}$$

Some run capacitors come with dual plates to operate a compressor and a fan motor from the same capacitor. These capacitors will come with three terminals on top of the capacitor. The terminals will be marked as "C" for common terminal, "**herm**" terminal for the compressor, and "**fan**" terminal for the fan motor. It is possible for one of the plates of the capacitor to be defective and the other plate to be OK. In this case the capacitor would have to be replaced with a new one.

Because run capacitors are left in the motor circuit during the motor operation, they have a tendency to overheat at times. This may cause a bulge in the casing of the capacitor. Although the capacitor may still operate OK, it is recommended that technicians replace these capacitors. This will avoid the possibility of an explosion of the capacitor while a technician is working on the system.

Practical Competency 38

Using an Ohmmeter to Check Run and Start Capacitors

SUGGESTED MATERIALS

Textbook

Refrigeration & Air Conditioning Technology, 5th Edition, Thomson Delmar Learning
Unit 17—*Types of Electrical Motors*
Unit 20—*Troubleshooting Electric Motors*

Review Topics
Checking Capacitors

COMPETENCY OBJECTIVE

The student will be able to check a capacitor with an analog meter.

OVERVIEW

A capacitor is a device that stores excess electrons on its plates. The excess electrons are propelled ahead of the applied voltage of a circuit once the motor circuit is turned on. Capacitance is the term used to measure the amount of electrical charge of the capacitor. The capacitance of the capacitor is marked on the side of the capacitor's case. The capacitance rating of the capacitor will be shown as MFD (microfarads) or μF, which also stands for microfarad. The Greek letter μ is the symbol for micro. Capacitors have the ability to store this energy over a long period of time. Care should be taken to make sure that the capacitor is discharged once it is removed from a motor circuit. Some capacitors will have a bleed resistor soldered across the capacitor terminals. This resistor is used to discharge the capacitor plates each time the circuit of the motor is shut down. Even with a bleed resistor on a capacitor, it is still recommended that the plates be manually discharged.

The applied voltage of the capacitor will also be marked on the capacitor case. No voltage higher than the listed voltage on the capacitor should be applied. A higher voltage will destroy the capacitor plates and could also cause the capacitor to explode.

There are two types of capacitors used in HVAC-R equipment—the run capacitor and the start capacitor. The run capacitor is referred to as the "tin can" because its case is made of tin. It is installed into a motor circuit to increase a motor's operating efficiency. The start capacitor comes in a plastic case and is installed in a motor circuit to give a motor a greater starting torque. There are times when both the run and start capacitors may be used in the same motor circuit. Both capacitors operate in the same manner. The only difference between them is the amount of capacitance energy the capacitor can store.

Run capacitors come in sizes from 1.5 microfarads to 60 microfarads and are wired parallel into the motor circuit with the run and start winding of the motor. Run capacitors are left in the motor circuit during the motor's complete operation. Start capacitors come in sizes from 75 microfarads to 600 microfarads. The start capacitor is wired in series with the start winding of the motor and must be removed from the motor circuit once the motor reaches 75% of its rated speed. This is accomplished by a centrifugal switch, or motor starting relays.

Motors that are equipped with capacitors are classified as PSC motors, CS motors, and CSCR motors. A PSC motor is equipped with a run capacitor to give it better running efficiency. A CS motor is equipped with a start capacitor to give it better starting torque. A CSCR motor is equipped with both a run and start capacitor to give it better running efficiency and greater starting torque.

In any of the motors, if the capacitor fails, the capacitor would have to be replaced with one that had the same MFD rating. The voltage rating should be the same, but a higher voltage rated capacitor could be used in the motor circuit as long as the capacitance of the capacitor was the same.

Some run capacitors come with dual plates to operate a compressor and a fan motor from the same capacitor. These capacitors will come with three terminals on top of the capacitor. The terminals will be marked "C," for common terminal, "herm" terminal for the compressor, and "fan" terminal for the fan motor. It is possible for one of the plates of the capacitor to be defective and the other plate to be OK. In this case the capacitor would have to be replaced with a new one.

Because run capacitors are left in the motor circuit during the motor operation, they have a tendency to overheat at times. This may cause a bulge in the casing of the capacitor. Although the capacitor may still operate OK, it is recommended that technicians replace these capacitors. This will avoid the possibility of an explosion of the capacitor while a technician is working on the system.

EQUIPMENT REQUIRED

1 Analog ohmmeter
1 Start capacitor
1 Run capacitor
1 Screwdriver with insulated handle

SAFETY PRACTICES

The capacitor should be removed from the motor circuit and discharged before checking with an ohmmeter. Be sure to use a screwdriver that has an insulated handle when discharging the capacitor.

COMPETENCY PROCEDURES   Checklist

1. Disconnect power to the equipment where the capacitor is being tested.   ❏
2. Remove the capacitor from the motor circuit.   ❏
3. Discharge the capacitor by holding the handle of the screwdriver and touch
   the screwdriver shaft across the capacitor terminal plates.   ❏

> NOTE: *There may be a loud snap when capacitor is discharged.*

> NOTE: *A 20,000-ohm resistor could also be used for discharging the capacitor, by placing the resistor terminal leads across the capacitor terminals.*

4. Discharge the capacitor again in the same manner listed in procedure 3.   ❏

> NOTE: *If the capacitor has a bleed resistor across the capacitor terminals the bleed resistor can be removed or left in on the capacitor. The test results will be different depending on whether the resistor is removed or left on the capacitor.*

CHECKING A CAPACITOR WITHOUT A BLEED RESISTOR

5. Set the ohmmeter range to the R ∞ 100 or R ∞ 1000 scale.   ❏
6. Touch the ohmmeter leads to the capacitor's terminals.   ❏

> **NOTE:** *The meter needle should rise fast and fall back slowly until the meter needle reads infinity.*

7. Reverse the meter leads and touch the capacitor terminals again. ❏

> **NOTE:** *The meter needle should rise fast again and fall slowly until meter needle reads infinity. The above test tells you that the capacitor is good and capable of taking a charge and discharging.*

> **NOTE:** *This test does not tell you if the capacitor is discharging at the correct capacitance. A capacitor tester would have to be used to check this.*

CHECKING A CAPACITOR WITH A BLEED RESISTOR

8. Set the ohmmeter range to the R ∞ 100 or R ∞ 1000 scale. ❏
9. Touch the ohmmeter leads to the capacitor's terminals. ❏

> **NOTE:** *The meter needle should rise fast and fall back slowly until the meter needle reads the bleed resistor value.*

10. Reverse the meter leads and touch the capacitor terminals again. ❏

> **NOTE:** *The meter needle should rise fast again and fall slowly until the meter needle reads the bleed resistor value. The above test tells you that the capacitor is good and capable of taking a charge and discharging.*

> **NOTE:** *This test does not tell you if the capacitor is discharging at the correct capacitance. A capacitor tester would have to be used to check this.*

11. Have your instructor check your work. ❏
12. Disconnect equipment, and return all tools and test equipment to their proper location. ❏

RESEARCH QUESTIONS

1. What capacitors are used in helping the motor to operate at better efficiency?

   _____

2. What capacitor is used in assisting a motor to have a better starting torque?

   _____

3. What is meant by the voltage range that is printed on the capacitor?

   _____

4. When replacing a capacitor in a motor circuit, you can never go higher in what rating of the capacitor?

_____

_____

5. The capacitance value of a capacitor is measured in what?

_____

_____

Passed Competency _____ **Failed Competency** _____

Instructor Signature _____ **Grade** _____

REFRIGERATION FUNDAMENTALS

Theory Lesson: ACR Tubing

SUGGESTED MATERIALS

Textbook
Refrigeration & Air Conditioning Technology, 5th Edition, Thomson Delmar Learning
Unit 7—Tubing and Piping

Review Topics
Types and Sizes of Tubing; Purpose of Tubing and Piping

Key Terms
ACR tubing • brazing • flaring • inside diameter (ID) • nominal size tubing • outside diameter (OD) • soft ACR tubing • soldering • swaging

OVERVIEW

ACR (air conditioning and refrigeration) tubing is classified as type K and type L. The type K tubing is a heavy-walled tubing, whereas type L tubing is medium-walled tubing. Type L tubing is the most common tubing used in HVAC-R sealed systems. Both types of tubing can be in soft- or hard-drawn form. All ACR tubing is measured by its **outside diameter (OD)** (**Figure 3–1**).

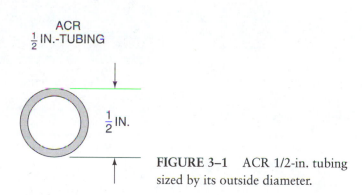

ACR
$\frac{1}{2}$ IN.-TUBING

$\frac{1}{2}$ IN.

FIGURE 3–1   ACR 1/2-in. tubing
sized by its outside diameter.

The tubing is very flexible and should be handled carefully so as not to bend or smash it.

The tubing end of the ACR tubing comes capped and should be recapped after being used to prevent moisture and noncondensables from entering the tubing. Never uncoil a roll of soft tubing from the inside of the roll (**Figure 3–2**).

ACR tubing can be cut and bent with special ACR tools such as bending springs, tubing benders, and tubing cutters (**Figure 3–3**). Joining ACR tubing can be accomplished by using flared fittings (**Figure 3–4**), soldering or brazing (**Figure 3–5**), or making a swage joint (**Figure 3–6**).

Flaring is the process of making a fabricated mechanical joint for sealing copper tubing in a system. **Swaging** of soft copper tubing is used when joining two pieces of copper together that are the same size without using fittings. **Soldering** is the process of fastening two base metals together by using a third filler

FIGURE 3–2 A roll of soft tubing. Place on a flat surface and unroll. (*Photo by Bill Johnson*)

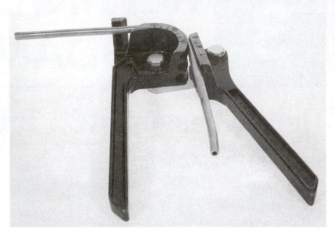

FIGURE 3–3 Use of a lever-type tube bender. (*Photo by Bill Johnson*)

FIGURE 3–4 Flared tubing and flare fitting. (*Photo by Bill Johnson*)

FIGURE 3–5 Soldering or brazing. (*Photo by Bill Johnson*)

FIGURE 3–6 Joining tubing by a swaged joint. (*Photo by Bill Johnson*)

metal that melts at a temperature below 800 degrees Fahrenheit. **Brazing** is the process of fastening two base metals together by using a third filler metal that melts at a temperature above 800 degrees (F).

ACR hard-drawn tubing is normally used in commercial refrigeration sealed systems and cannot be bent as easily as the soft tubing. This requires ACR fittings to be used for connection of tubing for making joints. The type of fitting to be used depends on the installation, and can be flared, soldered, or brazed into the system.

Tubing that is used for plumbing and heating is called **nominal size tubing**. This tubing also comes in hard- or soft-drawn form. There are three different styles of nominal size tubing: types L, M, and K. Type L is a medium-walled tubing and is the most common tubing used in plumbing and heating application. Plumbing and heating tubing is measured by its inside diameter (ID) (**Figure 3–7**).

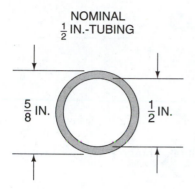

NOMINAL
$\frac{1}{2}$ IN.-TUBING

$\frac{5}{8}$ IN. $\frac{1}{2}$ IN.

FIGURE 3–7   Copper tubing used for plumbing and heating (nominal) is sized by its inside diameter.

This means that the outside diameter (OD) of the tubing is 1/8" thicker than the ID measurement.

Practical Competency 39

Using a Tube Cutter

SUGGESTED MATERIALS

Textbook
Refrigeration & Air Conditioning Technology, 5th Edition, Thomson Delmar Learning
Unit 7—Tubing and Piping

Review Topics
Types and Sizes of Tubing; Cutting Tubing

COMPETENCY OBJECTIVE

The student will be able to cut and ream copper tubing.

OVERVIEW

Soft and hard copper tubing can be cut with a tubing cutter or hacksaw. In most cases, soft copper tubing will be cut with a hand-held tube cutter. Hard-drawn copper can be cut with either a tube cutter or a hacksaw. Care should be taken not to smash the tubing from overtightening the tube cutter when making cuts. All cut areas should be reamed before connections are made to fitting or joints to be soldered. Care should be taken when reaming the tubing to allow the burrs to fall from the tubing and not inside the tubing. The ends of ACR tubing should be capped to prevent moisture from entering the tubing.

EQUIPMENT REQUIRED

1 Tubing cutter
1 12" Piece of 1/2" ACR tubing
1 Tape rule

SAFETY PRACTICES

Use care when cutting and reaming copper tubing.

COMPETENCY PROCEDURES

Checklist

1. Take the 12" piece of copper tubing and mark a line on the copper tube every 4". ❑
2. Open the tube cutter wide enough to fit over the copper tubing. ❑
3. Align the tube cutter wheel with one of the 4" marks on the 12" piece of copper tube (**Figure 3–8**). ❑
4. Tighten the tube cutter knob until the cutter exerts a moderate pressure on the tubing. (*Do not overtighten or the tube will be bent.*) ❑
5. Turn the tube cutter around the copper tubing, keeping a moderate pressure applied (**Figure 3–9**). ❑
6. With each revolution of the tube cutter, gradually turn the tube cutter adjustment knob. (*Do not overtighten.*) ❑

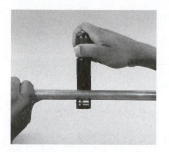

FIGURE 3–8 Aligning tube cutter wheel with cutting mark. (*Photo by Bill Johnson*)

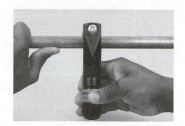

FIGURE 3–9 Revolve the cutter around the tubing. (*Photo by Bill Johnson*)

7. Continue until the final cut of the tube is made (**Figure 3–10**). ❑
8. Repeat the same procedures until the tube is cut at all 4" marks. ❑
9. Remove the tube cutter-reaming blade. ❑
10. Insert the reaming blade into the copper tubing (**Figure 3–11**). ❑

FIGURE 3–10 Completed cut. (*Photo by Bill Johnson*)

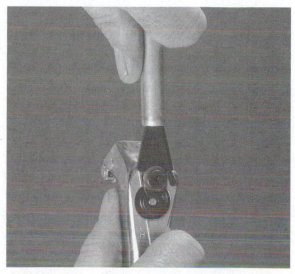

FIGURE 3–11 Reamer being used to deburr copper tubing.

11. Remove the burrs by rotating the reaming blade around the inner surface of the copper tubing. ❑
12. Take care not to allow burring chips to fall into the tubing. ❑
13. Ream both ends of all 4" pieces of copper tubing. ❑
14. Have your instructor check your work. ❑
15. Return all tools and equipment to their proper location. ❑

> **NOTE:** *Save the copper pieces for Practical Competencies 40 and 41—Swaging Copper Tubing and Flaring Copper Tubing.*

RESEARCH QUESTIONS

1. What happens when the tube cutter is tightened too tight on the copper during the cutting process?

_____

_____

2. Why should the copper tubing be reamed to remove the burrs from the cut?

_____

_____

3. Does its ID or OD measure ACR tubing?

_____

_____

4. What is annealing?

_____

_____

5. Does its ID or OD measure nominal-size tubing?

_____

_____

Passed Competency _____   Failed Competency _____

Instructor Signature _____   Grade _____

Practical Competency 40

Swaging Copper Tubing

SUGGESTED MATERIALS

Textbook
Refrigeration & Air Conditioning Technology, 5th Edition, Thomson Delmar Learning
Unit 7—Tubing and Piping

Review Topics
Swaging Techniques

COMPETENCY OBJECTIVE

The student will be able to make a swaged joint.

OVERVIEW

Swaging is the process of joining two pieces of copper tubing of the same size together by expanding or stretching the end of one of the pieces to fit over the other so that they may be joined and soldered or brazed together. The advantage of this is the elimination of possible leaks from two mechanically joined pieces of copper tubing.

The two common tools used to make swaging joints are the swaging punch and swaging block. **A rule to follow** is: *The length of the tubing that fits over the other is equal to the approximate OD (outside diameter) of the tubing being swaged.* Using a drop of refrigeration oil on the swaging tool will help the swaging tool slide easier. This oil must be cleaned off before the joint is soldered or brazed. Swaging joints should always be checked for cracks or other defects before soldering or brazing.

EQUIPMENT REQUIRED

1 Swaging block
4 4" Pieces of 1/2" copper tubing
1 1/2" Swaging tool
1 Ball peen hammer
Refrigerant oil

SAFETY PRACTICES

Use care when swaging copper tubing.

COMPETENCY PROCEDURES

Checklist

1. Take one of the 4" pieces of copper tubing and place it in the proper size hole of the swaging block (**Figure 3–12**). ❑
2. Adjust the height of the copper tubing by aligning the swaging tool next to the copper tubing and moving it to determine the depth of the swage (**Figure 3–13**). ❑
3. Tighten the copper tubing into place by adjusting the swaging block locking wing nuts. ❑
4. Place a film of refrigerant oil around the swaging tool. ❑

FIGURE 3–12 Tubing placed in swaging block. (*Photo by Bill Johnson*)

FIGURE 3–13 Aligning copper tube and swaging tool. (*Photo by Bill Johnson*)

5. Insert the swaging tool into the opening of the copper tubing. ❑

> **NOTE:** *If swaging tool will not slide into tube opening easily, make sure tubing is deburred and try using a little refrigerant oil on the swaging tool.*

6. Hold the swaging block and tubing in one hand. ❑
7. Use the ball peen hammer and tap the swaging tool lightly to start the swage (**Figure 3–14**). ❑

FIGURE 3–14 A swaging punch expanding metal. (*Photo by Bill Johnson*)

8. Continue to hammer the swaging tool into the copper tubing until it reaches the required depth of the swage. ❑
9. Remove the swaging tool from the swaged copper. ❑

> **NOTE:** *You may have to tap on the side of the swaging tool or use an adjustable wrench to remove the swaging tool from the copper tubing.*

10. Once swaging tool is removed, unscrew the swaging block locking nuts and remove the copper from the swaging block. ❏
11. Wipe the refrigerant oil from the swaging tool and copper tubing. ❏
12. Inspect the swaged copper for cracks or other defects. ❏
13. Insert one of the other 4" pieces of copper tubing into the swaged copper joint. ❏
14. *Use procedures 1–13 and swage another piece of copper tubing.* ❏
15. Have your instructor check your work. ❏
16. Return all tools and equipment to their proper location. ❏

RESEARCH QUESTIONS

1. What is the definition of swaging?

   _____

   _____

2. When using methods of joining copper, what is the advantage of making a swage joint over making a flared joint?

   _____

   _____

3. As a general rule, what should be the length of the overlap of the swage?

   _____

   _____

4. In what size rolls is soft copper tubing normally available?

   _____

   _____

5. What should you do if you see or suspect a crack in a flared joint?

   _____

   _____

Practical Competency 41

Flaring Copper Tubing

SUGGESTED MATERIALS

Textbook

Refrigeration & Air Conditioning Technology, 5th Edition, Thomson Delmar Learning
Unit 7—Tubing and Piping

Review Topics

Making Flare Joints

COMPETENCY OBJECTIVE

The student will be able to make a flared joint.

OVERVIEW

Flared fittings are made on soft copper tubing. This joint uses a flare at one end of the tubing fitted against an angle on a flared fitting. It is secured with a flare nut that fits behind the flare on the tubing. The flared fittings used to join flared tubing together are forged from brass (**Figure 3–15**).

FIGURE 3–15   Examples of flare fittings. (*Photo by Bill Johnson*)

It is recommended that when flared fittings are used in areas with *no extreme temperature changes, the long flare nuts should be used for all joint connections.* In areas where there will be *extreme temperature changes, the flared fitting connections should be made with short flare nuts.*

Using a few drops of refrigeration oil on joining flared fittings will prevent the tubing from twisting when the connections are made. When tightening flared joints, the joint should be hand tightened and the wrench tightened one full turn with a flare wrench. Overtightening any flare joint can cause the tubing or flare to split and cause leaks.

EQUIPMENT REQUIRED

1 Flaring block
1 Flaring yoke
2 4" Pieces of 1/2" copper tubing
2 1/2" Flare nuts
1 1/2" × 1/2" Flared union
Refrigeration oil

SAFETY PRACTICES

Use care when flaring copper tubing.

COMPETENCY PROCEDURES

1. Use two remaining 4" pieces of 1/2" copper tubing. ❑
2. Ream the copper tubing ends to remove the burrs. ❑
3. Place a flare nut over one of the pieces of copper, with the flare nut threaded end facing the end of the copper to be flared. ❑
4. Slide the tubing about 3/16" above the flaring block. ❑
5. Clamp the copper tubing into the flaring block (**Figure 3–16**). ❑
6. Place a film of refrigeration oil on the flaring yoke (**Figure 3–17**). ❑

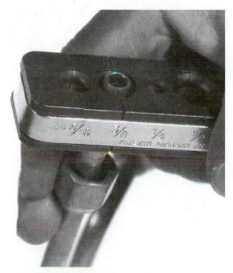

FIGURE 3–16   Placing tube in flaring block. (*Photo by Bill Johnson*)

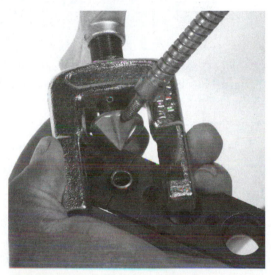

FIGURE 3–17   Adding refrigeration oil to flaring yoke. (*Photo by Bill Johnson*)

7. Place the flaring yoke on the flaring block with the tapered cone centered over the end of the copper tube (**Figure 3–18**). ❑
8. Turn the handle of the flaring yoke firmly until the yoke cone is seated the whole way down. ❑
9. Unscrew the flaring yoke and remove from the flaring block. ❑
10. Inspect the flare. Make sure the flare has seated the whole way in the flaring mold and has no cracks (**Figure 3–19**). ❑
11. Remove flared copper from the flaring block. ❑
12. Flare one end of the other piece of copper tubing. ❑
13. Use the 1/2" × 1/2" flaring union and join the two flared pieces of copper tubing together. ❑
14. Have your instructor check your work. ❑
15. Return all tools and equipment to their proper location. ❑

RESEARCH QUESTIONS

1. Overtightening the flaring yoke could cause what to happen to the flared joint?

_____

_____

FIGURE 3–18   Centering flaring yoke on flaring block and copper tubing. (*Photo by Bill Johnson*)

FIGURE 3–19   Inspecting flare. (*Photo by Bill Johnson*)

2. What could happen if the copper tubing were extended too far above the flaring block?

_____

_____

3. What is the standard degree angle of most flared joints?

_____

_____

4. What is a disadvantage of a flared fitting joint compared to a soldered joint?

_____

_____

5. What is a double-thickness flared joint?

_____

_____

Passed Competency _____   Failed Competency _____

Instructor Signature _____   Grade _____

Theory Lesson: Air-Acetylene (Figure 3–20)

FIGURE 3–20   A typical air–acetylene setup. (*Photo by Bill Johnson*)

SUGGESTED MATERIALS

Textbook
Refrigeration & Air Conditioning Technology, 5th Edition, Thomson Delmar Learning
Unit 7—Tubing and Piping

Review Topics
Heat Sources for Soldering and Brazing

Key Terms
acetone • acetylene • air–acetylene • brazing • B tank • fusible plug • left-hand threaded soldering • MC tank • pressure regulator • working pressure

OVERVIEW

In soldering and brazing a fuel gas must be used to cause the filler metal to melt and fill the space of the joint area. Higher temperatures can be reached by mixing certain gases with oxygen. In the **air–acetylene outfit**, regular atmospheric air is used because it contains enough oxygen to create a flame hot enough to melt **soft solders** below 800 degrees Fahrenheit. The type of gas used determines the flame temperature of a torch. **Acetylene** produces the highest temperatures compared to such gases as propane, butane, or natural gas.

The air–acetylene outfit combines gas from the cylinder with combustion air drawn from the atmosphere to produce the soldering flame. Atmospheric air contains 21% oxygen and is enough to create a flame hot enough to melt the lower temperature soldering alloys. **Acetylene gas** is a colorless gaseous hydrocarbon made from a chemical reaction of water and calcium carbide. When acetylene is burned with oxygen, flame temperatures as high as 5660 degrees Fahrenheit can be obtained.

The Department of Transportation (DOT) approves the bottles that are used to hold acetylene gas. Each cylinder is equipped with a **fusible plug** located at the bottom of the cylinder and is designed to melt at 212 degrees Fahrenheit, in case cylinder pressures reach a dangerous level. Acetylene cylinders should always be stored in the upright position. This is because acetylene is a very unstable, unsafe, and explosive gas at pressures above 15 pounds per square inch (psi). Because of acetylene's instability, it is illegal to operate a torch with acetylene pressure greater than 15 psi. Most acetylene gages have a **Red** mark indicating 15 psi.

Torch manufacturers have designed their equipment so that a torch will operate properly with acetylene pressures from 1 to 5 psi. This pressure is high enough for most air conditioning and refrigeration applications.

To make acetylene cylinders safe to store under high pressures, manufacturers fill the cylinders with a **porous substance** called acetone, which absorbs acetylene. When acetylene is added to the cylinder it is absorbed by the **acetone**, allowing it to be more stable under high pressures.

When cylinders are laid down, the valve can open and the acetone can leak out of the hose with the acetylene. The lower the amount of acetone in the cylinder the more dangerous the pressure in the cylinder can be with the acetylene. Each time the cylinder is refilled, it is checked for the correct amount of acetone.

Acetylene cylinders come in various cylinder sizes. The two most common cylinders used in HVAC-R service work are the **MC tanks** and **B tank**. The MC is referred to as the **"motor car"** tank and is the smallest tank available for acetylene. These cylinders are equipped with a valve and threaded outlet for installing a **pressure regulator** and torch hose assembly.

All acetylene fittings are **left-hand threaded**. Care should be taken to make sure that fittings are turned in the correct direction when making connections to the **acetylene cylinder** and torch handle.

The **pressure regulator** is used to reduce the acetylene cylinder pressure of 250 psi down to a working pressure. Normally the regulator **working pressure** is around 5 psi. Single-gage pressure regulators (**Figure 3–21**) are used to show the working pressure traveling through the torch hose and torch tip. The **working pressure** can be adjusted by turning the knob or handle located on the front of the regulator. Turning the adjustment knob *counterclockwise will close the pressure regulator,* and *turning the adjustment knob clockwise will open the pressure regulator.*

---

*NOTE: Adjusting working pressure requires the torch valve to be open when making adjustments. Some regulators will have two gages (**Figure 3–22**); one is for the working pressure and the other is used to show the level of acetylene in the cylinder.*

FIGURE 3–21   Single-gage pressure regulator.

FIGURE 3–22   Two-gage regulator. (*Photo by Bill Johnson*)

Practical Competency 42

Setting Up an Air–Acetylene Outfit

SUGGESTED MATERIALS

Textbook
Refrigeration & Air Conditioning Technology, 5th Edition, Thomson Delmar Learning
Unit 7—Tubing and Piping

Review Topics
Heat Sources for Soldering and Brazing

COMPETENCY OBJECTIVE

The student will be able to set up an air–acetylene outfit.

OVERVIEW

Acetylene cylinders come in various cylinder sizes. The two most common cylinders used in HVAC-R service work are the **MC** and **B tank**. The MC is referred to as the **"motor car"** tank and is the smallest tank available for acetylene. These cylinders are equipped with a valve and threaded outlet for installing a pressure regulator and torch hose assembly.

All acetylene fittings are left-hand threaded. Care should be taken to make sure that fitting is turned in the correct direction when making connections to the acetylene cylinder and torch handle.

The pressure regulator is used to reduce the acetylene cylinder pressure of 250 psi down to a working pressure. Normally the regulator working pressure is around 5 psi. Single-gage pressure regulators are used to show the working pressure traveling through the torch hose and torch tip. The working pressure can be adjusted by turning the knob or handle located on the front of the regulator. Turning the adjustment knob counterclockwise will close the pressure regulator, and turning the adjustment knob clockwise will open the pressure regulator. Because of acetylene's instability, it is illegal to operate a torch with acetylene pressure greater than 15 psi. Most acetylene gages have a **Red** mark indicating 15 psi.

Torch manufacturers have designed their equipment so that a torch will operate properly with acetylene pressures from 1 to 5 psi. This pressure is high enough for most air conditioning and refrigeration applications.

> NOTE: *Adjusting working pressure requires the torch valve to be open when making adjustments. Some regulators will have two gages; one is for the working pressure and the other is used to show the level of acetylene in the cylinder.*

EQUIPMENT REQUIRED

1 Acetylene tank
1 Acetylene regulator with acetylene hose
1 Valve stem wrench
1 12" Adjustable wrench

1 Acetylene torch handle
1 Torch tip
1 Striker
1 Soap bubble solution

SAFETY PRACTICES

Do not make a connection to the acetylene tank around an open flame. Do not use a match or lighter to light the torch. Keep torch cylinders secured.

All connections should be free of dirt, dust, grease, and oil. Oxygen can produce an explosion when in contact with grease or oil.

NOTE: *It is illegal to operate a torch with an acetylene pressure greater than 15 psig.*

COMPETENCY PROCEDURES

Checklist

1. Check the regulator and make sure it is free of dirt, grease, and oil. ❑
2. Use the valve stem wrench and crack the acetylene tank valve to clear the valve of any foreign matter (**Figure 3–23**). ❑

FIGURE 3–23   Clear valve of foreign matter. (*Photo by Bill Johnson*)

NOTE: *Do not vent acetylene near an open flame.*

---

3. Use the 12" adjustable wrench and attach the acetylene regulator to the acetylene tank (**Figure 3–24**). ❑

FIGURE 3–24   Attaching regulator to B-tank. (*Photo by Bill Johnson*)

4. Use the adjustable wrench to attach the acetylene hose to the pressure regulator. ❑

5. Attach the other end of the acetylene hose to the torch handle. ❑
6. Make sure the torch knob is closed off. ❑
7. Turn the adjustment knob on the pressure regulator counterclockwise until there is no pressure on the spring. ❑
8. Use the valve stem wrench and open the acetylene tank valve by turning it counterclockwise one full turn (**Figure 3–25**). ❑
9. Turn the pressure regulator adjustment knob clockwise until a pressure of 10 psig is reached on the regulator pressure gage. ❑
10. Use a soap bubble solution and check all connections for leaks (**Figure 3–26**). ❑

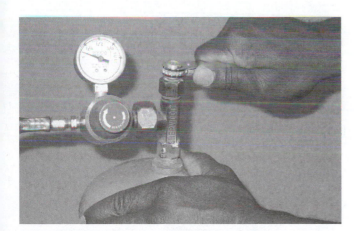

FIGURE 3–25   Opening acetylene valve. (*Photo by Bill Johnson*)

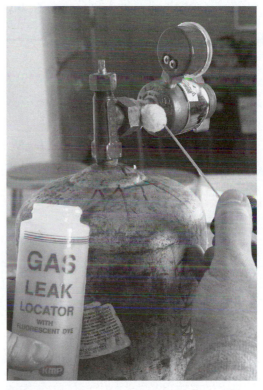

FIGURE 3–26   Checking for leaks at all connections.

11. Open the torch handle knob and use the sticker to light the torch. ❑
12. Adjust the torch tip (**Figure 3–27**). ❑
13. Turn off the flame by turning the torch handle knob clockwise. ❑
14. Use the valve stem wrench and close the acetylene tank valve by turning the valve stem clockwise. ❑
15. Turn the pressure regulator valve counterclockwise to release the pressure from the spring on the regulator. ❑

FIGURE 3–27   Adjusting torch flame.

16. Open the torch handle knob and bleed the hose and regulator to a zero pressure on the regulator. ❑
17. Once regulator pressure drops to zero, shut the torch handle off by turning it clockwise. ❑
18. Have your instructor check your work. ❑
19. Return all tools and equipment to their proper location. ❑

RESEARCH QUESTIONS

1. Most acetylene cylinders are pressurized with how much pressure?

   _____

   _____

2. What is the normal working pressure for acetylene?

   _____

   _____

3. What is meant by working pressure?

   _____

   _____

4. A single-gage pressure regulator is used to indicate what pressure?

   _____

   _____

5. What is the acetone used for in the acetylene cylinder?

   _____

   _____

Theory Lesson: Oxyacetylene Outfits (Figure 3–28)

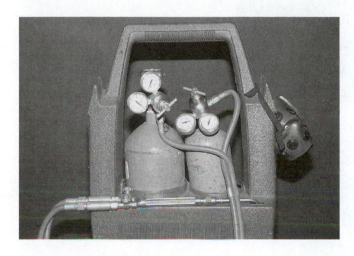

FIGURE 3–28   Oxyacetylene equipment. (*Photo by Bill Johnson*)

SUGGESTED MATERIALS

Textbook
Refrigeration & Air Conditioning Technology, 5th Edition, Thomson Delmar Learning
Unit 7—Tubing and Piping

Review Topics
Heat Sources for Soldering and Brazing

Key Terms
acetylene • backfire • flashback • left-hand threaded • oxyacetylene • oxygen regulator • pressure regulator • reverse flow valve working pressure • right-hand threaded operating pressures • safety relief device

OVERVIEW

Oxyacetylene outfits are used when higher temperatures for brazing joints are required. Brazing joints are much stronger than regular soft-soldered joints. The braze joint tolerates vibration without cracking, is a better leak-proof joint under high pressures, and is not corrosive. Some states require that all refrigeration joints be brazed rather than soldered.

Safety procedures should be followed when working with oxyacetylene outfits. At the higher temperatures, flying sparks can be produced during brazing; molten metal, gaseous fumes, and intense light rays are created, which can be harmful to the person using the outfit as well as those around the area (**Figure 3–29**).

The oxyacetylene outfit requires two cylinders of gas—one with oxygen and the other with acetylene gas. The **oxygen cylinder** has a pressure of about 2200 psi at 70 degrees Fahrenheit. The pressure in the cylinder will vary as the temperature of the cylinder changes. A **safety relief device** is built into the oxygen cylinder valves to protect against excessive pressure that could create an explosion. The **oxygen pressure regulator** is designed to apply a constant **working pressure** through the oxygen hose, no matter what the pressure is in the cylinder. (**Remember the cylinder pressure will vary because of ambient temperatures.**)

The **oxygen regulator** is connected directly to the oxygen cylinder valve (**Figure 3–30**).

Both regulators contain an outlet for the connection of the hoses leading to the torch handle. Regulators are also equipped with two gages; one is used to measure the **cylinder pressure** and the other is used to

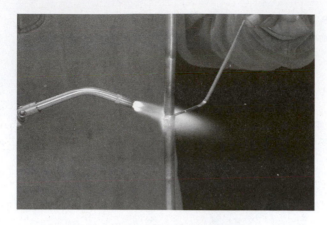

FIGURE 3-29   Brazing.

FIGURE 3–30   An oxygen gage and regulator.
(*Photo by Bill Johnson*)

indicate the **torch working pressure.** The regulators are equipped with a pressure adjustment handle to adjust the working pressure of the torch (**Figure 3–31**).

FIGURE 3–31   Oxygen regulator pressure adjustment handle.

Turning the adjustment handle **counterclockwise** will **close** the regulator valve, preventing gas from flowing to the torch handle. Turning the adjustment handle **clockwise** will **open** the valve to allow gas to flow to the torch handle. The pressure indicated on these regulators is in **pounds per square inch gage (psig).**

NOTE: The oxygen torch valve must be open when an adjustment is made to the working pressure.

Once the proper operating pressure is reached, the torch valve is closed, and the working operating pressure is set. When shutting the oxyacetylene outfit down, always turn both the oxygen and acetylene pressure adjustment knobs to the OFF position by turning them out counterclockwise the whole way.

Two regulators are required for oxyacetylene outfits, one for the **oxygen cylinder** and the other for the **acetylene cylinder**. These regulators, gages, and hoses are color-coded. The *oxygen regulator and hose is green* and the *acetylene regulator and hose is red*. The regulators and hoses for oxygen and acetylene outfits cannot be accidentally interchanged because of the manufactured threading system. All oxygen fittings are **right-hand threaded** and all acetylene fittings are **left-hand threaded**. Each regulator has two gages, one to register **tank pressure** and the other to register **pressure to the torch**.

A **reverse flow valve** should be used somewhere in the hoses. These valves allow the gas to flow in one direction only and prevent the two gases from mixing in the hoses.

> **NOTE:** *All connections should be free of dirt, dust, grease, and oil. Oxygen can produce an explosion when in contact with grease or oil.*

Operating pressures to the torch tips should be set to manufacturer recommended pressures (**Figure 3–32**). Tips from size 0 to 5 are designed to operate with a working pressure of 5 psig for acetylene and 10 psig for the oxygen regulators (**Figure 3–33** and **Figure 3–34**).

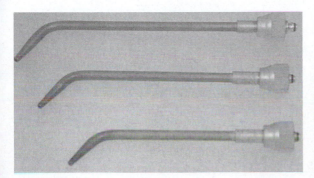

FIGURE 3–32   An assortment of oxyacetylene tips. (*Photo by Bill Johnson*)

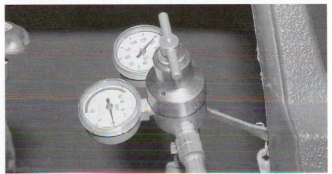

FIGURE 3–33   Acetylene regulator gage indicates 5 psig. (*Photo by Bill Johnson*)

FIGURE 3–34   Oxygen regulator gage indicates 10 psig. (*Photo by Bill Johnson*)

Because of acetylene's instability, it is illegal to operate a torch with acetylene pressure greater than 15 psig. Most acetylene gages have a **Red** mark indicating 15 psig. Torch manufacturers have designed their equipment so that a torch will operate properly with acetylene pressures from 1 to 5 psig. This pressure is high enough for most air conditioning and refrigeration applications.

Improper operation of the torch tip may create a condition called **backfire** in which the flame goes out with a loud cracking sound. This can be caused by the torch tip touching the piece being brazed or from flow pressures being too low.

Another condition that can develop is called **flashback**. In this case the flame burns back inside the torch tip and may even burn back through the hose and regulator. A shrill hissing or squealing sound may be created. This is a serious condition and an indication that something is wrong. When it occurs, the torch should be shut down immediately and allowed to cool before relighting the torch. This condition is normally caused by a clogged torch orifice, or, more likely, the working pressures of the oxygen and acetylene regulator are not set properly.

Practical Competency 43

Setting Up an Oxyacetylene Outfit

SUGGESTED MATERIALS

Textbook
Refrigeration & Air Conditioning Technology, 5th Edition, Thomson Delmar Learning
Unit 7—Tubing and Piping

Review Topics
Heat Sources for Soldering and Brazing

COMPETENCY OBJECTIVE

The student will be able to set up an oxyacetylene outfit.

OVERVIEW

The oxyacetylene outfit requires two cylinders of gases—one with oxygen and the other with acetylene gas. The oxygen cylinder has a pressure of about 2200 psi at 70 degrees Fahrenheit. The pressure in the cylinder will vary as the temperature of the cylinder changes. A safety relief device is built into the oxygen cylinder valves to protect against excessive pressure that could create an explosion. The oxygen pressure regulator is designed to apply a constant working pressure through the oxygen hose, no matter what the pressure is in the cylinder. **(Remember the cylinder pressure will vary because of ambient temperatures.)**

The oxygen regulator is connected directly to the oxygen cylinder valve. The regulator contains an outlet for the connection of the oxygen hose leading to the torch handle. The oxygen regulator is equipped with two gages; one is used to measure the cylinder pressure and the other is used to indicate the torch working pressure.

The regulator is equipped with a pressure adjustment handle to adjust the working pressure of the torch. Turning the adjustment handle counterclockwise will close the regulator valve and no oxygen gas will flow to the torch handle. Turning the adjustment handle clockwise will open the valve to allow oxygen gas to flow to the torch handle. Tips from size 0 to 5 are designed to operate with a working pressure of 5 psig for acetylene and 10 psig for the oxygen regulators.

NOTE: The oxygen torch valve must be open when an adjustment is made to the working pressure.

Once the proper operating pressure is reached, the torch valve is closed, and the working operating pressure is set. When shutting the oxyacetylene outfit down, always turn both the oxygen and acetylene pressure adjustment knobs to the OFF position by turning them out counterclockwise the whole way.

Two regulators are required for oxyacetylene outfits—one for the oxygen cylinder and the other for the acetylene cylinder. These regulators, gages, and hoses are color-coded. The oxygen regulator and hose is green and the acetylene regulator and hose is red. The regulators and hoses for oxygen and acetylene outfits cannot be accidentally interchanged because of the manufactured threading system. All oxygen fittings are right-hand threaded and all acetylene fittings are left-hand threaded.

EQUIPMENT REQUIRED

1 Oxygen tank
1 Acetylene tank
1 Acetylene regulator with acetylene hose
1 Oxygen regulator with hose
1 Valve stem wrench
1 12" adjustable wrench
1 Oxyacetylene torch handle
1 Torch tip
1 Striker
Soap bubble solution

SAFETY PRACTICES

Do not make a connection to the acetylene tank around an open flame. Do not use a match or lighter to light the torch. Keep torch cylinders secured.

NOTE: All connections should be free of dirt, dust, grease, and oil. Oxygen can produce an explosion when in contact with grease or oil.

---

*NOTE: It is illegal to operate a torch with acetylene pressure greater than 15 psig. Most acetylene gages have a **Red** mark indicating 15 psig.*

COMPETENCY PROCEDURES

Checklist

1. Check the regulators and make sure they are free of dirt, grease, and oil. ❏
2. Use the valve stem wrench and crack the acetylene tank valve to clear the valve of any foreign matter. ❏

NOTE: Do not vent acetylene near an open flame.

---

3. Crack the oxygen and acetylene tank valves by turning the tank handles to clear the valves of any foreign matter. ❏
4. Use the 12" adjustable wrench and attach the acetylene regulator to the acetylene tank. ❏
5. Use the 12" adjustable wrench and attach the oxygen regulator to the oxygen tank. ❏

NOTE: The acetylene and oxygen regulators are right-hand threaded.

---

6. Use the adjustable wrench to attach the red hose to the acetylene pressure regulator. ❏

NOTE: The acetylene hose connection is left-hand threaded.

---

7. Attach the other end of the red hose to the torch handle. ❑

8. Use the adjustable wrench to attach the green hose to the oxygen pressure regulator. ❑

> *NOTE: The oxygen hose connection is right-hand threaded.*

9. Attach the other end of the green hose to the torch handle. ❑

10. Make sure the torch knobs are closed off. ❑

11. Turn the adjustment knobs on the pressure regulators counterclockwise until there is no pressure on the regulator spring. (0 psig should be indicated on the regulator gages.) ❑

12. Use the valve stem wrench and open the acetylene tank valve by turning it counterclockwise one full turn. ❑

13. Open the oxygen tank valve completely by turning it counterclockwise. ❑

14. Turn the acetylene pressure regulator adjustment knob clockwise until a pressure of 5 psig is reached on the regulator pressure gage. ❑

15. Turn the oxygen pressure regulator adjustment knob clockwise until a pressure of 10 psig is reached on the regulator pressure gage. ❑

> *NOTE: These are the regulator pressure settings for normal brazing procedures.*

16. Use a soap bubble solution and check all connections for leaks. ❑

> *NOTE: Never open both the oxygen and acetylene torch valves at the same time and try to light the torch (Figure 3–35). Always open the acetylene torch valve first and light the torch, then mix the torch flame with oxygen until a neutral flame is obtained (Figure 3–36).*

FIGURE 3–35   An acetylene flame. (*Photo by Bill Johnson*)

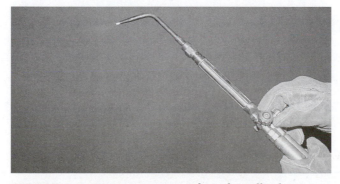

FIGURE 3–36 Mixing oxygen. (*Photo by Bill Johnson*)

CAUTION: Keep the torch pointed away from you and any combustible items.

17. Open the acetylene torch handle valve by turning it counterclockwise one-half turn. ❑
18. Use a striker and light the torch. ❑
19. Open the oxygen torch handle valve and adjust the oxygen to obtain a neutral flame (**Figure 3–37**). ❑

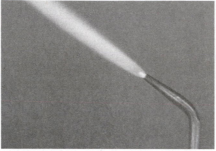

FIGURE 3–37 Neutral flame. (*Photo by Bill Johnson*)

Torch Shutdown Procedures

> **NOTE:** *With an oxyacetylene outfit, always shut the oxygen torch valve off first before closing the acetylene torch valve.*

20. Close the oxygen torch valve by turning the oxygen torch handle knob clockwise the whole way in. ❑
21. Close the acetylene torch valve by turning the acetylene torch handle knob clockwise the whole way in. ❑
22. Use the valve stem wrench and close the acetylene tank valve by turning the valve stem clockwise the whole way in. ❑
23. Close the oxygen tank valve by turning the knob clockwise the whole way in. ❑
24. Turn the pressure regulator valves counterclockwise to release the pressure from the spring on the regulator. ❑
25. Open the torch handle knobs and bleed the hoses and regulators till a zero pressure is indicated on the regulators. ❑
26. Once the regulator pressure drops to zero, shut the torch handles off by turning them clockwise. ❑
27. Have your instructor check your work. ❑
28. Return all tools and equipment to their proper location. ❑

RESEARCH QUESTIONS

1. Why is oxygen mixed with acetylene for soldering and brazing?

   _____

   _____

2. Why should oxygen or compressed air never be used to pressurize a system?

   _____

   _____

3. What is meant by reverse flow of gases from an oxyacetylene outfit?

_____

_____

4. What is backfire?

_____

_____

5. What is flashback?

_____

_____

Passed Competency _____ **Failed Competency** _____

Instructor Signature _____ **Grade** _____

Practical Competency 44

Lighting an Oxyacetylene Torch Outfit

SUGGESTED MATERIALS

Textbook
Refrigeration & Air Conditioning Technology, 5th Edition, Thomson Delmar Learning
Unit 7—Tubing and Piping

Review Topics
Heat Sources for Soldering and Brazing

COMPETENCY OBJECTIVE

The student will be able to light an oxyacetylene torch outfit.

OVERVIEW

The oxyacetylene outfit requires two cylinders of gases—one oxygen and one acetylene. The oxygen cylinder has a pressure of about 2200 psi at 70 degrees Fahrenheit. The pressure in the cylinder will vary as the temperature of the cylinder changes. A safety relief device is built into the oxygen cylinder valves to protect against excessive pressure that could create an explosion. The oxygen pressure regulator is designed to apply a constant working pressure through the oxygen hose, no matter what the pressure is in the cylinder. (**Remember the cylinder pressure will vary because of ambient temperatures.**)

The oxygen regulator is connected directly to the oxygen cylinder valve. The regulator contains an outlet for the connection of the oxygen hose leading to the torch handle. The oxygen regulator is equipped with two gages; one is used to measure the cylinder pressure and the other is used to indicate the torch working pressure.

The regulator is equipped with a pressure adjustment handle to adjust the working pressure of the torch. Turning the adjustment handle counterclockwise will close the regulator valve and no oxygen gas will flow to the torch handle. Turning the adjustment handle clockwise will open the valve to allow oxygen gas to flow to the torch handle. Tips from size 0 to 5 are designed to operate with a working pressure of 5 psig for acetylene and 10 psig for the oxygen regulators.

> **NOTE:** *The oxygen torch valve must be open when an adjustment is made to the working pressure.*

Once the proper operating pressure is reached, the torch valve is closed, and the working operating pressure is set. When shutting the oxyacetylene outfit down, always turn both the oxygen and acetylene pressure adjustment knobs to the OFF position by turning them out counterclockwise the whole way.

Two regulators are required for oxyacetylene outfits. These regulators, gages, and hoses are color-coded. The oxygen regulator and hose is green and the acetylene regulator and hose is red. The regulators and hoses for oxygen and acetylene outfits cannot be accidentally interchanged because of the manufactured threading system. All oxygen fittings are right-hand threaded and all acetylene fittings are left-hand threaded.

EQUIPMENT REQUIRED

Oxyacetylene torch outfit
1 Valve stem wrench
1 Torch tip
1 Striker

SAFETY PRACTICES

Do not make a connection to the acetylene tank around an open flame. Do not use a match or lighter to light the torch. Keep torch cylinders secured.

> *NOTE: All connections should be free of dirt, dust, grease, and oil. Oxygen can produce an explosion when in contact with grease or oil.*

> *NOTE: It is illegal to operate a torch with acetylene pressure greater than 15 psig. Most acetylene gages have a **Red** mark indicating 15 psig.*

COMPETENCY PROCEDURES

Checklist

1. Make sure the torch knobs are closed off. ❑
2. Turn the adjustment knobs on the pressure regulators counterclockwise until there is no pressure on the regulator spring. (0 psig should be indicated on the regulator gages.) ❑
3. Use the valve stem wrench and open the acetylene tank valve by turning it counterclockwise one full turn. ❑
4. Open the oxygen tank valve completely by turning it counterclockwise. ❑
5. Turn the acetylene pressure regulator adjustment knob clockwise until a pressure of 5 psig is reached on the regulator pressure gage. ❑
6. Turn the oxygen pressure regulator adjustment knob clockwise until a pressure of 10 psig is reached on the regulator pressure gage. ❑

> *NOTE: These are the regulator pressure settings for normal brazing procedures.*

7. Use a soap bubble solution and check all connections for leaks. ❑

> *NOTE: Never open both the oxygen and acetylene torch valves at the same time and try to light the torch. Always open the acetylene torch valve first and light the torch, then mix the torch flame with oxygen until a neutral flame is obtained.*

CAUTION: Keep the torch pointed away from you and any combustible items.

8. Open the acetylene torch handle valve by turning it counterclockwise one-half turn. ❑
9. Use a striker and light the torch. ❑
10. Open the oxygen torch handle valve and adjust the oxygen to obtain a neutral flame. ❑

Torch Shutdown Procedures

> *NOTE: With an oxyacetylene outfit, always shut the oxygen torch valve off first before closing the acetylene torch valve.*

11. Close the oxygen torch valve by turning the oxygen torch handle knob clockwise the whole way in. ❑

12. Close the acetylene torch valve by turning the acetylene torch handle knob clockwise the whole way in. ❑

13. Use the valve stem wrench and close the acetylene tank valve by turning the valve stem clockwise the whole way in. ❑

14. Close the oxygen tank valve by turning the knob clockwise the whole way in. ❑

15. Turn the pressure regulator valves counterclockwise to release the pressure from the spring on the regulator. ❑

16. Open the torch handle knobs and bleed the hoses and regulators until a zero pressure is indicated on the regulators. ❑

17. Once regulators pressures drops to zero, shut the torch handles off by turning them clockwise. ❑

18. Have your instructor check your work. ❑

19. Return all tools and equipment to their proper location. ❑

RESEARCH QUESTIONS

1. Brazing is done at what temperatures?

_____

_____

2. Soldering is done at what temperatures?

_____

_____

3. Why is flux used in soldering or brazing?

_____

_____

4. What type of soldering is suitable for moderate temperature and pressures?

_____

_____

5. What elements make up brazing filler metal alloys?

_____

_____

Passed Competency _____   Failed Competency _____

Instructor Signature _____   Grade _____

Theory Lesson: Proper Soldering Techniques (Figure 3–38)

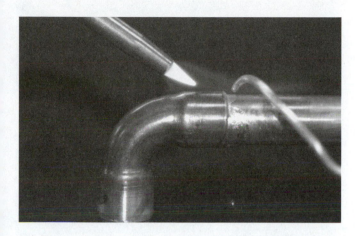

FIGURE 3–38   Soldering. (*Photo by Bill Johnson*)

SUGGESTED MATERIALS

Textbook
Refrigeration & Air Conditioning Technology, 5th Edition, Thomson Delmar Learning
Unit 7—Soldering and Piping

Review Topics
Soldering and Brazing Process; Soldering Techniques; Practical Soldering and Brazing Tips

Key Terms
brazing • capillary action • flux • solder • soldering

OVERVIEW

Soldering is the process of fastening two base metals together by using a third filler metal that melts at a temperature below 800 degrees Fahrenheit. **Brazing** is the process of fastening two base metals together by using a third filler metal that melts at a temperature above 800 degrees Fahrenheit. Good soldering techniques are important to ensure leak-proof joints in refrigeration sealed system work. Using the correct soldering equipment—solder, flux; cleaning the joint to be soldered; and applying the heat correctly are all important factors to be considered in good soldering techniques.

When preparing the joint to be soldered, make sure it is clean and free of dirt, oils, and oxidation (**Figure 3–39** and **Figure 3–40**).

Make sure that the joint is a tight fit. Select the correct soldering equipment for the job. Use the correct flux and solder for the job (**Figure 3–41**).

When applying the heat to the joint to be soldered, keep in mind that the solder will flow properly only when the pieces to be joined are heated equally and evenly. The soldering alloy will flow toward the heat with the greatest intensity. Understanding this principle allows solder to be moved so that the solder can be drawn into a joint to be soldered. Do not apply the flame directly to the solder (**Figure 3–42**).

Use the flame of the torch to heat the metals to a temperature hot enough to melt the solder. To see if the metal is hot enough to melt the solder, at intervals, touch the metal area being heated and see if the solder starts to melt (**Figure 3–43**). If it does, the metal is hot enough to melt the solder to complete the soldering joint. If the solder does not melt, move it away from the melt and continue to heat the joint area evenly and recheck with the solder.

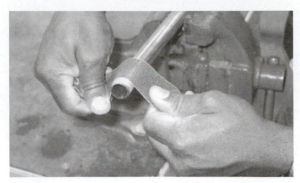

FIGURE 3–39   Cleaning tubing. (*Photo by Bill Johnson*)

FIGURE 3–40   Cleaning tubing. (*Photo by Bill Johnson*)

FIGURE 3–41   Applying flux. (*Photo by Bill Johnson*)

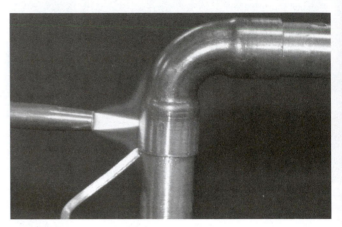

FIGURE 3–42   Applying heat to the metal. (*Photo by Bill Johnson*)

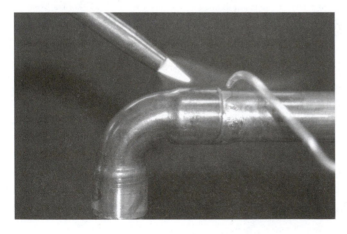

FIGURE 3–43   Testing metal with solder. (*Photo by Bill Johnson*)

Once the metal is hot enough to melt the solder, back off the flame just far enough to maintain the proper temperature until the solder fills the joint. Apply only enough solder to fill the joint. If the solder forms beads or balls of solder on the metal rather than flowing smoothly, the metal is probably dirty or not hot enough.

Solder will flow into the joint by a method called **capillary action**. This is the process by which the solder automatically fills the gap between the pieces of metal being soldered. For the best capillary action, the joint to be soldered should be a tight fit and supported to prevent movement. When the solder alloy enters the joint, the molecules of the solder have a greater attraction to the metal walls of the metals being joined together. This causes the solder to fill the joint completely.

Practical Competency 45

Proper Soldering and Brazing Techniques

SUGGESTED MATERIALS

Textbook
Refrigeration & Air Conditioning Technology, 5th Edition, Thomson Delmar Learning
Unit 7—Soldering and Piping

Review Topics
Soldering and Brazing Process; Soldering Techniques; Practical Soldering and Brazing Tips

COMPETENCY OBJECTIVE

The student will be able to properly solder and braze.

OVERVIEW

When applying the heat to the joint to be soldered, keep in mind that the solder will flow properly only when the pieces to be joined are heated equally and evenly. The soldering alloy will flow toward the heat with the greatest intensity. Understanding this principle allows solder to be moved so that the solder can be drawn into a joint to be soldered. Do not apply the flame directly to the solder. Use the flame of the torch to heat the metals to a temperature hot enough to melt the solder.

To see if the metal is hot enough to melt the solder, at intervals, touch the metal area being heated and see if the solder starts to melt. If it does, the metal is hot enough to melt the solder to complete the soldering joint. If the solder does not melt, move it away from the melt and continue to heat the joint area evenly and recheck with the solder.

Once the metal is hot enough to melt the solder, back off the flame just far enough to maintain the proper temperature until the solder fills the joint. Apply only enough solder to fill the joint. If the solder forms beads or balls of solder on the metal rather then flowing smoothly, the metal is probably dirty or the metal is not hot enough.

Solder will flow into the joint by a method called capillary action. This is the process by which the solder automatically fills the gap between the pieces of metal being soldered. For the best capillary action, the joint to be soldered should be a tight fit and supported to prevent movement. When the solder alloy enters the joint, the molecules of the solder have a greater attraction to the metal walls of the metals being joined together. This causes the solder to fill the joint completely.

EQUIPMENT REQUIRED

1 Soldering outfit unit
1 Striker
1 Sanding cloth or wire brush
1 Solder and flux
2 4" Pieces of 1/2" ACR copper tubing
1 1/2" ACR 90-degree elbow

SAFETY PRACTICES

Do not make a connection to the acetylene tank around an open flame. Do not use a match or lighter to light the torch. Keep the torch away from yourself and combustible items when lighting. Keep torch cylinders secured.

COMPETENCY PROCEDURES

Checklist

1. Use sand cloth or a wire brush to clean all mating joints to be soldered or brazed. ❑
2. Apply flux (if required) to all male copper connections. ❑
3. Insert the fluxed ends of the two 4" pieces of 1/2" copper into the 1/2" × 90 degree elbow. ❑
4. Support the tubing to be soldered or brazed. ❑
5. Set up the torch outfit for soldering or brazing the copper joint. ❑
6. Light the torch and adjust it to a neutral flame. ❑
7. Hold the torch so the inner cone of the flame just touches the metal near the vertical joint for a short time (**Figure 3–44**). ❑

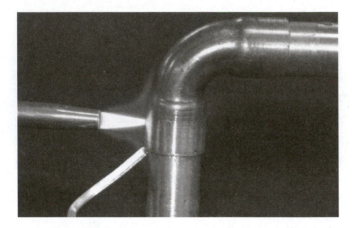

FIGURE 3–44　Heating tubing.
(*Photo by Bill Johnson*)

> **NOTE:** *The blue flame in the middle of the torch flame is the hottest part of the flame.*

8. Move the torch flame from the tubing to the fitting and heat the fitting area for a short time. ❑
9. Spread the heat evenly by moving the torch from the tubing to the fitting and from the fitting to the tubing. ❑
10. Test the copper to see if it is hot enough to melt the solder or brazing alloy. Do this by touching the joint with the solder or brazing rod and seeing if the metal melts the solder or brazing rod (**Figure 3–45**). ❑

> **NOTE:** *Do not melt the solder or brazing rod with the torch flame; the heat of the metal is used to melt the solder or brazing alloy.*

> **NOTE:** *If using a brazing alloy, the metal will become cherry red in color when the metal reaches the right temperature to melt the brazing rod.*

11. If the copper is not hot enough to melt the soldering alloy, continue to heat the copper surface evenly by moving the torch flame back and forth across the fitting and copper area. ❑
12. Continue to test the heat of the metal with the solder or brazing alloy. ❑

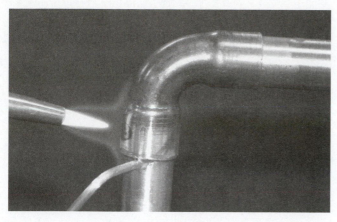

FIGURE 3–45 Testing the tubing. (*Photo by Bill Johnson*)

13. Once the soldering alloy flows freely from the heat of the metal, concentrate the torch flame a little above the soldering joint (**Figure 3–46**). ❑

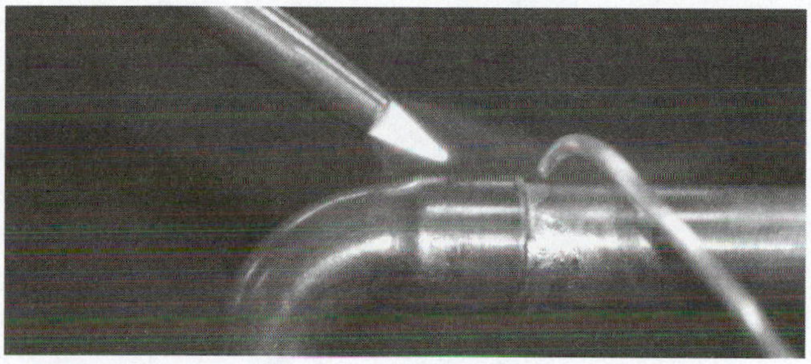

FIGURE 3–46 Soldering the joint. (*Photo by Bill Johnson*)

14. Allow enough soldering alloy to fill the vertical joint. ❑
15. Proceed to solder or braze the horizontal joint using the same heating techniques as before. ❑
16. Once the metal is hot enough to melt the soldering alloy, apply the soldering alloy to the bottom of the joint first. ❑
17. Then apply the soldering alloy to the sides. ❑
18. Then apply the soldering alloy to the top of the joint. ❑
19. Wipe the joints with a rag while they are still hot to remove any excess solder alloy. ❑

> **NOTE:** *You may want to practice a few joints to become more competent at soldering and brazing techniques.*

20. Have your instructor check your work. ❑
21. Return all tools and equipment to their proper location. ❑

RESEARCH QUESTIONS

1. What are some important factors in ensuring good solder joints?

_____

_____

2. It is important to apply the heat from the torch how?

_____

_____

3. During soldering, what should be used to melt the solder?

_____

_____

4. Solder will travel in which direction when applied to the joint to be soldered?

_____

_____

5. What is meant by capillary action during soldering?

_____

_____

Passed Competency _____ **Failed Competency** _____

Instructor Signature _____ **Grade** _____

Theory Lesson: The Vapor Compression Refrigeration Cycle (Figure 3–47)

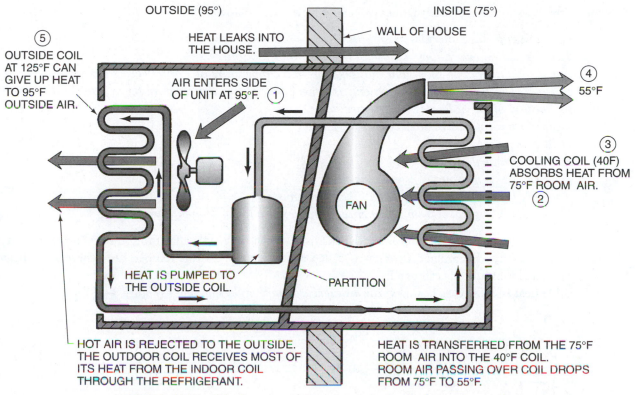

OUTSIDE (95°) INSIDE (75°)

HEAT LEAKS INTO THE HOUSE. WALL OF HOUSE

⑤ OUTSIDE COIL AT 125°F CAN GIVE UP HEAT TO 95°F OUTSIDE AIR.

AIR ENTERS SIDE OF UNIT AT 95°F. ①

④ 55°F

③ COOLING COIL (40F) ABSORBS HEAT FROM 75°F ROOM AIR.

②

FAN

HEAT IS PUMPED TO THE OUTSIDE COIL.

PARTITION

HOT AIR IS REJECTED TO THE OUTSIDE. THE OUTDOOR COIL RECEIVES MOST OF ITS HEAT FROM THE INDOOR COIL THROUGH THE REFRIGERANT.

HEAT IS TRANSFERRED FROM THE 75°F ROOM AIR INTO THE 40°F COIL. ROOM AIR PASSING OVER COIL DROPS FROM 75°F TO 55°F.

FIGURE 3–47 Window air conditioner mechanical refrigeration cycle.

SUGGESTED MATERIALS

Textbook
Refrigeration & Air Conditioning Technology, 5th Edition, Thomson Delmar Learning
Unit 3—Refrigeration and Refrigerants

Review Topics
Introduction to Refrigeration; Refrigeration; The Refrigeration Process; Refrigeration Components; The Evaporator; The Compressor; The Condenser; The Refrigerant Metering Device; Refrigeration System and Components

Key Terms
absolute zero • boiling • change of state • compressor • condensation • condenser • evaporation • evaporator • heat • heat of compression • high-side • latent heat • low-side • mechanical refrigeration • metering device • pressure • refrigerant • saturation • saturation temperature • sensible heat • sub-cooling • superheat • temperature difference • thermodynamics of heat • ton of refrigeration

OVERVIEW

The **refrigeration process** is defined as removing heat from an area or product where it is not wanted and transferring this heat to an area where it makes little or no difference.

Understanding what heat is and the types of heat involved in the refrigeration process need to be discussed to take hold of the concept of moving heat from one area to an area where it makes little difference.

There are different types of heat to be discussed, but most important is to understand what heat is and some of the laws of the thermodynamics of heat, which are as follows:

1. *Heat is defined as molecules in motion.*
2. *There is heat in everything because of the molecular motion.*
3. *Scientists state that all molecular motion stops at a temperature of −460 degrees Fahrenheit. This is considered to be **absolute zero**. Scientists believe that at this temperature there is no heat because the motion of all molecules has stopped.*
4. *Heat is not created; it is already here on Earth.*
5. *Heat comes from our sun.*
6. *Heat cannot be destroyed.*
7. *Heat can be turned into consumable energy.*
8. *The only thing that can be done with heat is to move it.*
9. *Heat will always travel from a higher temperature to a lower temperature.*

As defined, **heat is molecules in motion,** with scientists stating that molecular motion stops at a temperature of −460 degrees Fahrenheit, or what is called **absolute zero**. At this temperature, there is no heat because molecular motion has stopped.

Sensible heat is defined as heat *that can be measured with a thermometer* (**Figure 3–48**).

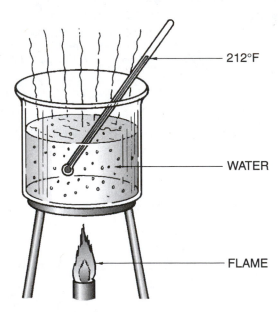

— 212°F

— WATER

— FLAME

FIGURE 3–48 Thermometer measuring the sensible heat temperature of water at 212 degrees (F).

Specific heat is defined as *the amount of heat (Btu's) required to raise or lower the temperature of 1 pound of a substance by 1 degree (F)* (**Figure 3–49**).

Notice the specific heat value of **water** is **1.00**. The chart shows that to raise or lower the temperature of 1 pound of water 1 degree (F) would require the addition or removal of the quantity of heat equal to one full Btu.

In addition, notice the specific heat value of **air** is **0.24** (*average*). The chart indicates that to raise or lower the temperature of 1 pound of air 1 degree (F) requires the addition or removal of the quantity of heat equal to a one fourth of a Btu of heat.

SUBSTANCE	SPECIFIC HEAT Btu/lb/°F	SUBSTANCE	SPECIFIC HEAT Btu/lb/°F
ALUMINUM	0.224	BEETS	0.90
BRICK	0.22	CUCUMBERS	0.97
CONCRETE	0.156	SPINACH	0.94
COPPER	0.092	BEEF, FRESH	
ICE	0.504	LEAN	0.77
IRON	0.129	FISH	0.76
MARBLE	0.21	PORK, FRESH	0.68
STEEL	0.116	SHRIMP	0.83
WATER	1.00	EGGS	0.76
SEA WATER	0.94	FLOUR	0.38
AIR	0.24 (AVERAGE)		

FIGURE 3–49   Specific heat table shows how much heat (Btu's) is required to raise or lower the temperature of 1 pound of several substances 1 degree (F).

If the quantity of heat equal to a **full Btu** were added or removed from air, its temperature would rise or fall by approximately 4 degrees. To cause water to change in temperature by 4 degrees would require the addition or removal of the quantity of heat generated by 4 full Btu's.

Btu stands for the **British Thermal Unit** and is defined as the amount (quantity) of heat required to raise the temperature of 1 pound of water, 1 degree (F).

The definition of the British Thermal Unit still leaves the unanswered question as to how much heat is equal to a Btu of heat. A **Btu of heat** is equal to the quantity of heat generated when one standard match is burned from beginning to end. The amount of **12,000 Btu's of heat removed or added to an area is equal to one ton of refrigeration (Figure 3–50).**

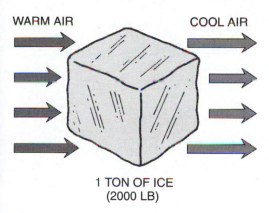

WARM AIR   COOL AIR

1 TON OF ICE
(2000 LB)

FIGURE 3–50   2000 lb of ice requires 144 Btu/lb to melt. 2000 lb × 144 Btu per pound = 288,000 Btu. When this is accomplished in a 24-hour period, it is known as a heat transfer rate of 1 ton of refrigeration. 1 ton = 12,000 Btu's per hour or 200 Btu's of heat per minute.

This means that a one-ton air conditioning unit has the ability to remove the quantity of heat generated by burning 12,000 matches in 1 hour.

Notice that Btu's have nothing to do with the temperature of a substance (**Figure 3–51**).

Temperature is defined as *the level of heat or molecular activity of a substance* (**Figure 3–52**). The quantity of heat in a substance is measured in Btu's.

Another type of heat to be discussed is **latent heat**. All **matter** can exist in three states: **solid, liquid,** and **vapor.** An example is water, which can be frozen to ice, melted to water, and turned to vapor. For any substance to change from one state of matter to another state of matter requires the addition or removal of what is referred to as **latent heat.** *This is heat that is required to cause a substance to change state.* A couple of facts must be understood relative to latent heat. This is heat that is also referred to as **"hidden heat."** During the latent heat process the temperature of the substance does not change, nor does the amount of pressure exerted on the substance. Latent heat only causes the substance to change state, and may be added or removed to cause the change of state to take place.

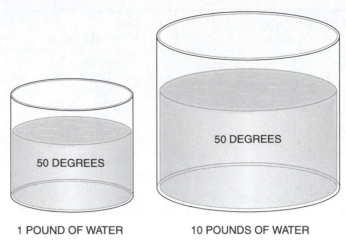

FIGURE 3–51   Two vessels at the same temperature, with one vessel having more Btu's (quantity of heat) than the other.

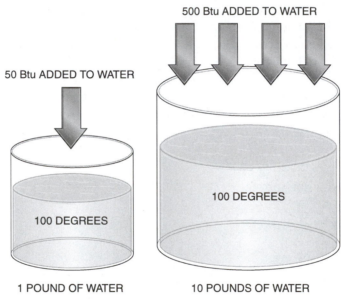

FIGURE 3–52   The Btu content of two vessels at the same temperature.

Superheat is defined *as heat required to raise the temperature of a vapor one or more degrees over the saturation temperature of an evaporating (boiling) liquid.* Superheating a vapor cannot happen as long as there is liquid in the mixture or the vapor has been moved far enough away from the evaporating liquid. Superheat does not change the pressure exerted on a substance.

Sub-cooling is defined *as lowering the temperature of liquid one or more degrees below its saturation temperature.* Sub-cooling does not change the amount of pressure exerted on a substance.

Take a look at the Heat/Temperature graph of water and the different types of heat discussed can be seen (**Figure 3–53**). Notice at **point 1** on the graph the water is in the solid state as ice at 32 degrees, with 16 Btu's of sensible heat. Between **points 2** and **3**, *latent heat is being added to the ice to cause it to melt. Notice that during this the latent heat process, the water and ice are still at 32 degrees.* Also notice that to completely melt all the ice to 100% water at 32 degrees requires the addition of **144 Btu's of latent heat**. From **points 3** and **4**, the water is being sensible heated to a temperature of 212 degrees (F). This requires the addition of 180 Btu's of heat.

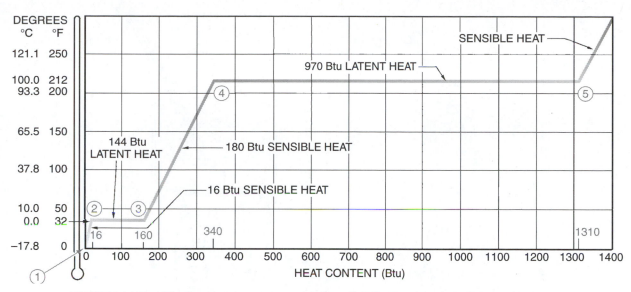

FIGURE 3–53   The heat/temperature graph for 1 lb of water at atmospheric pressure.

Water under atmospheric pressure will boil or evaporate at 212 degrees (F) if saturated with heat. This is referred to as its **saturation temperature**. Every substance has a saturation temperature, which is determined by the amount of pressure exerted on the substance (**Figure 3–54** and **Figure 3–55**).

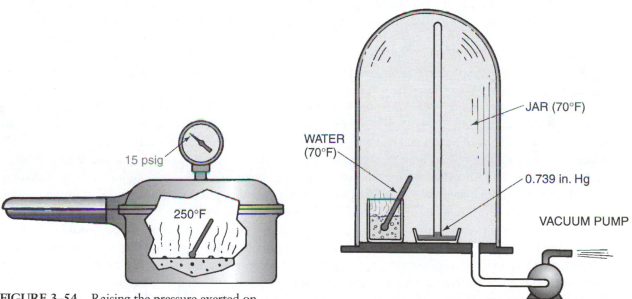

FIGURE 3–54   Raising the pressure exerted on water to 15 psig raises the water's saturation temperature to 250 degrees (F).

FIGURE 3–55   Lowering the pressure exerted on water will lower its saturation temperature.

At point 4, continuing to add heat to the water will not raise the water temperature; the additional heat will only saturate the water to a point that it will eventually start to boil or evaporate. It is important to note that during the boiling of the liquid at 212 degrees, the pot temperature is 212 degrees (F) no matter what the temperature is or how much heat is being used to heat the water (**Figure 3–56**).

In fact, the water is keeping the pot cool by throwing the additional heat out with the steam. **From points 4 and 5,** the water is boiling or evaporating until there is no longer any liquid left. Notice that to boil or evaporate 1 pound of water all the way to 100% vapor requires the addition or 970 Btu's of latent heat.

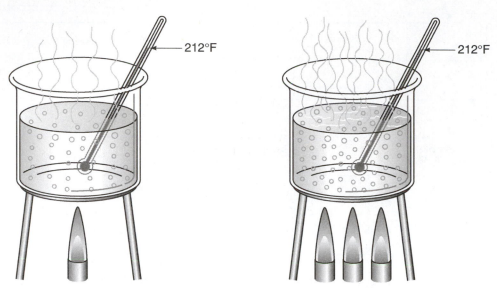

FIGURE 3-56 Adding three times as much heat only causes the water to boil faster; the water or pot does not increase in temperature.

Also notice that during the boiling or evaporation process, the temperature of the water and the vapor leaving the water did not change. The boiling liquid and steam vapor leaving the liquid were both at 212 degrees no matter how much heat was added.

At point 5, all the liquid has been boiled away and only 100% vapor exists. At this point if heat continues to be added to the mixture, the vapor temperature will rise. This is where superheating the vapor takes place. It is important to notice when the vapor became superheated; this happened only after the liquid was completely boiled away. This is referred to as superheated vapor.

In reviewing the graph there are a couple of things to consider. Notice how much heat needed to be added to the ice from point 1 to point 5. It took 1294 Btu's of heat to melt 1 pound of ice to water and then boil the water completely to 100% vapor. With the additional 16 Btu's the ice held at 32 degrees; this means the vapor contains 1310 Btu's of heat at point 5, before the vapor gets superheated. Taking into consideration the quantity of heat equal to 1 Btu, the water vapor now holds a tremendous amount of heat. In addition, notice that only 196 Btu's of this heat was **sensible heat** and 1114 Btu's of the heat was **latent heat. More heat was absorbed during the two changes of state** of the water than just heating the water from 32 degrees to 212 degrees (F). 144 Btu's of heat were absorbed by melting the ice, with the greatest amount of heat, 970 Btu's, absorbed by boiling the water to 100% vapor.

Since heat cannot be destroyed and can be accounted for, the same amount of heat would have to be removed to return the water back to 32-degree ice. The principle of boiling a liquid to a vapor is important to remember because the refrigeration process uses this principle to absorb heat from a conditioned space or product and reject the heat somewhere else by condensing a vapor back to a liquid.

It is important to understand the concepts discussed along with the most relevant law of the Thermodynamics of Heat that states: **"Heat always moves from a higher temperature to a lower temperature."** This law is the basis on which a vapor compression mechanical refrigeration system operates. The question might be asked as to why does *"heat of a higher temperature always travel towards heat of a lower temperature?"* In fact, the greater the temperature difference between two mediums, the faster the heat transfer. The simplest answer to this question is to remember that heat is defined as **molecules in motion.** *This means that molecules of a higher temperature are moving with faster and with greater force than heat molecules of a lower temperature.* When a temperature difference between two mediums does exist, molecules of the higher temperature medium will overcome molecules of a lower temperature until equilibrium of the two temperature differences is reached. At this point, all molecules of the two mediums are moving at the same speed, meaning that the two mediums have reached an equal temperature.

This law is the foundation on which vapor-compression refrigeration systems work. Mechanically a temperature difference where heat will move from a higher temperature to a lower temperature is established and maintained so that heat can be transferred from an area or product where it is not wanted and rejected to an area where it makes little or no difference.

To mechanically establish this concept requires that the mechanical refrigeration system consist of four main components: an evaporator, condenser, compressor, and metering device (**Figure 3–57**).

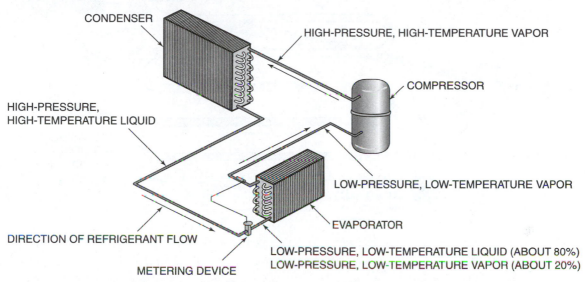

CONDENSER

HIGH-PRESSURE, HIGH-TEMPERATURE VAPOR

COMPRESSOR

HIGH-PRESSURE, HIGH-TEMPERATURE LIQUID

LOW-PRESSURE, LOW-TEMPERATURE VAPOR

EVAPORATOR

DIRECTION OF REFRIGERANT FLOW

LOW-PRESSURE, LOW-TEMPERATURE LIQUID (ABOUT 80%)
LOW-PRESSURE, LOW-TEMPERATURE VAPOR (ABOUT 20%)

METERING DEVICE

FIGURE 3–57   Four main components of a mechanical refrigeration system.

These four components are sealed and connected together by tubing, which could be copper, aluminum, or steel. Each of the four components is designed to work with any one of the others to accomplish the transfer of heat from a medium and reject this heat to another medium somewhere else.

The **evaporator's main function** is to **absorb heat** from a space to be conditioned or a product to be cooled or frozen. This component of the sealed system would be located somewhere in the area where heat is to be removed.

The **condenser's main function** is to **reject heat** that was absorbed by the evaporator into an area or medium where it makes little or no difference. This means that the condenser is going to be located in the area where rejecting the heat is not objectionable.

The **compressor's main function** is to **move heat**, which is absorbed by the evaporator, and pump it to the condenser so that it can be rejected somewhere else. The compressor's location will normally be where the condenser is located.

The **metering device's main function** is to **control the flow of refrigerant** into the evaporator. Normally the metering device will be located with the evaporator.

Each of the main four components of a vapor-compression refrigeration system has three jobs to perform. Gaining an understanding of the performance of each can assist the technician in evaluating sealed system performance and problems when they occur.

As mentioned, the evaporator's main function is to absorb heat from the conditioned space or product (**Figure 3–58**).
To perform this task, **the evaporator:**

Absorbs heat.
Boils (evaporates) a liquid to a vapor.
Superheats the vapor.

The condenser's main function is to reject heat from the refrigeration system (**Figure 3–59**).

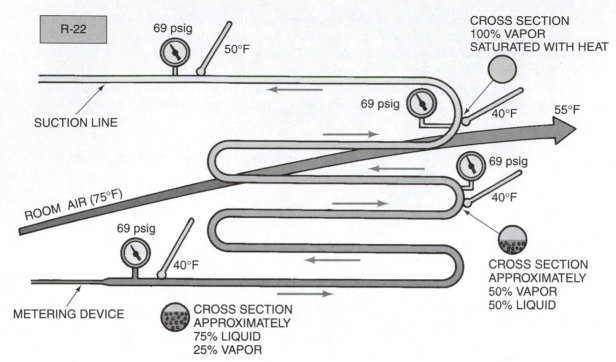

FIGURE 3–58 The evaporator absorbs heat in to the refrigeration system by boiling the refrigerant at a temperature lower than that of the room air passing over it.

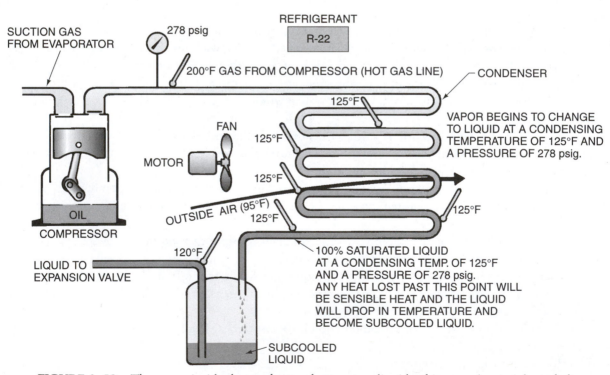

FIGURE 3–59 The vapor inside the condenser changes to a liquid refrigerant that is sub-cooled.

To perform this task, **the condenser:**

Rejects heat.
Condenses a vapor to a liquid.
Sub-cools the liquid refrigerant.

The compressor's main function is to move heat by pumping refrigerant vapor through the sealed system (**Figure 3–60**).

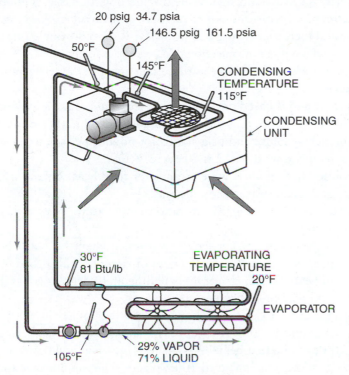

FIGURE 3–60 The compressor pumps heat through the sealed system by moving refrigerant vapor through the sealed system.

To perform this task, **the compressor:**

Sucks low temperature-superheated vapor from the evaporator.
Compresses the superheated vapor (**heat of compression**).
Discharges the heat of compression vapor to the condenser.

The main function of the metering device is to control the flow of liquid refrigerant into the evaporator (**Figure 3–61**).

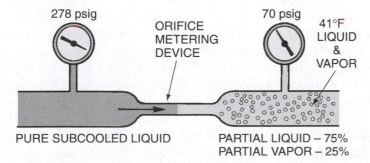

FIGURE 3–61 A fixed–orifice metering device.

To perform this task, the **metering device:**

Creates a pressure drop between the low and high side of the sealed system.
Controls the flow of liquid refrigerant into the evaporator.
Creates flash gas.

For the refrigeration process to take place efficiently, each of the four components is dependent on the others to perform their individual functions efficiently. Likewise, if one of the four main components fails in performing efficiently, the performance of the other three components will be affected, which will reduce the effectiveness of the refrigeration process.

Just having all four of these components joined together in a sealed circuit does not mean that heat can be absorbed or rejected. For these components to perform their function in the sealed system requires an additional element—**refrigerant**. The refrigerant of the system is what actually is used to absorb the heat from the conditioned area and is also used to reject the heat in an area where it is not objectionable. A **good refrigerant** is a liquid that *boils at low temperatures and pressures and condenses at high temperatures and pressures*. The concept of boiling a refrigerant and then condensing it is how heat is moved from an area where it is not wanted and then put into an area where it is not objectionable.

The principles that have been explained about water are the same as those that apply to the refrigerant used in a mechanical refrigeration system (**Figure 3–62**).

This chart is a portion of R-22 properties chart and is based on 1 pound of refrigerant. Take a look at 40 degrees down the Temperature column of the chart. Notice that the refrigerant is under 83.206 psia and 68.510 psig. **Psia** stands for **pounds per square inch abso**lute and **psig** stands for **pounds per square inch gage**. Because of these pressures, the saturation temperature of R-22 is 40 degrees (F).

Move across the top of the chart to the column to **Enthalpy**, which stands for **total heat content** of a substance. This includes the amount of **sensible** and **latent heat** in the substance.

Move down the Enthalpy Liquid column to the 40 degrees (F) saturation temperature. Notice that 40-degree liquid contains **21.422 Btu's of sensible heat.**

The next column over to the right is the **Latent column**. This shows how much latent heat will be absorbed by 40-degree R-22 if it is boiled or evaporated to a vapor. The amount of latent heat absorbed by boiling 1 pound of R-22 would be **86.720** Btu's of heat. As the refrigerant is boiling at 40 degrees (F) as it passes through the coil, the surface area of the coil will be 40 degrees.

Move to the Vapor column of the Enthalpy chart to see how much heat is contained in 100% 40-degree (F) vapor. The total heat contained in 1 pound of 40-degree vapor is **108.142 Btu's** of heat. This total is **21.422 Btu's of sensible heat** and **86.720 Btu's of latent heat**. As stated earlier, when boiling water, the greatest amount heat was absorbed when the water was boiled to 100% vapor. The same is true for boiling 40-degree (F) liquid refrigerant to 100% vapor. The greatest amount of heat was absorbed during the boiling process.

Taking this 40-degree (F) vapor to another location and using means to cool it down would eventually remove 86.720 Btu's of latent heat and return the refrigerant to 100%, 40-degree liquid.

The Vapor Compression Refrigeration Cycle

Observe **Figure 3–63**. Remembering the three jobs of metering device and evaporator, start reviewing the refrigeration process at the metering device. A **low-temperature, low-pressure refrigerant liquid is fed into the evaporator** of the sealed system by the metering device. Because the evaporator is attached to the suction side of the compressor, a low-pressure area exists in the evaporator coil. This causes the liquid refrigerant to boil (evaporate) at a low temperature and pressure, as it enters the evaporator.

The **metering device** also assists in creating this low-pressure area by reducing the flow of the liquid refrigerant from the condenser to the evaporator. The immediate boiling of the liquid refrigerant at **point 1** as it enters the evaporator is called **flash gas**. Under normal conditions, 25% of the 100% liquid entering the evaporator is used in this process. This 25% boiling liquid is used to cool down the 75% liquid to the saturation temperature of 40 degrees from **point 1 through point 5**. For the 75% low-temperature and low-pressure liquid to evaporate through the coil, it must be saturated with heat. The heat used to saturate the refrigerant comes from the 75-degree (F) air of the conditioned space. This means that the air is giving up heat to the refrigerant in the latent heat process. As the air gives up heat, causing the refrigerant to evaporate, its temperature will start to drop because it contains less heat. The leaving air temperature is approximately 55 degrees (F). This will continue to take place as long as the mechanical

TEMP.	PRESSURE		VOLUME cu ft/lb		DENSITY lb/cu ft		ENTHALPY Btu/lb			ENTROPY Btu/(lb)(°R)		TEMP.
°F	PSIA	PSIG	LIQUID v_f	VAPOR v_g	LIQUID $1/v_f$	VAPOR $1/v_g$	LIQUID h_f	LATENT h_{fg}	VAPOR h_g	LIQUID s_f	VAPOR s_g	°F
10	47.464	32.768	0.012088	1.1290	82.724	0.88571	13.104	92.338	105.442	0.02932	0.22592	10
11	48.423	33.727	0.012105	1.1077	82.612	0.90275	13.376	92.162	105.538	0.02990	0.22570	11
12	49.396	34.700	0.012121	1.0869	82.501	0.92005	13.648	91.986	105.633	0.03047	0.22548	12
13	50.384	35.688	0.012138	1.0665	82.389	0.93761	13.920	91.808	105.728	0.03104	0.22527	13
14	51.387	36.691	0.012154	1.0466	82.276	0.95544	14.193	91.630	105.823	0.03161	0.22505	14
15	52.405	37.709	0.012171	1.0272	82.164	0.97352	14.466	91.451	105.917	0.03218	0.22484	15
16	53.438	38.742	0.012188	1.0082	82.051	0.99188	14.739	91.272	106.011	0.03275	0.22463	16
17	54.487	39.791	0.012204	0.98961	81.938	1.0105	15.013	91.091	106.105	0.03332	0.22442	17
18	55.551	40.855	0.012221	0.97144	81.825	1.0294	15.288	90.910	106.198	0.03389	0.22421	18
19	56.631	41.935	0.012238	0.95368	81.711	1.0486	15.562	90.728	106.290	0.03446	0.22400	19
20	57.727	43.031	0.012255	0.93631	81.597	1.0680	15.837	90.545	106.383	0.03503	0.22379	20
21	58.839	44.143	0.012273	0.91932	81.483	1.0878	16.113	90.362	106.475	0.03560	0.22358	21
22	59.967	45.271	0.012290	0.90270	81.368	1.1078	16.389	90.178	106.566	0.03617	0.22338	22
23	61.111	46.415	0.012307	0.88645	81.253	1.1281	16.665	89.993	106.657	0.03674	0.22318	23
24	62.272	47.576	0.012325	0.87055	81.138	1.1487	16.942	89.807	106.748	0.03730	0.22297	24
25	63.450	48.754	0.012342	0.85500	81.023	1.1696	17.219	89.620	106.839	0.03787	0.22277	25
26	64.644	49.948	0.012360	0.83978	80.907	1.1908	17.496	89.433	106.928	0.03844	0.22257	26
27	65.855	51.159	0.012378	0.82488	80.791	1.2123	17.774	89.244	107.018	0.03900	0.22237	27
28	67.083	52.387	0.012395	0.81031	80.675	1.2341	18.052	89.055	107.107	0.03958	0.22217	28
29	68.328	53.632	0.012413	0.79604	80.558	1.2562	18.330	88.865	107.196	0.04013	0.22198	29
30	69.591	54.895	0.012431	0.78208	80.441	1.2786	18.609	88.674	107.284	0.04070	0.22178	30
31	70.871	56.175	0.012450	0.76842	80.324	1.3014	18.889	88.483	107.372	0.04126	0.22158	31
32	72.169	57.473	0.012468	0.75503	80.207	1.3244	19.169	88.290	107.459	0.04182	0.22139	32
33	73.485	58.789	0.012486	0.74194	80.089	1.3478	19.449	88.097	107.546	0.04239	0.22119	33
34	74.818	60.122	0.012505	0.72911	79.971	1.3715	19.729	87.903	107.632	0.04295	0.22100	34
35	76.170	61.474	0.012523	0.71655	79.852	1.3956	20.010	87.708	107.719	0.04351	0.22081	35
36	77.540	62.844	0.012542	0.70425	79.733	1.4199	20.292	87.512	107.804	0.04407	0.22062	36
37	78.929	64.233	0.012561	0.69221	79.614	1.4447	20.574	87.316	107.889	0.04464	0.22043	37
38	80.336	65.640	0.012579	0.68041	79.495	1.4697	20.856	87.118	107.974	0.04520	0.22024	38
39	81.761	67.065	0.012598	0.66885	79.375	1.4951	21.138	86.920	108.058	0.04576	0.22005	39
40	83.206	68.510	0.012618	0.65753	79.255	1.5208	21.422	86.720	108.142	0.04632	0.21986	40
41	84.670	69.974	0.012637	0.64643	79.134	1.5469	21.705	86.520	108.225	0.04688	0.21968	41
42	86.153	71.457	0.012656	0.63557	79.013	1.5734	21.989	86.319	108.308	0.04744	0.21949	42
43	87.655	72.959	0.012676	0.62492	78.892	1.6002	22.273	86.117	108.390	0.04800	0.21931	43
44	89.177	74.481	0.012695	0.61448	78.770	1.6274	22.558	85.914	108.472	0.04855	0.21912	44
45	90.719	76.023	0.012715	0.60425	78.648	1.6549	22.843	85.710	108.553	0.04911	0.21894	45
46	92.280	77.584	0.012735	0.59422	78.526	1.6829	23.129	85.506	108.634	0.04967	0.21876	46
47	93.861	79.165	0.012755	0.58440	78.403	1.7112	23.415	85.300	108.715	0.05023	0.21858	47
48	95.463	80.767	0.012775	0.57476	78.280	1.7398	23.701	85.094	108.795	0.05079	0.21839	48
49	97.085	82.389	0.012795	0.56532	78.157	1.7689	23.988	84.886	108.874	0.05134	0.21821	49
50	98.727	84.031	0.012815	0.55606	78.033	1.7984	24.275	84.678	108.953	0.05190	0.21803	50
51	100.39	85.69	0.012836	0.54698	77.909	1.8282	24.563	84.468	109.031	0.05245	0.21785	51
52	102.07	87.38	0.012856	0.53808	77.784	1.8585	24.851	84.258	109.109	0.05301	0.21768	52
53	103.78	89.08	0.012877	0.52934	77.659	1.8891	25.139	84.047	109.186	0.05357	0.21750	53
54	105.50	90.81	0.012898	0.52078	77.534	1.9202	25.429	83.834	109.263	0.05412	0.21732	54
55	107.25	92.56	0.012919	0.51238	77.408	1.9517	25.718	83.621	109.339	0.05468	0.21714	55
56	109.02	94.32	0.012940	0.50414	77.282	1.9836	26.008	83.407	109.415	0.05523	0.21697	56
57	110.81	96.11	0.012961	0.49606	77.155	2.0159	26.298	83.191	109.490	0.05579	0.21679	57
58	112.62	97.93	0.012982	0.48813	77.028	2.0486	26.589	82.975	109.564	0.05634	0.21662	58
59	114.46	99.76	0.013004	0.48035	76.900	2.0818	26.880	82.758	109.638	0.05689	0.21644	59
60	116.31	101.62	0.013025	0.46523	76.773	2.1154	27.172	82.540	109.712	0.05745	0.21627	60
61	118.19	103.49	0.013047	0.46523	76.644	2.1495	27.464	82.320	109.785	0.05800	0.21610	61
62	120.09	105.39	0.013069	0.45788	76.515	2.1840	27.757	82.100	109.857	0.05855	0.21592	62
63	122.01	107.32	0.013091	0.45066	76.386	2.2190	28.050	81.878	109.929	0.05910	0.21575	63
64	123.96	109.26	0.013114	0.44358	76.257	2.2544	28.344	81.656	110.000	0.05966	0.21558	64

FIGURE 3–62   A portion of properties of the R-22 Table. (*Courtesy of E.I. DuPont*)

refrigeration system is allowed to operate. At some point, so much heat is going to be used from the air in evaporating the liquid refrigerant in the evaporator that a control device will be used to shut the mechanical refrigeration system off once the space air temperature reaches the control set-point. The control device senses the sensible heat of the air and determines at which temperature the mechanical refrigeration system should be shut down.

Since the conditioned space air is giving up heat to help the refrigerant to evaporate in the evaporator, where does the heat from the air go? Most of the heat was used in the latent heat process (*boiling the liquid to vapor at 40 degrees*). This takes place to about **point 5** of **Figure 3–63**. Notice that between **point 5** and **point 7**, the low-temperature and low-pressure refrigerant vapor is warmed from 40 degrees (F) to 70 degrees (F). This is the area of the evaporator coil where the refrigerant vapor is being heated above the saturation temperature of 40 degrees (F), which is referred to as being **superheated**. The outlet tube of the evaporator coil is sealed to the suction inlet port of the compressor at **point 7**. The low-temperature and low-pressure superheated vapor is sucked out of the evaporator and into the suction side of the compressor, referred to as the low side of the sealed system. This low-temperature and low-pressure superheated vapor contains the heat absorbed in the evaporating process and superheating process in the evaporator. The removal of this heat by the evaporator causes the space temperature air to drop.

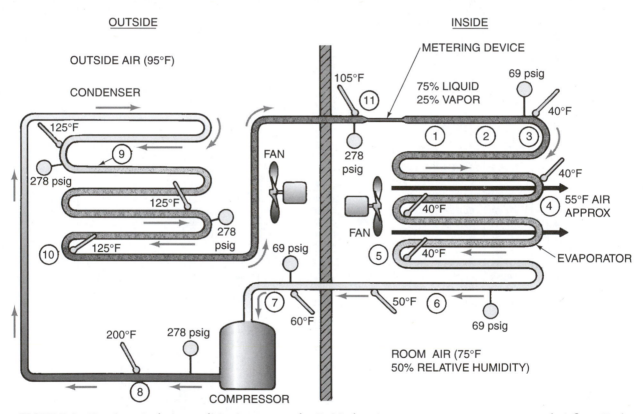

FIGURE 3–63 A typical air-conditioning system for R-22 showing temperatures, pressures, and airflow. Red indicates warm to hot; blue indicates cool to cold.

Now that the compressor has removed this low-temperature, low-pressure heat latent vapor from the evaporator, it has to move it somewhere else where it can be released. To be able to get the low-temperature, low-pressure superheated vapor to give up the heat it contains, the compressor must set up a temperature difference where this low-temperature, low-pressure superheated vapor will be higher in temperature than the area where it is to be released.

To accomplish this, the compressor compresses the low-temperature, low-pressure superheated vapor into a small space. This is called **heat of compression**. To understand this principle, think of heat of compression like the following example. Let's say that you were in a small room that had no way for heat to escape and a furnace was pumping 60-degree air into this room without having any control to shut the furnace OFF. Even though the furnace is **pumping only 60-degree** air into the room, at some point the room temperature will start to rise above 60 degrees. This is because the air and everything in the room cannot hold any more heat without the temperature rising.

This principle is basically what takes place in the compressor during the heat of compression. From **point 7 to point 8**, the compressor takes the heat that is already in the vapor from the evaporator at **60 degrees** and pushes it into a small space. By doing this, the temperature and pressure exerted on the vapor rise to a **high temperature of 200 degrees (F) superheated vapor**, with a **high pressure of 278 psig**. This makes the vapor hotter than the surrounding air so that heat can be rejected.

Once the compressor completes the heat of compression, it forces this **high-temperature, high-pressure superheated vapor** into the **condenser coil** at **point 8**. Remember the law of the thermodynamics of heat which states: *"Heat will always travel from a higher temperature to a lower temperature."* Since the **vapor being pumped into the condenser is at 200 degrees (F)** and the **outside air temperature is 95 degrees**, heat from the high-temperature, high-pressure superheated refrigerant vapor in the will naturally flow to the surrounding area temperature.

This is how the heat that was absorbed from the evaporator is moved and rejected to an area where it is not objectionable. The transfer of the heat from the high-temperature, high-pressure superheated vapor in the condenser is assisted with the use of a condenser fan. From **point 8 to point 9, sensible heat** from the high-temperature, high-pressure superheated refrigerant vapor is being rejected to the surrounding air. At **point 9**, notice that the refrigerant vapor has given up enough sensible heat to reach the **saturation temperature of 125 degrees (F)**. This saturation temperature is determined by the 278 psig being exerted on the vapor. At **point 9** to **point 10**, the 125-degree high-temperature, high-pressure superheated vapor continues to reject heat to the outside air temperature of 95 degrees in the form of latent heat. Rejecting latent heat from the vapor causes the vapor to condense back to a high-temperature, high-pressure liquid–vapor mixture. Eventually enough latent heat will be rejected from the vapor to cause the high-temperature, high-pressure liquid–vapor to be condensed back to 100% high-temperature high-pressure liquid at **point 10.**

Between point 10 and **point 11**, the high-temperature, high-pressure liquid continues to give up heat to the outside air temperature of 95 degrees. This is called **sub-cooling the liquid** below its saturation temperature. **At point 11**, the high-temperature, high-pressure liquid has been cooled to **105 degree (F)**, indicating that the refrigerant liquid has been cooled 20 degrees below the saturation temperature of 125 degrees (F).

Remember the three jobs of the condenser in any mechanical refrigeration system are to **reject heat, change the high-temperature, high-pressure vapor back to a high-temperature, high- pressure liquid**, and then **sub-cool the liquid** before it is fed by the metering device into the evaporator again.

High pressure exerted on the condenser forces the high-temperature, high-pressure sub-cooled liquid through the metering device and into the evaporator. Here, the metering device will create a pressure drop and feed low-temperature, low-pressure sub-cooled liquid refrigerant into the evaporator so that it can be evaporated to a vapor once again, by using heat from the air of the conditioned space. This process of evaporating, compressing heat, and rejecting heat will continue until some electrical control is used to turn the system OFF when a desired temperature has been reached.

Practical Competency 46

Learning the Vapor Compression Refrigeration Cycle

SUGGESTED MATERIALS

Textbook
Refrigeration & Air Conditioning Technology, 5th Edition, Thomson Delmar Learning
Unit 3—Refrigeration and Refrigerants

Review Topics
Refrigeration and Refrigerants; The Refrigeration Process; Refrigeration System and Components

COMPETENCY OBJECTIVE

The student will be able to explain the refrigeration cycle of a vapor compression sealed system.

OVERVIEW

The mechanical refrigeration system consists of components that are used with a refrigerant to set up the temperature difference needed to absorb the heat from the area where it is not wanted and to reject this heat into another area where it is not objectionable. The four main components of any refrigeration system are the evaporator, condenser, compressor, and the metering device. The evaporator of any refrigeration system is used to absorb the heat from the conditioned space or product. The evaporator of the sealed system is always located in the area where the heat is to be removed.

The condenser of the sealed system is so named because it is designed to condense a refrigerant vapor back to a liquid refrigerant. Understanding these principles can also help you to understand the state of the refrigerant in the evaporator and condenser of the sealed system.

The temperature at which evaporating and condensing takes place is determined by the pressure exerted on the refrigerant in the evaporator and condenser coils of the sealed system. The amount of pressure can be controlled in a few different ways. The amount of heat, amount of refrigerant, type of metering device used, and the compressor's compression ratio will affect the pressure exerted on the sealed system. All of these factors will affect the evaporating condensing temperature of the refrigerant in the sealed system.

EQUIPMENT REQUIRED

1 Light blue colored pencil
1 Dark blue pencil
1 Light red or pink colored pencil
1 Dark red pencil
1 No. 2 pencil

SAFETY PRACTICES

Technicians should follow all safety procedures and U.S. Environmental Protection Agency (EPA) regulations when working with refrigerants and refrigeration equipment.

COMPETENCY PROCEDURES

> NOTE: *Competency procedure will be used to complete the refrigeration cycle of the refrigerant in the sealed system components below (Figure 3–64).*

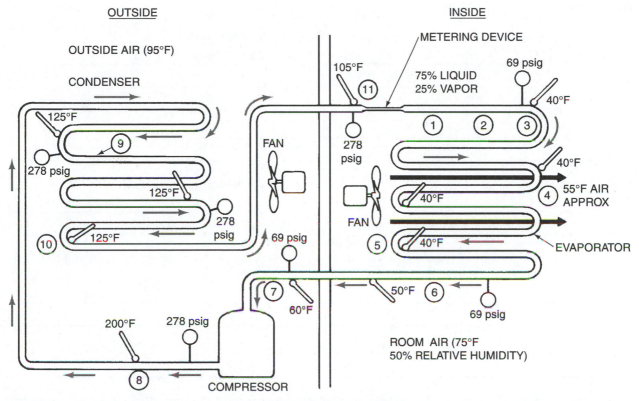

FIGURE 3–64   Refrigerant sealed system and components.

> NOTE: *This competency should be completed only after the student has a good understanding of the refrigeration cycle lesson. (Refer to "Theory Lesson: The Vapor Compression Refrigeration Cycle" and Figure 3–64.)*

1. Use the light red or pink pencil and color the correct area of the sealed system where a high-temperature, high-pressure superheated vapor refrigerant would exist.   ❑
2. Use the dark red pencil and color in the correct area of the sealed system where a high-temperature, high-pressure sub-cooled liquid refrigerant would exist.   ❑
3. Use the dark blue pencil and color in the correct area of the sealed system where a low-temperature, low-pressure liquid–vapor mixture would exist.   ❑
4. Use the light blue pencil and color in the correct area of the sealed system where a low-temperature, low-pressure superheated vapor would exist.   ❑
5. Using the No. 2 pencil, circle and label the areas of the sealed system where a low-temperature, low-pressure superheated vapor would exist.   ❑
6. Using the No. 2 pencil, circle and label the areas of the sealed system where there is a high-temperature, high-pressure sub-cooled liquid.   ❑
7. Using the No. 2 pencil, circle and label the area where flash gas exists.   ❑

8. Using the No. 2 pencil, circle and label the area where a low-temperature, low-pressure liquid–vapor mixture exists. ❏

9. Using the No. 2 pencil, circle and label the area where a high–temperature, high-pressure vapor–liquid mixture exists. ❏

10. Have your instructor check your work. ❏

RESEARCH QUESTIONS

1. List the three components that make up the low side of the sealed system.

_____

_____

2. List the three components that make up the high side of the sealed system.

_____

_____

3. List the component of the sealed system that moves heat up the temperature scale.

_____

_____

4. List the components of the sealed system that are the dividing points between the high and low sides of the sealed systems.

_____

_____

5. List the component that is used to absorb heat into the refrigerant.

_____

_____

6. List the component that is used to reject the heat from the refrigerant.

_____

_____

7. List the component of the sealed system where there is a change of state from a liquid to a vapor.

_____

_____

8. List the component of the sealed system where there is a change of state from a vapor to a liquid.

_____

_____

9. In which change of state of the refrigerant is heat absorbed?

_____

_____

10. In which change of state of the refrigerant is heat rejected?

_____

_____

11. List the component where latent heat of evaporation takes place.

_____

_____

12. List the component where latent heat of condensation takes place.

_____

_____

13. What component is used to control the flow of liquid refrigerant?

_____

_____

14. What type of heat is required to change the state of a substance?

_____

_____

15. What type of heat is required to raise the temperature of a substance?

_____

_____

Practical Competency 47

Interpreting a Pressure–Temperature Chart

SUGGESTED MATERIALS

Textbook
Refrigeration & Air Conditioning Technology, 5th Edition, Thomson Delmar Learning
Unit 3—Refrigeration and Refrigerants

Review Topics
Refrigeration and Refrigerants; Pressure and Temperature Relationship

COMPETENCY OBJECTIVE

The student will be able to interpret a pressure–temperature chart.

OVERVIEW

Water will **boil (evaporate)** at a temperature of 212 degrees (F) as long as it is under atmospheric pressure of 14.696 psi.

When heating a pot of water on the stove, the water temperature will rise until it reaches what is called its saturation temperature of 212 degrees (F). **Saturation temperature** is the point at which the substance, liquid, vapor, or solid, cannot **take on or lose any additional heat** and stay in the state that it is in.

Like any liquid, water can be made to boil (evaporate) at any temperature desired by changing the amount of pressure exerted on it. This means that there is a direct correlation between pressure and temperature.

If water is to be boiled at a temperature lower than 212 degrees (F), the pressure exerted on it must be reduced by some means.

If water is to boil at a temperature higher than 212 degree (F), the pressure exerted on it must be raised by some means.

Another point to understand is that the amount of heat added or rejected from a substance does not change the temperature at which the change of state will take place as long as the pressure exerted on the substance stays the same.

Having a pressure–temperature chart for a particular substance such as water or different refrigerants gives the individual the ability to determine the saturation temperature of a substance if the amount of pressure being exerted on it is known; or if the saturation temperature is known, the amount of pressure being exerted on it can be found.

The temperature at which evaporating and condensing of the refrigerant takes place in the vapor compression refrigeration system is determined by the amount of pressure exerted on the refrigerant in the evaporator or condenser coils of the sealed system. Pressure in a sealed system can be controlled in a few different ways. The amount of heat, amount of refrigerant, type of metering device being used, and the compressor's compression ratio will all have an effect on the amount of pressure exerted on both the low and high sides of a sealed system. In refrigeration, there is a direct relationship between temperature and pressure. If one factor is known (pressure or temperature), the other factor can be found by using a **pressure–temperature chart (P/T chart)**.

EQUIPMENT REQUIRED

Writing utensil
Pressure–temperature chart

SAFETY PRACTICES

None

COMPETENCY PROCEDURES

Checklist

> NOTE: *Using the pressure–temperature chart will complete all competency procedures.* **(Figure 3–65).**

1. *What do the columns of numbers across the top of the chart represent?* _____ ❑
2. *What six types of refrigerants are represented on this pressure–temperature chart?*
   _____ _____ _____ _____ _____ _____ ❑ ❑ ❑ ❑ ❑ ❑ ❑
3. *What do the numbers in the far left column represent?* _____
4. *What is the temperature range of the chart?* _____ to _____
5. *What do the numbers listed under the refrigerants represent?* _____
6. *What is the pressure range for R-12 on the chart?* _____ to _____
7. *What refrigerant has a pressure range of 0.3 to 616.2 psig?* _____
8. *What refrigerant has a saturation temperature of 40 degrees (F) at 35.1 psig?* _____ ❑
9. *What would be the pressure exerted on R-22 at a saturation temperature of 120 degrees?* _____ ❑
10. *What is the saturation temperature of R-134a at 11.9 psig?* _____ ❑
11. *What pressure is exerted on 404A at 60 degrees below zero?* _____ ❑
12. *What is the saturation temperature of R-502 with 432.9 psig exerted on it?* _____ ❑
13. *What refrigerant has the highest psig exerted on when it is at a temperature of 155 degrees (F)?* _____ ❑

You should have learned at this point that the P/T chart has a temperature range on the left-hand side of each of the columns of refrigerant scales. There are six different types of refrigerants listed with the pressure scales for each that go from inches of vacuum to higher-pressure scales.

In this example manifold gages are attached to both the low- and high-side service valves of an R-22 air conditioning system. The high-side pressure is 210 psig, and the low-side pressure is 68 psig. Using the P/T chart, the saturation temperature of the liquid refrigerant in the evaporator and the condensing temperature of the refrigerant vapor in the condenser can be determined. Finding 210 psig down the R-22 column, and reading the temperature to the left we can see that the vapor in condenser is condensing to a liquid when the vapor is cooled to 105 degrees (F). Find 68 psig down the R-22 column and look to the left at the temperature column. We can see that the liquid is boiling in the evaporator at 40 degrees (F).

The opposite is also true. If the evaporator coil surface temperature was determined to be 40 degrees (F) and the middle surface area of the condenser coil was determined to be 105 degrees (F), the pressure for both the low and high sides of the system could be determined. Find 40 degrees down the temperature scale and read to the right until the R-22 column is found. At this point you can see that a pressure of 68.5 psig is represented. Likewise, find 105 degrees (F) down the temperature column. Again read to the right until the R-22 column is located. At this point you can see that by knowing the temperature near the middle of the condenser coil surface, the high-side pressure would be approximately 210.8 psig.

TEMPERATURE

	REFRIGERANT					
°F	12	22	134a	502	404A	410A
−60	19.0	12.0		7.2	6.6	0.3
−55	17.3	9.2		3.8	3.1	2.6
−50	15.4	6.2		0.2	0.8	5.0
−45	13.3	2.7	14.7	1.9	2.5	7.8
−40	11.0	0.5	12.4	4.1	4.8	9.8
−35	8.4	2.6	9.7	6.5	7.4	14.2
−30	5.5	4.9	6.8	9.2	10.2	17.9
−25	2.3	7.4		12.1	13.3	21.9
−20	0.6	10.1	3.6	15.3	16.7	26.4
−18	1.3	11.3	2.2	16.7	18.2	28.2
−16	2.0	12.5	0.7	18.1	19.6	30.2
−14	2.8	13.8	0.3	19.5	21.1	32.2
−12	3.6	15.1	1.2	21.0	22.7	34.3
−10	4.5	16.5	2.0	22.6	24.3	36.4
−8	5.4	17.9	2.8	24.2	26.0	38.7
−6	6.3	19.3	3.7	25.8	27.8	40.9
−4	7.2	20.8	4.6	27.5	30.0	42.3
−2	8.2	22.4	5.5	29.3	31.4	45.8
0	9.2	24.0	6.5	31.1	33.3	48.3
1	9.7	24.8	7.0	32.0	34.3	49.6
2	10.2	25.6	7.5	32.9	35.3	50.9
3	10.7	26.4	8.0	33.9	36.4	52.3
4	11.2	27.3	8.6	34.9	37.4	53.6
5	11.8	28.2	9.1	35.8	38.4	55.0
6	12.3	29.1	9.7	36.8	39.5	56.4
7	12.9	30.0	10.2	37.9	40.6	57.8
8	13.5	30.9	10.8	38.9	41.7	59.3
9	14.0	31.8	11.4	39.9	42.8	60.7
10	14.6	32.8	11.9	41.0	43.9	62.2
11	15.2	33.7	12.5	42.1	45.0	63.7

TEMPERATURE

	REFRIGERANT					
°F	12	22	134a	502	404A	410A
12	15.8	34.7	13.2	43.2	46.2	65.3
13	16.4	35.7	13.8	44.3	47.4	66.8
14	17.1	36.7	14.4	45.4	48.6	68.4
15	17.7	37.7	15.1	46.5	49.8	70.0
16	18.4	38.7	15.7	47.7	51.0	71.6
17	19.0	39.8	16.4	48.8	52.3	73.2
18	19.7	40.8	17.1	50.0	53.5	75.0
19	20.4	41.9	17.7	51.2	54.8	76.7
20	21.0	43.0	18.4	52.4	56.1	78.4
21	21.7	44.1	19.2	53.7	57.4	80.1
22	22.4	45.3	19.9	54.9	58.8	81.9
23	23.2	46.4	20.6	56.2	60.1	83.7
24	23.9	47.6	21.4	57.5	61.5	85.5
25	24.6	48.8	22.0	58.8	62.9	87.3
26	25.4	49.9	22.9	60.1	64.3	90.2
27	26.1	51.2	23.7	61.5	65.8	91.1
28	26.9	52.4	24.5	62.8	67.2	93.0
29	27.7	53.6	25.3	64.2	68.7	95.0
30	28.4	54.9	26.1	65.6	70.2	97.0
31	29.2	56.2	26.9	67.0	71.7	99.0
32	30.1	57.5	27.8	68.4	73.2	101.0
33	30.9	58.8	28.7	69.9	74.8	103.1
34	31.7	60.1	29.5	71.3	76.4	105.1
35	32.6	61.5	30.4	72.8	78.0	107.3
36	33.4	62.8	31.3	74.3	79.6	108.4
37	34.3	64.2	32.2	75.8	81.2	111.6
38	35.2	65.6	33.2	77.4	82.9	113.8
39	36.1	67.1	34.1	79.0	84.6	116.0
40	37.0	68.5	35.1	80.5	86.3	118.3
41	37.9	70.0	36.0	82.1	88.0	120.5

TEMPERATURE

	REFRIGERANT					
°F	12	22	134a	502	404A	410A
42	38.8	71.4	37.0	83.8	89.7	122.9
43	39.8	73.0	38.0	85.4	91.5	125.2
44	40.7	74.5	39.0	87.0	93.3	127.6
45	41.7	76.0	40.1	88.7	95.1	130.0
46	42.6	77.6	41.1	90.4	97.0	132.4
47	43.6	79.2	42.2	92.1	98.8	134.9
48	44.6	80.8	43.3	93.9	100.7	136.4
49	45.7	82.4	44.4	95.6	102.6	139.9
50	46.7	84.0	45.5	97.4	104.5	142.5
55	52.0	92.6	51.3	106.6	114.6	156.0
60	57.7	101.6	57.3	116.4	125.2	170.0
65	63.8	111.2	64.1	126.7	136.5	185.0
70	70.2	121.4	71.2	137.6	148.5	200.8
75	77.0	132.2	78.7	149.1	161.1	217.6
80	84.2	143.6	86.8	161.2	174.5	235.4
85	91.8	155.7	95.3	174.0	188.6	254.2
90	99.8	168.4	104.4	187.4	203.5	274.1
95	108.2	181.8	114.0	201.4	219.2	295.0
100	117.2	195.9	124.2	216.2	235.7	317.1
105	126.6	210.8	135.0	231.7	253.1	340.3
110	136.4	226.4	146.4	247.9	271.4	364.8
115	146.8	242.7	158.5	264.9	290.6	390.5
120	157.6	259.9	171.2	282.7	310.7	417.4
125	169.1	277.9	184.6	301.4	331.8	445.8
130	181.0	296.8	198.7	320.8	354.0	475.4
135	193.5	316.6	213.5	341.2	377.1	506.5
140	206.6	337.2	229.1	362.6	401.4	539.1
145	220.3	358.9	245.5	385.9	426.8	573.2
150	234.6	381.5	262.7	408.4	453.3	608.9
155	249.5	405.1	280.7	432.9	479.8	616.2

VACUUM (in. Hg) – LIGHT FIGURES
GAGE PRESSURE (psig) – BOLD FIGURES

FIGURE 3–65   Pressure–temperature chart.

Conclusion

Knowing the type of refrigerant in a sealed system and the evaporating or condensing temperature of the refrigerant can be used to determine the pressure of either the low or high side of a sealed system. Likewise, knowing the type of refrigerant in the sealed system along with the low- or high-side operating pressure of the system can be used to determine the evaporation and condensing temperature of the refrigerant in a sealed system.

Use the pressure–temperature chart to complete the following procedures:

14. *At 155 degrees, which refrigerant on the P/T chart has the highest pressure exerted on it?* _____ ❑
15. *What is the temperature of R-12 with a pressure of 84.2 psig exerted on it?* _____ ❑
16. *What is the pressure exerted on R-502 at a temperature of 11 degrees?* _____ ❑
17. *What is the pressure of HFC-134 A at a temperature of 40 degrees?* _____ ❑
18. *What refrigerant is 130 degrees with a pressure of 296.8 psig exerted on it?* ❑
19. *What two refrigerants would have pressure below atmospheric pressure at a temperature of −25 degrees (F)* _____ _____ ❑
20. Have your instructor check your work. ❑

RESEARCH QUESTIONS

1. What is the boiling temperature of water at atmospheric pressure?

_____

_____

2. What is the freezing temperature of water?

_____

_____

3. What type of heat added or removed from a substance causes a change of state?

_____

_____

4. What is latent heat of evaporation?

_____

_____

5. How much latent heat is required to boil (evaporate) 1 pound of water to 100% vapor?

_____

_____

Passed Competency _____   Failed Competency _____

Instructor Signature _____   Grade _____

Theory Lesson: Refrigeration Manifold Gages (Figure 3–66)

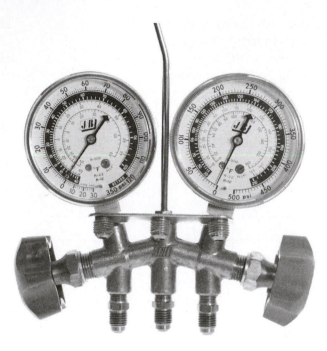

FIGURE 3–66 A gage manifold. (*Photo by Bill Johnson*)

SUGGESTED MATERIALS

Textbook

Refrigeration & Air Conditioning Technology, 5th Edition, Thomson Delmar Learning
Unit 3—Refrigeration and Refrigerants
Unit 8—System Evacuation
Unit 11—Calibrating Instruments
Unit 41—Troubleshooting

Review Topics

Pressure and Temperature Relationship; Gage Manifold Hoses; Pressure Test Instruments; Low-Side Gage Readings; High-Side Gage Readings

Key Terms

atmospheric pressure • compound gage • high-side gage • low-side gage • pressure–temperature • psig

OVERVIEW

One of the most important tools used by air conditioning and refrigeration technicians are refrigeration manifold gages. The gage manifold is used to measure actual pressures inside a vapor compression sealed system. These pressures can then be used to evaluate the operational performance of a sealed system, or used with a pressure–temperature chart to determine refrigerant saturation temperatures within the sealed system. There are two gages on the refrigeration manifold set. One gage is referred to as a compound gage. This is normally a **Blue gage** and is used to evaluate a system's **low-side operating and pressure (Figure 3–67).**

A **compound gage** has the ability to measure above and below atmospheric pressure. The outer large **Black numbers** on the gage face represent pressure in **psi** (*pounds per square inch*), and the outer scale **Red numbers** are used to measure the vacuum level during the evacuation of a sealed system.

The black number **zero (0)** on the gage represents atmospheric pressure of 14.696 psi. When the needle moves **above the (0)** gage pressure, it is measuring pressure above atmospheric pressure and is stated as **psig** (*pounds per square inch gage*).

When the needle of the gage is **below the (0)**, this is indicating pressure in **inches of mercury (Hg)** below atmospheric pressure. The low-side gage pressure in **Figure 3–94** ranges from **0 to 350 psig**. Some low-side gages will have a different range scale. Notice the inches of vacuum scale for the gage is from **0 to 30"** (inches) of Hg (mercury) vacuum.

The high-side gage of the manifold set is the **Red gage** and is used to measure the pressure of the high side of the sealed system (**Figure 3–68**).

FIGURE 3–67   Low-side gage reading 20 in. Hg vacuum. (*Photo by Bill Johnson*)

FIGURE 3–68   High-side gage indicating 226 psig. (*Photo by Eugene Silberstein*)

The pressure range of the high side gage is represented by the large black numbers just like on the low-side pressure gage. The high-side gage pressure scale ranges from **0 to 500 psig**. Notice that there is no pressure scale below the zero (0) on the gage. This gage does not have the ability to measure pressure below atmospheric pressure, so it is not referred to as a compound gage.

Look at the center number scales on each of the gages. You will also notice at the end of each of the center scales is a **refrigerant identification number**. In the example of the high-side gage there are three refrigerants listed. The **purple inner numbers** on the high-side gage represent temperatures of **R-502**. The **inner green numbers** represent temperatures for **R-22** and the **white numbers** represent temperatures for **R-12**. At one time these three refrigerants were the most popular refrigerants in use. With the number of new refrigerants in use today, there may be temperature scales for different types of refrigerants on the gages you may be using. Some gages may not have temperature scales listed for any refrigerant. The refrigerant temperature scales on the gage allow the pressure–temperature relationship of these particular refrigerants to be determined without the use of a P/T chart (for an **example**, refer to **Figure 3–69**).

Notice that the low-side gage reads 68.5 psig. If this were an **R-22 refrigerant system**, looking down the needle from the 68.5 psig scale to where it interconnects with the R-22 temperature scale (**Figure 3–70**), a temperature of **40 degrees (F)** is indicated. This shows that the **R-22** is saturating at **40 degrees (F)** based on the gage pressure of **68.5 psig**. Likewise, if the refrigerant sealed system contained **R-502**, the saturation

FIGURE 3–69 This gage reads 68.5 psig. (*Photo by Eugene Silberstein*)

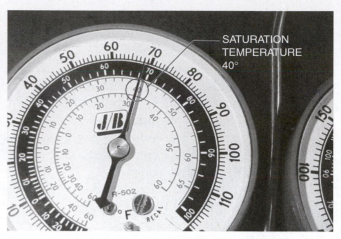

FIGURE 3–70 For R-22, a pressure of 68.5 psig corresponds to a saturation temperature of 40 degrees (F). (*Photo by Eugene Silberstein*)

temperature would be approximately **32 degrees**, or if **R-12**, the saturation temperature of the refrigerant would be approximately **68 degrees**.

If the sealed system contains a refrigerant other than that which is shown on a particular gage manifold set, the gages would be used to determine the **system's operating pressures**. A **pressure–temperature chart** for the particular refrigerant would have to be used to convert these gage pressures to saturation temperatures. For example, let's say that the system you are working on uses refrigerant **404A** and has a high-side pressure of **219 psig**. The manifold gage in use does not have a temperature scale for 404A, so a P/T chart would have to be used to determine the system's saturation temperature. Using the Lab Manual P/T chart, or P/T chart used in *Practical Competency 47*, find the pressure of 219 psig in the 404A refrigerant column, and read out to the temperature column; a saturation temperature of **95 degrees** is indicated. This procedure would have to be followed anytime a refrigerant other than that which is provided on the manifold gages is being used to check system pressures.

The hoses of a manifold gage set are normally colored to match the gages on the manifold. With a standard manifold gage set there will be three hoses about 36" long. Manifold gage hoses come as three different colors: red, blue, and yellow.

> **NOTE: Refrigerant hoses.** *Notice the ends of a refrigerant hose. One end has a shredder port stem and the other end does not. The end without the shredder port stem gets attached to the gage manifold. The end with the shredder stem is used to attach to a sealed system service valve. If the hoses are connected improperly to the manifold gage, pressures from a sealed system will not be indicated on the manifold gages.*

The **Red and Blue hoses** get attached to the corresponding gage color: **Blue hose** to the **Blue** low-side gage, and the **Red** hose to the **Red** high-side gage. The **Yellow hose** gets attached to the **center port** of the gage manifold.

Manifold gage hand valves are shown in **Figure 3–71**. *The hand valves of the manifold gage DO NOT have to be opened to measure the pressure of a sealed system.* The only time the manifold valves should be opened is during the following sealed system procedures:

Adding refrigerant
Recovering refrigerant
Leak testing a system
Evacuating a system

FIGURE 3–71   Manifold gage hand valves.

The *low-side* hand valve **opens clockwise** and closes **counterclockwise.** The *high-side* valve is **opened counterclockwise** and **closed counterclockwise.**

When closing the valves *DO NOT overtighten* the valves. The valves should be closed finger tight.

Most refrigerant manifold gages sets can be rebuilt. Seal kits for refrigerant hoses and hand valves can be purchased separately along with individual low- and high-side gages.

> **NOTE:** *Standard refrigerant manifold gages should not be used on heat pumps when checking the system's pressures during the heating cycle or used on a R-410A air conditioning system. Pressures on these systems can be higher than the gage pressure scale and cause damage to the pressure gage. Heat pump manifold gages can be purchased, as well as special gages for R-410A system.*

Practical Competency 48

Calibrating the Gage Manifold

SUGGESTED MATERIALS

Textbook

Refrigeration & Air Conditioning Technology, 5th Edition, Thomson Delmar Learning
Unit 3—Refrigeration and Refrigerants
Unit 8—System Evacuation
Unit 11—Calibrating Instruments
Unit 41—Troubleshooting

Review Topics

Pressure and Temperature Relationship; Gage Manifold Hoses; Pressure Test Instruments; Low-Side Gage Readings; High-Side Gage Readings

COMPETENCY OBJECTIVE

The student will be able to calibrate gage manifolds.

OVERVIEW

One of the most important tools used by air conditioning and refrigeration technicians are refrigeration manifold gages. The gage manifold is used to measure actual pressures inside a vapor-compression sealed system. These pressures can then be used to evaluate the operational performance of a sealed system, or used with a pressure–temperature chart to determine refrigerant saturation temperatures within the sealed system. There are two gages on the refrigeration manifold set. One gage is referred to as a compound gage. This is normally a **Blue gage** and is used to evaluate a system's **low-side operating and pressure.** (*Refer to Theory Lesson: Refrigeration Manifold Gage, Figure 3–67.*)

A **compound gage** has the ability to measure above and below atmospheric pressure. The outer large **Black numbers** on the gage face represent pressure in **psi** (*pounds per square inch*), and the outer scale **Red numbers** are used to measure the vacuum level during the evacuation of a sealed system.

The black number **zero (0)** on the gage represents atmospheric pressure of 14.696 psi. When the needle moves **above the (0)** gage pressure, it is measuring pressure above atmospheric pressure and is stated as **psig** (*pounds per square inch gage*).

When the needle of the gage is **below the (0),** this is indicating pressure in **inches of mercury (Hg)** below atmospheric pressure. The low-side gage pressure in (*Figure 3–68, Theory Lesson: Refrigeration Manifold Gage*) ranges from **0 to 350** psig. Some low-side gages will have a different range scale. Notice the inches of vacuum scale for the gage is from **0 to 30"** (inches) of Hg (mercury) vacuum.

The high-side gage of the manifold set is the **Red gage** and is used to measure the pressure of the high side of the sealed system. (*Refer to Figure 3–68, Theory Lesson: Refrigeration Manifold Gage.*)

The pressure range of the high-side gage is represented by the large black numbers just like on the low-side pressure gage. The high-side gage pressure scale ranges from **0 to 500 psig**. Notice that there is no pressure scale below the zero (0) on the gage. This gage does not have the ability to measure pressure below atmospheric pressure, so it is not referred to as a compound gage.

With a standard manifold gage set there will be three hoses about 36" long. Manifold gage hoses come as three different colors: red, blue, and yellow.

The **Red and Blue hoses** get attached to the corresponding gage color: **Blue hose** to the **Blue** low-side gage, and the **Red** hose to the **Red** high-side gage. The **Yellow hose** gets attached to the **center port** of the gage manifold.

The hand valves of the manifold gage **DO NOT** *have to be opened to measure the pressure of a sealed system.* The only time the manifold valves should be opened is during the following sealed system procedures:

Adding refrigerant
Recovering refrigerant
Leak testing a system
Evacuating a system

The *low-side* hand valve **opens clockwise** and closes **counterclockwise.** The *high-side* valve is **opened counterclockwise** and **closed counterclockwise.**

When closing the valves **DO NOT** *overtighten* the valves. The valves should be closed finger tight.

Most refrigerant manifold gages sets can be rebuilt. Seal kits for refrigerant hoses and hand valves can be purchased separately along with individual low- and high-side gages.

NOTE: *Standard refrigerant manifold gages should not be used on heat pumps when checking the system's pressures during the heating cycle or used on a R-410A air conditioning system. Pressures on these systems can be higher than the gage pressure scale and cause damage to the pressure gage. Heat pump manifold gages can be purchased, as well as special gages for R-410A systems.*

EQUIPMENT REQUIRED

Manifold gage set
Small standard screw driver

COMPETENCY PROCEDURES

Checklist

1. Remove the low-side gage hose from the blank port of the gage manifold. ❑
2. Remove the high-side gage hose from the blank port of the gage manifold. ❑
3. Look at the position of the gage manifold needle on both the high-side and low-side gage. ❑

NOTE: *If both gage needles point to 0 psig, the gages are properly calibrated and no calibration is required.*

If one or both of the gages are not aligned at 0 psig, remove the clear plastic cover from the gage that needs calibration.

4. Look at the bottom of the gage and notice a small adjustment screw. ❑
5. Using the small standard screwdriver, slowly turn the adjustment screw until the needle aligns with 0 psig on the gage (**Figure 3–72**). ❑

NOTE: *Tuning the screw clockwise will cause the needle to move counterclockwise. Turning the screw counterclockwise will cause the needle to move clockwise.*

FIGURE 3–72 Calibrating gage.

6. Once gage(s) have been calibrated, lightly tap the side of the gage and observe the gage needle. Make sure the needle stays in calibration. ❑
7. Once calibration is complete, replace plastic gage cover(s). ❑
8. Replace low-side and high-side hoses to the blank ports of the manifold gage set. ❑
9. Have your instructor check your work ❑
10. Replace tools and supplies in their proper location. ❑

RESEARCH QUESTIONS

1. 0 psig represents what pressure on the manifold gages?

_____

_____

2. What is a compound gage?

_____

_____

3. On the some high-side gage pressure scales, there is an area at the end of the pressure scale called "RETARD." What does this area represent?

_____

_____

4. What can happen to a gage if more pressure is applied than the gage pressure scale is designed to measure?

_____

_____

5. How is psia determined from gage pressure readings?

_____

_____

| Passed Competency _____ | Failed Competency _____ |
| Instructor Signature _____ | Grade _____ |

Practical Competency 49

Interrupting Refrigerant Gage Manifold Readings

SUGGESTED MATERIALS

Textbook

Refrigeration & Air Conditioning Technology, 5th Edition, Thomson Delmar Learning
Unit 3—Refrigeration and Refrigerates
Unit 8—System Evacuation
Unit 11—Calibrating Instruments
Unit 41—Troubleshooting

Review Topics

Pressure and Temperature Relationship; Gage Manifold Hoses; Pressure Test Instruments;
Low-Side Gage Readings; High-Side Gage Readings

COMPETENCY OBJECTIVE

The student will be able to interpret the pressure and temperature readings by using refrigerant manifold gages.

OVERVIEW

One of the most important tools used by air conditioning and refrigeration technicians are refrigeration manifold gages. The gage manifold is used to measure actual pressures inside a vapor compression sealed system. These pressures can then be used to evaluate the operational performance of a sealed system, or used with a pressure–temperature chart to determine refrigerant saturation temperatures within the sealed system. There are two gages on the refrigeration manifold set. One gage is referred to as a compound gage. This is normally a **Blue gage** and is used to evaluate a system's **low-side operating and pressure**. (*Refer to Figure 3–67, Theory Lesson: Refrigeration Manifold Gages.*)

A **compound gage** has the ability to measure above and below atmospheric pressure. The outer large **Black numbers** on the gage face represent pressure in **psi** (*pounds per square inch*), and the outer scale **Red numbers** are used to measure the vacuum level during the evacuation of a sealed system.

The black number **zero (0)** on the gage represents atmospheric pressure of 14.696 psi. When the needle moves **above the** (0) gage pressure, it is measuring pressure above atmospheric pressure and is stated as **psig** (*pounds per square inch gage*).

When the needle of the gage is **below the** (0), this is indicating pressure in **inches of mercury (Hg)** below atmospheric pressure.

The high-side gage of the manifold set is the **Red gage** and is used to measure the pressure of the high side of the sealed system. (*Refer to Figure 3–68, Theory Lesson: Refrigeration Manifold Gages.*)

The pressure range of the high-side gage is represented by the large black numbers just like on the low-side pressure gage. The high-side gage does not have the ability to measure pressure below atmospheric pressure, so it is not referred to as a compound gage.

The center-numbered scales on each of the gages represent temperature scales for the refrigerants listed at the end of each of the center temperature scales. Some gages may not have temperature scales listed for any refrigerant. The refrigerant temperature scales on the gage allow the pressure–temperature relationship of these particular refrigerants to be determined without the use of a P/T chart.

If the sealed system contains a refrigerant other than that which is shown on a particular gage manifold set, the gages would be used to determine the **system's operating pressures**. A **pressure–temperature chart** for the particular refrigerant would have to be used to convert these gage pressures to saturation temperatures. The hoses of a manifold gage set are normally colored to match the gages on the manifold. With a standard manifold gage set there will be three hoses about 36" long. Manifold gage hoses come as three different colors: red, blue, and yellow.

> *NOTE: Refrigerant hoses. Notice the ends of a refrigerant hose. One end has a shredder port stem and the other end does not. The end without the shredder port stem gets attached to the gage manifold. The end with the shredder stem is used to attach to a sealed system service valve. If the hoses are connected improperly to the manifold gage, pressures from a sealed system will not be indicated on the manifold gages.*

The **Red and Blue hoses** get attached to the corresponding gage color: **Blue hose** to the **Blue** low-side gage, and the **Red hose** to the **Red** high-side gage. The **Yellow hose** gets attached to the **center port** of the gage manifold.

The hand valves of the manifold gage DO NOT have to be opened to measure pressure of a sealed system. The only time the manifold valves should be opened is during the following sealed system procedures:

Adding refrigerant
Recovering refrigerant
Leak testing a system
Evacuating a system

The *low-side* hand valve **opens clockwise** and **closes counterclockwise**. The *high-side* valve is **opened counterclockwise** and **closed counterclockwise**.

When closing the valves *DO NOT* overtighten the valves. The valves should be closed finger tight.

> *NOTE: Standard refrigerant manifold gages should not be used on heat pumps when checking the system's pressures during the heating cycle or used on a R-410A air conditioning system. Pressures on these systems can be higher than the gage pressure scale and cause damage to the pressure gage. Heat pump manifold gages can be purchased, as well as special gages for R-410A systems.*

EQUIPMENT REQUIRED

3 Different refrigerant cylinders (*virgin or recovered cylinders*)
1 Manifold gage set
1 Temperature–pressure chart
1 Temperature recorder
Safety goggles
Safety gloves

SAFETY PRACTICES

The student should be competent in the use of refrigerant manifold gages. Follow all EPA rules and regulations when working with refrigerants.

COMPETENCY PROCEDURES

Checklist

> *NOTE: Using three different types of refrigerant cylinders will complete all competency procedures.*

1. Let the three refrigerant cylinders reach room temperature. ❏
2. *List the ambient temperature of the area where the cylinders are located.* _____ ❏

3. *Record the type of refrigerant of the first cylinder being tested.* _____ ❑
4. Use the pressure–temperature chart and locate the ambient temperature recorded in procedure 2. On the P/T chart find the column of the type of refrigerant cylinder being checked and correspond the temperature and pressure for this refrigerant. *Record what the cylinders pressure should be according to the P/T chart.* _____ ❑
5. Make sure the manifold gage valves are closed and the gage needles are zeroed in. ❑
6. Attach the low-side manifold gage to the refrigerant cylinder valve. ❑
7. Open the refrigerant cylinder valve. ❑
8. *Record the pressure of the cylinder.* _____ ❑
9. *Does this pressure correspond to your estimated pressure from procedure 4?* _____ ❑

NOTE: *This pressure should be within a couple of psig's of what you determined it should be according to the P/T chart. If the pressure is off by a great deal, either you did not use the proper temperature and pressure scales on the P/T chart, or the refrigerant may be contaminated.*

---

10. Close the refrigerant cylinder valve. ❑
11. Let the center hose of the manifold gage set hang down from the gage manifold. ❑
12. Slowly open the low-side manifold gage valve and purge refrigerant from the hose until the gage pressure reads zero. ❑
13. Close the low-side manifold gage valve. ❑
14. Remove the low-side hose from the refrigerant cylinder. ❑
15. *Record the second type of refrigerant cylinder being tested.* _____ ❑
16. *Follow the same procedures listed in step 4 and record what the cylinder's pressure should be according to the P/T chart.* _____ ❑
17. Make sure the manifold gage valves are closed and zeroed in. ❑
18. Attach the low-side manifold gage to the refrigerant cylinder valve. ❑
19. Open the refrigerant cylinder valve. ❑
20. *Record the pressure of the cylinder.* _____ ❑
21. *Does this pressure correspond to your estimated pressure from procedure 16?* _____ ❑

NOTE: *This pressure should be within a couple of psig's of what you determined it should be. If the pressure is off by a great deal, either you did not use the proper temperature and pressure scales on the P/T chart, or the refrigerant may be contaminated.*

---

22. Close the refrigerant cylinder valve. ❑
23. Let the center hose of the manifold gage set hang down from the gage manifold. ❑
24. Slowly open the low-side manifold gage valve and purge refrigerant from the hose until the gage indicates zero pressure. ❑
25. Close the low-side manifold gage valve. ❑
26. Remove the low-side hose from the refrigerant cylinder. ❑
27. *Record the third type of refrigerant cylinder being tested.* _____ ❑
28. *Follow the same procedures listed in step 4 and record what the cylinder's pressure should be according to the P/T chart.* _____ ❑
29. Make sure the manifold gage valves are closed and zeroed in. ❑
30. Attach the low-side manifold gage to the refrigerant cylinder valve. ❑
31. Open the refrigerant cylinder valve. ❑
32. *Record the pressure of the cylinder.* _____ ❑
33. *Does this pressure correspond to your estimated pressure from procedure 28?* _____ ❑

> **NOTE:** *This pressure should be within a couple of psig's of what you determined it should be. If the pressure is off by a great deal, either you did not use the proper temperature and pressure scales on the P/T chart, or the refrigerant may be contaminated.*

34. Close the refrigerant cylinder valve. ❏
35. Let the center hose of the manifold gage set hang down from the gage manifold. ❏
36. Open the low-side manifold gage valve and purge the refrigerant from the hose until the gage pressure indicates zero. ❏
37. Remove the low-side hose from the refrigerant cylinder. ❏
38. Have your instructor check your work. ❏
39. Return all equipment to its proper location. ❏

RESEARCH QUESTIONS

1. What is the saturation temperature of R-410A at 38 psig?

   _____

   _____

2. What is the psig of R-502 at 100 degrees (F)?

   _____

   _____

3. What is the psig of R-134A at 155 degrees (F)?

   _____

   _____

4. What is the saturation temperature of R-22 at 10 psig?

   _____

   _____

5. What is the saturation temperature of R-12 at 63 psig?

   _____

   _____

Passed Competency _____   Failed Competency _____

Instructor Signature _____ Grade _____

Practical Competency 50

Learning Valve Stem Service Valve Operation (Figure 3–73)

FIGURE 3–73   A service valve. (*Photo by Bill Johnson*)

SUGGESTED MATERIALS

Textbook

Refrigeration & Air Conditioning Technology, 5th Edition, Thomson Delmar Learning
Unit 25—Special Refrigeration System Components

Review Topics

Suction Service Valves; Discharge Service Valves

COMPETENCY OBJECTIVE

The student will be able to operate valve stem service valves.

OVERVIEW

Valve stem serve valves give the technician the ability to access the sealed system without having to add a saddle type service valve to the system. The suction valve is always located on the compressor of the sealed system and is normally larger than the discharge service valve. The discharge valve is either located on the compressor or on a liquid receiver. This type of valve is normally found on refrigeration equipment. The valve consists of a **valve cap, valve stem, packing gland, inlet, outlet**, and **valve body**. These valves are used as a gage port to obtain system pressures, throttle refrigerant gas flow to the compressor, valve off the compressor from the evaporator or condenser for service, and to charge the system with refrigerant. The valve stem has a square head and requires a valve wrench (**Figure 3–74**).

Technicians must become familiar with the operation of these valves and terms used with these valves. The following practical competency is designed to teach technicians the proper operation and terms associated with these valves.

EQUIPMENT REQUIRED

1 Manifold gage set
1 Refrigeration service wrench
1 Adjustable wrench
Safety glasses
Gloves

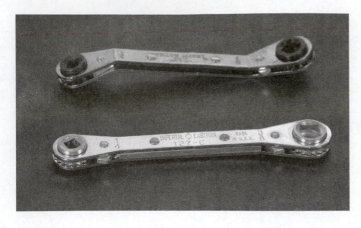

FIGURE 3–74 Refrigeration service wrench. (*Photo by Bill Johnson*)

SAFETY PRACTICES

The student should have experience using refrigerant manifold gages, adjustable wrench, and refrigeration wrench. Follow all EPA rules and regulations when working with refrigerants.

COMPETENCY PROCEDURES   **Checklist**

> *NOTE: This competency is to be performed on an operational sealed system with valve stem service valves.*

1. Turn the power to the **system OFF**.   ❑
2. Use the adjustable wrench and remove both the high-side and low-side service valve caps and gage port caps.   ❑
3. Make sure that manifold gage valves are closed.   ❑
4. Attach the low-side manifold gage hose to the low-side service valve.   ❑
5. Attach the high-side manifold gage hose to the high-side service valve.   ❑
6. Turn the system ON.   ❑
7. Use the refrigeration wrench and **mid-seat** the high-side service valve by **turning the valve clockwise** one half turn (**Figure 3–75**).   ❑

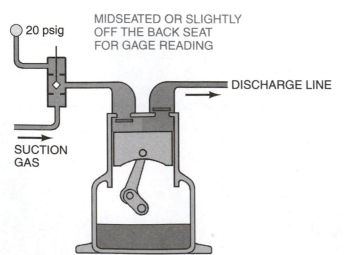

20 psig

MIDSEATED OR SLIGHTLY
OFF THE BACK SEAT
FOR GAGE READING

DISCHARGE LINE

SUCTION
GAS

FIGURE 3–75 Mid-seated valve for gage readings.

8. *Record the pressure indicated on the manifold high-side gage.* _____ ❑
9. Use the refrigeration wrench and **mid-seat** the low-side service valve by **turning the valve clockwise** one half turn. ❑
10. *Record the pressure indicated on the manifold low-side gage.* _____ ❑

NOTE: *In the above procedure you should have learned that to obtain a gage reading on a sealed system with valve stem service valves, you need to **mid-seat** the valve stem(s) one-half or one full turn only. This will gain access to the sealed system, allowing system pressure monitoring, recovery, evacuation, and recharge of a sealed system.*

*The valves at this point are said to be **cracked or at mid-position**. A service valve that is in the **cracked position** allows refrigerant pressure to the system to be read at the gage service port of the valve, and also allows refrigerant to pass from compressor valves out to the refrigeration lines. You also should have learned that the term **mid-seat** means to turn the valve in a **clockwise direction**.*

---

11. Use the refrigeration wrench and turn the high-side valve **counterclockwise** until it is snug. This is called **back-seating** the high-side valve. ❑
12. Use the refrigeration wrench and turn the low-side valve **counterclockwise** until it is snug. This is called **back-seating** the low-side valve (**Figure 3–76**). ❑

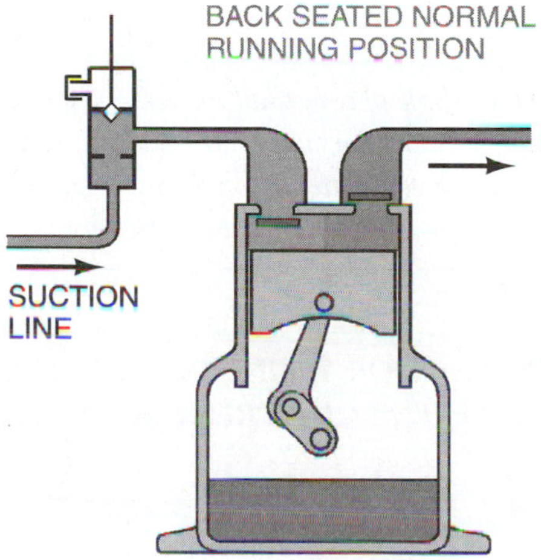

BACK SEATED NORMAL
RUNNING POSITION

DISCHARGE LINE

SUCTION
LINE

FIGURE 3–76 Valves are back-seated during normal running position.

13. Slowly remove the refrigeration hose from the low-side service valve port. ❑

NOTE: *There will be a loss of a small amount of refrigerant during this procedure unless the refrigerant hoses are low-loss fittings. Once the refrigerant has leaked from the hose, you should notice that there is no refrigerant leaking from the low-side service valve. This is because the service valve stem has been **back-seated** the whole way out. You should also realize that when service work is performed on a sealed system with valve stem service valves, the valve stem must be **back-seated** the whole way out **before removing refrigerant hoses**. If not, refrigerant will be lost from the system when manifold gage hoses are removed. **Back-seating the valves** requires the valve stem to be turned in a **counterclockwise position**.*

---

14. Reattach the low-side refrigeration hose to the low-side service valve. ❑
15. With the system operating, turn the low-side valve stem in the whole way until it becomes snug. This is called **front-seating the valve** (**Figure 3–77**). ❑

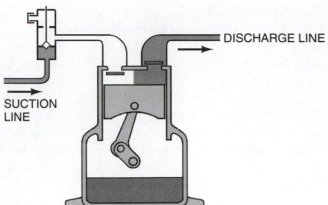

DISCHARGE LINE

SUCTION LINE

FIGURE 3–77 Front-seated. The compressor is isolated from the suction port.

16. Watch the low-side manifold gage pressure. ❏

> **NOTE:** *Systems you are using for this competency may have a low-pressure cut-out control. In this case shut the system down once the cut-out switch has turned the system OFF.*

17. If the system does not have a low-pressure cut-out control, wait for the low-side pressure to go into a vacuum and then **turn the system** OFF. ❏

> **NOTE: Front-seating** *a valve means to turn the valve clockwise the whole way in until the valve seats or becomes snug with the valve seat. Front-seating the low-side valve isolates the evaporator and low-side components from the compressor. Front-seating the high-side service valve isolates the high-side from the compressor. Front-seating both valves isolates the compressor from the evaporator and condenser.*

> **NOTE: NEVER RUN A SEALED SYSTEM WITH A COMPRESSOR HIGH-SIDE SERVICE VALVE FRONT SEATED. VERY DANGEROUS! CAN CAUSE THE COMPRESSOR CYLINDER HEADS TO EXPLODE.**

18. *Watch the low-side pressure and explain what happened.* ❏

> **NOTE:** *The low-side pressure should have stayed the same as long as the valve is front-seated.*

19. **Back-seat** the low-side valve one full turn counterclockwise. ❏
20. *Explain what happened to the low-side pressure.* ❏

> **NOTE:** *The low-side pressure should have risen in pressure because the service valve was opened during the back seating of the valve, allowing refrigerant to move through the low-side sealed system components.*

21. Back-seat the low-side valve the whole way out. ❏
22. Back-seat the high-side valve the whole way out. ❏
23. Slowly remove the low-side refrigerant hose. ❏

24. Slowly remove the high-side refrigerant hose. ❏
25. Replace the service valve caps. ❏
26. Replace the service valve gage port caps. ❏
27. Have your instructor check your work. ❏
28. Return all tools and test equipment to their proper location. ❏

RESEARCH QUESTIONS

1. Do sealed system service valves leak at times?

_____

_____

2. If service valves leak, in what areas can they leak?

_____

_____

3. Do all sealed system service valves have packing nuts?

_____

_____

4. Can sealed service valves be replaced?

_____

_____

5. Can a high-side service valve located on a liquid receiver be front-seated while the system is operational?

_____

_____

Passed Competency _____ Failed Competency _____

Instructor Signature _____ Grade _____

Practical Competency 51

Using Manifold Gages to Interpret Sealed System Pressures

SUGGESTED MATERIALS

Textbook

Refrigeration & Air Conditioning Technology, 5th Edition, Thomson Delmar Learning
Unit 3—Refrigeration and Refrigerates
Unit 8—System Evacuation
Unit 11—Calibrating Instruments
Unit 41—Troubleshooting

Review Topics

Pressure and Temperature Relationship; Gage Manifold Hoses; Pressure Test Instruments; Low-Side Gage
Readings; High-Side Gage Readings

COMPETENCY OBJECTIVE

Using refrigerant manifold gages, the student will be able to interpret sealed systems' pressures and saturation temperatures of the evaporator and condenser.

OVERVIEW

One of the most important tools used by air conditioning and refrigeration technicians are refrigeration
manifold gages. The gage manifold is used to measure actual pressures inside a vapor-compression sealed
system. These pressures can then be used to evaluate the operational performance of a sealed system, or
used with a pressure–temperature chart to determine refrigerant saturation temperatures within the sealed
system. There are two gages on the refrigeration manifold set. One gage is referred to as a compound gage.
This is normally a **Blue gage** and is used to evaluate a system's **low-side operating and pressure.** (Refer to
Figure 3–67, Theory Lesson: Refrigeration Manifold Gages.)

A **compound gage** has the ability to measure above and below atmospheric pressure. The outer large
Black numbers on the gage face represent pressure in **psi** (*pounds per square inch*), and the outer scale **Red
numbers** are used to measure the vacuum level during the evacuation of a sealed system.

The black number **zero (0)** on the gage represents atmospheric pressure of 14.696 psi. When the needle moves **above the (0)** gage pressure, it is measuring pressure above atmospheric pressure and is stated
as **psig** (*pounds per square inch gage*).

When the needle of the gage is **below the (0)**, this is indicating pressure in **inches of mercury (in. Hg)**
below atmospheric pressure.

The high-side gage of the manifold set is the **Red gage** and is used to measure the pressure of the high
side of the sealed system. (Refer to *Figure 3–68, Theory Lesson: Refrigeration Manifold Gages.*)

The pressure range of the high-side gage is represented by the large black numbers just like on the low-
side pressure gage. The high-side gage does not have the ability to measure pressure below atmospheric
pressure, so it is not referred to as a compound gage.

The center numbered scales on each of the gages represent temperature scales for the refrigerants listed
at the end of each of the center temperature scales. Some gages may not have temperature scales listed for
any refrigerant. The refrigerant temperature scales on the gage allow the pressure–temperature relationship
of these particular refrigerants to be determined without the use of a P/T chart.

If the sealed system contains a refrigerant other than that which is shown on a particular gage manifold set, the gages would be used to determine the **system's operating pressures**. A **pressure–temperature chart** for the particular refrigerant would have to be used to convert these gage pressures to saturation temperatures. The hoses of a manifold gage set are normally colored to match the gages on the manifold. With a standard manifold gage set there will be three hoses about 36" long. Manifold gage hoses come as three different colors: red, blue, and yellow.

> NOTE: *Refrigerant hoses. Notice the ends of a refrigerant hose. One end has a shredder port stem and the other end does not. The end without the shredder port stem gets attached to the gage manifold. The end with the shredder stem is used to attach to a sealed system service valve. If the hoses are connected improperly to the manifold gage, pressures from a sealed system will not be indicated on the manifold gages.*

The **Red and Blue hoses** get attached to the corresponding gage color: **Blue hose** to the **Blue** low-side gage and the **Red hose** to the **Red** high-side gage. The **Yellow hose** gets attached to the **center port** of the gage manifold.

The hand valves of the manifold gage DO NOT have to be opened to measure the pressure of a sealed system. The only time the manifold valves should be opened is during the following sealed system procedures:

Adding refrigerant
Recovering refrigerant
Leak testing a system
Evacuating a system

The *low-side* hand valve **opens clockwise** and **closes counterclockwise**. The *high-side* valve is **opened counterclockwise** and **closed counterclockwise**.

When closing the valves **DO NOT** overtighten the valves. The valves should be closed finger tight.

> NOTE: *Standard refrigerant manifold gages should not be used on heat pumps when checking the system's pressures during the heating cycle or used on a R-410A air conditioning system. Pressures on these systems can be higher than the gage pressure scale and cause damage to the pressure gage. Heat pump manifold gages can be purchased, as well as special gages for R-410A systems.*

EQUIPMENT REQUIRED

2 Operational sealed systems with different refrigerant charges
1 Manifold gage set
1 Temperature–pressure chart
1 Temperature recorder
1 Refrigeration wrench (*if applicable*)
Safety goggles
Safety gloves

SAFETY PRACTICES

The student should be competent in the use of refrigerant manifold gages. Follow all EPA rules and regulations when working with refrigerants.

COMPETENCY PROCEDURES   Checklist

CHECKING THE TEMPERATURE AND PRESSURE OF A SEALED SYSTEM 1
  1. Turn power to the unit OFF.   ❑
  2. Make sure the manifold gage valves are closed.   ❑

3. Connect the low-side manifold gage hose to the low-side service valve of the system. ❑
4. Connect the high-side manifold gage hose to the high-side service valve of the system. ❑
5. Turn the system ON. ❑
6. *List the type of refrigerant in the sealed system being checked.* _____ ❑
7. *Record the low-side operating pressure.* _____ ❑
8. *Use the P/T chart and determine the saturation temperature of the refrigerant corresponding to the low-side operating pressure.* _____ ❑
9. *Record the high-side operating pressure.* _____ ❑
10. *Use the P/T chart and determine the saturation temperature of the refrigerant corresponding to the high-side operating pressure.* _____ ❑
11. Use safety glasses and gloves and remove the manifold gage hoses from the sealed system. ❑
12. Turn the power to the unit OFF. ❑

CHECKING THE TEMPERATURE AND PRESSURE OF A SEALED SYSTEM 2

13. Turn the power to the unit OFF. ❑
14. Make sure the manifold gage valves are closed. ❑
15. Connect the low-side manifold gage hose to the low-side service valve of the system. ❑
16. Connect the high-side manifold gage hose to the high-side service valve of the system. ❑
17. Turn the system ON. ❑
18. *Record the type of refrigerant in the sealed system being checked.* _____ ❑
19. *Record the low-side operating pressure.* _____ ❑
20. *Use the P/T chart and determine the saturation temperature of the refrigerant corresponding to the low-side operating pressure.* _____ ❑
21. *Record the high-side operating pressure.* _____ ❑
22. *Use the P/T chart and determine the saturation temperature of the refrigerant according to the high-side operating pressure.* _____ ❑
23. Use safety glasses and gloves and remove the manifold gage hoses from the sealed system. ❑
24. Turn the power to the unit OFF. ❑
25. Have your instructor check your work. ❑
26. Disconnect equipment and return all tools and test equipment to their proper location. ❑

RESEARCH QUESTIONS

1. Which refrigerants have O ODP?

   A. HFCs
   B. HCFCs
   C. CFCs
   D. Both A and B

2. CFCs are more harmful to the stratospheric ozone layer than HCFCs because they contain _____.

3. What agency of the federal government is charged with implementing the United States Clean Air Act Amendments of 1990?

   _____

   _____

4. Why does the refrigerant pressure decrease in a refrigerant cylinder while charging with vapor?

   _____

   _____

5. What gas is commonly used to sweep a refrigeration system to push out any air in the system?

_____

_____

Passed Competency _____ Failed Competency _____

Instructor Signature _____ Grade _____

Practical Competency 52

Using a System's High-Side Pressure and Valve Stem Service Valves to Purge Air from the Manifold Gage

SUGGESTED MATERIALS

Textbook

Refrigeration & Air Conditioning Technology, 5th Edition, Thomson Delmar Learning
Unit 3—Refrigeration and Refrigerants
Unit 8—System Evacuation
Unit 11—Calibrating Instruments
Unit 41—Troubleshooting

Review Topics

Pressure and Temperature Relationship; Gage Manifold Hoses; Pressure Test Instruments; Low-Side Gage Readings; High-Side Gage Readings

COMPETENCY OBJECTIVE

The student will be able to install refrigeration manifold gages to a sealed system with valve stem service valves and use the high-side pressure of the system to purge air from the manifold gage set.

OVERVIEW

One of the most important tools used by air conditioning and refrigeration technicians are refrigeration manifold gages. The gage manifold is used to measure actual pressures inside a vapor-compression sealed system. These pressures can then be used to evaluate the operational performance of a sealed system, or used with a pressure–temperature chart to determine refrigerant saturation temperatures within the sealed system. There are two gages on the refrigeration manifold set. One gage is referred to as a compound gage. This is normally a **Blue gage** and is used to evaluate a system's **low-side operating and pressure.** (Refer to *Figure 3–67, Theory Lesson: Refrigeration Manifold Gages.*)

The high-side gage of the manifold set is the **Red gage** and is used to measure the pressure of the high side of the sealed system. (Refer to *Figure 3–68, Theory Lesson: Refrigeration Manifold Gages.*)

The pressure range of the high-side gage is represented by the large black numbers just like on the low-side pressure gage. The high-side gage does not have the ability to measure pressure below atmospheric pressure, so it is not referred to as a compound gage.

The hoses of a manifold gage set are normally colored to match the gages on the manifold. With a standard manifold gage set there will be three hoses about 36" long. Manifold gage hoses come as three different colors: red, blue, and yellow.

> NOTE: *Refrigerant hoses. Notice the ends of a refrigerant hose. One end has a shredder port stem and the other end does not. The end without the shredder port stem gets attached to the gage manifold. The end with the shredder stem is used to attach to a sealed system service valve. If the hoses are connected improperly to the manifold gage, pressures from a sealed system will not be indicated on the manifold gages.*

The **Red and Blue hoses** get attached to the corresponding gage color: **Blue hose** to the **Blue** low-side gage and the **Red hose** to the **Red** high-side gage. The **Yellow hose** gets attached to the **center port** of the gage manifold.

The hand valves of the manifold gage DO NOT have to be opened to measure pressure of a sealed system. The only time the manifold valves should be opened is during the following sealed system procedures:

Adding refrigerant
Recovering refrigerant
Leak testing a system
Evacuating a system

The *low-side* hand valve **opens clockwise** and **closes counterclockwise**. The *high-side* valve is **opened counterclockwise** and **closed counterclockwise**.

When closing the valves *DO NOT overtighten* the valves. The valves should be closed finger tight.

Manifold gage hoses contain air when they are attached to a sealed system. If air is allowed to enter the sealed system, moisture and noncondensables will enter the sealed system and contaminate the refrigeration system. In most cases the pressure inside the sealed system is greater than the atmospheric pressure and would prevent air from entering the sealed system when attached to the system just to obtain system pressures. If refrigerant is to be added, purging air from the hoses can be done by bleeding the hoses for a second at the hose manifold connection. This would normally be the center manifold hose and low-side hose connection. If servicing a refrigeration system that has valve stem serve valves, air can be purged from the manifold hoses following the procedures listed in this competency.

EQUIPMENT REQUIRED

1 Sealed system with valve stem service valves
1 Manifold gage set
1 Refrigeration wrench
1 Adjustable wrench
Liquid search-leak detector solution
Safety glasses

SAFETY PRACTICES

The student should be knowledgeable of the use of manifold gages and valve stem service valves. Follow all EPA rules and regulations when working with refrigerants.

COMPETENCY PROCEDURES   **Checklist**

> NOTE: *This competency is to be performed on an operational sealed system with valve stem service valves.*

1. Turn the power to the system OFF.   ❏
2. Gain access to the sealed system's service valves.   ❏
3. Use the adjustable wrench and remove both the high-side and low-side service valve caps
   and gage port caps.   ❏
4. Make sure that the refrigeration service valves are back seated.   ❏
5. Make sure that manifold gage valves are closed.   ❏
6. Attach the low-side manifold gage hose to the low-side service valve.   ❏
7. Attach the high-side manifold gage hose to the high-side service valve.   ❏
8. Attach the center hose of the manifold gage set to the manifold bar screws for securing the
   hoses of the gage set.   ❏
9. Mid-seat the **high-side service valve.**   ❏

> **NOTE:** *To mid-seat a valve, you must turn the valve stem one-half turn clockwise.*

10. *Record the pressure indicated on the manifold high-side gage.* _____ ❏
11. *Record the pressure indicated on the manifold low-side gage.* _____ ❏

> **NOTE:** *There should be a zero (0) gage reading on the low-side manifold gage set in the above procedure.*

12. Open the high-side valve on the **manifold gage** set. ❏

> **NOTE:** *This will force all the air in the hoses and manifold bar to collect at the center hose connection on the manifold bar, and also at the suction service valve.*

13. Loosen the center hose connection at the **manifold bar** and let it purge refrigerant and air from the manifold bar and hoses for a couple of seconds; then close the hose connection. ❏
14. Open the low-side **manifold gage valve.** ❏
15. *Record the low-side gage pressure.* _____ ❏
16. Close the low-side **manifold gage valve.** ❏
17. Close the high-side **manifold gage valve.** ❏
18. Loosen the low-side hose connection at the compressor **low-side service valve** and allow the refrigerant and air to purge from the hose for a couple of seconds. ❏

> **NOTE:** *With this process you have purged the air from the manifold and hoses so that no air will enter the sealed system when the suction valve is opened.*

19. Mid-seat the low-side service valve. ❏
20. Mid-seat the high-side service valve. ❏
21. *Record the low-side pressure.* _____ ❏
22. *Record the high-side pressure.* _____ ❏

> **NOTE:** *The pressures should be equal at this point.*

23. Turn the power to the system ON and start the system. ❏
24. Let the system operate for 5 minutes. ❏
25. *Record the operating low-side pressure.* _____ ❏
26. *Record the operating high-side pressure.* _____ ❏
27. Turn the power to the unit OFF. ❏
28. Back-seat the low-side service valve the whole way by turning the valve stem counterclockwise. ❏
29. Back-seat the high-side service valve the whole way by turning the valve stem counterclockwise. ❏
30. Open the low-side manifold gage valve. ❏
31. Open the high-side manifold gage valve. ❏
32. Slowly loosen the center hose connection at the manifold bar and allow the pressure from the gages to purge. ❏
33. Once gage pressures are at zero (0), close both manifold gage valves. ❏

34. Remove the hose connection from the system's service valves. ☐
35. Replace sealed system service valve service port caps. ☐
36. Check for leaks around the valves. ☐
37. Have your instructor check your work. ☐
38. Disconnect the equipment and return all tools and test equipment to their proper location. ☐

RESEARCH QUESTIONS

1. Why is it important to remove the air from the hoses and manifold gages before opening the valves to the sealed system?

_____

_____

2. How could the air and the contaminants from the manifold gages be drawn into a sealed system if there is refrigerant in the system?

_____

_____

3. What happens to the noncondensable gas once it is drawn into the sealed system?

_____

_____

4. What else could be contained in the air that could be harmful to the sealed system?

_____

_____

5. What effect will moisture in a sealed system cause?

_____

_____

Passed Competency _____ Failed Competency _____

Instructor Signature _____ Grade _____

Practical Competency 53

Proper Procedures for Installing Manifold Gages and Purging Air from Hoses on a Sealed System Equipped with Schrader Valves

SUGGESTED MATERIALS

Textbook

Refrigeration & Air Conditioning Technology, 5th Edition, Thomson Delmar Learning
Unit 5—Tools and Equipment
Unit 8—System Evacuation
Unit 9—Recovery, Recycling, Reclaiming, and Retrofitting
Unit 41—Troubleshooting

Review Topics

Low-Loss Fittings; Schrader Valves; Piercing Valves; System Valves; Systems with Schrader Valves

COMPETENCY OBJECTIVE

The student will be able to install refrigeration manifold gages on a sealed system equipped with Schrader valves and measure system operational pressures.

OVERVIEW

There are many types of service valve adapters available on the market. Piercing valves are sometimes called line tap valves.

In addition, service valve adapters are used on sealed systems where no service valves or ports are available to gain entry to the sealed system.

Clamp-on type piercing valves are used for temporary access to the sealed system during recovery or charging process. According to EPA, clamp-on type valves must be removed from the sealed system when service is complete (**Figure 3–78**).

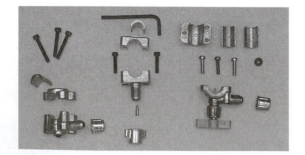

FIGURE 3–78   Tube piercing valves. (*Photo by Bill Johnson*)

These valves have a sharp pin, which is driven through the system's tubing. When the needle is retracted, access is gained to the sealed system.

Valves used for permanent access to the sealed system are the Schrader valves (**Figure 3–79** and **Figure 3–80**).

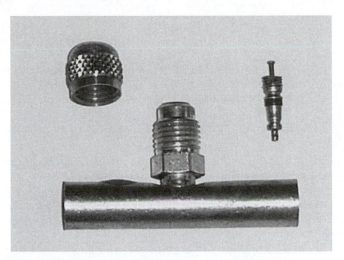

FIGURE 3–79   A Schrader valve with core and cap. (*Photo by John Tomczyk*)

FIGURE 3–80   Solder on type Schrader valve. (*Photo by Bill Johnson*)

If these valves leak, the valve core can be replaced. Schrader valve caps should be placed finger tight on the valve once service work is complete. This assists in preventing leaks along with preventing accidental depression of the valve core stem.

These valves are made of brass, which is a soft metal. Overtightening the cap can distort the top of the valve and prevent easy replacement of the valve core stem. The valve cap has a neoprene gasket, which helps in sealing the valve and also protects the valve top. Caps should be put on the valve finger tight.

EQUIPMENT REQUIRED

1 Manifold gage set
1 Flared plug 3/8 " (*if required*)
Safety glasses
Gloves

SAFETY PRACTICES

The student should be knowledgeable in the use of refrigerant manifold gages and interpretation of the pressure–temperature relationship. Follow all EPA rules and regulations when working with refrigerants.

COMPETENCY PROCEDURES   Checklist

> NOTE: *This competency is to be performed on an operational sealed system with Schrader valves.*

1. Turn the power to the system OFF.   ❑
2. Make sure that manifold gage valves are closed.   ❑
3. Remove the caps to the low-side and high-side Schrader valves.   ❑

> NOTE: *Make sure not to lose the O-ring in the cap.*

4. Attach the high-side manifold gage hose to the high-side Schrader valve. ❑
5. *Record the high-side gage standing pressure.* _____ ❑
6. Attach the low-side manifold gage hose to the low-side Schrader valve. ❑
7. *Record the low-side gage standing pressure.* _____ ❑
8. Attach the center hose of the manifold to the manifold bar or use a 3/8" flared plug to seal center hose end. ❑
9. Open the high-side manifold gage valve. ❑
10. Loosen the center hose connection at the manifold bar or flared plug and let it purge air and refrigerant from the high-side hose for a couple of seconds; then close the center hose connection. ❑

> **NOTE:** *In this procedure you have removed the air from the manifold and high-side hose.*

11. Close the high-side manifold gage valve. ❑
12. Open the low-side manifold gage valve. ❑
13. *Record the low-side gage standing pressure.* _____ ❑
14. Loosen the center hose connection at the manifold bar or flared plug and purge air and refrigerant from the low-side hose for a couple of seconds; then close the center hose connection. ❑

> **NOTE:** *In this procedure you have removed the air from the manifold and low-side hose.*

15. Close the low-side manifold gage valve. ❑

> **NOTE:** *You have completed purging the air and contaminants from the manifold bar and service hoses so the sealed system can be serviced.*

16. If manifold gage hoses do not have low-loss fittings, make sure that you have safety gloves on for the removal of hoses from the service valves. ❑
17. Loosen and remove the low-side manifold gage hose from the low-side Schrader valve. ❑
18. Loosen and remove the high-side manifold gage hose from the high-side Schrader valve. ❑
19. Check for leaks at the Schrader valve cores. ❑
20. Make sure that the Schrader valve caps contain the sealing O-rings. ❑
21. Replace the Schrader valve caps. ❑
22. Have your instructor check your work. ❑
23. Disconnect equipment and return all tools and test equipment to their proper location. ❑

RESEARCH QUESTIONS

1. Why is it important to purge air from hoses before opening the valves to the sealed system?

_____

_____

2. What else could be contained in the air that could be harmful to the sealed system?

_____

_____

3. What effect can moisture in a sealed system cause?

_____

_____

4. Where do most Schrader valves leak?

_____

_____

5. Do you have to recover the refrigerant from the sealed system to replace a Schrader valve core? (Explain your answer).

_____

_____

Passed Competency _____   Failed Competency _____

Instructor Signature _____   Grade _____

Theory Lesson: Refrigerant Recovery, Recycling, and Reclaim (Figure 3–81)

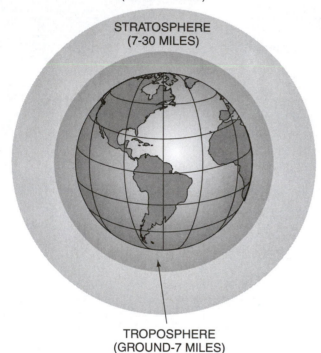

ATMOSPHERIC REGIONS

IONOSPHERE
(30-300 MILES)

STRATOSPHERE
(7-30 MILES)

TROPOSPHERE
(GROUND-7 MILES)

FIGURE 3–81   Atmospheric regions.

SUGGESTED MATERIALS

Textbook
Refrigeration & Air Conditioning Technology, 5th Edition, Thomson Delmar Learning
Unit 9—Refrigerant and Oil Chemistry and Management—Recovery, Recycling, Reclaiming, and Retrofitting

Review Topics
Recovery, Recycle, or Reclaim

Key Terms
active method • ARI Standard 700 • Clean Air Act • De minimums • DOT (Department of Transportation) • liquid recovery • passive method • reclaim • recover • recycle • system-dependent • vapor recovery

OVERVIEW

As of July 1, 1992, the EPA has mandated that it is illegal to release refrigerants into the atmosphere during the maintenance, service, and disposal of refrigeration and air conditioning equipment. According to the Clean Air Act of 1990, technicians may not knowingly vent or otherwise release or dispose of any substance used as a refrigerant in such appliances in a manner that permits refrigerants to enter the environment. Technicians are required to contain all refrigerants except those used to purge lines and tools of the trade, such as the gage manifold or any device used to capture and save refrigerants.

De minimus releases associated with good faith attempts to recapture and recycle or safely dispose of any such substance shall not be subject to prohibition of the Clean Air Act of 1990.

Severe fines and penalties are provided for, including prison terms for violations of the Clean Air Act. The first level of penalty is the power of the EPA to obtain an injunction against the offending party, prohibiting the discharge of refrigerant to the atmosphere. The second level of penalty is a $25,000 fine per day and a prison term not exceeding 5 years. To help in catching violators of the Clean Air Act of 1990, a $10,000 bounty is offered and may be given to any person who furnishes information leading to the conviction of a person willfully venting refrigerant into the environment.

As of November 14, 1994, the EPA has mandated that all refrigeration and air conditioning technicians must be certified in order to purchase refrigerants for the service and repair of refrigeration and air conditioning equipment. There are four different certification areas for different types of service work.

To become certified a technician must take an approved EPA Proctored test and pass the core section and either Type 1, Type II, or Type III certification sections to become certified. A technician who passes the core and Type 1, Type II, and Type III is automatically Universally Certified. Each level consists of 25 questions and a technician must receive a 70% or better to gain certification in the area(s) being tested.

> **NOTE:** *Technician certification is under Section 608 of the Clean Air Act. Gaining certification in any Section 608 area does not permit technicians to work on auto air conditioning. Auto air conditioning technicians are certified under Section 609 of the Clean Air Act.*

Certification areas:

Type I Certification—Small Appliances: *Manufactured, charged and hermetically sealed with 5 lb or less of refrigerant. Includes refrigerators, freezers, room air conditioners, packaged terminal heat pumps, dehumidifiers, under-the-counter ice makers, vending machines, and drinking water coolers.*

Type II Certification—High-Pressure Appliance: *Uses refrigerant with a boiling point between –50 degrees (C) (–58 degrees F) and 10 degrees (C) (50 degrees F) at atmospheric pressure. Includes 12, 22, 114, 500, and 502 refrigerants. Replacement refrigerants for these refrigerants are also included.*

Type III Certification—Low-Pressure Appliance: *Uses refrigerant with a boiling point above 10 (C) (50 degrees F) at atmospheric pressure. Includes 11, 113, 123, and their replacement refrigerants.*

Universal Certification—*Certified in Type I, II, III.*

> **NOTE:** *It is important to note that technicians are responsible to follow any future changes in the law(s) relative to Section 608 Certification or changes in the Clean Air Act relative to the recovery, recycling, and reclamation of refrigerants.*

As of July 1, 1992, the EPA has mandated that it is illegal to release refrigerants into the atmosphere during the maintenance, service, and disposal of refrigeration and air conditioning equipment. According to the Clean Air Act of 1990, technicians may not knowingly vent or otherwise release or dispose of any substance used as a refrigerant in such appliances in a manner that permits refrigerants to enter the environment. Technicians are required to contain all refrigerants except those used to purge lines and tools of the trade, such as the gage manifold or any device used to capture and save refrigerants.

The laws governing the release of CFCs and HCFCs into the atmosphere are very strict and can result in heavy penalties if they are violated. This has led to the development of specific procedures that must be followed to **recover, recycle,** and **reclaim** refrigerants. During the recovery of refrigerant only DOT (**Department of Transportation**) approved cylinders are to be used.

These cylinders are painted yellow on the shoulder area and 12 inches down the side. The body of the cylinders is painted gray. These cylinders are to be tested 5 years after the test date, which is stamped on the shoulder of the handle of the recovery cylinder, along with the tank's weight and fill capacity (**Figure 3–82**).

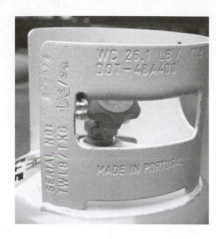

FIGURE 3–82   Recovery cylinder tank weight, capacity, and test date.

The EPA mandates that recovery cylinders not be filled to more than 80% of their liquid weight.

Recovery is the process of removing refrigerant from a sealed system and storing it into an external cylinder without necessarily testing or reprocessing it in any way. The recovery of refrigerant from a sealed system can be accomplished by using several methods, either **passive recovery** or **system-dependent,** and **active recovery.** With these procedures the refrigerant within a sealed system can be removed in the **vapor state,** referred to as **vapor recovery** or **liquid state,** referred to as **liquid recovery.**

The **passive method or system-dependent** recovery process uses the internal pressure of the system or system's compressor to assist in the removal of refrigerant from the sealed system. If the refrigerant system has an operational compressor, this is often the best method for recovery and recommended when the total system refrigerant charge is less than 15 lb. With the passive method of recovery, EPA mandates that 90% of the refrigerant be removed from the system (**Figure 3–83**).

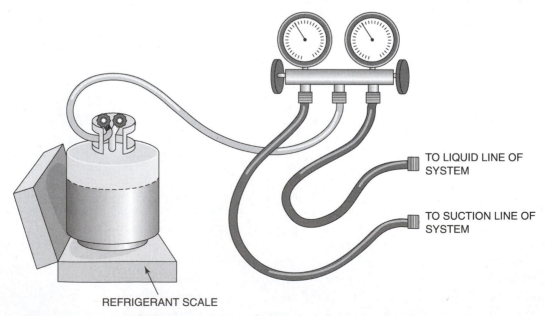

TO LIQUID LINE OF SYSTEM

TO SUCTION LINE OF SYSTEM

REFRIGERANT SCALE

FIGURE 3–83   Setup for passive refrigerant recovery.

Setup for passive recovery is demonstrated in **Figure 3–83.** Recovering liquid would require the **high-side valve** on the gage manifold and the **vapor valve** on the recovery cylinder to be in the **opened position.** The **low-side valve** on the manifold would be in the **closed** position. When refrigerant stops flowing into the recovery tank, the manifold gage **high-side valve** would be **closed** and the manifold gage **low-side valve** would be **opened.** This allows refrigerant from recovery cylinder to flow back into the sealed system, causing the recovery tank pressure to drop.

> **NOTE:** *Packing the recovery cylinder in cold water or ice will also assist in dropping the recovery cylinder's pressure and speed up the recovery process.*

After a period of time the manifold gage **low-side valve** should be **closed** and the manifold gage **high-side valve opened** again to allow more liquid to be forced into the recovery cylinder.

Another method of recovery is called the **active method** of recovery. This method is to be used on refrigeration systems with a charge of 15 lb of refrigerant or more, or if the refrigeration system has an inoperative compressor. The active method of recovery requires the use of a self-contained recovery unit that has been certified for use by the Department of Transportation (DOT), Air Conditioning and Refrigeration Institute (ARI), and the Environmental Protection Agency (EPA).

EPA requires that 80% of the refrigerant be removed from the system with the active method of recovery.

There are many different types of recovery units on the market today and they operate in the same way. Always follow the guidelines and operation instructions for the recovery unit being used. Most recovery units today have **common features** that include inlet and outlet ports, inlet and outlet valves, gages, and a means to purge the unit of refrigerant once recovery from the sealed system is complete. The feature of **"purging"** aids in preventing mixing or cross contaminating of refrigerates.

The idea of vapor recovery is to use the recovery unit to suck vapor out of the sealed system's low side and allow the recovery unit to discharge the refrigerant vapor into the vapor valve of the recovery bottle. This requires a hose connection between the system's low-side valve and recovery unit's **IN valve.** Another hose is attached between the recovery unit's **OUT valve** and **recovery bottle's vapor valve.**

The recovery process is shown in **Figure 3–84.**

In most cases the recovery process begins with recovering as much liquid from the system as possible and completing the recovery process by switching over to the vapor recovery. The reason for this procedure is that liquid recovery is much quicker than vapor recovery.

During the **liquid recovery process,** the manifold gage **high-side valve is opened** because it is attached to the liquid line of the refrigeration sealed system. Once liquid recovery is complete, the manifold gage high-side valve is **closed** and the manifold gage **low-side valve** is **opened** to recover refrigerant in the vapor state from the sealed system.

Adding a filter drier and sight glass/moisture indicator installed between in the hose between the refrigeration system and recovery unit protects the recovery unit from particulate matter and allows for monitoring refrigerant state and moisture content as it passes to the recovery unit (**Figure 3–85**).

The liquid recovery process is shown in **Figure 3–86.**

Some recovery units use the push–pull method for recovery refrigerant in the liquid state from a sealed system. This is accomplished by connecting a refrigerant hose between the **high-side valve** of the sealed system and the **liquid valve** of the recovery cylinder. Another refrigerant hose is connected from the recovery unit's **INLET valve** and the refrigerant cylinder's **Vapor Valve.** The final connection is a refrigerant hose from the recovery unit's **OUTLET valve** and the refrigerant system's **low-side valve.** All valves are opened and when the recovery unit is started, refrigerant vapor is pulled out the recovery cylinder's vapor vale and condensed by the recovery unit. A small amount of liquid is pushed by the recovery unit into the refrigeration system's low-side valve where it flashes to a vapor, building pressure through the refrigeration systems, forcing liquid to flow from the refrigeration sealed system's high-side valve to the recovery cylinder's liquid valve. A sight glass and filter drier should be in the line connection between the refrigeration system's high-side

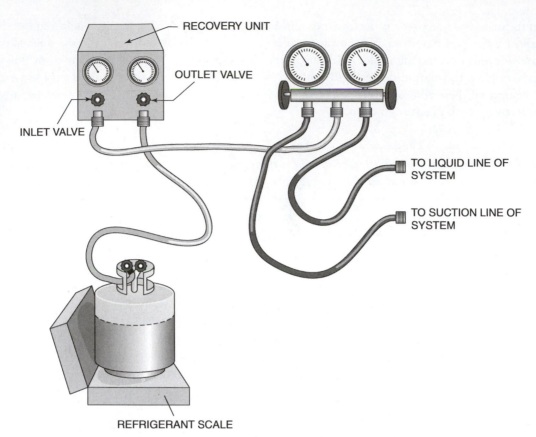

RECOVERY UNIT

OUTLET VALVE

INLET VALVE

TO LIQUID LINE OF SYSTEM

TO SUCTION LINE OF SYSTEM

REFRIGERANT SCALE

FIGURE 3–84 Setup for active refrigerant recovery.

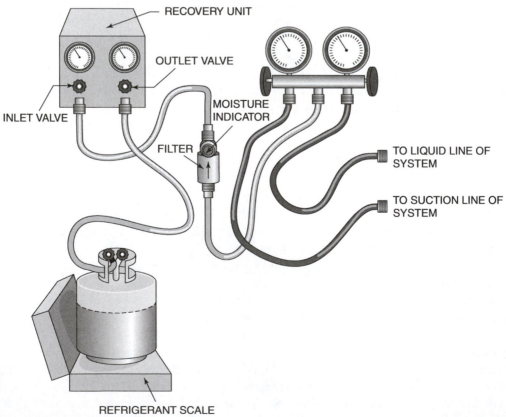

RECOVERY UNIT

OUTLET VALVE

INLET VALVE

MOISTURE INDICATOR

FILTER

TO LIQUID LINE OF SYSTEM

TO SUCTION LINE OF SYSTEM

REFRIGERANT SCALE

FIGURE 3–85 Active recovery setup with moisture indicator and filter installed.

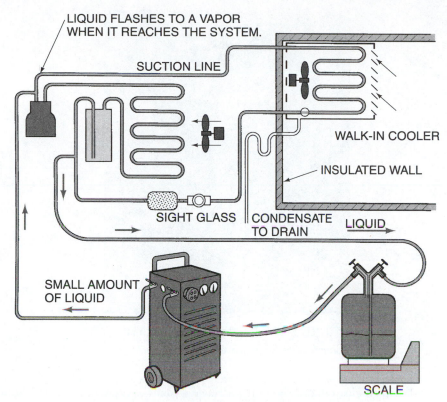

FIGURE 3–86   Liquid recovery push–pull method.

valve and recovery cylinder's liquid valve. This permits observation and filtering of refrigerant as it is forced from the sealed system into the recovery cylinder. Once visual inspection of the sight glass indicates that liquid is no longer flowing from the sealed system to the recovery cylinder, the liquid recovery process should be stopped and connections for vapor recovery should be made to complete the recovery process of the refrigeration sealed system.

Universal guidelines to be followed when using recovery units include:

- *Make sure the recovery unit is purged of any residue refrigerant left in the unit from previous use.*
- *Make sure the recovery bottle contains the same type of refrigerant that is to be recovered from a sealed system.*
- *Make sure that all hoses are purged to remove air from them prior to attachment to the recovery unit and refrigeration system.*
- *Warmer temperatures will increase the speed of recovery.*
- *Use filter drier and sight glass during recovery to protect the recovery unit and assist in monitoring refrigerant condition and state.*
- *Follow the recovery unit's maintenance guidelines as established by the manufacturers.*
- *Except for oil-less recovery units, change oil and filters often to reduce the chance of refrigerant contamination.*
- *Make sure that only qualified individuals use the recovery unit.*
- *Wear safety glasses and gloves during the recovery process.*
- *Recover refrigerant in a well ventilated area.*
- *Empty recovery bottles should be evacuated and cleaned before being used in the recovery process.*
- *Only DOT-approved recover cylinders should be used for refrigerant storage.*

Recycling is the process of removing refrigerant from a sealed system into an external cylinder and then cleaning the refrigerant for reuse by using a recovery/recycling unit (**Figure 3–87**).

This is accomplished by passing the refrigerant through an oil separator and using single and multiple passes through devices such as replaceable core filter driers to reduce moisture, acid, and particle matter.

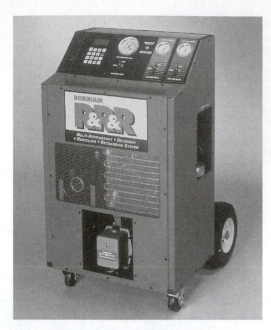

FIGURE 3–87 Recovery/recycling units.
(*Courtesy of Robinair Corporation*)

These units are designed to clean the refrigerant for reuse under the guidelines of ARI Standard 700. This process can be completed at the job site or at the service center.

> **NOTE:** *Extreme care should be taken to avoid the cross contamination of refrigerants in the process of recovery, and recycling. Technicians should make sure that manifold hoses with quick connect fittings are bled before attaching to another sealed system with a different refrigerant.*

Reclaiming is the process of cleaning refrigerant to new product specification by means that may require distillation. This process requires chemical analysis of the refrigerant to determine that proper specifications are met. This process can be accomplished only at a reprocessing or manufacturing facility.

There are instances when refrigerants will be so badly contaminated that the reclaim facility will not be able to clean the refrigerant to new product specifications. This could include refrigerants from a burnout or refrigerants that are mixed. In either case the only option for the reclaim facility is to destroy the refrigerant. Incinerating the refrigerant at a temperature of approximately 1200 degrees (F) does this. This is very expensive and the cost to complete this process is passed on to the technician.

EPA has set standards for evacuation levels for recovery equipment to ensure that the proper amount of refrigerant is removed from the refrigeration system.

Small appliances that are identified as refrigeration systems are manufactured, charged, and hermetically sealed with 5 lb or less of refrigerant. This includes refrigerators, freezers, room air conditioners, packaged heat pumps, dehumidifiers, vending machines, drinking water coolers, and under-the-counter ice makers (**Figure 3–88**).

For all other refrigeration and air conditioning equipment, EPA has established Levels of Evacuation for Recovery/Recycling Equipment (**Figure 3–89**).

Recovery Efficiency Requirements for Small Appliances*		
Recovery Efficiencies Required	Recovered Percentages	Inches of Mercury Vacuum
For active and passive equipment manufactured after November 15, 1993, for service or disposal of small appliances with an operative compressor on the small appliance.	90%	4*
For active and passive equipment manufactured after November 15, 1993, for service or disposal with an inoperative compressor on the small appliance.	80%	4*
For grandfathered active and passive equipment manufactured before November 15, 1993, for service or disposal with or without an operating compressor on the small appliance.	80%	4*

*ARI 740-Standards
*NOTE: Small appliances are products that are fully manufactured, charged, and hermetically sealed in a factory with 5 lb or less of refrigerant.

FIGURE 3–88   Small appliances recovery requirements.

Required Levels of Evacuation for Air-Conditioning, Refrigeration, and Recovery/Recycling Equipment (Except for small appliances, MVACs, and MVAC-like equipment) Inches of Hg Vacuum		
Type of Air-Conditioning or Refrigeration Equipment	Using Recovery or Recycling Equipment Manufactured before November 15, 1993	Using Recovery or Recycling Equipment Manufactured on or after November 15, 1993
HCFC-22 equipment, or isolated component of such equipment, normally containing less than 200 pounds of refrigerant.	0	0
HCFC-22 equipment, or isolated component of such equipment, normally containing 200 pounds or more of refrigerant.	4	10
Other high pressure equipment, or isolated component of such equipment, normally containing less than 200 pounds of refrigerant.	4	10
Other high pressure equipment, or isolated component of such equipment, normally containing 200 pounds or more of refrigerant.	4	15
Very high pressure equipment.	0	0
Low pressure equipment.	25	29

NOTE: MVAC = Motor Vehicle Air Conditioning

FIGURE 3–89   Required levels of evacuation for air conditioning, refrigeration, and recovery/recycling equipment. (*Courtesy of U.S. EPA*)

Practical Competency 54

Recovering Refrigerant from a System with an Operating Compressor

SUGGESTED MATERIALS

Textbook
Refrigeration & Air Conditioning Technology, 5th Edition, Thomson Delmar Learning
Unit 9—Refrigerant and Oil Chemistry and Management—Recovery, Recycling, Reclaiming, and Retrofitting

Review Topics
Recovery, Recycle, or Reclaim

COMPETENCY OBJECTIVE

The student will be able to recover refrigerant from a sealed system with an operating compressor.

OVERVIEW

As of July 1, 1992, the EPA has mandated that it is illegal to release refrigerants into the atmosphere during the maintenance, service, and disposal of refrigeration and air conditioning equipment. According to the Clean Air Act of 1990, technicians may not knowingly vent or otherwise release or dispose of any substance used as a refrigerant in such appliances in a manner that permits refrigerants to enter the environment. Technicians are required to contain all refrigerants except those used to purge lines and tools of the trade, such as the gage manifold or any device used to capture and save refrigerants.

The laws governing the release of CFCs and HCFCs into the atmosphere are very strict and can result in heavy penalties if they are violated. This has led to the development of specific procedures that must be followed to **recover, recycle,** and **reclaim** refrigerants. During the recovery of refrigerant only DOT-approved cylinders are to be used.

These cylinders are painted yellow on the shoulder area and 12 inches down the side. The body of the cylinders is painted gray.

EPA mandates that recovery cylinders not be filled to more then 80% of their liquid weight.

Recovery is the process of removing refrigerant from a sealed system and storing it into an external cylinder without necessarily testing or reprocessing it in any way. The recovery of refrigerant from a sealed system can be accomplished by using several methods, either **passive recovery** or **system–dependent**, and **active recovery**. With these procedures the refrigerant within a sealed system can be removed in the **vapor state**, referred to as **vapor recovery** or **liquid state**, referred to as **liquid recovery**.

The **passive method or system-dependent** recovery process uses the internal pressure of the system or system's compressor to assist in the removal of refrigerant from the sealed system. If the refrigerant system has an operational compressor, this is often the best method for recovery and recommended when the total system refrigerant charge is less than 15 lb. With the passive method of recovery, EPA mandates that 90% of the refrigerant be removed from the system.

The setup for passive recovery is demonstrated in *Figure 3–83, Theory Lesson: Refrigerant Recovery, Recycling, and Reclaim.*

Recovering liquid would require the **high-side valve** on the gage manifold and the **vapor valve** on the recovery cylinder to be in the **opened position**. The **low-side valve** on the manifold would be in the **closed** position. When refrigerant stops flowing into the recovery tank, the manifold gage **high-side valve** would be **closed** and the manifold gage **low-side valve** would be **opened**. This allows refrigerant from recovery cylinder to flow back into the sealed system, causing the recovery tank pressure to drop.

> **NOTE:** *Packing the recovery cylinder in cold water or ice will also assist in dropping the recovery cylinder's pressure and speed up the recovery process.*

After a period of time the manifold gage **low-side valve** should be **closed** and the manifold gage **high-side valve opened** again to allow more liquid to be forced into the recovery cylinder.

There are many different types of recovery units on the market today and they operate in the same way. Always follow the guidelines and operation instructions for the recovery unit being used. Most recovery units today have **common features** that include inlet and outlet ports, inlet and outlet valves, gages, and a means to purge the unit of refrigerant once recovery from the sealed system is complete. The feature of **"purging"** aids in preventing mixing or cross contaminating of refrigerates.

In most cases the recovery process begins with recovering as much liquid from the system as possible and completing the recovery process by switching over to the vapor recovery. The reason for this procedure is that liquid recovery is much quicker than vapor recovery.

During the **liquid recovery process**, the manifold gage **high-side valve** is **opened** because it is attached to the liquid line of the refrigeration sealed system. Once liquid recovery is complete, the manifold gage high side valve is **closed** and the manifold gage **low-side valve** is **opened** to recover refrigerant in the vapor state from the sealed system.

Adding a filter drier and sight glass/moisture indicator installed between in the hose between the refrigeration system and recovery unit, protects the recovery unit from particulate matter, and allows for monitoring refrigerant state and moisture content as it passes to the recovery unit. (Refer to *Figure 3–85, Theory Lesson: Refrigerant Recovery, Recycling, and Reclaim.*)

Some recovery units use the push–pull method for recovery refrigerant in the liquid state from a sealed system. This is accomplished by connecting a refrigerant hose between the **high-side valve** of the sealed system and the **Liquid Valve** of the recovery cylinder. Another refrigerant hose is connected from the recovery unit's **INLET valve** and the refrigerant cylinder's **Vapor Valve**. The final connection is a refrigerant hose from the recovery unit's **OUTLET valve** and the refrigerant system's **low-side valve**. All valves are opened and when the recovery unit is started, refrigerant vapor is pulled out the recovery cylinders vapor vale and condensed by the recovery unit. A small amount of liquid is pushed by the recovery unit into the refrigeration system's low-side valve where it flashes to a vapor, building pressure through the refrigeration systems, forcing liquid to flow from the refrigeration sealed system's high-side valve to the recovery cylinder's liquid valve. A sight glass and filter drier should be in the line connection between the refrigeration system's high-side valve and recovery cylinder's liquid valve. This permits observation and filtering of refrigerant as it is forced from the sealed system into the recovery cylinder. Once visual inspection of the sight glass indicates that liquid is no longer flowing from the sealed system to the recovery cylinder, the liquid recovery process should be stopped and connections for vapor recovery should be made to complete the recovery process of the refrigeration sealed system.

Universal guidelines to be followed when using recovery units include:

- *Make sure the recovery unit is purged of any residue refrigerant left in the unit from previous use.*
- *Make sure the recovery bottle contains the same type of refrigerant that is to be recovered from a sealed system.*
- *Make sure that all hoses are purged to remove air from them prior to attachment to the recovery unit and refrigeration system.*
- *Warmer temperatures will increase the speed of recovery.*
- *Use filter drier and sight glass during recovery to protect the recovery unit and assist in monitoring refrigerant condition and state.*

- *Follow the recovery unit's maintenance guidelines as established by the manufacturers.*
- *Except for oil-less recovery units, change oil and filters often to reduce the chance of refrigerant contamination.*
- *Make sure that only qualified individuals use the recovery unit.*
- *Wear safety glasses and gloves during the recovery process.*
- *Recover the refrigerant in a well ventilated area.*
- *Empty recovery bottles should be evacuated and cleaned before being used in the recovery process.*
- *Only DOT-approved recover cylinders should be used for refrigerant storage.*

NOTE: Extreme care should be taken to avoid the cross contamination of refrigerants in the process of recovery and recycling. Technicians should make sure that manifold hoses with quick connect fittings are bled before attaching to another sealed system with a different refrigerant.

EPA has set standards for evacuation levels for recovery equipment to ensure that the proper amount of refrigerant is removed from the refrigeration system.

Small appliances that are identified as refrigeration systems are manufactured, charged, and hermetically sealed with 5 lb or less of refrigerant. This includes refrigerators, freezers, room air conditioners, packaged heat pumps, dehumidifiers, vending machines, drinking water coolers, and under-the-counter ice makers.

For all other refrigeration and air conditioning equipment, EPA has established Levels of Evacuation for Recovery/Recycling Equipment (**Figure 3–90**).

Required Levels of Evacuation for Air-Conditioning, Refrigeration, and Recovery/Recycling Equipment (Except for small appliances, MVACs, and MVAC-like equipment) Inches of Hg Vacuum		
Type of Air-Conditioning or Refrigeration Equipment	**Using Recovery or Recycling Equipment Manufactured before November 15, 1993**	**Using Recovery or Recycling Equipment Manufactured on or after November 15, 1993**
HCFC-22 equipment, or isolated component of such equipment, normally containing less than 200 pounds of refrigerant.	0	0
HCFC-22 equipment, or isolated component of such equipment, normally containing 200 pounds or more of refrigerant.	4	10
Other high pressure equipment, or isolated component of such equipment, normally containing less than 200 pounds of refrigerant.	4	10
Other high pressure equipment, or isolated component of such equipment, normally containing 200 pounds or more of refrigerant.	4	15
Very high pressure equipment.	0	0
Low pressure equipment.	25	29

NOTE: MVAC = Motor Vehicle Air Conditioning

FIGURE 3–90 Required levels of evacuation for air conditioning, refrigeration, and recovery/recycling equipment. (*Courtesy of U.S. EPA*)

EQUIPMENT REQUIRED

1 Operational sealed system
1 Sight glass with 1/4" flared fittings

1 Filter drier 1/4" × 1/4" flare
2 Access valves (*if required*)
1 DOT recovery cylinder
1 Refrigerant scales
1 Manifold gage set
1 Refrigerant wrench (*if required*)
1 Adjustable wrench (*if required*)
2 Additional refrigerant hose (*if required*)

SAFETY PRACTICES

Wear goggles or safety glasses and gloves when working with refrigerants in sealed systems. Beware of high pressures and follow all electrical safety rules. Follow all EPA rules and regulations while working with refrigerants.

COMPETENCY PROCEDURES   Checklist

1. Turn the refrigerant sealed system OFF (*if applicable*). ❑
2. *If applicable*, remove valve stem service valve caps and service port caps. ❑
3. Make sure that the recovery cylinder is evacuated or has the same refrigerant in it as is in the system being recovered. ❑
4. Set the recovery cylinder on the refrigerant scale. ❑
5. If required, attach line valves to the suction and liquid lines of the sealed system. ❑
6. Determine the recovery cylinder's 80% fill rate. ❑
7. *Record the cylinder's 80% fill rate.* _____ ❑
8. *Record the recovery units DOT test date.* _____ ❑
9. *Record the type of refrigerant being recovered.* _____ ❑
10. *List the type of refrigeration equipment refrigerant is being recovered from.* _____ ❑
11. *Review the EPA "Required Levels of Evacuation for Air Conditioning, and Refrigeration" or "Recovery Efficiency Requirements for Small Appliances" charts and record the required recovery level for the equipment that the refrigerant is being recovered from.* _____ ❑
12. **Close** manifold **gage valves.** ❑
13. Install the manifold high-side hose to the sealed system's high-side valve. ❑
14. Install the manifold low-side hose to the sealed system's low-side valve. ❑
15. Install the manifold center hose to the filter drier and sight glass. (*Use additional refrigerant hose if necessary.*) ❑
16. Connect another hose from the filter drier and sight glass to the recovery cylinder's vapor valve. (*Use additional refrigerant hose if necessary.*) ❑
17. *If applicable,* mid-seat the valve stem service valves. ❑
18. **Open** the manifold **high-side valve.** ❑
19. **Open** the manifold **low-side valve.** ❑
20. Loosen the hose connection on the recovery cylinder for 1 second and then retighten the hose. ❑

NOTE: This will purge any air from the gage hoses.

---

21. **Close** the **high-side** manifold valve. ❑
22. **Close** the **low-side** manifold valve. ❑
23. **Start** the refrigeration sealed system. ❑
24. *If applicable,* turn on the refrigerant scale. ❑
25. Zero the display on the refrigerant scale. ❑
26. **Open** the manifold **high-side valve.** ❑
27. **Open** the **vapor valve** on the recovery cylinder. ❑

> **NOTE:** *Refrigerant will flow from the sealed system, through the manifold gage and into the recovery cylinder.*

28. Monitor the system and the display on the refrigerant scale. ❏
29. When the reading on the refrigerant scale's display indicates that no more refrigerant is being introduced into the recovery cylinder, **close the manifold high-side valve.** ❏
30. **Open** the **low-side** manifold valve. ❏

> **NOTE:** *This will allow vapor refrigerant to flow back into the refrigeration sealed system.*

31. Allow the pressure in the recovery cylinder to stabilize. ❏
32. **Close** the manifold **low-side valve.** ❏
33. **Open** the manifold **high-side valve** to allow liquid recovery to resume. ❏
34. Repeat Steps 31 through 35 until the proper recovery level is reached. ❏

> **NOTE:** *If the entire refrigerant cannot be recovered using this method, or if the high-side pressure rises, close valves on the recovery cylinder and manifold gages. Replace the recovery cylinder with another evacuated cylinder, and continue the recovery process until the required recovery level is reached.*

35. When recovery level is reached, **close** the manifold **high-side valve.** ❏
36. **Close** the manifold **low-side valve.** ❏
37. **Close** the recovery cylinder **vapor valve.** ❏
38. Turn the refrigeration system OFF. ❏
39. Slowly remove the manifold center hose from the recovery cylinder. ❏
40. Slowly remove the manifold high-side and low-side hoses from the sealed system. ❏
41. Tag the refrigerant cylinder and label with the type of refrigerant, amount of refrigerant, owner of refrigerant, and recovery date. ❏

> **NOTE:** *If manifold gage hoses are low-loss fittings, be sure to purge refrigerant from hoses.*

42. Have your instructor check your work. ❏
43. Disconnect the equipment and return all tools and equipment to their proper location. ❏

RESEARCH QUESTIONS

1. Why is liquid recovery a faster method of recovery than vapor recovery?

   _____

   _____

2. With the passive method of recovery, the system's compressor is used to do what?

   _____

   _____

3. Who is the approval agency for recovery cylinders?

   _____

   _____

4. How often should recovery cylinders be tested?

_____

_____

5. Why should a recovery cylinder be filled to only 80% of its total weight?

_____

_____

Passed Competency _____   Failed Competency _____

Instructor Signature _____   Grade _____

Practical Competency 55

Active Method of Recovery – Recovering Refrigerant from a System with a Self-Contained Recovery Unit

SUGGESTED MATERIALS

Textbook

Refrigeration & Air Conditioning Technology, 5th Edition, Thomson Delmar Learning
Unit 9—Refrigerant and Oil Chemistry and Management—Recovery, Recycling, Reclaiming, and Retrofitting

Review Topics

Recovery, Recycle, or Reclaim

COMPETENCY OBJECTIVE

The student will be able to use the Active Method of Recovery and recover refrigerant from a sealed system.

OVERVIEW

As of July 1, 1992, the EPA has mandated that it is illegal to release refrigerants into the atmosphere during the maintenance, service, and disposal of refrigeration and air conditioning equipment. According to the Clean Air Act of 1990, technicians may not knowingly vent or otherwise release or dispose of any substance used as a refrigerant in such appliances in a manner that permits refrigerants to enter the environment. Technicians are required to contain all refrigerants except those used to purge lines and tools of the trade, such as the gage manifold or any device used to capture and save refrigerants.

The laws governing the release of CFCs and HCFCs into the atmosphere are very strict and can result in heavy penalties if they are violated. This has led to the development of specific procedures that must be followed to **recover, recycle,** and **reclaim** refrigerants. During the recovery of refrigerant only DOT-approved cylinders are to be used.

These cylinders are painted yellow on the shoulder area and 12 inches down the side. The body of the cylinders is painted gray.

EPA mandates that recovery cylinders not be filled to more then 80% of their liquid weight.

Recovery is the process of removing refrigerant from a sealed system and storing it into an external cylinder without necessarily testing or reprocessing it in any way. The recovery of refrigerant from a sealed system can be accomplished by using several methods, either **Passive Recovery** or **System–Dependent,** and **Active Recovery.** With these procedures the refrigerant within a sealed system can be removed in the **Vapor state,** referred to as **Vapor Recovery** or **Liquid state,** referred to as **Liquid Recovery.**

The **active method** of recovery is to be used on refrigeration systems with a charge of 15 lb of refrigerant or more, or if the refrigeration system has an inoperative compressor. The active method of recovery requires the use of a self-contained recovery unit that has been certified for use by the Department of Transportation (DOT), Air Conditioning and Refrigeration Institute (ARI), and the Environmental Protection Agency (EPA).

EPA requires that 80% of the refrigerant be removed from the system with the active method of recovery.

There are many different types of recovery units on the market today and they operate in the same way. Always follow the guidelines and operation instructions for the recovery unit being used. Most recovery units today have **common features** that include inlet and outlet ports, inlet and outlet valves, gages, and a means to purge the unit of refrigerant once recovery from the sealed system is complete. The feature of **"purging"** aids in preventing mixing or cross contaminating of refrigerants.

The idea of **vapor recovery** is to use the recovery unit to suck vapor out of the sealed system's low side and allow the recovery unit to discharge the refrigerant vapor into the vapor valve of the recovery bottle. This requires a hose connection between the system's low-side valve and the recovery unit's **IN valve**. Another hose is attached between the recovery unit's **OUT valve** and **recovery bottle's vapor valve**.

In most cases the recovery process begins with recovering as much liquid from the system as possible and completing the recovery process by switching over to the vapor recovery. The reason for this procedure is that liquid recovery is much quicker than vapor recovery.

During the **liquid recovery process**, the manifold gage **high-side valve** is **opened** because it is attached to the liquid line of the refrigeration sealed system. Once liquid recovery is complete, the manifold gage high-side valve is **closed** and the manifold gage **low-side valve** is **opened** to recover refrigerant in the vapor state from the sealed system.

> **NOTE:** *Packing the recovery cylinder in cold water or ice will also assist in dropping the recovery cylinder's pressure and speed up the recovery process.*

After a period of time the manifold gage **low-side valve** should be **closed** and the manifold gage **high-side valve opened** again to allow more liquid to be forced into the recovery cylinder.

Adding a filter drier and sight glass/moisture indicator installed between in the hose between the refrigeration system and recovery unit protects the recovery unit from particulate matter, and allows for monitoring refrigerant state and moisture content as it passes to the recovery unit (Refer to *Figure 3–85, Theory Lesson: Refrigerant Recovery, Recycling, and Reclaim.*)

Some recovery units use the push–pull method for recovery refrigerant in the liquid state from a sealed system. This is accomplished by connect a refrigerant hose between to the **high-side valve** of the sealed system and the **Liquid Valve** of the recovery cylinder. Another refrigerant hose is connected from the recovery unit's **INLET valve** and the refrigerant cylinder's **Vapor Valve**. The final connection is a refrigerant hose from the recovery unit's **OUTLET valve** and the refrigerant system's **low-side valve**. All valves are opened and when the recovery unit is started, refrigerant vapor is pulled out the recovery cylinder's vapor vale and condensed by the recovery unit. A small amount of liquid is pushed by the recovery unit into the refrigeration system's low-side valve where it flashes to a vapor, building pressure through the refrigeration systems, forcing liquid to flow from the refrigeration sealed system's high-side valve to the recovery cylinder's liquid valve. A sight glass and filter drier should be in the line connection between the refrigeration system's high-side valve and recovery cylinder's liquid valve. This permits observation and filtering of refrigerant as it is forced from the sealed system into the recovery cylinder. Once visual inspection of the sight glass indicates that liquid is no longer flowing from the sealed system to the recovery cylinder, the liquid recovery process should be stopped and connections for vapor recovery should be made to complete the recovery process of the refrigeration sealed system.

Universal guidelines to be followed when using recovery units include:

- *Make sure the recovery unit is purged of any residue refrigerant left in the unit from previous use.*
- *Make sure the recovery bottle contains the same type of refrigerant that is to be recovered from a sealed system.*
- *Make sure that all hoses are purged to remove air from them prior to attachment to the recovery unit and refrigeration system.*
- *Warmer temperatures will increase the speed of recovery.*
- *Use filter drier and sight glass during recovery to protect the recovery unit and assist in monitoring refrigerant condition and state.*
- *Follow the recovery unit's maintenance guidelines as established by the manufacturers.*
- *Except for oil-less recovery units, change oil and filters often to reduce the chance of refrigerant contamination.*
- *Make sure that only qualified individuals use the recovery unit.*

- *Wear safety glasses and gloves during the recovery process.*
- *Recover refrigerant in a well ventilated area.*
- *Empty recovery bottles should be evacuated and cleaned before being used in the recovery process.*
- *Only DOT-approved recovery cylinders should be used for refrigerant storage.*

> **NOTE:** *Extreme care should be taken to avoid the cross contamination of refrigerants in the process of recovery and recycling. Technicians should make sure that manifold hoses with quick connect fittings are bled before attaching to another sealed system with a different refrigerant.*

EPA has set standards for evacuation levels for recovery equipment to ensure that the proper amount of refrigerant is removed from the refrigeration system.

Small appliances that are identified as refrigeration systems are manufactured, charged, and hermetically sealed with 5 lb or less of refrigerant. This includes refrigerators, freezers, room air conditioners, packaged heat pumps, dehumidifiers, vending machines, drinking water coolers, and under-the-counter ice makers.

For all other refrigeration and air conditioning equipment, EPA has established Levels of Evacuation for Recovery/Recycling Equipment.

EQUIPMENT REQUIRED

1 Recovery unit
2 Access valves (*if required*)
1 DOT recovery cylinder
1 Refrigerant scale
1 Manifold gage set
1 Refrigerant wrench (*if required*)
1 Adjustable wrench (*if required*)
2 Additional refrigerant hose (*if required*)

SAFETY PRACTICES

Wear goggles or safety glasses and gloves when working with refrigerant in sealed systems. Beware of high pressures and follow all electrical safety rules. Follow all EPA rules and regulations while working with refrigerants.

COMPETENCY PROCEDURES   Checklist

1. If required, check the oil and filters of the recovery unit being used. ❑
2. Turn the refrigeration or air conditioning system OFF. ❑
3. Make sure the recovery unit is purged of all refrigerant. ❑
4. Make sure the recovery cylinder is evacuated or has the same type of refrigerant in it that is to be recovered. ❑
5. If the recovery unit has an electrical hookup for an 80% stop-fill switch, complete the hook-up to the recovery tank as required by the manufacturer. ❑
6. Set the recovery cylinder on refrigerant scales. ❑
7. Make sure the recovery unit's inlet and outlet valves are closed. ❑
8. *If required*, attach refrigerant access valve to the low side of the sealed system. ❑
9. *If required*, attach the refrigerant access valve to the high side of the sealed system. ❑
10. Determine the recovery cylinder's 80% fill rate. ❑
11. *Record the cylinder's 80% fill rate.* _____ ❑
12. *Record the recovery unit's DOT test date.* _____ ❑
13. *Record the type of refrigerant being recovered.* _____ ❑
14. *List the type of refrigeration equipment refrigerant is being recovered from.* _____ ❑

15. *Review the EPA "Required Levels of Evacuation for Air Conditioning, and Refrigeration"* *or "Recovery Efficiency Requirements for Small Appliances" charts and record the required* *recovery level for the equipment refrigerant is being recovered from.* _____ ❑
16. Attach a manifold low-side gage hose to the sealed system low-side access valve. ❑
17. Attach a manifold high-side gage hose to the sealed system's high-side access valve. ❑
18. Attach a manifold gage center hose to the recovery unit's INLET valve. ❑
19. Connect an additional refrigerant hose to the recovery unit's OUTLET valve. ❑
20. Connect the opposite end of the hose from the recovery unit outlet valve to the recovery cylinder's VAPOR valve. ❑
21. **Close** both INLET and OUTLET valves or recovery unit. ❑
22. **Open** the manifold gage **high-side valve.** ❑
23. Loosen the hose connection at the recovery cylinder to purge air from the hose connected between the recovery unit and the recovery cylinder. ❑
24. Turn on the refrigerant scale. ❑
25. Zero the display on the refrigerant scale. ❑
26. Open the INLET valve on the recovery unit. ❑
27. Open the OUTLET valve of the recovery unit. ❑
28. Allow the pressure in the recovery unit to equalize. ❑
29. Plug the recovery unit into a power supply. ❑
30. Turn the recovery unit ON. ❑
31. To **recover liquid, open** the manifold gage **high-side valve.** ❑
32. **Open** the **vapor valve** on the recovery cylinder. ❑
33. When the liquid refrigerant has been recovered, **open** the manifold gage **low-side valve.** ❑
34. Continue recovery until the proper recovery level is reached. ❑
35. **Close** the manifold gage **low-side valve.** ❑
36. **Close** the manifold gage **high-side valve.** ❑
37. Follow the recovery unit's instructions to **purge** refrigerant from the **recovery unit.** ❑
38. **Close** the recovery unit's **INLET valve.** ❑
39. **Close** the recovery unit's **OUTLET valve.** ❑
40. **Close** the recovery cylinder's **vapor valve.** ❑
41. Turn the recovery unit OFF. ❑
42. Disconnect the manifold gage hose from the sealed system. ❑
43. Disconnect the manifold gage center hose from the recovery unit's INLET valve. ❑
44. Disconnect the hose connection between the recovery unit and the recovery cylinder. ❑
45. Have your instructor check your work. ❑
46. Return all tools and equipment to their proper location. ❑

RESEARCH QUESTIONS

1. When using the active method of recovery, the EPA requires what percentage of refrigerant recovery for small appliances?

_____

_____

2. Technicians certified in which to areas of certification can work on small appliances?

_____

_____

3. According to EPA, what percentage of refrigerant must be removed from a sealed system when using the passive method of refrigerant?

_____

_____

4. What method of recovery would be used on a sealed system with an inoperative compressor?

_____

_____

5. What year did technician certification become mandated?

_____

_____

Passed Competency _____   **Failed Competency** _____

Instructor Signature _____   **Grade** _____

Practical Competency 56

System-Dependent Recovery from a Small Appliance (Figure 3–91)

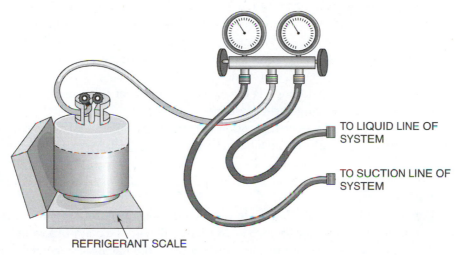

TO LIQUID LINE OF SYSTEM

TO SUCTION LINE OF SYSTEM

REFRIGERANT SCALE

FIGURE 3–91   System-dependent recovery connection.

SUGGESTED MATERIALS

Textbook
Refrigeration & Air Conditioning Technology, 5th Edition, Thomson Delmar Learning
Unit 9—Refrigerant and Oil Chemistry and Management—Recovery, Recycling, Reclaiming, and Retrofitting

Review Topics
Recovery, Recycle, or Reclaim

COMPETENCY OBJECTIVE

The student will be able to use the system-dependent method of recovery and recover refrigerant from a small appliance with an operating compressor.

OVERVIEW

As of July 1, 1992, the EPA has mandated that it is illegal to release refrigerants into the atmosphere during the maintenance, service, and disposal of refrigeration and air conditioning equipment. According to the Clean Air Act of 1990, technicians may not knowingly vent or otherwise release or dispose of any substance used as a refrigerant in such appliances in a manner that permits refrigerants to enter the environment. Technicians are required to contain all refrigerants except those used to purge lines and tools of the trade, such as the gage manifold or any device used to capture and save refrigerants.

The laws governing the release of CFCs and HCFCs into the atmosphere are very strict and can result in heavy penalties if they are violated. This has led to the development of specific procedures that must be followed to **recover, recycle,** and **reclaim** refrigerants. During the recovery of refrigerant only DOT-approved cylinders are to be used.

These cylinders are painted yellow on the shoulder area and 12 inches down the side. The body of the cylinders is painted gray.

EPA mandates that recovery cylinders not be filled to more then 80% of their liquid weight.

Recovery is the process of removing refrigerant from a sealed system and storing it into an external cylinder without necessarily testing or reprocessing it in any way. The recovery of refrigerant from a sealed system can be accomplished by using several methods, either **passive recovery** or **system-dependent,** and **active recovery.** With these procedures the refrigerant within a sealed system can be removed in the **vapor state,** referred to as **vapor recovery** or **liquid state,** referred to as **liquid recovery.**

The **passive method or system-dependent** recovery process uses the internal pressure of the system or system's compressor to assist in the removal of refrigerant from the sealed system. If the refrigerant system has an operational compressor, this is often the best method for recovery and recommended when the total system refrigerant charge is less than 15 lb. With the passive method of recovery, EPA mandates that 90% of the refrigerant be removed from the system. Systems with an inoperative compressor, EPA mandates 80% recovery of refrigerant from the sealed system.

EPA has set standards for evacuation levels for recovery equipment to ensure that the proper amount of refrigerant is removed from the refrigeration system.

Small appliances that are identified as refrigeration systems are manufactured, charged, and hermetically sealed with 5 lb or less of refrigerant. This includes refrigerators, freezers, room air conditioners, packaged heat pumps, dehumidifiers, vending machines, drinking water coolers, and under-the-counter ice makers.

The setup for passive recovery is demonstrated in *Figure 3–83, Theory Lesson: Refrigerant Recovery, Recycling, and Reclaim.*

Recovering liquid would require the **high-side valve** on the gage manifold and the **vapor valve** on the recovery cylinder to be in the **opened position.** The **low-side valve** on the manifold would be in the **closed** position. When refrigerant stops flowing into the recovery tank, the manifold gage **high-side valve** would be **closed** and the manifold gage **low-side valve** would be **opened.** This allows refrigerant from recovery cylinder to flow back into the sealed system, causing the recovery tank pressure to drop.

NOTE: *Packing the recovery cylinder in cold water or ice will also assist in dropping the recovery cylinder's pressure and speed up the recovery process.*

After a period of time the manifold gage **low-side valve** should be **closed** and the manifold gage **high-side valve opened** again to allow more liquid to be forced into the recovery cylinder.

There are many different types of recovery units on the market today and they operate in the same way. Always follow the guidelines and operation instructions for the recovery unit being used. Most recovery units today have **common features** that include inlet and outlet ports, inlet and outlet valves, gages, and a means to purge the unit of refrigerant once recovery from the sealed system is complete. The feature of "purging" aids in preventing mixing or cross contaminating of refrigerates.

In most cases the recovery process begins with recovering as much liquid from the system as possible and completing the recovery process by switching over to the vapor recovery. The reason for this procedure is because liquid recovery is much quicker than vapor recovery.

During the **liquid recovery process,** the manifold gage **high-side valve** is **opened** because it is attached to the liquid line of the refrigeration sealed system. Once liquid recovery is complete, the manifold gage high side valve is **closed** and the manifold gage **low-side valve** is **opened** to recover refrigerant in the vapor state from the sealed system.

Universal guidelines to be followed when using recovery units include:

- *Make sure the recovery unit is purged of any residue refrigerant left in the unit from previous use.*
- *Make sure the recovery bottle contains the same type of refrigerant that is to be recovered from a sealed system.*

- Make sure that all hoses are purged to remove air from them prior to attachment to the recovery unit and refrigeration system.
- Warmer temperatures will increase the speed of recovery.
- Use filter drier and sight glass during recovery to protect recovery unit and assist in monitoring refrigerant condition and state.
- Follow the recovery unit's maintenance guidelines as established by the manufacturers.
- Except for oil-less recovery units, change oil and filters often to reduce the chance of refrigerant contamination.
- Make sure that only qualified individuals use the recovery unit.
- Wear safety glasses and gloves during the recovery process.
- Recover refrigerant in a well ventilated area.
- Empty recovery bottles should be evacuated and cleaned before being used in the recovery process.
- Only DOT-approved recovery cylinders should be used for refrigerant storage.

> **NOTE:** Extreme care should be taken to avoid the cross contamination of refrigerants in the process of recovery and recycling. Technicians should make sure that manifold hoses with quick connect fittings are bled before attaching to another sealed system with a different refrigerant.

EQUIPMENT REQUIRED

1 Operational small appliance
2 Access valves (*if required*)
1 DOT recovery cylinder
1 Refrigerant scales
1 Manifold gage set
1 Refrigerant wrench (*if required*)
1 Adjustable wrench (*if required*)
2 Additional refrigerant hose (*if required*)
1 Heat gun—heat lamps—heat blankets (*if required*)
1 Filter drier 1/4" × 1/4" flare (*if required*)
1 Refrigerant sight glass—1/4" × 1/4" flare (*if required*)
2 Additional refrigerant hoses (*if required*)

SAFETY PRACTICES

Wear goggles or safety glasses and gloves when working with refrigerants in sealed systems. Beware of high pressures and follow all electrical safety rules. Follow all EPA rules and regulations while working with refrigerants.

COMPETENCY PROCEDURES

Checklist

1. Turn small appliance OFF (*if applicable*). ❏
2. *If applicable*, remove valve stem service valve caps and service port caps. ❏
3. Make sure that the recovery cylinder is evacuated or has the same refrigerant in it as is in the system being recovered. ❏
4. Set the recovery cylinder on a refrigerant scale. ❏
5. If required, attach line valves to the suction and liquid lines of the sealed system. ❏
6. Determine the recovery cylinder's 80% fill rate. ❏
7. *Record the cylinder's 80% fill rate.* _____ ❏
8. *Record the recovery unit's DOT test date.* _____ ❏
9. *Record the type of refrigerant being recovered.* _____ ❏
10. *List the type of refrigeration equipment refrigerant is being recovered from.* _____ ❏
11. *Review the EPA "Required Levels of Evacuation for Air Conditioning, and Refrigeration" or "Recovery Efficiency Requirements for Small Appliances" charts and record the required recovery level for the equipment refrigerant is being recovered from.* _____ ❏

12. **Close** the manifold **gage valves.** ❑
13. Install the manifold high-side hose to the sealed system's high-side valve. ❑
14. Install the manifold low-side hose to the sealed system's low-side valve. ❑
15. Install the manifold center hose to the filter drier and sight glass. (*Use additional refrigerant hose if necessary.*) ❑
16. Connect another hose from the filter drier and sight glass to the recovery cylinder's vapor valve. (*Use additional refrigerant hose if necessary.*) ❑
17. *If applicable,* mid-seat valve stem service valves. ❑
18. **Open** the manifold **high-side valve.** ❑
19. **Open** the manifold **low-side valve.** ❑
20. Loosen the hose connection on the recovery cylinder for 1 second and then retighten the hose. ❑

> *NOTE: This will purge any air from the gage hoses.*

21. **Close** the **high-side** manifold valve. ❑
22. **Close** the **low-side** manifold valve. ❑
23. START the refrigeration sealed system. ❑
24. *If applicable,* turn on the refrigerant scale. ❑
25. Zero the display on the refrigerant scale. ❑
26. **Open** the manifold **high-side valve.** ❑
27. **Open** the **vapor valve** on the recovery cylinder. ❑

> *NOTE: Refrigerant will flow from the sealed system, through the manifold gage and into the recovery cylinder.*

28. Monitor the system and the display on the refrigerant scale. ❑
29. When the reading on the refrigerant scale's display indicates that no more refrigerant is being introduced into the recovery cylinder, **close the manifold high-side valve.** ❑
30. **Open** the **low-side** manifold valve. ❑

> *NOTE: This will allow vapor refrigerant to flow back into the refrigeration sealed system.*

31. Allow the pressure in the recovery cylinder to stabilize. ❑
32. **Close** the manifold **low-side valve.** ❑
33. **Open** the manifold **high-side valve** to allow liquid recovery to resume. ❑
34. Repeat Steps 30 through 33 until the proper recovery level is reached. ❑

> *NOTE: If the entire refrigerant cannot be recovered using this method, or if the high-side pressure starts to rise, the cylinder if full. Close valves on the recovery cylinder and manifold gages. Replace the recovery cylinder with another evacuated cylinder, and continue the recovery process until the required recovery level is reached.*

> *NOTE: During the process, if the high-side pressure starts to rise, the cylinder is full. An addition external cylinder needs to be used to capture more of the refrigerant.*

35. When recovery level is reached, **close** the manifold **high-side valve.** ❑
36. **Close** the manifold **low-side valve.** ❑
37. **Close the** recovery cylinder **vapor valve.** ❑
38. Turn the refrigeration system OFF. ❑
39. Slowly remove manifold center hose from recovery cylinder. ❑
40. Slowly remove manifold high-side and low-side hoses from the sealed system. ❑
41. Tag the refrigerant cylinder and label with the type of refrigerant, amount of refrigerant, owner of refrigerant, and recovery date. ❑

NOTE: If manifold gage hoses are low-loss fittings, be sure to purge refrigerant from hoses.

---

42. Have your instructor check your work. ❑
43. Disconnect equipment and return all tools and equipment to their proper location. ❑

RESEARCH QUESTIONS

1. According to the EPA, recovering refrigerant with the assistance of the unit's compressor is called what method of recovery?

   _____

   _____

2. In the answer to question 1, the EPA requires what percentage of refrigerant recovery from the system?

   _____

   _____

3. According to EPA, why should external recovery cylinders not be filled more than 80% of their liquid weight?

   _____

   _____

4. According to EPA, how often should refrigerant recovery cylinders be tested?

   _____

   _____

5. Who is the approving agency for recovery cylinders?

   _____

   _____

Passed Competency _____ Failed Competency _____

Instructor Signature _____ Grade _____

Theory Lesson: Leak Testing Refrigerant Sealed Systems (Figure 3–92)

FIGURE 3–92   Using a electronic leak detector to locate refrigerant leaks. (*Photo by Eugene Silberstein*)

SUGGESTED MATERIALS

Textbook
Refrigeration & Air Conditioning Technology, 5th Edition, Thomson Delmar Learning
Unit 8—System Evacuation

Review Topics
Standing Pressure Test; Leak Detection Methods; Leak Detection Tips

Key Terms
dry nitrogen • R-22 • standing pressure test • trace gas

OVERVIEW

The *Standing Pressure Test* is the method most widely accepted in the field for leak-checking a sealed system. This is the process of pressurizing a sealed system with dry nitrogen (**Figure 3–93**) to a set pressure and marking the manifold gage where the needle indicates pressure exerted on the system (**Figure 3–94**).

In addition, letting the system stand pressurized for 5 to 10 minutes to see if there is any movement of the gage needle. Normally the low side can be pressurized to 150 psig and the high side of the sealed system can be pressurized to 250 psig if it is isolated from the total system. Total system pressurization is normally done at 150 psig. In either case, when pressure is introduced into a sealed system, the manufacturer's test pressures should never be exceeded. Once a system is pressurized, any movement of the needle below the gage marking is an indication of a sealed system leak. No movement of the needle is an indication of a tight sealed system with no leaks. EPA has approved small amounts of R-22 as the trace gas (**Figure 3–95**), which can be added to CFC and HCFC systems for leak detections. EPA also mandates that HFC refrigerate systems be leak tested by dry nitrogen only. Traces of R-22 and nitrogen used for leak testing CFC and HCFC sealed systems may be vented to the atmosphere under Section 608 of the EPA guidelines.

FIGURE 3–93   Dry nitrogen with pressure regulator.

FIGURE 3–94   Marking a gage for pressure testing.

FIGURE 3–95   R-22 cylinder.

Before performing a standing pressure test on a sealed system, always do a visual inspection of the sealed system. Look for fresh oil or dust spots where dust has collected on the oil spots. This is a good indication of a leak or the area where there is a leak. A visual inspection would include all refrigerant lines, soldered joints, flared fittings, evaporator coil, condenser coil, compressor, and any other special sealed system components. Technicians should also listen for leaking refrigerant from the sealed system. It is important to note that a large number of leaks on split system air conditioning and heat pumps are found at the system gage connections ports. These port connections should be checked for leaking refrigerant before manifold gages are attached to the system. If a leak is not visually noticeable or leaks cannot be heard, the standing pressure test will have to be performed on the sealed system.

Before performing a leak test, refrigerant that may still be within the system must be recovered. It is unlawful to use dry nitrogen and the existing system refrigerant for leak testing.

Once a sealed system is pressurized with dry nitrogen and R-22 trace gas, halide torch leak detectors (**Figure 3–96**), ultrasonic leak detetors (**Figure 3–97**), ultraviolet leak detectors (**Figure 3–98**), electronic leak detectors (**Figure 3–99**), and leak detection solutions (**Figure 3–100**) can be used to locate the leak.

EPA states the ultrasonic and electronic leak detectors are good for locating general area leaks, while leak detections solutions are best for pinpointing system leaks. Once the leak is located, mixtures of dry nitrogen and R-22, or dry nitrogen only may be vented to the atmosphere. Because of the sensitivity of

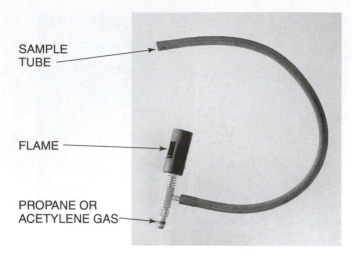

SAMPLE TUBE

FLAME

PROPANE OR ACETYLENE GAS

FIGURE 3–96   Halide torch for leak detection. (*Photo by Bill Johnson*)

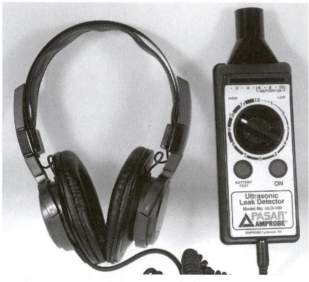

FIGURE 3–97   An ultrasonic leak detector that listens for leaks. (*Photo by Bill Johnson*)

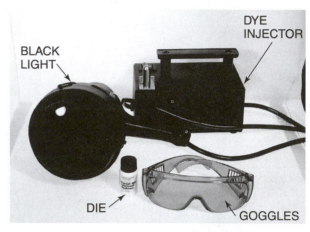

BLACK LIGHT

DYE INJECTOR

DIE

GOGGLES

FIGURE 3–98   An ultraviolet leak detector that uses black light and a special die. (*Photo by Bill Johnson*)

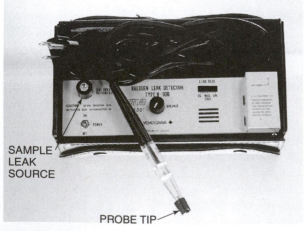

SAMPLE LEAK SOURCE

PROBE TIP

FIGURE 3–99   An electronic leak detector. (*Photo by Bill Johnson*)

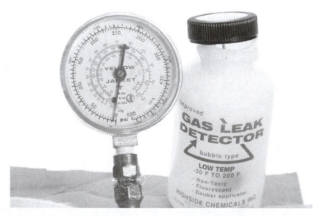

FIGURE 3–100   Gas leak detector solution. (*Photo by Bill Johnson*)

some leak detectors, attempting to locate a leak with these instruments should not be rushed. Leak detection with these devices is a slow process. Probes of leak detectors instruments should be moved slowly, normally about 1 inch every 2 seconds. Very small leaks would require the instrument probe to be directly over the leak for instrument indication of a leak.

NOTE: Never use pressurized air to leak-check a sealed system because some refrigerants can react with pressurized air and cause an explosion. Also never start the compressor on a sealed system that is pressurized with dry nitrogen.

Practical Competency 57

Using Nitrogen and R-22 for Leak-Checking a CFC or HCFC Sealed System (Figure 3–101)

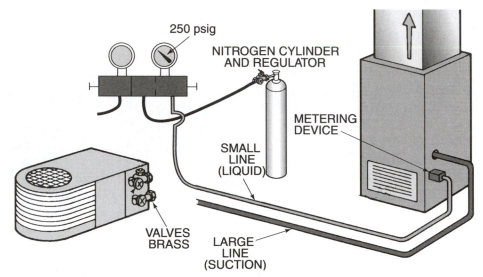

250 psig

NITROGEN CYLINDER
AND REGULATOR

METERING
DEVICE

SMALL
LINE
(LIQUID)

VALVES
BRASS

LARGE
LINE
(SUCTION)

FIGURE 3–101 Using nitrogen and R-22 to pressurize isolated sealed system components for leak testing.

SUGGESTED MATERIALS

Textbook
Refrigeration & Air Conditioning Technology, 5th Edition, Thomson Delmar Learning
Unit 8—System Evacuation

Review Topics
Standing Pressure Test; Leak Detection Methods; Leak Detection Tips

COMPETENCY OBJECTIVE

The student will be able to use nitrogen and a trace amount of R-22 to leak-check a CFC or HCFC sealed system.

OVERVIEW

The *Standing Pressure Test* is the method most widely accepted in the field for leak-checking a sealed system. This is the process of pressurizing a sealed system with dry nitrogen to a set pressure, marking the manifold gage where the needle indicates pressure exerted on the system, and letting the system stand pressurized for 5 to 10 minutes to see if there is any movement of the gage needle. Normally the low side can be pressurized to 150 psig and the high side can be pressurized to 250 psig. In either case, when pressure is introduced into a sealed system, the manufacturer's test pressures should never be exceeded. Once a system is pressurized, any movement of the needle below the gage marking is an indication of a sealed system leak. No movement of the needle is an indication of a tight sealed system with no leaks. EPA has approved small amounts of R-22 as the trace gas, which can be added to CFC and HCFC systems for leak detections. EPA also mandates that HFC refrigerate systems be leak-tested by dry nitrogen only. Traces of R-22 and nitrogen used for leak-testing CFC and HCFC sealed systems may be vented to the atmosphere under Section 608 of the EPA guidelines. Before performing a standing pressure test on a sealed system, always do a visual inspection of the sealed system. Look for fresh oil or dust spots where dust collects on the oil. A visual inspection would include all refrigerant lines, soldered joints, flared fittings, evaporator coil, condenser coil, compressor, and any other special sealed system components. Technicians should also listen for leaking refrigerant from the sealed system. It's important to note that a lot of leaks on split system air conditioning and heat pumps are found at the system gage connections ports. Theses port connections should be checked for leaking refrigerant before manifold gages are attached to the system. If leak is not visually noticeable or leaks cannot be heard, using the standing pressure test on the sealed system will have to be performed. Before performing a leak test, refrigerant, which may still be within the system, must be recovered. It is unlawful to use dry nitrogen and the existing system refrigerant for leak testing. Once a standing pressure test is performed on a sealed system, halide torch leak detectors, ultrasonic leak detectors, ultraviolet leak detectors, electronic leak detectors, and leak detection solutions can be used to locate the leak.

EPA states the ultrasonic leak detectors are good for locating general area leaks, while leak detections solutions are best for pinpointing system leaks. Once the leak is located, mixtures of dry nitrogen and R-22, or dry nitrogen only may be vented to the atmosphere.

> NOTE: Never use pressurized air for leak-checking a sealed system, because some refrigerants can react with pressurized air and cause an explosion. Also never start the compressor on a sealed system that is pressurized with nitrogen.

EQUIPMENT REQUIRED

Sealed system to be leak-checked
R-22 cylinder
Nitrogen tank with regulator
Manifold gage set
Refrigeration wrench (*if applicable*)
Adjustable wrench (*if applicable*)
Liquid leak detector solution
Leak detector instrument

SAFETY PRACTICES

Wear goggles or safety glasses and gloves when working with refrigerant in sealed systems. Beware of high pressures and follow all electrical safety rules. Follow all EPA rules and regulations while working with refrigerants.

COMPETENCY PROCEDURES   Checklist

1. Make sure that refrigerant is recovered from the sealed system being leak tested.   ❏
2. Make sure the sealed system is turned OFF.   ❏
3. Remove sealed system service valve caps.   ❏
4. *If applicable*, remove sealed system valve stem caps.   ❏
5. Make sure that manifold gage vales are closed.   ❏
6. Attach the manifold low-side gage hose to the sealed system low-side gage connection port.   ❏

7. Attach the manifold high-side gage hose to the sealed system high-side gage connection port. ❑
8. Attach the manifold gage center hose the R-22 refrigerant cylinder. ❑
9. Open the R-22 refrigerant cylinder valve. ❑
10. Slowly loosen the center hose connection at the manifold valve body until a small amount of refrigerant leaks out, immediately retighten the hose connection. ❑
11. *If applicable*, mid-seat the sealed system low and high-side valve stem valves one-half turn. ❑
12. Slowly open the manifold gage low-side valve and allow R-22 to enter the system until about 10 psig is indicated on the manifold gages. ❑
13. Once about 10 psig is indicated on the manifold gages, close the low-side manifold gage. ❑
14. Turn the R-22 cylinder valve OFF. ❑
15. Slowly loosen the center hose from the R-22 cylinder and allow the refrigerant in the hose to bleed out. ❑
16. Make sure that the nitrogen cylinder regulator valve stem is back seated so that there is no pressure being felt against the valve stem. ❑
17. Attach the manifold gage center hose to the nitrogen tank regulator connection port. ❑
18. Open the nitrogen tank valve. ❑
19. Check the nitrogen tank pressure gage. (Pressure should be indicated on only one of the regulator gages.) This is the nitrogen cylinder pressure. ❑
20. Open the manifold's low and high-side gage valves. ❑
21. Slowly turn the nitrogen regulator valve stem to force dry nitrogen into the sealed system. ❑
22. Add dry nitrogen to the system until 150 psig is indicated on the high-side manifold gage. ❑
23. Close the nitrogen tank valve. ❑
24. Close the manifold gage valves. ❑
25. Let system stand pressurized for 5 to 10 minutes. ❑
26. Use a leak detector or liquid solution to check for sealed system leaks. ❑
27. If pressure holds or a leak area is identified, slowly back seat the regulator valve stem until there is no opposing pressure being felt on the regulator stem. ❑
28. Slowly loosen the manifold center hose from the nitrogen regulator connection port. ❑
29. Slowly open the low and high-side gage manifold valves and allow mixture of R-22 and nitrogen to vent to the atmosphere. ❑
30. Once nitrogen is completely emptied from the sealed system, repair leak if required. ❑
31. If additional work is not required on the sealed system, remove manifold gage hoses from the sealed system. ❑
32. *If applicable*, back seat the sealed system low and high-side valves. ❑
33. *If applicable*, replace valve stem caps. ❑
34. Replace sealed system service port connection caps. ❑
35. Have instructor check your work. ❑
36. Return nitrogen tank, R-22 cylinder and tools and equipment to their proper location. ❑

RESEARCH QUESTIONS

1. Why is pulling a vacuum NOT the best method for determining if a system is leaking or not?

_____

_____

2. Why should the low-side of a sealed system not be pressurized more than 150 psig?

_____

_____

3. Why should small leaks within a system be found and repaired?

_____

_____

4. Because of high efficiency air conditioning and heat pump systems, how much refrigerant leaked from the system will affect the system's efficiency?

_____

_____

5. Why should oxygen or compressed air never be used to pressurize a system?

_____

_____

Passed Competency _____   Failed Competency _____

Instructor Signature _____   Grade _____

Practical Competency 58

Using Nitrogen to Check for Leaks in an HFC Sealed System

SUGGESTED MATERIALS

Textbook

Refrigeration & Air Conditioning Technology, 5th Edition, Thomson Delmar Learning
Unit 8—System Evacuation

Review Topics

Standing Pressure Test; Leak Detection Methods; Leak Detection Tips

COMPETENCY OBJECTIVE

The student will be able to use nitrogen to leak-check an HFC sealed system.

OVERVIEW

The *Standing Pressure Test* is the method most widely accepted in the field for leak-checking a sealed system. This is the process of pressurizing a sealed system with dry nitrogen to a set pressure, marking the manifold gage where the needle indicates pressure exerted on the system, and letting the system stand pressurized for 5 to 10 minutes to see if there is any movement of the gage needle. Normally the low-side can be pressurized to 150 psig and the high-side can be pressurized to 250 psig. In either case, when pressure is introduced into a sealed system, the manufacturer's test pressures should never be exceeded. Once a system is pressurized, any movement of the needle below the gage marking is an indication of a sealed system leak. No movement of the needle is an indication of a tight sealed system with no leaks. **EPA mandates that HFC refrigerate systems be leak-tested by dry nitrogen only.** Before performing a standing pressure test on a sealed system, always do a visual inspection of the sealed system. Look for fresh oil or dust spots where dust collects on the oil. A visual inspection would include all refrigerant lines, soldered joints, flared fittings, evaporator coil, condenser coil, compressor, and any other special sealed system components. Technicians should also listen for leaking refrigerant from the sealed system. It is important to note that a lot of leaks on split system air conditioning and heat pumps are found at the system gage connections ports. Theses port connections should be checked for leaking refrigerant before manifold gages are attached to the system. If a leak is not visually noticeable or leaks cannot be heard, using the standing pressure test on the sealed system will have to be performed.

Before performing a leak test, refrigerant that may still be within the system must be recovered. It is unlawful to use dry nitrogen and the existing system refrigerant for leak-testing. Once a standing pressure test is performed on a sealed system, halide torch leak detectors, ultrasonic leak detectors, ultraviolet leak detectors, electronic leak detectors, and leak detection solutions can be used to locate the leak.

EPA states the ultrasonic leak detectors are good for locating general area leaks, while leak detections solutions are best for pinpointing system leaks. Once the leak is located, mixtures of dry nitrogen and R-22 or dry nitrogen only may be vented to the atmosphere.

> **NOTE:** *Never use pressurized air for leak-checking a sealed system, because some refrigerants can react with pressurized air and cause an explosion. Also never start the compressor on a sealed system that is pressurized with nitrogen.*

EQUIPMENT REQUIRED

Sealed system to be leak-checked
Nitrogen tank with regulator
Manifold gage set
Refrigeration wrench (*if applicable*)
Adjustable wrench (*if applicable*)
Liquid leak detector solution
Leak detector instrument

SAFETY PRACTICES

Wear goggles or safety glasses and gloves when working with refrigerant in sealed systems. Beware of high pressures and follow all electrical safety rules. Follow all EPA rules and regulations while working with refrigerants.

COMPETENCY PROCEDURES   Checklist

1. Make sure that refrigerant is recovered from the sealed system being leak tested. ❏
2. Make sure the sealed system is turned OFF. ❏
3. Remove sealed system service valve caps. ❏
4. *If applicable*, remove sealed system valve stem caps. ❏
5. Make sure that manifold gage vales are closed. ❏
6. Attach the manifold low-side gage hose to the sealed system low-side gage connection port. ❏
7. Attach the manifold high-side gage hose to the sealed system high-side gage connection port. ❏
8. *If applicable*, mid-seat the sealed system low- and high-side valve stem valves one-half turn. ❏
9. Make sure that the nitrogen cylinder regulator valve stem is back seated so that there is no pressure being felt against the valve stem. ❏
10. Attach the manifold gage center hose to the nitrogen tank regulator connection port. ❏
11. Open the nitrogen tank valve. ❏
12. Check the nitrogen tank pressure gage. (*Pressure should be indicated on only one of the regulator gages.*) This is the nitrogen cylinder pressure. ❏
13. Open the manifold's low and high-side gage valves. ❏
14. Slowly turn the nitrogen regulator valve stem to force dry nitrogen into the sealed system. ❏
15. Add dry nitrogen to the system until 150 psig is indicated on the high-side manifold gage. ❏
16. Close the nitrogen tank valve. ❏
17. Close the manifold gage valves. ❏
18. Let the system stand pressurized for 5 to 10 minutes. ❏
19. Use a leak detector or liquid solution to check for sealed system leaks. ❏
20. If pressure holds or a leak area is identified, slowly back-seat the regulator valve stem until there is no opposing pressure being felt on the regulator stem. ❏
21. Slowly loosen the manifold center hose from the nitrogen regulator connection port. ❏
22. Slowly open the low- and high-side gage manifold valves and allow nitrogen to vent to the atmosphere. ❏
23. Once nitrogen is completely emptied from the sealed system, repair the leak if required. ❏
24. If additional work is not required on the sealed system, remove manifold gage hoses from the sealed system. ❏
25. *If applicable*, back seat the sealed system low- and high-side valves. ❏
26. *If applicable*, replace valve stem caps. ❏
27. Replace sealed system service port connection caps. ❏
28. Have your instructor check your work. ❏
29. Return nitrogen tank, tools, and equipment to their proper location. ❏

RESEARCH QUESTIONS

1. Why is nitrogen ONLY to be used for leak-testing HFC sealed systems?

   _____

   _____

2. HFC refrigerants contain which three elements?

   _____

   _____

3. List a noncondensable gas.

   _____

   _____

4. How many refrigerants are used to make a binary refrigerant?

   _____

   _____

5. What is meant by a refrigerant's temperature glide?

   _____

   _____

Passed Competency _____ Failed Competency _____

Instructor Signature _____ Grade _____

Theory Lesson: Evacuation (Figure 3–102)

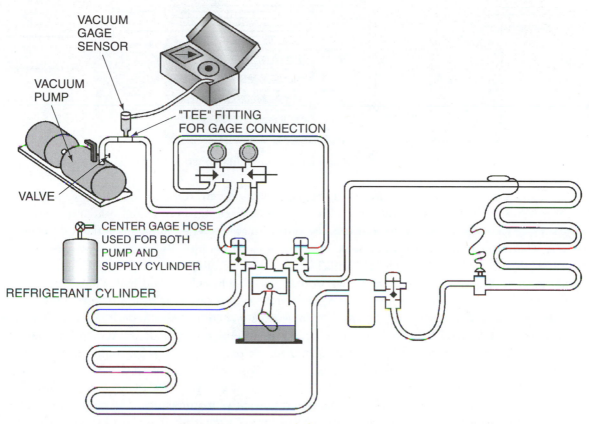

FIGURE 3–102   Sealed system evacuation.

SUGGESTED MATERIALS

Textbook
Refrigeration & Air Conditioning Technology, 5th Edition, Thomson Delmar Learning
Unit 8—System Evacuation

Review Topics
Purpose of System Evacuation; Theory Involved with Evacuation

Key Terms
air • atmospheric pressure • compound gage • fozene gas • hydrochloric acids • hydrofluoric acids • hydrogen • in. Hg • moisture • nitrogen • noncondensable gas • oxygen • psia • psig • sludge • vacuum

OVERVIEW

Removing air and other noncondensables, along with water vapor, from a refrigeration sealed system is known as **evacuation**. Air and moisture inside of a refrigeration system can create many problems that are detrimental to the efficiency of the equipment and its components. Air contains **oxygen, nitrogen, hydrogen,** and water in the form of water vapor. **Nitrogen** is a noncondensable gas, which will occupy space at the top of the condenser and cause higher than normal operating high-side pressures, higher discharge temperatures, and higher compressor/ compression ratios, which will affect the efficiency of the refrigeration process (**Figure 3–103**).

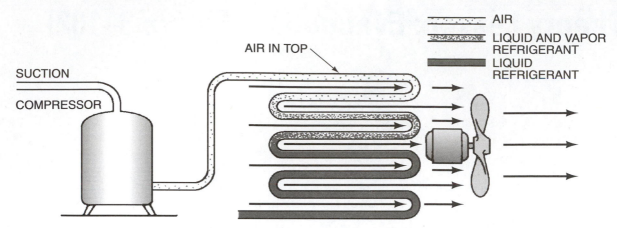

FIGURE 3–103 A condenser containing noncondensable gases.

Other gases and water vapor cause a chemical reaction with refrigerant and refrigerant oil within the refrigeration sealed system, producing acids that cause deterioration of sealed system components, copper plating of running gears, and breakdown of motor insulation. When moisture in the form of water vapor, heat, and refrigerant is present in a sealed system, acids will eventually form as a result of the presence of chemicals such as **hydrogen, fluorine,** and **chlorine,** which are used to manufacture **CFC, HCFC,** and **HFC** refrigerants. When heated and mixed with other gases, **hydrochloric** and **hydrofluoric acids** are formed. These acids can deteriorate compressor motor windings and cause corrosion of metal alloys used in the makeup of sealed refrigeration systems. The longer the system operates under these conditions, the more likely sludge is to from. **Sludge** is a tightly bound mixture of water, acid, and oil. Under these conditions the compressor will eventually burn out. If moisture is present in the refrigerant sealed system during a compressor burnout, fozene gas will also be formed. **Fozene gas** is very harmful to humans when burned and can be deadly if released in an area where poor ventilation exists. Avoiding corrosion and sludge problems is dependent on good service procedures and preventive maintenance. The only way to rid the sealed system of moisture is to use good evacuation procedures. Evacuation does not remove sludge; this requires cleanup procedures in which oversized driers specified for sludge removal must be used.

Acid test kits are available so that sealed systems can be checked for acid content.

Sight glass can be used to indicate moisture content in a system (**Figure 3–104**).

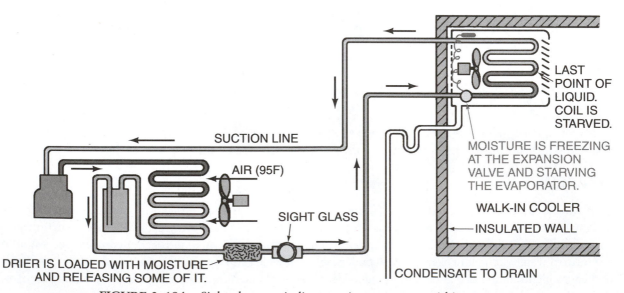

FIGURE 3–104 Sight glass can indicate moisture content within a system.

These compounds should not be allowed to remain in the sealed system.

A vacuum must be applied to a sealed system to remove these contaminants. Lowering the pressure in a sealed system to below atmospheric pressure creates a vacuum. Atmospheric pressure is 14.696 **psia** (pounds per square inch absolute) or 29.92 in. Hg (inches of mercury). On a manifold gage, the pressure is scaled in **psig**. This is pounds per square inch gage.

To convert the gage pressure to psia, add 14.696 to the actual gage pressure. To measure a vacuum with a gage you need a gage that can read below atmospheric pressure. The low-side gage of the manifold gage set is referred to as a **compound gage** and can read above and below atmospheric pressure.

When using the manifold gage to measure the level of vacuum, the system has to be evacuated until the compound gage indicates a 29.92 in. Hg vacuum. The vacuum must be pulled to this level to remove atmospheric pressure.

For years, sealed systems have been evacuated and leak-checked by pulling 29 in. Hg vacuum for a certain amount of time. It was assumed that this removed all the contaminants and noncondensable gases from the sealed system. The 29.92 in. Hg vacuum was also used to determine if a system had a leak. If a system could not be pulled to 29.92 in. Hg vacuum, it was a good indicator of a leak. Technicians have learned over the last few years that the above method is not the best for removing all noncondensables and moisture, or for leak-checking a sealed system. It has been found that even when a vacuum is measured by time and pulling to 29 in. Hg, there could still be a leak in the system and there could still be contaminants in the system. Other methods of measuring a system's vacuum are now preferred and recommended by EPA.

Micron vacuum gages, thermistor vacuum gages, and U-tube manometers are better tools for determining the vacuum levels of a sealed system (**Figure 3–105**).

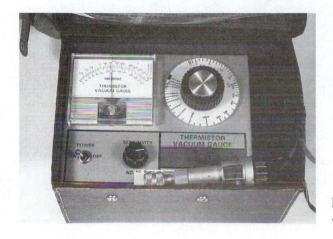

FIGURE 3–105   An analog micron gage for deep vacuum measurements. (*Photo by Bill Johnson*)

With the new EPA regulations on refrigerants, the EPA has recommended that systems be evacuated to at least a 500-micron vacuum or equivalent. At this level the chances of noncondensables and a leak in the system are unlikely.

Practical Competency 59

Draining and Replacing Vacuum Pump Oil (Figure 3–106)

FIGURE 3–106   A two-stage rotary vacuum pump and vacuum pump oil.

SUGGESTED MATERIALS

Textbook
Refrigeration & Air Conditioning Technology, 5th Edition, Thomson Delmar Learning
Unit 8—System Evacuation

Review Topics
The Vacuum Pump

COMPETENCY OBJECTIVE

The student will be able to drain and change the oil in a vacuum pump.

OVERVIEW

Most vacuum pumps used in the refrigeration industry are rotary compressors, with two-stage rotary vacuum pumps being cable of pulling the lowest vacuum on refrigeration sealed systems. The EPA requires that refrigeration sealed system be evacuated down to a minimum of 500 microns. The vacuum pump pulls vapor from refrigeration sealed systems and discharges this vapor into the atmosphere. As the pressure within the sealed system drops, moisture in the system boils off into a vapor, which is pulled out by the vacuum pump. The efficiency of a vacuum pump depends on the correct vacuum pump oil being used and regular oil changes. Vacuum pumps are rated in CFMs (*cubic feet per minute*). Residential air conditioning and light commercial refrigeration equipment can be effectively evacuated with a two-stage 4 CFM pump.

When pumps are used to evacuate a system, the vacuum pump oil can become contaminated with sealed system impurities. This could be moisture, acids, or other impurities. If the oil becomes saturated with these impurities, the vacuum pump will not operate effectively and will not be able to pull a sealed system down to the required micron level. For this reason the vacuum pump oil should be changed regularly, and before each use when installing new refrigeration or air conditioning equipment. Make sure that the proper oil level is maintained in the pump during usage. For most vacuum pumps the oil level is maintained at the middle of the pump's sight glass (**Figure 3–107**).

FIGURE 3–107 Oil level at the middle of the vacuum pump sight glass.

Underfilling a pump can cause poor pump operation and overfilling the pump can result in vacuum pump oil being blown from the vacuum pump's exhaust port. Always follow the manufacturer's guidelines for proper pump operation, vacuum pump oil type, and maintenance procedures.

EQUIPMENT REQUIRED

Vacuum pump
Vacuum pump oil (*manufacturer's rated oil for vacuum pump*)
Oil container
Paper towel or rag (*if applicable*)
Funnel (*if applicable*)

SAFETY PRACTICES

Wear goggles or safety glasses and gloves.

COMPETENCY PROCEDURES

Checklist

1. Turn the vacuum pump ON and let it run for a couple of minutes to allow the vacuum pump oil to warm up.   ❏
2. Turn the pump OFF once it has warmed up the oil.   ❏

3. Place the vacuum pump so that the pumps drain plug is over the oil container.
4. Remove the vacuum pump oil drain plug (**Figure 3–108**).

❑
❑

FIGURE 3–108   Removal of oil drain plug.

NOTE: The drain plug is normally at the bottom of the pump.

---

5. Allow vacuum pump oil to drain in the oil container.
6. Once the oil is completely drained, replace the vacuum pump drain plug.
7. Check the manufacturer's instructions and locate the oil fill cap.
8. Remove the vacuum pump oil fill cap.
9. Check the manufacturer's instruction manual and add the recommended vacuum pump oil.
10. Fill the pump until the oil level reaches the middle of the pump's sight glass (**Figure 3–109**).
11. Replace the oil fill cap.
12. Cap the inlet port on the pump (**Figure 3–110**).

❑
❑
❑
❑
❑
❑
❑
❑

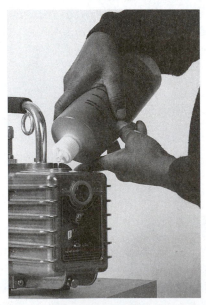

FIGURE 3–109   Adding oil to vacuum pump.

FIGURE 3–110   Cap the vacuum pump's inlet port.

13. Turn the pump ON and allow it to run for 1 minute. ❑
14. Check the oil level in the sight glass while the pump is running. ❑

> **NOTE:** If the oil level is **below the level mark** on the sight glass, remove the oil fill cap and slowly add more oil into the pump until the oil reaches the oil level mark in the sight glass.
>
> If the oil level is **over the oil level mark**, turn the vacuum pump OFF and remove the oil drain plug and drain oil until it reaches the oil level mark on the sight glass.

15. Replace the oil fill cap turn the vacuum pump OFF. ❑
16. Close the oil container. ❑
17. Properly dispose of the used oil. ❑
18. Wipe up any excess oil on vacuum pump or surrounding area. ❑
19. Have your instructor check your work. ❑
20. Return all materials and supplies to their proper location. ❑

RESEARCH QUESTIONS

1. Vacuum levels are accurately measured in what?

   _____

   _____

2. System evacuation is used to remove what from a sealed system?

   _____

   _____

3. How often should vacuum pump oil be changed?

   _____

   _____

4. What impurities can contaminate a sealed system?

   _____

   _____

5. Vacuum pump oil has a _____ boiling point.

Passed Competency _____   **Failed Competency** _____

Instructor Signature _____   **Grade** _____

Practical Competency 60

Evacuating an Air Conditioning System (Figure 3–111)

FIGURE 3–111   Sealed system evacuation.

SUGGESTED MATERIALS

Textbook
Refrigeration & Air Conditioning Technology, 5th Edition, Thomson Delmar Learning
Unit 8—System Evacuation

Review Topics
Purpose of System Evacuation; Theory Involved with Evacuation

COMPETENCY OBJECTIVE

The student will be able to remove moisture and noncondensables from an air conditioning system with a vacuum pump.

OVERVIEW

Removing air and other noncondensables, along with water vapor, from a refrigeration sealed system is known as **evacuation**. Air and moisture inside of a refrigeration system can create many problems that are detrimental to the efficiency of the equipment and its components. Air contains **oxygen, nitrogen, hydrogen,** and water in the form of water vapor. **Nitrogen** is a noncondensable gas, which will occupy space at the top of the condenser and cause higher than normal operating high-side pressures, higher discharge temperatures, and higher compressor compression ratios, which will affect the efficiency of the refrigeration process. (Refer to *Theory Lesson: Evacuation, Figure 3–103.*)

Other gases and water vapor cause a chemical reaction with refrigerant and refrigerant oil within the refrigeration sealed system, producing acids that cause deterioration of sealed system components, copper plating of running gears, and breakdown of motor insulation. When moisture in the form of water vapor, heat, and refrigerant is present in a sealed system, acids will eventually form because of the presence of chemicals such as **hydrogen, fluorine,** and **chlorine,** which are used to manufacture **CFC, HCFC,** and **HFC** refrigerants. When heated and mixed with other gases, **hydrochloric** and **hydrofluoric acids** are formed. These acids can deteriorate compressor motor windings, and cause corrosion of metal alloys used in the makeup of sealed refrigeration systems. The longer the system operates under these conditions the more likely sludge is to form. **Sludge** is a tightly bound mixture of water, acid, and oil. Under these conditions the compressor will eventually burn out. If moisture is present in the refrigerant sealed system during a compressor burnout, fozene gas will also be formed. **Fozene gas** is very harmful to humans when burned and can be deadly if released in an area where poor ventilation exists. Avoiding corrosion and sludge problems is dependent on good service procedures and preventive maintenance. The only way to rid the sealed system of moisture is to use good evacuation procedures. Evacuation does not remove sludge; this requires cleanup procedures in which oversized driers specified for sludge removal must be used.

Acid test kits are available so that sealed systems can be checked for acid content.

Sight glass can be used to indicate moisture content in a system. (Refer to *Theory Lesson: Evacuation, Figure 3–104*).

These compounds should not be allowed to remain in the sealed system. A vacuum must be applied to a sealed system to remove these contaminants.

With the new EPA regulations on refrigerants, the EPA has recommended that systems be evacuated to at least a 500-micron vacuum or equivalent. At this level the chances of noncondensables and a leak in the system are unlikely.

EQUIPMENT REQUIRED

Air conditioning system
1 Vacuum pump
1 Vacuum pump three-way valve or flare tee (*used for attachment of micron gage*)
1 Manifold gage
1 Additional refrigerant hose (*if required*)
1 Micron gage

SAFETY PRACTICES

Wear goggles or safety glasses and gloves when working with refrigerant in sealed systems. Beware of high pressures and follow all electrical safety rules. Follow all EPA rules and regulations while working with refrigerants and evacuating sealed systems.

COMPETENCY PROCEDURES   Checklist

> **NOTE:** *Refrigerant should be recovered from the air conditioning system before proceeding with system evacuation.*

1. Remove high- and low-side manifold hoses from the blank ports on the gage manifold. ❑
2. Calibrate manifold gages if needed. ❑
3. Make sure gage manifold service valves are closed. ❑
4. Make sure vacuum pump oil level is correct. ❑
5. Remove condensing unit service valve caps. ❑
6. Attach high-side gage manifold hose to condensing unit high-side valve (**Figure 3–112**). ❑
7. Attach low-side gage manifold hose to the condensing unit low-side service valve. ❑
8. Connect the center hose from the gage manifold to the inlet port of the vacuum pump. ❑
9. Set up micron gage so as to enable measurement of vacuum level of sealed system. ❑

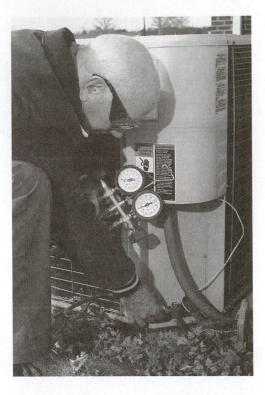

FIGURE 3–112 Attaching high-side manifold hose to condensing unit high-side valve.

NOTE: *Setup for micron gage measurement of sealed system vacuum will vary dependent on type of micron gage and vacuum pump service valve.*

---

10. Turn on the vacuum pump. ❑
11. Slowly open the valve on the vacuum pump. ❑
12. Open the high-side gage manifold valve. ❑
13. Open the low-side manifold gage. ❑
14. Turn on the micron gage. ❑
15. Calibrate the micron gage (if required). ❑

NOTE: *EPA mandates that a vacuum be pulled on a sealed system to a level of at least 500 microns or lower. The amount of time required to reach the required vacuum level will vary depending on the size of the sealed system and the amount of contaminates within the system.*

---

16. Let system evacuation continue until a vacuum level of 500 microns is reached. ❑
17. Close the vacuum pump valve. ❑
18. Turn the vacuum pump OFF. ❑
19. Monitor the micron gage level for a minimum of 5 minutes. ❑

NOTE: *Any rise in micron gage reading above 500 microns indicates that the system may still hold contaminants or possibly has a leak. This would require you to continue to evacuate the system.*

If a 500-micron vacuum cannot be obtained, check system for leaks, change vacuum pump oil, and evacuate the system again.

For larger sealed systems, a much longer time for the standing test may be required and recommended.

20. Once proper vacuum is reached, close the vacuum pump valve. ❑
21. Turn the vacuum pump OFF. ❑
22. Close the low-side manifold gage valve. ❑
23. Close the high-side manifold gage valve. ❑
24. Turn OFF the micron gage. ❑
25. Disconnect the micron gage. ❑
26. Disconnect the manifold center hose from the vacuum pump. ❑
27. Leave the manifold gage set attached to the air conditioning sealed system. ❑

> **NOTE:** *The system is ready to be charged when instructed to do so by your instructor.*

28. Have your instructor check your work. ❑
29. Return the vacuum pump and micron gage to their proper location. ❑

RESEARCH QUESTIONS

1. What problem can noncondensables cause in a sealed system?

_____

_____

2. What is meant by pulling a vacuum?

_____

_____

3. What is a compound gage?

_____

_____

4. What three things does a vacuum of 500 microns assure asssume the technician of?

_____

_____

5. What is the best way of checking a system for leaks?

_____

_____

Passed Competency _____ Failed Competency _____

Instructor Signature _____ Grade _____

Practical Competency 61

Evacuating a Sealed System That Has Valve Stem Service Valves (Figure 3–113)

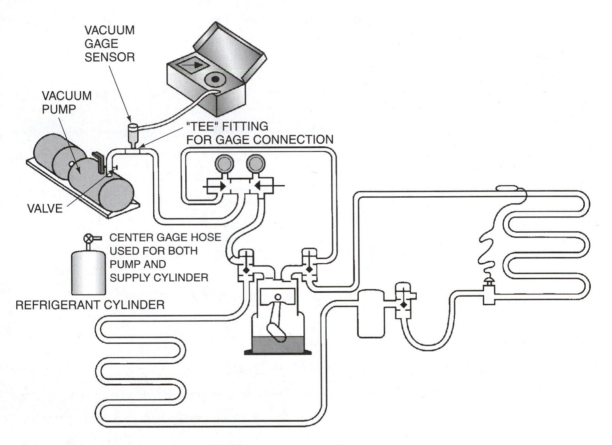

FIGURE 3–113   Sealed system evacuation.

SUGGESTED MATERIALS

Textbook
Refrigeration & Air Conditioning Technology, 5th Edition, Thomson Delmar Learning
Unit 8—System Evacuation

Review Topics
Purpose of System Evacuation; Theory Involved with Evacuation

COMPETENCY OBJECTIVE

The student will be able to use a vacuum pump and remove moisture and non-condensables from a sealed system with Valve Stem Service Valves.

OVERVIEW

NOTE: *Refer to Practical Competency 50—"Learning Valve Stem Service Valve Operation" to learn proper procedures for accessing sealed systems with Valve Stem Service Valves.*

Valve Stem Service Valves

Valve stem serve valves give the technician the ability to access the sealed system without having to add a saddle type service valve to the system. The suction valve is always located on the compressor of the sealed system and is normally larger than the discharge service valve. The discharge valve is ocated either lon the compressor or on a liquid receiver. This type of valve is normally found on refrigeration equipment. The valve consists of a **valve cap, valve stem, packing gland, inlet, outlet,** and **valve body.** These valves are used as a gage port to obtain system pressures, throttle refrigerant gas flow to the compressor, valve off the compressor from the evaporator or condenser for service, to charge the system with refrigerant. The valve stem has a square head and requires a valve wrench.

Evacuation

Removing air and other noncondensables, along with water vapor from a refrigeration-sealed system is known as **evacuation.** Air and moisture inside of a refrigeration system can create many problems that are detrimental to the efficiency of the equipment and its components. Air contains **oxygen, nitrogen, hydrogen,** and water in the form of water vapor. **Nitrogen** is a noncondensable gas, which will occupy space at the top of the condenser and cause higher than normal operating high-side pressures, higher discharge temperatures, and higher compressor compression ratios, which will affect the efficiency of the refrigeration process. (Refer to *Theory Lesson: Evacuation, Figure 3–103*).

Other gases and water vapor cause a chemical reaction with refrigerant and refrigerant oil within the refrigeration sealed system, producing acids that cause deterioration of sealed system components, copper plating of running gears, and breakdown of motor insulation. When moisture in the from of water vapor, heat, and refrigerant is present in a sealed system, acids will eventually form because of the presence of chemicals such as **hydrogen, fluorine,** and **chlorine,** which are used to manufacture **CFC, HCFC,** and **HFC** refrigerants. When heated and mixed with other gases, **hydrochloric** and **hydrofluoric acids** are formed. These acids can deteriorate compressor motor windings and cause corrosion of metal alloys used in the makeup of sealed refrigeration systems. The longer the system operates under these conditions the more likely sludge is to form. **Sludge** is a tightly bound mixture of water, acid, and oil. Under these conditions the compressor will eventually burn out. If moisture is present in the refrigerant sealed system during a compressor burnout, fozene gas will also be formed. **Fozene gas** is very harmful to humans when burned and can be deadly if released in an area where poor ventilation exists. Avoiding corrosion and sludge problems is dependent on good service procedures and preventive maintenance. The only way to rid the sealed system of moisture is to use good evacuation procedures. Evacuation does not remove sludge; this requires cleanup procedures in which oversized driers specified for sludge removal must be used.

Acid test kits are available so that sealed systems can be checked for acid content. Sight glass can be used to indicate moisture content gin a system. (Refer to *Theory Lesson: Evacuation, Figure 3–104*.)

These compounds should not be allowed to remain in the sealed system. A vacuum must be applied to a sealed system to remove these contaminants. Lowering the pressure in a sealed system to below atmospheric pressure creates a vacuum.

With the new EPA regulations on refrigerants, the EPA has recommended that systems be evacuated to at least a 500-micron vacuum or equivalent. At this level the chances of noncondensables and a leak in the system are unlikely.

EQUIPMENT REQUIRED

Sealed system with valve stem service valves
1 Vacuum pump
1 Vacuum pump three-way valve or flared tee (used for attachment of micron gage)

1 Manifold gage
1 Additional refrigerant hose (*if required*)
1 Micron gage
Adjustable wrench
Refrigeration wrench
Heat gun or heat lamp

SAFETY PRACTICES

Wear goggles or safety glasses and gloves when working with refrigerant in sealed systems. Beware of high pressures and follow all electrical safety rules. Follow all EPA rules and regulations while working with refrigerants and evacuating sealed systems.

> **NOTE:** *NEVER RUN A SEALED SYSTEM WITH A COMPRESSOR HIGH-SIDE SERVICE VALVE FRONT SEATED. VERY DANGEROUS! CAN CAUSE THE COMPRESSOR CYLINDER HEADS TO EXPLODE.*

COMPETENCY PROCEDURES

Checklist

> **NOTE:** *Refrigerant should be recovered from the air conditioning system before proceeding with system evacuation.*

1. Remove high- and low-side manifold hoses from the blank ports on the gage manifold. ❏
2. Calibrate manifold gages if needed. ❏
3. Make sure gage manifold service valves are closed. ❏
4. Make sure vacuum pump oil level is correct. ❏
5. Remove high- and low-side valve stems caps. ❏
6. Make sure that high- and low-side valve stems are completely back seated. ❏
7. Remove high- and low-side service valve caps. ❏
8. Attach low-side manifold gage hose to the sealed system low-side service valve port. ❏
9. Attach the high-side manifold gage hose to the sealed system high-side service valve port. ❏
10. Connect the center hose from the gage manifold to the inlet port of the vacuum pump. ❏
11. Set up a micron gage so as to enable measurement of the vacuum level of the sealed system. ❏

> **NOTE:** *Setup for micron gage measurement of the sealed system vacuum will vary depending on type of micron gage and vacuum pump service valve.*

> **NOTE:** *If the system being evacuated has a liquid receiver with a valve stem service valve, the micron gage could be attached at this location to enable evacuation level.*

12. Turn on the vacuum pump. ❏
13. Slowly open the valve on the vacuum pump. ❏
14. Mid-seat the sealed system low-side valve stem one-half turn. ❏
15. Mid-seat the sealed system high-side valve stem one-half turn. ❏
16. Open the high-side gage manifold valve. ❏
17. Open the low-side manifold gage. ❏

18. Turn on micron gage. ❑
19. Calibrate micron gage (*if required*). ❑
20. *Mid-seat the liquid receiver valve stem service valve one-half turn (if used).* ❑

> **NOTE:** *EPA mandates that a vacuum be pulled on a sealed system to a level of at least 500 microns or lower. The amount of time required to reach the required vacuum level will vary depending on the size of the sealed system and the amount of contaminants within the system.*

21. Let system evacuation continue until a vacuum level of 500 microns is reached. ❑

> **NOTE:** *In larger systems heat should be added to areas of the sealed system where liquid refrigerant and moisture can collect. Adding heat to these areas helps in the evacuation procedure by causing these forms of matter to boil off so that they can be removed from the system by the vacuum pump (**Figure 3–114**).*

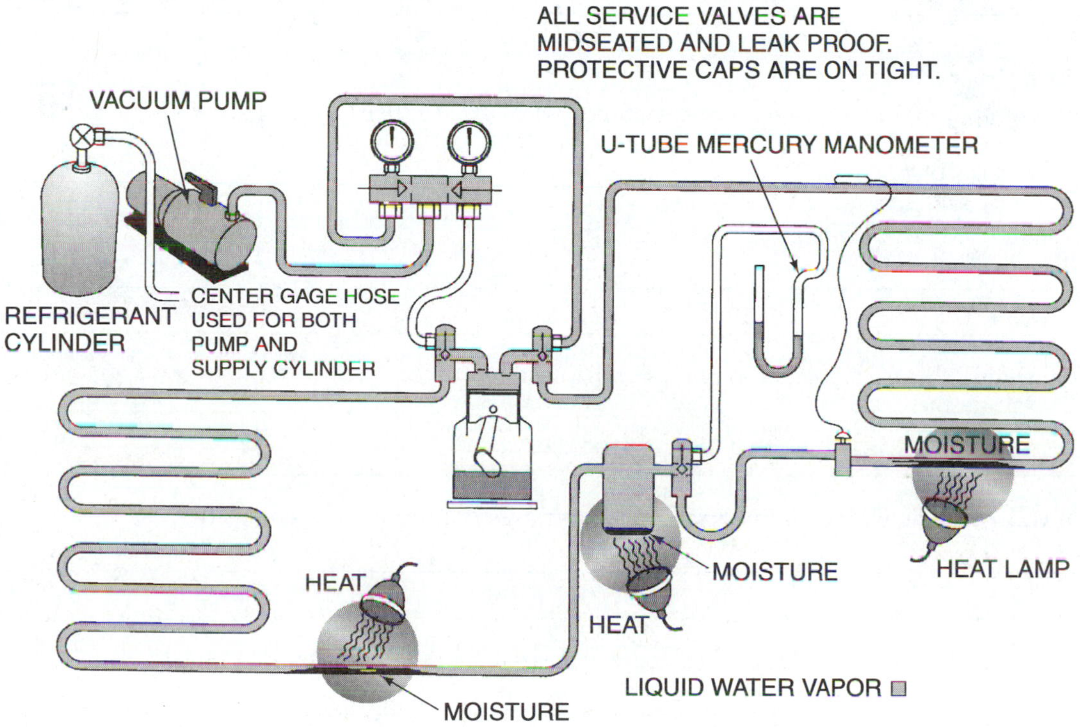

FIGURE 3–114 Heat applied to key areas of the sealed system so that moisture and liquid refrigerant can boil off and be removed from the sealed system by the vacuum pump.

22. Close the vacuum pump valve. ❑
23. Turn the vacuum pump OFF. ❑
24. Monitor the micron gage level for a minimum of 5 minutes. ❑

> **NOTE:** *Any rise in micron gage reading above 500 microns indicates that the system may still hold contaminants or possibly has a leak. This would require you to continue to evacuate the system.*

    If a 500-micron vacuum cannot be obtained, check system for leaks, change vacuum pump oil, and evacuate the system again.

For larger sealed systems, a much longer time for the standing test may be required and recommended.

25. Once proper vacuum is reached, close the vacuum pump valve. ❏
26. Turn the vacuum pump OFF. ❏
27. Close the low-side manifold gage valve. ❏
28. Close the high-side manifold gage valve. ❏
29. Turn OFF the micron gage. ❏
30. Back seat the high- and low-side sealed system valve stem. ❏
31. Back seat the liquid receiver valve stem (*if used*). ❏
32. Disconnect the micron gage. ❏
33. Replace the liquid receiver service valve cap. ❏
34. Replace the liquid receiver valve stem cap. ❏
35. Disconnect the manifold center hose from the vacuum pump. ❏
36. Replace the high- and low-side sealed system valve stem caps. ❏
37. Leave manifold gage set attached to sealed system. ❏

> **NOTE:** *System is ready to be charged when instructed to do so by your instructor.*

38. Have your instructor check your work. ❏
39. Return vacuum pump, micron gage, heating lamp, or heat gun to their proper location. ❏

RESEARCH QUESTIONS

1. What is the best way of checking a system for leaks?

   _____

   _____

2. How often should the oil in a vacuum pump be changed?

   _____

   _____

3. What is a way of checking the oil in a vacuum pump to see if it needs to be changed?

   _____

   _____

4. Pulling a vacuum down to 50 microns can cause what to happen to the vacuum pump?

   _____

   _____

5. What is a Triple Evacuation?

   _____

   _____

Passed Competency _____ Failed Competency _____

Instructor Signature _____ Grade _____

Student Name _____ Grade _____ Date _____

Practical Competency 62

Multiple Evacuation of a Sealed System

NOTE: Multiple evacuation procedures can be performed on all refrigeration sealed systems. The steps listed in Competency 62 include steps to set up for the Multiple Evacuation procedures on a sealed system, which has valve stem service valves. Sealed systems without valve stem service valves would not require all the procedures listed in this competency. What is important to identify is the actual setup of the equipment to be able to perform a multiple evacuation on a sealed system.

SUGGESTED MATERIALS

Textbook

Refrigeration & Air Conditioning Technology, 5th Edition, Thomson Delmar Learning
Unit 8—System Evacuation

Review Topics

Purpose of System Evacuation; Theory Involved with Evacuation; Multiple Evacuations

COMPETENCY OBJECTIVE

The student will be able to use a vacuum pump and perform multiple evacuations on a sealed system.

OVERVIEW

Removing air and other noncondensables, along with water vapor, from a refrigeration sealed system is known as **evacuation**. Air and moisture inside of a refrigeration system can create many problems that are detrimental to the efficiency of the equipment and its components. Air contains **oxygen, nitrogen, hydrogen**, and water in the form of water vapor. **Nitrogen** is a noncondensable gas, which will occupy space at the top of the condenser and cause higher than normal operating high-side pressures, higher discharge temperatures, and higher compressor gcompression ratios, which will affect the efficiency of the refrigeration process. (Refer to *Theory Lesson: Evacuation, Figure 3–103*.)

Other gases and water vapor cause a chemical reaction with refrigerant and refrigerant oil within the refrigeration sealed system, producing acids that cause deterioration of sealed system components, copper plating of running gears and breakdown of motor insulation. When moisture in the form of water vapor, heat, and refrigerant is present in a sealed system, acids will eventually form as a result of the presence of chemicals such as **hydrogen, fluorine**, and **chlorine**, which are used to manufacture **CFC, HCFC**, and **HFC** refrigerants. When heated and mixed with other gases, **hydrochloric** and **hydrofluoric acids** are formed. These acids can deteriorate compressor motor windings and cause corrosion of metal alloys used in the makeup of sealed refrigeration systems. The longer the system operates under these conditions the more likely sludge is to form. **Sludge** is a tightly bound mixture of water, acid, and oil. Under these conditions the compressor will eventually burnout. If moisture is present in the refrigerant sealed system during a compressor burnout, fozene gas will also be formed. **Fozene gas** is very harmful to humans when burned and can be deadly if released in an area where poor ventilation exists. Avoiding corrosion and sludge problems is dependent on good service procedures and preventive maintenance. The only way to rid the sealed system of moisture is to use good evacuation procedures. Evacuation does not remove sludge; this requires cleanup procedures where oversized driers specified for sludge removal must be used.

Acid test kits are available so that sealed systems can be checked for acid content. Sight glass can be used to indicate moisture content in a system. (Refer to *Theory Lesson: Evacuation, Figure 3–104.*)

These compounds should not be allowed to remain in the sealed system. A vacuum must be applied to a sealed system to remove these contaminants. Lowering the pressure in a sealed system to below atmospheric pressure creates a vacuum.

With the new EPA regulations on refrigerants, the EPA has recommended that systems be evacuated to at least a 500-micron vacuum or equivalent. At this level the chances of non-condensables and a leak in the system are unlikely.

Multiple Evacuation

Technicians at times will perform multiple evacuations on the same sealed system. This process is used to remove the atmosphere from a sealed system to the lowest level of contamination. This type of evacuation is accomplished by evacuating a system to a low vacuum, about 1 or 2 mm Hg (1000 to 2000 microns) and then introduce a small amount of refrigerant back into the sealed system, then evacuated to a 1 or 2 mm Hg vacuum again. Repeating this process three times is referred to as a triple evacuation. On the final vacuum the system would be charged with the manufacturer's recommended charge.

EQUIPMENT REQUIRED

Refrigerant sealed system
Vacuum pump
Flared tee 1/4" (*if applicable*)
Vacuum pump three-way valve (*if applicable*)
Refrigerant manifold gage set
Additional refrigerant hose
Electronic micron gage (*if applicable*)
U-tube manometer (*if applicable*)
Torpedo level (*if applicable*)
Cylinder of R-22 (*used for CFC or HCFC systems only*)
Nitrogen tank with regulator (*if applicable*)

> NOTE: *Dry nitrogen can be used on CFC, HCFC, instead of the trace gas R-22. For HFC systems, dry nitrogen only must be used.*

Electronic charging scale
Refrigeration wrench (*if applicable*)
Adjustable wrench (*if applicable*)
Cylinder of refrigerant required for system charge

SAFETY PRACTICES

Wear goggles or safety glasses and gloves when working with refrigerant in sealed systems. Beware of high pressures and follow all electrical safety rules. Follow all EPA rules and regulations while working with refrigerants.

COMPETENCY PROCEDURES

Checklist

1. Make sure that the system refrigerant has been recovered from the system. ❑
2. Make sure power to sealed system is turned OFF. ❑
3. Remove the sealed system high and low-side valve stem valve covers (*if applicable*). ❑
4. Remove the sealed system high- and low-side valve stem gage port caps. ❑

5. Attach the low-side manifold hose to the sealed system's low-side gage port. ❑
6. Connect the high-side manifold hose to the sealed system's high-side gage port. ❑
7. Attach the center manifold gage hose to the vacuum pump port. ❑
8. Attach a refrigerant hose to the micron gage or U-tube manometer. ❑
9. If the sealed system is equipped with a receiver that has a valve stem valve attached to it, attach the other end of the hose from the micron gage or U-tube manometer there. If the system does not have a receiver, *proceed to step 11.* ❑
10. On sealed systems with a receiver and valve stem valve, mid-seat the service valve (*if applicable*). ❑
11. Attach the other end of the hose from the micron gage or U-tube manometer to the vacuum pump flared tee or vacuum pump three-way valve. ❑
12. *If applicable*, mid-seat the sealed system low- and high-side valves one-half turn. ❑
13. Open the manifold gage high- and low-side valves. ❑
14. Open the vacuum pump valve. ❑
15. Turn on the micron gage (*if applicable*). ❑
16. If using a U-tube manometer, position and level the U-tube manometer (*if applicable*). ❑
17. Turn ON the vacuum pump. ❑
18. Open the manifold gage high- and low-side valves. ❑
19. Let the vacuum pump operate until a 1 or 2 mm Hg or 1000-to 2000-micron vacuum is indicated. ❑
20. Close the manifold gage high- and low-side valves. ❑
21. Turn the vacuum pump OFF. ❑
22. Remove the manifold gage center hose from the vacuum pump. ❑
23. Attach the manifold center hose to a cylinder of R-22 or dry nitrogen cylinder. ❑
24. Open the manifold gage high- and low-side valves and allow a small amount of refrigerant or dry nitrogen to enter the system until the vacuum rises to about 20 in. Hg on the low-side manifold gage. ❑
25. Close manifold gage high- and low-side valves. ❑
26. Reattach the manifold gage center hose the vacuum pump. ❑
27. Start the vacuum pump. ❑
28. Open manifold gage high- and low-side valves. ❑
29. Once again let the vacuum pump operate until a 1- or 2-mm Hg or 1000- to 2000-micron vacuum is indicated. ❑
30. At this point, *repeat Steps 20 through 26.* ❑
31. Start the vacuum pump to evacuate the sealed system for the third time. ❑
32. Allow the vacuum pump to run until a 500-micron vacuum is reached, or if measuring a vacuum with the U-tube manometer all the mercury columns to "Flat Out." This is where the mercury columns are equal. ❑
33. Close the manifold gage high- and low-side valves. ❑
34. Turn the vacuum pump OFF. ❑

PROCEDURES FOR CHARGING A SEALED SYSTEM WITHOUT VALVE STEM SERVICE VALVES

35. Connect the manifold center hose to the refrigerant cylinder required to charge the sealed system. ❑
36. Open the refrigerant cylinder valve. ❑
37. Loosen the center hose at the manifold bar to purge (bleed) air from the hose, and then tighten hose. ❑
38. Set refrigerant cylinder on the charge scales in such a position that the system will be charged with liquid refrigerant. ❑
39. Turn ON the refrigerant charging scales (*if applicable*). ❑
40. Check the manufacturer's charging plate and set the charging scales to the correct amount of charge required for the sealed system (*if applicable*). ❑

41. Open the manifold gage high-side valve. ❑
42. Allow the correct amount of charge to enter the sealed system. ❑

> **NOTE:** *If total charge cannot be added to the sealed system, set up charging procedures so that additional charge can be added while the sealed system is operating.*

43. Once system is completely charged, stop the charging procedure. In addition, take appropriate steps to remove refrigerant cylinder, charging scales, and manifold gages from the sealed system. ❑
44. Replace service port caps. ❑
45. Turn sealed system ON and allow the unit to operate. ❑

PROCEDURES FOR CHARGING A SEALED SYSTEM WITH VALVE STEM SERVICE VALVES AND LIQUID RECEIVER

46. Back-seat the receiver service valve. ❑
47. Remove the micron gage or U-tube manometer (*if applicable*). ❑
48. Connect a refrigerant hose to the refrigerant cylinder required to charge the sealed system. ❑
49. Connect the other end of the refrigerant cylinder hose to the receiver service valve port. ❑
50. Open the refrigerant cylinder valve. ❑
51. Loosen the hose end at the receiver and purge (bleed) air from the hose, then tighten hose. ❑
52. Set refrigerant cylinder on the charge scales in such a position that the system will be charged with liquid refrigerant. ❑
53. Turn ON the refrigerant charging scales (*if applicable*). ❑
54. Check the manufacturer's charging plate and set the charging scales to the correct amount of charge required for the sealed system (*if applicable*). ❑
55. Mid-seat the receiver service valve stem and allow the correct amount of charge to enter the sealed system. ❑

> **NOTE:** *If total charge cannot be added to the sealed system, set up charging procedures so that additional charge can be added while the sealed system is operating.*

56. Once the system is completely charged, stop the charging procedure In addition, take appropriate steps to remove refrigerant cylinder, charging scales, and manifold gages from the sealed system. ❑
57. Replace service port caps and service valve stem covers. ❑
58. Turn sealed system ON and allow unit to operate. ❑
59. Have your instructor check your work. ❑

RESEARCH QUESTIONS

1. What is the advantage of performing a triple evacuation on a system?

   _____

   _____

2. What is the purpose of adding refrigerant to the system and then circulating it through the system?

   _____

   _____

3. If a U-tube manometer is used to measure the vacuum level, what level in (mm Hg) is recommended for a proper vacuum?

_____

_____

4. A vacuum measurement of 5 mm Hg is equal to how many microns?

_____

_____

5. How many microns are equal to 1 in. Hg?

_____

_____

Theory Lesson: Sealed System Charging with Azeotropic and Near-Azeotropic Refrigerants (Figure 3–115)

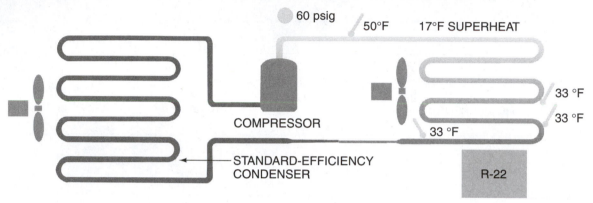

FIGURE 3–115 The evaporator is starved for refrigerant because of a low refrigerant charge.

SUGGESTED MATERIALS

Textbook

Refrigeration & Air Conditioning Technology, 5th Edition, Thomson Delmar Learning
Unit 10—*System Charging*
Unit 9—*Refrigerant and Oil Chemistry and Management—Recovery, Recycling, Reclaiming, and Retrofitting*

Review Topics

Charging a Refrigeration System; Vapor Refrigerant Charging; Liquid Refrigerant Charging; Weighing Refrigerant; Using Charging Devices; Using Charging Charts; Sub-cooling Method of Charging for TXV Systems; Refrigerant Blends

Key Terms

azeotrope/azeotropic • dry-bulb temperature • fractionation • graduated cylinders • holding charge • latent heat • liquid refrigerant charging • line set • manufacturer's sub-cooling tables • manufacturer's superheat tables • near-azeotropic refrigerants • overcharged • refrigerant • sensible heat • sling psychrometer • sub-cooling • superheat • temperature glide • total load • undercharged • vapor refrigerant charging • weighing refrigerant • wet-bulb temperature

OVERVIEW

Air conditioning, heat pumps, and refrigeration equipment must have the correct amount of refrigerant charge for the equipment to operate and perform as efficiently as it was designed to. With today's high-efficiency equipment, ensuring that a system is charged with the correct amount of refrigerant is one of the most important procedures a technician is required to perform. This requires that technicians have a thorough understanding of the system operation and operating conditions that indicate whether a system is properly *charged, overcharged,* or *undercharged*. Refrigerant can be added to a sealed system in the vapor or liquid state, which can be accomplished by *weighing the refrigerant charge* into the sealed system, *measuring the charge* into the sealed system (**Figure 3–116**), using *system operating charts* (**Figure 3–117**), and using manufacturer's *superheat charging tables* (**Figure 3–118**) and *suction line temperature tables* (**Figure 3–119**) or, alternatively, manufacturer's *sub-cooling charging tables* for systems with a TXV metering device (**Figure 3–120**).

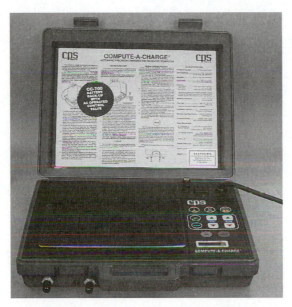

FIGURE 3–116   A programmable electronic scale. (*Photo by Bill Johnson*)

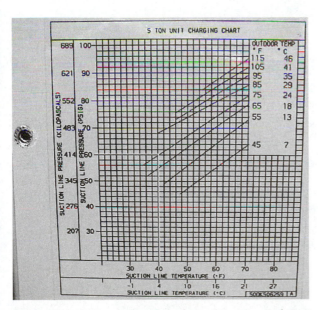

FIGURE 3–117   A charging chart. (*Courtesy of Ferris State University, Photo by John Tomczyk*)

Outdoor Temp (F)	Indoor Coil Entering Air (F) WB													
	50	52	54	56	58	60	62	64	66	68	70	72	74	76
55	9	12	14	17	20	23	26	29	32	35	37	40	42	45
60	7	10	12	15	18	21	24	27	30	33	35	38	40	43
65	—	6	10	13	16	19	21	24	27	30	33	36	38	41
70	—	—	7	10	13	16	19	21	24	27	30	33	36	39
75	—	—	—	6	9	12	15	18	21	24	28	31	34	37
80	—	—	—	—	5	8	12	15	18	21	25	28	31	35
85	—	—	—	—	—	—	8	11	15	19	22	26	30	33
90	—	—	—	—	—	—	5	9	13	16	20	24	27	31
95	—	—	—	—	—	—	—	6	10	14	18	22	25	29
100	—	—	—	—	—	—	—	—	8	12	15	20	23	27
105	—	—	—	—	—	—	—	—	5	9	13	17	22	26
110	—	—	—	—	—	—	—	—	—	6	11	15	20	25
115	—	—	—	—	—	—	—	—	—	—	8	14	18	23

FIGURE 3–118   Superheat charging table for fixed-bore metering device systems. (*Courtesy of Carrier Corporation*)

Superheat Temp (F)	Suction Pressure at Service Port (psig)								
	61.5	64.2	67.1	70.0	73.0	76.0	79.2	82.4	85.7
0	35	37	39	41	43	45	47	49	51
2	37	39	41	43	45	47	49	51	53
4	39	41	43	45	47	49	51	53	55
6	41	43	45	47	49	51	53	55	57
8	43	45	47	49	51	53	55	57	59
10	45	47	49	51	53	55	57	59	61
12	47	49	51	53	55	57	59	61	63
14	49	51	53	55	57	59	61	63	65
16	51	53	55	57	59	61	63	65	67
18	53	55	57	59	61	63	65	67	69
20	55	57	59	61	63	65	67	69	71
22	57	59	61	63	65	67	69	71	73
24	59	61	63	65	67	69	71	73	75
26	61	63	65	67	69	71	73	75	77
28	63	65	67	69	71	73	75	77	79
30	65	67	69	71	73	75	77	79	81
32	67	69	71	73	75	77	79	81	83
34	69	71	73	75	77	79	81	83	85
36	71	73	75	77	79	81	83	85	87
38	73	75	77	79	81	83	85	87	89
40	75	77	79	81	83	85	87	89	91

FIGURE 3–119 Suction line temperature table. (*Courtesy of Carrier Corporation*)

Pressure (psig) at Service Fitting	Required Subcooling Temperature (F)					
	0	5	10	15	20	25
134	76	71	66	61	56	51
141	79	74	69	64	59	54
148	82	77	72	67	62	57
156	85	80	75	70	65	60
163	88	83	78	73	68	63
171	91	86	81	76	71	66
179	94	89	84	79	74	69
187	97	92	87	82	77	72
196	100	95	90	85	80	75
205	103	98	93	88	83	78
214	106	101	96	91	86	81
223	109	104	99	94	89	84
233	112	107	102	97	92	87
243	115	110	105	100	95	90
253	118	113	108	103	98	93
264	121	116	111	106	101	96
274	124	119	114	109	104	99
285	127	122	117	112	107	102
297	130	125	120	115	110	105
309	133	128	123	118	113	108
321	136	131	126	121	116	111
331	139	134	129	124	119	114
346	142	137	132	127	122	117
359	145	140	135	130	125	120

FIGURE 3–120 Sub-cooling chart for R-22. (*Courtesy of Carrier Corporation*)

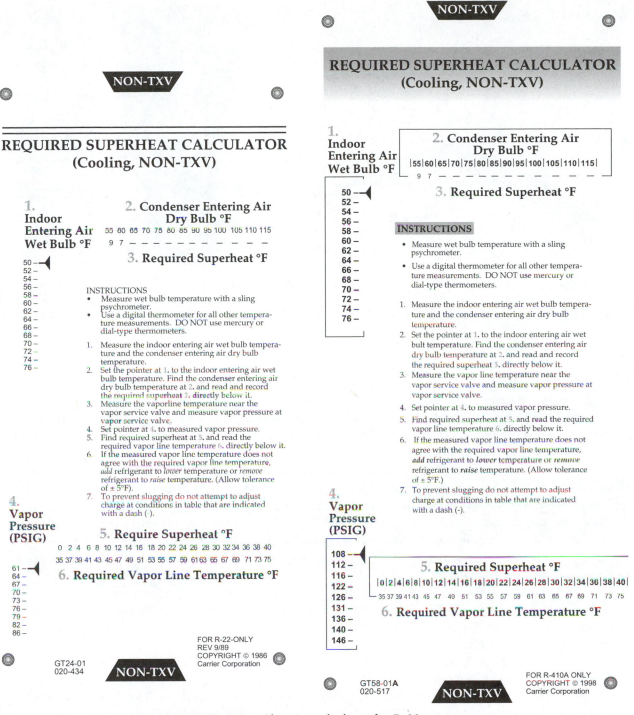

FIGURE 3–121   Charging calculator for R-22 systems.

Charging calculators are also available for use with R-22 and R-410A split system air conditioning and heat pumps (**Figure 3–121**). These calculators can be used to charge, check charge, and adjust system charge, along with the ability to check the system for proper airflow. To use these calculators requires that technicians be able to determine the total load, which the sealed system is operating in at the time the system is being charged, or charge is being checked. *Total load* refers to the amount of *sensible heat* and *latent heat* in the air the system is trying to condition. Determining the total load of the air requires the use of a *sling psychrometer* or other device that can measure *wet-bulb* and *dry-bulb* temperature of the air being conditioned by the air conditioning system (**Figure 3–122** and **Figure 3–123**).

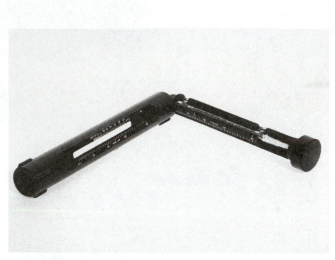

FIGURE 3–122 Sling psychrometer. (*Photo by Bill Johnson*)

FIGURE 3–123 Two styles of digital sling psychrometers. (*A, Courtesy of Amprobe*; B, *Courtesy of UEi*)

The use of these calculators and sling psychrometers will be discussed in greater detail in future Theory Lessons and Practical Competencies.

Most split-system air conditioning and split system heat pumps come precharged from the factory. This is referred to as the *holding charge* and in most cases the proper amount of refrigerant for the system components and refrigerant piping between the condensing unit and the system air handler or evaporator coil. Manufacturers calculate the *holding charge* based on either 15 or 25 feet of refrigerant piping (*line set*) between the condenser and the system air handler or evaporator. Always check the manufacturer's installation instructions for the line set length in which the holding charge was calculated. In addition, listed in the system installation instructions are directions for the addition or removal of refrigerant for line sets that are shorter or longer than those on which the system charge was based. Regardless of the system's holding charge, it is highly recommended that the system's charged be checked for correctness once installation is complete.

Some manufacturers ship condensing units without a holding charge. These systems are normally pressurized with dry nitrogen. With these units the dry nitrogen must be released and the system evacuated and charged according to the manufacturer's recommended charge amount, which is normally listed on the condensing unit nameplate.

Because of ozone depletion and global warming issues, new refrigerants have been introduced to the industry to replace the traditional ozone depleting refrigerants, which were used in the industry for years. These new refrigerants require different charging procedures owing to their chemical makeup. The characteristics of these refrigerants during the evaporation and condensing process are different compared to traditional refrigerants used in the industry. The different characteristic of these new refrigerants also requires different calculations for checking system *superheat* and *sub-cool* readings.

Traditional refrigerants of CFCs and HCFCs are known as *azeotropes* or *azeotropic* mixtures. These mixtures are blended refrigerants made up of two or more liquids. When mixed together these *refrigerant blends* form a new pure compound that has its own temperature and pressure relationship. In other words, the different refrigerants in the mixture do not separate within the mixture and retain some of their individual characteristics at a given temperature and pressure (**Figure 3–124**).

With these refrigerants only one boiling point (evaporating temperature) and one condensing point (condensing temperature) exists for each given system pressure.

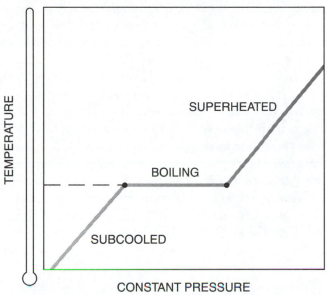

FIGURE 3–124   An azeotropic refrigerant blend showing only one temperature for a given pressure as it boils (evaporates).

Many of the new refrigerants are referred to as *near-azeotropic* or zeotropic refrigerants and do not behave like azeotropic refrigerants. These refrigerants are blends, but referred to as *refrigerant mixtures* because refrigerants used within the blend can still separate into individual refrigerants, like water and oil, which can be mixed, but if allowed to sit for any period of time the water and oil will separate. This is where a difference in the pressure–temperature relationship arises versus the pressure–temperature relationship of azeotropic refrigerants.

Near-azeotropic blends experience a *temperature glide* (**Figure 3–125**).

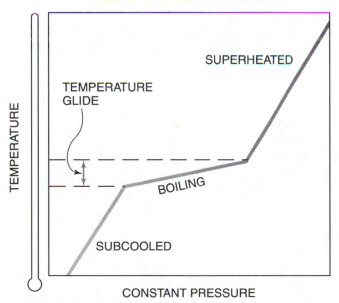

FIGURE 3–125   A near-azeotropic refrigerant blend showing temperature glide as it boils at a constant pressure.

This temperature glide occurs when the blend has many temperatures as it evaporates and condenses at a given pressure. This means that refrigerants used in the mixture will change phase from liquid to vapor or vapor back to a liquid faster than other refrigerants used to make the mixture. Therefore, it is possible with these refrigerant mixtures for one refrigerant in the mixture to be boiled away to 100% vapor and get *superheated* while other refrigerants in the mixture are still boiling from liquid to vapor. Likewise it is possible for one or more of the refrigerants to be condensed from a vapor to 100% liquid and start to *sub-cool*, while other refrigerants within the mixture are still being condensed from vapor back to a liquid.

Near-azeotropic refrigerants may also experience what is referred to as *fractionation*. This occurs when one or more of the refrigerants in the blend will condense or evaporate at different rates than other refrigerants making up the mixture. Fractionation happens only when the leak occurs as a vapor leak within the sealed system; it will not occur when the system is leaking pure liquid. This means that during a vapor leak within a system charged with near-azeotropic refrigerant, refrigerants with in the mixture will leak out at different rates, which can have an effect on sealed system efficiency if system charge is *"topped off"* during a leak situation, rather than being completely removed from the system and a new charge weighed into the system. Fractionation also occurs if refrigerant is removed from a near-azeotropic refrigerant cylinder. For this reason, near-azeotropic or zeotropic blends should be *charged as a liquid* into a sealed system.

Practical Competency 63

Measuring the Dry-Bulb and Wet-Bulb Temperature and Relative Humidity of Air Using a Sling Psychrometer

SUGGESTED MATERIALS

Textbook
Refrigeration & Air Conditioning Technology, 5th Edition, Thomson Delmar Learning
Unit 5—Tools and Equipment

Review Topics
Sling Psychrometer

COMPETENCY OBJECTIVE

The student will be able to use a sling psychrometer determine the dry-bulb and wet-bulb temperature and relative humidity percentage of air by using a sling psychrometer.

OVERVIEW

The main purpose of air conditioning is to provide a comfortable and healthy environment in a conditioned space or building for humans. Comfort is obtained by controlling the total heat load of the air within these spaces. Total heat is the amount of sensible and latent heat within the air being conditioned. In most cases humans feel comfortable in air temperatures where the sensible heat temperature is approximately 75 degree (F) with a latent heat factor of 50% humidity. Sling psychrometers are used to evaluate the percentage of relative humidity in the air along with the sensible heat intensity. Sling pshychrometers are used to measure the wet-bulb and dry-bulb temperature of air. A sling psychrometer consists of two different thermometers. One of the thermometers is kept dry, and is referred to as the dry-bulb. The other thermometer has a sock or cloth wrapped around and is dampened for testing purposes. This thermometer is referred to as the wet-bulb.

The dry thermometer (dry bulb) is used to measure the sensible heat temperature of the air being tested. The wet-bulb thermometer measures the temperature taking the humidity level into account. The two bulbs are whirled together in the air being tested. This causes evaporation to occur at the wick of the wet-bulb thermometer, giving it a lower temperature reading. The difference in temperature will depend on the amount of humidity in the air. The drier the air, the greater the difference in the temperature readings between the dry-bulb and wet-bulb due to the fact that drier air can absorb and hold more moisture. During periods of low humidity, the water from the sock on the wet-bulb will evaporate quickly, removing heat from the wet-bulb thermometer. This will cause the wet-bulb reading to drop. During periods of high humidity, the moisture will not evaporate and the wet-bulb temperature reading will be very close to that of the dry-bulb. Once the dry-bulb and wet-bulb temperatures are measured, these readings can be used to determine the percentage of humidity within the air where the temperatures where measured. In cases where technicians are using manufacturer's superheat, sub-cooling charts for checking, adjusting, or charging air conditioning equipment, the dry-bulb and wet-bulb temperatures of the air will have to be known. This is also true when using superheat and sub-cooling calculators.

EQUIPMENT REQUIRED

Sling psychrometer

SAFETY PRACTICES

Care should be taken to avoid hitting any person or solid object when swinging the sling psychrometer.

COMPETENCY PROCEDURES

Checklist

1. Remove end cap of sling psychrometer. ❑
2. Fill psychrometer reservoir with cool water. ❑
3. Replace psychrometer end cap. ❑
4. Be sure that wick from the reservoir covers the end of the wet-bulb. ❑
5. Make sure that the wick around the set-bulb is dampened, not soaked with water. ❑
6. Pull psychrometer body from the psychrometer handle and let it hang free. ❑
7. Use the tube of the psychrometer as a handle and whirl the psychrometer body in the air being tested for 1 1/2 minutes. ❑

NOTE: *Swing psychrometer body at a steady even speed. Do not swing to psychrometer too fast or too slow.*

8. After swing the psychrometer for a minute and a half, read the temperatures of the wet-bulb and dry-bulb. ❑
9. *Record the dry-bulb temperature.* _____ ❑
10. *Record the wet-bulb temperature.* _____ ❑

NOTE: *Notice that the wet-bulb temperatures are marked on both sides of the handle section of the sling psychrometer. The dry-bulb temperatures are marked on both sides of the sling psychrometer body. The percentage of relative humidity is also marked on the psychrometer body.*

11. Push the psychrometer body back in to the psychrometer handle. ❑

DETERMINING THE PERCENTAGE OF RELATIVE HUMIDITY

12. Move the psychrometer body and align the recorded dry-bulb temperature underneath the recorded wet-bulb temperature. ❑
13. Read the percentage of relative humidity as indicated by the arrow at the bottom relative humidity percentage. ❑
14. *Record the percentage of relative humidity in the air tested?* _____ ❑

NOTE: *Have your instructor assign three other locations where the air can be tested and the relative humidity can be determined.*

RECORD THE FOLLOWING INFORMATION:

AREA 1 (LOCATION) _____
Measured wet-bulb temperature _____
Measured dry-bulb temperature _____
Record the percentage of relative humidity _____

AREA 2 (LOCATION) _____
Measured wet-bulb temperature _____
Measured dry-bulb temperature _____
Record the percentage of relative humidity _____

AREA 3 (LOCATION) _____
Measured wet-bulb temperature _____
Measured dry-bulb temperature _____
Record the percentage of relative humidity _____

AREA 4 (LOCATION) *OUTSIDE AIR*_____
Measured wet-bulb temperature _____
Measured dry-bulb temperature _____
Record the percentage of relative humidity _____

ANSWER THE FOLLOWING QUESTIONS:

1. *What area had the highest humidity?*

   _____
   _____

2. *Out of all the different relative humidity reading, why do you think the area listed in* **Question 1** *one had the highest percentage of humidity?*

   _____
   _____

3. *What area tested had the lowest percentage of relative humidity?*

   _____
   _____

4. *Why do you think the area listed in* **Question 3** *had the lowest percentage of relative humidity?*

   _____
   _____

5. *What area had the highest ambient temperature?*

   _____
   _____

6. *Why do you think the area listed in* **Question 5** *had the highest temperature?*

   _____
   _____

7. *What area had the lowest ambient temperature?*

   _____
   _____

8. *Why do you think this area had the lowest ambient temperature?*

   _____
   _____

9. *What area do you think could be cooled down the fastest?*

   _____
   _____

10. *Why do you think the area listed in* **Question 10** *can be cooled down the fastest?*

   _____
   _____

11. Have your instructor check your work.

Practical Competency 64

Liquid Charging a Split System Air Conditioning Unit Using Charging Scales (Figure 3–126)

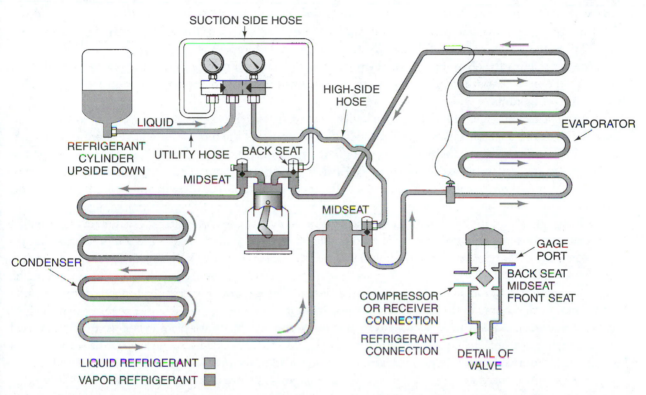

FIGURE 3–126   This system is being charged while it has no refrigerant in it. The liquid refrigerant moves toward the evaporator and the condenser when doing this. No liquid refrigerant will enter the compressor.

SUGGESTED MATERIALS

Textbook

Refrigeration & Air Conditioning Technology, 5th Edition, Thomson Delmar Learning
Unit 10—System Charging
Unit 9—Refrigerant and Oil Chemistry and Management—Recovery, Recycling, Reclaiming, and Retrofitting

Review Topics

Liquid Refrigerant Charging; Weighing Refrigerant; Using Charging Devices; Azeotropic Refrigerants

COMPETENCY OBJECTIVE

The student will be able to liquid charge a central air conditioning system using charging scales.

OVERVIEW

Adding refrigerant to refrigeration sealed system is referred to as charging. Air conditioning, heat pumps, and refrigeration equipment must have the correct amount of refrigerant charge for the equipment to operate and perform as efficiently as it was designed to. With today's high-efficiency equipment, ensuring that a system is charged with the correct amount of refrigerant is one of the most important procedures a technician is required to perform. This requires that technicians have a thorough understanding of the system operation and operating conditions which indicate whether a system is properly *charged*, *overcharged*, or *undercharged*. Refrigerant can be added to a sealed system in the vapor or liquid state by *weighing the refrigerant charge* into the sealed system, measuring the charge into the sealed system (*Theory Lesson: Sealed System Charging, Figure 3–116*), using *system operating charts* (*Theory Lesson: Sealed System Charging, Figure 3–117*), manufacturer's superheat charging tables (*Theory Lesson: Sealed System Charging, Figure 3–118*), suction line temperature tables (*Theory Lesson: Sealed System Charging, Figure 3–119*), or manufacturer's sub-cooling charging tables for systems with a TXV metering device (*Theory Lesson: Sealed System Charging, Figure 3–120*).

Charging calculators are also available for use with R-22 and R-410A split-system air conditioning and heat pumps (*Theory Lesson: Sealed System Charging, Figure 3–121*).

The procedures for charging refrigeration systems with either vapor or liquid refrigerant are different, and with the new near-azeotrope refrigerants, technicians must follow proper procedures when working with these refrigerants.

The most accurate method of charging a sealed system is to weigh the manufacturer's required charge into the system. Before a technician considers the method and state that the refrigerant will be charged into a system, technicians must be familiar with the two different types of refrigerant blends. The azeotrope refrigerant blends can be charged into a sealed system in either the vapor or liquid state. These refrigerant blends are combined in a chemical process in such a manner that the total refrigerant mixture will not separate into the individual refrigerants within the mixture. These refrigerants are considered to be pure compounds.

Adding refrigerant in the liquid state to a sealed system is normally done after a vacuum has been pulled on the system. It is important to note that liquid should never be permitted to enter the compressor of the system. When liquid charging a system, the liquid is normally measured or weighed by some means to determine the correct amount of charge into the system. Sometimes the full amount of refrigerant charge cannot be accomplished in the liquid state. At this point the system's charge would have to be topped off by charging the balance of the charge in the vapor state. To get liquid from the refrigerant cylinder requires that the bottle be turned upside down for azeotrope refrigerants. Near-azeotrope refrigerant cylinders are always used in the upright position. The liquid refrigerant would be added to the sealed system through the sealed system liquid line or liquid receiver if applicable. Liquid can be added to the low side of a system, but special restrictive devices must be used to allow refrigerant liquid to vaporize before entering the sealed system's low side.

EQUIPMENT REQUIRED

1 Manifold gage set
1 Vacuum pump (*if applicable*)
1 Micron gage (*if applicable*)
1 Charging scale
1 Clamp-on ammeter
1 Refrigerant cylinder—sealed system type
2 Tap line valve (*if required*)

SAFETY PRACTICES

Wear goggles or safety glasses and gloves when working with refrigerants in sealed systems. Beware of high pressures and follow all electrical safety rules. Follow all EPA rules and regulations while working with refrigerants.

> NOTE: *Competency Procedures could be used to weigh the charge into any sealed system.*

COMPETENCY PROCEDURES

> *NOTE: The system to be charged should be evacuated and ready to charge. If not, use proper procedures to evacuate system, and proceed with system charging, Practical Competency 60.*

1. *Review unit nameplate and record the type of refrigerant required for the unit.* _____ ❑
2. *Review unit nameplate and record the manufacturer's charge amount.* _____ ❑
3. Set charging cylinder on charging scales. ❑
4. Connect the center hose of the manifold gage set to the valve port of the refrigerant cylinder. ❑

> *NOTE: If using a recovery bottle of refrigerant, connect manifold center hose to the Liquid Valve of Recovery Cylinder.*

5. Open the refrigerant cylinder valve. ❑
6. Slowly loosen center hose connection on the gage manifold for about 2 seconds to purge air from the center hose, and then retighten it. ❑
7. If charging cylinder is not a recovery cylinder, place charging cylinder upside down on the charging scale. ❑
8. Follow the manufacturer's instructions for the charging scales you are using and prepare charging scales for charging. ❑
9. Open the manifold high-side gage valve and allow the required amount of liquid refrigerant to enter the system. ❑

> *NOTE: Depending on the amount of refrigerant designed for the system you are charging, you may not be able to complete the system's total charge with liquid. The charge for the system may have to be completed by topping off the charge in the vapor state.*

10. Close the manifold high-side gage valve once the charge is complete. ❑
11. Turn the system ON. ❑
12. *If applicable,* adjust charge by opening the low-side manifold gage valve and add refrigerant in the vapor state to the sealed system's low side. ❑
13. Allow the system to run once the required system charge is correct. ❑
14. *After 10 minutes of operation, record the system's low-side operating pressure.* _____ ❑
15. *Record the low-side saturation temperature.* _____ ❑
16. *Record the system's high-side operating pressure.* _____ ❑
17. *Record the high-side saturation temperature.* _____ ❑
18. *Check the unit manufacturer's nameplate and record the manufacturer's RLA (Run Load Amperage).* _____ ❑
19. Use the clamp-on ammeter and measure the system's RLA amperage at the unit disconnect (**Figure 3–127**). ❑
20. *Record the system's operating RLA.* _____ ❑

> *NOTE: The system's operating RLA should be at or close to the manufacturer's rated RLA if the system's charge is correct.*

21. Close the charging cylinder valve. ❑
22. Slowly loosen and remove the hose from the charging cylinder. ❑

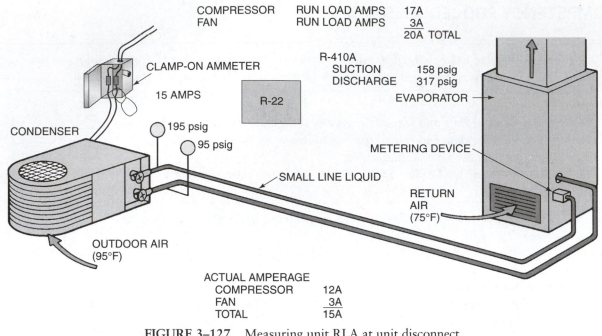

FIGURE 3–127 Measuring unit RLA at unit disconnect.

23. Slowly remove the low-side and high-side hoses from the system's service valves. ❑
24. Replace system service valve caps. ❑
25. Have your instructor check your work. ❑
26. At the direction of your instructor, turn unit OFF, disconnect all equipment, and return all tools and test equipment to their proper location. ❑

RESEARCH QUESTIONS

1. Why is weighing a liquid charge into a system better than vapor charging a system?

_____

_____

2. When charging with liquid, what are a couple of ways of keeping the cylinder's pressure up so that the liquid will be forced into the high side of the system?

_____

_____

3. Is it possible to charge a system with liquid on the low side of the system?

_____

_____

4. Why is it important that liquid refrigerant be allowed to enter the compressor during operation?

_____

_____

5. How can liquid refrigerant safely be charged through the low side of a sealed system?

_____

_____

Passed Competency _____   Failed Competency _____

Instructor Signature _____   Grade _____

Practical Competency 65

Liquid Charging Central Air Conditioning System with Near-Azeotropic–Zeotropic Refrigerant Blends (R-410A/Puron) (Figure 3–128)

DATA PLATE

SPECIFICATIONS
R-410A
HIGH EFFICIENCY
FLA 10
LRA 55
VOLTS 208/230
SINGLE PHASE
SUBCOOLING 15

R-410A PRESSURE/TEMP CHART

F	PRESSURE (psig)
95	295 psig
100	317 psig
105	340 psig
110	365 psig
115	390 psig
120	417 psig
125	446 psig
130	475 psig
135	506 psig
140	539 psig

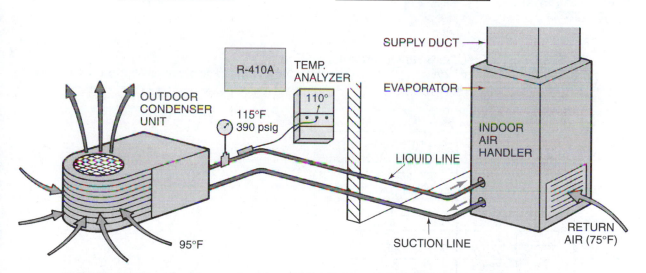

FIGURE 3–128   Charging method for a high-efficiency split R-410A air conditioning system.

SUGGESTED MATERIALS

Textbook

Refrigeration & Air Conditioning Technology, 5th Edition, Thomson Delmar Learning
Unit 10—System Charging
Unit 9—Refrigerant and Oil Chemistry and Management—Recovery, Recycling, Reclaiming, and Retrofitting

Review Topics

Liquid Refrigerant Charging; Weighing Refrigerant; Using Charging Devices; Near-Azeotropic–(Zeotropic) Refrigerants

COMPETENCY OBJECTIVE

The student will be able to charge central air conditioning systems that use near-azeotropic refrigerants.

OVERVIEW

The procedures for charging refrigeration systems with either vapor or liquid refrigerant are different. With the new near-azeotrope refrigerants, proper procedures for charging and checking the charge of these systems are different and must be learned by technicians.

Many of the new refrigerants are referred to as *near-azeotropic* or zeotropic refrigerants. These refrigerants do not behave like traditional azeotropic refrigerants because refrigerants used within these blends can still separate into individual refrigerants and also can retain some of their individual characteristics under a given pressure, like water and oil, which can be mixed, but if allowed to sit for any period of time the water and oil will separate. For this reason near-azeotropic or zeotropic refrigerants are referred to as *refrigerant mixtures*. This is where a difference in pressure–temperature relationship arises versus the pressure–temperature relationship of azeotropic refrigerants. Unlike azeotropic refrigerants, near-azeotropic blends experience what is called *temperature glide*.

Temperature glide occurs when the blend has many temperatures as it evaporates and condenses at a given pressure. This means that refrigerants used in the mixture will change phase from liquid to vapor or vapor back to a liquid faster than other refrigerants used to make the mixture (**Figure 3–129**).

Therefore, it possible with these refrigerant mixtures for one refrigerant in the mixture to be boiled away to 100% vapor and get *superheated* while other refrigerants in the mixture are still boiling from liquid to vapor (**Figure 3–130**).

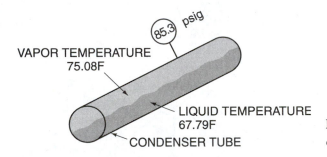

FIGURE 3–129 The pressure–temperature relationship of R-401A as it phase changes in the condenser tube.

	°F	PRESSURE PER SQUARE INCH, (psig)
V	75.08	85.3
L	67.79	85.3
V	76.30	87.3
L	69.04	87.3
V	77.51	89.3
L	70.27	89.3
V	78.69	91.3
L	71.48	91.3
V	79.86	93.3
L	72.67	93.3
V	81.02	95.3
L	73.85	95.3

R-401 A

V = VAPOR
L = LIQUID

FIGURE 3–130 The pressure–temperature relationship of R-401A showing the liquid and vapor at different temperatures at the same pressure.

Likewise it is possible for one or more of the refrigerants to be condensed from a vapor to 100% liquid and starting to *sub-cool* while other refrigerants within the mixture are still being condensed from a vapor back to a liquid. One temperature no longer corresponds to one pressure as with azeotropic refrigerants. Temperature glide can range from 2 to 14 degrees (F) in refrigeration and air conditioning applications depending on the specific blend of refrigerant being used. Because of temperature glide different ways of calculating superheat and sub-cooling are required.

When working on systems with near-azeotropic blends that have temperature glide, new pressure–temperature charts are available for technicians to use when checking a system for the correct amount of refrigerant charge (**Figure 3–131**).

These pressure–temperature charts instruct service technicians who are checking system superheat to use the *dew point valve* shown on the pressure–temperature chart. Dew point is the temperature at which saturated vapor first starts to condense. When checking the system sub-cooling value, technicians are instructed to use the *bubble point* values listed on the chart. Buddle point is the temperature at which the saturated liquid starts to boil off its first bubble of vapor.

Near-azeotropic refrigerants may also experience what is referred to as *fractionation*. This occurs when one or more of the refrigerants in the blend will condense or evaporate at different rates than other refrigerants making up the mixture. Fractionation happens only when the leak occurs as a vapor leak within the sealed system; it will not occur when the system is leaking pure liquid. This means that during a vapor leak within a system charged with near-azeotropic refrigerant, refrigerants with in the mixture will leak out at different rates, which can have an effect on sealed system efficiency if system charge is *"topped off"* during a leak situation, rather than being completely removed from the system and a new charge weighed into the system. Fractionation also occurs if refrigerant is removed from a near-azeotropic refrigerant cylinder. For this reason, near-azeotropic or zeotropic blends should be *charged as a liquid* into a sealed system. When adding liquid to the suction line of a system that is running, has to be restricted and vaporized into the system to avoid any damage to the compressor (**Figure 3–132**).

Addition of refrigerant in the liquid state to a sealed system is normally done after a vacuum has been pulled on the system. It is important to note that liquid should never be permitted to enter the compressor of the system. When liquid charging a system, the liquid is normally measured or weighed by some means to determine the correct amount of charge into the system. To get liquid from a near-azeotropic or zeotropic refrigerant cylinder requires that the cylinder always be used in the upright position. Liquid refrigerant can added to a sealed system through the liquid line (high side) or liquid receiver if applicable. Liquid can be added to the low side of a system, but special restrictive devices must be used to allow refrigerant liquid to vaporize before entering the sealed system's low side.

EQUIPMENT REQUIRED

1 Manifold gage set
1 Vacuum pump (*if applicable*)
1 Micron gage (*if applicable*)
1 Charging scale
1 Clamp-on ammeter
1 Refrigerant cylinder—near-azeotropic–zeotropic refrigerant

SAFETY PRACTICES

Wear goggles or safety glasses and gloves when working with refrigerant in sealed systems. Beware of high pressures and follow all electrical safety rules. Follow all EPA rules and regulations while working with refrigerants.

> *NOTE: Competency Procedures could be used to weigh the charge into any sealed system.*

COMPETENCY PROCEDURES

Checklist

> *NOTE: The system to be charged should be evacuated and ready to charge. If not, use proper procedures to evacuate system, and proceed with system charging, Practical Competency 60.*

PRESSURE-TEMPERATURE CHART

PSIG	Temperature, °F Pink — MP39 or 401A (X)	Sand — HP80 or 402A (L)	Orange — HP62 or 404A (S)	Green — KLEA 60 or 407A	Lt. Brown — 9000 or KLEA 66 407C (N)	Brown — FX-56 or 409A
5*	−23	−59	−57	−45	−40	−22
4*	−22	−58	−56	−43	−39	−20
3*	−20	−56	−54	−42	−37	−19
2*	−19	−55	−53	−41	−36	−17
1*	−17	−54	−52	−39	−35	−16
0	−16	−53	−51	−38	−34	−15
1	−13	−50	−48	−36	−31	−12
2	−11	−48	−46	−33	−29	−9
3	−9	−45	−43	−31	−27	−7
4	−6	−43	−41	−29	−24	−5
5	−4	−41	−39	−27	−22	−2
6	−2	−39	−37	−25	−20	0
7	0	−37	−35	−23	−18	2
8	2	−36	−33	−21	−17	4
9	4	−34	−32	−20	−15	6
10	6	−32	−30	−18	−13	8
11	8	−30	−28	−16	−12	9
12	9	−29	−27	−15	−10	11
13	11	−27	−25	−13	−8	13
14	13	−26	−23	−12	−7	14
15	14	−24	−22	−10	−5	16
16	16	−23	−20	−9	−4	17
17	17	−21	−19	−8	−3	19
18	19	−20	−18	−6	−1	20
19	20	−19	−16	−5	0	22
20	21	−17	−15	−4	1	23
21	23	−16	−14	−2	3	25
22	24	−15	−12	−1	4	26
23	25	−14	−11	0	5	27
24	27	−12	−10	1	6	29
25	28	−11	−9	2	8	30
26	29	−10	−8	4	9	31
27	30	−9	−6	5	10	32
28	32	−8	−5	6	11	34
29	33	−7	−4	7	12	35
30	34	−6	−3	8	13	36
31	35	−5	−2	9	14	37
32	36	−4	−1	10	15	38
33	37	−2	0	11	16	39
34	38	−1	1	12	17	40
35	39	0	2	13	18	41
36	40 30	0	3	14	19	43
37	42 31	1	4	15	20	44
38	43 32	2	5	16	21	45 30
39	44 33	3	6	17	22	46 31
40	45 34	4	7	18	23	47 32
42	46 36	6	10	19	25	48 34
44	48 38	8	12	21	26	50 36
46	50 40	10	14	23	28	38
48	42	11	16	24	30	39
50	44	13	16	26	31	41
52	45	14	17	28	33	43
54	47	16	19	29	34	45
56	49	18	20	31	36	46
58	50	19	22	32	37	48
60	52	20	23	33	39	50
62	53	22	25	35	40 30	51
64	55	23	26	36	42 32	53
66	56	25	27	37	43 32	54
68	58	26	29	39	44 33	56
70	59	27	30 29	40 30	46 34	57
72	61	29	32 31	41 31	47 36	58
74	62	30	33 32	43 32	48 37	60
76	64	31	34 33	44 34	49 38	61
78	65	32 30	35 34	45 35	39	63
80	66	34 31	37 36	46 36	41	64
85	69	37 34	40 39	49 39	44	67
90	73	40 37	42 42	42	46	70
95	76	42 40	45 44	45	49	73
100	78	45 43	48 47	47	52	76
105	81	48 45	50	50	54	79
110	84	50 48	52	53	57	82
115	87	50	55	55	59	84
120	89	53	57	57	62	87
125	92	55	59	60	64	89
130	94	57	62	62	66	92
135	96	60	64	64	69	94
140	99	62	66	66	71	96
145	101	64	68	68	73	99
150	103	66	70	70	75	101
155	105	68	72	72	77	103
160	108	70	74	74	79	105
165	110	72	76	76	81	107
170	112	74	78	78	82	109
175	114	75	80	80	84	111
180	116	77	82	81	86	113
185	117	79	83	83	88	115
190	119	81	85	85	90	117
195	121	82	87	87	91	119
200	123	84	88	88	93	121
205	125	86	90	90	95	123
210	127	87	92	91	96	124
220	130	91	95	94	99	128
230	133	94	98	97	102	131
240	136	97	101	100	105	134
250	140	99	104	103	108	137
260	143	102	107	106	111	141
275	147	106	111	110	115	145
290	151	110	115	114	119	149
305	155	114	118	117	123	153
320	159	118	122	121	126	157
335	163	121	126	124	130	161
350	167	125	129	128	133	165
365	170	128	132	131	137	169

Header: REFRIGERANT – (Sporlan Code). The chart is annotated with "BUBBLE POINT" and "DEW POINT" regions (with arrows) for each refrigerant blend; where two values appear in a cell they correspond to the bubble point and dew point temperatures.

*Inches mercury below one atmosphere

FIGURE 3–131 A pressure–temperature chart for refrigerant blends with a temperature glide. (*Courtesy of Sporlan Valve Company*)

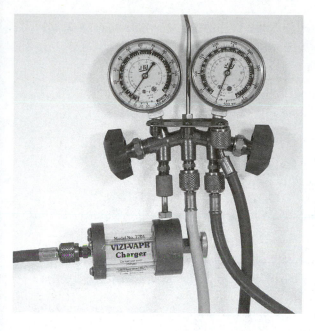

FIGURE 3–132 A charging device in the gage line between the liquid refrigerant in the cylinder and the suction line of the system. (*Photo by John Tomczyk*)

1. *Review unit nameplate and record the type of refrigerant required for the unit.* _____ ❏
2. *Review unit nameplate and record the manufacturer's charge amount.* _____ ❏
3. Set the charging cylinder on charging scales. ❏
4. Connect the center hose of the manifold gage set to the valve port of the refrigerant cylinder. ❏

NOTE: *If using a recovery bottle of refrigerant, connect manifold center hose to the Liquid Valve of Recovery Cylinder.*

---

5. Open the refrigerant cylinder valve. ❏
6. Slowly loosen the center hose connection on the gage manifold for about 2 seconds to purge air from the center hose, and then retighten it. ❏
7. If charging cylinder is not a recovery cylinder, place the charging cylinder upright on the charging scale. ❏
8. Follow the manufacturer's instructions for the charging scales you are using and prepare charging scales for charging. ❏
9. Open the manifold high-side gage valve and allow the required amount of liquid refrigerant to enter the system. ❏

NOTE: *Depending on the amount of refrigerant designed for the system you are charging, you may not be able to complete the system's total charge with liquid entering the high side of the system. To complete the charge for the system technicians may have to complete the charging process by running the unit and adding the balance of the liquid charge through the low side of the system.*

---

10. Close the manifold high-side gage valve once charge is complete. ❏
11. Turn the system ON. ❏
12. *If applicable*, adjust charge by opening the low-side manifold gage valve and add refrigerant in the liquid state to the sealed system's low side. ❏

> **NOTE:** *When adding liquid to the suction line of a system that is running, has to be restricted and vaporized into the system to avoid any damage to the compressor. Refer to Figure 3–132.*

13. Allow system to run once the required system charge is correct. ❑
14. *After 10 minutes of operation, record the system's low-side operating pressure.* _____ ❑
15. *Record the low-side saturation temperature.* _____ ❑
16. *Record the system's high-side operating pressure.* _____ ❑
17. *Record the high-side saturation temperature.* _____ ❑
18. *Check the unit manufacturer's nameplate and record the manufacturer's RLA (Run Load Amperage).* _____ ❑
19. Use the clamp-on ammeter and measure the system's RLA amperage at the unit disconnect. ❑
20. *Record the system's operating RLA.* _____ ❑

> **NOTE:** *The system's operating RLA should be at or close to the manufacturer's rated RLA if the system's charge is correct.*

21. Close the charging cylinder valve. ❑
22. Slowly loosen and remove the hose from the charging cylinder. ❑
23. Slowly remove the low-side and high-side hoses from the system's service valves. ❑
24. Replace the system service valve caps. ❑
25. Have your instructor check your work. ❑
26. At the direction of your instructor, turn unit OFF, disconnect all equipment, and return all tools and test equipment to their proper location. ❑

RESEARCH QUESTIONS

1. How is refrigerant cylinder pressure kept above the system pressure when charging with vapor from the cylinder?

   _____

   _____

2. Why does the refrigerant pressure decrease in the refrigerant cylinder while charging with vapor?

   _____

   _____

3. What methods besides weighing and measuring are used for charging systems?

   _____

   _____

4. What is the main difference between zeotropic and azeotropic refrigerant blends?

   _____

   _____

5. How many refrigerants are used to make up a ternary refrigerant blend?

   _____

   _____

Passed Competency _____   Failed Competency _____

Instructor Signature _____   Grade _____

Practical Competency 66

Adjusting the Charge of a Central Air Conditioning System by Adding Refrigerant Vapor

> NOTE: *The competency procedures listed are those for adding refrigerant vapor to a central air conditioning system that is low on charge.*

SUGGESTED MATERIALS

Textbook
Refrigeration & Air Conditioning Technology, 5th Edition, Thomson Delmar Learning
Unit 10—System Charging
Unit 9—Refrigerant and Oil Chemistry and Management—Recovery, Recycling, Reclaiming, and Retrofitting

Review Topics
Liquid Refrigerant Charging; Weighing Refrigerant; Using Charging Devices;
Azeotropic Refrigerants; Vapor Refrigerant Charging

COMPETENCY OBJECTIVE

The student will be able to charge or adjust the charge of an air conditioning system by using refrigerant vapor.

OVERVIEW

Adding refrigerant to refrigeration sealed system is referred to as charging. Air conditioning, heat pumps, and refrigeration equipment must have the correct amount of refrigerant charge for the equipment to operate and perform as efficiently as it was designed to. With today's high-efficiency equipment, ensuring that a system is charged with the correct amount of refrigerant is one of the most important procedures a technician is required to perform. With today's high-efficiency equipment, ensuring that a system is charged with the correct amount of refrigerant is one of the most important procedures a technician is required to perform. This requires that technicians have a thorough understanding of the system operation and operating conditions which indicate whether a system is properly *charged*, *overcharged*, or *undercharged*. Refrigerant can be added to a sealed system in the vapor or liquid state by *weighing the refrigerant charge* into the sealed system and measuring the charge into the sealed system (*Theory Lesson: Sealed System Charging, Figure 3–116*), using *system operating charts* (*Theory Lesson: Sealed System Charging, Figure 3–117*), manufacturer's superheat charging tables (*Theory Lesson: Sealed System Charging, Figure 3–118*), suction line temperature tables (*Theory Lesson: Sealed System Charging, Figure 3–119*), or manufacturer's sub-cooling charging tables for systems with a TXV metering device (*Theory Lesson: Sealed System Charging, Figure 3–120*).

Charging calculators are also available for use with R-22 and R-410A split-system air conditioning and heat pumps (*Theory Lesson: Sealed System Charging, Figure 3–121*).

Vapor refrigerant charging of a system is accomplished by allowing refrigerant vapor to move out of the refrigerant cylinder and into the low-pressure side of the refrigerant system. If the sealed system is in a vacuum or if the system is out of refrigerant, refrigerant vapor can be added to the low and high sides of the sealed system. When the system is operational, however, refrigerant vapor is always added through the low side of the system.

Liquid can also be added to the low side of a system, but special restrictive devices must be used to allow refrigerant liquid to vaporize before entering the sealed system's low side.

EQUIPMENT REQUIRED

1 Manifold gage set
1 Charging scale
1 Clamp-on ammeter
1 Refrigerant cylinder R-22

SAFETY PRACTICES

Wear goggles or safety glasses and gloves when working with refrigerant in sealed systems. Beware of high pressures and follow all electrical safety rules. Follow all EPA rules and regulations while working with refrigerants.

NOTE: *Competency Procedures could be used to adjust or charge any sealed system by adding refrigerant to the system in the vapor state.*

The following procedures can be used to add refrigerant vapor to a sealed system for charging purposes or adjusting the charge of a sealed system.

The unit should be operating in designed load conditions. If the system has a low-pressure cut-out control, it may have to be bypassed to get the system to operate until proper operating pressure is reached.

If charging a system that has a fixed-bore metering device, the student will have to know what operating saturation temperature is required for the system he is charging. (Check with your instructor if you are not sure what the operating low-side pressure should be for the system you are working on.)

If the system has a TXV (thermostatic expansion valve) with a sight glass on the liquid line, charge in the vapor state while observing the sight glass. Once the sight glass is clear of liquid, charging is complete.

COMPETENCY PROCEDURES
Checklist

1. Check to see that both the high-side and low-side service valves are closed on the manifold gage set. ☐
2. Remove the system's service valve caps and connect the low-side and high-side manifold gage hoses to the system's low- and high-side service valves. ☐
3. Connect the center hose of the manifold set to the valve port of the refrigerant cylinder. ☐
4. Set the refrigerant cylinder on the charging scales. ☐
5. Open the refrigerant cylinder valve. ☐
6. Loosen the center hose connection on the gage manifold for about 2 seconds to purge any air from the center hose and then tighten. ☐
7. Turn the refrigerant charging scale ON. ☐
8. Follow charging scale procedures and reset or zero the charging scale. ☐
9. Start the refrigeration system and let the unit operate for 10 minutes. ☐

NOTE: *Under standard conditions of a 75-degree (F) inside air temperature with 50% humidity and an outside temperature of 95 degrees (F), the refrigerant in the evaporator should be evaporating at approximately 40 degrees. For Refrigerant 22, this equals a low-side operating pressure of approximately 68.5 psig.*

---

10. *After 10 minutes of run time, record the system's low-side-operating pressure.* _____ ❏

11. *Convert the system's low-side operating pressure to saturation temperature and record.* _____ ❏

12. Open the manifold low-side valve and introduce refrigerant vapor into the system. ❏

13. Monitor the refrigerant scale and when 4 ounces of refrigerant vapor has been weighed into the system, close the low-side valve. ❏

14. *Allow the system to operate for 5 minutes and record the system's low-side pressure.* _____ ❏

15. *Convert low-side pressure to saturation.* _____ ❏

16. *If the system is still undercharged, introduce another 4 ounces of refrigerant into the system.* ❏

17. *Allow the system to operate for 5 minutes and record the system's low-side pressure.* _____ ❏

18. *Convert low-side pressure to saturation.* _____ ❏

19. Continue adding refrigerant to the system until the desired operating pressure and evaporator saturation temperature are reached. ❏

20. Once the system is charged to the desired operating pressure, record the following: ❏

    A. *Low-side operating psig.* _____ ❏
    B. *Low-side saturation temperature.* _____ ❏
    C. *High-side operating psig.* _____ ❏
    D. *High-side saturation temperature.* _____ ❏

21. *Check the unit's nameplate and record the manufacturer's required RLA (Run Load Amperage).* _____ ❏

22. *Use the clamp-on ammeter and record the system's operating RLA once the charge is complete.* _____ ❏

NOTE: *The system's operating RLA should be the required manufacturer's rated RLA, or close to it if system's charge is correct. Adjust the charge if needed.*

---

23. If the charge is complete, close the refrigerant cylinder valve. ❏

24. Have your instructor check your work. ❏

25. Once the work has been checked, slowly loosen and remove the manifold center hose from the refrigerant cylinder. ❏

26. Slowly remove the low- and high-side manifold hoses from the system's low-and high-side service valves. ❏

27. Replace the system service valve caps. ❏

28. Turn the charging scales OFF. ❏

29. Turn the sealed system OFF. ❏

30. Return all equipment and supplies to their proper location. ❏

RESEARCH QUESTIONS

1. High-temperature refrigeration is referred to as what type of refrigeration?

_____

_____

2. Medium-temperature refrigeration is referred to as what type of refrigeration?

_____

_____

3. Low-temperature refrigeration is referred to as what type of refrigeration?

_____

_____

4. What creates the pressure on the low side of a refrigeration system?

_____

_____

5. What determines the operating temperature of the evaporator coil?

_____

_____

Passed Competency _____   Failed Competency _____

Instructor Signature _____   Grade _____

Theory Lesson: Evaporators and Evaporator Superheat (Figure 3–133)

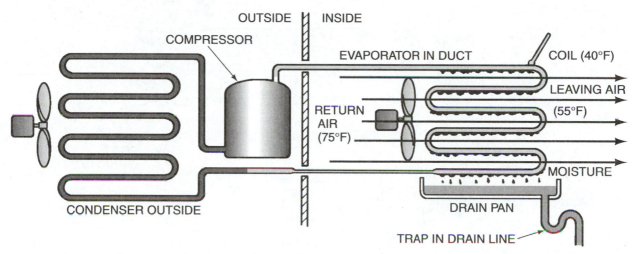

FIGURE 3–133   The evaporator absorbs heat from the air and causes moisture from the air to condense on the coil's cold surface.

SUGGESTED MATERIALS

Textbook

Refrigeration & Air Conditioning Technology, 5th Edition, Thomson Delmar Learning
Unit 21—Evaporators and Refrigeration Systems
Unit 36—Refrigeration Applied to Air Conditioning
Unit 40—Typical Operating Conditions

Review Topics

The Evaporator; The Evaporator and Boiling Temperature; Removing Moisture; Evaporator Evaluation; The Flooded Evaporator; Dry-Type Evaporator Performance; The A Coil; The Slant Coil; The H Coil

Key Terms

A coil • direct expansion evaporators • dry type evaporators • flood-back • flooded evaporator • H coil • hidden heat superheat • latent heat • latent heat of evaporation process • saturation temperature • sensible heat • slant coil • standard conditions • starved evaporator • standard operating conditions

OVERVIEW

The evaporator of refrigeration sealed system is the main component for absorbing heat from the conditioned medium (*air or water*). The heat absorbing process is accomplished by maintaining the evaporator coil surface area at a lower temperature than that of the air coming across the coil (**Figure 3–134**).

    The boiling temperature of liquid refrigerant in the coil determines the evaporator's surface operating temperature. Refer to *Figure 3–134* and you can see that the 75-degree (F) air is being passed over a 40-degree (F) evaporator coil, gives up heat to the evaporator, and leaves the coil at 55 degrees (F). As the 75-degree (F) air passes over the evaporator cold surface, two types of heat are absorbed from the air, *sensible heat* and *latent heat*. *Sensible heat* is referred to as heat that can be measured with a thermometer and can be seen and measured by the drop in air temperature as it leaves the coil at 55 degrees (F). The other

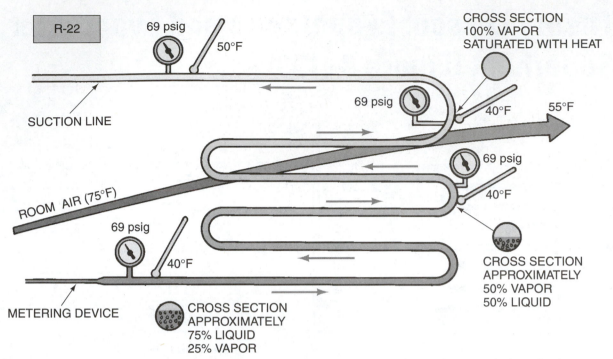

FIGURE 3–134   The evaporator coil surface area is maintained at 40 degrees because the liquid refrigerant passing through the coil is boiling at 40 degrees.

type of heat is referred to as latent heat. By definition, *latent heat* is heat that is required to cause a substance to change from one state of matter to another state of matter. All matter can exist in three states: *solid, liquid, or vapor*; further, the state of matter can be changed by the addition or removal of *latent heat*. *For example,* **ice to water** would require the *addition of latent heat*; **water to ice** would require the *removal of latent heat*. Controlling the amount of pressure exerted on the substance will also determine the actual temperature at which a substance changes state (**Figure 3–135**).

In the evaporator in *Figure 3–134*, the liquid refrigerant is boiling, or changing state from a liquid to a vapor at 40 degrees (F). This is referred to as the latent heat of evaporation process. By definition, *latent*

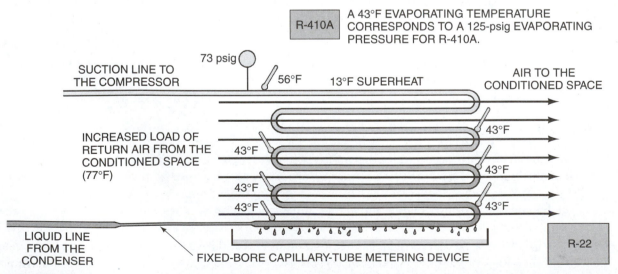

FIGURE 3–135   The low-side pressure on the evaporator has changed due to an increase in the air temperature coming across the evaporator coil. In turn the refrigerant boiling temperature in the evaporator has also risen.

heat of evaporation is the amount of heat required to cause a liquid to boil (evaporate) to a vapor. In reality, heat from the 75-degree air passing over the evaporator surface is being used in the *latent heat of evaporation* process, causing the liquid refrigerant to boil from liquid to vapor at 40 degrees (F). The absorption of heat from the 75-degree air causing the latent heat of evaporation process to take place is reflected by the change in the *sensible heat temperature* of the air as it leaves the evaporator coil at 55 degrees (F). It is important to understand that *latent heat does not cause a change in the temperature of a substance; latent heat only causes a substance to change state.* This is why latent heat is also referred to as "*hidden heat.*" Anytime a substance changes from one state of matter to another, a large amount of latent heat will have to be added or removed for the change of state to take place.

The design of mechanical refrigeration systems allows a liquid to boil in the evaporator at low temperatures and pressures so that heat can be absorbed from the medium being cooled (**Figure 3–136**).

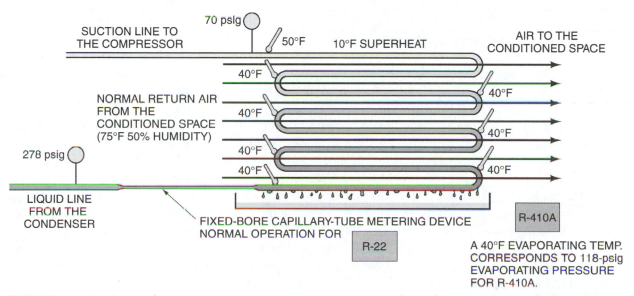

FIGURE 3–136 Air conditioning evaporators operate at approximately 40 degrees under standard conditions.

Not only does the cool surface area of the evaporator remove sensible heat from the air, but it is also used to cool moisture vapor that is normally mixed with the air. The cooling of the water vapor causes water droplets to form and collect on the evaporator coil. The water droplets drain from the coil surface and collect in a condensate pan located somewhere in an area outside of the conditioned space (**Figure 3–137**).

In refrigeration applied to *air conditioning, liquid refrigerant is designed to boil in the evaporator at range of approximately 35 to 55 degrees (F), with 40-degree (F) saturation temperature being the norm under standard operating conditions. Standard operating conditions are referenced as conditions in which the inside air temperature is 75 degrees (F) with 50% humidity and the outside air temperature is 95 degrees (F). Under these conditions, the refrigerant inside of the air conditioner evaporator will be boiling at approximately 40 degrees (F).* Whenever the system is operating outside of these conditions, the boiling temperature of the refrigerant inside the evaporator will change, affecting the evaporator's surface temperature. *By design, the air conditioning and heat pump evaporator's surface temperature will be approximately 30 to 35 degrees colder than the air temperature being drawn across the coil.* This is due to the boiling refrigerant temperature within the coil.

This knowledge can be helpful to technicians in evaluating what a fixed-bore metering device sealed system's operating low-side pressure should be under most load conditions. *For example*, if the air temperature coming across the evaporator coil is 75 degrees, subtract 30 and 35 degrees (F) from the air temperature of 75 degrees to determine the boiling temperature of the refrigerant within the evaporator. This would mean that the refrigerant in the evaporator coil should be boiling at approximately 40 to 45 degrees (F). Next, determine system refrigerant type and convert the evaporator coil temperature of 40 to 45 degrees (F) to psig. This would tell a technician what range the system's low-side operating pressure should be under these conditions. This

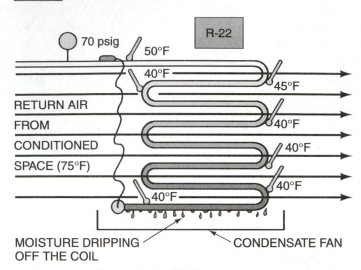

FIGURE 3–137   Humidity in the room air being condensed on the evaporator coil surface as the air is cooled.

information can be used to evaluate actual system operating low-side pressures. Pressure readings outside of this range (*plus or minus 10% to allow for the percentage of relative humidity within the air being conditioned*) is an indication that the system in not operating in the design range, and technicians should look for reasons as to why and what corrective actions could be taken to bring system pressures more into the proper range.

The amount of heat absorbed from the medium to be cooled is based on a number of factors. The temperature and percentage of the humidity in the air, how fast and the amount of air moved across the evaporator coil, along with the amount of liquid refrigerant fed into the evaporator coil by the system's metering device are just a few of these factors. Ideally, liquid refrigerant should be fed into the evaporator coil surface area without overfeeding the coil to a point that liquid refrigerant would pass through the whole coil and eventually enter the system's compressor. Referring to *Figure 3–136,* under standard conditions you can see that refrigerant liquid is fed and boiled at 40 degrees (F) through the evaporator about 90% of the coil surface area. Through this area of the evaporator coil, the latent heat process of evaporation is taking place, removing sensible heat from the room air and latent heat from the water vapor, causing water droplets to condense on the coil surface and drain to the condensate pan. Removal of both sensible and latent heat will cause the room air temperature to drop. At about the 90% area of the evaporator coil, the refrigerant is at the 100% vapor state. This means that no longer is any liquid refrigerant left in the refrigerant mixture. At this point, heat from the room air heats the vapor above the 40-degree boiling temperature of the refrigerant. This is referred to as *superheating the vapor.* In other words, *superheating a vapor means* raising the temperature of the vapor one or more degrees (F) over the saturation temperature of the boiling liquid refrigerant. By design, under standard conditions, heating the vapor approximately 10 degrees over the refrigerant saturation temperature of 40 degrees is normal. This means that the refrigerant entering the suction side of the compressor will be approximately 50 degrees (F), or superheated by 10 degrees (F) over the saturation temperature of the boiling liquid at 40 degrees (F).

Any change in the outside or inside air temperature, along with the percentage of relative humidity above or below those listed as standard conditions, will affect the temperature of the evaporator coil surface area and also the amount of evaporator superheat (**Figure 3–138** and **Figure 3–139**).

Evaporator superheat readings above what is expected under standard conditions are referred to as a *starved evaporator.*

Evaporator superheat readings below what is expected under standard conditions are referred to as a *flooded evaporator.*

The amount of evaporator superheat can be used to evaluate the evaporator's efficiency. Remember that the purpose of the evaporator is to absorb heat from the conditioned space or product, and that the greatest amount of heat absorbed by the evaporator is during the boiling process of liquid to vapor. How efficiently the evaporator is performing is based on how far through the coil liquid refrigerant is allowed to

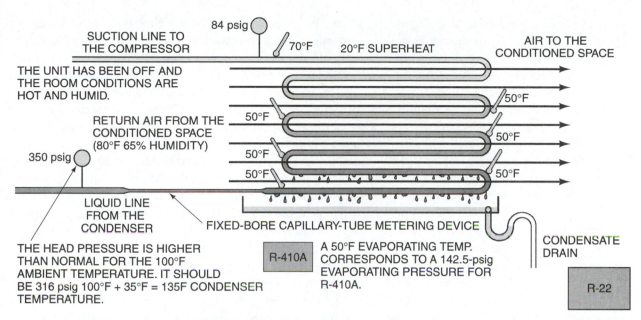

FIGURE 3–138   This is a fixed-bore metering device when an increase in load has caused the evaporator saturation temperature to rise and the amount of evaporator superheat to rise.

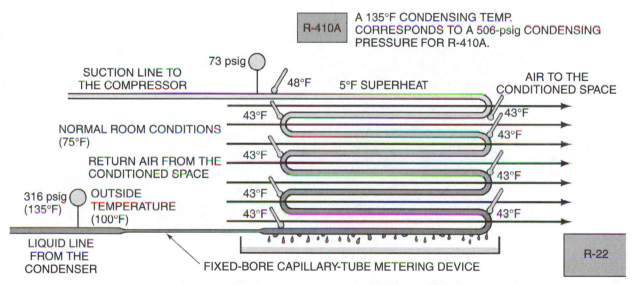

FIGURE 3–139   Increase in outside air temperature has caused an increase in system high-side pressure. The increased high-side pressure has increased the flow of refrigerant through the fixed-bore metering device, decreasing the amount of evaporator superheat.

boil before superheating the vapor begins. The ideal situation is to boil liquid refrigerant through the coil as much as possible without flooding liquid back to the compressor.

Because it is impossible to visually see inside the coils of the evaporator, checking the amount of evaporator superheat is a method of evaluating the efficiency of the latent heat of evaporation process of the evaporator.

The amount of evaporator superheat can be determined by knowing the saturation temperature of the liquid in the evaporator. This can be found by obtaining the sealed system's low-side operating psig, and using a pressure–temperature chart to convert this pressure to saturation temperature based on the type of refrigerant being used in the system. Next, the end-run temperature of the evaporator will have to be measured. This is normally done by placing a temperature probe at the suction line of the condensing unit and allowing the unit to operate for at least 15 minutes before recording suction line temperature. The amount of evaporator superheat can than be determined by subtracting the evaporator's saturation temperature from the evaporator's end-run temperature. All dry type or direct expansion evaporators are designed to operate with a superheat range of 8 to 12 degrees under standard operating conditions.

In air conditioning and heat pump equipment, three basic types of evaporators are used: the *A coil*, the *slant coil*, and the *H coil*. All are *dry-type of direct expansion evaporators*. This is because these evaporators operate with a required amount of superheat at the end of the coils by design.

The A coil evaporator (**Figure 3–140**) is used for up flow, down flow, and horizontal flow applications. It consists of two coils with their circuits side by side and spread apart at the bottom, which forms the letter A. The condensate pan is located at the bottom of the A pattern for up flow or down flow applications. In horizontal flow applications, the condensate pan is placed at the bottom of the coil and the coil is turned on its side. Airflow through the A coil is through the core of the coil.

The slant coil (**Figure 3–141**) can be used for up flow, down flow, or horizontal flow when designed for these applications. It is a one-piece coil that is mounted in the duct on an angle (usually 60 degrees) or slant to give the coil more surface area. The condensate pan is located at the bottom of the slant.

The H coil (**Figure 3–142**) is normally used in horizontal applications, but can be adapted to vertical applications by using a special drain pan design. The condensate drain pan is normally located at the bottom of the H pattern.

FIGURE 3–140   The A coil evaporator. (*Courtesy of Carrier Corporation*)

FIGURE 3–141   A slant coil. (*Courtesy of BDP Company*)

FIGURE 3–142   An H coil. (*Courtesy of BDP Company*)

Practical Competency 67

Measuring the Amount of Superheat on an Air Conditioner or Heat Pump Evaporator Coil

SUGGESTED MATERIALS

Textbook

Refrigeration & Air Conditioning Technology, 5th Edition, Thomson Delmar Learning
Unit 21—*Evaporators and Refrigeration Systems*
Unit 36—*Refrigeration Applied to Air Conditioning*
Unit 40—*Typical Operating Conditions*

Review Topics

The Evaporator; The Evaporator and Boiling Temperature; Removing Moisture; Evaporator Evaluation; The Flooded Evaporator; Dry-Type Evaporator Performance

COMPETENCY OBJECTIVE

The student will be able to measure the amount of superheat on an evaporator coil.

OVERVIEW

Superheating a vapor means to raise the temperature of the vapor one or more degrees (F) over the saturation temperature of the boiling liquid refrigerant. In refrigeration applied to *air conditioning, liquid refrigerant is designed to boil in the evaporator at range of approximately 35 to 55 degrees (F), with a 40-degree (F) saturation temperature being the norm under standard operating conditions. Standard operating conditions* are referenced as conditions in which the *inside air temperature is 75 degrees (F) with 50% humidity and the outside air temperature is 95 degrees (F). Under these conditions, the refrigerant inside of the air conditioner evaporator will be boiling at approximately 40 degrees (F). By design, under standard conditions, heating the vapor approximately 10 degrees over the refrigerant saturation temperature of 40 degrees is normal. Whenever the system is operating outside of these conditions, the boiling temperature of the refrigerant inside the evaporator will change, affecting the evaporator's surface temperature.

Evaporator superheat readings above what is expected under standard conditions are referred to as a *starved evaporator*.

Evaporator superheat readings below what is expected under standard conditions are referred to as a *flooded evaporator*. (Refer to *Theory Lesson: Evaporators and Evaporator Superheat, Figure 3–138 and Figure 3–139*.)

The amount of evaporator superheat can be used to evaluate the evaporator's efficiency. Remember that the purpose of the evaporator is to absorb heat from the conditioned space or product, and that the greatest amount of heat absorbed by the evaporator is during the boiling process of liquid to vapor. How efficient the evaporator is performing is based on how far through the coil liquid refrigerant is allowed to boil before superheating the vapor begins. The ideal situation is to boil liquid refrigerant through the coil as much as possible without flooding liquid back to the compressor.

Because it is impossible to visually see inside the coils of the evaporator, checking the amount of evaporator superheat is a method of evaluating the efficiency of the latent heat of evaporation process of the evaporator.

The amount of evaporator superheat can be determined by knowing the saturation temperature of the liquid in the evaporator. This can be found by obtaining the sealed system's low-side operating psig, and using a pressure–temperature chart to convert this pressure to saturation temperature based on the type of refrigerant being used in the system. Next, the end-run temperature of the evaporator will have to be measured. This is normally done by placing a temperature probe at the suction line of the condensing unit and allowing the unit to operate for at least 15 minutes before recording suction line temperature. The amount of evaporator superheat can than be determined by subtracting the evaporator's saturation temperature from the evaporator's end-run temperature. All dry-type or direct expansion evaporators are designed to operate with a superheat range of 8 to 12 degrees under standard operating conditions.

EQUIPMENT REQUIRED

1 Refrigeration manifold gage set
1 Temperature meter (*with temperature probes*)
1 Refrigeration wrench (*if applicable*)
1 Adjustable wrench (*if applicable*)
Safety glasses
Insulated gloves
Insulation tape

SAFETY PRACTICES

The student should be knowledgeable in the use of tools and testing equipment. Follow all EPA rules and regulations when working with refrigerants.

COMPETENCY PROCEDURES   Checklist

> **NOTE:** *If using a heat pump for completion of this competency, place the heat pump in the cooling mode.*

1. Turn the sealed system ON.   ❑
2. Make sure manifold gage valves are closed.   ❑
3. Remove the system suction service valve cap.   ❑
4. Connect the low-side manifold hose to the sealed system suction service valve. ❑
5. Attach a temperature probe near the suction line service valve (**Figure 3–143**). ❑
6. Insulate the temperature probe.   ❑
7. *Record the type of system refrigerant.* _____   ❑
8. Let the system operate for at least 15 minutes.   ❑
9. *Record the system's suction line temperature.* _____   ❑
10. *Record the low-side suction pressure of the system (**Figure 3–144**).*_____ ❑
11. Use the P/T chart or manifold gage (**Figure 3–144**) scale and convert low-side operating pressure to temperature.   ❑
12. *Record the saturation temperature of refrigerant in the evaporator.* _____ ❑
13. Determine amount of superheat by the following formula (**Figure 3–145**):   ❑

    Suction line temperature _____
    MINUS
    Saturation temperature _____
    EQUALS evaporator superheat _____

14. *Record the amount of superheat.* _____   ❑
15. *Is the recorded superheat higher or lower than that required for air conditioners according to required under standard conditions?* _____   ❑

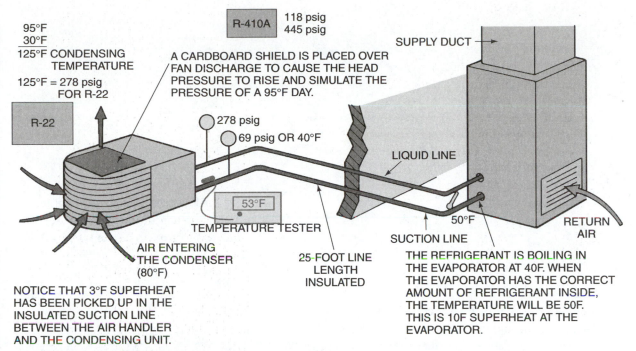

95°F
30°F
125°F CONDENSING
TEMPERATURE

125°F = 278 psig
FOR R-22

R-22

R-410A   118 psig
         445 psig

A CARDBOARD SHIELD IS PLACED OVER
FAN DISCHARGE TO CAUSE THE HEAD
PRESSURE TO RISE AND SIMULATE THE
PRESSURE OF A 95°F DAY.

278 psig
69 psig OR 40°F

53°F

TEMPERATURE TESTER

AIR ENTERING
THE CONDENSER
(80°F)

25-FOOT LINE
LENGTH
INSULATED

SUPPLY DUCT

LIQUID LINE

SUCTION LINE

50°F

RETURN
AIR

THE REFRIGERANT IS BOILING IN
THE EVAPORATOR AT 40F. WHEN
THE EVAPORATOR HAS THE CORRECT
AMOUNT OF REFRIGERANT INSIDE,
THE TEMPERATURE WILL BE 50F.
THIS IS 10F SUPERHEAT AT THE
EVAPORATOR.

NOTICE THAT 3°F SUPERHEAT
HAS BEEN PICKED UP IN THE
INSULATED SUCTION LINE
BETWEEN THE AIR HANDLER
AND THE CONDENSING UNIT.

FIGURE 3–143   Attaching temperature probe to sealed system suction line.

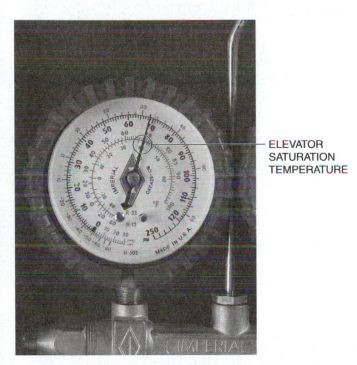

ELEVATOR
SATURATION
TEMPERATURE

FIGURE 3–144   Obtaining system operating
low-side pressure.

16. *Using a temperature probe, measure and record the temperature of the air being cooled
    by the system being checked.* _____   ❑

17. *Using a temperature probe, measure and record the temperature of the air after it has
    been passed over the evaporator coil.* _____   ❑

18. Determine the air temperature drop across the coil by: ❑

    Return air temperature _____

    MINUS

    Supply air temperature _____

    EQUALS *Temperature drop* _____

19. *Record the amount of temperature drop.* _____   ❑

20. *Is the amount of temperature drop higher or lower than that required under standard
    operating conditions?* _____   ❑

FIGURE 3–145 Subtract evaporator saturation temperature from the suction line temperature to obtain system superheat.

21. Turn the unit OFF. ❑
22. Remove suction line temperature probe. ❑
23. Remove manifold gage hose from the sealed system suction service valve. ❑
24. Replace the sealed system suction service valve cap. ❑
25. Have your instructor check your work. ❑
26. Return equipment and all tools to their proper location. ❑

RESEARCH QUESTIONS

1. What is a starved evaporator?

_____

_____

2. Superheat is required on dry-type or direct expansion evaporator to protect which component of the sealed system?

_____

_____

3. Can a fixed-bore metering device maintain constant evaporator superheat?

_____

_____

4. What metering device is designed to maintain constant evaporator superheat within a range of 8 to 12 degrees?

_____

_____

5. If the indoor blower stopped running on an air conditioning system with a fixed-bore metering device, would the evaporator's superheat go up or down?

_____

_____

Passed Competency _____ Failed Competency _____

Instructor Signature _____ Grade _____

Practical Competency 68

Measuring the Amount of Evaporator Superheat of an Air Conditioning or Heat Pump System Under Different Load Conditions

SUGGESTED MATERIALS

Textbook

Refrigeration & Air Conditioning Technology, 5th Edition, Thomson Delmar Learning
Unit 21—Evaporators and Refrigeration Systems
Unit 36—Refrigeration Applied to Air Conditioning
Unit 40—Typical Operating Conditions

Review Topics

The Evaporator; The Evaporator and Boiling Temperature; Removing Moisture; Evaporator Evaluation; The Flooded Evaporator; Dry-Type Evaporator Performance

COMPETENCY OBJECTIVE

The student will be able to measure the amount of evaporator superheat under different load conditions.

OVERVIEW

Superheating a vapor means raising the temperature of the vapor one or more degrees (F) over the saturation temperature of the boiling liquid refrigerant. In refrigeration applied to *air conditioning, liquid refrigerant is designed to boil in the evaporator at range of approximately 35 to 55 degrees (F), with 40-degree (F) saturation temperature being the norm under standard operating conditions. Standard operating conditions* are referenced as conditions in which the inside air temperature is *75 degrees (F) with 50% humidity and the outside air temperature is 95 degrees (F). Under these conditions, the refrigerant inside of the air conditioner evaporator will be boiling at approximately 40 degrees (F).* By design, under standard conditions, heating the vapor approximately 10 degrees over the refrigerant saturation temperature of 40 degrees is normal. Whenever the system is operating outside of these conditions, the boiling temperature of the refrigerant inside the evaporator will change, affecting the evaporator's surface temperature.

Evaporator superheat readings above what is expected under standard conditions are referred to as a *starved evaporator.*

Evaporator superheat readings below what is expected under standard conditions are referred to as a *flooded evaporator.* (Refer to *Theory Lesson: Evaporators and Evaporator Superheat, Figure 3–138 and 3–139.*)

The amount of evaporator superheat can be used to evaluate the evaporator's efficiency. Remember that the purpose of the evaporator is to absorb heat from the conditioned space or product, and that the greatest amount of heat absorbed by the evaporator is during the boiling process of liquid to vapor. How efficient the evaporator is performing is based on how far through the coil liquid refrigerant is allowed to boil before superheating the vapor begins. The ideal situation is to boil liquid refrigerant through the coil as much as possible without flooding liquid back to the compressor.

Because it is impossible to visually see inside the coils of the evaporator, checking the amount of evaporator superheat is a method of evaluating the efficiency of the latent heat of evaporation process of the evaporator.

The amount of evaporator superheat can be determined by knowing the saturation temperature of the liquid in the evaporator. This can be found by obtaining the sealed system's low-side operating psig, and using a pressure–temperature chart to convert this pressure to saturation temperature based on the type of refrigerant being used in the system. Next, the end-run temperature of the evaporator will have to be measured. This is normally done by placing a temperature probe at the suction line of the condensing unit and allowing the unit to operate for at least 15 minutes before recording suction line temperature. The amount of evaporator superheat can than be determined by subtracting the evaporator's saturation temperature from the evaporator's end-run temperature. All dry-type or direct expansion evaporators are designed to operate with a superheat range of 8 to 12 degrees under standard operating conditions.

EQUIPMENT REQUIRED

1 Refrigeration manifold gage set
1 Temperature meter (*with temperature probes*)
1 Refrigeration wrench (*if applicable*)
1 Adjustable wrench (*if applicable*)
Safety glasses
Insulated gloves
Insulation tape
Heat gun
Piece of cardboard (*big enough to block evaporator surface area*)

SAFETY PRACTICES

The student should be knowledgeable in the use of tools and testing equipment. Follow all EPA rules and regulations when working with refrigerants.

COMPETENCY PROCEDURES

Checklist

> *NOTE:* If performing Competency on a heat pump, place heat pump in the cooling mode to perform competency.

1. Turn the sealed system ON. ❏
2. Make sure manifold gage valves are closed. ❏
3. Remove the system suction service valve cap. ❏
4. Connect the low-side manifold gage hose to the sealed system suction service valves. ❏
5. Attach a temperature probe near the suction line service valve (**Figure 3–146**). ❏
6. Insulate the temperature probe. ❏
7. *Record the type of system refrigerant.* _____ ❏
8. Let the system operate for at least 15 minutes. ❏
9. *Record the system's suction line temperature.* _____ ❏
10. *Record the low-side suction pressure of the system.* _____ ❏
11. Use the P/T chart or manifold gage scale and convert low-side operating pressure to temperature. ❏
12. *Record the saturation temperature of the refrigerant in the evaporator.* _____ ❏
13. Determine the amount of superheat by the following formula: ❏

Suction line temperature _____

MINUS

Saturation temperature _____

EQUALS Evaporator superheat _____

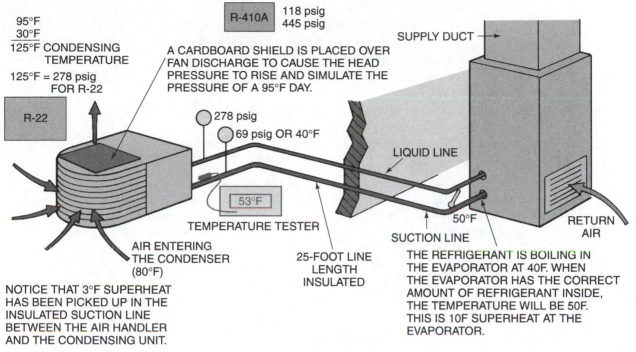

95°F
30°F
125°F CONDENSING
TEMPERATURE

125°F = 278 psig
FOR R-22

R-22

R-410A   118 psig
         445 psig

A CARDBOARD SHIELD IS PLACED OVER
FAN DISCHARGE TO CAUSE THE HEAD
PRESSURE TO RISE AND SIMULATE THE
PRESSURE OF A 95°F DAY.

278 psig
69 psig OR 40°F

SUPPLY DUCT →

LIQUID LINE

53°F

TEMPERATURE TESTER

50°F

SUCTION LINE

RETURN
AIR

AIR ENTERING
THE CONDENSER
(80°F)

25-FOOT LINE
LENGTH
INSULATED

THE REFRIGERANT IS BOILING IN
THE EVAPORATOR AT 40F. WHEN
THE EVAPORATOR HAS THE CORRECT
AMOUNT OF REFRIGERANT INSIDE,
THE TEMPERATURE WILL BE 50F.
THIS IS 10F SUPERHEAT AT THE
EVAPORATOR.

NOTICE THAT 3°F SUPERHEAT
HAS BEEN PICKED UP IN THE
INSULATED SUCTION LINE
BETWEEN THE AIR HANDLER
AND THE CONDENSING UNIT.

FIGURE 3–146   Attaching temperature probe to sealed system suction line.

14. *Record the amount of superheat.* _____   ❑

EVALUATING EVAPORATOR SUPERHEAT UNDER REDUCED LOAD

15. Use a piece of paper or cardboard and block the evaporator surface by 50%.   ❑
16. Allow the unit to operate for 10 minutes.   ❑
17. *Record the system's suction line temperature.* _____   ❑
18. *Record the low-side suction pressure of the system.* _____   ❑
19. Use the P/T chart or manifold gage scale and convert low-side operating pressure to
    temperature.   ❑
20. *Record the saturation temperature of refrigerant in the evaporator.* _____   ❑
21. Determine amount of superheat by the following formula: ❑

    Suction line temperature _____
    ### MINUS
    Saturation temperature _____
    EQUALS *Evaporator superheat* _____

22. *Record the amount of superheat.* _____   ❑
23. *With the reduced load on the evaporator, did the evaporator superheat go higher, lower,*
    or stay the same as the amount recorded in Step 14? _____   ❑
24. *Explain why you think the amount of superheat in Step 23 was higher, lower,*
    or the same as the amount recorded in Step 14. _____
    _____
    _____   ❑

EVALUATING EVAPORATOR SUPERHEAT UNDER INCREASED LOAD

25. Use a heat gun or another source of heat to increase the heat load on the evaporator coil.   ❑
26. Allow the unit to operate for 10 minutes.   ❑
27. *Record the system's suction line temperature.* _____   ❑
28. *Record the low-side suction pressure of the system.* _____   ❑
29. Use the P/T chart or manifold gage scale and convert low-side operating pressure to
    temperature.   ❑
30. *Record the saturation temperature of the refrigerant in the evaporator.* _____   ❑

31. *Determine the amount of superheat by the following formula:*
   Suction line temperature _____
   MINUS
   Saturation temperature _____
   EQUALS *Evaporator superheat* _____   ❏

32. *Record the amount of superheat.* _____   ❏

33. *With the increased load on the evaporator, did the evaporator superheat go higher, lower or stay the same as the amount recorded in Step 14?* _____   ❏

34. *Explain why you think the amount of superheat in Step 33 was higher, lower, or the same as the amount recorded in Step 14.* _____
   _____
   _____   ❏

35. Have your instructor check your work.   ❏

36. Disconnect equipment and return all tools and test equipment to their proper location.   ❏

RESEARCH QUESTIONS

1. What would happen to the evaporator superheat if the evaporator fan stopped working?
   _____
   _____

2. What would happen to the evaporator superheat if the condenser fan stopped working?
   _____
   _____

3. What would happen to the evaporator superheat if the system were low on refrigerant charge?
   _____
   _____

4. What would happen to the evaporator superheat if the sealed system were overcharged?
   _____
   _____

5. What effect would high evaporator superheat have on the compressor of the sealed system?
   _____
   _____

Passed Competency _____   Failed Competency _____

Instructor Signature _____   Grade _____

Practical Competency 69

Checking an Evaporator for Flood-Back Using the Hand-Touch Method (Figure 3–147)

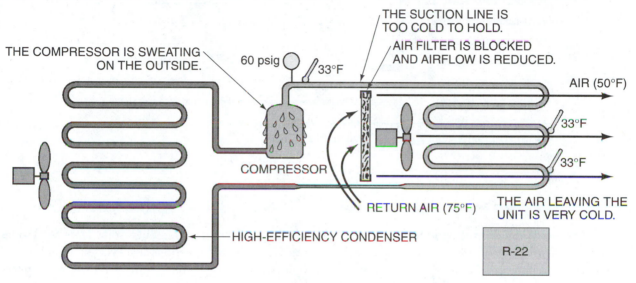

THE SUCTION LINE IS TOO COLD TO HOLD.

AIR FILTER IS BLOCKED AND AIRFLOW IS REDUCED.

THE COMPRESSOR IS SWEATING ON THE OUTSIDE.

60 psig   33°F

AIR (50°F)

33°F

33°F

COMPRESSOR

RETURN AIR (75°F)

THE AIR LEAVING THE UNIT IS VERY COLD.

HIGH-EFFICIENCY CONDENSER

R-22

FIGURE 3-147   An evaporator flooded with refrigerant.

SUGGESTED MATERIALS

Textbook

Refrigeration & Air Conditioning Technology, 5th Edition, Thomson Delmar Learning
Unit 21—Evaporators and Refrigeration Systems
Unit 36—Refrigeration Applied to Air Conditioning
Unit 40—Typical Operating Conditions

Review Topics

The Evaporator; The Evaporator and Boiling Temperature; Removing Moisture; Evaporator Evaluation; The Flooded Evaporator; Dry-Type Evaporator Performance

COMPETENCY OBJECTIVE

The student will be able to determine if a refrigeration system is flooding back to the compressor by using the hand-touch method.

OVERVIEW

An evaporator that has very little or no superheat is referred to as a flooded evaporator. This is a situation in which liquid refrigerant is boiling too far through the coil and possibly enters the compressor on the low side. Liquid flood-back can occur for many reasons, such as a refrigerant overcharge, low load condition, restricted airflow, dirty coil, or a metering device that is overfeeding the coil. Flood-back in most cases is not caused by a system overcharge or an overfeeding metering device. In most cases, a flood-back situation is due to reduced airflow, reduced load, dirty filters, loose evaporator fan motor belt, and so forth. There are those in the field who state that a frosted suction line or a compressor that is "sweating" is a good indication of evaporator flood-back.

Although flood-back could be at fault for these conditions, especially in air conditioning and heat pump equipment, it is not the only reason that could cause such conditions. In refrigeration equipment, in addition to the situations stated above, a low-temperature superheated vapor could also cause suction line frosting or compressor sweating. Evaporator flood-back can be checked in a couple of ways; one is to evaluate evaporator superheat (**Figure 2–148**).

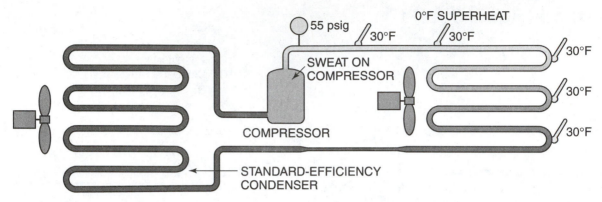

FIGURE 3-148   When the suction pressure is too low and the superheat is low, the unit is not boiling the refrigerant in the evaporator. The coil is flooded with refrigerant.

Another method is the hand-touch method, which is simple to perform, although some facts need to be understood about refrigerant in the liquid and vapor state. Whenever liquid refrigerant is boiling, heat is being absorbed from the conditioned air. The refrigerant vapor leaving the boiling refrigerant is at the same temperature as the liquid refrigerant. Remember that refrigerant vapor cannot be superheated until there no longer is any liquid refrigerant left boiling in the mixture. At this point, the refrigerant vapor can be heated (superheated) above the saturation temperature of the boiling liquid. The human body's average temperature is 98.6 degrees (F). If a technician places a hand on the suction line of an operational sealed system, the heat from the hand will be transferred to the refrigerant vapor or boiling liquid. If superheated vapor is passing through the suction line where the hand is placed, the technician's hand would feel the coolness of the suction line and then feel the line get warmer because of the heat from the hand. This is an indication that superheat vapor refrigerant is passing through the suction line where the hand is placed. The additional heat from the hand is superheating the suction vapor to a higher temperature. If liquid refrigerant were passing through the suction line where the hand were placed, the technician would feel the coolness of the suction line and then feel the hand getting colder, even to the point that the hand would have to be removed because of feeling pain. This is an indication of liquid flood-back. The heat from the technician's hand is causing the liquid refrigerant in the suction line to boil faster where the hand is placed; in turn heat is being removed from the technician's hand and eventually frostbite could occur if the hand were not removed from the suction line. In this situation, liquid is flooding back to the compressor and if not corrected, the compressor could be damaged.

EQUIPMENT REQUIRED

Something to restrict evaporator airflow

SAFETY PRACTICES

The student should be knowledgeable in the use of tools and testing equipment and follow all safety procedures when working with live voltage. The student should also follow all EPA rules and regulations when working with refrigerants.

COMPETENCY PROCEDURES

> **NOTE:** *This competency could be performed on any operational air conditioning, heat pump, or refrigeration equipment.*

1. Turn the power to the unit ON. ❏
2. Let the system operate for 10 minutes. ❏
3. Wrap your hand around the suction line of the sealed system and leave it there a few seconds. ❏
4. *Explain what you felt happened.* ❏

   _____

> **NOTE:** *If the system is not flooding back, your hand should feel cold when you first wrap your hand around the suction line and then continue to get warmer. This tells you that there is superheated vapor coming back to the system.*

5. Use some means and restrict the airflow to the evaporator coil. ❏
6. Let the system operate for 10 minutes. ❏

> **NOTE:** *If the system is equipped with a low-pressure cut-out control, you may have to jumper the control to keep the system operational for this test.*

7. Wrap your hand around the suction line and leave it there a few seconds. ❏
8. *Explain what response you felt with your hand.* ❏

   _____
   _____

9. *Was the system flooding back?* _____ ❏

> **NOTE:** *If liquid refrigerant is flooding back through the suction line, the technician should have felt the coldness of the suction line and then feel the heat being draw from the hand by the boiling refrigerant.*

10. Correct the restriction to system airflow. ❏
11. Allow system to operate until system flood-back has cleared up. ❏
12. Turn the power to the unit OFF. ❏
13. Replace all panels (*if applicable*). ❏
14. Have your instructor check your work. ❏
15. Disconnect equipment and return all tools and test equipment to their proper location. ❏

RESEARCH QUESTIONS

1. What is flood-back?

   _____

   _____

2. Are some evaporators flooded by design?

   _____

   _____

3. On a system that has liquid refrigerant flood-back, why does your hand continue to get colder when you use the hand-touch method to check for flood-back?

_____

_____

4. When using the hand-touch method to check for liquid flood-back, why does your hand first feel cold and then continue to warm up when the system is not flooding back?

_____

_____

5. If a system is flooding back, does this mean that the system is charged incorrectly?

_____

_____

Passed Competency _____   Failed Competency _____

Instructor Signature _____   Grade _____

Practical Competency 70

Evaluating Evaporator Efficiency

SUGGESTED MATERIALS

Textbook

Refrigeration & Air Conditioning Technology, 5th Edition, Thomson Delmar Learning
Unit 21—Evaporators and Refrigeration Systems
Unit 36—Refrigeration Applied to Air Conditioning
Unit 40—Typical Operating Conditions

Review Topics

The Evaporator; The Evaporator and Boiling Temperature; Removing Moisture; Evaporator Evaluation; The Flooded Evaporator; Dry-Type Evaporator Performance

COMPETENCY OBJECTIVE

The student will be able to evaluate the evaporator efficiency of an air conditioner or heat pump.

OVERVIEW

The amount of evaporator superheat can be used to evaluate the evaporator's efficiency. Remember that the purpose of the evaporator is to absorb heat from the conditioned space or product, and that the greatest amount of heat absorbed by the evaporator is during the boiling process of liquid to vapor. How efficient the evaporator is performing is based on how far through the coil liquid refrigerant is allowed to boil before superheating the vapor begins. The ideal situation is to boil liquid refrigerant through the coil as much as possible without flooding liquid back to the compressor.

Because it is impossible to visually see inside the coils of the evaporator, checking the amount of evaporator superheat is a method of evaluating the efficiency of the latent heat of evaporation process of the evaporator.

The amount of evaporator superheat can be determined by knowing the saturation temperature of the liquid in the evaporator. This can be found by obtaining the sealed system's low-side operating psig, and using a pressure–temperature chart to convert this pressure to saturation temperature based on the type of refrigerant being used in the system. Next, the end-run temperature of the evaporator will have to be measured. This is normally done by placing a temperature probe at the suction line of the condensing unit and allowing the unit to operate for at least 15 minutes before recording suction line temperature. The amount of evaporator superheat can than be determined by subtracting the evaporator's saturation temperature from the evaporators end-run temperature. All dry-type or direct expansion evaporators are designed to operate with a superheat range of 8 to 12 degrees under standard operating conditions.

EQUIPMENT REQUIRED

1 Manifold gage set
1 Temperature tester (*with probes*)

1 Refrigeration wrench
1 Refrigeration saddle service valve (*if applicable*)
Insulation tape
Assorted hand tools

SAFETY PRACTICES

The student should be knowledgeable in the use of tools and testing equipment and follow all safety procedures when working with live voltage. The student should also follow all EPA rules and regulations when working with refrigerants.

COMPETENCY PROCEDURES

Checklist

> **NOTE:** *This competency is designed to evaluate the efficiency of the actual evaporator of a sealed system, and can be performed on any operational air conditioner or heat pump system.*

1. Turn the sealed system ON. ❏
2. Make sure manifold gage valves are closed. ❏
3. Remove the system suction service valve cap. ❏
4. Connect the low-side manifold hose to sealed system low-side service valve. ❏
5. Attach a temperature probe near the suction line service valve. ❏
6. Insulate the temperature probe. ❏
7. *Record the type of system refrigerant.* _____ ❏
8. Let the system operate for at least 15 minutes. ❏
9. *Record the system's suction line temperature.* _____ ❏
10. *Record the low-side suction pressure of the system.* _____ ❏
11. Use the P/T chart or manifold gage scale and convert low-side operating pressure to temperature. ❏
12. *Record the saturation temperature of refrigerant in the evaporator.* _____ ❏
13. *Determine the amount of superheat by the following formula:* ❏
$$\text{Suction line temperature } \underline{\hspace{2cm}}$$
$$\textbf{\textit{MINUS}}$$
$$\textit{Saturation temperature } \underline{\hspace{2cm}}$$
$$\textbf{\textit{EQUALS}} \textit{ Evaporator superheat } \underline{\hspace{1.5cm}}$$
14. *Record the amount of superheat.* _____ ❏
15. *Using a temperature probe, measure and record the temperature of the air being cooled by the system being checked.* _____ ❏
16. *Using a temperature probe, measure and record the temperature of the air after it has been passed over the evaporator coil.* _____ ❏
17. *Determine the air temperature drop across the coil by:* ❏
$$\textit{Return air temperature } \underline{\hspace{2cm}}$$
$$\textbf{\textit{MINUS}}$$
$$\textit{Supply air temperature } \underline{\hspace{2cm}}$$
$$\textbf{\textit{EQUALS}} \textit{ Temperature drop } \underline{\hspace{1.5cm}}$$
18. *Record the amount of temperature drop.* _____ ❏
19. *Is the amount of temperature drop higher or lower than that required under standard operating conditions?* _____ ❏

> **NOTE:** *Based on standard operating conditions, air conditioning equipment should be operating with in a range of 8 to12 degrees of evaporator superheat. Remember that standard operating conditions are stated as:* **an inside air temperature of 75 degrees, with 50% humidity, and an outside air temperature of 95 degrees (F)**. *If the evaporator were operating with no superheat, it could be stated that the evaporator was operating at 100% efficiency in absorbing latent heat from the air passing over the evaporator. With the evaporator operating with 8 to 12 degrees of superheat, it could be stated that the evaporator's efficiency is about 88% to 92%, which is normal if operating under standard conditions. Air temperature above or below standard conditions will change the efficiency percentage of the evaporator in absorbing latent heat from the air passing over the coil.*

20. *Is the temperature of the air passing over the evaporator coil higher or lower than that listed under standard operating conditions?* _____ ❑
21. *Is recorded superheat in Step 14 higher or lower than that required for air conditioners under standard operating conditions?* _____ ❑
22. *Based on the temperature of the air passing over the evaporator and the amount of superheat recorded in Step 14, what percentage of the evaporator coil's surface is being used for the latent heat process of absorbing heat from the air?* _____ ❑
23. *Do you feel that the evaporator is operating within the efficiency range based on current load conditions?* _____ ❑
24. *Explain your reasoning for your answer in Step 23.* _____ ❑
_____

25. Turn the unit OFF. ❑
26. Remove the suction line temperature probe. ❑
27. Remove the manifold gage hose from the sealed system suction service valve. ❑
28. Replace the sealed system suction service valve cap. ❑
29. Have your instructor check your work. ❑
30. Return equipment and all tools to their proper location. ❑

RESEARCH QUESTIONS

1. If a system is charged correctly, what other conditions could cause a flooded evaporator?
_____
_____

2. Besides high-load conditions, what other conditions could cause higher than normal evaporator superheat?
_____
_____

3. What effect would a blocked air filter have on evaporator superheat?
_____
_____

4. Most heat is absorbed by an evaporator during what refrigerant process?
_____
_____

5. What factors determine a system's low-side operating pressure?

_____

_____

Passed Competency _____   Failed Competency _____

Instructor Signature _____   Grade _____

Theory Lesson: Superheat and Sub-cooling Calculations for Systems with Near-Azeotropic (Zeotropic) Refrigerant Blends

SUGGESTED MATERIALS

Textbook

Refrigeration & Air Conditioning Technology, 5th Edition, Thomson Delmar Learning
Unit 10—System Charging

Review Topics

Charging Near-Azeotropic (Zeotropic) Refrigerant Blends; Superheat Calculations; Sub-cool Calculations

Key Terms

bubble point • dew point • fractionation • near-azeotropic refrigerants • sub-cooling • superheat • temperature glide

OVERVIEW

Many of the new refrigerants are referred to as *near-azeotropic* or zeotropic refrigerants and do not behave like azeotropic refrigerants. These refrigerants are blends, but are referred to as *refrigerant mixtures* because refrigerants used within the blend can still separate into individual refrigerants, like water and oil, which can be mixed, but if allowed to sit for any period of time the water and oil will separate. This is where a difference in the pressure–temperature relationship arises versus the pressure–temperature relationship of azeotropic refrigerants.

Near-azeotropic blends experience a *temperature glide* (**Figure 3–149**).

This temperature glide occurs when the blend has many temperatures as it evaporates and condenses at a given pressure. This means that refrigerants used in the mixture will change phase from liquid to vapor or vapor back to a liquid faster than other refrigerants used to make the mixture. Therefore, it is possible with these refrigerant mixtures for one refrigerant in the mixture to be boiled away to 100% vapor and get *superheated* while other refrigerants in the mixture are still boiling from liquid to vapor (**Figure 3–150**).

Likewise it is possible for one or more of the refrigerants to be condensed from a vapor to 100% liquid and start to *sub-cool* while other refrigerants within the mixture are still being condensed from vapor back to a liquid.

Near-azeotropic refrigerants may also experience what is referred to as *fractionation*. This occurs when one or more of the refrigerants in the blend will condense or evaporate at different rates than other refrigerants making up the mixture. Fractionation happens only when the leak occurs as a vapor leak within the sealed system; it will not occur when the system is leaking pure liquid. This means that during a vapor leak within a system charged with near-azeotropic refrigerant, refrigerants within the mixture will leak out at different rates, which can have an effect on sealed system efficiency if system charge is *"topped off"* during a leak situation, rather than being completely removed from the system and a new charge weighed into the system. Fractionation also occurs if refrigerant is removed from a near-azeotropic refrigerant cylinder. For this reason, near-azeotropic or zeotropic blends should be *charged as a liquid* into a sealed system.

New pressure–temperature charts have been designed and make it easier for technicians to check, charge, and perform superheat and sub-cool calculations for equipment charged with a near-azeotopic refrigerant blend.

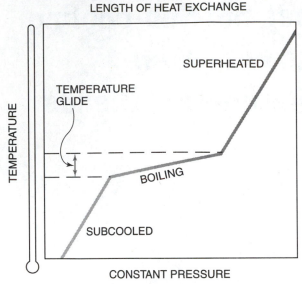

LENGTH OF HEAT EXCHANGE

TEMPERATURE

SUPERHEATED

TEMPERATURE GLIDE

BOILING

SUBCOOLED

CONSTANT PRESSURE

FIGURE 3–149 A near-azeotropic refrigerant blend showing temperature glide as it boils at a constant pressure.

	F	PRESSURE PER SQUARE INCH, (psig)
V	75.08	85.3
L	67.79	85.3
V	76.30	87.3
L	69.04	87.3
V	77.51	89.3
L	70.27	89.3
V	78.69	91.3
L	71.48	91.3
V	79.86	93.3
L	72.67	93.3
V	81.02	95.3
L	73.85	95.3

R-401 A

V = VAPOR
L = LIQUID

(A)

85.3 psig

VAPOR TEMPERATURE
75.08°F

LIQUID TEMPERATURE
67.79°F

CONDENSER TUBE

(B)

FIGURE 3–150 Difference in refrigerant vapor and refrigerant liquid.

For system superheat calculations, technicians are instructed to use the **dew point** values of the pressure–temperature chart. *Dew point is the temperature at which saturated vapor first starts to condense.* For sub-cool calculations, technicians are instructed to use the **bubble point** values of the pressure–temperature chart. *Bubble point is the temperature where the saturated liquid starts to boil off its first bubble of vapor.*

Example: Superheat Calculation

Refrigerant 404A
System suction pressure = 33 psig
Evaporator outlet temperature = 10 degrees (F)
Refer to the pressure–temperature chart in **Figure 3–151**.

A. Locate **33 psig** on the P/T chart.
B. Find the corresponding evaporator temperature by reading from 33 psig over to the R-404A Refrigerant column. The saturation temperature is **0 degrees (F)**.
C. **Zero (0) degrees** is the **dew point temperature** for R-404A.
D. The evaporator outlet temperature is **10 degrees (F)**.
E. Superheat equals:

Evaporator outlet temperature = 10 degrees (F)
MINUS
Evaporating temperature = 0 Degrees (F)
Evaporator superheat = 10 Degrees (F)

Example: Sub-cool Calculation

Refrigerant 404A
High-side pressure = 250 psig
Condenser outlet temperature = 90 degrees (F)
Refer to pressure–temperature chart in **Figure 3–151**.

PRESSURE-TEMPERATURE CHART

Temperature, °F

REFRIGERANT – (Sporlan Code)

PSIG	Pink MP39 or 401A (X)	Sand HP80 or 402A (L)	Orange HP62 or 404A (S)	Green KLEA 60 or 407A	Lt. Brown 9000 or KLEA 66 407C (N)	Brown FX-56 or 409A
5*	−23	−59	−57	−45	−40	−22
4*	−22	−58	−56	−43	−39	−20
3*	−20	−56	−54	−42	−37	−19
2*	−19	−55	−53	−41	−36	−17
1*	−17	−54	−52	−39	−35	−16
0	−16	−53	−51	−38	−34	−15
1	−13	−50	−48	−36	−31	−12
2	−11	−48	−46	−33	−29	−9
3	−9	−45	−43	−31	−27	−7
4	−6	−43	−41	−29	−24	−5
5	−4	−41	−39	−27	−22	−2
6	−2	−39	−37	−25	−20	0
7	0	−37	−35	−23	−18	2
8	2	−36	−33	−21	−17	4
9	4	−34	−32	−20	−15	6
10	6	−32	−30	−18	−13	8
11	8	−30	−28	−16	−12	9
12	9	−29	−27	−15	−10	11
13	11	−27	−25	−13	−8	13
14	13	−26	−23	−12	−7	14
15	14	−24	−22	−10	−5	16
16	16	−23	−20	−9	−4	17
17	17	−21	−19	−8	−3	19
18	19	−20	−18	−6	−1	20
19	20	−19	−16	−5	0	22
20	21	−17	−15	−4	1	23
21	23	−16	−14	−2	3	25
22	24	−15	−12	−1	4	26
23	25	−14	−11	0	5	27
24	27	−12	−10	1	6	29
25	28	−11	−9	2	8	30
26	29	−10	−8	4	9	31
27	30	−9	−6	5	10	32
28	32	−8	−5	6	11	34
29	33	−7	−4	7	12	35
30	34	−6	−3	8	13	36
31	35	−5	−2	9	14	37
32	36	−4	−1	10	15	38
33	37	−2	0	11	16	39
34	38	−1	1	12	17	40
35	39	0	2	13	18	41
36	40	0	3	14	19	43
37	42	30 / 1	4	15	20	44
38	43	31 / 2	5	16	21	45 / 30
39	44	32 / 3	6	17	22	46 / 31
40	45	33 / 4	7	18	23	47 / 32
42	46	34 / 6	8	19	25	48 / 34
44	48	36 / 8	10	21	26	50 / 36
46	50	38 / 10	12	23	28	38
48		40 / 11	14	24	30	39
50	44	13 / 16	26	31		41
52	45	14 / 17	28	33		43
54	47	16 / 19	29	34		45
56	49	18 / 20	31	36		46
58	50	19 / 22	32	37		48
60	52	20 / 23	33	39		60
62	53	22 / 25	35	40		51
64	55	23 / 26	36	42 / 30		53
66	56	25 / 27	38	43 / 32		54
68	58	26 / 29	39	44 / 33		56
70	59	27 / 30	29 / 40	30 / 46	34	57
72	61	29 / 32	31 / 41	31 / 47	36	58
74	62	30 / 33	32 / 43	32 / 48	37	60
76	64	31 / 34	33 / 44	34 / 49	38	61
78	65	32 / 30	35 / 34	45 / 35	39	63
80	66	34 / 31	37 / 36	46 / 36	41	64
85	69	37 / 34	40 / 39	49 / 39	44	67
90	73	40 / 37	42 / 42		42 / 46	70
95	76	42 / 40	45 / 44		45 / 49	73
100	78	45 / 43	48 / 47		47 / 52	76
105	81	48 / 45	50		50 / 54	79
110	84	50 / 48	52		53 / 57	82
115	87	50	55		55 / 59	84
120	89	53	57		57 / 62	87
125	92	55	59		60 / 64	89
130	94	57	62		62 / 66	92
135	96	60	64		64 / 69	94
140	99	62	66		66 / 71	96
145	101	64	68		68 / 73	99
150	103	66	70		70 / 75	101
155	105	68	72		72 / 77	103
160	108	70	74		74 / 79	105
165	110	72	76		76 / 81	107
170	112	74	78		78 / 82	109
175	114	75	80		80 / 84	111
180	116	77	82		81 / 86	113
185	117	79	83		83 / 88	115
190	119	81	85		85 / 90	117
195	121	82	87		87 / 91	119
200	123	84	88		88 / 93	121
205	125	86	90		90 / 95	123
210	127	87	92		91 / 96	124
220	130	91	95		94 / 99	128
230	133	94	98		97 / 102	131
240	136	97	101		100 / 105	134
250	140	99	104		103 / 108	137
260	143	102	107		106 / 111	141
275	147	106	111		110 / 115	145
290	151	110	115		114 / 119	149
305	155	114	118		117 / 123	153
320	159	118	122		121 / 126	157
335	163	121	126		124 / 130	161
350	167	125	129		128 / 133	165
365	170	128	132		131 / 137	169

Vertical labels in the chart columns: BUBBLE POINT and DEW POINT regions are indicated for each refrigerant.

*Inches mercury below one atmosphere

FIGURE 3–151 A pressure–temperature chart for refrigerant blends with dew point and bubble point temperature.

F. Locate **250 psig** on the P/T chart.
G. Find the corresponding evaporator temperature by reading from 250 psig over to the R-404A refrigerant column. The saturation temperature is **104 degrees (F)**.
H. **104 degrees** is the **bubble point temperature** for R-404A.
I. The condenser outlet temperature is **90 degrees (F)**.
J. Sub-cool equals:

Condensing temperature = 104 degrees (F)
MINUS
Condenser outlet temperature = 90 degrees (F)
Sub-cool equals = 14 degrees (F)

Notice that the pressure–temperature chart is laid out in such a way as to avoid confusion between liquid or vapor temperature when calculating superheat and sub-cooling values.

Practical Competency 71

Measuring Superheat of a Sealed System with a Near-Azeotropic (Zeotropic) Refrigerant Blend

SUGGESTED MATERIALS

Textbook

Refrigeration & Air Conditioning Technology, 5th Edition, Thomson Delmar Learning
Unit 10—System Charging
Unit 21—Evaporators and Refrigeration Systems
Unit 36—Refrigeration Applied to Air Conditioning
Unit 40—Typical Operating Conditions

Review Topics

Charging Near-Azeotropic (Zeotropic) Refrigerant Blends; The Evaporator; The Evaporator and Boiling Temperature; Evaporator Evaluation; The Flooded Evaporator; Dry-Type Evaporator Performance

COMPETENCY OBJECTIVE

The student will be able to measure system superheat of an air conditioner or heat pump charged with a near-azeotropic (zeotropic) refrigerant blend.

OVERVIEW

Many of the new refrigerants are referred to as *near-azeotropic* or zeotropic refrigerants and do not behave like azeotropic refrigerants. These refrigerants are blends, but referred to as *refrigerant mixtures* because refrigerants used within the blend can still separate into individual refrigerants, like water and oil, which can be mixed, but if allowed to sit for any period of time the water and oil will separate. This is where a difference in the pressure–temperature relationship arises versus the pressure–temperature relationship of azeotropic refrigerants.

Near-azeotropic blends experience a *temperature glide*. (Refer to *Figure 3–149, Theory Lesson: Superheat and Sub-cooling Calculations for Systems with Near-Azeotropic [Zeotropic] Refrigerant Blends*.)

This temperature glide occurs when the blend has many temperatures as it evaporates and condenses at a given pressure. This means that refrigerants used in the mixture will change phase from liquid to vapor or vapor back to a liquid faster than other refrigerants used to make the mixture. Therefore, it is possible with these refrigerant mixtures for one refrigerant in the mixture to be boiled away to 100% vapor and get *superheated* while other refrigerants in the mixture are still boiling from liquid to vapor. (Refer to *Figure 3–150, Theory Lesson: Superheat and Sub-cooling Calculations for Systems with Near-Azeotropic [Zeotropic] Refrigerant Blends*.)

Likewise it is possible for one or more of the refrigerants to be condensed from a vapor to 100% liquid and start to *sub-cool* while other refrigerants within the mixture are still being condensed from vapor back to a liquid.

Near-azeotropic refrigerants may also experience what is referred to as *fractionation*. This occurs when one or more of the refrigerants in the blend will condense or evaporate at different rates than other refrigerants making up the mixture. Fractionation happens only when the leak occurs as a vapor leak within the sealed system; it will not occur when the system is leaking pure liquid. This means that during a vapor leak within a system charged with near-azeotropic refrigerant, refrigerants within the mixture will leak out at different rates, which can have an effect on sealed system efficiency if system charge is *"topped off"* during a leak situation, rather than being completely removed from the system and a new charge weighed into the system. Fractionation also occurs if refrigerant is removed from a near-azeotropic refrigerant cylinder. For this reason, near-azeotropic or zeotropic blends should be *charged as a liquid* into a sealed system.

New pressure–temperature charts have been designed and make it easier for technicians to check, charge, and perform superheat and sub-cool calculations for equipment charged with near-azeotropic refrigerant blends. (Refer to *Figure 3–151, Theory Lesson—Superheat and Sub-cooling Calculations for Systems with Near Azeotropic [Zeotropic] Refrigerant Blends.*)

For system superheat calculations, technicians are instructed to use the **dew point** values of the pressure–temperature chart. *Dew point is the temperature at which saturated vapor first starts to condense.* For sub-cool calculations, technicians are instructed to use the **bubble point** values of the pressure–temperature chart. *Bubble point is the temperature at which the saturated liquid starts to boil off its first bubble of vapor.*

EQUIPMENT REQUIRED

1 Manifold gage set (R-410A)
1 Pressure–temperature chart with R-410A refrigerant scale
1 Temperature tester (*with probes*)
Insulation tape
Assorted hand tools

SAFETY PRACTICES

The student should be knowledgeable in the use of tools and testing equipment and follow all safety procedures when working with live voltage. The student should have training when working with near-azeotropic refrigerants. The student should also beware of high pressures and follow all EPA rules and regulations when working with refrigerants.

COMPETENCY PROCEDURES   Checklist

1. Turn the sealed system ON.   ❑
2. Make sure manifold gage valves are closed.   ❑
3. Remove the system suction service valve cap.   ❑
4. Connect the low-side manifold hose to the sealed system suction service valve.   ❑
5. Attach a temperature probe near the suction line service valve.   ❑
6. Insulate the temperature probe.   ❑
7. *Record the type of system refrigerant.* _____   ❑
8. Let the system operate for at least 15 minutes.   ❑
9. *Record the system's suction line temperature.* _____   ❑
10. *Record the low-side suction pressure of the system.* _____   ❑
11. Use the P/T chart or manifold gage scale and convert low-side operating pressure to
    temperature.   ❑
12. *Record the saturation temperature of refrigerant in the dew point temperature.* _____   ❑
13. Determine the amount of superheat by the following formula   ❑

                    Suction line temperature _____
                              MINUS
                    Dew point temperature _____
                    EQUALS *Evaporator superheat* _____

14. *Record the amount of superheat.* _____   ❑
15. *Is the recorded superheat within the range as determined by standard operating
    conditions?* _____   ❑

16. *If recorded superheat is not within design range, explain why you think it isn't.* ❏

_____

_____

17. *Using a temperature probe, measure and record the temperature of the air being cooled by the system being checked.* _____ ❏

18. *Using a temperature probe, measure and record the temperature of the air after it has been passed over the evaporator coil.* _____ ❏

19. Determine the air temperature drop across the coil by: ❏

Return air temperature _____

MINUS

Supply air temperature _____

EQUALS *Temperature drop*

20. *Record the amount of temperature drop.* _____ ❏

21. *Is the amount of temperature drop higher or lower than that required under standard operating conditions?* _____ ❏

22. *If the temperature drop is not within design range as set under standard operating conditions, explain why you think it isn't.* ❏

_____

_____

23. Turn the system off. ❏

24. Have your instructor check your work. ❏

25. Disconnect all equipment and return materials to their proper location. ❏

RESEARCH QUESTIONS

1. What is the average operating high-side pressure for a R-410A air conditioning system?

_____

_____

2. What is the average operating low-side pressure for a R-410A air conditioning system?

_____

_____

3. R-410A air conditioning systems have high-efficiency air-cooled condensers; by design, the condensing temperature for these units is how many degrees above ambient temperature?

_____

_____

4. Under standard conditions, R-410A evaporators are designed to operate how many degrees cooler than the air coming across the evaporator?

_____

_____

5. What is the average sub-cool range for condensers used in R-410A air conditioning systems operating under standard conditions?

_____

_____

Passed Competency _____   Failed Competency _____

Instructor Signature _____   Grade _____

Practical Competency 72

Using Manufacturer's Charging Chart to Check or Charge an Air Conditioning or Heat Pump Unit (Figure 3–152)

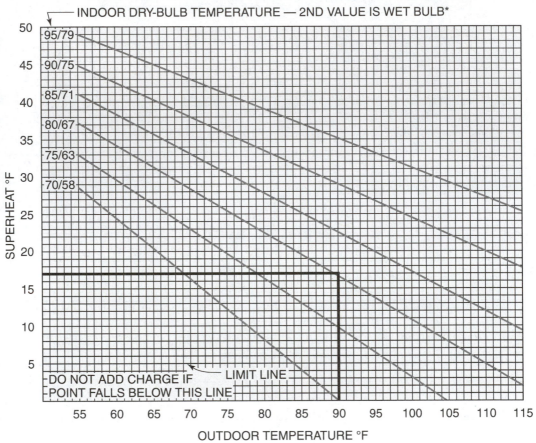

Chart based on 400 cfm/ton indoor airflow and 50% relative humidity.

FIGURE 3–152   A charging curve used for charging a split air-conditioning system incorporating a capillary tube or fixed-bore metering device.

SUGGESTED MATERIALS

Textbook
Refrigeration & Air Conditioning Technology, 5th Edition, Thomson Delmar Learning
Unit 10—System Charging
Unit 9—Refrigerant and Oil Chemistry and Management—Recovery, Recycling, Reclaiming, and Retrofitting

Review Topics
Liquid Refrigerant Charging; Weighing Refrigerant; Using Charging Devices; Using Charging Charts; Using Charging Curves; Azeotropic Refrigerants; Vapor Refrigerant Charging

COMPETENCY OBJECTIVE

The student will be able to charge or adjust the charge of an air conditioning or heat pump system by using a manufacturer's charging chart.

OVERVIEW

Adding refrigerant to refrigeration sealed system is referred to as charging. Air conditioning, heat pumps, and refrigeration equipment must have the correct amount of refrigerant charge for the equipment to operate and perform as efficiently as it was designed to. With today's high-efficiency equipment, ensuring that a system is charged with the correct amount of refrigerant is one of the most important procedures a technician is required to perform. This requires that technicians have a thorough understanding of the system operation and operating conditions which indicate whether a system is properly *charged, overcharged*, or *undercharged*. Refrigerant can be added to a sealed system in the vapor or liquid state, which can be accomplished by *weighing the refrigerant charge* into the sealed system and measuring the charge into the sealed system (*Theory Lesson: Sealed System Charging, Figure 3–116*), using *system operating charts (Theory Lesson: Sealed System Charging, Figure 3–117*), manufacturer's superheat charging tables (*Theory Lesson: Sealed System Charging, Figure 3–118*), suction line temperature tables (*Theory Lesson: Sealed System Charging, Figure 3–119*), or manufacturer's sub-cooling charging tables for systems with a TXV metering device (*Theory Lesson: Sealed System Charging, Figure 3–120*).

Manufacturers will often supply charts or curves to assistant in correctly checking or charging air-conditioning and heat pumps. The charts are referred to as charging charts or charging curves. These charts are based on superheat or sub-cool readings under certain load conditions in which the air conditioning or heat pump may be operating. The type of metering device is also taken into consideration by the manufacturer when supplying these charts with the equipment. Most air conditioning and heat pump equipment is designed with capillary tube, fixed orifice, or piston type metering devices. For this type of metering device, the charts or curves will be based on system **superheat values**. For systems with thermostatic expansion devices (TXV), charts and cures will be based on system **sub-cooling values**. The use of the charts requires technicians to evaluate the load conditions the air conditioning or heat pump is operating in. Evaluating the load refers to the amount of sensible and latent heat in the air being conditioned. The procedures to use manufacturer's charging charts or charging curves are universal, although the safest and most accurate way to charge any unit is to follow the manufacturer's charging instructions that come attached to the unit or supplied in the unit installation manual.

EQUIPMENT REQUIRED

1 Manufacturer's charging chart or charging chart (*Figure 3–152*)
Air conditioning or heat pump unit
1 Sling psychrometer
1 Temperature tester
1 Manifold gage set
1 Refrigerant cylinder that matches system refrigerant

SAFETY PRACTICES

Wear goggles or safety glasses and gloves when working with refrigerant in sealed systems. Beware of high pressures and follow all electrical safety rules. Follow all EPA rules and regulations while working with refrigerants.

> **NOTE:** *Competency can be performed on an air conditioning unit or a heat pump operating in the cooling mode. The charging chart shown in Figure 3–133 is based on 400 CFMs of air per ton or airflow with 50% relative humidity across the evaporator coil.*

COMPETENCY PROCEDURES

1. Turn the system ON. ❑
2. *Measure and record the indoor return air dry-bulb temperature.* _____ ❑

NOTE: *Use the indoor air wet-bulb temperature if the percentage of relative humidity in the air is above 70% or below 20%.*

---

3. *Measure and record the outdoor dry-bulb temperature at the condensing unit.* _____ ❑
4. Make sure the manifold gage valves are closed. ❑
5. Purge any refrigerant from refrigerant hoses. ❑
6. Attach the manifold low-side gage hose to the sealed system low-side valve. ❑
7. Attach the manifold high-side gage hose to the sealed system high-side valve. ❑
8. Attach and insulate a temperature probe lead at or near the suction line service valve of the condensing unit. ❑
9. Let operate for 15 minutes. ❑
10. *Record the type of system refrigerant.* _____ ❑
11. *Record the temperature of the suction line.* _____ ❑
12. *Record the system's low-side operating pressure (psig).* _____ ❑
13. *Use a pressure–temperature chart and convert and record the system's low-side pressure to refrigerant saturation temperature.* _____ ❑
14. Determine the system superheat by subtracting the system's low-side saturation temperature from the system suction line temperature. ❑

Suction line temperature _____

MINUS

Low-side saturation temperature = **System superheat**

15. *Record the amount of superheat.* _____ ❑
16. *Refer to the system charging chart or charging chart shown in Figure 3–152.* ❑
17. Locate the indoor dry-bulb temperature lines on the charging chart. ❑
18. Locate the outdoor dry-bulb temperature lines on the charging chart. ❑
19. Locate where the indoor dry-bulb temperature and outdoor dry-bulb temperature intersect on the charging chart. ❑
20. Follow the intersecting line over to the superheat values. ❑
21. *Record the required superheat according to the charging chart.* _____ ❑
22. *Does the recorded superheat value recorded in Step 15 match the charging chart required superheat value recorded in Step 21?* _____ ❑
23. *Is the recorded superheat value in Step 15 higher or lower than the required superheat listed in Step 21?* _____ ❑

NOTE: *If the superheat of the system is more than 5 degrees (F) **above** what the charging chart requires, add refrigerant through the system's low side until the system's superheat is within 5 degrees (F) of the required superheat according to the charging chart.*

*If the superheat of the system is more than 5 degrees (F) **below** what the charging charts calls for, use a recovery unit and recover some refrigerant from the system until the superheat is within 5 degrees of what is required according to the charging chart.*

NOTE: *Make adjustments to system refrigerant system charge slowly and let system operate for an additional 15 minutes before recalculating system superheat.*

> **NOTE:** *Any adjustments to system charge will require recalculation of the system's superheat.*

ONCE SUPERHEAT IS WITHIN RANGE, RECALCULATE SYSTEM SUPERHEAT.

24. *Record the temperature of the suction line.* _____ ❑
25. *Record the system's low-side operating pressure (psig).* _____ ❑
26. *Use a pressure–temperature chart and convert and record the system's low-side pressure to refrigerant saturation temperature.* _____ ❑
27. Determine system superheat by subtracting the system's low-side saturation temperature from the system suction line temperature: ❑

Suction line temperature _____

MINUS

Low-side saturation temperature = **System superheat**

28. *Record the amount of superheat.* _____ ❑
29. *Once the system charge is within required range, have your instructor check your work.* ❑
30. With the instructor's approval, turn the unit OFF. ❑
31. Remove the temperature probe from the system suction line. ❑
32. Remove low- and high-side manifold hoses from system service valves. ❑
33. Replace system service valve caps. ❑
34. Return equipment to its proper location. ❑

RESEARCH QUESTIONS

1. How is liquid refrigerant added to the refrigeration system when the system is out of refrigerant?

_____

_____

2. How is the refrigerant cylinder pressure kept above the system pressure when charging with vapor from a cylinder?

_____

_____

3. Why does the refrigerant pressure decrease in a refrigerant cylinder while charging with vapor?

_____

_____

4. What other methods besides weighing and measuring are used for charging systems?

_____

_____

5. What is the main difference between zeotropic and azeotropic refrigerant blends?

_____

_____

Passed Competency _____ **Failed Competency** _____

Instructor Signature _____ **Grade** _____

Practical Competency 73

Using the Sub-cool Method of Charging R-22 Air Conditioning or Heat Pump Equipment Supplied with a Thermostatic Expansion Device (TXV)

SUGGESTED MATERIALS

Textbook

Refrigeration & Air Conditioning Technology, 5th Edition, Thomson Delmar Learning
Unit 10—System Charging
Unit 21—Evaporators and Refrigeration Systems
Unit 24—Expansion Devices
Unit 36—Refrigeration Applied to Air Conditioning
Unit 40—Typical Operating Conditions

Review Topics

TXV Metering Devices; Air-Cooled Condensers; Condenser Sub-cooling; Standard Air-Cooled Condensers; High-Efficiency Air-Cooled Condensers

COMPETENCY OBJECTIVE

The student will be able to properly charge or check the charge of an air conditioning or heat pump equipped with a TXV metering device by using the sub-cool method.

OVERVIEW

Some air conditioning and heat pump equipment are equipped with a Thermostatic Expansion Valve (TXV). These types of metering devices are designed to maintain a constant amount of superheat on the evaporator under any load conditions, including changing outdoor ambient temperatures. TXV are provided mainly on high-efficiency equipment. When used, a refrigerant charging method known as the "Sub-cooling Method" must be used to charge or to check the charge on air conditioning or heat pump equipment so equipped. This method of charging or checking a system's charge involves measuring the amount of liquid line sub-cooling exiting the condenser. Because of the different SEER ratings on the market today, the actual sub-cooling range for particular equipment varies. Manufacturers have made an effort to provide the required sub-cool amount for air cool condensers on the system data plate or technical manual. In the absence of such information, use a sub-cooling value of 10 to 15 degrees (F).

EQUIPMENT REQUIRED

Manufacturer's sub-cooling chart or *Figure 3–153*
1 Temperature tester (with temperature leads)
1 Manifold gage set
1 Refrigerant cylinder that matches system refrigerant
Insulation tape
Assorted hand tools

SAFETY PRACTICES

Wear goggles or safety glasses and gloves when working with refrigerant in sealed systems. Beware of high pressures and follow all electrical safety rules. Follow all EPA rules and regulations while working with refrigerants.

> **NOTE:** *If competency is being performed on a heat pump, perform procedures with the heat pump operating in the air conditioning mode.*

COMPETENCY PROCEDURES

Checklist

1. Turn the unit ON. ❏
2. Allow the unit to operate for 15 minutes. ❏
3. Make sure manifold gage valves are closed. ❏
4. Remove the system's high-side service valve cap. ❏
5. Attach the manifold high-side hose to the system's high-side service valve. ❏
6. Attach a temperature probe to the condenser's liquid line. ❏
7. Insulate the temperature probe. ❏
8. *After 15 minutes of operation, record the system's high-side pressure. _____* ❏
9. Use a gage manifold or a pressure temperature chart and convert high-side pressure to saturation temperature. ❏
10. *Record the high-side saturation temperature. _____* ❏
11. *Record the temperature of the condenser's liquid line. _____* ❏
12. *Determine condenser sub-cool amount by the following formula:* ❏

   High-side saturation temperature _____
   ### MINUS
   Condenser liquid line temperature _____
   EQUALS *Condenser sub-cooling _____*

13. *Record the amount of condenser sub-cooling. _____* ❏

> **NOTE:** *Required sub-cooling used for this competency is based on 15 degrees. If required sub-cooling is known for the equipment competency is being performed on, use the manufacturer's required superheat rather than 15 degrees.*

14. Refer to **Figure 3–153.** ❏
15. Find the approximate high-side pressure recorded in **Step 8** on the Sub-cooling Chart in **Figure 3–153.** ❏
16. Follow the high-side pressure recorded in Step 8 across the sub-cooling chart to the 15-degree (F) required sub-cooling column. ❏
17. *Record the required liquid line temperature according to the sub-cooling chart. _____* ❏

> **NOTE:** *According to the sub-cooling chart, the liquid line temperature recorded in Step 17 is the required liquid line temperature for this unit if the system is charged correctly.*

18. *Record the actual liquid line temperature recorded in Step 11. _____* ❏
19. *Was the actual recorded liquid line temperature within 3 to 4 degrees of the temperature required according to the sub-cooling chart? _____* ❏

Pressure (psig) at Service Fitting	Required Subcooling Temperature (F)					
	0	5	10	15	20	25
134	76	71	66	61	56	51
141	79	74	69	64	59	54
148	82	77	72	67	62	57
156	85	80	75	70	65	60
163	88	83	78	73	68	63
171	91	86	81	76	71	66
179	94	89	84	79	74	69
187	97	92	87	82	77	72
196	100	95	90	85	80	75
205	103	98	93	88	83	78
214	106	101	96	91	86	81
223	109	104	99	94	89	84
233	112	107	102	97	92	87
243	115	110	105	100	95	90
253	118	113	108	103	98	93
264	121	116	111	106	101	96
274	124	119	114	109	104	99
285	127	122	117	112	107	102
297	130	125	120	115	110	105
309	133	128	123	118	113	108
321	136	131	126	121	116	111
331	139	134	129	124	119	114
346	142	137	132	127	122	117
359	145	140	135	130	125	120

FIGURE 3–153 Manufacturer's charging chart. (*Courtesy of Carrier Corporation*)

> **NOTE:** *Allow a 3- to 4-degree (F) temperature tolerance between required and actual liquid line temperatures.*
>
> *If the measured liquid line temperature recorded in Step 11 is **higher** than the desired liquid line temperature recorded in Step 17, **add refrigerant** to the unit.*
>
> *If the measured liquid line temperature recorded in Step 11 is **lower** than the desired liquid line temperature recorded in Step 17, **recover refrigerant** from the unit.*

MAKING CHARGE ADJUSTMENTS

When adding or removing refrigerant from the system, it should be done in small increments. Allow at least 15 minutes of system operation between refrigerant adjustments before retaking system pressures and temperatures.

20. Adjust the system charge if required. ❑
21. Once the system is within tolerance according to the sub-cooling chart, have your instructor check your work. ❑
22. Turn the system OFF. ❑
23. Remove all equipment and return materials to their proper location. ❑

RESEARCH QUESTIONS

1. What is the difference between standard and high-efficiency air-cooled condensers?

2. Is the designed sub-cooling range for high-efficiency air-cooled condensers higher or lower than that which is established for standard air-cooled condensers?

_____

_____

3. The SEER rating of air conditioning and heat pump equipment is based on what factors?

_____

_____

4. What factors could cause condenser sub-cooling to be higher than the designed range?

_____

_____

5. What factors could cause condenser sub-cooling to be lower than the designed range?

_____

_____

Passed Competency _____ **Failed Competency** _____

Instructor Signature _____ **Grade** _____

Theory Lesson: The "Big Question"— Charging and Checking the Charge of Air Conditioning and Heat Pump Equipment

SUGGESTED MATERIALS

Textbook
Refrigeration & Air Conditioning Technology, 5th Edition, Thomson Delmar Learning
Unit 10—System Charging

Review Topics
Superheat Charging Charts; Sub-cool Charging Charts; Charging Charts; Charging Tables; Charging Curves; Charging Calculators

Key Terms
airflow • charging calculators • charging charts • efficiency • frost-line • load • sub-cooling • superheat • system amperage • system pressures

OVERVIEW

It is important for HVAC technicians to understand that systems that are undercharged or overcharged will remove heat from a conditioned area, although they will not perform efficiently doing it!

It has been reported by national air conditioning associations that more than 70% of air conditioning and heat pump equipment in use in the United States is either *overcharged* or *undercharged*. This is a staggering number when efficiency performance of air conditioning and heat pump equipment relies on operating with the correct refrigerant charge. Additional factors that affect the efficiency of such equipment are making sure that the system is moving the correct airflow in CFMs (*cubic feet per minute*) for both the evaporator and condenser. For forced air systems, making sure that the ductwork is sized correctly and operating with the correct static pressure is important. Making sure that the evaporator and condenser coils are clean and that the equipment is sized properly are all important to the efficient performance of air conditioning and heat pump equipment. Yet while having all these additional factors correct, having the correct refrigerant charge is paramount in comparison, especially with the new higher rated SEER equipment. *Remember, it is the refrigerant within the mechanical refrigeration system that actually absorbs heat and rejects heat from the conditioned space.* Having too much or too little refrigerant within the system affects the equipment's ability and efficient performance in absorbing and rejecting heat within an area where comfort is trying to be sustained, let alone shortening the life of the equipment. Charging and checking the charge on air conditioning and heat pump equipment has become one of the most important aspects of the HVAC industry and a certified technician's duties.

Many methods of charging and checking the charge of systems are used in the industry, and most work if done properly. Weight charging to the manufacturer's system charge requirements, superheat charging, and sub-cool charging are probably the most accurate, if all other conditions mentioned above are correct, especially making sure that the system is operating with the correct airflow.

Charging or checking a system's charge by *temperature drop, system amperage, system pressures*, and *frost-line* are other methods of charging where most questions arise regarding the accuracy of the system charge through the use of these procedures and practices. Unfortunately, using these procedures seems to be the most well known by technicians in the field today. Used on their own, these procedures leave too many variables that affect their validity, and in turn cause a technician to unknowingly overcharge or undercharge a system. Procedures such as these can be validated by technicians if they would take the time to compare and confirm the system's load and proper airflow against a manufacturer's charging chart or the use of a charging calculator. Correct airflow once established stays constant. The load that the system deals with can change often. This factor alone has the greatest impact on nullifying those practices used by technicians as stated earlier. The load is referred to as the air temperature, and the percentage of humidity within the air that the evaporator is absorbing, heat. The actual temperature of the air along with the percentage of humidity in the air *affect the system's operating pressures, the temperature drop across the evaporator coil, the system's operating amperage, and suction line temperature.* Any change in *"the load"* will affect every one of the factors that are being used to charge or check a system's charge. Understanding this fact, you can see that without knowing the *"actual load"* the system is dealing with at the time the system is being charged or checked, the validation of using such methods as *temperature drop, system amperage, system pressures,* and *frost-line procedures is inaccurate for the most part.* In reality, *overcharging and undercharging a system using these procedures on their own is unavoidable. Using these practices may provide the system with a charge that is close to what is recommended, but without validation by other means, the system's charge is called into question.*

Another issue technicians seem to take for granted is manufactured "pre-charged equipment." Years ago, all equipment was charged in the field by the technician. In most cases, these units were also overcharged or undercharged by technicians in the field. To correct this problem, manufacturers started manufacturing equipment with the correct charge in the system. This was to make the technician's job easier and to make the equipment operate more efficiently. Although this equipment is charged at the factory for a certain line set length, technicians avoid taking the time to confirm the system's charge once the installation is complete. *Personal experience has proven that more then half of the "pre-charged equipment" installed required charge adjustment on completion of the installation.*

To correct the staggering amount of equipment that is not charged correctly, manufacturers started providing charging tables and charts with their equipment over the years. Unfortunately, most technicians avoided using these tables for whatever reasons and depended more on past practices or procedures as stated above. For the most part the equipment works, but not as efficient as it was designed to. *"Efficiency" is a key word for the HVAC industry today. New EPA mandates for higher efficiency equipment and ozone-free refrigerants are just a couple of issues that the HVAC industry must deal with. Future years will bring stricter regulations on HVAC equipment performance and higher efficiency levels for such equipment. These are not only requirements for industry manufacturers; they are also requirements for HVAC technicians.*

Technicians need to *learn proper procedures for system charging and checking the charge of manufacturer's equipment. In all cases, manufacturers provide the correct charging procedures and procedures for checking the charge of their equipment in the unit installation manual.* Technicians need to reference this information and follow the procedures listed to ensure that the equipment is charged correctly.

If the unit installation manual is not available, technicians should become familiar with using a universal charging calculator. These calculators can be used to charge or check a system's charge under most load conditions. They take the *"guess work"* out of the equation when it comes to charging practices.

Theory Lesson: Procedures for Using an R-22 Superheat/Sub-cooling Charging Calculator

Using and understanding how to use a **superheat/sub-cooling calculator** takes the *"guess work"* out of checking or charging air conditioning and heat pump equipment along with checking for proper system airflow. The calculator is designed for use on TXV and non-TXV equipment. Technicians should use the **Required Superheat (Cooling, Non-TXV) side for systems with fixed-bore metering devices,** and the **Sub-cooling (Cooling, TXV)** for systems with a TXV metering device. The calculator can also be used to check a system's airflow by using the **Proper Airflow Range (Cooling)** portion of the calculator.

PROCEDURES FOR USING REQUIRED SUPERHEAT CALCULATOR (COOLING, NON-TXV)

> *NOTE: The temperature and pressures used in this example are only being used to assist the technician in becoming familiar with proper usage of the calculator.*

Actual readings would have to be taken for each system being charged or checked. Technicians would be required to use a sling psychrometer to measure indoor wet-bulb temperature, access the sealed system's low-side pressure, and use a temperature probe to measure the system's suction line temperature and outdoor condensing temperature.

Example Readings
1. Indoor Wet-Bulb—**66 degrees (F)**
2. Condensing Entering Air Dry-Bulb Temperature—**95 degrees (F)**
3. Required Superheat—**10 degrees (F)**
4. Vapor Pressure (psig)—**70 psig**

PROCEDURES: REQUIRED SUPERHEAT METHOD

STEP 1
A. Use a sling psychrometer and take wet-bulb and dry-bulb readings of the inside air being conditioned.
B. Example indoor air wet-bulb temperature is 66 degrees (F).
C. Slide the calculator slide rule down until the arrow lines up with 66 degrees at **Number 1** (*Indoor Entering Air Wet-Bulb Temperature*) on the charging calculator (**Figure 3–154**).

STEP 2
A. Find the *Condenser Entering Air Dry-Bulb Temperature* of *95 degrees* at **Number 2** on the charging calculator (refer to **Figure 3–154**).

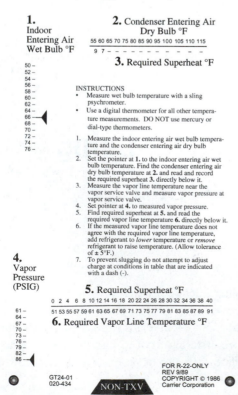

FIGURE 3–154 Setting calculator at point **1** for 66-degree wet bulb.

STEP 3

A. Find the ***Required Superheat Temperature*** of 10 degrees at **Number 3** by reading the ***Superheat Value*** directly below the ***95-Degree Condenser Entering Air Dry-Bulb Temperature*** (refer to **Figure 3–154**).

> NOTE: *Required superheat of 10 degrees (F) will be referenced again at **Number 5** on the charging calculator.*

STEP 4

A. Obtain the system's low-side operating pressure (vapor pressure – psig).

B. Using the example low-side pressure of 70 psig, slide the slide rule of the charging calculator to 70 psig at **Number 4** (**Figure 3–155**).

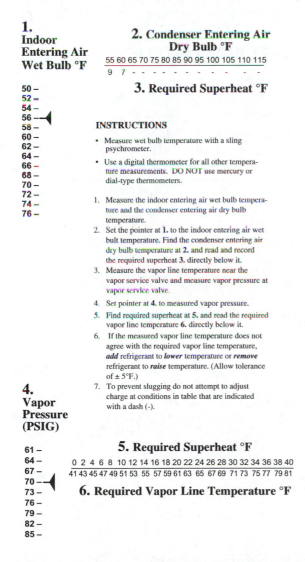

REQUIRED SUPERHEAT CALCULATOR
(Cooling, NON-TXV)

1.
Indoor
Entering Air
Wet Bulb °F

2. Condenser Entering Air Dry Bulb °F

55 60 65 70 75 80 85 90 95 100 105 110 115
9 7 - - - - - - - - - - -

3. Required Superheat °F

50 –
52 –
54 –
56 –
58 –
60 –
62 –
64 –
66 –
68 –
70 –
72 –
74 –
76 –

INSTRUCTIONS

- Measure wet bulb temperature with a sling psychrometer.
- Use a digital thermometer for all other temperature measurements. DO NOT use mercury or dial-type thermometers.

1. Measure the indoor entering air wet bulb temperature and the condenser entering air dry bulb temperature.
2. Set the pointer at **1.** to the indoor entering air wet bult temperature. Find the condenser entering air dry bulb temperature at **2.** and read and record the required superheat **3.** directly below it.
3. Measure the vapor line temperature near the vapor service valve and measure vapor pressure at vapor service valve.
4. Set pointer at **4.** to measured vapor pressure.
5. Find required superheat at **5.** and read the required vapor line temperature **6.** directly below it.
6. If the measured vapor line temperature does not agree with the required vapor line temperature, ***add*** refrigerant to ***lower*** temperature or ***remove*** refrigerant to ***raise*** temperature. (Allow tolerance of ± 5°F.)
7. To prevent slugging do not attempt to adjust charge at conditions in table that are indicated with a dash (-).

4.
Vapor
Pressure
(PSIG)

61 –
64 –
67 –
70 –
73 –
76 –
79 –
82 –
85 –

5. Required Superheat °F

0 2 4 6 8 10 12 14 16 18 20 22 24 26 28 30 32 34 36 38 40
41 43 45 47 49 51 53 55 57 59 61 63 65 67 69 71 73 75 77 79 81

6. Required Vapor Line Temperature °F

GT58-01**A**
020-517

FOR R-410A ONLY
COPYRIGHT © 1998
Carrier Corporation

 NON-TXV

FIGURE 3–155   70 psig at point 4.

STEP 5

A. Find the *Required Superheat Temperature* of *10 degrees* at **Number 5** of the charging calculator (refer to **Figure 3–155**).

STEP 6

A. At **Number 6** of the charging calculator find the *Required Vapor Line Temperature* directly below the *Required Superheat Temperature* of *10 degrees (F)* (refer to **Figure 3–155**).

B. The charging calculator shows a *Required Vapor Line Temperature of 51 Degrees (F)* at **Number 6** (refer to **Figure 3–155**).

NOTE: Underneath the Charging Calculator Instructions at Number 6 it states the following: If the measured vapor line temperature does not agree with the required vapor line temperature, ADD refrigerant to LOWER temperature or REMOVE refrigerant to RAISE temperature. (Allow a tolerance of + or –5 degrees F.)

Instruction Number 7 states the following: To prevent slugging do not attempt to adjust charge at conditions in the table that are indicated with a dash (–).

PROCEDURES FOR USING SUB-COOLING CALCULATOR (COOLING, TXV)

NOTE: The temperature and pressures used in this example are only being used to assist the technician in becoming familiar with proper usage of the calculator.

Actual readings would have to be taken for each system being charged or checked. Technicians would be required to access the sealed system's high-side pressure, and use a temperature probe to measure the system's liquid line temperature.

Example Readings

1. Required Sub-cooling Temperature—**15 degrees (F)**
2. Liquid Pressure (psig) at Service Valve—**223 psig**
3. Required Liquid Line Temperature—**94 degrees (F)**

PROCEDURES: SUB-COOLING METHOD

STEP 1

A. Check the system's nameplate or service literature and determine the sub-cooling range for the air cooled condenser.

B. In this example, a sub-cooling range of **15 degrees** is being used.

C. At **Number 1**, slide the calculator's slide rule down until the arrow aligns with 15 degrees *Required Sub-cooling (F)* (**Figure 3–156**).

STEP 2

A. Determine the system's operating high-side pressure.

B. In the example *223 psig* is being used.

C. Find *223 psig* at **Number 2** of the calculator (refer to **Figure 3–156**).

STEP 3

A. Determine the liquid line temperature by looking directly underneath *223 psig* at **Number 2**.

B. *94 degrees* is the *Required Liquid Line Temperature* at **Number 3**.

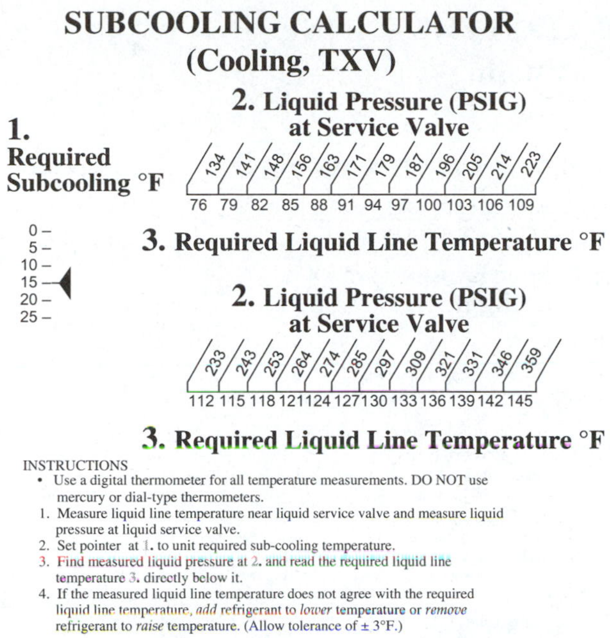

SUBCOOLING CALCULATOR
(Cooling, TXV)

1.
Required
Subcooling °F

2. Liquid Pressure (PSIG)
at Service Valve

134 141 148 156 163 171 179 187 196 205 214 223

76 79 82 85 88 91 94 97 100 103 106 109

0 –
5 –
10 –
15 –
20 –
25 –

3. Required Liquid Line Temperature °F

2. Liquid Pressure (PSIG)
at Service Valve

233 243 253 264 274 285 297 309 321 331 346 359

112 115 118 121 124 127 130 133 136 139 142 145

3. Required Liquid Line Temperature °F

INSTRUCTIONS
• Use a digital thermometer for all temperature measurements. DO NOT use
  mercury or dial-type thermometers.
1. Measure liquid line temperature near liquid service valve and measure liquid
   pressure at liquid service valve.
2. Set pointer at 1. to unit required sub-cooling temperature.
3. Find measured liquid pressure at 2. and read the required liquid line
   temperature 3. directly below it.
4. If the measured liquid line temperature does not agree with the required
   liquid line temperature, *add* refrigerant to *lower* temperature or *remove*
   refrigerant to *raise* temperature. (Allow tolerance of ± 3°F.)

FIGURE 3–156 Setting sub-cool
temperature at 15.

NOTE: *Under the Charging Calculator Instructions at Number 4 it states the following:*
    If the measured liquid line temperature does not agree with the required liquid line temperature,
ADD *refrigerant to* LOWER *temperature or* REMOVE *refrigerant to* RAISE *temperature. (Allow a*
tolerance of + or –3 degrees F.)

PROCEDURES: PROPER AIRFLOW RANGE—COOLING

NOTE: *The temperatures in this example are only being used to assist the technician in becoming familiar with proper usage of the calculator.*
    Actual dry-bulb and wet-bulb temperatures would be required to calculate the Required Airflow Range.

Example Readings
1. Indoor Entering Air Dry-Bulb Temperature—**76 degrees** (F)
2. Indoor Entering Air Wet-Bulb Temperature—**64 degrees** (F)
3. Proper Evaporator Coil Leaving Air Dry-Bulb Temperature—**57 degrees** (F)

Use a temperature tester and measure the *Indoor Entering Air Dry-Bulb Temperature. Example temperature is 76 degrees.*

STEP 1
A. Slide the calculator slide rule down to *76 degrees* at **Number 1** (**Figure 3–157**).
B. Use a sling psychrometer and measure the *Indoor Entering Air Wet-Bulb Temperature. Example temperature is 64 degrees.*

STEP 2
A. Find *64-Degree (F) Wet-Bulb Temperature* at **Number 2** (refer to **Figure 3–157**).
B. Measure the leaving air temperature (supply air).

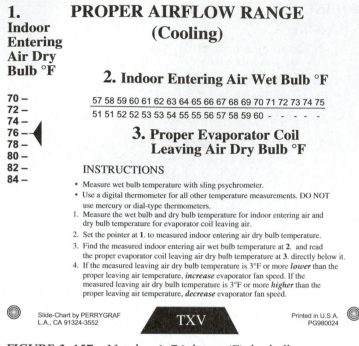

FIGURE 3–157 Number 1–76-degree (F) dry bulb.

STEP 3

A. Read the proper *Evaporator Leaving Air Dry-Bulb Temperature* directly below the **64-Degree (F) Indoor Entering Air Wet-Bulb Temperature. An example is 57 degrees (F).**

B. Compare the actual leaving air temperature to the *Proper Evaporator Coil Leaving Air Dry-Bulb Temperature of 57 Degrees (F).*

NOTE: Under the Charging Calculator Instructions at Number 4 it states the following:

*If the measured leaving air dry-bulb temperature is 3 degrees (F) or more **LOWER** than the proper leaving air temperature, INCREASE evaporator fan speed. If the measured leaving air-dry-bulb temperature is 3 degrees (F) or more **HIGHER** than the proper leaving air temperature, DECREASE evaporator fan speed.*

Practical Competency 74

Charging an R-22 Air Conditioning or Heat Pump by Using a Superheat/Sub-cooling Calculator—Superheat Method

SUGGESTED MATERIALS

Textbook
Refrigeration & Air Conditioning Technology, 5th Edition, Thomson Delmar Learning
Unit 10—System Charging
Unit 40—Typical Operating Conditions

Review Topics
Vapor Refrigerant Charging; Liquid Refrigerant Charging;

COMPETENCY OBJECTIVE

The student will be able to properly charge an air conditioning or heat pump unit using a charging calculator and the superheat method.

OVERVIEW

It is important for HVAC technicians to understand that systems that are undercharged or overcharged will remove heat from a conditioned area, although they will not perform efficiently doing it!
    It has been reported by national air conditioning associations that more than 70% of air conditioning and heat pump equipment in use in the United States is either *overcharged* or *undercharged*. This is a staggering number when efficient performance of air conditioning and heat pump equipment relies on operating with the correct refrigerant charge. Additional factors that affect the efficiency of such equipment are making sure that the system is moving the correct airflow in CFMs (*cubic feet per minute*) for both the evaporator and condenser. For forced air systems, making sure that the ductwork is sized correctly and operating with the correct static pressure is important. Making sure that the evaporator and condenser coils are clean and that the equipment is sized properly are all important to the efficient performance of air conditioning and heat pump equipment. Yet while having all these additional factors correct, having the correct refrigerant charge is paramount in comparison, especially with the new higher rated SEER equipment. *Remember, it is the refrigerant within the mechanical refrigeration system that actually absorbs heat and rejects heat from the conditioned space.* Having too much or too little refrigerant within the system affects the equipment's ability and efficiency performance in absorbing and rejecting heat within an area where comfort is trying to be sustained, let alone shortening the life of the equipment.
    Charging and checking the charge on air conditioning and heat pump equipment has become one of the most important aspects of the HVAC industry, and a certified technician's duties.
    Using and understanding how to use a **superheat/sub-cooling calculator** takes the *"guess work"* out of checking or charging air conditioning and heat pump equipment along with checking for proper system airflow. The calculator is designed for use on TXV and non-TXV equipment. Technicians should use the **Required Superheat (Cooling, Non-TXV) side for systems with fixed-bore metering devices,** and the **Sub-cooling (Cooling, TXV)** for systems with a TXV metering device. The calculator can also be used to check the system's airflow by using the **Proper Airflow Range (Cooling)** portion of the calculator.

EQUIPMENT REQUIRED

Superheat/sub-cooling charging calculator
1 Sling psychrometer
1 Temperature tester (with temperature leads)
1 Manifold gage set
1 Refrigerant cylinder that matches system refrigerant
Insulation tape
Assorted hand tools

SAFETY PRACTICES

Wear goggles or safety glasses and gloves when working with refrigerant in sealed systems. Beware of high pressures and follow all electrical safety rules. Follow all EPA rules and regulations while working with refrigerants.

> NOTE: *If competency is being performed on a heat pump, perform procedures with the heat pump operating in the air conditioning mode.*

> NOTE: *System being charged should have been leak-checked and evacuated prior to performing this competency.*

COMPETENCY PROCEDURES

Checklist

1. The sealed system should be OFF. ❏
2. Use the sling psychrometer and measure the Indoor Entering Air Wet-Bulb Temperature. ❏
3. *Record the Indoor Entering Air Wet-Bulb Temperature.* _____ ❏
4. Use the required superheat calculator (*Cooling, NON-TXV*) and slide the slide rule of the charging calculator down to the recorded Indoor Entering Air Wet-Bulb Temperature at **Number 1** of the charging calculator. ❏
5. Use the temperature tester and measure the Condenser Entering Air Dry-Bulb Temperature. ❏
6. *Record the Condenser Entering Air Dry-Bulb Temperature.* _____ ❏
7. Find the Condenser Entering Air Dry-Bulb Temperature at **Number 2** of the charging calculator. ❏
8. Directly below the Condenser Entering Air Dry-Bulb Temperature is the Required Superheat Temperature at **Number 3** of the charging calculator. ❏
9. *Record the Required Superheat Temperature.* _____ ❏
10. Attach and insulate a temperature probe near the system's suction line service valve. ❏
11. Add refrigerant through the system's low-side valve in the vapor state until the system's low-side pressure is brought out of a vacuum. ❏
12. Turn the sealed system ON. ❏
13. Continue to add refrigerant in the vapor state until the system's low-side pressure is operating at least **64 psig**. ❏
14. Once the system's operating low-side pressure is at least *64 psig*, stop charging and allow the system to operate for 10 minutes. ❏
15. At **Number 5** of the charging calculator, find the Required Superheat Temperature as recorded in **Step 9.** ❏
16. Directly below the Required Superheat at **Number 5** is the Required Vapor Line Temperature at **Number 6.** ❏

17. *Record the Required Vapor Line Temperature as indicated by the charging calculator at* **Number 6.** _____ ❑

18. *Check the system's suction line temperature and record the actual suction line temperature.* _____ ❑

> **NOTE:** *Underneath the Instructions of the Charging Calculator at Number 6 it states the following: If the measured vapor line temperature does not agree with the required vapor line temperature,* **ADD** *refrigerant to* **LOWER temperature** *or* **REMOVE** *refrigerant to* **RAISE temperature.** *(Allow tolerance of + or –5 degrees F.)*

19. *Add or remove refrigerant to the system in small amounts until the system's suction line temperature falls within (+ or –5 degrees) of the Required Vapor Line Temperature as indicated by the charging calculator.* ❑

> **NOTE:** *Any change in refrigerant charge to the system will change the system's low-side operating pressure. The vapor pressure at Number 4 will have to be readjusted as the system's pressures change as a result of the addition or removal of refrigerant to the system. The same required superheat temperature would continue to be used to determine Required Vapor Line Temperature unless there has been a considerable time lapse or temperature change from the original temperature readings taken at Numbers 1 and 2.*

20. Once the system charge is correct, have your instructor check your work. ❑

21. On completion of system charge, remove all equipment and return to proper location. ❑

RESEARCH QUESTIONS

1. If oil is used in a compressor for lubrication, what is used to cool the compressor motor windings?

   _____

   _____

2. What is a starved evaporator?

   _____

   _____

3. What causes refrigerant inside the evaporator to evaporate?

   _____

   _____

4. What is a flooded evaporator?

   _____

   _____

5. How many fins are there per inch for evaporators used in air conditioning equipment?

   _____

   _____

Passed Competency _____   Failed Competency _____

Instructor Signature _____   Grade _____

Practical Competency 75

Charging an R-22 Air Conditioning or Heat Pump by Using a Superheat/Sub-cooling Calculator—Sub-cool Method

SUGGESTED MATERIALS

Textbook

Refrigeration & Air Conditioning Technology, 5th Edition, Thomson Delmar Learning
Unit 10—System Charging
Unit 40—Typical Operating Conditions

Review Topics

Vapor Refrigerant Charging; Liquid Refrigerant Charging

COMPETENCY OBJECTIVE

The student will be able to properly charge an air conditioning or heat pump unit equipped with a TXV metering device by using a charging calculator and the sub-cool method.

OVERVIEW

It is important for HVAC technicians to understand that systems that are undercharged or overcharged will remove heat from a conditioned area, although they will not perform efficiently doing it!

It has been reported by national air conditioning associations that more than 70% of air conditioning and heat pump equipment in use in the United States is either *overcharged* or *undercharged*. This is a staggering number when efficiency performance of air conditioning and heat pump equipment relies on operating with the correct refrigerant charge. Additional factors that affect the efficiency of such equipment are making sure that the system is moving the correct airflow in CFMs (*cubic feet per minute*) for both the evaporator and condenser. For forced air systems, making sure that the ductwork is sized correctly and operating with the correct static pressure is important. Making sure that the evaporator and condenser coils are clean and that the equipment is sized properly are all important to the efficient performance of air conditioning and heat pump equipment. Yet while having all these additional factors correct, having the correct refrigerant charge is paramount in comparison, especially with the new higher rated SEER equipment. *Remember, it is the refrigerant within the mechanical refrigeration system that actually absorbs heat and rejects heat from the conditioned space.* Having too much or too little refrigerant within the system affects the equipment's ability and efficient performance in absorbing and rejecting heat within an area where comfort is trying to be sustained, let alone shortening the life of the equipment.

Charging and checking the charge on air conditioning and heat pump equipment has become one of the most important aspects of the HVAC industry, and a certified technician's duties. Using and understanding how to use a **superheat/sub-cooling calculator** takes the *"guess work"* out of checking or charging air conditioning and heat pump equipment along with checking for proper system airflow. The calculator is designed for use on TXV and non-TXV equipment. Technicians should use the **Required Superheat (Cooling, Non-TXV) side for systems with fixed-bore metering devices,** and the **Sub-cooling (Cooling, TXV) for systems with a TXV metering device.** The calculator can also be used to check the system's airflow by using the **Proper Airflow Range (Cooling)** portion of the calculator.

EQUIPMENT REQUIRED

Superheat/sub-cooling charging calculator
1 Temperature tester (with temperature leads)
1 Manifold gage set
1 Refrigerant cylinder that matches system refrigerant
Insulation tape
Assorted hand tools

SAFETY PRACTICES

Wear goggles or safety glasses and gloves when working with refrigerant in sealed systems. Beware of high pressures and follow all electrical safety rules. Follow all EPA rules and regulations while working with refrigerants.

> NOTE: *If competency is being performed on a heat pump, perform procedures with the heat pump operating in the air conditioning mode.*

> NOTE: *System being charged should have been leak-checked and evacuated prior to performing this competency.*

COMPETENCY PROCEDURES
Checklist

1. The sealed system should be OFF. ❑
2. Check the system's nameplate or service literature and determine the sub-cooling range for the air cooled condenser. ❑

> NOTE: *If the system sub-cooling range is unavailable use 10 or 15 degrees of sub-cooling.*

3. At **Number 1,** slide the calculator's slide rule down until the arrow aligns with the desired sub-cool temperature. ❑
4. Attach and insulate a temperature probe to the system liquid line just ahead of the system's high-side service valve. ❑
5. Add refrigerant to the system in either the vapor or liquid state and bring the system's low-side pressure out of a vacuum. ❑
6. Turn the system ON. ❑
7. Continue to charge the system until a low-side operating pressure of 62 psig is established. ❑
8. Let the system operate for 10 minutes. ❑
9. *Check and record the system's operating pressure.* _____ ❑
10. Find the system's high-side pressure at **Number 2** of the charging calculator. ❑
11. Determine the liquid line temperature by looking directly underneath the system's operating high-side pressure at **Number 2.** ❑
12. *Record the required Liquid Line Temperature according to the charging calculator.* _____ ❑

> NOTE: *Under the Charging Calculator Instructions at Number 4 it states the following:*
> *If the measured liquid line temperature does not agree with the required liquid line temperature,* **ADD refrigerant to LOWER temperature** *or* **REMOVE refrigerant to RAISE temperature.** *(Allow a tolerance of + or −3 degrees(F.)*

> **NOTE:** *Any change in refrigerant charge to the system will change the systems high-side operating pressure. The Liquid Pressure (psig) at Number 2 will have to be readjusted as systems pressures change due to the addition or removal of refrigerant to the system. The same required sub-cool temperature would continue to be used to determine.*

13. Once the system charge is correct, have your instructor check your work. ❑
14. On completion of system charge, remove all equipment and return to its proper location. ❑

RESEARCH QUESTIONS

1. What is the definition of sub-cooling?

   _____

   _____

2. What would happen to a system's operating head pressure if the condenser were operating in low ambient temperature conditions?

   _____

   _____

3. What are a couple of methods of maintaining a system's operating head pressure during low ambient temperature conditions?

   _____

   _____

4. What are the three functions of an air-cooled condenser?

   _____

   _____

5. Piston type metering devices for split-system air conditioning or heat pump systems gets sized to which component of the sealed system?

   _____

   _____

Passed Competency _____   Failed Competency _____

Instructor Signature _____   Grade _____

Practical Competency 76

Checking for Proper Airflow Range of a Split-System Air Conditioner or Heat Pump Using a Charging Calculator

SUGGESTED MATERIALS

Textbook

Refrigeration & Air Conditioning Technology, 5th Edition, Thomson Delmar Learning
Unit 10—System Charging
Unit 40—Typical Operating Conditions

Review Topics

Vapor Refrigerant Charging; Liquid Refrigerant Charging

COMPETENCY OBJECTIVE

The student will be able check a split-system air conditioner or heat pump for proper airflow by using a charging calculator.

OVERVIEW

One of the additional factors that affect the efficiency of air conditioning and heat pump equipment is making sure that the system is moving the correct airflow in CFMs (*cubic feet per minute*) for both the evaporator and condenser. For most installations 400 to 450 CFMs of air are required per ton of cooling.

EQUIPMENT REQUIRED

Superheat/sub-cooling charging calculator
1 Sling psychrometer
1 Temperature tester (with temperature leads)
1 Manifold gage set
1 Refrigerant cylinder that matches system refrigerant
Insulation tape
Assorted hand tools

SAFETY PRACTICES

Wear goggles or safety glasses and gloves when working with refrigerant in sealed systems. Beware of high pressures and follow all electrical safety rules. Follow all EPA rules and regulations while working with refrigerants.

COMPETENCY PROCEDURES

Checklist

1. Turn the system ON. ❏
2. Allow system to operate for 10 minutes. ❏
3. Use the Proper Airflow Range (Cooling) of the charging calculator. ❏
4. Use a temperature tester and measure the Indoor Entering Air Dry-Bulb Temperature. ❏
5. *Record the Indoor Entering Air Temperature.* _____ ❏

6. Slide the calculator slide rule to the proper indoor entering air temperature at **Number 1** of the Proper Airflow Range. ❏
7. Use the sling psychrometer and measure the Indoor Entering Air Wet-Bulb Temperature. ❏
8. *Record the Indoor Entering Air Wet-Bulb Temperature._____* ❏
9. Find the Indoor Entering Air Wet-Bulb Temperature at **Number 2** of the charging calculator. ❏
10. *Read the **Proper Evaporator Leaving Air Dry-Bulb Temperature** directly below the measured Indoor Entering Air Wet-Bulb Temperature recorded at Step 8.* ❏
11. *Record the Proper Evaporator Leaving Air Dry-Bulb Temperature as established by the charging calculator. _____* ❏
12. *Measure the actual leaving evaporator air temperature._____* ❏
13. *Record the actual leaving evaporator air temperature._____* ❏
14. Compare the actual leaving evaporator air temperature to the Proper Evaporator Coil Leaving Air Dry-Bulb Temperature established by the charging calculator. ❏

NOTE: *Under the Charging Calculator Instructions at Number 4 it states the following:*
    *If the measured leaving air-dry-bulb temperature is 3 Degrees (F) or more **LOWER** than the proper leaving air temperature, **INCREASE** evaporator fan speed. If the measured leaving air dry-bulb temperature is 3 degrees (F) or more **HIGHER** than the proper leaving air temperature, **DECREASE** evaporator fan speed.*

---

15. If airflow is not correct, make adjustments as needed to bring the system airflow into proper range. ❏
16. Have your instructor check your work. ❏
17. Turn the system OFF. ❏
18. Return all supplies and equipment to their proper location. ❏

RESEARCH QUESTIONS

1. What effect would restricted airflow have on a system's evaporator?
   _____
   _____

2. What effect would a blocked air filter have on a system's evaporator?
   _____
   _____

3. What effect would restricted airflow have on an air-cooled condenser?
   _____
   _____

4. Air conditioning and furnaces with air conditioning that have a multispeed blower should have the blower operating at which speed for the cooling cycle?
   _____
   _____

5. Furnaces with air conditioning that have a multispeed blower should have the blower operating at which speed for the heating cycle?
   _____
   _____

Passed Competency _____   Failed Competency _____

Instructor Signature _____   Grade _____

Theory Lesson: Procedures for Using an R-410A Superheat/Sub-cooling Charging Calculator

SUGGESTED MATERIALS

Textbook
Refrigeration & Air Conditioning Technology, 5th Edition, Thomson Delmar Learning
Unit 10—System Charging

Review Topics
Charging Near-Azeotropic (Zeotropic) Refrigerant Blends; Superheat Calculations; Sub-cool Calculations

Key Terms
bubble point • fractionation dew point • near-azeotropic refrigerants • sub-cooling • superheat • temperature glide

OVERVIEW

Many of the new refrigerants are referred to as *near-azeotropic* or zeotropic refrigerants and do not behave like azeotropic refrigerants. These refrigerants are blends, but referred to as *refrigerant mixtures* because refrigerants used within the blend can still separate into individual refrigerants, like water and oil, which can be mixed, but if allowed to sit for any period of time the water and oil will separate. This is where a difference in the pressure–temperature relationship arises versus the pressure–temperature relationship of azeotropic refrigerants.

Near-azeotropic blends experience a *temperature glide*. (Refer to *Figure 3–149, Theory Lesson: Superheat and Sub-cooling Calculations for Systems with Near-Azeotropic [Zeotropic] Refrigerant Blends.*)

This temperature glide occurs when the blend has many temperatures as it evaporates and condenses at a given pressure. This means that refrigerants used in the mixture will change phase from liquid to vapor or vapor back to a liquid faster than other refrigerants used to make the mixture. Therefore, it is possible with these refrigerant mixtures for one refrigerant in the mixture to be boiled away to 100% vapor and get *superheated* while other refrigerants in the mixture are still boiling from liquid to vapor. (Refer to *Figure 3–150, Theory Lesson: Superheat and Sub-cooling Calculations for Systems with Near-Azeotropic [Zeotropic] Refrigerant Blends.*)

Likewise it is possible for one or more of the refrigerants to be condensed from a vapor to 100% liquid and start to *sub-cool* while other refrigerants within the mixture are still being condensed from vapor back to a liquid.

Near-azeotropic refrigerants may also experience what is referred to as *fractionation*. This occurs when one or more of the refrigerants in the blend will condense or evaporate at different rates than other refrigerants making up the mixture. Fractionation happens only when the leak occurs as a vapor leak within the sealed system; it will not occur when the system is leaking pure liquid. This means that during a vapor leak within a system charged with near-azeotropic refrigerant, refrigerants within the mixture will leak out at different rates, which can have an effect on sealed system efficiency if system charge is *"topped off"* during a leak situation, rather than being completely removed from the system and a new charge weighed into the system. Fractionation also occurs if refrigerant is removed from a near-azeotropic refrigerant cylinder. For this reason, near-azeotropic or zeotropic blends should be *charged as a liquid* into a sealed system.

New pressure–temperature charts have been designed and make it easier for technicians to check, charge, and perform superheat and sub-cool calculations for equipment charged with near-azeotropic refrigerant blends. (Refer to *Figure 3–151, Theory Lesson: Superheat and Sub-cooling Calculations for Systems with Near-Azeotropic [Zeotropic] Refrigerant Blends.*)

For system superheat calculations, technicians are instructed to use the **dew point** values of the pressure temperature chart. *Dew point is the temperature at which saturated vapor first starts to condense.* For sub-cool calculations, technicians are instructed to use the **bubble point** values of the pressure–temperature chart. *Bubble point is the temperature at which the saturated liquid starts to boil off its first bubble of vapor.*

Using and understanding how to use a **superheat/sub-cooling calculator** takes the *"guess work"* out of checking or charging air conditioning and heat pump equipment along with checking for proper system airflow. The calculator is designed for use on TXV and non-TXV equipment. Technicians should use the **Required Superheat (Cooling, Non-TXV) side for systems with fixed-bore metering devices,** and the **Sub-cooling (Cooling, TXV) for systems with a TXV metering device.** The calculator can also be used to check the system's airflow by using the **Proper Airflow Range (Cooling)** portion of the calculator.

PROCEDURES FOR USING REQUIRED SUPERHEAT CALCULATOR (COOLING, NON-TXV)

> **NOTE:** *The temperature and pressures used in this example are only being used to assist the technician in becoming familiar with proper usage of the calculator.*
>
> *Actual readings would have to be taken for each system being charged or checked. Technicians would be required to use a sling psychrometer to measure indoor wet-bulb temperature, access the sealed system's low-side pressure, and use a temperature probe to measure the system's suction line temperature and outdoor condensing temperature.*

Example Readings
1. Indoor Wet-Bulb—**66 degrees (F)**
2. Condensing Entering Air Dry-Bulb Temperature—**95 degrees (F)**
3. Required Superheat—**10 degrees (F)**
4. Vapor Pressure (psig)—**70 psig**

PROCEDURES: REQUIRED SUPERHEAT METHOD

STEP 1
A. Use a sling psychrometer and take wet-bulb and dry-bulb readings of the inside air being conditioned.
B. Example indoor air wet-bulb temperature is 66 degrees (F).
C. Slide the calculator slide rule down until the arrow lines up with 66 degrees at **Number 1** (*Indoor Entering Air Wet-Bulb Temperature*) on the charging calculator (**Figure 3–158**).

STEP 2
A. Find the *Condenser Entering Air Dry-Bulb Temperature* of *95 degrees* at **Number 2** on the charging calculator (refer to **Figure 3–158**).

STEP 3
A. Find the *Required Superheat Temperature* of **10 degrees** at **Number 3** by reading the *Superheat Value* directly below the *95-degree Condenser Entering Air Dry-Bulb Temperature* (refer to **Figure 3–158**).

> **NOTE:** *Required superheat of 10 degrees (F) will be referenced again at **Number 5** on the charging calculator.*

STEP 4
A. Obtain the system's low-side operating pressure (vapor pressure – psig).
B. Using the example low-side pressure of 70 psig, slide the slide rule of the charging calculator to 70 psig at **Number 4** (**Figure 3–159**).

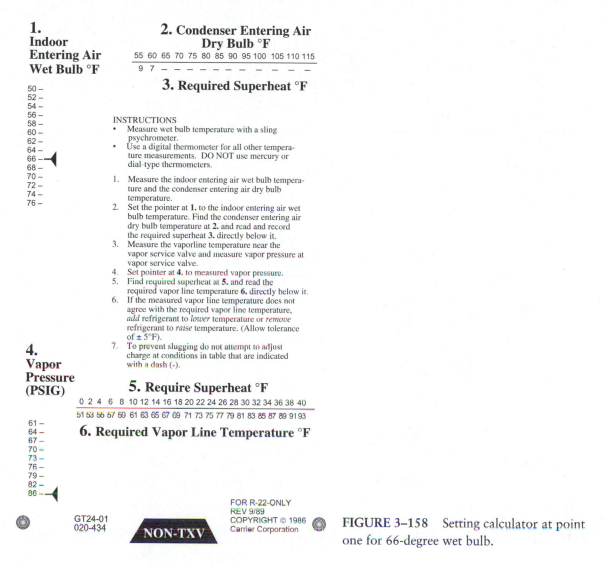

REQUIRED SUPERHEAT CALCULATOR
(Cooling, NON-TXV)

1.
Indoor Entering Air Wet Bulb °F

2. Condenser Entering Air Dry Bulb °F

55 60 65 70 75 80 85 90 95 100 105 110 115
9 7 – – – – – – – – – –

3. Required Superheat °F

50 –
52 –
54 –
56 –
58 –
60 –
62 –
64 –
66 –
68 –
70 –
72 –
74 –
76 –

INSTRUCTIONS
- Measure wet bulb temperature with a sling psychrometer.
- Use a digital thermometer for all other temperature measurements. DO NOT use mercury or dial-type thermometers.

1. Measure the indoor entering air wet bulb temperature and the condenser entering air dry bulb temperature.
2. Set the pointer at **1.** to the indoor entering air wet bulb temperature. Find the condenser entering air dry bulb temperature at **2.** and read and record the required superheat **3.** directly below it.
3. Measure the vaporline temperature near the vapor service valve and measure vapor pressure at vapor service valve.
4. Set pointer at **4.** to measured vapor pressure.
5. Find required superheat at **5.** and read the required vapor line temperature **6.** directly below it.
6. If the measured vapor line temperature does not agree with the required vapor line temperature, *add* refrigerant to *lower* temperature or *remove* refrigerant to *raise* temperature. (Allow tolerance of ± 5°F).
7. To prevent slugging do not attempt to adjust charge at conditions in table that are indicated with a dash (-).

4.
Vapor Pressure (PSIG)

5. Require Superheat °F

0 2 4 6 8 10 12 14 16 18 20 22 24 26 28 30 32 34 36 38 40
51 53 55 57 59 61 63 65 67 69 71 73 75 77 79 81 83 85 87 89 91 93

6. Required Vapor Line Temperature °F

61 –
64 –
67 –
70 –
73 –
76 –
79 –
82 –
86 –

GT24-01
020-434

NON-TXV

FOR R-22-ONLY
REV 9/89
COPYRIGHT © 1986
Carrier Corporation

FIGURE 3–158 Setting calculator at point one for 66-degree wet bulb.

STEP 5

A. Find the ***Required Superheat Temperature*** of **10 degrees** at **Number 5** of the charging calculator (refer to **Figure 3–159**).

STEP 6

A. At **Number 6** of the charging calculator find the ***Required Vapor Line Temperature*** directly below the ***Required Superheat Temperature of 10 degrees (F)*** (refer to **Figure 3–159**).

B. The charging calculator shows a ***Required Vapor Line Temperature of 51 Degrees (F)*** at **Number 6** (refer to ***Figure 3–159***).

> NOTE: *Underneath the Charging Calculator Instructions at Number 6 it states the following: If the measured vapor line temperature does not agree with the required vapor line temperature, ADD refrigerant to LOWER temperature or REMOVE refrigerant to RAISE temperature. (Allow a tolerance of + or –5 degrees F.)*
>
> *Instruction Number 7 states the following: To prevent slugging do not attempt to adjust charge at conditions in the table that are indicated with a dash (–).*

REQUIRED SUPERHEAT CALCULATOR
(Cooling, NON-TXV)

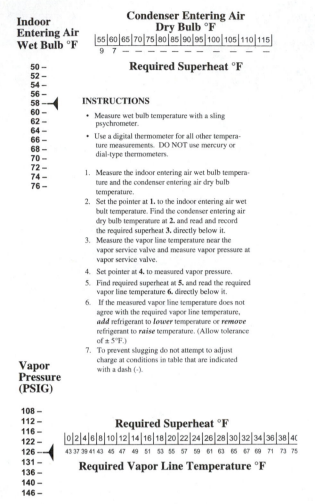

Indoor Entering Air Wet Bulb °F

Condenser Entering Air Dry Bulb °F

55	60	65	70	75	80	85	90	95	100	105	110	115
9	7	—	—	—	—	—	—	—	—	—	—	—

Required Superheat °F

50 –
52 –
54 –
56 –
58 –◄
60 –
62 –
64 –
66 –
68 –
70 –
72 –
74 –
76 –

INSTRUCTIONS

- Measure wet bulb temperature with a sling psychrometer.

- Use a digital thermometer for all other temperature measurements. DO NOT use mercury or dial-type thermometers.

1. Measure the indoor entering air wet bulb temperature and the condenser entering air dry bulb temperature.
2. Set the pointer at **1.** to the indoor entering air wet bult temperature. Find the condenser entering air dry bulb temperature at **2.** and read and record the required superheat **3.** directly below it.
3. Measure the vapor line temperature near the vapor service valve and measure vapor pressure at vapor service valve.
4. Set pointer at **4.** to measured vapor pressure.
5. Find required superheat at **5.** and read the required vapor line temperature **6.** directly below it.
6. If the measured vapor line temperature does not agree with the required vapor line temperature, **add** refrigerant to **lower** temperature or **remove** refrigerant to **raise** temperature. (Allow tolerance of ± 5°F.)
7. To prevent slugging do not attempt to adjust charge at conditions in table that are indicated with a dash (-).

Vapor Pressure (PSIG)

108 –
112 –
116 –
122 –
126 –◄
131 –
136 –
140 –
146 –

Required Superheat °F

0	2	4	6	8	10	12	14	16	18	20	22	24	26	28	30	32	34	36	38	4(
43	37	39	41	43	45	47	49	51	53	55	57	59	61	63	65	67	69	71	73	75

Required Vapor Line Temperature °F

FIGURE 3–159   126 psig at point 4.

PROCEDURES FOR USING SUB-COOLING CALCULATOR (COOLING, TXV)

> NOTE: *The temperature and pressures used in this example are only being used to assist the technician in becoming familiar with proper usage of the calculator.*
>
> *Actual readings would have to be taken for each system being charged or checked. Technicians would be required to access the sealed system's high-side pressure, and use a temperature probe to measure the system's liquid line temperature.*

Example Readings
1. Required Sub-cooling Temperature—**15 degrees (F)**
2. Liquid Pressure (psig) at Service Valve—**223 psig**
3. Required Liquid Line Temperature—**94 degrees (F)**

PROCEDURES: SUB-COOLING METHOD

STEP 1
A. Check the system's nameplate or service literature and determine the sub-cooling range for the air cooled condenser.

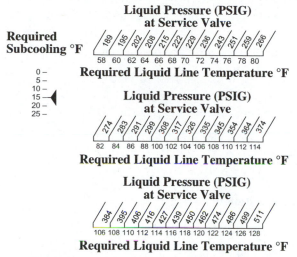

SUBCOOLING CALCULATOR
(Cooling, TXV)

Liquid Pressure (PSIG)
at Service Valve

Required
Subcooling °F

189 195 202 208 215 222 229 236 243 251 259 266
58   60   62   64   66   68   70   72   74   76   78   80

Required Liquid Line Temperature °F

0 –
5 –
10 –
15 –◀
20 –
25 –

Liquid Pressure (PSIG)
at Service Valve

274 283 291 299 308 317 326 335 345 354 364 374
82   84   86   88   100 102 104 106 108 110 112 114

Required Liquid Line Temperature °F

Liquid Pressure (PSIG)
at Service Valve

384 395 406 416 427 439 450 462 474 486 499 511
106 108 110 112 114 116 118 120 122 124 126 128

Required Liquid Line Temperature °F

INSTRUCTIONS
- Use a digital thermometer for all temperture measurement. DO NOT use mercury or dial-type thermometers.
1. Measure liquid line temperature near liquid service valve and measure liquid pressure at liquid service valve.
2. Set pointer at **1.** to unit required sub-cooling temperature.
3. Find measured liquid pressure at **2.** and read the required liquid line temperature **3.** directly below it.
4. If the measured liquid line temperature does not agree with the required liquid line temperature, **add** refrigerant to **lower** temperature or **remove** refrigerant to **raise** temperature. (Allow tolerance of ± 3°F.)

FIGURE 3–160   Setting sub-cool temperature at 16.

B. In this example, a sub-cooling range of **15 degrees** is being used.
C. At **Number 1**, slide the calculator's slide rule down until the arrow aligns with 15 degrees *Required Sub-cooling (F)* (**Figure 3–160**).

STEP 2
A. Determine the system's operating high-side pressure.
B. In the example *364 psig* is being used.
C. Find *364 psig* at **Number 2** of the calculator (refer to **Figure 3–160**).

STEP 3
A. Determine the liquid line temperature by looking directly underneath *364 psig* at **Number 2**.
B. *94 degrees* is the *Required Liquid Line Temperature* at **Number 3**.

NOTE: *Under the Charging Calculator Instructions at Number 4 it states the following:*
*   If the measured liquid line temperature does not agree with the required liquid line temperature, ADD refrigerant to LOWER temperature or REMOVE refrigerant to RAISE temperature. (Allow a tolerance of + or –3 degrees F.)*

PROCEDURES: PROPER AIRFLOW RANGE—COOLING

NOTE: *The temperatures in this example are only being used to assist the technician in becoming familiar with proper usage of the calculator.*
*   Actual dry-bulb and wet-bulb temperatures would be required to calculate the Required Airflow Range.*

Example Readings

1. Indoor Entering Air Dry-Bulb Temperature—**76 degrees (F)**
2. Indoor Entering Air Wet-Bulb Temperature—**64 degrees (F)**
3. Proper Evaporator Coil Leaving Air Dry-Bulb Temperature—**57 degrees (F)**

Use a temperature tester and measure the *Indoor Entering Air Dry-Bulb Temperature. Example temperature is 76 degrees.*

STEP 1

A. Slide the calculator slide rule down to **76 degrees** at **Number 1 (Figure 3–161).**

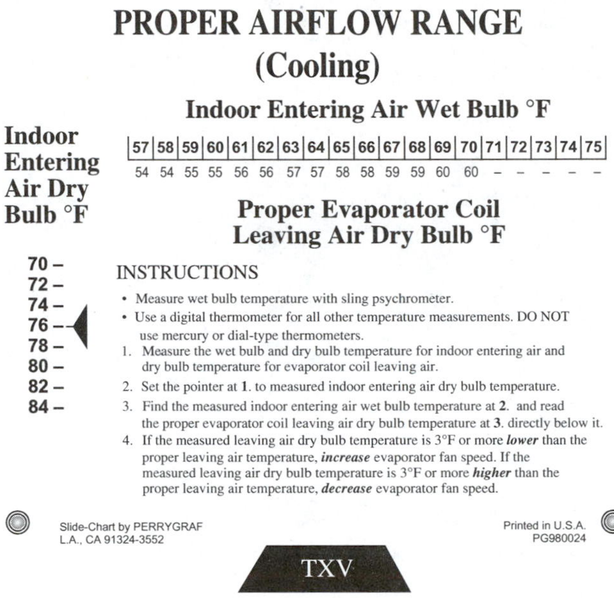

PROPER AIRFLOW RANGE
(Cooling)

Indoor Entering Air Wet Bulb °F

Indoor Entering Air Dry Bulb °F

57	58	59	60	61	62	63	64	65	66	67	68	69	70	71	72	73	74	75
54	54	55	55	56	56	57	57	58	58	59	59	60	60	–	–	–	–	–

Proper Evaporator Coil Leaving Air Dry Bulb °F

70 –
72 –
74 –
76 – ◄
78 –
80 –
82 –
84 –

INSTRUCTIONS

• Measure wet bulb temperature with sling psychrometer.
• Use a digital thermometer for all other temperature measurements. DO NOT use mercury or dial-type thermometers.
1. Measure the wet bulb and dry bulb temperature for indoor entering air and dry bulb temperature for evaporator coil leaving air.
2. Set the pointer at **1.** to measured indoor entering air dry bulb temperature.
3. Find the measured indoor entering air wet bulb temperature at **2.** and read the proper evaporator coil leaving air dry bulb temperature at **3.** directly below it.
4. If the measured leaving air dry bulb temperature is 3°F or more *lower* than the proper leaving air temperature, *increase* evaporator fan speed. If the measured leaving air dry bulb temperature is 3°F or more *higher* than the proper leaving air temperature, *decrease* evaporator fan speed.

Slide-Chart by PERRYGRAF
L.A., CA 91324-3552

Printed in U.S.A.
PG980024

TXV

FIGURE 3–161   Number 1–76-degrees (F) dry bulb.

B. Use a sling psychrometer and measure the *Indoor Entering Air Wet-Bulb Temperature. Example temperature is* **64 degrees.**

STEP 2

A. Find *64-Degree (F) Wet-Bulb Temperature* at **Number 2** (refer to **Figure 3–161**).
B. Measure the leaving air temperature (supply air).

STEP 3

A. Read the proper *Evaporator Leaving Air Dry-Bulb Temperature* directly below the *64-Degree (F) Indoor Entering Air Wet-Bulb Temperature. An example is 57 degrees (F).*
B. Compare the actual leaving air temperature to the *Proper Evaporator Coil Leaving Air Dry-Bulb Temperature of 57 Degrees (F).*

NOTE: Under the Charging Calculator Instructions at Number 4 it states the following:
If the measured leaving air dry-bulb temperature is 3 degrees (F) or more **LOWER** *than the proper leaving air temperature,* **INCREASE** *evaporator fan speed. If the measured leaving air-dry-bulb temperature is 3 degrees (F) or more* **HIGHER** *than the proper leaving air temperature,* **DECREASE** *evaporator fan speed.*

Practical Competency 77

Charging an R-410A Air Conditioning or Heat Pump by Using a Superheat/Sub-cooling Calculator—Superheat Method

SUGGESTED MATERIALS

Textbook

Refrigeration & Air Conditioning Technology, 5th Edition, Thomson Delmar Learning
Unit 10—System Charging
Unit 40—Typical Operating Conditions

Review Topics

Vapor Refrigerant Charging; Liquid Refrigerant Charging

COMPETENCY OBJECTIVE

The student will be able to properly charge an air conditioning or heat pump unit using a charging calculator and the superheat method.

OVERVIEW

It is important for HVAC technicians to understand that systems that are undercharged or overcharged will remove heat from a conditioned area, although they will not perform efficiently doing it!

It has been reported by national air conditioning associations that more than 70% of air conditioning and heat pump equipment in use in the United States is either *overcharged* or *undercharged*. This is a staggering number when efficiency performance of air conditioning and heat pump equipment relies on operating with the correct refrigerant charge. Many of the new refrigerants are referred to as *near-azeotropic* or zeotropic refrigerants and do not behave like azeotropic refrigerants. These refrigerants are blends, but referred to as *refrigerant mixtures* because refrigerants used within the blend can still separate into individual refrigerants, like water and oil, which can be mixed, but if allowed to sit for any period of time the water and oil will separate. This is where a difference in the pressure–temperature relationship arises versus the pressure–temperature relationship of azeotropic refrigerants.

Near-azeotropic blends experience a *temperature glide*. (Refer to *Figure 3–149, Theory Lesson: Superheat and Sub-cooling Calculations for Systems with Near-Azeotropic [Zeotropic] Refrigerant Blends.*)

This temperature glide occurs when the blend has many temperatures as it evaporates and condenses at a given pressure. This means that refrigerants used in the mixture will change phase from liquid to vapor or vapor back to a liquid faster than other refrigerants used to make the mixture. Therefore, it is possible with these refrigerant mixtures for one refrigerant in the mixture to be boiled away to 100% vapor and get *superheated* while other refrigerants in the mixture are still boiling from liquid to vapor. (Refer to *Figure 3–150, Theory Lesson: Superheat and Sub-cooling Calculations for Systems with Near-Azeotropic [Zeotropic] Refrigerant Blends.*)

Likewise it is possible for one or more of the refrigerants to be condensed from a vapor to 100% liquid and start to *sub-cool* while other refrigerants within the mixture are still being condensed from vapor back to a liquid. Near-azeotropic refrigerants may also experience what is referred to as *fractionation*. This occurs when one or more of the refrigerants in the blend will condense or evaporate at different rates than other refrigerants making up the mixture. Fractionation happens only when the leak occurs as a vapor leak

409

within the sealed system; it will not occur when the system is leaking pure liquid. This means that during a vapor leak within a system charged with near-azeotropic refrigerant, refrigerants within the mixture will leak out at different rates, which can have an effect on sealed system efficiency if system charge is *"topped off"* during a leak situation, rather than being completely removed from the system and a new charge weighed into the system. Fractionation also occurs if refrigerant is removed from a near-azeotropic refrigerant cylinder. For this reason, near-azeotropic or zeotropic blends should be *charged as a liquid* into a sealed system. New pressure–temperature charts have been designed and make it easier for technicians to check, charge, and perform superheat and sub-cool calculations for equipment charged with near-azeotropic refrigerant blends. (Refer to *Figure 3–151, Theory Lesson: Superheat and Sub-cooling Calculations for Systems with Near-Azeotropic [Zeotropic] Refrigerant Blends.*)

For system superheat calculations, technicians are instructed to use the **dew point** values of the pressure–temperature chart. *Dew point is the temperature at which saturated vapor first starts to condense.* For sub-cool calculations, technicians are instructed to use the **bubble point** values of the pressure–temperature chart. *Bubble point is the temperatureat which the saturated liquid starts to boil off its first bubble of vapor.*

Using and understanding how to use a **superheat/sub-cooling calculator** takes the *"guess work"* out of checking or charging air conditioning and heat pump equipment along with checking for proper system airflow. The calculator is designed for use on TXV and non-TXV equipment. Technicians should use the **Required Superheat (Cooling, Non-TXV) side for systems with fixed-bore metering devices,** and the **Subcooling (Cooling, TXV) for systems with a TXV metering device.** The calculator can also be used to the check system's airflow by using the **Proper Airflow Range (Cooling)** portion of the calculator.

EQUIPMENT REQUIRED

R-410A Superheat/sub-cooling charging calculator
1 Sling psychrometer
1 Temperature tester (with temperature leads)
1 Manifold gage set
1 Refrigerant cylinder that matches system refrigerant
Insulation tape
Assorted hand tools

SAFETY PRACTICES

Wear goggles or safety glasses and gloves when working with refrigerant in sealed systems. Beware of high pressures and follow all electrical safety rules. Follow all EPA rules and regulations while working with refrigerants.

> NOTE: *If competency is being performed on a heat pump, perform procedures with the heat pump operating in the air conditioning mode.*

> NOTE: *The system being charged should have been leak-checked and evacuated prior to performing this competency.*

COMPETENCY PROCEDURES   Checklist

1. The sealed system should be OFF.   ☐
2. Use the sling psychrometer and measure the Indoor Entering Air Wet-Bulb Temperature.   ☐
3. *Record the Indoor Entering Air Wet-Bulb Temperature.* _____   ☐
4. Use the Required Superheat Calculator (*Cooling, NON-TXV*) and slide the slide rule of the charging calculator down to the recorded Indoor Entering Air Wet-bulb Temperature at **Number 1.**   ☐

5. Use the temperature tester and measure the Condenser Entering Air Dry-Bulb Temperature. ❑
6. *Record the Condenser Entering Air Dry-Bulb Temperature.* _____ ❑
7. Find the Condenser Entering Air Dry-Bulb Temperature at **Number 2** of the charging calculator. ❑
8. Under directly below the Condenser Entering Air Dry-Bulb Temperature is the Required Superheat Temperature at **Number 3** of the charging calculator. ❑
9. *Record the Required Superheat Temperature.* _____ ❑
10. Attach and insulate a temperature probe near the system's suction line service valve. ❑
11. Add refrigerant through the system's low-side valve in the vapor state until the system's low-side pressure is brought out of a vacuum. ❑
12. Turn the sealed system ON. ❑
13. Continue to add refrigerant in the vapor state until the system's low-side pressure is operating at least **64 psig.** ❑
14. Once the system's operating low-side pressure is at least *64 psig,* stop charging and allow the system to operate for 10 minutes. ❑
15. At **Number 5** of the charging calculator, find the Required Superheat Temperature as recorded in **Step 9.** ❑
16. Directly below the Required Superheat at **Number 5** is the Required Vapor Line Temperature at **Number 6.** ❑
17. *Record the Required Vapor Line Temperature as indicated by the charging calculator at Number 6.* _____ ❑
18. *Check the system's suction line temperature and record actual suction line temperature.* _____ ❑

> **NOTE:** *Underneath the Charging Calculator Instructions at Number 6 it states the following:*
> *If the measured vapor line temperature does not agree with the required vapor line temperature,* **ADD** *refrigerant to* **LOWER** *temperature or* **REMOVE** *refrigerant to* **RAISE** *temperature. (Allow tolerance of + or −5 degrees F.)*

19. Add or remove refrigerant to the system in small amounts until the system's suction line temperature falls within (+ *or* −5 degrees) of the Required Vapor Line Temperature as indicated by the charging calculator. ❑

> **NOTE:** *Any change in refrigerant charge to the system will change the system's low-side operating pressure. The Vapor Pressure at Number 4 will have to be readjusted as the system's pressures change due to the addition or removal of refrigerant to the system. The same required superheat temperature would continue to be used to determine Required Vapor Line Temperature unless there has been a considerable time lapse or temperature change from the original temperature readings taken at Numbers 1 and 2.*

20. Once the system charge is correct, have your instructor check your work. ❑
21. On completion of the system charge, remove all equipment and return to its proper location. ❑

RESEARCH QUESTIONS

1. What is the design superheat range for an air conditioning system operating under standard conditions?

   _____

   _____

2. What type of equipment normally has the refrigerant charge printed on the nameplate?

_____

_____

3. What is the main difference between a zeotropic and azeotropic refrigerant blend?

_____

_____

4. When should a service technician use the sub-cooling method of charging air conditioning and heat pumps systems?

_____

_____

5. What causes fractionation to happen in certain blends of refrigerant?

_____

_____

Passed Competency _____   **Failed Competency** _____

Instructor Signature _____   **Grade** _____

Practical Competency 78

Charging an R-410A Air Conditioning or Heat Pump by Using a Superheat/Sub-cooling Calculator—Sub-cool Method

SUGGESTED MATERIALS

Textbook

Refrigeration & Air Conditioning Technology, 5th Edition, Thomson Delmar Learning
Unit 10—System Charging
Unit 40—Typical Operating Conditions

Review Topics

Vapor Refrigerant Charging; Liquid Refrigerant Charging

COMPETENCY OBJECTIVE

The student will be able to properly charge an air conditioning or heat pump unit equipped with a TXV metering device by using a charging calculator and the sub-cool method.

OVERVIEW

It is important for HVAC technicians to understand that systems that are undercharged or overcharged will remove heat from a conditioned area, although they will not perform efficiently doing it!

It has been reported by national air conditioning associations that more than 70% of air conditioning and heat pump equipment in use in the United States is either *overcharged* or *undercharged*. This is a staggering number when efficiency performance of air conditioning and heat pump equipment relies on operating with the correct refrigerant charge. Many of the new refrigerants are referred to as *near-azeotropic* or zeotropic refrigerants and do not behave like azeotropic refrigerants. These refrigerants are blends, but referred to as *refrigerant mixtures* because refrigerants used within the blend can still separate into individual refrigerants, like water and oil, which can be mixed, but if allowed to sit for any period of time the water and oil will separate. This is where a difference in the pressure–temperature relationship arises versus the pressure–temperature relationship of azeotropic refrigerants.

Near-azeotropic blends experience a *temperature glide*. (Refer to *Figure 3–149, Theory Lesson: Superheat and Sub-cooling Calculations for Systems with Near-Azeotropic [Zeotropic] Refrigerant Blends*.)

This temperature glide occurs when the blend has many temperatures as it evaporates and condenses at a given pressure. This means that refrigerants used in the mixture will change phase from liquid to vapor or vapor back to a liquid faster than other refrigerants used to make the mixture. Therefore, it is possible with these refrigerant mixtures for one refrigerant in the mixture to be boiled away to 100% vapor and get *superheated* while other refrigerants in the mixture are still boiling from liquid to vapor. (Refer to *Figure 3–150, Theory Lesson: Superheat and Sub-cooling Calculations for Systems with Near Azeotropic [Zeotropic] Refrigerant Blends*.)

Likewise it is possible for one or more of the refrigerants to be condensed from a vapor to 100% liquid and start to *sub-cool*, while other refrigerants within the mixture are still being condensed from vapor back to a liquid. Near-azeotropic refrigerants may also experience what is referred to as *fractionation*. This occurs when one or more of the refrigerants in the blend will condense or evaporate at different rates than other refrigerants making up the mixture. Fractionation happens only when the leak occurs as a vapor leak within the sealed system; it will not occur when the system is leaking pure liquid. This means that during a vapor leak within a system charged with near-azeotropic refrigerant, refrigerants within the mixture will leak out at different rates, which can have an effect on sealed system efficiency if the system charge is *"topped off"* during a leak situation, rather than being completely removed from the system and a new charge weighed into the system. Fractionation also occurs if refrigerant is removed from a near-azeotropic refrigerant cylinder. For this reason, near-azeotropic or zeotropic blends should be *charged as a liquid* into a sealed system. New pressure–temperature charts have been designed and make it easier for technicians to check, charge, and perform superheat and sub-cool calculations for equipment charged with near-azeotropic refrigerant blends. (Refer to *Figure 3–151, Theory Lesson: Superheat and Sub-cooling Calculations for Systems with Near Azeotropic [Zeotropic] Refrigerant Blends*.)

For system superheat calculations, technicians are instructed to use the **dew point** values of the pressure–temperature chart. *Dew point is the temperature at which saturated vapor first starts to condense.* For sub-cool calculations, technicians are instructed to use the **bubble point** values of the pressure temperature chart. *Bubble point is the temperature at which the saturated liquid starts to boil off its first bubble of vapor.*

Using and understanding how to use a **superheat/sub-cooling calculator** takes the *"guess work"* out of checking or charging air conditioning and heat pump equipment along with checking for proper system airflow. The calculator is designed for use on TXV and non-TXV equipment. Technicians should use the **Required Superheat (Cooling, Non-TXV) side for systems with fixed-bore metering devices, and the Sub-cooling (Cooling, TXV)** for systems with a TXV metering device. The calculator can also be used to check the system's airflow by using the **Proper Airflow Range (Cooling)** portion of the calculator.

EQUIPMENT REQUIRED

R-410A Superheat/sub-cooling charging calculator
1 Temperature tester (with temperature leads)
1 Manifold gage set
1 Refrigerant cylinder that matches system refrigerant
Insulation tape
Assorted hand tools

SAFETY PRACTICES

Wear goggles or safety glasses and gloves when working with refrigerant in sealed systems. Beware of high pressures and follow all electrical safety rules. Follow all EPA rules and regulations while working with refrigerants.

NOTE: If competency is being performed on a heat pump, perform procedures with the heat pump operating in the air conditioning mode.

NOTE: System being charged should have been leak-checked and evacuated prior to performing this competency.

COMPETENCY PROCEDURES   Checklist

1. The sealed system should be OFF.   ❑
2. Check the system's nameplate or service literature and determine the sub-cooling range for the
   air cooled condenser.   ❑

NOTE: If the system's sub-cooling range is unavailable use 10 or 15 degrees of sub-cooling.

3. At **Number 1,** slide the calculators slide rule down until the arrow aligns with the desired sub-cool temperature. ❏

4. Attach and insulate a temperature probe to the system liquid line just ahead of the system's high-side service valve. ❏

5. Add refrigerant to the system in either the vapor or liquid state and bring the system's low-side pressure out of a vacuum. ❏

6. Turn the system ON. ❏

7. Continue to charge system until a low-side operating pressure of 62 psig is established. ❏

8. Let the system operate for 10 minutes. ❏

9. *Check and record the system's operating pressure.* _____ ❏

10. Find the system's high-side pressure at **Number 2** of the charging calculator. ❏

11. Determine the liquid line temperature by looking directly underneath the system's operating high-side pressure at **Number 2.** ❏

12. *Record the required Liquid Line Temperature according to the charging calculator.* _____ ❏

NOTE: *Under the Charging Calculator Instructions at Number 4 it states the following:*
    If the measured liquid line temperature does not agree with the required liquid line temperature, **ADD** *refrigerant to* **LOWER** *temperature or* **REMOVE** *refrigerant to* **RAISE** *temperature. (Allow a tolerance of + or −3 degrees (F).*

NOTE: *Any change in refrigerant charge to the system will change the system's high-side operating pressure. The Liquid Pressure (psig) at Number 2 will have to be readjusted as the system's pressures change as a result of the addition or removal of refrigerant to the system. The same required sub-cool temperature would continue to be used to determine.*

---

13. Once the system charge is correct, have your instructor check your work. ❏

14. On completion of the system charge, remove all equipment and return to its proper location. ❏

RESEARCH QUESTIONS

1. How is condenser high efficiency obtained?

    _____

    _____

2. The evaporator design temperature may in some cases operate at a slightly _____ (higher or lower) pressure and temperature on high-efficiency equipment because the evaporator is larger.

3. The three major power-consuming devices on an air conditioning system that may have to be analyzed are the _____, _____, and the _____.

4. Air conditioning systems normally move about _____ CFM of air per ton.

5. The compressor amperage of a 3-ton system operating on 230 volts is approximately _____ A.

Passed Competency _____   Failed Competency _____

Instructor Signature _____   Grade _____

Practical Competency 79

Checking for Proper Airflow Range of a Split-System Air Conditioner or Heat Pump Using a R-410A Charging Calculator

SUGGESTED MATERIALS

Textbook
Refrigeration & Air Conditioning Technology, 5th Edition, Thomson Delmar Learning
Unit 10—System Charging
Unit 40—Typical Operating Conditions

Review Topics
Vapor Refrigerant Charging; Liquid Refrigerant Charging

COMPETENCY OBJECTIVE

The student will be able check a split-system air conditioner or heat pump for proper airflow by using a charging calculator.

OVERVIEW

One of the additional factors that affect the efficiency of air conditioning and heat pump equipment is making sure that the system is moving the correct airflow in CFMs (*cubic feet per minute*) for both the evaporator and condenser. For most installations 400 to 450 CFMs of air are required per ton of cooling.

EQUIPMENT REQUIRED

R-410A Superheat/sub-cooling charging calculator
1 Sling psychrometer
1 Temperature tester (with temperature leads)
1 Manifold gage set
1 Refrigerant cylinder that matches system refrigerant
Insulation tape
Assorted hand tools

SAFETY PRACTICES

Wear goggles or safety glasses and gloves when working with refrigerant in sealed systems. Beware of high pressures and follow all electrical safety rules. Follow all EPA rules and regulations while working with refrigerants.

COMPETENCY PROCEDURES   Checklist

1. Turn the system ON.   ❏
2. Allow the system to operate for 10 minutes.   ❏
3. Use the Proper Airflow Range (Cooling) of the charging calculator.   ❏
4. Use a temperature tester and measure the Indoor Entering Air Dry-Bulb Temperature.   ❏

5. *Record the Indoor Entering Air Temperature.* _____ ❏
6. Slide the calculator slide rule to the proper indoor entering air temperature at Number 1 of the Proper Airflow Range. ❏
7. Use the sling psychrometer and measure the Indoor Entering Air Wet-Bulb Temperature. ❏
8. *Record the Indoor Entering Air Wet-Bulb Temperature.* _____ ❏
9. Find the Indoor Entering Air Wet-Bulb Temperature at **Number 2** of the charging calculator. ❏
10. *Read the **Proper Evaporator Leaving Air Dry-Bulb Temperature** directly below the measured Indoor Entering Air Wet-Bulb Temperature recorded at Step 8.* ❏
11. *Record the Proper Evaporator Leaving Air Dry-Bulb Temperature as established by the charging calculator.* _____ ❏
12. *Measure the actual leaving evaporator air temperature.* _____ ❏
13. *Record the actual leaving evaporator air temperature.* _____ ❏
14. Compare the actual leaving evaporator air temperature to the Proper Evaporator Coil Leaving Air Dry-bulb Temperature established by the charging calculator. ❏

NOTE: *Under the Charging Calculator Instructions at Number 4 it states the following:*
*If the measured leaving air-dry-bulb temperature is 3 degrees (F) or more **LOWER** than the proper leaving air temperature, **INCREASE** evaporator fan speed. If the measured leaving air dry-bulb temperature is 3 degrees (F) or more **HIGHER** than the proper leaving air temperature, **DECREASE** evaporator fan speed.*

---

15. If airflow is not correct, make adjustments as needed to bring the system airflow into the proper range. ❏
16. Have your instructor check your work. ❏
17. Turn the system OFF. ❏
18. Return all supplies and equipment to their proper location. ❏

RESEARCH QUESTIONS

1. Why do most air conditioning systems not need a defrost system?

   _____

   _____

2. The typical temperature difference between the entering air and the boiling refrigerant on a standard air conditioning evaporator is _____ (F).

3. **True or False:** When troubleshooting a small system, the first step is to attach the high and low pressure gages.

   _____

4. The suction gas may have a high superheat if the unit has a _____ charge.

5. What are two types of fixed-bore metering devices? _____ and _____

Passed Competency _____   Failed Competency _____

Instructor Signature _____   Grade _____

Theory Lesson: Charging Procedures in the Field

SUGGESTED MATERIALS

Textbook
Refrigeration & Air Conditioning Technology, 5th Edition, Thomson Delmar Learning
Unit 41—Troubleshooting

Review Topics
Charging Procedures in the Field; Fixed-Bore Metering Devices; Capillary Tube and Orifice Type; Field Charging the TXV System

Key Terms
condenser head pressure • condensing temperature • design condenser sub-cooling • design evaporator superheat • liquid line • standard operating conditions • suction line • typical reference points • typical sub-cooling • typical superheat

OVERVIEW

When charging or checking the charge of any air conditioning or heat pump system it's important that technicians follow manufacturer's guidelines when available. There are times when the technician needs to check or make system charge adjustments and the manufacturer's guidelines are not available. Besides using a charging calculator, charging charts, or charging curves, technicians are left with making personal judgments as to a system's charge condition. Technicians can accurately determine or adjust a system's charge without the aid of such tools, but to do so requires an understanding of refrigeration fundamentals, the manufacturer's design of such equipment, and how the equipment should perform under certain conditions.

Field charging procedures are best established by knowing how the equipment is to perform under *Standard Operating Conditions*. When a system is charged correctly and operating within Standard Conditions, a prescribed amount of refrigerant should be in the condenser, the evaporator, and the system's liquid line. **It's important that technicians understand the following facts about air conditioning and heat pump equipment:**

1. *The amount of refrigerant in the evaporator can be measured by the Superheat Method.*
2. *The amount of refrigerant in the condenser can be measured by the Sub-cooling Method.*
3. *The amount of refrigerant in the liquid line may be determined by measuring the length and calculating the refrigerant charge.*
4. *If the evaporator is performing correctly, the liquid line has the correct charge.*

When system charge adjustments are required, technicians need to establish charging procedures in the field for all types of equipment. The following are a couple of methods used for different types of equipment.

PROCEDURES FOR SYSTEMS WITH FIXED-BORE METERING DEVICES: CAPILLARY TUBE AND PISTON (ORIFICE) TYPE

Fixed-bore metering devices allow refrigerant flow based on the difference in the inlet (high-side) and outlet (low-side) pressures. Ideally, the best situation would be to check for a system's correct charge under Standard Operating Conditions, 75-degree (F) return air with 50% humidity, and an outside temperature of 95 degrees. Unfortunately, rarely do these exact conditions exist. When a system is operating outside of these conditions, different system pressures superheat readings, and sub-cooling temperatures will occur. The condition that most affects system readings is the outside ambient air temperature. In lower than normal outside air temperatures, the condenser will become more efficient in rejecting heat from the refrigerant vapor, causing condensing of vapor to liquid to happen sooner, raising the condenser's sub-cooling temperature, and in turn lowering the system's high-side operating pressure.

Since fixed-bore metering devices rely on a pressure difference between the system's high-side and low-side pressures to feed the correct amount of refrigerant into the evaporator, lower outside temperature can create the effect of partially starving the evaporator for refrigerant. In other words, because the condenser is operating more efficiently, there is more liquid refrigerant in the condenser than that which is being fed into the evaporator by the metering device, in turn creating a situation in which the evaporator becomes starved.

The lower than normal high-side pressure under this condition is what makes it hard to evaluate a system's charge, although typical operating conditions **can be simulated by reducing the airflow across the condenser, which will cause the system's high-side pressure to rise (Figure 3–162).**

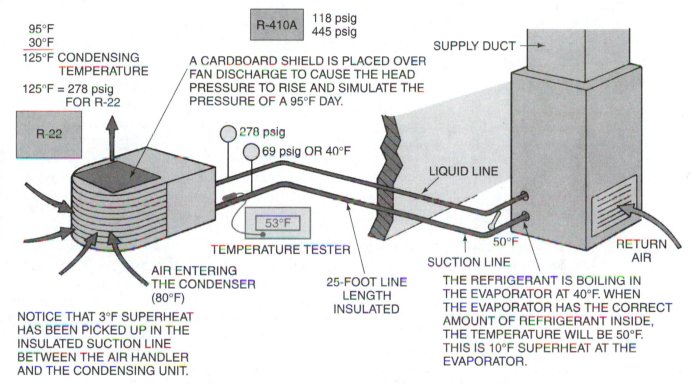

FIGURE 3–162   Blocking condenser airflow to simulate typical outside air temperature of 95 degrees (F), increasing system head pressure to correlate with established conditions, so that the Superheat Method can be used to check the charge of a system equipped with a fixed-bore metering device.

On a 95-degree (F) day the highest condenser head pressure for **R-22 systems is 278 psig**, which equals to a condensing temperature of 125 degrees for standard air-cooled condensers. Remember that standard air-cooled condensers condense the refrigerant vapor to liquid at approximately 30 degrees (F) above the outside air temperature. (*95-degree day plus 30 degrees equals 125 degrees saturation temperature.*) *R-410A systems would operate with a high-side pressure of approximately **445 psig** under standard conditions. For high-efficiency equipment, a head pressure of **250 psig** would be sufficient for an **R-22 system**, and **390 psig** for an **R-410A system.***

The higher high-side pressure pushes the refrigerant through the metering device at the correct rate, rather than allowing refrigerant liquid buildup in the condenser as with lower ambient conditions.

> *NOTE: Once the condenser's airflow is restricted, allow time for the system's high-side pressure to build. If the correct high-side pressure is not in range as that which should be under standard conditions, refrigerant will have to added or removed from the system to bring the high-side pressure into range. Once the correct pressure is established, the following additional procedures can be used to bring the system's charge into range.*

Using this procedure allows the technician to perform a superheat check on the system's evaporator, although a superheat check at the actual evaporator is not always easy with split- systems, so a superheat check at the condensing unit suction line can be made. When using the system's suction line, the length of the line set would need to be taken into consideration. Remember that under standard conditions, 75-degree (F) inside air with 50% humidity, and an outside air temperature of 95 degrees (F), the evaporator superheat should be around 10 degrees. Since the superheat is actually measured at the system's suction line, the length of the system's line set needs to be taken into consideration.

With the proper high-side pressure and a properly insulated suction line, a **line set length of 10 to 30 feet should indicate a superheat value of 10 to 15 degrees (F) if the system's charge is close to being correct. Line sets of 30 to 50 feet should indicate a superheat value of 15 to 18 degrees.**

Procedures for Field Charging a TXV System

A TXV metering device is designed to maintain a constant superheat on an evaporator of 8 to 12 degrees under all load conditions. Depending on the load on the evaporator, the TXV may not need to feed the system's total charge to the evaporator, so it needs space to store the additional refrigerant when not in use. Liquid receivers are equipped with TXV systems for this very reason, and work in conjunction with the system's condenser to store liquid refrigerant in low ambient temperature conditions. This process always ensures enough liquid refrigerant is available to the TXV for feeding the evaporator. Because of this, a TXV system will not be affected as much as a system with a fixed-bore metering device during low ambient temperatures. In fact a TXV system could be undercharged or overcharged, yet have enough refrigerant to maintain designed superheat on the evaporator. For this reason, checking or adjusting the charge of a system with a TXV metering device cannot be performed by using the Superheat Method.

The charge of air conditioning and heat systems equipped with a TXV metering device can be checked in much the same way as a system with a fixed-bore metering device during low ambient temperature conditions. The only difference is that once pressures are established to represent a 95-degree (F) day, the sub-cool method of checking or charging the system would be used.

Once again, typical operating conditions **would need to be simulated by reducing the airflow across the condenser, which will cause the system's high-side pressure to rise.**

On a 95-degree (F) day the highest condenser **head pressure for R-22 systems is 278 psig**, which equals to a condensing temperature of 125 degrees for standard air-cooled condensers. Remember that standard air-cooled condensers condense the refrigerant vapor to liquid at approximately 30 degrees (F) above the outside air temperature. (*95-degree day plus 30 degrees equals 125 degrees saturation temperature*). *R-410A systems would operate with a high-side pressure of approximately **445 psig** under standard conditions. For **high-efficiency equipment**, a head pressure of **250 psig** would be sufficient for an R-22 system, and **390 psig** for an R-410A system.*

Once the correct high-side pressure for the type of refrigerant is simulated, checking the condensers sub-cooling can be used to evaluate the system's charge. A **typical sub-cooling** circuit will sub-cool the liquid refrigerant **from 10 to 20 degrees (F)** than the condenser's condensing temperature. Under the simulated conditions, a temperature lead should be attached to the condenser's **liquid line and a temperature of 105 to 115 degrees (F)**, equivalent to 10 to 20 degrees (F) of sub-cooling, should be established.

NOTE: *Establishing the proper sub-cooling range under these simulated conditions means that the system's charge is quite close to accurate.*

NOTE: (OVERCHARGED SYSTEM) *A sub-cooling temperature higher than 20 degrees (F) is an indication of an overcharged system. This will require that some of the refrigerant from the system be recovered until the sub-cooling temperature is brought within the design range.*

NOTE: (UNDERCHARGED SYSTEM) *Sub-cooling temperatures lower than the 10- to 20-degree (F) range are an indication that the system is undercharged. Refrigerant would have to be added to the system until the sub-cooling temperature is brought within the design range.*

NOTE: *Refrigerant charge adjustments should be made guardedly. Once an adjustment is made, allow system to stabilize before making additional adjustments.*

Practical Competency 80

Field Charging or Checking the Charge of an Air Conditioner or Heat Pump System with a Fixed-Bore Metering Device

SUGGESTED MATERIALS

Textbook
Refrigeration & Air Conditioning Technology, 5th Edition, Thomson Delmar Learning
Unit 41—Troubleshooting
Unit 22—Condensers

Review Topics
Charging Procedures in the Field; Fixed-Bore Metering Devices; Capillary Tube and Orifice Type

COMPETENCY OBJECTIVE

The student will be able charge or check the charge of an air conditioner or heat pump during low ambient temperature by using field service procedures.

OVERVIEW

When charging or checking the charge of any air conditioning or heat pump system it's important that technicians follow manufacturer's guidelines when available. There are times when the needs to check or make system charge adjustments and the manufacturer's guidelines are not available. Besides using a charging calculator, charging charts, or charging curves, technicians are left with making personal judgments as to a system's charge condition. Technicians can accurately determine or adjust a system's charge without the aid of such tools, but to do so requires an understanding of refrigeration fundamentals, the manufacturer's design of such equipment, and how the equipment should perform under certain conditions.

Field charging procedures are best established by knowing how the equipment is to perform under *Standard Operating Conditions*. When a system is charged correctly and operating within Standard Conditions, a prescribed amount of refrigerant should be in the condenser, the evaporator, and the system's liquid line. **It's important that technicians understand the following facts about air conditioning and heat pump equipment:**

1. *The amount of refrigerant in the evaporator can be measured by the Superheat Method.*
2. *The amount of refrigerant in the condenser can be measured by the sub-cooling method.*
3. *The amount of refrigerant in the liquid line may be determined by measuring the length and calculating the refrigerant charge.*
4. *If the evaporator is performing correctly, the liquid line has the correct charge.*

When system charge adjustments are required, technicians need to establish charging procedures in the field for all types of equipment. The following are a couple of methods used for different types of equipment.

Procedures for Systems with Fixed-Bore Metering Devices: Capillary Tube and Piston (Orifice) Type

Fixed-bore metering devices allow refrigerant flow based on the difference in the inlet (high-side) and outlet (low-side) pressures. Ideally, the best situation would be to check for a system's correct charge under

Standard Operating Conditions, 75-degree (F) return air with 50% humidity, and an outside temperature of 95 degrees. Unfortunately, rarely do these exact conditions exist. When a system is operating outside of these conditions, different system pressures superheat readings, and sub-cooling temperatures will occur. The condition that most affects system readings is the outside ambient air temperature. In lower than normal outside air temperatures, the condenser will become more efficient in rejecting heat from the refrigerant vapor, causing condensing of vapor to liquid to occur sooner, raising the condenser's sub-cooling temperature, and in turn lowering the system's high-side operating pressure.

Since fixed-bore metering devices rely on a pressure difference between the system's high-side and low-side pressures to feed the correct amount of refrigerant into the evaporator, lower outside temperature can create the effect of partially starving the evaporator for refrigerant. In other words, because the condenser is operating more efficiently, there is more liquid refrigerant in the condenser than that which is being fed into the evaporator by the metering device, in turn creating a situation in which the evaporator becomes starved.

The lower than normal high-side pressure under this condition is what makes checking a system's charge hard to evaluate, although typical operating conditions **can be simulated by reducing the airflow across the condenser, which will cause the system's high-side pressure to rise.**

On a 95-degree (F) day the highest condenser head pressure for **R-22 systems is 278 psig**, which equals to a condensing temperature of 125 degrees for standard air-cooled condensers. Remember that standard air-cooled condensers condense the refrigerant vapor to liquid at approximately 30 degrees (F) above the outside air temperature. (*95-degree day plus 30 degrees equals 125 degrees saturation temperature*). *R-410A systems would operate with a high-side pressure of approximately **445 psig** under standard conditions. For **high-efficiency equipment**, a head pressure of **250 psig** would be sufficient for an **R-22 system**, and **390 psig** for an **R-410A system.***

The higher high-side pressure pushes the refrigerant through the metering device at the correct rate, rather than allowing refrigerant liquid buildup in the condenser as with lower ambient conditions.

> **NOTE:** *Once the condenser's airflow is restricted, allow time for the system's high-side pressure to build. If the correct high-side pressure is not in range as that which should be under standard conditions, refrigerant will have to added or removed from the system to bring the high-side pressure into range. Once the correct pressure is established, the following additional procedures can be used to bring the system's charge into range.*

Using this procedure allows the technician to perform a superheat check on the system's evaporator, although a superheat check at the actual evaporator is not always easy with split- systems, so a superheat check at the condensing unit suction line can be made. When using the system's suction line, the length of the line set would need to be taken into consideration. Remember that under standard conditions, 75 degree (F) inside air with 50% humidity and an outside air temperature of 95 degrees (F), the evaporator superheat should be around 10 degrees. Since the superheat is actually measured at the system's suction line, the length of the system's line set needs to be taken into consideration.

With the proper high-side pressure and a properly insulated suction line, a *line set length of 10 to 30 feet should indicate a superheat value of 10 to 15 degrees (F) if the system's charge is close to being correct. Line sets of 30 to 50 feet should indicate a superheat value of 15 to 18 degrees.*

EQUIPMENT REQUIRED

Air conditioner or heat pump system
1 Piece of cardboard/wood shield large enough to restrict condenser airflow
1 Temperature tester (with temperature leads)
1 Manifold gage set
1 Refrigerant cylinder that matches system refrigerant
Insulation tape
Assorted hand tools

SAFETY PRACTICES

Wear goggles or safety glasses and gloves when working with refrigerant in sealed systems. Beware of high pressures and follow all electrical safety rules. Follow all EPA rules and regulations while working with refrigerants.

> **NOTE:** *This competency is best performed on a system operating with an outside air temperature below 95 degrees (F).*

COMPETENCY PROCEDURES

Checklist

1. Turn the system OFF. ☐
2. Remove high- and low-side service valve caps. ☐
3. Make sure that manifold gage valves are closed. ☐
4. Attach the manifold high-side hose to the system's high-side valve service port. ☐
5. Attach the manifold low-side hose to the system's low-side valve service port. ☐
6. Attach a temperature lead to the system's suction line, near the system's low-side service valve. ☐
7. Insulate the temperature lead. ☐
8. *Estimate and record the systems line set length.* _____ ☐
9. *According to the Competency Overview and considering the approximate length of systems line set, what should be the approximate amount of superheat if the system's charge is correct?* _____ ☐
10. *Record the type of refrigerant used in the system.* _____ ☐
11. Turn the system ON. ☐
12. Allow system to operate for 10 to 15 minutes. ☐
13. *Record the systems operating low-side pressure.* _____ ☐
14. *Use a P/T chart and record the system's low-side saturation temperature.* _____ ☐
15. *Record the suction line temperature.* _____ ☐
16. *Determine the amount of the system's superheat:* ☐

Suction line temperature _____

MINUS

Low-side saturation temperature _____

EQUALS *Superheat amount* _____

17. *Record the system's superheat.* _____ ☐
18. *Does the superheat amount fall in range with that which is stated in the Overview based on a 95-degree (F) outside temperature and length of the system's line set?* _____ ☐
19. *Record the systems operating high-side pressure.* _____ ☐
20. *Does the system's high-side pressure fall in line with an outside temperature of a 95-degree (F) day as described in the Competency Overview?* _____ ☐

> **NOTE:** *If the system's superheat and high-side pressure* **are within range** *as described in the Competency Overview, the system's charge is correct.* **Do not proceed;** *have your Instructor check your work.*
>
> *If the system's superheat and high-side pressure* **are not within range** *as described in the Competency Overview,* **proceed to the next steps** *of the Competency.*

21. Place the cardboard/wood shield over the top of the condenser to restrict condenser airflow. (Refer to **Figure 2-268.**) ☐
22. Allow the system to operate for 10 to 15 minutes with restricted condenser airflow. ☐
23. *Record the system's high-side pressure.* _____ ☐

24. *Does the system's high-side pressure fall in line for the Type of Refrigerant as described in the Competency Overview, based on a 95-degree day?* _____ ❏

> **NOTE:** *With condenser air restricted, if the high-side pressure is higher or lower than that which is expected on a 95 degree (F) day according to the Competency Overview,* **ADD** *Refrigerant to* **RAISE** *high-side pressure, or* **REMOVE** *Refrigerant to* **LOWER** *high-side pressure.*

25. Adjust the system's charge as necessary to bring the system's operating high-side pressure into line based on a 95-degree (F) day. ❏
26. *Record the system's operating high-side pressure once charge adjustment is complete.*
    _____ ❏
27. *Record the system's operating low-side pressure.* _____ ❏
28. *Use a P/T chart and record the system's low-side saturation temperature.* _____ ❏
29. *Record the suction line temperature.* _____ ❏
30. Determine the amount of the system's superheat: ❏
                    Suction line temperature _____
                              MINUS
              Low-side saturation temperature _____
              EQUALS *Superheat amount* _____
31. *Record the system's superheat.* _____ ❏
32. *Does the superheat amount fall in range with that which is stated in the Overview based on a 95-degree (F) outside temperature and length of the system's line set?* _____ ❏
33. Make an additional adjustment to system charge if need be. ❏
34. Once system charge is in line based on a 95-degree day, have your instructor check your work. ❏
35. Remove the condenser airflow restrictor. ❏
36. Let the system operate for 10 minutes. ❏
37. *Record the system's operating low-side pressure.* _____ ❏
38. *Use a P/T chart and record the system's low-side saturation temperature.* _____ ❏
39. *Record the suction line temperature.* _____ ❏
40. *Determine the amount of the system's superheat:* ❏
                    Suction line temperature _____
                              MINUS
              Low-side saturation temperature _____
              EQUALS *Superheat amount* _____
41. *Record the system's superheat.* _____ ❏
42. *Does the superheat amount fall in range with that which is stated in the Overview based on a 95-degree (F) outside temperature and length of the system's line set?* _____ ❏
43. *Record the system's operating high-side pressure.* _____ ❏
44. Have your instructor check your work. ❏
45. Turn the system OFF. ❏
46. Remove manifold gage hoses from high- and low-system service valves. ❏
47. Replace the service valve caps. ❏
48. Remove the temperature lead probe from the suction line. ❏
49. Return all supplies and equipment to their proper location. ❏

RESEARCH QUESTIONS

1. The three materials of which condensers are normally made are _____, _____,
   and _____.

2. When is the most heat removed from the refrigerant in the condensing process?

   _____

   _____

3. After heat is absorbed into a condenser medium in a water-cooled condenser, the heat can be deposited in one of two places. What are they? _____ and _____

4. When a standard efficiency air-cooled condenser is used, the condensing refrigerant will normally be _____ (F) higher in temperature than the entering air temperature.

5. Four methods for controlling head pressure in an air-cooled condenser are _____, _____, _____, and _____.

Practical Competency 81

Field Charging or Checking the Charge of an Air Conditioner or Heat Pump System with a TXV Metering Device

SUGGESTED MATERIALS

Textbook
Refrigeration & Air Conditioning Technology, 5th Edition, Thomson Delmar Learning
Unit 41—Troubleshooting
Unit 22—Condensers

Review Topics
Charging Procedures in the Field; Fixed-Bore Metering Devices; Capillary Tube and Orifice Type

COMPETENCY OBJECTIVE

The student will be able charge or check the charge of an air conditioner or heat pump during low ambient temperature by using field service procedures.

OVERVIEW

When charging or checking the charge of any air conditioning or heat pump system it's important that technicians follow manufacturer's guidelines when available. There are times when the technician needs to check or make system charge adjustments and the manufacturer's guidelines are not available. Besides using a charging calculator, charging charts, or charging curves, technicians are left with making personal judgments as to a system's charge condition. Technicians can accurately determine or adjust a system's charge without the aid of such tools, but to do so requires an understanding of refrigeration fundamentals, the manufacturer's design of such equipment, and how the equipment should perform under certain conditions.

Field charging procedures are best established by knowing how the equipment is to perform under *Standard Operating Conditions*. When a system is charged correctly and operating within Standard Conditions, a prescribed amount of refrigerant should be in the condenser, the evaporator, and the system's liquid line. **It's important that technicians understand the following facts about air conditioning and heat pump equipment:**

1. *The amount of refrigerant in the evaporator can be measured by the superheat method.*
2. *The amount of refrigerant in the condenser can be measured by the sub-cooling method.*
3. *The amount of refrigerant in the liquid line may be determined by measuring the length and calculating the refrigerant charge.*
4. *If the evaporator is performing correctly, the liquid line has the correct charge.*

When system charge adjustments are required, technicians need to establish charging procedures in the field for all types of equipment. The following are a couple of methods used for different types of equipment.

Procedures for Field Charging a TXV System

A TXV metering device is designed to maintain a constant superheat on an evaporator of 8 to 12 degrees under all load conditions. Depending on the load on the evaporator, the TXV may not need to feed the system's total charge to the evaporator, so it needs space to store the additional refrigerant when not in use. Liquid receivers are equipped with TXV systems for this very reason, and work in conjunction with the system's condenser to store liquid refrigerant in low ambient temperature conditions. This process always ensures enough liquid refrigerant is available to the TXV for feeding the evaporator. Because of this, a TXV system will not be affected as much as a system with a fixed-bore metering device during low ambient temperatures. In fact a TXV system could be undercharged or overcharged, yet have enough refrigerant to maintain designed superheat on the evaporator. For this reason, checking or adjusting the charge of a system with a TXV metering device cannot be performed by using the Superheat Method.

The charge of air conditioning and heat systems equipped with a TXV metering device can be checked in much the same way as a system with a fixed-bore metering device during low ambient temperature conditions. The only difference is that once pressures are established to represent a 95-degree (F) day, the Subcool Method of checking or charging the system would be used.

Once again, typical operating conditions **would need to be simulated by reducing the airflow across the condenser, which will cause the system's high-side pressure to rise.**

On a 95-degree (F) day, the highest condenser **head pressure for R-22 systems is 278 psig**, which equals to a condensing temperature of 125 degrees for standard air-cooled condensers. Remember that standard air-cooled condensers condense the refrigerant vapor to liquid at approximately 30 degrees (F) above the outside air temperature. (*95-degree day plus 30 degrees equals 125 degrees saturation temperature*). *R-410A systems would operate with a high-side pressure of approximately **445 psig** under standard conditions. For **high-efficiency equipment**, a head pressure of **250 psig** would be sufficient for an R-22 system, and **390 psig** for an R-410A system.*

Once the correct high-side pressure for the type of refrigerant is simulated, checking the condenser's sub-cooling can be used to evaluate the system's charge. A **typical sub-cooling** circuit will sub-cool the liquid refrigerant **from 10 to 20 degrees (F)** than the condenser's condensing temperature. Under the simulated conditions, a temperature lead should be attached to the condenser's **liquid line and a temperature of 105 to 115 degrees (F)**, equivalent to 10 to 20 degrees (F) of sub-cooling, should be established.

> *NOTE: Establishing the proper sub-cooling range under these simulated conditions means that the system's charge is quite close to accurate.*

> *NOTE: (OVERCHARGED SYSTEM) A sub-cooling temperature higher than 20 degrees (F) is an indication of an overcharged system. This will require that some of the refrigerant from the system be recovered until the sub-cooling temperature is brought within the design range.*

> *NOTE: (UNDERCHARGED SYSTEM) Sub-cooling temperatures lower than the 10- to 20-degree (F) range are an indication that the system is undercharged. Refrigerant would have to be added to the system until the sub-cooling temperature is brought within the design range.*

> *NOTE: Refrigerant charge adjustments should be made guardedly. Once an adjustment is made, allow the system to stabilize before making additional adjustments.*

EQUIPMENT REQUIRED

Air conditioner or heat pump system
1 Piece of cardboard/wood shield large enough to restrict condenser airflow
1 Temperature tester (with temperature leads)
1 Manifold gage set
1 Refrigerant cylinder that matches system refrigerant
Insulation tape
Assorted hand tools

SAFETY PRACTICES

Wear goggles or safety glasses and gloves when working with refrigerant in sealed systems. Beware of high pressures and follow all electrical safety rules. Follow all EPA rules and regulations while working with refrigerants.

COMPETENCY PROCEDURES

Checklist

1. Turn the system OFF. ❑
2. Remove the high service valve cap. ❑
3. Make sure that manifold gage valves are closed. ❑
4. Attach the manifold high-side hose to the system's high-side valve service port. ❑
5. Attach a temperature lead to the system's liquid line, near the system's high-side service valve. ❑
6. Insulate the temperature lead. ❑
7. *Record the type of refrigerant used in the system.* _____ ❑
8. Turn the system ON. ❑
9. Allow system to operate for 10 to 15 minutes. ❑
10. *Record the system's operating high-side pressure.* _____ ❑
11. *Use a P/T chart and record the system's high-side saturation temperature.* _____ ❑
12. *Record the liquid line temperature.* _____ ❑
13. *Determine the amount of the system's sub-cool amount:* ❑

High-side saturation temperature _____

MINUS

Liquid line temperature _____

EQUALS Sub-cool amount _____

14. *Record the system's sub-cool.* _____ ❑
15. *Does the sub-cool amount fall in range with that which is stated in the Overview based on a 95-degree (F) outside temperature?* _____ ❑
16. *Record the system's operating high-side pressure.* _____ ❑
17. *Does the system's high-side pressure fall in line with an outside temperature of a 95-degree (F) day as described in the Competency Overview?* _____ ❑

> **NOTE:** *If the system's sub-cool and high-side pressure **are within range** as described in the Competency Overview, the system's charge is correct.* **Do not proceed**; *have your instructor check your work.*

*If the system's sub-cool and high-side pressure **are not within range** as described in the Competency Overview, **proceed to the next steps** of the Competency.*

18. Place the cardboard/wood shield over the top of the condenser to restrict condenser airflow. ❑
19. Allow system to operate for 10 to 15 minutes with restricted condenser airflow. ❑
20. *Record the system's high-side pressure.* _____ ❑
21. *Does the system's high-side pressure fall in line for the type of refrigerant as described in the Competency Overview, based on a 95-degree day?* _____ ❑

> **NOTE:** *With condenser air restricted, if the high-side pressure is higher or lower than that which is expected on a 95-degree (F) day according to the Competency Overview,* **ADD** *Refrigerant to* **RAISE** *high-side pressure, or* **REMOVE** *Refrigerant to* **LOWER** *high-side pressure.*

22. Adjust the system's charge as necessary to bring the system's operating high-side pressure into line based on a 95-degree (F) day. ❑
23. *Record the system's operating high-side pressure once charge adjustment is complete.* _____ ❑
24. *Use a P/T chart and record the system's high-side saturation temperature.* _____ ❑

25. *Record the liquid line temperature.* _____ ❑

26. Determine the amount of the system's sub-cool: ❑

High-side saturation temperature _____

MINUS

Liquid line temperature _____

EQUALS Sub-cool amount _____

27. *Record the system's sub-cool amount.* _____ ❑

28. *Does the sub-cool amount fall in range with that which is stated in the Overview based on a 95-degree (F) outside temperature and length of the system's line set?* _____ ❑

29. Make an additional adjustment to system charge if need be. ❑

30. Once system charge is in line based on a 95-degree day, have your instructor check your work. ❑

31. Remove the condenser airflow restrictor. ❑

32. Let the system operate for 10 minutes. ❑

33. *Record the system's operating high-side pressure.* _____ ❑

34. *Use a P/T chart and record the system's high-side saturation temperature.* _____ ❑

35. *Record the liquid lines temperature.* _____ ❑

36. *Determine the amount of the system's sub-cool:* ❑

High-side saturation temperature _____

MINUS

Liquid line temperature _____

EQUALS Sub-cool amount _____

37. *Record the system's sub-cool.* _____ ❑

38. *Does the sub-cool amount fall in range with that which is stated in the Overview based on a 95-degree (F) outside temperature and length of the system's line set?* _____ ❑

39. Have your instructor check your work. ❑

40. Turn the system OFF. ❑

41. Remove the manifold gage hose from the high system service valves. ❑

42. Replace the service valve cap. ❑

43. Remove the temperature lead probe from the liquid line. ❑

44. Return all supplies and equipment to their proper location. ❑

RESEARCH QUESTIONS

1. The prevailing winds can affect which of the air-cooled condensers?

_____

_____

2. **True or False:** Air-cooled condensers are much more efficient than water-cooled condensers.

_____

_____

3. The type of condenser that has the lowest operating head pressure is the _____ cooled condenser.

4. Name two ways in which the heat from an air-cooled condenser can be used for heat reclaim _____ and _____.

5. What is the approximate condensing temperature for high-efficiency condensers?

_____

_____

Passed Competency _____   Failed Competency _____

Instructor Signature _____   Grade _____

Theory Lesson: Electric Motors (Figure 3–163)

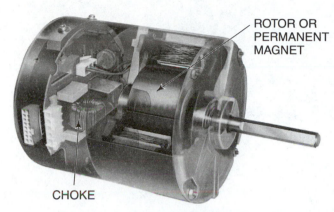

ROTOR OR
PERMANENT
MAGNET

CHOKE

FIGURE 3–163   An electronically commutated motor.
(*Courtesy of General Electric*)

SUGGESTED MATERIALS

Textbook

Refrigeration & Air Conditioning Technology, 5th Edition, Thomson Delmar Learning
Unit 17—Types of Electric Motors
Unit 18—Application of Motors
Unit 19—Motor Controls
Unit 20—Troubleshooting Electric Motors

Review Topics

Electric Motors and Magnetism; Single-Phase Open Motors; Split-Phase Motors; Capacitor-Start Motors; Capacitor-Start, Capacitor Run Motors; Permanent Split-Capacitor Motors; Shaded-Pole Motors; Three-Phase Motors; Single-Phase Hermetic Motors; Two-Speed Compressor Motors; Three-Phase Motor Compressors; Variable-Speed Motors

Key Terms

capacitor-start, capacitor run motors • capacitor-start motors • common winding • dual-voltage three-phase motors • high-voltage connection • low-voltage connection • permanent split-capacitor motors • run winding • single-phase open motors • split-phase motors • start winding • three-phase delta connection • three-phase motors • WYE connection

OVERVIEW

There are many types of motors used for different applications in the HVAC industry. Motors work on the principle of electrical magnetism and consist of a stator (**Figure 3–164**), a rotor (**Figure 3–165**), motor housing, and end bell caps (**Figure 3–166**).

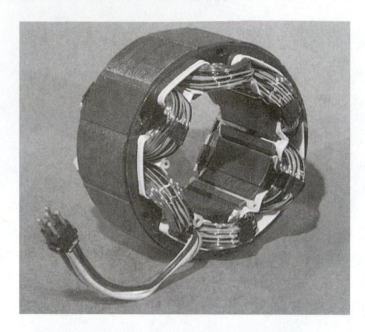

FIGURE 3–164 Stator made up of start and run windings that do not rotate. (*Courtesy of A.O. Smith*)

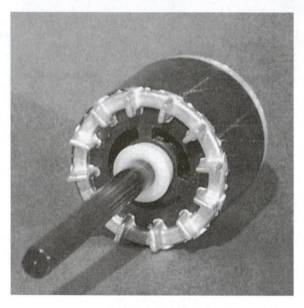

FIGURE 3–165 Squirrel-cage rotor. (*Courtesy of A.O. Smith*)

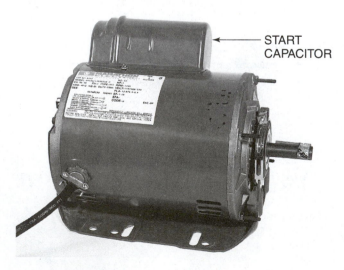

START CAPACITOR

FIGURE 3–166 Motor housing and end bell caps. (*Courtesy of W.W. Grainger, Inc.*)

There are three motor types:

1. *Open Motors* (**Figure 3–167**)

FIGURE 3–167   Open type motor. (*Courtesy of Universal Electric Company*)

These motors have open slots in the motor shell for the purpose of moving air across the motor windings for cooling. These motors are not designed to be used in dirty and wet locations.

2. *Enclosed Motors* (**Figure 3–168**)

FIGURE 3–168   A totally enclosed motor. (*Photo by Bill Johnson*)

Enclosed motors are designed to be used in dirty conditions. The motor windings of these motors are not designed to be cooled by air but depend on the type of Motor Insulation Classification as a method of determining the motor's operating temperature range (**Figure 3–169**).

Class A	221° F	(105° C)
Class B	266° F	(130° C)
Class F	311° F	(155° C)
Class H	356° F	(180° C)

FIGURE 3–169 Temperature classification of typical motors.

3. *Drip-Proof Motors* (**Figure 3–170**)

FIGURE 3–170 A drip-proof motor. (*Photo by Bill Johnson*)

These motors are designed to be used in wet locations where water can fall on them.

DIFFERENT MOTORS

Single-Phase Motors (Figure 3–171)
The voltage supplied for single-phase motors will be either 115 volts or 208 to 230 volts for most residential applications, and 230 volts to 460 volts for commercial applications. Some of these motors can be dual voltage, which can be wired for the correct voltage by the technician in the field. 115-volt motors will be powered by one hot leg, and contain one start and one run winding (**Figure 3–172**), except for shaded-pole motors, which have one winding (**Figure 3–173**).

In addition, two hot legs will power 208/230-volt motors. (Refer to **Figure 3–171.**) Dual-voltage motors will contain two run windings and one start winding. The run windings operate on 115 volts no matter what voltage the motor is wired to operate on.

Shaded-Pole Motors (Figure 3–174 A–C)
These motors have very little starting torque and are used mainly for light duty application, specifically the movement of air.

Split-Phase Motors, also referred to as *induction-start-induction run* (*ISIR*) *motors,* have medium starting torque and are normally used for operating fans in the fractional horsepower range. These motors have two motor windings, the start and the run winding.

These motors are referred to as *split-phase* because a single power (voltage) supply is split between two individual windings, the RUN and the START (**Figure 3–175**). Both windings are used in the starting application of

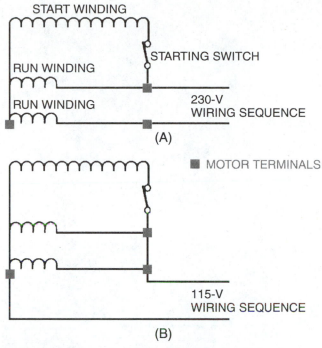

FIGURE 3–171 The wiring diagram of a dual-voltage motor. It is made to operate using 115 volts or 230 volts, depending on how the motor is wired. (A) 230-volt wiring sequence. (B) 115-volt wiring sequence.

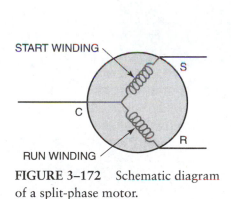

FIGURE 3–172 Schematic diagram of a split-phase motor.

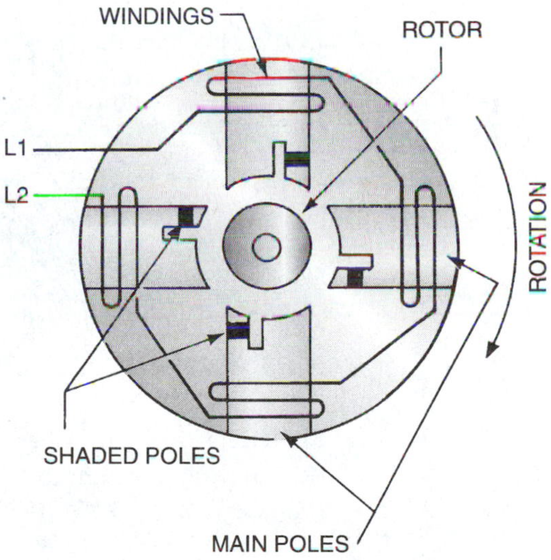

FIGURE 3–173 Shaded-pole motor winding.

these motors, although once the motor reaches three fourths of its rated speed some means of removing the start winding from the circuit must be employed, such as a *centrifugal switch* (**Figure 3–176**).

On start-up, voltage is applied to both the start and run windings of the motor which causes the motor to start to turn. Once the motor reaches 75% of its rated speed, the contacts of the centrifugal switch open, removing the start winding from the motor circuit.

(A)

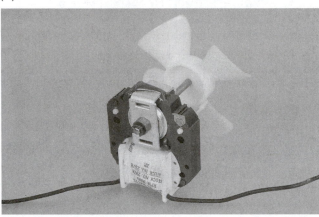

(B)

(C)

FIGURE 3–174   (A) Shaded-pole blower motor.
(B) Evaporator fan motor. (C) Condenser fan motor.
(*Photos by Bill Johnson*)

The starting torque and running efficiency of split-phase motors can be enhanced by the addition of a run or start capacitor, or both (**Figure 3–177**).

Run capacitors increase the motor's efficiency and are left in the motor circuit during operation. Run capacitors are wired parallel between the run and start winding of the motor (**Figure 3–178**).

The start winding of the motor need not be removed from the motor circuit in this type of application because the run capacitor limits the amount of current draw through the start winding while it is in operation.

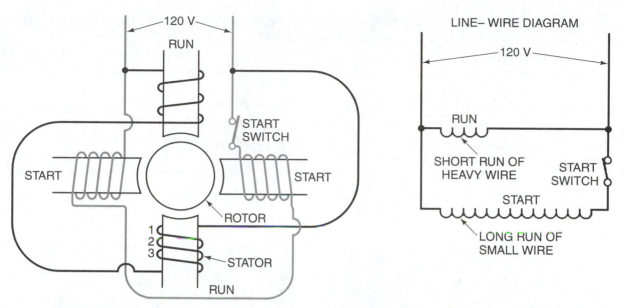

FIGURE 3–175 Power applied to start and run windings through a common connection.

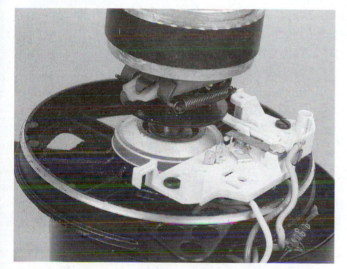

FIGURE 3–176 The centrifugal switch located at the end of the motor. (*Photo by Bill Johnson*)

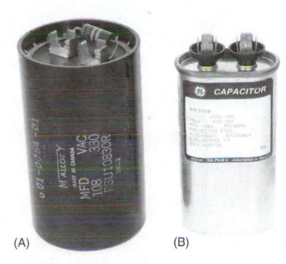

FIGURE 3–177 (A) Start capacitor. (B) Run capacitor. (*Photo by Bill Johnson*)

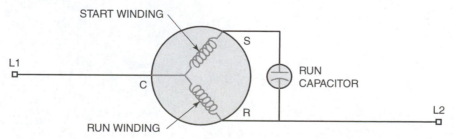

FIGURE 3–178 Diagram of run capacitor wired in parallel across the run and start windings of the motor.

Motors with run capacitors are called **permanent split-capacitor motors,** or **PSC motors** (**Figure 3–179**). PSC motors can be single-speed or multispeed (**Figure 3–180**).

Multispeed motors are easy to recognize by the number of muticolored lead wires. Each wire lead with the common motor connection provides a different speed. Multispeed motors have multiple taps or poles for each speed (**Figure 3–181**).

Motors equipped with a start capacitor are called **capacitor-start motors or CS motors** (**Figure 3–182**).

CS motors have a very high starting torque and work on the same principle as a split-phase motor. Once these motors reach 75% of their rated speed, both the start winding and start capacitor are removed from the motor circuit (**Figure 3–183**).

Capacitor-start, capacitor run motors (CSCR motors) incorporate both a run and start capacitor. These motors offer both running efficiency and a high starting torque (**Figure 3–184**).

FIGURE 3–179   A permanent split-capacitor motor. (*Courtesy of Universal Electric Company*)

FIGURE 3–180   A multispeed PSC motor. (*Photo by Bill Johnson*)

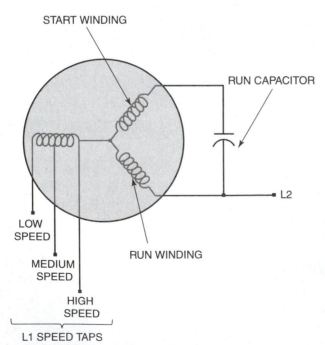

FIGURE 3–181   Multispeed motor schematic.

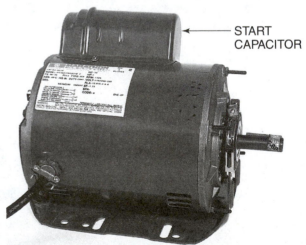

FIGURE 3–182   Capacitor start (CS) motor. (*Courtesy of W.W. Grainger, Inc.*)

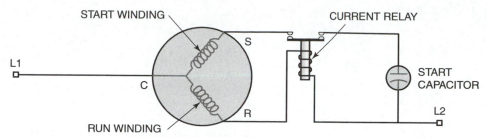

FIGURE 3–183 A wiring diagram of a capacitor start motor with a current relay used to remove both the start winding and start capacitor once the motor reaches 75% of its rated speed.

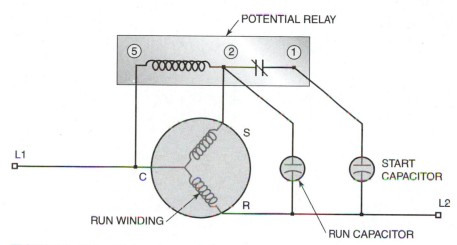

FIGURE 3–184 Wiring diagram of a CSCR motor using a potential relay. The potential relay removes the start capacitor from the motor's circuit once the relay's pick-up voltage is created across the motor windings.

Because CSCR motors are equipped with a run capacitor, along with a start capacitor, the start winding is not removed from the motor circuit, although some means must be employed to remove the start capacitor once the motor reaches 75% of its rated speed.

Three-phase motors do not have run, start, and common windings as in a split-phase motor. These motors have the highest starting torque and best running efficiency without incorporating a run capacitor or start capacitor. There are three individual "hot legs" or motor windings that are 120 degrees out of phase with the next winding. Three phase motors and equipment can be used only where three-phase voltage is supplied (**Figure 3–185**).

Three hot lines identified as L1, L2, and L3 feed the power to the motor. For a 240-volt, three–phase power supply, the voltage measured between any two of the hot legs is 240 volts (**Figure 3–186**).

The windings of three-phase motors can be connected in a couple of different ways. The **WYE configuration** three-phase motor has three motor terminals marked **T1, T2,** and **T3.** WYE connection motors are designed to operate at a specific voltage and are easily identifiable by the three designated motor terminal connections or motor leads with the motor windings displayed on a diagram in the form of a **Y** (**Figure 3–187**).

The directional rotation of these motors can be easily reversed by switching any two of the power lead connections or motor terminal connections (**Figure 3–188**).

There are also three-phase motors for dual-voltage motors and require different motor connection configurations for high-voltage and low-voltage power supply. In dual-voltage motors there are nine different motor leads. The three main motor windings are divided into two smaller windings, making a total of six (**Figure 3–189**).

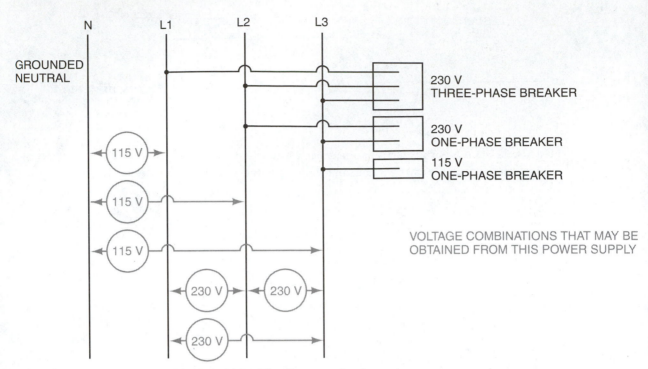

FIGURE 3–185 The diagram of a three-phase power supply.

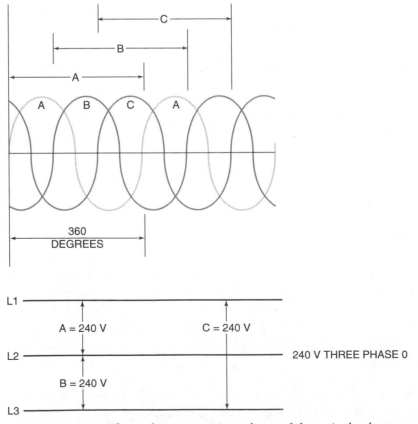

FIGURE 3–186 Three-phase power is made up of three single-phase supplied, each of which is 120 degrees out of phase with the next.

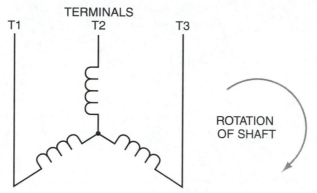

FIGURE 3–187   Three-phase motor windings.

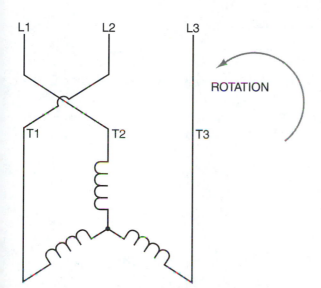

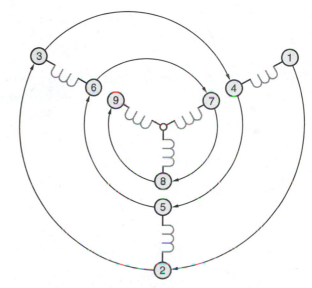

FIGURE 3–188   Switching any two power or motor leads will switch the rotation of the motor.

FIGURE 3–189   Method used to number the wires in a dual-voltage motor.

Each of the motor leads is numbered and care must be taken when wiring the motor leads for different voltage applications or serious damage to the motor could occur. For **high-voltage connection**, the windings are connected in series with each other, forming a large Y (**Figure 3–190** and **Figure 3–191**).

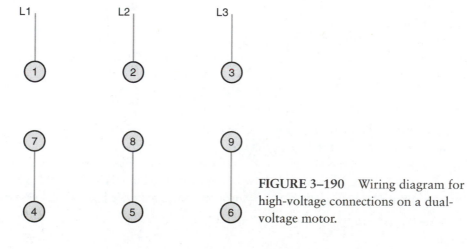

FIGURE 3–190   Wiring diagram for high-voltage connections on a dual-voltage motor.

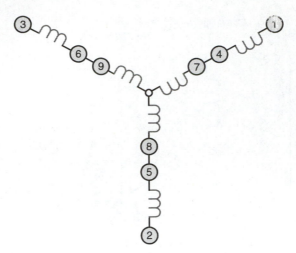

FIGURE 3–191   Windings connected in series.

When a dual-voltage motor is operating at the higher voltage, the current draw of the motor is one half that of the same motor operating at one half the voltage.

Low-voltage connections for a dual-voltage three-phase motor have the motor windings connected in parallel with each other, forming two smaller Y configurations. (**Figure 3–192** and **Figure 3–193**).

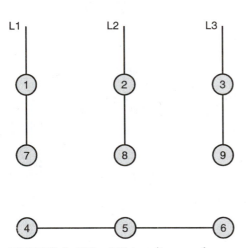

FIGURE 3–192   Wiring diagram for the low-voltage connections on a dual-voltage motor.

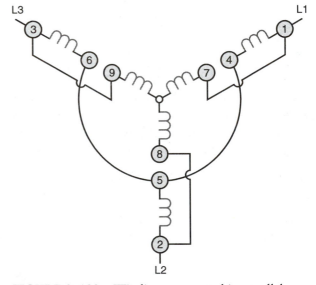

FIGURE 3–193   Windings connected in parallel.

The amperage draw of the dual-voltage motor operating at the low voltage will be twice that of an identical motor wired to operate at the higher voltage.

The **DELTA configuration** three-phase motor will have 12 motor lead wires and can be dual-voltage motors. For high-voltage applications, the motor windings are wired in series (**Figure 3–194**).

In addition, for low-voltage applications, the motor windings are wired in parallel (**Figure 3–195**).

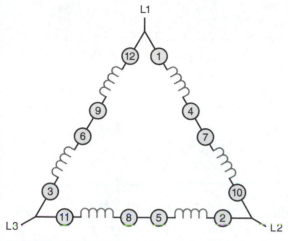

FIGURE 3–194   High-voltage delta configuration.

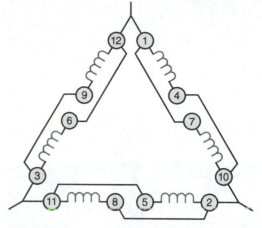

FIGURE 3–195   Low-voltage delta connection.

Electric motors used in compressor applications are normally split-phase or three-phase motors. Split-phase compressor motors can also employ the benefits of the addition of a run, start, or both types of capacitors.

Theory Lesson: Air Conditioning and Heat Pump Compressors and Motor Troubleshooting

SUGGESTED MATERIALS

Textbook
Refrigeration & Air Conditioning Technology, 5th Edition, Thomson Delmar Learning
Unit 20—Troubleshooting Electric Motors
Unit 23—Compressors
Unit 29—Troubleshooting and Typical Operating Conditions for Commercial Refrigeration
Unit 40—Typical Operating Conditions
Unit 41—Troubleshooting

Review Topics
The Function of the Compressor; Types of Compressors; Compressor Vacuum Test; Closed Loop Compressor Running Bench Test; Closed-Loop Compressor Running Field Test; Compressor Running Test in the System; Open Motor Windings; Shorted Motor Winding

Key Terms
absolute discharge pressure • absolute pressure • absolute suction pressure • closed-loop compressor bench test • closed-loop compressor test • compression ratio • compressor types • compressor vacuum test • full-load amperage (FLA – RLA) • heat of compression • lock rotor test • open winding test • pumping capacity • running field test • shorted winding test • short to ground test • single-phase hermetic motors • three-phase motor compressors • two-speed compressor motors • variable-speed motors

OVERVIEW

There are five types of compressors used in the refrigeration and air conditioning industry. They are the reciprocating (**Figure 3–196**), the screw (**Figure 3–197**), the rotary (**Figure 3–198**), the scroll (**Figure 3–199**), and the centrifugal (**Figure 3–200**).

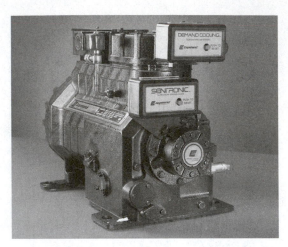

FIGURE 3–196   A reciprocating compressor. (*Courtesy of Copeland Corporation*)

FIGURE 3–197   A screw compressor. (*Courtesy of Frick Company*)

All of these compressors are designed as refrigerant vapor pumps (**Figure 3–201**).

All types of compressors serve two functions in the refrigeration system. One of their tasks is to remove the low-temperature, low-pressure refrigerant vapor from the evaporator, while reducing the evaporator pressure to a point where the desired evaporator temperature can be maintained. The second task of the compressor is to raise the pressure of the low-temperature, low-pressure suction vapor to a level where its saturation temperature is higher than the temperature of the condenser-cooling medium (**Figure 3–202**).

Compression ratio is a technical expression of the pressure difference between the compressor's low-side pressure and the high-side pressure (**Figure 3–203**) and is used to compare pumping conditions of a compressor.

The **compressor compression ratio** is determined by the dividing the compressor's *absolute discharge pressure* by the compressor's *absolute suction pressure*.

$$\text{Compression ratio} = \frac{\textit{Absolute discharge psia}}{\textit{Absolute suction psia}}$$

Absolute pressures are found by adding atmospheric pressure of 14.7 psi to the system's operating low-side and high-side gage pressures (psig). Compression ratios of 3:1 are considered normal, although any situation that could cause an increase in a system's head pressure or a decrease in a system's suction pressure will cause higher compression ratios. When the compression ratio gets above the 12:1 range, the refrigerant gas temperature leaving the compressor can rise to a point at which the compressor's lubricating oil becomes overheated. Refrigerant oil that becomes overheated can turn into carbon and create acids within the system.

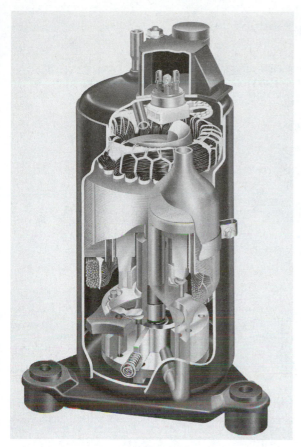

FIGURE 3–198   A rotary compressor. (*Reprinted with permission of Motors and Armatures. Inc.*)

FIGURE 3–199   A scroll compressor. (*Courtesy of Copeland Corporation*)

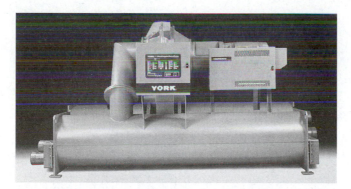

FIGURE 3–200   A centrifugal compressor. (*Courtesy of York International Corp.*)

Because of the compressor compression ratio, *heat is pumped up the temperature scale.* During the compression of the low-temperature vapor from the evaporator, the vapor is compressed into a smaller area. This causes the temperature and pressure to rise. This is called **heat of compression.** The compressor takes the heat that is already in the vapor from the evaporator (**Figure 3–204**) and pushes it into a small space (**Figure 3–205**).

| R-12 | LOW-TEMPERATURE APPLICATION |

THE REFRIGERANT ON THE LOW-PRESSURE SIDE OF THE SYSTEM IS EVAPORATING AT −13°F AND 3 psig.

THE REFRIGERANT ON THE HIGH-PRESSURE SIDE OF THE SYSTEM IS CONDENSING AT 125°F AND 169 psig.

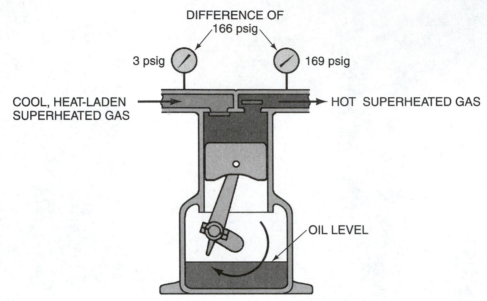

DIFFERENCE OF 166 psig

3 psig   169 psig

COOL, HEAT-LADEN SUPERHEATED GAS   HOT SUPERHEATED GAS

OIL LEVEL

FIGURE 3–201   Cool, heat-laden superheated gas is sucked in the low side and hot superheated gas is discharged out the high side.

LOW-TEMPERATURE

| R-12 |

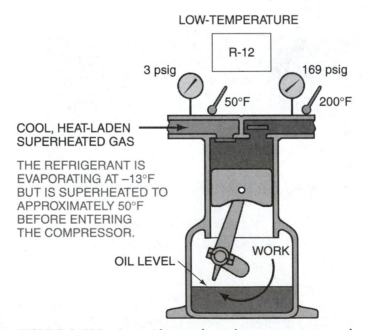

3 psig   169 psig

50°F   200°F

COOL, HEAT-LADEN SUPERHEATED GAS

THE REFRIGERANT IS EVAPORATING AT −13°F BUT IS SUPERHEATED TO APPROXIMATELY 50°F BEFORE ENTERING THE COMPRESSOR.

OIL LEVEL   WORK

FIGURE 3–202   Low-side superheated gas temperature and pressure and high-side superheated gas temperature and pressure.

THE REFRIGERANT
IS EVAPORATING AT
20°F AND 21 psig.

NOTE THAT THE REFRIGERANT IN THIS
MEDIUM-TEMPERATURE APPLICATION
CONDENSES AT THE SAME CONDITIONS
AS THE PREVIOUS LOW-TEMPERATURE
APPLICATION.

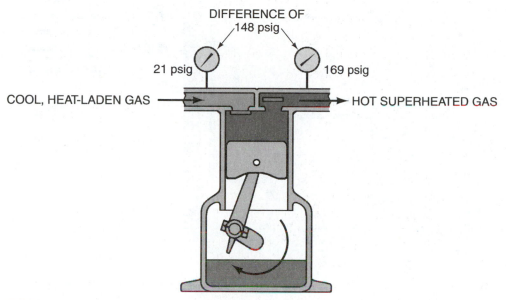

DIFFERENCE OF
148 psig

21 psig   169 psig

COOL, HEAT-LADEN GAS →   → HOT SUPERHEATED GAS

FIGURE 3–203   Compression ratio is the technical expression of pressure difference between the low side and high side of the compressor.

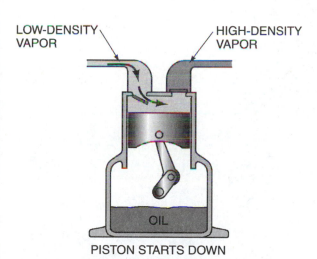

LOW-DENSITY
VAPOR

HIGH-DENSITY
VAPOR

OIL

PISTON STARTS DOWN

FIGURE 3–204   On the down stroke of the compressor, low-temperature low-pressure low-density vapor fills the compressor cylinder.

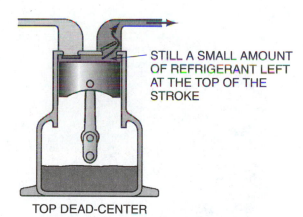

STILL A SMALL AMOUNT
OF REFRIGERANT LEFT
AT THE TOP OF THE
STROKE

TOP DEAD-CENTER

FIGURE 3–205   Compressing the refrigerant vapor into a smaller area, causing the refrigerant vapor temperature, density, and pressure to increase.

By doing this, the temperature and pressure exerted on the vapor rise to a high temperature and high-pressure vapor that is hotter than the area in which it is to be released. Once the compressor completes the heat of compression, it forces this high-temperature, high-pressure vapor into the condenser coil, which is normally located in an area where the heat from the evaporator is to be released (**Figure 3–206**).

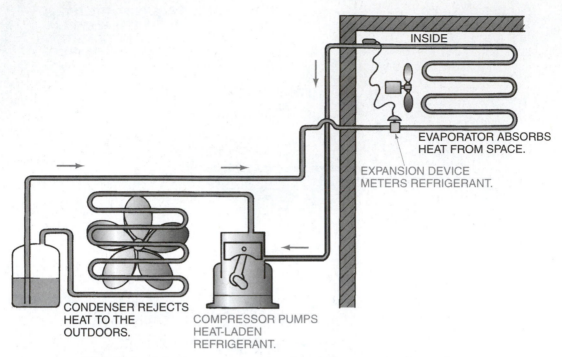

INSIDE

EVAPORATOR ABSORBS
HEAT FROM SPACE.

EXPANSION DEVICE
METERS REFRIGERANT.

CONDENSER REJECTS
HEAT TO THE
OUTDOORS.

COMPRESSOR PUMPS
HEAT-LADEN
REFRIGERANT.

FIGURE 3–206   Compressor pumps heat-laden vapor to the condenser.

The electrical motors of the compressors can be contained in the compressor housing or can be separate from the compressor such as an open-drive compressors (**Figure 3–207**).

FIGURE 3–207   Open drive compressor and drive motor. (*Courtesy of Carrier Corporation*)

These systems can be either belt-driven or direct-drive type.

The compressor compression ratio can be used to evaluate compressor efficiency, although inefficient compressor operation can be one of the most difficult problems to find. Remember that compressors are vapor pumps and should be able to create a pressure from the low side of the system to the high side under design conditions. The efficiency of a compressor is determined by its ability to pump refrigerant through the system at designed pumping capacities. The pumping capacity for all compressors is not the same. Compressors used in high- temperature, medium-temperature, and low-temperature refrigeration application are designed with different pumping capabilities because of the operating temperature ranges of the evaporator coils in the three different refrigeration applications. Using the standard operating conditions for the type of refrigeration equipment can aid the technician in checking the pumping capacity of that system's compressor. The following test can be used to evaluate and identify the inefficient compressor's problem.

The **Compressor Vacuum Test** can be performed on a reciprocating compressor to see if the suction valves are seating correctly on at least one cylinder (**Figure 3–208**).

The **Closed-Loop Compressor Running Bench Test** can be used to determine whether the compressor can pump or not (**Figure 3–209**).

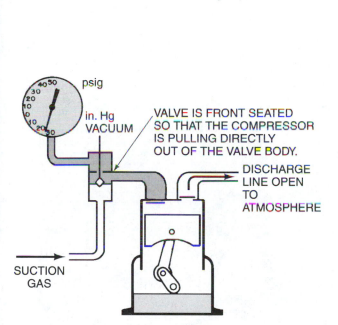

THE TYPICAL SUCTION AND DISCHARGE PRESSURE
FOR THIS MEDIUM-TEMPERATURE R-12 SYSTEM

20 psig   180 psig

THIS VALVE IS USED TO
THROTTLE THE SYSTEM.

VALVE IS CLOSED
DURING TEST.
REFRIGERANT OR
NITROGEN DRUM
IT CAN BE OPENED
SLIGHTLY FROM TIME
TO TIME TO ALLOW
REFRIGERANT OR
NITROGEN INTO SYSTEM.

SUCTION LINE
CAPPED

COMMON
MOTOR LEAD

230 V

11 A

CLAMP-ON
AMMETER

COMPRESSOR HAS A FULL-LOAD RATING
OF 11.1 A AT 230 V.

FIGURE 3–209   Closed-loop compressor running bench test.

psig

in. Hg
VACUUM

VALVE IS FRONT SEATED
SO THAT THE COMPRESSOR
IS PULLING DIRECTLY
OUT OF THE VALVE BODY.

DISCHARGE
LINE OPEN
TO
ATMOSPHERE

SUCTION
GAS

FIGURE 3–208   A compressor vacuum test.

The **Closed-Loop Compressor Running Field Test** is used to determine if a compressor can pump or not, but can be performed in the field if the compressor has suction and discharge service valves (**Figure 3–210**).

The **Compressor Running Test in the System** can be used by creating typical design conditions and checking the pumping capacity of a compressor (**Figure 3–211**).

An **Open Winding Test** (**Figure 3–212**) can evaluate the motor windings of compressors
Shorted Winding Test (**Figure 3–213**)

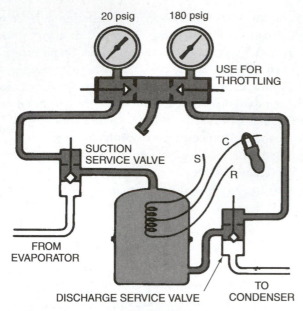

20 psig   180 psig

USE FOR THROTTLING

SUCTION SERVICE VALVE

C

S

R

FROM EVAPORATOR

DISCHARGE SERVICE VALVE

TO CONDENSER

THE SUCTION SERVICE VALVE MAY BE CRACKED FROM TIME TO TIME TO ALLOW SMALL AMOUNTS OF REFRIGERANT TO ENTER COMPRESSOR UNTIL TEST PRESSURES ARE REACHED.

NOTE: START THIS TEST WITH GAGE MANIFOLD VALVES WIDE OPEN. THEN USE THE HIGH SIDE VALVE FOR THROTTLING.

FIGURE 3–210   Closed-loop compressor running field test.

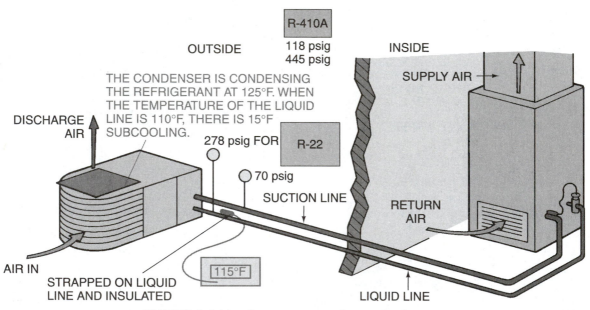

OUTSIDE

R-410A
118 psig
445 psig

INSIDE

SUPPLY AIR →

THE CONDENSER IS CONDENSING THE REFRIGERANT AT 125°F. WHEN THE TEMPERATURE OF THE LIQUID LINE IS 110°F, THERE IS 15°F SUBCOOLING.

DISCHARGE AIR

278 psig FOR   R-22

70 psig

SUCTION LINE

RETURN AIR

AIR IN

STRAPPED ON LIQUID LINE AND INSULATED

115°F

LIQUID LINE

FIGURE 3–211   Compressor running test in the system.

Short to Ground Test (Figure 3–214) or **Lock Rotor Test (Figure 3–215).**

A compressor that is operating in designed load conditions and pumping at design capacity will draw at or near the manufacturer's full-load amperage.

These tests should be performed by qualified technicians or under close supervision of a qualified instructor.

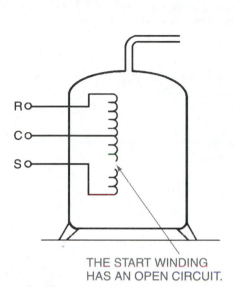

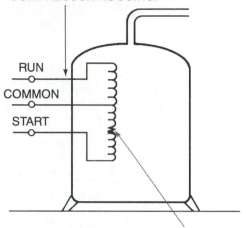

THE MOTOR TERMINALS ARE INSULATED WHERE THEY PASS THROUGH THE COMPRESSOR HOUSING.

RUN

COMMON

START

SOME OF THE WINDINGS ARE TOUCHING EACH OTHER AND REDUCING THE RESISTANCE IN THE START WINDING.

THE START WINDING HAS AN OPEN CIRCUIT.

FIGURE 3–212   A compressor with open start winding.

FIGURE 3–213   A compressor with shorted winding.

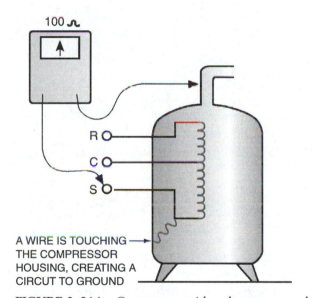

A WIRE IS TOUCHING THE COMPRESSOR HOUSING, CREATING A CIRCUT TO GROUND

FIGURE 3–214   Compressor with a short to ground.

The compressor vacuum test is used to evaluate the suction valve operation of reciprocating compressors. All reciprocating compressors with a good suction valve should be able to pull a 26 to 28 in. Hg vacuum using atmosphere as the discharge pressure. Checking the discharge pressure of a compressor can be simulated by correct adjustments of the compressor's service valves and using a refrigerant or nitrogen as the compression gas. Once the gas is compressed and the compressor is stopped, there should be no equalization of the suction and discharge pressures.

NOTE: This test cannot be used to evaluate multi-cylinder compressors.

WHEN THE MOTOR HAS CORRECT VOLTAGE TO ALL
WINDINGS DURING A START ATTEMPT AND DRAWS
LOCK ROTOR AMPERES, THE MOTOR SHOULD
START. THIS TEST CAN BE PERFORMED WITH
1 METER BUT REQUIRES 3 START ATTEMPTS.

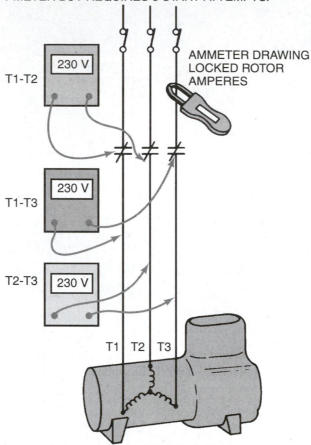

FIGURE 3–215 Compressor locked rotor test.

Good Motor Winding Resistance Check—Split-Phase Motors

NOTE: *When checking **split-phase motor windings**, good motor windings would be found by a resistance check between the motor's winding terminals. If the motor windings are good, a resistance check across the motor terminals should indicate three separate and different resistance values, with one of those resistance values totaling the sum of the other two (**Figure 3–216**).*

If the motor windings are good, the measured resistance values can also be used to determine the common, run, and start terminals of the motor (*Figure 3–217*).

NOTE: *A split-phase motor could have good motor windings and still not operate. In most cases, this is due to the motor being in locked rotor.*

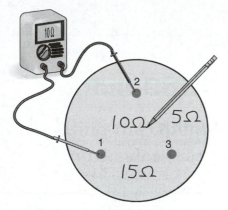

FIGURE 3–216 Good split-phase motor windings as indicated by measured resistance values.

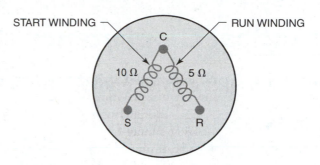

FIGURE 3–217 Good resistance measurements on a split-phase motor can also be used to determine common, run, and start terminals.

Good Motor Winding Resistance Check—Three-Phase Motors

Three-phase motors do not have common, run, or start windings. The motor windings are of the same resistance and depend on line voltage phasing to operate. A resistance check of these motors should indicate the same resistance across all motor terminals to reflect good motor windings (**Figure 3–218**).

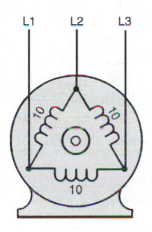

FIGURE 3–218 Good three-phase motor windings are indicated by the same resistance values across each winding.

Practical Competency 82

Checking Split-Phase Compressor Motor Windings

SUGGESTED MATERIALS

Textbook

Refrigeration & Air Conditioning Technology, 5th Edition, Thomson Delmar Learning
Unit 20—Troubleshooting Electric Motors
Unit 23—Compressors
Unit 41—Troubleshooting

Review Topics
Split-Phase Motors

COMPETENCY OBJECTIVE

The student will be able to check the motor windings of a split-phase motor.

OVERVIEW

There are many different types of motors used in the air conditioning, refrigeration, and heating industry. Most of these motors perform many different applications and most work by the principle of magnetic induction. Technicians must gain an understanding of the following motors to successfully troubleshoot and work with them in the field: *the split-phase motor, CS motor, CSCR motor, PSC motor, shaded-pole motor, and three-phase motor.*

The method of operation of each of these motors is similar but each serves a different purpose. All of these motors have an electrical winding or windings that are used to create the magnetic induction for motor operation. **Split-phase motors,** also referred to as *induction-start-induction run (ISIR) motors,* have medium starting torque and are normally used for operating fans in the fractional horsepower range. These motors have two motor windings, the start and the run winding.

These motors are referred to as *split-phase* because a single power (voltage) supply is split between two individual windings, the RUN and the START.

Both windings are used in the starting application of these motors, although once the motor reaches three fourths of its rated speed some means of removing the start winding from the circuit must be employed, such as a *centrifugal switch.*

On start-up, voltage is applied to both the start and run windings of the motor which causes the motor to start to turn. Once the motor reaches 75% of its rated speed, the contacts of the centrifugal switch open, removing the start winding from the motor circuit.

The starting torque and running efficiency of split-phase motors can be enhanced by the addition of a run or start capacitor, or both.

Run capacitors increase the motor's efficiency and are left in the motor circuit during operation. Run capacitors are wired parallel between the run and start winding of the motor.

The start winding of the motor need not be removed from the motor circuit in this type of application because the run capacitor limits the amount of current draw through the start winding while it is in operation. Motors with run capacitors are called **permanent split-capacitor motors,** or **PSC motors.**

454

Good Motor Winding Resistance Check—Split-Phase Motors

> *NOTE: When checking **split-phase motor windings**, good motor windings would be found by a resistance check between the motor's winding terminals. If the motor windings are good, a resistance check across the motor terminals should indicate three separate and different resistance values, with one of those resistance values totaling the sum of the other two.*

If the motor windings are good, the measured resistance values can also be used to determine the common, run, and start terminals of the motor.

> *NOTE: A split-phase motor could have good motor windings and still not operate. In most cases, this is due to the motor being in locked rotor.*

EQUIPMENT REQUIRED

Compressor (*in or out of system*)
Ohmmeter or VOM meter
Assorted hand tools (*if applicable*)

SAFETY PRACTICES

Make sure meter is set to the proper function and the electrical power is disconnected from the motor. Discharge capacitors used in compressor motor circuit before removing motor terminal wires (if applicable).

> *NOTE: For safety, motor winding value readings should be obtained through the motor's terminal wiring leads when possible. If not applicable, readings can be taken at the compressor motor terminals, but should be obtained with care.*

COMPETENCY PROCEDURES

Checklist

1. Make sure power has been removed from the system or compressor. ☐
2. Gain access to the compressor motor terminals or motor terminal wire leads (**Figure 3–219**). ☐

FIGURE 3–219   Compressor motor terminals. (*Photo by Bill Johnson*)

3. Discharge all motor circuit capacitors before removing compressor terminal wires (*if applicable*). ❑

4. Remove compressor motor wiring leads from associated components (*relays, capacitors, etc.*) (*if applicable*). ❑

NOTE: *If removing motor wires from associated components in equipment, make notes for rewiring motor lead wires back into the equipment.*

---

5. Draw a picture of the compressor's terminal layout below (**Figure 3–220**). ❑

6. Number the compressor terminal or compressor wire leads from the terminal layout above as "**1,**" "**2,**" and "**3.**" (Refer to **Figure 3–220**.) ❑

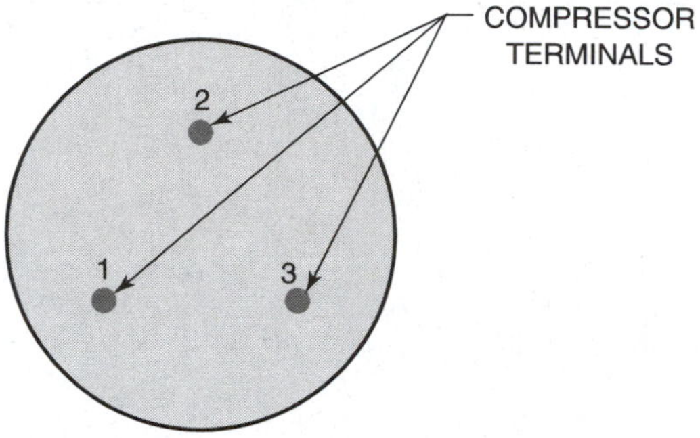

COMPRESSOR
TERMINALS

2

1   3

FIGURE 3–220   Drawing a picture of compressor terminal layout.

NOTE: *If testing compressor motor terminal wires, number wires and wire colors so that identification and resistance readings can be recorded and identified (for example, Yellow Wire 1, Black Wire 2, etc.)*

---

7. Set the ohmmeter the lowest ohms value scale. (*R × 1, R × 200, etc.*) ❑

8. Zero meter *(if applicable)*. ❑

9. Place ohmmeter leads on compressor terminal or wire leads 1 and 2 and check for a resistance value reading on the meter (**Figure 3–221**). ❑

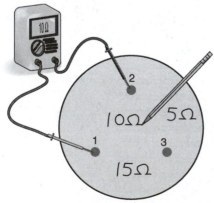

FIGURE 3–221   Measuring resistance across compressor terminals of wire leads.

10. *Record your ohms value reading at compressor terminals or wire leads 1 and 2.* _____ ❑

11. Place ohmmeter leads on compressor terminal or wire leads 1 and 3 and check for a resistance value reading on the meter. ❑

12. *Record your ohms value reading at compressor terminals or wire leads 1 and 3.* _____ ❑
13. Place ohmmeter leads on compressor terminals or wire leads 2 and 3 and check for a resistance value reading on meter. ❑
14. *Record your ohms value reading at compressor terminals or wire leads 2 and 3.* _____ ❑

GOOD MOTOR WINDING RESISTANCE CHECK—SPLIT-PHASE MOTORS

> *NOTE: When checking **split-phase motor windings**, good motor windings would be found by a resistance check between the motor's winding terminals. If the motor windings are good, a resistance check across the motor terminals should indicate three separate and different resistance values, with one of those resistance values totaling the sum of the other two. (Refer to **Figure 3–221**.)*

> *NOTE: An infinity reading (No Ohms Value Reading) is an indication of an open motor winding. A compressor or any motor with an open motor winding would have to be replaced.*

15. *Were the motor windings of the split-phase motor OK?* _____ ❑
16. Have your instructor check your work. ❑
17. Return all equipment and supplies to their original location. ❑

RESEARCH QUESTIONS

1. Reciprocating compressors are used in what type of refrigeration application?

   _____

   _____

2. Screw compressors are used in what type of refrigeration application?

   _____

   _____

3. Rotary compressors are used in what type of refrigeration application?

   _____

   _____

4. Scroll compressors are used in what type of refrigeration application?

   _____

   _____

5. Centrifugal compressors are used in what type of refrigeration applications?

   _____

   _____

Passed Competency _____ Failed Competency _____

Instructor Signature _____ Grade _____

Practical Competency 83

Identifying Common, Run, and Start Terminals of a Split-Phase Compressor Motor

SUGGESTED MATERIALS

Textbook
Refrigeration & Air Conditioning Technology, 5th Edition, Thomson Delmar Learning
Unit 20—Troubleshooting Electric Motors
Unit 23—Compressors
Unit 41—Troubleshooting

Review Topics
Split-Phase Motors

COMPETENCY OBJECTIVE

The student will be able to identify the Common, Run, and Start terminals of a split-phase compressor.

OVERVIEW

There are many different types of motors used in the air conditioning, refrigeration, and heating industry. Most of these motors perform many different applications and most work by the principle of magnetic induction. Technicians must gain an understanding of the following motors to successfully troubleshoot and work with them in the field: *the split-phase motor, CS motor, CSCR motor, PSC motor, shaded-pole motor, and three-phase motor.*

The method of operation of each of these motors is similar but each serves a different purpose. All of these motors have an electrical winding or windings that are used to create the magnetic induction for motor operation. **Split-phase motors,** also referred to as *induction-start-induction run (ISIR) motors,* have medium starting torque and are normally used for operating fans in the fractional horsepower range. These motors have two motor windings, the start and the run winding.

These motors are referred to as *split-phase* because a single power (voltage) supply is split between two individual windings, the RUN and the START.

Both windings are used in the starting application of these motors, although once the motor reaches three-fourths of its rated speed some means of removing the start winding from the circuit must be employed, such as a *centrifugal switch.*

On start-up, voltage is applied to both the start and run windings of the motor, which causes the motor to start to turn. Once the motor reaches 75% of its rated speed, the contacts of the centrifugal switch open, removing the start winding from the motor circuit.

The starting torque and running efficiency of split-phase motors can be enhanced by the addition of a run or start capacitor, or both.

Run capacitors increase the motor's efficiency and are left in the motor circuit during operation. Run capacitors are wired parallel between the run and start winding of the motor.

The start winding of the motor need not be removed from the motor circuit in this type of application because the run capacitor limits the amount of current draw through the start winding while it is in operation. Motors with run capacitors are called **permanent split-capacitor motors,** or **PSC motors.**

Good Motor Winding Resistance Check—Split-Phase Motors

> *NOTE: When checking **split-phase motor windings**, good motor windings would be found by a resistance check between the motor's winding terminals. If the motor windings are good, a resistance check across the motor terminals should indicate three separate and different resistance values, with one of those resistance values totaling the sum of the other two.*

If the motor windings are good, the measured resistance values can also be used to determine the common, run, and start terminals of the motor.

The **highest** resistance from Common is identified as the **START** terminal and the **lowest** resistance from the Common terminal is the **RUN** terminal.

EQUIPMENT REQUIRED

Compressor (*in or out of system*)
Ohmmeter or VOM meter
Assorted hand tools (*if applicable*)

SAFETY PRACTICES

Make sure the meter is set to the proper function and the electrical power is disconnected from the motor. **Discharge capacitors used in compressor motor circuit before removing motor terminal wires (if applicable).**

> *NOTE: For safety, motor winding value readings should be obtained through the motor's terminal wiring leads when possible. If not applicable, readings can be taken at the compressor motor terminals, but should be obtained with care.*

COMPETENCY PROCEDURES   Checklist

1. Make sure power has been removed from the system or compressor.   ❑
2. Gain access to the compressor motor terminals or motor terminal wire leads.   ❑
3. Discharge all motor circuit capacitors before removing compressor terminal wires (*if applicable*).   ❑
4. Remove compressor motor wiring leads from associated components (*relays, capacitors, etc.*) (*if applicable*).   ❑

> *NOTE: If removing motor wires from associated components in equipment, make notes for rewiring motor lead wires back into equipment.*

5. Draw a picture of the compressor's terminal layout.   ❑
6. Number the compressor terminal or compressor wire leads from the terminal layout above as "1," "2," and "3."   ❑

> *NOTE: If testing compressor motor terminal wires, number wires and wire colors so that identification and resistance readings can be recorded and identified (for example, Yellow Wire 1, Black Wire 2, etc.).*

7. Set the ohmmeter the lowest ohms value scale (*R* × *1*, *R* × *200*, etc.). ❑
8. Zero meter (*if applicable*). ❑
9. *Place ohmmeter leads on compressor terminal or wire leads 1 and 2 and check for a resistance value reading on meter.* ❑
10. *Record your ohms value reading at compressor terminals or wire leads 1 and 2.* _____ ❑
11. *Place ohmmeter leads on compressor terminal or wire leads 1 and 3 and check for a resistance value reading on meter.* ❑
12. *Record your ohms value reading at compressor terminals or wire leads 1 and 3.* _____ ❑
13. *Place ohmmeter leads on compressor terminals or wire leads 2 and 3 and check for a resistance value reading on meter.* ❑
14. *Record your ohms value reading at compressor terminals or wire leads 2 and 3.* _____ ❑

GOOD MOTOR WINDING RESISTANCE CHECK—SPLIT-PHASE MOTORS

> **NOTE:** *When checking* **split-phase motor windings**, *good motor windings would be found by a resistance check between the motor's winding terminals. If the motor windings are good, a resistance check across the motor terminals should indicate three separate and different resistance values, with one of those resistance values totaling the sum of the other two.*

15. *Were the motor windings of the split-phase motor OK?* _____ ❑

IDENTIFYING THE COMMON—RUN—START MOTOR TERMINALS

16. *Between which two terminal numbers did you get the highest resistance value? Terminals* _____ *and* _____ ❑
17. The two terminals listed in *Step 16* are the RUN and START terminals, the terminal left over is the COMMON TERMINAL. ❑
18. *Record the COMMON terminal number.* _____ ❑
19. *From the COMMON terminal, which terminal has the highest resistance value?* _____ ❑
20. Terminal number listed in *Step 19* is the START *winding.* ❑
21. *Record the START terminal.* _____ ❑
22. *From the COMMON terminal, which terminal has the lowest resistance value?* _____ ❑
23. The terminal with the lowest resistance from the common terminal is the RUN Winding. ❑
24. *Record the RUN terminal.* _____ ❑

> **NOTE:** *Common terminal to the highest resistance is START. Common terminal to the lowest resistance is RUN.*

25. Have your instructor check your work. ❑
26. Return all equipment and supplies to their original location. ❑

RESEARCH QUESTIONS

1. The two types of motors used for hermetic compressors are _____ and _____.
2. Define BEMF.

   _____

   _____

3. The two types of relays used to start hermetic compressors are the _____ relay and the _____.

4. **True or False:** All large motors are three-phase.

_____

_____

5. **True or False:** Three-phase motors have low starting torque.

_____

_____

Passed Competency _____   **Failed Competency** _____

Instructor Signature _____   **Grade** _____

Practical Competency 84

Checking a Compressor for Open Motor Windings (Figure 3–222)

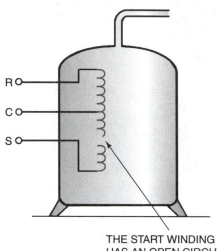

THE START WINDING
HAS AN OPEN CIRCUT   **FIGURE 3–222**   Open motor winding test.

SUGGESTED MATERIALS

Textbook
Refrigeration & Air Conditioning Technology, 5th Edition, Thomson Delmar Learning
Unit 20—Troubleshooting Electric Motors
Unit 23—Compressors
Unit 41—Troubleshooting

Review Topics
Open Motor Test

COMPETENCY OBJECTIVE

The student will be able to check a compressor for an open motor winding by using an ohmmeter.

OVERVIEW

There are many different types of motors used in the air conditioning, refrigeration, and heating industry. Most of these motors perform many different applications and most work by the principle of magnetic induction. Technicians must gain an understanding of the following motors to successfully troubleshoot and work with them in the field: *the split-phase motor, CS motor, CSCR motor, PSC motor, shaded-pole motor, and three-phase motor.*

The method of operation of each of these motors is similar but each serves a different purpose. All of these motors have an electrical winding or windings that are used to create the magnetic induction for motor operation. An open motor winding can be caused by an internal motor overload protector that has opened due to compressor overheating. Before condemning a compressor or motor, feel the compressor or motor housing and see if it is hot. If the motor housing feels the same temperature as the surrounding ambient temperature, it probably has an actual open motor circuit. If the motor housing is hot to the touch, allow time for the motor to cool down and see if the internal overload protector resets. The cool down period can take a long time. Running cold water over the compressor housing can shorten the time. Once the motor has cooled down, recheck the motor windings. If an infinity reading is still indicated, the motor has an open winding and will need to be replaced. If the compressor motor windings resets, the windings are OK and indicates that the motor overheated for some reason. Technicians should take the time to determine what caused the compressor to overheat, and make corrections.

Good Motor Winding Resistance Check—Split-Phase Motors

> NOTE: *When checking* **split-phase motor windings**, *good motor windings would be found by a resistance check between the motor's winding terminals. If the motor windings are good, a resistance check across the motor terminals should indicate three separate and different resistance values, with one of those resistance values totaling the sum of the other two.*

If the motor windings are good, the measured resistance values can also be used to determine the common, run, and start terminals of the motor.

> NOTE: *A split-phase motor could have good motor windings and still not operate. In most cases, this is due to the motor being in locked rotor.*

EQUIPMENT REQUIRED

Compressor (*in or out of system*)
Ohmmeter or VOM meter
Assorted hand tools (*if applicable*)

SAFETY PRACTICES

Make sure meter is set to the proper function and the electrical power is disconnected from the motor. Discharge capacitors used in compressor motor circuit before removing motor terminal wires (if applicable).

> NOTE: *For safety, motor winding value readings should be obtained through the motor's terminal wiring leads when possible. If not applicable, readings can be taken at the compressor motor terminals, but should be obtained with care.*

COMPETENCY PROCEDURES

Checklist

1. Make sure power has been removed from the system or compressor. ❏
2. Gain access to the compressor motor terminals or motor terminal wire leads. ❏
3. Discharge all motor circuit capacitors before removing compressor terminal wires (*if applicable*). ❏
4. Remove compressor motor wiring leads from associated components (*relays, capacitors, etc.*) (*if applicable*). ❏

> NOTE: *If removing motor wires from associated components in equipment, make notes for rewiring motor lead wires back into equipment.*

5. Draw a picture of the compressor's terminal layout. ❑
6. Number the compressor terminal or compressor wire leads from the terminal layout above as "*1*," "*2*," and "*3*." ❑

> **NOTE:** *If testing compressor motor terminal wires, number wires and wire colors so that identification and resistance readings can be recorded and identified (for example, Yellow Wire 1, Black Wire 2, etc.).*

7. Set the ohmmeter the lowest ohms value scale ($R \times 1$, $R \times 200$, etc.). ❑
8. *Zero meter (if applicable).* ❑
9. *Place ohmmeter leads on compressor terminal or wire leads 1 and 2 and check for a resistance value reading on meter.* ❑
10. *Record your ohms value reading at compressor terminals or wire leads 1 and 2.* _____ ❑
11. *Place ohmmeter leads on compressor terminal or wire leads 1 and 3 and check for a resistance value reading on meter.* ❑
12. *Record your ohms value reading at compressor terminals or wire leads 1 and 3.* _____ ❑
13. *Place ohmmeter leads on compressor terminals or wire leads 2 and 3 and check for a resistance value reading on meter.* ❑
14. *Record your ohms value reading at compressor terminals or wire leads 2 and 3.* _____ ❑

> **NOTE:** *An infinity reading (No Ohms Value Reading) is an indication of an open motor winding. A compressor or any motor with an open motor winding would have to be replaced.*

CAUTION: Figure 3–223

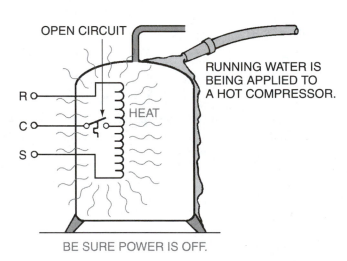

FIGURE 3–223   Internal open motor overload protector.

15. *Did the motor you checked have an open winding?* _____ ❑
16. Have your instructor check your work. ❑
17. Return all equipment and supplies to their original location. ❑

RESEARCH QUESTIONS

1. **True or False:** All electric motors must be cooled, or they will overheat.

2. When an open motor gets up to speed, the _____ switch takes the start winding out of the circuit.

3. Is the resistance in the start winding greater or less than the resistance in the run winding?

   _____

   _____

4. The two popular operating voltages for residences are _____ volts and _____ volts.

5. The two types of bearings commonly used on small motors are _____ and _____.

Passed Competency _____   **Failed Competency** _____

Instructor Signature _____   **Grade** _____

Practical Competency 85

Checking a Compressor for a Shorted Motor Winding (Figure 3–224)

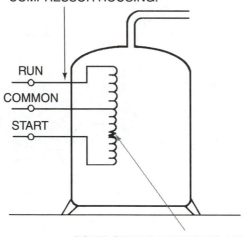

THE MOTOR TERMINALS ARE INSULATED WHERE THEY PASS THROUGH THE COMPRESSOR HOUSING.

RUN

COMMON

START

SOME OF THE WINDINGS ARE TOUCHING EACH OTHER AND REDUCING THE RESISTANCE IN THE START WINDING.

FIGURE 3–224   A compressor with a shorted winding.

SUGGESTED MATERIALS

Textbook
Refrigeration & Air Conditioning Technology, 5th Edition, Thomson Delmar Learning
Unit 20—Troubleshooting Electric Motors
Unit 23—Compressors
Unit 41—Troubleshooting

Review Topics
Shorted Motor Windings; Compressor Electrical Checkup

COMPETENCY OBJECTIVE

The student will be able to check the compressor windings and determine if the motor has a shorted motor winding.

OVERVIEW

There are many different types of motors used in the air conditioning, refrigeration, and heating industry. Most of these motors perform many different applications and most work by the principle of magnetic induction. Technicians must gain an understanding of the following motors to successfully troubleshoot and work with them in the field: *the split-phase Motor, CS motor, CSCR motor, PSC motor, shaded-pole motor, and three-phase motor.*

The method of operation of each of these motors is similar but each serves a different purpose. All of these motors have an electrical winding or windings that are used to create the magnetic induction for motor operation. There are also special electrical controls that may be added to the motor or into the motor circuit to assist the motor in starting and running, along with providing motor protection.

Motor windings are normally made from bare copper wire and are coated with an insulator to keep the copper wires that make up the motor winding from touching each other.

Short circuits in a motor winding occur when the wire used to make up the motor windings touch each other where the insulation used to protect the wire becomes worn or burned off due to the winding over-heating. This creates a short path through which electrical current can flow. This path has a lower resistance and increases the current flow in the winding(s). The best method for checking a motor for shorted windings is to know what the actual motor winding resistance values should be according to the manufacturer. Some manufacturers provide **resistance charts** for some motors that list what the actual, *run winding and start winding resistance values* should be (**Figure 3–225**).

Compressor Model	Voltage	MOTOR AMPS				FUSE SIZE		Winding Resistance in Ohms
		Full Winding		1/2 Winding		Recommended Max		
		Rated Load	Locked Rotor	Rated Load	Locked Rotor	Fusetron	Std.	
9RA - 0500 - CFB	230/1/60	27.5	125.0			FRN-40	50	Start 1.5 Run 0.40
9RB TFC	208-230/3/60	22.0	115.0			FRN-25	40	0.51-0.61
9RJ TFD	460/3/60	12.1	53.0			FRS-15	15	2.22-2.78
9TK TFE	575/3/60	7.8	42.0			FRS-10	15	3.40-3.96
MRA FSR	200-240/3/50	17.0	90.0	8.5	58.0	FRN-25	35	0.58-0.69
MRB FSM	380-420/3/50	9.5	50.0	4.8	32.5	FRS-15	20	1.80-2.15
MRF								

FIGURE 3–225   Resistances for some typical hermetic compressor. (*Courtesy Copeland Corporation*)

Motors with shorted windings will have resistance values that are lower than the rated value. In most cases, these charts are not available to technicians when checking motors in the field, which makes for difficult diagnosis. Without resistance charts, a helpful fact for technicians to take into consideration to assist in the diagnostic procedure is to remember that when evaluating the motor windings of a split-phase motor, *the start winding has a higher resistance value than the motor's run winding.* When the *run winding's resistance* is higher than the motor's *start winding resistance,* this is a good indication of a shorted start winding. A resistance reading of 0 ohms across any motor winding indicates a complete short.

*When checking **split-phase motor windings**, good motor windings would be found by a resistance check between the motor's winding terminals. If the motor windings are good, a resistance check across the motor terminals should indicate three separate and different resistance values, with one of those resistance values totaling the sum of the other two.*

If the motor windings are good, the measured resistance values can also be used to determine the common, run, and start terminals of the motor.

> NOTE: *A split-phase motor could have good motor windings and still not operate. In most cases, this is due to the motor being in locked rotor.*

> NOTE: *For this competency, compare the actual motor windings measured resistance with the manufacturer's resistance chart if available. If a resistance chart for the motor being tested is not available, compare actual readings based on the fact that the motor start winding should have a higher resistance value than the motors run winding.*

EQUIPMENT REQUIRED

Split-phase motor (*in or out of system*)
1 Ohmmeter or VOM
Assorted hand tools (*if applicable*)

SAFETY PRACTICES

Make sure meter is set to the proper function and the electrical power is disconnected from the motor. Discharge capacitors used in compressor motor circuit before removing motor terminal wires.

> *NOTE: For safety, motor winding value readings should be obtained through the motor's terminal wiring leads when possible. If not applicable, readings can be taken at the compressor motor terminals, but should be obtained with care.*

COMPETENCY PROCEDURES

Checklist

1. Make sure power has been removed from the system or compressor. ❑
2. Gain access to the compressor motor terminals or motor terminal wire leads. ❑
3. Discharge all motor circuit capacitors before removing compressor terminal wires (*if applicable*). ❑
4. Remove compressor motor wiring leads from associated components (*relays, capacitors, etc.*) (*if applicable*). ❑

> *NOTE: If removing motor wires from associated components in equipment, make notes for rewiring motor lead wires back into equipment.*

5. Draw a picture of the compressor's terminal layout. ❑
6. Number the compressor terminal or compressor wire leads from the terminal layout above as "1," "2," and "3." ❑

> *NOTE: If testing compressor motor terminal wires, number wires and wire colors so that identification and resistance readings can be recorded and identified (for example, Yellow Wire 1, Black Wire 2, etc.).*

7. *If available, record the actual motor winding resistance values from the motor or manufacturer's* **resistance chart.** *If a resistance chart for the motor being tested is not available, compare actual readings based on the fact that the motor's start winding should have a higher resistance value than the motor's run winding.* ❑
   Terminals 1 and 2 _____ ❑
   Terminals 1 and 3 _____ ❑
   Terminals 2 and 3 _____ ❑
8. Set the ohmmeter the lowest ohms value scale (*R × 1, R × 200, etc.*). ❑
9. *Zero meter (if applicable).* ❑
10. *Place ohmmeter leads on compressor terminal or wire leads 1 and 2 and check for a resistance value reading on meter.* ❑
11. *Record your ohms value reading at compressor terminals or wire leads 1 and 2.* _____ ❑
12. *Place ohmmeter leads on compressor terminal or wire leads 1 and 3 and check for a resistance value reading on the meter.* ❑
13. *Record your ohms value reading at compressor terminals or wire leads 1 and 3.* _____ ❑
14. *Place ohmmeter leads on compressor terminals or wire leads 2 and 3 and check for a resistance value reading on the meter.* ❑

15. *Record your ohms value reading at compressor terminals or wire leads 2 and 3.* _____ ❏

NOTE: *Motors with shorted windings will have resistance values that are lower than the rated value. In most cases, these charts are not available to technicians when checking motors in the field, which makes for difficult diagnosis. Without* **resistance charts,** *a helpful fact for technicians to take into consideration to assist in the diagnostic procedure is to remember that when evaluating the motor windings of a split-phase motor, the start winding has a higher resistance value than the motor's run winding. When the run winding's resistance is higher than the motor's start winding resistance, this is a good indication of a shorted start winding. A resistance reading of 0 ohms across any motor winding indicates a complete short.*

---

16. *Did the motor being checked have shorted windings according to recorded resistance values?* _____ ❏
17. Have your instructor check your work. ❏
18. Return all equipment and supplies to their original location. ❏

RESEARCH QUESTIONS

1. What is a shorted motor winding?

   _____

   _____

2. What is a "shunted winding"?

   _____

   _____

3. Could a motor that has a shorted motor winding operate?

   _____

   _____

4. If a motor has shorted windings and operates, what will happen to the motor's RLA?

   _____

   _____

5. If a motor has shorted windings and operates, will the motor operate at the designed RPMs?

   _____

   _____

Passed Competency _____ Failed Competency _____

Instructor Signature _____ Grade _____

Practical Competency 86

Short-to-Ground Compressor Motor Test (Figure 3–226)

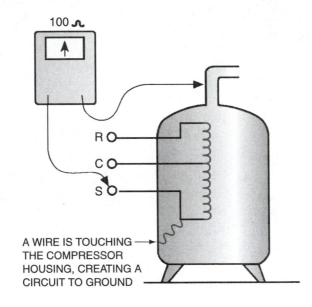

100 Ω

R○

C○

S○

A WIRE IS TOUCHING
THE COMPRESSOR
HOUSING, CREATING A
CIRCUIT TO GROUND

FIGURE 3–226 A compressor with a
short to ground.

SUGGESTED MATERIALS

Textbook
Refrigeration & Air Conditioning Technology, 5th Edition, Thomson Delmar Learning
Unit 20—Troubleshooting Electric Motors
Unit 23—Compressors
Unit 41—Troubleshooting

Review Topics
Short-to-Ground Motor Windings; Compressor Electrical Checkup

COMPETENCY OBJECTIVE

The student will be able to check the compressor windings and determine if the motor has a short to ground.

OVERVIEW

There are many different types of motors used in the air conditioning, refrigeration, and heating industry. Most of these motors perform many different applications and most work by the principle of magnetic induction. Technicians must gain an understanding of the following motors to successfully troubleshoot and work with them in the field: *the split-phase motor, CS motor, CSCR motor, PSC motor, shaded-pole motor, and three-phase motor.*

The method of operation of each of these motors is similar but each serves a different purpose. All of these motors have an electrical winding or windings that are used to create the magnetic induction for motor operation. There are also special electrical controls that may be added to the motor or into the motor circuit to assist the motor in starting and running, along with providing motor protection.

Motor windings are normally made from bare copper wire and are coated with an insulator to keep the copper wires that make up the motor winding from touching each other.

Short to ground occurs when a current-carrying wire makes contact with a ground or zero potential metallic surface such as the compressor shell, the refrigerant-carrying copper tubing or piping, the evaporator and condenser coils, or the casing of the condensing unit or air handler. A short circuit from a motor winding to ground or the frame of a motor can be detected with an ohmmeter using the R × 10,000 ohm scale. Any measurable resistance from the motor windings terminals and the compressor shell, suction line, or discharge line is an indication that the motor is grounded and needs to be replaced or repaired. When a meter reads a very slight resistance to ground, the windings may be dirty and damp if it is an open motor. Cleaning these types of motors will eliminate the short-to-ground in most cases. Compressor oil or liquid refrigerant splashing on the motor windings could also indicate a slight resistance to ground for some compressors. If a motor shows a slight short-to-ground, run the motor for a little while and perform a short-to-ground test again. If a slight short-to-ground is still reflected, it is an indication that the motor is probably going to fail soon if the system is not cleaned. For compressors, the addition or changing of a suction line drier may help remove particles that are circulating in the system and causing the slight ground.

*When checking **split-phase motor windings**, good motor windings would be found by a resistance check between the motor's winding terminals. If the motor windings are good, a resistance check across the motor terminals should indicate three separate and different resistance values, with one of those resistance values totaling the sum of the other two.*

If the motor windings are good, the measured resistance values can also be used to determine the common, run, and start terminals of the motor.

> NOTE: *A split-phase motor could have good motor windings and still not operate. In most cases, this is due to the motor being in locked rotor.*

EQUIPMENT REQUIRED

Split-phase motor (*in or out of system*)
1 ohmmeter or VOM
Assorted hand tools (*if applicable*)

SAFETY PRACTICES

Make sure meter is set to the proper function and the electrical power is disconnected from the motor. Discharge capacitors used in compressor motor circuit before removing motor terminal wires.

> NOTE: *For safety, motor winding value readings should be obtained through the motor's terminal wiring leads when possible. If not applicable, readings can be taken at the compressor motor terminals, but should be obtained with care.*

COMPETENCY PROCEDURES

Checklist

1. Make sure power has been removed from the system or compressor. ❑
2. Gain access to the compressor motor terminals or motor terminal wire leads. ❑
3. Discharge all motor circuit capacitors before removing compressor terminal wires (*if applicable*). ❑
4. Remove compressor motor wiring leads from associated components (*relays, capacitors, etc.*) (*if applicable*). ❑

5. Draw a picture of the compressor's terminal layout. ❑
6. Number the compressor terminal or compressor wire leads from the terminal layout above as "**1,**" "**2,**" and "**3.**" ❑

7. Set the ohmmeter to the ohms value scale of $R \times 10,000$. ❑
8. *Zero meter (if applicable).* ❑
9. Take one of the ohmmeter leads and touch the top of the compressor. Leave this meter lead at this location through this testing procedure. ❑

10. *Place opposite meter leads on compressor terminal wire lead 1 and record the ohms value reading from the ohmmeter. Ohms value reading at terminal 1.* _____ ❑
11. *Once again place the ohmmeter's lead on compressor terminal wire lead 2 and record the ohms value reading from the ohmmeter.*
    Ohms value reading at terminal 2. _____ ❑
12. *Again, place the ohmmeter lead on compressor terminal wire lead 3 and record ohms value reading from the ohmmeter.*
    Ohms value reading at terminal 3. _____ ❑

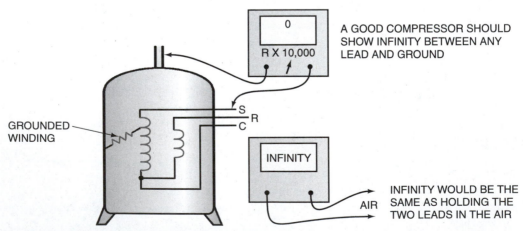

FIGURE 3–227 A good compressor should show infinity between any lead and ground. Infinity would be the same as holding the two leads in the air.

13. *Did any of the motor leads tested indicate a short-to-ground reading?* _____ ❏

14. *If a short-to-ground reading was indicated, which winding or wire number(s) indicated a short-to-ground?*

      Terminal 1 _____ ❏
      Terminal 2 _____ ❏
      Terminal 3 _____ ❏

15. *If a short-to-ground reading was indicated on any of the motor windings, did the resistance reading indicate a SLIGHT or a COMPLETE short to ground?* _____ ❏

16. *If a short-to-ground reading was indicated, could anything be done to eliminate the short to ground (i.e., **running the motor for awhile or using a suction line drier**)?*
_____ ❏

17. Have your instructor check your work. ❏

18. Return all equipment and supplies to their original location. ❏

RESEARCH QUESTIONS

1. Could a motor that has a winding grounded to the motor case run?
_____
_____

2. An open winding or shorted winding can be determined with which of the following instruments:
   A. Voltmeter
   B. Ammeter
   C. Ohmmeter
   D. All the above

3. A decrease in resistance of a compressor's motor winding will cause the motor amperage to:
   A. Decrease
   B. Increase
   C. Stay at the rated amperage
   D. None of the above

4. What size of resistor is recommended for shorting across an electrical motor capacitor before checking with an ohmmeter?
_____
_____

5. What is the physical difference between a run and a start capacitor?
   Run capacitor _____
   Start capacitor _____

Passed Competency _____   Failed Competency _____

Instructor Signature _____   Grade _____

Practical Competency 87

Checking the Windings of a Three-Phase Compressor

SUGGESTED MATERIALS

Textbook

Refrigeration & Air Conditioning Technology, 5th Edition, Thomson Delmar Learning
Unit 18—*Application of Motors*
Unit 20—*Troubleshooting Electric Motors*
Unit 23—*Compressors*
Unit 41—*Troubleshooting*

Review Topics

Short-to-Ground Motor Windings; Compressor Electrical Checkup

COMPETENCY OBJECTIVE

The student will be able to check the motor windings of a three-phase motor.

OVERVIEW

Three-phase motors do not have a run, start, and common windings as in a split-phase motor. These motors have the highest starting torque and best running efficiency without incorporating a run capacitor or start capacitor. There are three individual "hot legs" or motor windings that are 120 degrees out of phase with the next winding. Three-phase motors and equipment can be used only where three-phase voltage is supplied.

Three hot lines identified as L1, L2, and L3 feed the power to the motor. For a 240-volt, three-phase power supply, the voltage measured between any two of the hot legs is 240 volts.

The windings of three-phase motors can be connected in a couple of different ways. The **WYE configuration** three-phase motor has three motor terminals marked **T1, T2,** and **T3**. WYE connection motors are designed to operate at a specific voltage and are easily identifiable by the three designated motor terminal connections or motor leads with the motor windings displayed on a diagram in the form of a **Y**.

The directional rotation of these motors can be easily reversed by switching any two of the power lead connections or motor terminal connections.

There are also three-phase motors for dual-voltage motors and require different motor connection configurations for high-voltage and low-voltage power supply. In dual-voltage motors there are nine different motors leads. The three main motor windings are divided into two smaller windings, making a total of six.

Each of the motor leads is number and care must be taken when wiring the motor leads for different voltage applications or serious damage to the motor could occur. For **high-voltage connection,** the windings are connected in series with each other, forming a large Y.

When a dual-voltage motor is operating at the higher voltage, the current draw of the motor is one half that of the same motor operating at one-half the voltage.

Low-voltage connections for a dual-voltage three-phase motor have the motor windings connected in parallel with each other, forming two smaller Y configurations.

The amperage draw of the dual-voltage motor operating at the low voltage will be twice that of an identical motor wired to operate at the higher voltage.

The **delta configuration** three-phase motor will have twelve motor lead wires and can be dual-voltage motors. For high-voltage applications, the motor windings are wired in series, and for low-voltage applications, the motor windings are wired in parallel.

Electric motors used in compressor applications are normally split-phase or three-phase motors. Split-phase compressor motors can also employ the benefits of the addition of a run, start, or both types of capacitors. Since three-phase motors do not have a common, run, or start winding, a resistance reading from motor terminal to motor terminal should indicate the same resistance value across all terminals (**Figure 3–228**).

Good Motor Winding Resistance Check—Three-Phase Motors

Three-phase motors do not have common, run, or start windings. The motor windings are of the same resistance and depend on line voltage phasing to operate. A resistance check of these motors should indicate the same resistance across all motor terminals to reflect good motor windings (**Figure 3–229**).

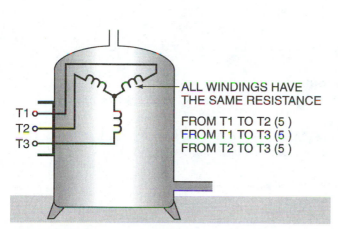

FROM T1 TO T2 (5)
FROM T1 TO T3 (5)
FROM T2 TO T3 (5)

ALL WINDINGS HAVE THE SAME RESISTANCE

FIGURE 3–228 A three-phase compressor with three leads for the three windings. The windings all have the same resistance.

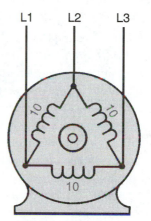

FIGURE 3–229 Good three-phase motor windings are indicated by the same resistance values across each winding.

EQUIPMENT REQUIRED

1 Three-phase compressor
1 Ohmmeter or VOM
Assorted hand tools (*if applicable*)

SAFETY PRACTICES

Make sure the meter is set to the proper function and the electrical power is disconnected from the motor.

> *NOTE: It is recommended that the ohms readings should not be taken at the compressor winding terminals. Motor winding value readings should be obtained through the motor's terminal wiring leads.*

COMPETENCY PROCEDURES

Checklist

1. Make sure power has been removed from the system or compressor. ❑
2. Gain access to the compressor motor terminals or motor terminal wire leads. ❑

3. Remove compressor motor wiring leads from associated components (*relay, contactor, electronic board, etc.*) (*if applicable*). ❑

> **NOTE:** *If removing motor wires from associated components in equipment, make notes for rewiring motor lead wires back into equipment.*

4. Draw a picture of the compressor's terminal layout. ❑
5. Number the compressor terminal or compressor wire leads from the terminal layout above as "**1**," "**2**," and "**3**." ❑

> **NOTE:** *If testing compressor motor terminal wires, number wires and wire colors so that identification and resistance readings can be recorded and identified (for example, Yellow Wire 1, Black Wire 2, etc.)*

6. Set the ohmmeter the lowest ohms value scale ($R \times 1$, $R \times 200$, etc). ❑
7. Zero meter (*if applicable*). ❑
8. Place ohmmeter leads on compressor terminal or wire leads 1 and 2 and check for a resistance value reading on meter. ❑
9. Record your ohms value reading at compressor terminals or wire leads 1 and 2. _____ ❑
10. Place ohmmeter leads on compressor terminal or wire leads 1 and 3 and check for a resistance value reading on meter. ❑
11. Record your ohms value reading at compressor terminals or wire leads 1 and 3. _____ ❑
12. Place ohmmeter leads on compressor terminals or wire leads 2 and 3 and check for a resistance value reading on meter. ❑
13. Record your ohms value reading at compressor terminals or wire leads 2 and 3. _____ ❑
14. Record resistance values: ❑
    Terminals 1 and 2 _____ ❑
    Terminals 1 and 3 _____ ❑
    Terminals 2 and 3 _____ ❑

GOOD MOTOR WINDING RESISTANCE CHECK—THREE-PHASE MOTORS

> **NOTE:** *Three-phase motors do not have a common, run, or start winding. The motor windings are of the same resistance and depend on line voltage phasing to operate. A resistance check of these motors should indicate the same resistance across all motor terminals to reflect good motor windings.*

> **NOTE:** *If an open motor winding were indicated on the motor winding test, it is important to remember that three-phase motors have internal over load protection on each winding. When checking motor windings, be sure to allow the motor to cool to ambient temperature before making a final decision on motor winding failure (**Figure 3–230 A and B**)*

15. *From resistance values taken, were the motor windings of the motor tested OK?* _____ ❑
16. Have your instructor check your work. ❑
17. Return all equipment and supplies to their original location. ❑

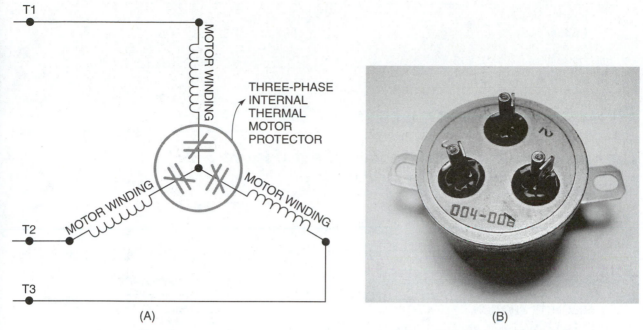

FIGURE 3–230 (A) Three-phase internal thermal overload protection. (B) A three-phase internal overload protector. (*Courtesy Ferris State University. Photo by John Tomczyk*)

RESEARCH QUESTIONS

1. **True or False:** Three-phase motors are often used in air conditioning systems in residences.

   _____

   _____

2. What is a matched set of belts?

   _____

   _____

3. Two types of motor drives are _____ and _____ drives.

4. Why must direct-drive couplings be aligned?

   _____

   _____

5. Why must resilient-mount motors have a ground strap?

   _____

   _____

Passed Competency _____ Failed Competency _____

Instructor Signature _____ Grade _____

Practical Competency 88

Compressor Compression Ratio Test (Figure 3–231)

COMPRESSION RATIO

USING 1 STAGE OF COMPRESSION

$$CR = \frac{\text{ABSOLUTE DISCHARGE}}{\text{ABSOLUTE SUCTION}}$$

$$= \frac{291 \text{ psig} + 15 \text{ (ATMOSPHERE)}}{11.48 \text{ psia}}$$

$$= \frac{306 \text{ psia}}{11.48 \text{ psia}}$$

$$= 26.65 \text{ TO } 1$$

USING 2 STAGES OF COMPRESSION

$$CR \text{ (1ST STAGE)} = \frac{\text{ABSOLUTE DISCHARGE}}{\text{ABSOLUTE SUCTION}}$$

$$= \frac{55 \text{ (40 psig} + 15 \text{ ATM)}}{11.48}$$

$$= 4.79 : 1$$

$$CR \text{ (2ND STAGE)} = \frac{\text{ABSOLUTE DISCHARGE}}{\text{ABSOLUTE SUCTION}}$$

$$= \frac{306 \text{ psia (291 psig} + 15 \text{ ATM)}}{55 \text{ psia}}$$

$$= 5.56 \text{ TO } 1$$

FIGURE 3–231   The compression ratio and two-stage refrigeration.

SUGGESTED MATERIALS

Textbook
Refrigeration & Air Conditioning Technology, 5th Edition, Thomson Delmar Learning
Unit 23—Compressors
Unit 41—Troubleshooting

Review Topics
The Function of the Compressor

COMPETENCY OBJECTIVE

The student will be able to check and evaluate the compression ratio of a compressor.

OVERVIEW

All types of compressors serve two functions in the refrigeration system. One of its tasks is to remove the low-temperature, low-pressure refrigerant vapor from the evaporator, while reducing the evaporator pressure to a point at which the desired evaporator temperature can be maintained. The second task of the compressor is to raise the pressure of the low-temperature, low-pressure suction vapor to a level where its saturation temperature is higher than the temperature of the condenser-cooling medium (**Figure 3–232**).

Compression ratio is a technical expression of the pressure difference between the compressor's low-side pressure and the high-side pressure (**Figure 3–233**) and is used to compare pumping conditions of a compressor.

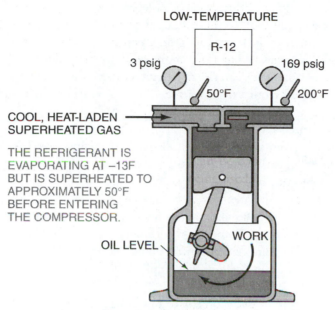

FIGURE 3–232 Low-side superheated gas temperature and pressure and high-side superheated gas temperature and pressure.

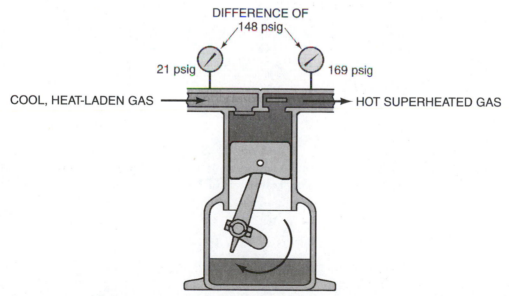

FIGURE 3–233 Compression ratio is the technical expression of pressure difference between the low side and high side of the compressor.

The **compressor compression ratio** is determined by the dividing the compressor's *absolute discharge pressure* by the compressor's *absolute suction pressure*.

$$\text{Compression ratio} = \frac{\textit{Absolute discharge psia}}{\textit{Absolute suction psia}}$$

Absolute pressures are found by adding atmospheric pressure of 14.7 psi to the system's operating low-side and high-side gage pressures (psig). Compression ratios of 3:1 are considered normal, although any situation that could cause an increase in a system's head pressure or a decrease in a system's suction pressure will cause higher compression ratios. When the compression ratio gets above the 12:1 range, the refrigerant gas temperature leaving the compressor can rise to a point at which the compressor's lubricating oil becomes overheated. Refrigerant oil that becomes overheated can turn into carbon and create acids within the system.

Because of the compressor compression ratio, *heat is pumped up the temperature scale.* During the compression of the low-temperature vapor from the evaporator, the vapor is compressed into a smaller area. This causes the temperature and pressure to rise. This is called **heat of compression.** The compressor takes the heat that is already in the vapor from the evaporator and pushes it into a small space.

By doing this, the temperature and pressure exerted on the vapor rise to a high temperature and high-pressure vapor that is hotter than the area in which it is to be released. Once the compressor completes the heat of compression, it forces this high-temperature, high-pressure vapor into the condenser coil, which is normally located in an area where the heat from the evaporator is to be released.

The electrical motors of the compressors can be contained in the compressor housing or can be separate from the compressor such as an open-drive compressors. These systems can be either belt-driven or direct-drive type.

The compressor's compression ratio can be used to evaluate compressor efficiency, although inefficient compressor operation can be one of the most difficult problems to find. Remember that compressors are vapor pumps and should be able to create a pressure from the low side of the system to the high side under design conditions. The efficiency of a compressor is determined by its ability to pump refrigerant through the system at designed pumping capacities. The pumping capacity for all compressors is not the same. Compressors used in high-temperature, medium-temperature, and low-temperature refrigeration applications are designed with different pumping capabilities because of the operating temperature ranges of the evaporator coils in the three different refrigeration applications. Using the standard operating conditions for the type of refrigeration equipment can aid the technician in checking the pumping capacity of that system's compressor. The following test can be used to evaluate and identify the inefficient compressor's problem.

EQUIPMENT REQUIRED

Operation air conditioning or heat pump
Manifold gage set
Refrigeration wrench (*if applicable*)
Assorted hand tools (*if applicable*)

SAFETY PRACTICES

Wear goggles or safety glasses and gloves when working with refrigerant in sealed systems. Beware of high pressures and follow all electrical safety rules. Follow all EPA rules and regulations while working with refrigerants.

COMPETENCY PROCEDURES

Checklist

1. Turn air conditioning or heat pump unit ON. ❑
2. Remove the low- and high-side service valve caps. ❑
3. Make sure manifold gage valves are closed. ❑
4. Attach low- and high-side manifold gage hoses to low- and high-side service valves of the operating unit. ❑

5. Allow the unit to operate for 10 minutes. ❏

DETERMINING COMPRESSOR COMPRESSION RATIO

6. *Record the system's operating high-side psig.* _____ ❏
7. *Determine high-side psia.*
   High side psig _____ *+ 15 psi =* _____ *psia* ❏
8. *Record high-side psig.* _____ ❏
9. *Record the system's low-side psig.* _____ ❏
10. *Determine low-side psia.*
    Low-side psig _____ *+ 15 psi =* _____ *psia* ❏
11. *Record high-side psia.* _____ ❏
12. *Determine compressor **compression ratio**.*

$$\textbf{Compression ratio} = \frac{\textit{High-side psia}}{\textit{Low-side psia}}$$ ❏

13. **Compression Ratio** $= \dfrac{\textit{High-side psia}}{\textit{Low-side psia}}$ ❏

14. *Record compressor compression ratio.* _____ ❏

NOTE: *Air conditioning compression ratios of 3:1 are considered normal, although any situation that could cause an increase in a system's head pressure or a decrease in a system's suction pressure will cause higher compression ratios. When the compression ratio gets above the 12:1 range, the refrigerant gas temperature leaving the compressor can rise to a point at which the compressor's lubricating oil becomes overheated. Refrigerant oil that becomes overheated can turn into carbon and create acids within the system.*

---

15. *Would you consider the compression ratio in proper design range? (**Explain your reasoning.**)*
    _____ ❏
16. Have your instructor check your work. ❏
17. Turn the unit OFF. ❏
18. Carefully remove the system's high-side and low-side manifold gage hoses. ❏
19. Replace unit service valve caps. ❏
20. Return all equipment and supplies to their original location. ❏

RESEARCH QUESTIONS

1. **True or False:** A compressor can compress a liquid.

   _____

   _____

2. Which compressor type uses pistons to compress the gas?
   A. Scroll
   B. Reciprocating
   C. Rotary
   D. Screw

3. Which style of compressor uses belts to turn the compressor?
   A. Open drive
   B. Closed drive
   C. Chassis drive
   D. Hermetic drive

4. At what speeds in rpm does a hermetic compressor normally turn?

    A. 1750

    B. 3450

    C. 4500

    D. Both A and B

5. What lubricates the refrigeration compressor?

    A. Refrigerant

    B. Moisture

    C. Desiccants

    D. Oil

Passed Competency _____   **Failed Competency** _____

Instructor Signature _____   **Grade** _____

Practical Competency 89

Reciprocating Compressor Vacuum Test (Atmosphere as Head Pressure)

NOTE: *This test is to be performed on reciprocating compressors with valve stem service valves installed and can be performed in or out of a sealed system as long as the compressor has valve stem service valves.*

NOTE: *If using atmosphere as the head pressure for this test, refrigerant should be recovered from the system before performing a vacuum test on a reciprocating compressor in a sealed system (if applicable).*

SUGGESTED MATERIALS

Textbook
Refrigeration & Air Conditioning Technology, 5th Edition
Unit 29—Troubleshooting and Typical Operating Conditions for Commercial Refrigeration

Review Topics
Compressor Vacuum Test

COMPETENCY OBJECTIVE

The student will be able to check for inefficient compressor suction valve by performing a compressor vacuum test.

OVERVIEW

The compressor vacuum test is used to evaluate the suction valve operation of reciprocating compressors. All reciprocating compressors with a good suction valve should be able to pull a 26- to 28-in. Hg vacuum, using atmosphere as the discharge pressure (**Figure 3–234**).

Checking the discharge pressure of a compressor can be simulated by correctly adjusting the compressor's service valves and using a *refrigerant or nitrogen* as the compression gas. Under these conditions, a reciprocating compressor should pull about 24 in. Hg vacuum, against 100 psig discharge pressure (**Figure 3–235**).

Once the compressor has reached a 24 in. Hg vacuum and the differential pressure of 100 psig is reached, stop the compressor and let the pressures sit. There should be no equalization of the suction and discharge pressures. When refrigerant is used for this pumping test, the 100 psig may drop some because of the condensing refrigerant.

NOTE: *This test **cannot be used to evaluate multicylinder compressors** because if one of the compressors cylinders is good, a vacuum will be pulled.*

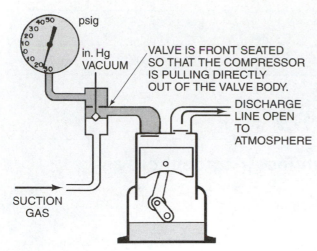

FIGURE 3-234 Most reciprocating compressors can pull 26 in. to 28 in. Hg vacuum with the atmosphere as the discharge pressure.

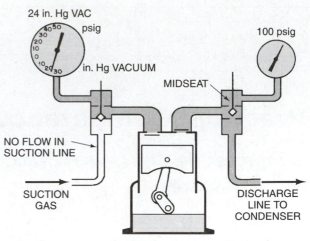

FIGURE 3-235 Perform a compressor check by pulling a vacuum while using refrigerant or nitrogen to build the system head pressure to 100 psig.

EQUIPMENT REQUIRED

Reciprocating compressor (*in or out of sealed system*) with low- and high-side valve stem service valves
1 Refrigeration service wrench
1 Refrigeration manifold gage with low-loss fittings

SAFETY PRACTICES

Wear goggles or safety glasses and gloves when working with refrigerant in sealed systems. Beware of high pressures and follow all electrical safety rules. Follow all EPA rules and regulations while working with refrigerants.

COMPETENCY PROCEDURES

Checklist

> NOTE: *Most compressor motors are cooled with suction gas and will get hot if operated for any length of time without being cooled. The vacuum test should not take more than 3 to 5 minutes to avoid damage to the compressor.*

1. Turn the refrigeration sealed system off (*if applicable*). ❑
2. Recover refrigerant (*if applicable*). ❑
3. Gain access to the refrigeration system's service valves. ❑
4. Make sure manifold gage valves are closed. ❑
5. Remove compressor service valve caps. ❑
6. Attach the low-side manifold gage hose to the suction service valve of the refrigeration sealed system. ❑
7. Attach the high-side manifold gage hose to the discharge service valve of the refrigeration system. ❑
8. Front-seat the compressor's suction service valve the whole way. ❑
9. Mid-seat the discharge service valve by turning it clockwise one full turn. ❑
10. Turn the refrigeration system ON and allow it to operate for 2 to 3 minutes. (**Compressor should be operating.**) ❑

> NOTE: *If vacuum test is being performed on a reciprocating compressor in a sealed system that has a low-pressure cut-out control, the control will have to jumped to allow the compressor to operate for the test.*

11. Observe the suction pressure while the compressor is operating. ❏

12. *After the compressor has operated for 2 to 3 minutes, record the suction pressure.* _____ ❏

13. **Shut OFF** the compressor after 2 to 3 minutes of operation and observe the compressors suction pressure for an additional 2 to 3 minutes. ❏

14. *Record the suction pressure of the compressor after it has been OFF for about 2 to 3 minutes.* _____ ❏

NOTE: *The compressor should have been able to pull a 26- to 28-inch vacuum after about 2 to 3 minutes of operation and hold the vacuum level for 2 to 3 minutes after the compressor has been shut off. This indicates that the compressor's suction valve is working OK.*

If the 26- to 28-inch vacuum cannot be reached in the 2- to 3-minute period, shut the system off for about 3 minutes and try the test again. In any case, if the compressor cannot pull a 26- to 28-inch vacuum in about 2 minutes, the suction valve has to be replaced. If the compressor is a welded hermetic compressor, it has to be replaced.

---

15. Was the suction valve of the compressor you tested able to pull to a 26- to 28-inch vacuum? (**Yes or No**) _____ ❏

16. Leave the refrigeration system turned OFF. ❏

17. Backseat the suction service valve by turning the service valve counterclockwise until the valve cannot be turned any further. ❏

NOTE: *Do not overtighten the valve.*

---

18. Back-seat the discharge service valve by turning the service valve counterclockwise until the valve cannot be turned any further. ❏

19. Remove low- and high-gage hoses. ❏

20. Replace the service valve cap. ❏

21. Return all equipment to its original condition and return all materials and supplies to their proper location. ❏

22. Have your instructor check your work. ❏

RESEARCH QUESTIONS

1. What symptoms could indicate a leaking compressor suction service valve in a refrigeration system?
   _____
   _____

2. What symptoms could indicate a broken compressor suction valve in a refrigeration system?
   _____
   _____

3. How can a hermetic compressor be tested for efficiency?
   _____
   _____

4. How are the motor windings of compressor cooled?
   _____
   _____

5. What compression ratio is considered too high?

_____

_____

Passed Competency _____ Failed Competency _____

Instructor Signature _____ Grade _____

Practical Competency 90

Closed-Loop Compressor Running Bench Test

SUGGESTED MATERIALS

Textbook
Refrigeration & Air Conditioning Technology, 5th Edition, Thomson Delmar Learning

Review Topics
Unit 23—Compressors
Unit 29—Troubleshooting and Typical Operating Conditions for Commercial Refrigeration

COMPETENCY OBJECTIVE

The student will be able to check a compressor's pumping capacity by using the closed-loop method.

OVERVIEW

The efficiency of a compressor is its ability to pump refrigerant through the system at designed pumping capacities. The pumping capacity is not the same for all compressors because of the operating temperature range of the evaporator coils used in high-temperature, medium-temperature, and low-temperature refrigeration applications.

The design operating evaporator coil temperature range for **high-temperature refrigeration** *is 35 to 55 degrees (F) with 40 degrees being the norm for standard equipment operating under standard operating conditions.*

Medium-temperature refrigeration *standard operating evaporator coil temperature range is 10 to 45 degrees (F), with 20 degrees (F) being the norm under standard design conditions.*

Low-temperature refrigeration *operating evaporator coil temperature range is 5 to –40 degrees (F) with –15 degrees (F) being the norm under design operating conditions.*

Another fact to be considered when evaluating a compressor pumping capacity is the type of condenser being used on the system. There are two types of air-cooled condensers, *standard air-cooled* and *high-efficiency* air-cooled condensers. Each has a different design saturation temperature.

The saturation temperature range for **standard air-cooled condensers** *used in refrigeration applications is normally 30 degrees above the ambient temperature (**Figure 3–236**).*

The design saturation temperature of **high-efficiency air-cooled condensers** *is between 15 and 20 degrees above the ambient temperature (**Figure 3–237**).*

The designed saturation temperature for **water-cooled condensers** is 10 degrees above the leaving water temperature (**Figure 3–238**).

Using the standard operating conditions for the type of refrigeration equipment can aid the technician in checking the pumping capacity of a system's compressor in or out of the refrigeration system by simulating the system's standard operating conditions and checking the manufacturer's rated full-load amperage for the compressor.

Run Testing (Figure 3–239)

A compressor running test can be accomplished by connecting a line from the discharge to the suction of the compressor and operating the compressor in a closed loop. A difference in pressure can be obtained with a valve arrangement or gage manifold. This will prove that the compressor can pump.

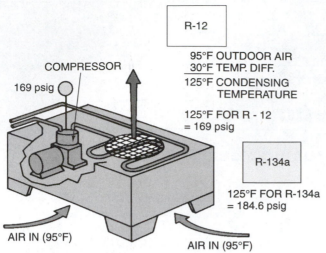

R-12

95°F OUTDOOR AIR
30°F TEMP. DIFF.
125°F CONDENSING
TEMPERATURE

125°F FOR R - 12
= 169 psig

R-134a

125°F FOR R-134a
= 184.6 psig

COMPRESSOR

169 psig

AIR IN (95°F)

AIR IN (95°F)

STANDARD EFFICIENCY AIR-COOLED CONDENSER
CONDITIONS WITH A TYPICAL LIGHT FILM OF
AIRBORNE DIRT CONDENSING AT 30°F ABOVE
THE AMBIENT ENTERING AIR TEMPERATURE.

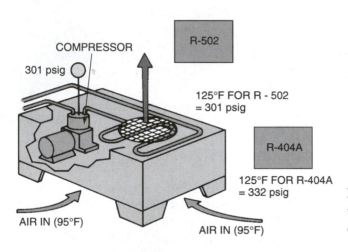

COMPRESSOR

301 psig

R-502

125°F FOR R - 502
= 301 psig

R-404A

125°F FOR R-404A
= 332 psig

AIR IN (95°F)

AIR IN (95°F)

FIGURE 3–236 Typical saturation temperatures and pressures for standard air-cooled condensers.

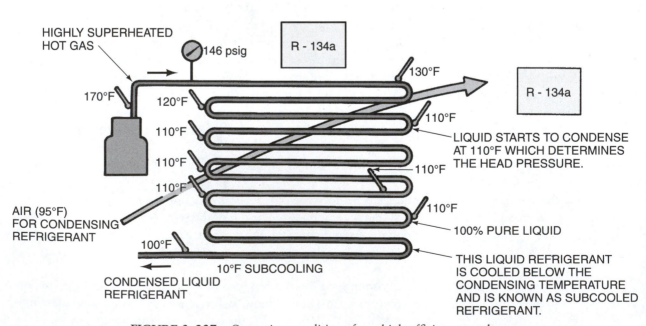

HIGHLY SUPERHEATED
HOT GAS

146 psig

R - 134a

130°F

R - 134a

170°F

120°F

110°F

110°F

110°F

110°F

110°F

110°F

110°F

LIQUID STARTS TO CONDENSE
AT 110°F WHICH DETERMINES
THE HEAD PRESSURE.

100% PURE LIQUID

AIR (95°F)
FOR CONDENSING
REFRIGERANT

100°F

10°F SUBCOOLING

CONDENSED LIQUID
REFRIGERANT

THIS LIQUID REFRIGERANT
IS COOLED BELOW THE
CONDENSING TEMPERATURE
AND IS KNOWN AS SUBCOOLED
REFRIGERANT.

FIGURE 3–237 Operating conditions for a high-efficiency condenser.

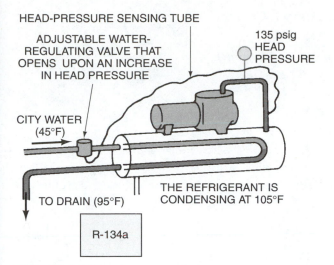

HEAD-PRESSURE SENSING TUBE

ADJUSTABLE WATER-REGULATING VALVE THAT OPENS UPON AN INCREASE IN HEAD PRESSURE

135 psig HEAD PRESSURE

CITY WATER (45°F)

TO DRAIN (95°F)

THE REFRIGERANT IS CONDENSING AT 105°F

R-134a

FIGURE 3–238 A water-cooled waste water system. The saturation temperature is 10 degrees above the leaving water temperature.

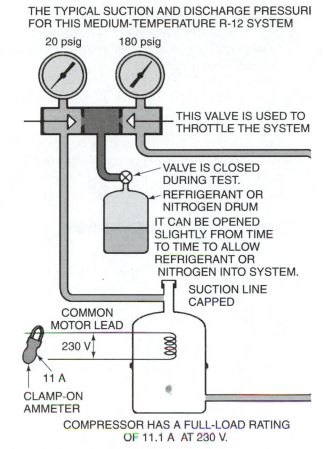

THE TYPICAL SUCTION AND DISCHARGE PRESSURI FOR THIS MEDIUM-TEMPERATURE R-12 SYSTEM

20 psig   180 psig

THIS VALVE IS USED TO THROTTLE THE SYSTEM

VALVE IS CLOSED DURING TEST.

REFRIGERANT OR NITROGEN DRUM IT CAN BE OPENED SLIGHTLY FROM TIME TO TIME TO ALLOW REFRIGERANT OR NITROGEN INTO SYSTEM.

SUCTION LINE CAPPED

COMMON MOTOR LEAD

230 V

11 A

CLAMP-ON AMMETER

COMPRESSOR HAS A FULL-LOAD RATING OF 11.1 A AT 230 V.

FIGURE 3–239 Compressor running test.

NOTE: *When the compressor is hermetic, it should operate at close to full-load current in the closed loop when design pressures are duplicated. Nitrogen or refrigerant can be used as a gas for the compressor to compress. Typical operating pressures will have to be duplicated for the compressor to operate near the manufacturer's rated full-load amperage.*

NOTE: *A compressor that is operating in designed load conditions and pumping at design capacity will draw at or near the manufacturer's full-load amperage.*

SAFETY PRECAUTION: *This test should be performed only under the close supervision of an instructor or by a qualified person and on equipment fewer than 3 hp. Safety goggles must be worn during this test.*

EQUIPMENT REQUIRED

NOTE: *The compressor used to complete this competency must have suction and discharge service valves so the compressor's discharge pressure gas can pass through the gage manifold and back into the suction service valve.*

1 Compressor (*3 hp or less*)
1 Clamp-on ammeter
1 Refrigeration manifold gage with low-loss fittings

1 Refrigeration wrench or service valve Allen wrench

Nitrogen tank with regulator (*if applicable*)

Refrigerant (*type used for the sealed system application*) (*if applicable*)

Pressure—temperature chart

Safety goggles

SAFETY PRACTICES

This test should be performed by experienced technicians or with the assistance of an experienced technician or instructor. Wear goggles or safety glasses and gloves when working with refrigerant in sealed systems. Beware of high pressures and follow all electrical safety rules. Follow all EPA rules and regulations while working with refrigerants.

COMPETENCY PROCEDURES   Checklist

> **NOTE:** *Compressor should be wired so that it can be electrically operated.*

1. The compressor should be OFF.   ❏
2. Remove the compressor service valve caps.   ❏
3. Make sure manifold gage valves are closed.   ❏
4. Use the gage manifold and fasten the suction line to the low-pressure gage line in such a manner that the compressor is pumping only from the gage line. (Refer to *Figure 3–239.*)   ❏
5. Fasten the discharge gage line to the discharge valve port in such a manner that the compressor is pumping only into the gage manifold. (Refer to *Figure 3–239.*)   ❏
6. Attach the manifold center hose to the refrigerant cylinder or nitrogen cylinder. (Refer to *Figure 3–239.*)   ❏

DO NOT OPEN CYLINDER VALVE AT THIS TIME!

7. Record the refrigerant application in which the compressor or sealed system being tested (*High Temperature—Medium Temperature—Low Temperature*). _____   ❏
8. *If compressor is being benched tested with refrigerant, record the refrigerant type.*

_____   ❏

DETERMINE DESIGN LOW-SIDE OPERATING PSIG

9. *Referring to the Competency Overview for the refrigeration application of the compressor, what is the design saturation temperature for the evaporator of the compressor being tested?*

_____   ❏

10. *Use a pressure–temperature chart and convert the saturation temperature to psig for the refrigeration application the compressor is designed to operate with.* _____   ❏

> **NOTE:** *This will be the low-side pressure being simulated to confirm the compressor's pumping capacity.*

DETERMINE HIGH-SIDE OPERATING PSIG (AIR-COOLED CONDENSER)

11. *What type of condenser would be used with the compressor being tested (Standard Air-Cooled—High-Efficiency—Water-Cooled)?* _____   ❏
12. *Referring to the Competency Overview, what is the design saturation temperature range for the condenser that is or would be used with the compressor being tested?* _____   ❏
13. *Determine and record the high-side saturation temperature by adding the design saturation temperature range to a standard 95-degree (F) outside air temperature.* _____   ❏
14. *Use a pressure–temperature chart and convert saturation temperature to psig for the refrigerant being used or simulated for the compressor being tested.* _____   ❏

> **NOTE:** *This will be the high-side pressure being simulated in this test to evaluate the compressor's pumping capacity.*

DETERMINE HIGH-SIDE OPERATING PSIG (WATER-COOLED)

15. *Referring to the Competency Overview, what is the design saturation temperature range for water-cooled condensers?* _____ ❑
16. *Determine and record the high-side saturation temperature by adding the design saturation temperature to the designed leaving water temperature.* _____ ❑
17. *Use a pressure–temperature chart and convert saturation temperature to psig for the refrigerant being used or simulated based on the compressor being tested.* _____ ❑

> **NOTE:** *This will be the high-side pressure being simulated in this test to evaluate the compressor's pumping capacity.*

18. *Record what low-side pressure will be simulated to check the compressor's pumping capacity.* _____ ❑
19. *Record what high-side pressure will be simulated to check the compressor's pumping capacity.* _____ ❑
20. *FIND and RECORD the manufacturer's FULL-LOAD AMPERAGE for the compressor being tested.* _____ ❑

> **NOTE:** *When the desired low-side and high-side pressure are simulated, the amperage reading on the compressor motor should compare closely to the manufacturer's rated full-load amperage for the compressor being tested.*

21. Place a clamp-on ammeter around the compressor's common motor lead wire. (Refer to Figure 3–239.) ❑
22. Completely open both manifold gage valves. ❑
23. Start the compressor and observe the compressor low- and high-side pressures. ❑

> **NOTE:** *The compressor should now be pumping out the discharge line back into the suction line.*

> **NOTE:** *If the correct pressures cannot be reached with the amount of refrigerant in the compressor, a small amount of refrigerant or nitrogen may be added to the loop system by slightly opening the refrigerant or nitrogen cylinder valve.*

24. *When the desired low-side and high-side pressures are reached, record the compressor's full-load amperage.* _____ ❑
25. *Once the compressor's full-load amperage reading has been obtained, turn the compressor OFF.* ❑

> **NOTE:** *The compressor should operate at near full-load amperage when the design voltage is supplied to the motor. The amperage may vary slightly from the full-load amperage because of the input voltage. For example, a voltage above the nameplate voltage for the compressor will cause amperage below full load and vice versa.*

> **NOTE:** *If designed simulated pressures cannot be obtained with the compressor, or the required full-load amperage, or close to full load, amperage cannot be obtained, the compressor is inefficient in its pumping capacity and should be replaced or rebuilt.*

26. *Were simulated pressures obtained in the test?* _____ ❏
27. *Was the compressor's full-load amperage within the manufacturer's rated full-load amperage?*
    _____ ❏
28. Have your instructor check your work. ❏
29. Disconnect equipment, return all tools, and test equipment to their proper location. ❏

RESEARCH QUESTIONS

1. What is the purpose of the compressor in a refrigeration system?

   _____

   _____

2. What formula is used to determine a compressor's compression ratio?

   _____

   _____

3. What will happen to the refrigerant discharge temperature if there is an increase of superheat on the system's low side?

   _____

   _____

4. Why must discharge gas temperatures be considered too high?

   _____

   _____

5. What is used to cool a compressor's motor windings?

   _____

   _____

Passed Competency _____ Failed Competency _____

Instructor Signature _____ Grade _____

Practical Competency 91

Closed-Loop Compressor Running Field Test

SUGGESTED MATERIALS

Textbook
Refrigeration & Air Conditioning Technology, 5th Edition, Thomson Delmar Learning

Review Topics
Unit 23—Compressors
Unit 29—Troubleshooting and Typical Operating Conditions for Commercial Refrigeration

COMPETENCY OBJECTIVE

The student will be able to check a compressor's pumping capacity by using the closed-loop method in the field.

OVERVIEW

The efficiency of a compressor is its ability to pump refrigerant through the system at designed pumping capacities. The pumping capacity is not the same for all compressors because of the operating temperature range of the evaporator coils used in high-temperature, medium-temperature, and low-temperature refrigeration applications.

The design operating evaporator coil temperature range for **high-temperature refrigeration** *is 35 to 55 degrees (F), with 40 degrees being the norm for standard equipment operating under standard operating conditions.*

Medium-temperature refrigeration *standard operating evaporator coil temperature range is 10 to 45 degrees (F), with 20 degrees (F) being the norm under standard design conditions.*

Low-temperature refrigeration *operating evaporator coil temperature range is 5 to –40 degrees (F), with –15 degrees (F) being the norm under design operating conditions.*

Another fact to be considered when evaluating a compressor pumping capacity is the type of condenser being used on the system. There are two types of air-cooled condensers, *standard air-cooled* and *high-efficiency* air-cooled condensers. Each has a different design saturation temperature.

*The saturation temperature range for **standard air-cooled condensers** used in refrigeration applications is normally 30 degrees above the ambient temperature.*

*The design saturation temperature of **high-efficiency air-cooled condensers** is between 15 and 20 degrees above the ambient temperature.*

The designed saturation temperature for water-cooled condensers is 10 degrees above the leaving water temperature.

Using the standard operating conditions for the type of refrigeration equipment can aid the technician in checking the pumping capacity of a system's compressor in or out of the refrigeration system by simulating the system's standard operating conditions and checking the manufacturer's rated full-load amperage for the compressor.

Run Field Testing
When a compressor has suction and discharge service valves, this test can be performed in place in the system using a gage manifold as the loop.

> *NOTE: When the compressor is hermetic, it should operate at close to full-load current in the closed loop when design pressures are duplicated.*

> *NOTE: A compressor that is operating in designed load conditions and pumping at design capacity will draw at or near the manufacturer's full-load amperage.*

SAFETY PRECAUTION: This test should be performed only under the close supervision of an instructor or by a qualified person and on equipment fewer than 3 hp. Safety goggles must be worn during this test.

EQUIPMENT REQUIRED

> *NOTE: The compressor used to complete this competency must have suction and discharge service valves so the compressor's discharge pressure gas can pass through the gage manifold and back into the suction service valve.*

1 Sealed system compressor (*3 hp or less*)
1 Clamp-on ammeter
1 Refrigeration manifold gage with low-loss fittings
1 Refrigeration wrench or service valve Allen wrench
Nitrogen tank with regulator (*if applicable*)
Refrigerant (*type used for the sealed system application*) (*if applicable*)
3/8" Flared plug (*if applicable*)
Pressure–temperature chart
Safety goggles

SAFETY PRACTICES

This test should be performed by experienced technicians or with the assistance of an experienced technician or instructor. Wear goggles or safety glasses and gloves when working with refrigerant in sealed systems. Beware of high pressures and follow all electrical safety rules. Follow all EPA rules and regulations while working with refrigerants.

COMPETENCY PROCEDURES   Checklist

1. Turn the refrigeration system OFF.   ❏
2. Remove service caps from service valves.   ❏
3. Attach the manifold low-side hose to the compressor low-side service valve.   ❏
4. Attach the manifold high-side hose to the compressor high-side service valve.   ❏
5. Use some means to restrict the manifold center hose.   ❏
6. Front seat (turn clockwise) the suction service valve the whole way.   ❏
7. Front seat (turn clockwise) the high-side service valve the whole way.   ❏
8. *Record the refrigerant application in which the compressor or sealed system being tested (High Temperature—Medium Temperature—Low Temperature).* _____   ❏

DETERMINE DESIGN LOW-SIDE OPERATING PSIG

9. *Referring to the Competency Overview for the refrigeration application of the compressor, what is the design saturation temperature for the evaporator of the compressor being tested?* _____   ❏

10. *Use a pressure–temperature chart and convert the saturation temperature to psig for the refrigeration application the compressor is designed to operate with.* _____   ❏

> **NOTE:** *This will be the low-side pressure being simulated to confirm the compressor's pumping capacity.*

DETERMINE HIGH-SIDE OPERATING PSIG (AIR-COOLED CONDENSER)

11. *What type of condenser would be used with the compressor being tested (Standard Air-Cooled—High-Efficiency—Water-Cooled)?* _____ ❑
12. *Referring to the Competency Overview, what is the design saturation temperature range for the condenser that is or would be used with the compressor being tested?* _____ ❑
13. *Determine and record the high-side saturation temperature by adding the design saturation temperature range to a standard 95-degree (F) outside air temperature.* _____ ❑
14. *Use a pressure–temperature chart and convert saturation temperature to psig for the refrigerant being used or simulated for the compressor being tested.* _____ ❑

> **NOTE:** *This will be the high-side pressure being simulated in this test to evaluate the compressor's pumping capacity.*

DETERMINE HIGH-SIDE OPERATING PSIG (WATER-COOLED CONDENSER)

15. *Referring to the Competency Overview, what is the design saturation temperature range for water-cooled condensers?* _____ ❑
16. *Determine and record the high-side saturation temperature by adding the design saturation temperature to the designed leaving water temperature.* _____ ❑
17. *Use a pressure–temperature chart and convert saturation temperature to psig for the refrigerant being used or simulated based on the compressor being tested.* _____ ❑

> **NOTE:** *This will be the high-side pressure being simulated in this test to evaluate the compressor's pumping capacity.*

18. *Record what low-side pressure will be simulated to check the compressor's pumping capacity.* _____ ❑
19. *Record what high-side pressure will be simulated to check the compressor's pumping capacity.* _____ ❑
20. *FIND and RECORD the manufacturer's FULL-LOAD AMPERAGE for the compressor being tested.* _____ ❑

> **NOTE:** *When the desired low-side and high-side pressure are simulated, the amperage reading on the compressor motor should compare closely to the manufacturer's rated full-load amperage for the compressor being tested.*

21. Place a clamp-on ammeter around the compressor's common motor lead wire. ❑
22. **SAFETY PRECAUTION:** OPEN BOTH GAGE MANIFOLD VALVES ALL THE WAY. ❑
23. **SAFETY PRECAUTION:** START THE COMPRESSOR AND KEEP YOUR HAND ON THE POWER SWITCH. ❑

> **NOTE:** *The compressor should be pumping out through the discharge port, through the gage manifold, and back into the suction port.*

> **NOTE:** *If the correct pressures cannot be reached the manifold gage high-side valve can be throttled toward the seat to restrict the flow of refrigerant and create a differential in pressure.* **DO NOT CLOSE THE VALVE ENTIRELY;** *only throttle until the desired high-side and low-side pressure are obtained.*

24. *When the desired low-side and high-side pressures are reached, record the compressor's full-load amperage.* ____ ❑
25. *Once the compressor's full-load amperage reading has been obtained, turn the compressor OFF.* ❑

> **NOTE:** *The compressor should operate at near full-load amperage when the design voltage is supplied to the motor. The amperage may vary slightly from the full-load amperage because of the input voltage. For example, a voltage above the nameplate voltage for the compressor will cause amperage below full load and vice versa.*

> **NOTE:** *If designed simulated pressures cannot be obtained with the compressor, or the required full-load amperage, or close to full load, amperage cannot be obtained, the compressor is inefficient in its pumping capacity and should be replaced or rebuilt.*

26. *Were simulated pressures obtained in the test?* _____ ❑
27. *Was the compressor's full-load amperage within the manufacturer's rated full-load amperage?* _____ ❑
28. Have your instructor check your work. ❑
29. Back-seat the suction valve the whole way out on the unit. (**Turn valve counterclockwise.**) ❑
30. Back-seat the discharge valve the whole way out on the unit. (**Turn valve counterclockwise.**) ❑
31. Close the manifold gage valves. ❑
32. Disconnect equipment and return all tools and test equipment to their proper location. ❑

RESEARCH QUESTIONS

1. Reciprocating compressors are what type of displacement pumps?

   _____

   _____

2. Which compressor type uses pistons to compress the gas?
   A. Scroll
   B. Reciprocating
   C. Rotary
   D. Screw

3. **True or False:** A compressor can compress a liquid.

   _____

   _____

4. At what speeds in rpm does a hermetic compressor normally turn?
   A. 1750
   B. 3450
   C. 4500
   D. Both A and B

5. What effect does a slight amount of liquid refrigerant have on a compressor over a long period of time?

_____

_____

Passed Competency _____ **Failed Competency** _____

Instructor Signature _____ **Grade** _____

Practical Competency 92

Compressor Running Test in the System

SUGGESTED MATERIALS

Textbook
Refrigeration & Air Conditioning Technology, 5th Edition, Thomson Delmar Learning

Review Topics
Unit 23—Compressors
Unit 29—Troubleshooting and Typical Operating Conditions for Commercial Refrigeration

COMPETENCY OBJECTIVE

The student will be able to perform a compressor running test while in the system.

OVERVIEW

The efficiency of a compressor is its ability to pump refrigerant through the system at designed pumping capacities. The pumping capacity is not the same for all compressors because of the operating temperature range of the evaporator coils used in high-temperature, medium-temperature, and low-temperature refrigeration applications.

The design operating evaporator coil temperature range for **high-temperature refrigeration** *is 35 to 55 degrees (F), with 40 degrees being the norm for standard equipment operating under standard operating conditions.*

Medium-temperature refrigeration *standard operating evaporator coil temperature range is 10 to 45 degrees (F), with 20 degrees (F) being the norm under standard design conditions.*

Low-temperature refrigeration *operating evaporator coil temperature range is 5 to –40 degrees (F), with –15 degrees (F) being the norm under design operating conditions.*

Another fact to be considered when evaluating a compressor pumping capacity is the type of condenser being used on the system. There are two types of air-cooled condensers, *standard air-cooled* and *high-efficiency* air-cooled condensers. Each has a different design saturation temperature.

*The saturation temperature range for **standard air-cooled condensers** used in refrigeration applications is normally 30 degrees above the ambient temperature.*

*The design saturation temperature of **high-efficiency air-cooled condensers** is between 15 and 20 degrees above the ambient temperature.*

The designed saturation temperature for **water-cooled condensers** is 10 degrees above the leaving water temperature.

Using the standard operating conditions for the type of refrigeration equipment can aid the technician in checking the pumping capacity of a system's compressor in or out of the refrigeration system by simulating the system's standard operating conditions and checking the manufacturer's rated full-load amperage for the compressor.

Running Test in the System

A system running test in the system can be performed by creating typical design conditions in the system.

> **NOTE:** *When the compressor is hermetic, it should operate at close to full-load current in the closed loop when design pressures are duplicated.*

> **NOTE:** *A compressor that is operating in designed load conditions and pumping at design capacity will draw at or near the manufacturer's full-load amperage.*

SAFETY PRECAUTION: This test should be performed only under the close supervision of an instructor or by a qualified person and on equipment fewer than 3 hp. Safety goggles must be worn during this test.

EQUIPMENT REQUIRED

1 Sealed system compressor (*3 hp or less*)
1 Clamp-on ammeter
1 Refrigeration manifold gage with low-loss fittings
1 Refrigeration wrench or service valve Allen wrench (*if applicable*)
Pressure–temperature chart
Safety goggles

SAFETY PRACTICES

Wear goggles or safety glasses and gloves when working with refrigerant in sealed systems. Beware of high pressures and follow all electrical safety rules. Follow all EPA rules and regulations while working with refrigerants.

COMPETENCY PROCEDURES   **Checklist**

> **NOTE:** *Make sure that the charge is correct (not over or under).*

1. Turn the refrigeration system OFF. ❑
2. Remove service caps from service valves. ❑
3. Attach manifold low-side hose to compressor low-side service valve. ❑
4. Attach manifold high-side hose to compressor high-side service valve. ❑
5. Record the refrigerant application in which the compressor or sealed system being tested (*High Temperature—Medium Temperature—Low Temperature*). _____ ❑

DETERMINE DESIGN LOW-SIDE OPERATING PSIG
6. *Referring to the Competency Overview for the refrigeration application of the compressor, what is the design saturation temperature for the evaporator of the compressor being tested?*
   _____ ❑
7. *Use a pressure–temperature chart and convert the saturation temperature to psig for the refrigeration application the compressor is designed to operate with.* _____ ❑

> **NOTE:** *This will be the low-side pressure being simulated to confirm the compressor's pumping capacity.*

DETERMINE HIGH-SIDE OPERATING PSIG (AIR-COOLED CONDENSER)
8. *What type of condenser would be used with the compressor being tested (Standard Air-Cooled—High-Efficiency—Water-Cooled)?* _____ ❑

9. *Referring to the Competency Overview, what is the design saturation temperature range for the condenser that is or would be used with the compressor being tested?* _____ ❏

10. *Determine and record the high-side saturation temperature by adding the design saturation temperature range to a standard 95-degree (F) outside air temperature.* _____ ❏

11. *Use a pressure–temperature chart and convert saturation temperature to psig for the refrigerant being used or simulated for the compressor being tested.* _____ ❏

NOTE: *This will be the high-side pressure being simulated in this test to evaluate the compressor's pumping capacity.*

DETERMINE HIGH-SIDE OPERATING PSIG (WATER-COOLED CONDENSER)

12. Referring to the Competency Overview, what is the design saturation temperature range for water cooled condensers? _____ ❏

13. Determine and record the high-side saturation temperature by adding the design saturation temperature to the designed leaving water temperature. _____ ❏

14. Use a pressure–temperature chart and convert saturation temperature to psig for the refrigerant being used or simulated based on the compressor being tested. _____ ❏

NOTE: *This will be the high-side pressure being simulated in this test to evaluate the compressor's pumping capacity.*

---

15. *Record what low-side pressure will be simulated to check the compressor's pumping capacity.* _____ ❏

16. *Record what high-side pressure will be simulated to check the compressor's pumping capacity.* _____ ❏

17. *FIND and RECORD the manufacturer's FULL-LOAD AMPERAGE for the compressor being tested.* _____ ❏

NOTE: *When the desired low-side and high-side pressure are simulated, the amperage reading on the compressor motor should compare closely to the manufacturer's rated full-load amperage for the compressor being tested.*

---

18. Place a clamp-on ammeter around the compressor's common motor lead wire. ❏
19. Turn the system ON. ❏
20. Allow system to operate for 10 minutes. ❏
21. *Record the system's high-side operating psig.* _____ ❏

NOTE: *If high-side pressure is not in range as that which was determined in Step 11 or Step 16, restrict the condenser airflow enough to cause the system's high-side pressure to rise high enough to reach what would be required under standard conditions for the type of system condenser.*

---

22. *When the desired high-side pressures is reached, record the compressor's full-load amperage.* ____ ❏
23. *Once the compressor's full-load amperage reading has been obtained, turn the system OFF.* ❏
24. *Was the compressor operating at the manufacturer's rated full-load amperage once the correct high-side pressure was reached?* _____ ❏

> **NOTE:** *If designed simulated pressures cannot be obtained with the compressor, or the required full-load amperage, or close to full load, amperage cannot be obtained, the compressor is inefficient in its pumping capacity and should be replaced or rebuilt.*

25. *Was simulated high-side pressure obtained in the test?* _____ ❏
26. Have your instructor check your work. ❏
27. Disconnect the manifold gage hoses from the sealed system service valves. ❏
28. Replace sealed system service caps (*if applicable*). ❏
29. Return all tools and test equipment to their proper location. ❏

RESEARCH QUESTIONS

1. Why do some condensers have to be cleaned with brushes and others with chemicals?

   _____

   _____

2. When is the most heat removed from the refrigerant in the condensing process?

   _____

   _____

3. **True or False:** Air-cooled condensers are much more efficient than water-cooled condensers.

   _____

   _____

4. The type of condenser that has the lowest operating head pressure is the _____ cooled condenser.

5. The three materials of which condensers are normally made are _____, _____, and _____.

Passed Competency _____   Failed Competency _____

Instructor Signature _____   Grade _____

Practical Competency 93

Cleaning a Dirty Sealed System (Figure 3–240)

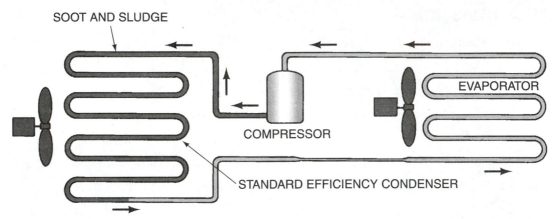

FIGURE 3–240   Soot and sludge move into the condenser when a motor burn occurs while the compressor is running.

SUGGESTED MATERIALS

Textbook
Refrigeration & Air Conditioning Technology, 5th Edition, Thomson Delmar Learning
Unit 8—System Evacuation

Review Topics
Cleaning a Dirty System

COMPETENCY OBJECTIVE

The student will be able to clean and remove contaminants from a sealed system.

NOTE: *The refrigerant must be recovered from the sealed system prior to performing the following procedures.*

OVERVIEW

It is important that refrigeration sealed system be kept clean. Moisture, air, acids, and particulate matter within a sealed system can be very damaging to the sealed system components. Hermetic motors inside a sealed system are the source of heat in a motor burn circumstance. This heat source can heat the refrigerant and oil to a temperature that will break down the oil and refrigerants to acids, soot, and sludge that cannot be removed with a vacuum pump. When a motor burn occurs while the compressor is running, the soot and sludge from the hot oil move into the condenser. (Refer to *Figure 3–240*.) Proper procedures should be used when removing and cleaning a refrigeration system of contaminants.

EQUIPMENT REQUIRED

Safety goggles
Gloves (appropriate for working with refrigerants)
Vacuum pump
Micron gage
Nitrogen bottle with regulator
Refrigeration manifold gage set
Refrigeration wrench (*if applicable*)
Suction line filter drier (*sized for unit*)
Liquid line filter drier (*sized for the unit*)
System refrigerant type
Leak detector
Solder and soldering outfit
Temperature tester

SAFETY PRACTICES

Wear goggles or safety glasses and gloves when working with refrigerant in sealed systems. Beware of high pressures and follow all electrical safety rules. Follow all EPA rules and regulations while working with refrigerants.

COMPETENCY PROCEDURES

Checklist

> *NOTE: Whenever working with a burned-out system, extreme caution must be used when handling and or operating the system. Adequate ventilation, butyle gloves, and safety glasses are required to protect the technician from acids. The oil from a burnout can cause serious skin irritation and possible burns.*

1. Shut power OFF to the unit. ❏
2. Recover refrigerant from sealed system. ❏
3. Open or mid-seat the suction line service valve to the atmosphere. ❏
4. Open or mid-seat the discharge line service valve to the atmosphere. ❏

> *NOTE: On small appliances without access valves, you will have to add some type of permanent service valve to the suction and discharge line of the sealed system.*

5. Attach the high-side manifold gage hose to the discharge service valve port. ❏
6. Open the high-side manifold gage valve. ❏
7. Attach the center hose from the manifold gage set to the nitrogen regulator. ❏
8. Close the suction valve of the gage manifold set (**Figure 3–241**). ❏

> *NOTE: Make sure the nitrogen regulator valve is back-seated the whole way out before turning on the nitrogen bottle tank valve.*

9. Turn ON the nitrogen bottle tank valve. ❏
10. Cover the system's low-side valve with a cloth. (This will catch solid contaminants as they are blown through the system.) ❏

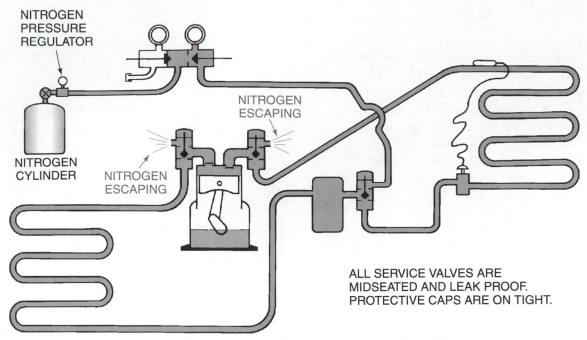

FIGURE 3–241 Setup for using dry nitrogen to sweep the system before evacuation.

11. Slowly turn the nitrogen regulator valve and build a pressure of 150 psig on the regulator. ❏

> *NOTE: Do **not** exceed the system working pressure with nitrogen.*

---

12. Let nitrogen circulate through the system until contaminants are no longer being discharged from the system. ❏
13. Once contaminants are no longer being discharged from the system, back-seat the nitrogen regulator valve the whole way out to stop the flow of nitrogen through the sealed system. ❏

> *NOTE: Once the nitrogen regulator valve has been shut, nitrogen gas may still be released through the suction valve.*

---

14. Disconnect the high-side manifold gage hose from the high-side service valve. ❏
15. Leave the high-side service valve open. ❏
16. Connect the high-side hose from the manifold gage to the suction service valve gage port. ❏
17. Make sure the high-side manifold gage valve is open. ❏
18. Cover the system's high-side valve with a cloth. (*This will catch solid contaminants as they are blown through the system.*) ❏
19. Slowly front-seat the nitrogen valve and build a pressure of 150 psig on the sealed system. ❏
20. Let nitrogen circulate through the system until contaminants are no longer being discharged from the system. ❏
21. Once contaminants are no longer being discharged from the system, back-seat the nitrogen regulator valve the whole way out to stop the flow of nitrogen through the sealed system. ❏

> **NOTE:** *Once the nitrogen regulator valve has been shut, nitrogen gas could still be released through the suction valve.*

> **NOTE:** *Depending on the level of contaminants in the system, monitor what condition the nitrogen gas is in when it is blown out of the system's service valves. If a great deal of dirty oil or debris is being bled from the system as nitrogen continues to discharge once the nitrogen tank has been turned OFF, repeat the procedures over again until the nitrogen gas discharge seems clean. You may also use a Total Test to evaluate the system condition.*

22. Once all nitrogen has leaked from the system, remove the center hose from the nitrogen regulator. ❑
23. Remove manifold gage hose from the high-side service valve of the system. ❑

> **NOTE:** *If necessary, provide means to protect service valves from excessive heat (if applicable).*

24. Use the soldering outfit and install a suction line on the system. ❑
25. Use the soldering outfit and install a liquid line drier on system (**Figure 3–242**). ❑

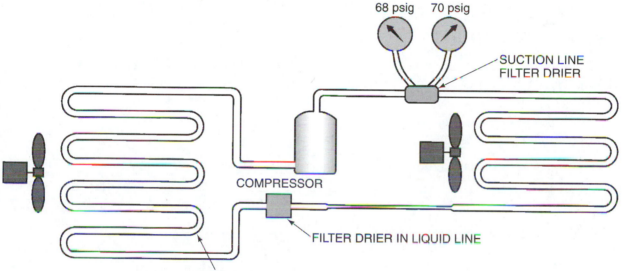

FIGURE 3–242   Adding suction line and liquid line driers to sealed system and checking pressure drop across suction line drier.

26. Pressurize systems with nitrogen and leak-check solder joints at driers. ❑
27. Use the vacuum pump and micron gage and evacuate the system to 500 microns. ❑
28. Recharge the system with the proper refrigerant. ❑
29. Turn the unit ON. ❑
30. Operate the system for at least 15 minutes to allow refrigerant to circulate through system filter driers. ❑
31. After system operation for 15 minutes, check for a temperature drop across the liquid line drier by measuring the in and out temperature of the drier (**Figure 3–243**). ❑
32. *Record the temperature of liquid line entering the drier.* _____ ❑
33. *Record the temperature of liquid line leaving the drier.* _____ ❑
34. *Record the temperature difference.* _____ ❑

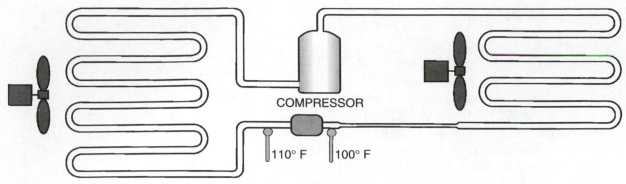

FIGURE 3–243 A 10-degree (F) difference between the IN and OUT of drier indicates a pressure drop; the drier should be replaced.

> **NOTE:** *If there is more than a 10-degree drop across the drier, the temperature drop is too much and the drier should be replaced.*
>
> *If the drier requires replacement, perform a pump down or recover the refrigerant from the system and replace the drier. Evacuate the system and recharge the system. Let the system operate for 15 minutes and check the temperature drop across the liquid line drier again.*

35. Once the system is clean, have your work checked by your instructor. ❑
36. Disconnect equipment and return all tools and test equipment to their proper location. ❑

RESEARCH QUESTIONS

1. The low-pressure side of the system must not be pressurized to more than 150 psig because of:
   A. The type of refrigerant
   B. Using dry nitrogen
   C. The low-side test pressure
   D. Atmospheric pressure

2. **True or False:** The high-pressure side working pressure for most systems is 700 psig.

   _____

   _____

3. Why should oxygen or compressed air never be used to pressurize a system?

   _____

   _____

4. **True or False:** A vacuum pump can remove any type of foreign matter that enters a refrigeration system.

   _____

   _____

5. Are the best vacuum pumps one stage or two stage?

   _____

   _____

Passed Competency _____	**Failed Competency** _____
Instructor Signature _____	**Grade** _____

Practical Competency 94

Procedures for Checking for Noncondensables Within a Sealed System (Figure 3–244)

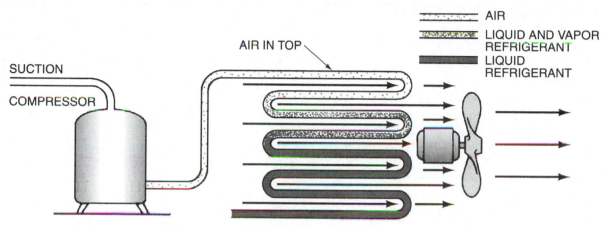

FIGURE 3–244   A condenser containing noncondensable gases.

SUGGESTED MATERIALS

Textbook
Refrigeration & Air Conditioning Technology, 5th Edition, Thomson Delmar Learning
Unit 8—System Evacuation

Review Topics
Purpose of System Evacuation

COMPETENCY OBJECTIVE

The student will be able to check for noncondensables in a sealed system.

OVERVIEW

Noncondensables trapped within refrigeration sealed systems can cause numerous problems in the operation and efficiency of refrigeration equipment. Air is a noncondensable gas that will occupy space in the condenser of the sealed system and cause an increase in the system's operating high-side pressure along with causing a decrease in the system's cooling efficiency.

Air allowed to enter a sealed system can also contain moisture, which can cause a freeze-up in the sealed system or create fozene gas during a compressor burnout. If air or other noncondensables are detected within a sealed system, proper procedures should be followed to remove such contaminants from the system. Noncondensables can be detected within a sealed system by letting the system stand and reach surrounding air temperature and using the pressure–temperature relationship for the type of refrigerant being used within the system. The following competency explains the procedures to check for air and other noncondensables within a sealed system.

EQUIPMENT REQUIRED

Manifold gage set
Pressure–temperature chart
Temperature recorder
Refrigeration wrench (*if applicable*)
Permanent line tap valve (*if applicable*)

SAFETY PRACTICES

Wear goggles or safety glasses and gloves when working with refrigerant in sealed systems. Beware of high pressures and follow all electrical safety rules. Follow all EPA rules and regulations while working with refrigerants.

COMPETENCY PROCEDURES

Checklist

> *NOTE: Manifold gages should be checked for proper calibration to "0" before performing this competency.*

1. Turn the refrigeration system OFF. ❑
2. If need be, add a permanent line valve to the high-side line of the sealed system. ❑

> *NOTE: If system is equipped with valve stem service valves, once manifold gages have been attached; mid-seat the high-side service valve one fourth of a turn to gain access to the system's high-side pressure.*

3. Make sure low-side and high-side manifold gage valves are closed. ❑
4. Attach the manifold high-side hose to the system's high-side line service valve. ❑
5. Allow the system's high-side pressure to stabilize. ❑
6. Allow the system to sit idle long enough for the system's condenser to reach the surrounding air temperature. ❑
7. *Record the type of refrigerant used in the system.* _____ ❑
8. *Record the condenser surrounding ambient air temperature.* _____ ❑
9. *Record the system's stabilized high-side pressure.* _____ ❑
10. *From the manifold gage or pressure–temperature chart, convert the system's high-side pressure to saturation temperature based on the type of refrigerant used in the system.* _____ ❑
11. *Record the air temperature around the condenser of the sealed system.* _____ ❑

> *NOTE: The recorded high-side saturation temperature should be the same as that of the recorded condenser surrounding air temperature.*

12. *Was the sealed systems high-side saturation temperature the same as the surrounding air temperature?* _____ ❑
13. *If the high-side saturation temperature **was not** the same as the surrounding air temperature, record the temperature difference between the two temperatures.* _____ ❑
14. *Record the psig difference between what the high-side psig should be based on the surrounding air temperature and what the actual system's high-side pressure is.* _____ ❑

> **NOTE:** *If the standing high-side pressure is a minimum of 5 to 10 psig higher than it should be based on pressure–temperature relationship of the surrounding air temperature the system has air in it and needs to have the refrigerant recovered, driers added or changed, system evacuated and recharged.*

15. *According to the temperature–pressure relationship and recorded high-side system pressure, would you say that the system (does or does not) contain noncondensables?* _____   ❏
16. Have your instructor check your work.   ❏
17. Disconnect equipment and return all tools and test equipment to their proper location.   ❏

RESEARCH QUESTIONS

1. What are noncondensable gases?

   _____

   _____

2. In most cases, where do noncondensable gases collect in a sealed system?

   _____

   _____

3. What effect will noncondensable gases have on the operation of the sealed system?

   _____

   _____

4. What is one way of determining that noncondensables are removed from a system during evacuation?

   _____

   _____

5. The only two products that should be circulating in a refrigeration system are _____ and _____.

Passed Competency _____	**Failed Competency** _____
Instructor Signature _____	**Grade** _____

Theory Lesson: Expansion Devices

SUGGESTED MATERIALS

Textbook
Refrigeration & Air Conditioning Technology, 5th Edition, Thomson Delmar Learning
Unit 24—Expansion Devices

Key Terms
bulb pressure • capillary tube metering device • evaporator pressure • external equalizer • metering device • orifice type metering device • spring pressure • thermostatic expansion valve (TXV, TEV) • valve system

OVERVIEW

There are four main components to every mechanical refrigeration sealed system: *the compressor, condenser, evaporator, and expansion device*. The **expansion device** is also referred to as the **metering device**, which is responsible to control the flow of liquid refrigerant into the evaporator, along with creating a pressure drop between the low-side and high-side of the sealed system (**Figure 3–245**).

The three most widely used expansion devices in air conditioning and heat pump equipment will be discussed in this lesson, although there are other types that are used for other special refrigeration applications. Expansion devices discussed in this lesson are the **capillary tube, orifice**, and **thermostatic expansion valve**. The purpose of these expansion devices is similar, although the operation of each is very different.

Capillary Tube Metering Device (Figure 3–246)
Capillary tube metering devices are made of a length of copper tube with a fixed-bore diameter. The bore size and the length of the copper tube are designed by the manufacturers to feed the correct amount of refrigerant flow into the system evaporator, based on the difference in the inlet and outlet pressure (**Figure 3–247**).

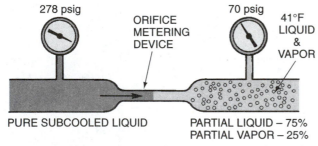

FIGURE 3–245 Metering device control the flow of liquid refrigerant into the evaporator and creates a pressure drop between the high side and low side of the sealed system.

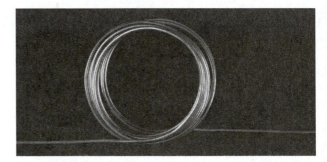

FIGURE 3–246 Capillary tube metering device.

   This type of metering device allows the total system charge to move through the system anytime the refrigeration system is in operation, although the capillary tube metering device does not control evaporator superheat or pressure and is used on systems that have a critical charge. When the refrigerant charge is analyzed and unit is operating at the design conditions, a specific amount of refrigerant is in the evaporator, and a specific amount of refrigerant in the condenser. This is the amount of refrigerant required for proper refrigeration. Any other refrigerant that is in the system is in the pipes for circulating purposes only. In most capillary tube systems, the charge is printed on the nameplate of the unit. Critically charged equipment is the type of

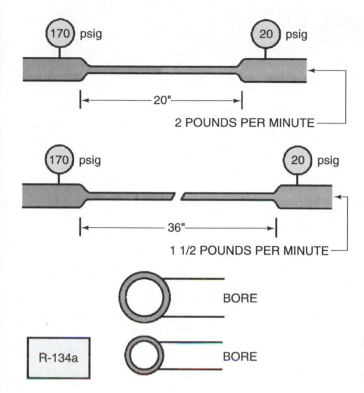

FIGURE 3–247 The length of the capillary tube as well as the bore determines the flow rate of the refrigerant, which is rated in pounds per minute.

refrigerated equipment used where the load is relatively constant with no large fluctuations. During any sealed system repair, technicians should measure the charge into the system during recharging situations. A properly charged system should maintain about 10 degrees (F) of superheat when system is operating in design conditions. A range of 8 to 12 degrees of superheat with these types of systems is acceptable. With capillary tube metering devices, if the system is operating in loads that are higher than design conditions, the evaporator superheat will be higher, but will fall into design range once load conditions fall into design range. The opposite conditions are true if load conditions are lower than design conditions. Low loads would create low evaporator superheat, and as the load increases on the system, so will evaporator superheat. During the OFF cycle of such equipment, the capillary tube allows for equalization of the system's low-side and high-side pressures.

Orifice-Type Metering Device (Figure 3–248)

This type of metering device is very popular in split-system air conditioning and heat pump installations. They are referred to as a *metering orifice, piston, or check valve.* They operate similar to the capillary tube metering device although they are made of brass and have a fixed-bore diameter through the center of the orifice. The size of the orifice is marked on the brass body and is matched to the tonnage of the condensing unit. The orifice is normally fixed into an actuator and sealed in the liquid line at the evaporator by the use of flared fittings. This allows a technician in the field to change the orifice size if need be. *During new installation of a split-system air conditioning unit or system heat pump, technicians must make sure that the correct orifice size is matched to the tonnage of the condensing unit.* Technicians will find the correct orifice required for the system attached to or in the installation manual of the condensing unit. *Technicians must check the evaporator coil metering orifice size and confirm that it is the same size as the one equipped with the condensing unit. If the orifice supplied in the evaporator coil is not the same size as that which came with the condensing unit, it must be replaced. The rule of thumb is that the correct orifice for the system is the one that is supplied with the condenser or outdoor unit.* Failure by a technician to confirm the correct orifice size for an installation will lead to system inefficiency.

The orifice-type metering device does not control evaporator superheat or system pressure either, although it allows full liquid flow in one direction (**Figure 3–249**) and restricted flow in the other direction

FIGURE 3–248   Orifice or piston metering device.

(**Figure 3–250**), making it an ideal metering device for heat pump applications. This type of metering device also allows for system pressure equalization during the system's OFF cycle.

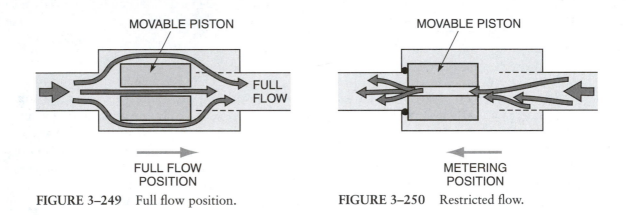

FIGURE 3–249   Full flow position.   **FIGURE 3–250**   Restricted flow.

Thermostatic Expansion Valve (TXV) (Figure 3–251)

Thermostatic expansion valve are also known as TEV or TXV metering devices. These metering devices are designed to maintain a constant superheat of 8 to 12 degrees (F) on an evaporator under any load condition (**Figure 3–252**).

FIGURE 3–251   A solder-type TXV. (*Courtesy of Parker Hannifin Corp.*)

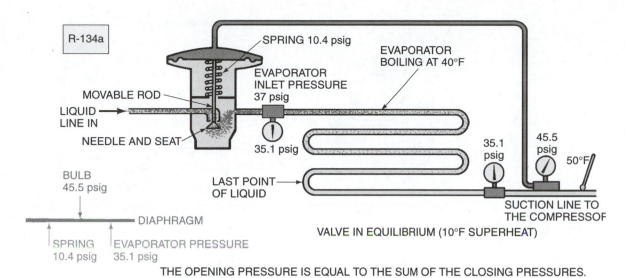

FIGURE 3–252   An ideal evaporator coil with no pressure drop across it and 10 degrees (F) of superheat.

The TXV is well suited for use on systems that experience a wide range of load changes on the evaporator.

Three pressures control the operation of the TXV (**Figure 3–253**). These are the expansion valve spring pressure (**F3**), evaporator pressure (**F2**), and the bulb pressure (**F1**). The **evaporator pressure** and **spring pressure** work together as the *closing force* underneath the diaphragm of the valve, restricting refrigerant flow into the evaporator (**Figure 3–254**).

The **bulb pressure** is transmitted to the top of the valve diaphragm through the capillary tube and works against the evaporator and spring pressure to *open the valve* to allow more refrigerant flow into the evaporator (**Figure 3–255**).

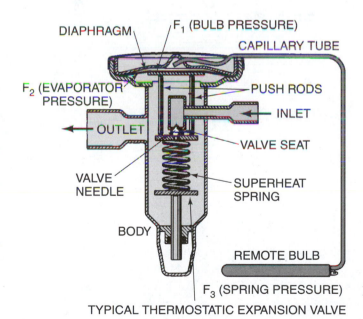

TYPICAL THERMOSTATIC EXPANSION VALVE

FIGURE 3–253   TXV operating pressures.
(*Courtesy of Singer Controls Division*)

The degree to which the refrigerant vapor is superheated at the end run of the evaporator depends on the load on the evaporator and the amount of liquid refrigerant being fed into the coil. Because of the wide range of refrigeration equipment used in the industry, manufacturers have developed different types of thermostatic expansion valves to provide maximum system performance in each application, although all TXV valves are designed to maintain a constant design superheat on the evaporator for maximum efficiency.

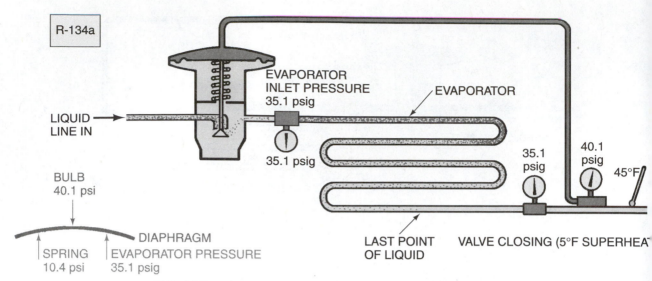

FIGURE 3–254 Evaporator and spring pressures work together to close the feeding of refrigerant of the valve.

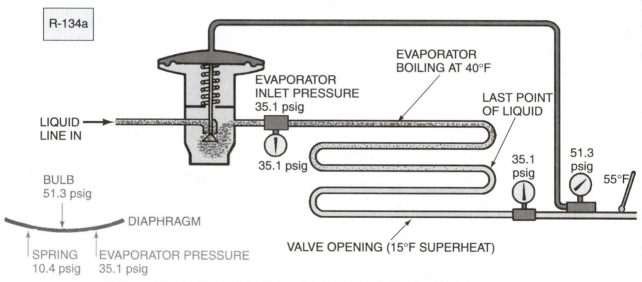

FIGURE 3–255 Bulb pressure is greater than evaporator and spring pressure, opening the valve to allow more refrigerant flow to evaporator.

The feeler bulb should be located as close as possible to the outlet of the evaporator on a horizontal line, making sure that the sensing bulb has good thermal contact with the suction line (**Figure 3–256**).

Using two straps can assist in ensuring that a good contact is made. The bulb should also be insulated to prevent false temperature reading from ambient air. On suction lines that are 5/8" or larger, the sensing bulb should be located at a 4 o'clock or 8 o'clock position. On suction lines that are 7/8" or larger, the bulb should be located at a 10 o'clock or 2 o'clock position. Some TXV valves have a third connection called an **external equalizer** (**Figure 3–257**).

This type of TXV is used on systems where there is a pressure drop on the evaporator. This normally occurs on evaporators with very long circuits. The pressure drop occurs between the inlet and outlet pressures of the evaporator.

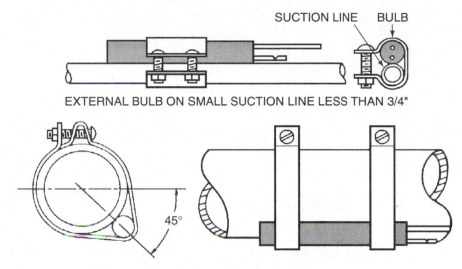

SUCTION LINE   BULB

EXTERNAL BULB ON SMALL SUCTION LINE LESS THAN 3/4"

45°

EXTERNAL BULB ON LARGE SUCTION LINE OVER 7/8"

FIGURE 3–256   TXV bulb location and good thermal contact with suction line.

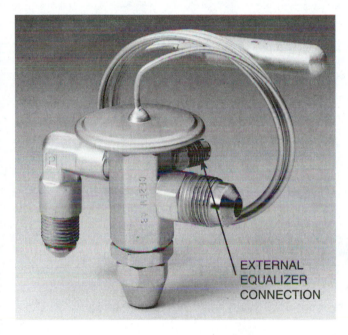

EXTERNAL
EQUALIZER
CONNECTION

FIGURE 3–257   The third connection on this expansion valve is called the external equalizer. (*Courtesy of Parker Hannifin Corp.*)

Evaporators that have a 5 psig or more drop from inlet pressure to outlet pressure require a TXV with an external equalizer. Excess pressure drop in an evaporator coil where a regular TXV is used will cause the valve to starve the coil of refrigerant. The external equalizer line should be connected downstream of the feeler bulb to prevent liquid refrigerant from entering the equalizer line (**Figure 3–258**).

TXV valves are set by the manufacturers to maintain a superheat range of 8 to 12 degrees for evaporators. In most cases, it is not necessary to change the superheat setting. Superheat ranges for the TXV can be changed by making adjustments with the TXV valve stem. Adjustments can be made by turning the TXV valve stem (**Figure 3–259**).

It is important to understand that superheat adjustments on the TXV are not recommended by manufacturers because they are preset to maintain a superheat range of 8 to 12 degrees. In most cases, adjustments

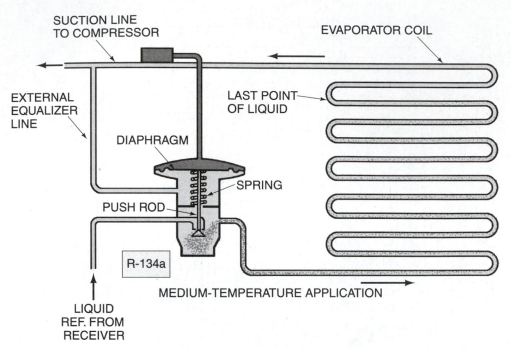

FIGURE 3–258 Correct method for connecting an external equalizer line.

FIGURE 3–259 Superheat adjustment stem. (*Photo by Bill Johnson*)

for superheat on an evaporator will not have to be made and usually other factors will affect the evaporator's performance more than the TXV.

If superheat adjustments are required for whatever reason, you should turn the valve stem only one-quarter turn at a time, and then allow the system to operate until system pressures and temperatures have stabilized before making another adjustment. One-quarter turn of the adjustment stem can change the superheat about one degree Fahrenheit. Turning the valve adjustment stem *clockwise will increase the amount of superheat* on the evaporator. Turning the valve stem *counterclockwise will decrease the amount of superheat* on the system.

Practical Competency 95

Evaluating the Operation of a Sealed System with a Capillary Tube Metering Device (Figure 3–260)

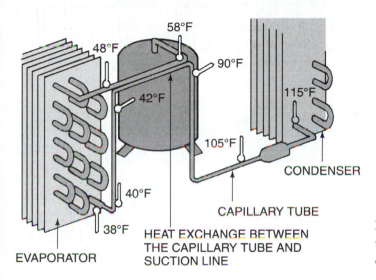

FIGURE 3–260   Typical temperatures along the capillary tube and suction-line heat exchanger.

SUGGESTED MATERIALS

Textbook
Refrigeration & Air Conditioning Technology, 5th Edition, Thomson Delmar Learning
Unit 24—*Expansion Devices*
Unit 41—*Troubleshooting*

Review Topics
The Capillary Tube Metering Device; Operating Charge for the Capillary Tube System; Fixed-Bore Metering Devices; Capillary Tube and Orifice Type

COMPETENCY OBJECTIVE

The student will be able to evaluate the operation of refrigeration sealed system utilizing a capillary tube-metering device.

OVERVIEW

Capillary tube metering devices (**Figure 3–246**) are made of a length of copper tube with a fixed-bore diameter. The bore size and the length of the copper tube are designed by the manufacturers to feed the correct amount of refrigerant flow into the system evaporator, based on the difference in the inlet and outlet pressure. (Refer to *Figure 3–247*.) This type of metering device allows the total system charge to move through the system anytime the refrigeration system is in operation, although the capillary tube metering device does not control evaporator superheat or pressure and is used on systems that have a critical charge. When the

517

refrigerant charge is analyzed and unit is operating at the design conditions, a specific amount of refrigerant is in the evaporator, and a specific amount of refrigerant in the condenser. This is the amount of refrigerant required for proper refrigeration. Any other refrigerant that is in the system is in the pipes for circulating purposes only. In most capillary tube systems, the charge is printed on the nameplate of the unit. Critically charged equipment is the type of refrigerated equipment used where the load is relatively constant with no large fluctuations. During any sealed system repair, technicians should measure the charge into the system during recharging situations. A properly charged system should maintain about 10 degrees (F) of superheat when the system is operating in design conditions. A range of 8 to 12 degrees of superheat with these types of systems is acceptable. With capillary tube metering devices, if the system is operating in loads that are higher than design conditions, the evaporator superheat will be higher, but will fall into design range once load conditions fall into design range. The opposite conditions are true if load conditions are lower than design conditions. Low loads would create low evaporator superheat, and as the load increases on the system, so will evaporator superheat. During the OFF cycle of such equipment, the capillary tube allows for equalization of the system's low-side and high-side pressures.

EQUIPMENT REQUIRED

Manifold gage set
Temperature meter
Refrigeration wrench (*if applicable*)
Permanent line Schrader valves (*if applicable*)
Insulation tape
Pressure–temperature chart
Safety gloves
Safety goggles

SAFETY PRACTICES

The student should be knowledgeable in the use of tools and testing equipment. Follow all EPA rules and regulations when working with refrigerants.

COMPETENCY PROCEDURES   Checklist

> **NOTE:** *This competency can be completed on any refrigeration system that has capillary tube or orifice type metering device.*

1. Turn OFF the power to the unit. ❏
2. Make sure the manifold gage valves are closed. ❏
3. Install access valves to system low and high-side lines (*if applicable*). ❏
4. Connect the high-side manifold gage hose to the refrigeration system's high-side service valve. ❏
5. Connect the low-side manifold gage hose to the refrigeration system's low-side service valve. ❏
6. Connect a temperature probe to the system's suction line near the compressor. ❏
7. Insulate the temperature probe. ❏
8. Turn the system ON and let it operate for 10 minutes. ❏
9. *Measure and record the ambient air temperature in which the evaporator is removing heat from.* _____ ❏
10. *Measure and record the ambient air temperature in which the system condenser is rejecting heat to.* _____ ❏
11. *Is system operating within standard operating conditions?* _____ ❏

Standard Operating Conditions:

75-degree inside air temperature with 50% humidity

95-degree outside air temperature

12. *If the system is not operating within design conditions, is it operating at below or above standard conditions?* _____ ❏

13. *Based on the load conditions the unit is operating in, would you expect the evaporator superheat to be* **higher** *or* **lower** *than the designed superheat range of 8 to 12 degrees (F).* _____ ❏

14. *List the type of refrigerant used in the system being tested.* _____ ❏

15. *Record the low-side operating pressure.* _____ ❏

16. *Determine the system's low-side saturation temperature by converting the system's low-side pressure to temperature based on system refrigerant type.* _____ ❏

17. *Record the suction line temperature.* _____ ❏

18. *Determine the system's superheat by using the following formula:* ❏

Suction line temperature _____

MINUS

Saturation temperature _____

EQUALS *Evaporator superheat* _____

19. *Record the amount of system superheat.* _____ ❏

20. *Understanding the conditions, the system is being operated in, would you say that the recorded superheat is in range of what you would expect based on these operating conditions?* _____ ❏

21. *Record the high-side operating pressure.* _____ ❏

22. *Use the pressure-temperature chart and convert the high-side pressure to saturation temperature.* _____ ❏

23. *Use some means to restrict the airflow over the system's condensers and allow the systems high-side pressure to increase approximately 25 to 30 psig higher than that which was recorded in Step 21.* ❏

NOTE: Do not allow high-side pressure to rise and operate over 300 psig for any length of time.

---

24. *Record the new high-side operating pressure.* _____ ❏

25. *Use the pressure–temperature chart and convert the new high-side pressure to saturation temperature.* _____ ❏

26. Allow system to operate for 10 minutes at these conditions. ❏

27. *After 10 minutes of operation, record the systems low-side operating pressure.* _____ ❏

28. *Did the system's low-side operating pressure increase, decrease, or stay the same with an increase in the system's high-side pressure?* _____ ❏

29. *Explain why you think the system's low-side pressure increased, decreased, or stayed the same with an increase of the system's high-side pressure.* ❏

_____

_____

30. *Determine the system's low-side saturation temperature by converting the system's low-side pressure to temperature based on system refrigerant type.* _____ ❏

31. *Record the suction line temperature.* _____ ❏

32. *Determine the system's superheat by using the following formula:* ❏

Suction line temperature _____

MINUS

Saturation temperature _____

EQUALS *Evaporator superheat* _____

33. *Record the amount of system superheat.* _____ ❏

34. *With the condenser airflow restricted, did the evaporator superheat decrease, increase, or stay the same?* _____ ❑

35. *Explain why you think the evaporator superheat decreased, increased, or stayed the same with restricted airflow across the condenser.* ❑

_____

_____

36. Remove the restriction at the condenser. ❑
37. Allow the system to operate for 5 minutes. ❑
38. Use some means and restrict the airflow across the evaporator by 50%. ❑
39. Allow the system to operate with restricted airflow across the evaporator for 10 minutes. ❑
40. *After 10 minutes of operation, record the system's high-side operating pressure.*
_____ ❑
41. *Use the pressure–temperature chart and convert the high-side pressure to saturation temperature.* _____ ❑
42. *With restricted airflow across the evaporator, did the system's high-side pressure and saturation temperature increase, decrease, or stay the same?* _____ ❑
43. *Explain why you think the system's high-side pressure and saturation temperature decreased, increased, or stayed the same with restricted airflow across the evaporator.* ❑

_____

_____

44. *After 10 minutes of operation, record the system's low-side operating pressure.*
_____ ❑

45. *Determine the system's low-side saturation temperature by converting systems low-side pressure to temperature based on system refrigerant type.* _____ ❑
46. *Record the suction line temperature.* _____ ❑
47. *Determine the system's superheat by using the following formula:* ❑

Suction line temperature _____

MINUS

Saturation temperature _____

EQUALS *Evaporator superheat* _____

48. *Record the amount of system superheat.* _____ ❑
49. *With the evaporator airflow restricted by 50%, did the evaporator superheat decrease, increase, or stay the same?* _____ ❑
50. *Explain why you think the evaporator's superheat decreased, increased, or stayed the same with restricted airflow across the evaporator.* ❑

_____

_____

51. *With the evaporator airflow restricted by 50%, did the system's operating low-side pressure increase, decrease, or stay the same?* _____ ❑
52. *Explain why you think the low-side pressure increased, decreased, or stayed the same with the evaporator airflow restricted by 50%.* ❑

_____

_____

NOTE: *What you should have learned in this competency is that capillary tube or orifice type metering devices do not control or maintain evaporator superheat or systems' operating pressures. As the load that the system is dealing with increases, the system's operating pressures and amount of superheat increase. As the load that the system is dealing with decreases, the system's operating pressures and system superheat will decrease.*

53. Remove evaporator airflow restriction. ❏
54. Turn the power OFF to the unit. ❏
55. Have your instructor check your work ❏
56. Disconnect equipment and return all tools and test equipment to their proper location. ❏

RESEARCH QUESTIONS

1. What is the purpose of the metering device in a sealed system?

   _____

   _____

2. What causes the refrigerant to flow through the capillary tube or orifice type metering devices?

   _____

   _____

3. What would happen to the superheat of an undercharged refrigeration system that has a capillary tube metering device?

   _____

   _____

4. Do the fixed-orifice type metering devices operate the same as a capillary tube metering device?

   _____

   _____

5. How are capillary tube metering devices sized for refrigeration units?

   _____

   _____

Passed Competency _____   Failed Competency _____

Instructor Signature _____   Grade _____

Practical Competency 96

Adjusting the Superheat Setting on a TXV Metering Device

SUGGESTED MATERIALS

Textbook
Refrigeration & Air Conditioning Technology, 5th Edition, Thomson Delmar Learning
Unit 24—Expansion Devices

Review Topics
TXV Components; Thermostatic Expansion Valve; Needle and Seat; The Spring

COMPETENCY OBJECTIVE

The student will be able to evaluate the sealed system response while making superheat adjustments on a refrigeration system with a TXV metering device.

OVERVIEW

TXV valves are set by the manufacturers to maintain a superheat range of 8 to 12 degrees for evaporators. In most cases, it is not necessary to change the superheat setting. Superheat ranges for the TXV can be changed by making adjustments with the TXV valve stem. Adjustments are made by turning the TXV valve stem. It is important to understand that superheat adjustments on the TXV are not recommended. In most cases adjustments for superheat on an evaporator will not have to be made. Usually other factors will affect the evaporator's performance more than the TXV.

When making superheat adjustments on the TXV, you should turn the valve stem only one-quarter turn at a time and then allow the system to operate until system pressures and temperatures have stabilized before making other another adjustment. One-quarter turn of the adjustment stem can change the superheat about one degree Fahrenheit. Turning the valve adjustment stem clockwise will increase the amount of superheat on the evaporator. Turning the valve stem counterclockwise will decrease the amount of superheat on the system.

EQUIPMENT REQUIRED

Refrigeration system with a TXV metering device
Manifold gage set
Temperature meter
Heat gun
Refrigeration wrench
Adjustable or box end wrenches
Safety glasses
Insulated gloves

SAFETY PRACTICES

The student should be knowledgeable in the use of tools and testing equipment. Follow all EPA rules and regulations when working with refrigerants.

COMPETENCY PROCEDURES

> **NOTE:** *This competency requires a system that has a TXV metering device charged properly and working in design operating pressure.*

1. Turn the power to the unit OFF. ❏
2. Make sure manifold gage valves are closed. ❏
3. Connect the low-side manifold gage hose to the suction line service valve. ❏
4. Mid-seat the system suction service valves one-quarter turn to obtain system pressure (*if applicable*). ❏
5. Attach a temperature probe next to the TXV metering device bulb element. ❏
6. Insulate the temperature probe. ❏
7. Turn system ON. ❏
8. Let unit operate for 15 minutes. ❏
9. *Record the operating low-side suction pressure.* _____ ❏
10. Use the pressure–temperature chart and convert pressure to saturation temperature for the refrigerant being used in the system. ❏
11. *Record the saturation temperature of refrigerant in evaporator.* _____ ❏
12. *Record the temperature of the suction line at the TXV sensing bulb.* _____ ❏
13. *Determine amount of superheat by the following formula:* ❏

Suction line temperature _____

MINUS

Saturation temperature _____

EQUALS *Evaporator superheat* _____

14. *Record the amount of superheat.* _____ ❏
15. *Is the system superheat in the design range?* _____ ❏
16. Let the unit continue to operate. ❏
17. Remove the TXV adjustment screw cap. ❏
18. Use the refrigeration wrench and turn the TXV adjustment screw clockwise one complete turn. ❏

> **NOTE:** *Superheat adjustments made on TXV metering devices are done by making adjustments with the metering device adjustment stem. Turning the adjustment screw clockwise will increase the superheat on the system. Turning the adjustment screw counterclockwise will lower the super-heat on the system.*

Superheat adjustments should be made when system is operating at or near design conditions. After making an adjustment, allow the system to operate for 15 minutes before rechecking the superheat of the system.

19. Let the unit operate for 15 minutes. ❏
20. *After 15 minutes of operation, record the system's low side operating pressure.* _____ ❏
21. Use the pressure–temperature chart and convert pressure to temperature for the refrigerant used in system. ❏
22. *Record the saturation temperature of refrigerant in evaporator.* _____ ❏
23. *Record the temperature of the suction line at the TXV sensing bulb.* _____ ❏
24. Determine the amount of superheat by the following formula: ❏

Suction line temperature _____

MINUS

Saturation temperature _____

EQUALS *Evaporator superheat* _____

25. *Record the amount of superheat.* _____ ❏
26. *Was there an increase in the system superheat?* _____ ❏
27. Use the refrigeration wrench and turn the TXV adjustment screw counterclockwise two complete turns. ❏
28. Let the unit operate for 15 minutes. ❏
29. *After 15 minutes of operation, record the system's low side operating pressure.* _____ ❏
30. Use the pressure–temperature chart and convert pressure to temperature for the refrigerant used in system. ❏
31. *Record the saturation temperature of refrigerant in evaporator.* _____ ❏
32. *Record the temperature of the suction line at the TXV sensing bulb.* _____ ❏
33. *Determine amount of superheat by the following formula:* ❏

Suction line temperature _____

MINUS

Saturation temperature _____

EQUALS *Evaporator superheat* _____

34. *Record the amount of superheat.* _____ ❏
35. *Was there a decrease in the evaporator superheat?* _____ ❏
36. *Adjust the TXV valve stem clockwise one full turn.* ❏
37. Turn unit OFF. ❏
38. Backseat the system's low-side service valve. ❏
39. Replace the TXV valve stem cap. ❏
40. Slowly remove the manifold gages. ❏
41. Have your instructor check your work. ❏
42. Disconnect equipment and return all tools and test equipment to their proper location. ❏

RESEARCH QUESTIONS

1. TXV valves are preset to maintain a superheat range of how much?

   _____

   _____

2. In most cases, is there a need for superheat adjustments on a system with a TXV metering device?

   _____

   _____

3. Which direction should the TXV valve stem be turned to decrease the amount of superheat on a evaporator?

   _____

   _____

4. The sensing bulb of the TXV should be located where on the refrigeration system?

   _____

   _____

5. The bulb location for suctions up to 5/8" should be located at which clock locations?

   _____

   _____

Passed Competency _____   Failed Competency _____

Instructor Signature _____   Grade _____

Theory Lesson: Condensers (Figure 3–261 and 3–262)

FIGURE 3–261   Air-cooled condenser. (*Courtesy of Carrier Corporation*)

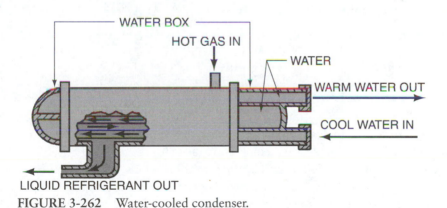

FIGURE 3-262   Water-cooled condenser.

SUGGESTED MATERIALS

Textbook

Refrigeration & Air Conditioning Technology, 5th Edition, Thomson Delmar Learning
Unit 22—Condensers

Review Topics

Theory of Heat; Refrigeration and Refrigerants; Condensers; Water-Cooled Condensers; Air-Cooled Condensers; Standard Efficiency; High Efficiency

Key Terms

design saturation temperatures • design sub-cooling temperatures • high-efficiency air-cooled condensers • high-side operating pressure • latent heat • saturation temperature • standard air-cooled condensers • sub-cooling • water-cooled condenser

OVERVIEW

The condenser of refrigeration sealed systems performs three functions in the refrigeration process. It rejects heat from the refrigerant vapor, condenses vapor to liquid, and sub-cools liquid refrigerant (**Figure 3–263**).

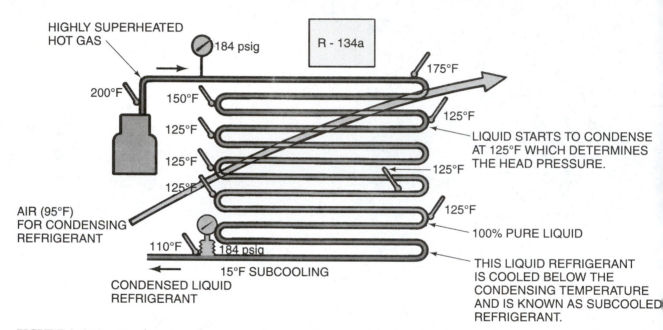

FIGURE 3–263 Condenser performance rejects sensible and latent heat condenses refrigerant vapor to liquid sub-cools the liquid.

How fast a condenser can perform these functions can be used to evaluate the condenser's performance and efficiency.

During rejection of **latent heat** from the refrigerant vapor, the vapor reaches a point of saturation and starts to condense back to the liquid state. The temperature at which this process happens is determined by the pressure being exerted on the vapor. The **saturation temperature** of a condenser can be determined by using the manufacturer's design condensing temperature range for the different type of condensers. In residential and light commercial air conditioning and heat pump applications, two types of condensers are used: air-cooled and water-cooled condensers. There are two types of air-cooled condensers used, standard efficiency and high efficiency. Manufacturers have designed all **standard air-cooled condensers** to condense the refrigerant vapor to a liquid at a temperature of about 30 to 35 degrees above the outside ambient air temperature (**Figure 3–264**).

New **high-efficiency air-cooled condensers** condense the refrigerant vapor back to a liquid at 15 to 25 degrees above the outside ambient air temperature based on the SEER rating of the equipment. The higher the SEER rating of the equipment, the lower the saturation temperature of the refrigerant (**Figure 3–265**).

Water-cooled condensers are designed to condense refrigerant vapor to liquid at 10 degrees above the leaving water temperature in summer applications (**Figure 3–266**) and approximately 20 degrees above the leaving water temperature in winter applications (**Figure 3–267**).

Knowing this information can assist technicians in evaluating a system's **high-side operating pressure** and **saturation temperature**. A system's **correct saturation temperature** can be determined by measuring the

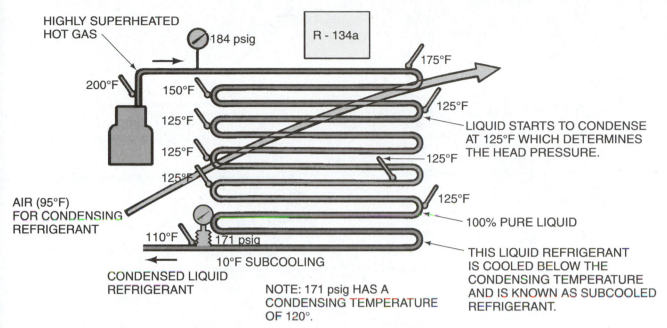

FIGURE 3–264   Operating conditions for standard air-cooled condenser.

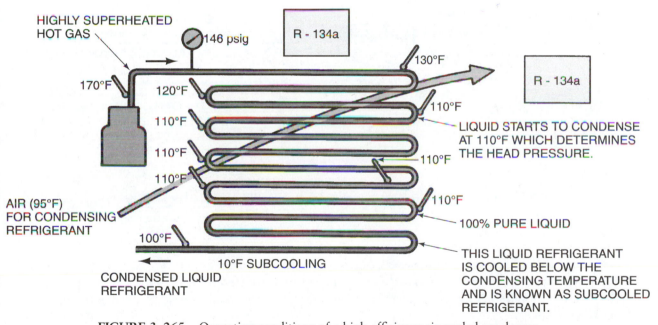

FIGURE 3–265   Operating conditions of a high-efficiency air-cooled condenser.

outside ambient air temperature and adding the manufacturer's designed condensing temperature to it, based on the type of condenser being used in the system. This tells the technician what the saturation temperature of the refrigerant should be for the condenser under current conditions. Using the saturation temperature can assist the technician in determining what the system's **high-side operating pressure** should be under current conditions. The technician would convert the saturation temperature to pressure for the type of system refrigerant being used. This would be what the system's operating head pressure should be under current conditions. This information can then be used to evaluate actual system operating high-side pressure and saturation temperature. An actual system pressure and saturation temperature of **plus or minus 10%**

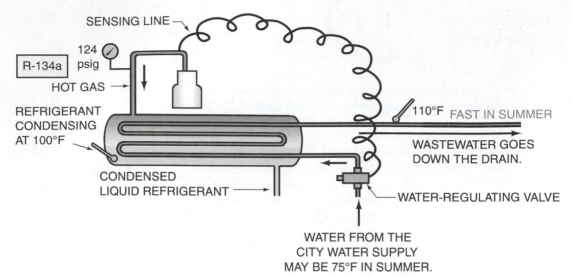

FIGURE 3–266 Operating conditions for water-cooled condensers in summer operation.

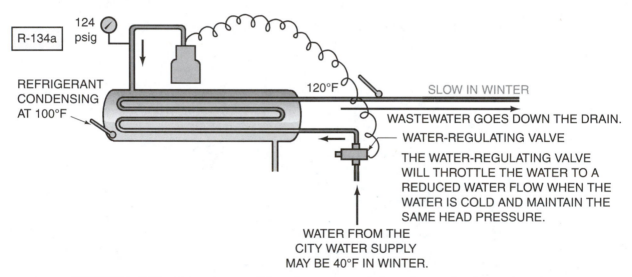

FIGURE 3–267 Operating conditions for water-cooled condensers in winter operation.

would be acceptable. Any pressure and saturation temperature higher or lower than the allowable 10% range is an indication to the technician that the condenser is working too hard or below an acceptable range. Technicians should look for conditions that could cause the condenser to perform outside the determined acceptable range.

Condenser efficiency can be evaluated by measuring the **condenser's sub-cooling**. Remember that **sub-cooling** is lowering the temperature of a liquid one degree or more below the saturation temperature. Manufacturers have also established **designed condenser sub-cooling ranges**. Knowing these designed sub-cooling temperatures is a way to evaluate the condenser's efficiency. Sub-cooling tells how fast the heat is being rejected from the vapor, along with how fast the change of state from a vapor to a liquid is taking place.

All **standard air-cooled condensers** are designed to **sub-cool the liquid a minimum of 10 degrees** below the saturation temperature. **High-efficiency air-cooled condensers** are designed to **sub-cool the liquid by a minimum of 20 degrees** below the saturation temperature. The sub-cooling range for **water-cooled condensers** is **approximately 5 to 10 degrees** (F) below the saturation temperature.

If the condenser is **operating in the designed minimum sub-cooling range**, the **condenser is working efficiently**. At times, the sub-cooling temperatures may be higher than the designed sub-cooling range. This can happen because of lower medium temperatures and due to lower compressor loads and other factors. *Higher sub-cooling temperatures will add 1% to 2% efficiency to the metering device function.*

A **low sub-cooling or no sub-cooling** at all means that the condenser is having problems performing its three designed functions. Low sub-cooling or none at all affects the performance of the metering device and affects the evaporator's ability to absorb heat from the conditioned space. **Sub-cooling checks on a condenser tells technician three things:**

1. *If the condenser is rejecting heat efficiently.*
2. *If the vapor is condensing back to 100% liquid at the end-run of the condenser.*
3. *How efficiently the condenser is operating in the sealed system.*

The **amount of condenser sub-cooling** can be determined by converting system operating high-side pressure to saturation temperature based on system refrigerant, and then measuring the end-run temperature of the condenser. Subtracting the condenser liquid line temperature from the condenser's saturation temperature will equal the amount of condenser sub-cooling.

Sub-cooling formula:

Condenser saturation temperature _____

MINUS

Condenser liquid line temperature _____

EQUALS Condenser sub-cooling _____

Practical Competency 97

Measuring the Sub-Cooling of an Air-Cooled Condenser (Figure 3–268)

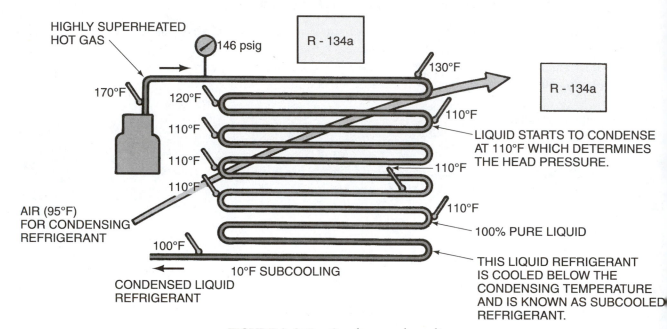

FIGURE 3–268   Condenser sub-cooling.

SUGGESTED MATERIALS

Textbook
Refrigeration & Air Conditioning Technology, 5th Edition, Thomson Delmar Learning
Unit 22—Condensers
Unit 36—Refrigeration Applied to Air Conditioning

Review Topics
The Condenser; Condenser Evaluation; Water-Cooled Condensers; Air-Cooled Condensers; Standard Efficiency; High Efficiency

COMPETENCY OBJECTIVE

The student will be able to evaluate the condenser's operating efficiency by measuring the sub-cooling range of an air-cooled condenser.

OVERVIEW

Condenser efficiency can be evaluated by measuring the **condenser's sub-cooling**. Remember that **sub-cooling** is lowering the temperature of a liquid one degree or more below the saturation temperature.

Manufacturers have also established **designed condenser sub-cooling ranges**. Knowing these designed sub-cooling temperatures is a way to evaluate the condenser's efficiency. Sub-cooling tells how fast the heat is being rejected from the vapor, along with how fast the change of state from a vapor to a liquid is taking place.

All **standard air-cooled condensers** are designed to **sub-cool the liquid a minimum of 10 degrees** below the saturation temperature. **High-efficiency air-cooled condensers** are designed to **sub-cool the liquid by a minimum of 20 degrees** below the saturation temperature. The sub-cooling range for **water-cooled condensers** is **approximately 5 to 10 degrees** (F) below the saturation temperature.

If the condenser is **operating in the designed minimum sub-cooling range**, the **condenser is working efficiently**. At times, the sub-cooling temperatures may be higher than the designed sub-cooling range. This can happen because of lower medium temperatures and due to lower compressor loads and other factors. *Higher sub-cooling temperatures will add 1% to 2% efficiency to the metering device function.*

A **low sub-cooling or no sub-cooling** at all means that the condenser is having problems performing its three designed functions. Low sub-cooling or none at all affects the performance of the metering device and affects the evaporator's ability to absorb heat from the conditioned space. **Sub-cooling checks on a condenser tells technician three things:**

1. *If the condenser is rejecting heat efficiently.*
2. *If the vapor is condensing back to 100% liquid at the end-run of the condenser.*
3. *How efficiently the condenser is operating in the sealed system.*

The **amount of condenser sub-cooling** can be determined by converting system operating high-side pressure to saturation temperature based on system refrigerant, and then measuring the end-run temperature of the condenser. Subtracting the condenser liquid line temperature from the condenser's saturation temperature will equal the amount of condenser sub-cooling.

Sub-cooling formula:

Condenser saturation temperature _____
MINUS
Condenser liquid line temperature _____
EQUALS *Condenser Sub-cooling* _____

EQUIPMENT REQUIRED

Manifold gage set
Temperature tester
Refrigeration wrench (*if applicable*)
Heat gun
Insulation tape
Safety goggles
Safety gloves

SAFETY PRACTICES

The student should be knowledgeable in the use of tools and testing equipment. Follow all EPA rules and regulations when working with refrigerants.

COMPETENCY PROCEDURES   Checklist

1. Turn the system OFF.   ❏
2. Gain access to the system's high-side service valve.   ❏
3. Attach high-side manifold gage to high-side service valve port.   ❏
4. Attach a temperature probe as close as possible to the condensers liquid line.
   (*Use insulation tape to insulate temperature probe lead.*)   ❏
5. Turn the system ON.   ❏
6. Let the system operate for approximately 10 minutes.   ❏

Sub-cooling Check 1

> **NOTE:** *Be careful of high-side temperatures and pressures.*

7. *List the type of system condenser.* _____ ❑
8. *Record the system refrigerant type.* _____ ❑
9. *Record the manufacturer's design sub-cool range.* _____ ❑
10. *Record the system's operating high-side pressure.* ____ ❑
11. *Record the high-side saturation temperature of the refrigerant.* _____ ❑
12. *Record condenser's liquid line temperature.* _____ ❑
13. *Determine the amount of condenser sub-cooling by using the following formula:* ❑
    Condenser saturation temperature _____
    MINUS
    Condenser liquid line temperature _____
    EQUALS *Condenser sub-cooling* _____
14. *Record the amount of sub-cooling temperature.* _____ ❑
15. *Was the amount of sub-cooling in the manufacturer's design range?* _____ ❑

Sub-cooling Check 2 (Condenser Restricted by 50%)

16. Block 50% of the surface area of the condenser coil. ❑
17. *Record the system's operating high-side pressure.* ____ ❑
18. *Record the high-side saturation temperature of the refrigerant.* _____ ❑
19. *Record the condenser's liquid line temperature.* _____ ❑
20. Determine the amount of condenser sub-cooling by using the following formula: ❑
    Condenser saturation temperature _____
    MINUS
    Condenser liquid line temperature _____
    EQUALS *Condenser sub-cooling* _____
21. *Record the amount of sub-cooling temperature.* _____ ❑
22. *Did the condenser sub-cool increase, decrease, or stay the same with the*
    condenser airflow restricted by 50%? _____ ❑
23. *Explain why you think the condenser's sub-cool increased, decreased, or stayed the same*
    with the condenser airflow restricted by 50%. ❑

    _____

    _____

Sub-cooling Check 3 (Totally Restricted Condenser)

24. Totally restrict condenser airflow. ❑
25. Let the system run for about 3 to 5 minutes. ❑

> **NOTE:** *Do not let the high-side pressure reach higher than 325 psig.*

26. *Record the system's operating high-side pressure.* ____ ❑
27. *Record the high-side saturation temperature of the refrigerant.* _____ ❑
28. *Record the condenser's liquid line temperature.* _____ ❑
29. Remove the blockage from the condenser. ❑
30. Determine the amount of condenser sub-cooling by using the following formula: ❑
    Condenser saturation temperature _____
    MINUS
    Condenser liquid line temperature _____
    EQUALS *Condenser sub-cooling* _____

31. *Record the amount of sub-cooling.* _____ ❏
32. *Did the condenser sub-cool increase, decrease, or stay the same with the condenser airflow fully restricted?* _____ ❏
33. *Explain why you think the condenser's sub-cool increased, decreased, or stayed the same with the condenser airflow fully restricted.* ❏

_____

_____

34. Turn the unit OFF. ❏
35. Have your instructor check your work. ❏
36. Remove gages and temperature probes. ❏
37. Return all tools and test equipment to their proper location. ❏

RESEARCH QUESTIONS

1. When is the most heat removed from the refrigerant in the condensing process?

_____

_____

2. What is latent heat of condensation?

_____

_____

3. What makes high-efficiency condensers high efficient?

_____

_____

4. What is heat reclaim?

_____

_____

5. Name two ways in which the heat from an air-cooled condenser can be used for heat reclaim.

_____

_____

Passed Competency _____   Failed Competency _____

Instructor Signature _____   Grade _____

Practical Competency 98

Standard Efficiency Air-Cooled Condenser Performance Evaluation

SUGGESTED MATERIALS

Textbook

Refrigeration & Air Conditioning Technology, 5th Edition, Thomson Delmar Learning
Unit 22—Condensers
Unit 40—Typical Operating Conditions

Review Topics

Air-Cooled Condensers; Standard Efficiency; High Efficiency; Documentation with the Unit

COMPETENCY OBJECTIVE

The student will be able to evaluate the performance of standard air-cooled condenser.

OVERVIEW

The condenser of refrigeration sealed systems performs three functions in the refrigeration process. It rejects heat from the refrigerant vapor, condenses vapor to liquid, and sub-cools liquid refrigerant. (*Theory Lesson: Condensers, Figure 3–263.*)

How fast a condenser can perform these functions can be used to evaluate the condenser's performance and efficiency.

During rejection of **latent heat** from the refrigerant vapor, the vapor reaches a point of saturation and starts to condense back to the liquid state. The temperature at which this process happens is determined by the pressure being exerted on the vapor. The **saturation temperature** of a condenser can be determined by using the manufacturer's design condensing temperature range for the different type of condensers. In residential and light commercial air conditioning and heat pump applications, two types of condensers are used: air-cooled and water-cooled condensers. There are two types of air-cooled condensers used, standard efficiency and high efficiency. Manufacturers have designed all **standard air-cooled condensers** to condense the refrigerant vapor to a liquid at a temperature of about 30 to 35 degrees above the outside ambient air temperature. (*Theory Lesson: Condensers, Figure 3–264.*)

Knowing this information can assist technicians in evaluating a system's **high-side operating pressure** and **saturation temperature**. A system's **correct saturation temperature** can be determined by measuring the outside ambient air temperature and adding the manufacturer's designed condensing temperature to it, based on the type of condenser being used in the system. This tells the technician what the saturation temperature of the refrigerant should be for the condenser under current conditions. Using the saturation temperature can assist the technician in determining what the system's **high-side operating pressure** should be under current conditions. The technician would convert the saturation temperature to pressure for the type of system refrigerant being used. This would be what the system's operating head pressure should be under current conditions.

This information can then be used to evaluate actual system operating high-side pressure and saturation temperature. An actual system pressure and saturation temperature of **plus or minus 10%** would be acceptable. Any pressure and saturation temperature higher or lower than the allowable 10% range is an indication to the technician that the condenser is working too hard or below an acceptable range. Technicians should look for conditions that could cause the condenser to perform outside the determined acceptable range.

Condenser efficiency can be evaluated by measuring the **condenser's sub-cooling**. Remember that **sub-cooling** is lowering the temperature of a liquid one degree or more below the saturation temperature. Manufacturers have also established **designed condenser sub-cooling ranges**. Knowing these designed sub-cooling temperatures is a way to evaluate the condenser's efficiency. Sub-cooling tells how fast the heat is being rejected from the vapor, along with how fast the change of state from a vapor to a liquid is taking place.

All **standard air-cooled condensers** are designed to **sub-cool the liquid a minimum of 10 degrees** below the saturation temperature.

If the condenser is **operating in the designed minimum sub-cooling range**, the **condenser is working efficiently**. At times, the sub-cooling temperatures may be higher than the designed sub-cooling range. This can happen because of lower medium temperatures and due to lower compressor loads and other factors. *Higher sub-cooling temperatures will add 1% to 2% efficiency to the metering device function.*

A **low sub-cooling or no sub-cooling** at all means that the condenser is having problems performing its three designed functions. Low sub-cooling or none at all affects the performance of the metering device and affects the evaporator's ability to absorb heat from the conditioned space.

The **amount of condenser sub-cooling** can be determined by converting system operating high-side pressure to saturation temperature based on system refrigerant, and then measuring the end- run temperature of the condenser. Subtracting the condenser liquid line temperature from the condensers saturation temperature will equal the amount of condenser sub-cooling.

EQUIPMENT REQUIRED

Manifold gage set
Temperature meter
Insulation tape
Permanent line tap vales (*if applicable*)
Pressure–temperature chart
Safety goggles
Safety gloves
Refrigeration wrench (*if applicable*)

SAFETY PRACTICES

The student should be knowledgeable in the use of tools and testing equipment. Follow all EPA rules and regulations when working with refrigerants.

COMPETENCY PROCEDURES   Checklist

> **NOTE:** *This competency can be performed on a central air conditioning unit with a standard air-cooled condenser or a window air conditioner that is charged properly.*

1. Turn the system OFF.   ❑
2. Make sure the manifold gage valves are closed.   ❑
3. Install permanent high and low-side service valves (*if applicable*).   ❑
4. Attach the low-side manifold gage hose to the system's low-side service valve port.   ❑
5. Attach the high-side manifold gage hose to the system's high-side service valve port.   ❑

6. Use the temperature meter and locate a temperature probe to measure outside ambient air temperature around the condenser. ❑

7. Locate another temperature probe in the condenser's liquid line. ❑

8. Turn the unit ON and allow unit to operate for 5 to 10 minutes. ❑

9. *After run time of 5 to 10 minutes, record the system's low-side operating pressure.* _____ ❑

10. *Use a pressure–temperature chart and convert low-side psig to saturation temperature.* _____ ❑

EVALUATING THE CONDENSER'S OPERATING HIGH-SIDE PRESSURE

11. *Record the system's operating high-side pressure.* _____ ❑

12. *Record the condenser's entering air temperature.* _____ ❑

13. *Is the system's operating high-side pressure in the design range based on the manufacturer's design condensing temperature of 30 to 35 degrees above ambient temperature for standard air-cooled condensers?* _____ ❑

EVALUATING THE CONDENSER'S SATURATION TEMPERATURE RANGE

14. *Use a pressure–temperature chart and convert high-side psig to saturation temperature.* _____ ❑

15. *Is the condenser's saturation temperature in the design range based on the manufacturer's design condensing temperature of 30 to 35 degrees above ambient temperature for standard air-cooled condensers?* _____ ❑

EVALUATING THE CONDENSER'S SUB-COOLING RANGE

16. *Record the condenser's liquid line temperature.* _____ ❑

17. *Determine condenser sub-cooling:* ❑

Condenser saturation temperature _____
MINUS
Condenser liquid line temperature _____
EQUALS Condenser sub-cooling _____

18. *Record the condenser sub-cooling.* _____ ❑

19. *Is the condenser's sub-cooling temperature within design range as that which is specified by the manufacturer for standard air-cooled condensers?* _____ ❑

EVALUATING THE CONDENSER'S TEMPERATURE DIFFERENCE

20. *Record the condenser entering air temperature.* _____ ❑

21. *Determine entering air and saturation temperature difference using the formula:* ❑

Condensing temperature _____
MINUS
Entering air temperature _____
EQUALS Temperature difference _____

22. *Record temperature difference.* _____ ❑

23. *Is the condenser operating at less than a 30-degree (F) temperature difference?* _____ ❑

NOTE: *If the condenser is operating at less than a 30-degree (F) temperature difference, it is probably operating efficiently. If the temperature difference is more than 30 degrees (F), the condenser is having trouble rejecting heat and may need to be cleaned or evaluated for other conditions, which could prevent it from rejecting heat in the proper range.*

---

24. Turn the unit OFF. ❑

25. Have your instructor check your work. ❑

26. Use proper procedures and remove manifold gages and temperature tester. ❑

27. Return all tools and test equipment to their proper location. ❑

RESEARCH QUESTIONS (CONDENSER PERFORMANCE EVALUATION RESULTS)

1. Was the condenser's operating high-side pressure within design range? _____ (*If not, explain what you think could be the reason(s) as to why the high-side pressure is not within design range.*)

   _____

   _____

2. Was the condenser's saturation temperature within design range? _____ (*If not, explain what you think could be the reason(s) as to why the condenser's saturation temperature was not within design range.*)

   _____

   _____

3. Was the condenser's sub-cooling within design range? _____ (*If not, explain what you think could be the reason(s) as to why the condenser's sub-cool temperature was not within design range.*)

   _____

   _____

4. Was the condenser's temperature difference within the design range? _____ (*If not, explain what you think could be the reason(s) as to why the entering air temperature and saturation temperature difference was not within design range.*)

   _____

   _____

5. How would you rate this condenser's performance?

   _____

   _____

Passed Competency _____   **Failed Competency** _____

Instructor Signature _____   **Grade** _____

Practical Competency 99

High-Efficiency Air-Cooled Condenser Performance Evaluation

SUGGESTED MATERIALS

Textbook

Refrigeration & Air Conditioning Technology, 5th Edition, Thomson Delmar Learning
Unit 22—*Condensers*
Unit 40—*Typical Operating Conditions*

Review Topics

Air-Cooled Condensers; Standard Efficiency; High Efficiency; Documentation with the Unit

COMPETENCY OBJECTIVE

The student will be able to evaluate the performance of a high-efficiency air-cooled condenser.

OVERVIEW

The condenser of refrigeration sealed systems performs three functions in the refrigeration process. It rejects heat from the refrigerant vapor, condenses vapor to liquid, and sub-cools liquid refrigerant. (*Theory Lesson: Condensers, Figure 3–263.*)

How fast a condenser can perform these functions can be used to evaluate the condenser's performance and efficiency.

During rejection of **latent heat** from the refrigerant vapor, the vapor reaches a point of saturation and starts to condense back to the liquid state. The temperature at which this process happens is determined by the pressure being exerted on the vapor. The **saturation temperature** of a condenser can be determined by using the manufacturer's design condensing temperature range for the different type of condensers. In residential and light commercial air conditioning and heat pump applications, two types of condensers are used: air cooled and water cooled condensers. There are two types of air-cooled condensers used, standard efficiency and high efficiency.

New **high-efficiency air-cooled condensers** condense the refrigerant vapor back to a liquid at 15 to 25 degrees above the outside ambient air temperature based on the SEER rating of the equipment. The higher the SEER rating of the equipment, the lower the saturation temperature of the refrigerant. (*Theory Lesson: Condensers, Figure 3–265.*)

Knowing this information can assist technicians in evaluating a system's **high-side operating pressure** and **saturation temperature**. Determining a system's **correct saturation temperature** can be found by measuring the outside ambient air temperature and adding the manufacturer's designed condensing temperature to it, based on the type of condenser being used in the system. This tells the technician what the saturation temperature of the refrigerant should be for the condenser under current conditions.

Using the saturation temperature can assist the technician in determining what the system's **high-side operating pressure** should be under current conditions. The technician would convert the saturation temperature to pressure for the type of system refrigerant being used. This would be what the system's operating head pressure should be under current conditions. This information can then be used to evaluate actual system operating high-side pressure and saturation temperature. An actual system pressure and saturation temperature of **plus or minus 10%** would be acceptable. Any pressure and saturation temperature higher or lower than the allowable 10% range is an indication to the technician that the condenser is working too hard or below an acceptable range. Technicians should look for conditions that could cause the condenser to perform outside the determined acceptable range.

Condenser efficiency can be evaluated by measuring the **condenser's sub-cooling**. Remember that **sub-cooling** is lowering the temperature of a liquid one degree or more below the saturation temperature. Manufacturers have also established **designed condenser sub-cooling ranges**. Knowing these designed sub-cooling temperatures is a way to evaluate the condenser's efficiency. Sub-cooling tells how fast the heat is being rejected from the vapor, along with how fast the change of state from a vapor to a liquid is taking place.

High-efficiency air-cooled condensers are designed to **sub-cool the liquid by a minimum of 20 degrees** below the saturation temperature. If the condenser is **operating in the designed minimum sub cooling range, the condenser is working efficiently**. At times, the sub-cooling temperatures may be higher than the designed sub-cooling range. This can happen because of lower medium temperatures and due to lower compressor loads and other factors. *Higher sub-cooling temperatures will add 1% to 2% efficiency to the metering device function.*

A **low sub-cooling or no sub-cooling** at all means that the condenser is having problems performing its three designed functions. Low sub-cooling or none at all affects the performance of the metering device and affects the evaporator's ability to absorb heat from the conditioned space.

The **amount of condenser sub-cooling** can be determined by converting system operating high-side pressure to saturation temperature based on system refrigerant, and then measuring the end-run temperature of the condenser. Subtracting the condenser liquid line temperature from the condensers saturation temperature will equal the amount of condenser sub-cooling.

EQUIPMENT REQUIRED

Manifold gage set
Temperature meter
Insulation tape
Permanent line tap vales (*if applicable*)
Pressure–temperature chart
Safety goggles
Safety gloves
Refrigeration wrench (*if applicable*)

SAFETY PRACTICES

The student should be knowledgeable in the use of tools and testing equipment. Follow all EPA rules and regulations when working with refrigerants.

COMPETENCY PROCEDURES   Checklist

> NOTE: *This competency can be performed on a central air conditioning unit with a high-efficiency air-cooled condenser that is charged properly.*

1. Turn the system OFF.   ❏
2. Make sure the manifold gage valves are closed.   ❏
3. Attach the low-side manifold gage hose to the system's low-side service valve port.   ❏
4. Attach the high-side manifold gage hose to the system's high-side service valve port.   ❏
5. Use the temperature meter and locate a temperature probe to measure outside ambient air temperature around the condenser.   ❏
6. Locate another temperature probe in the condenser's liquid line.   ❏
7. Turn the unit ON and allow the unit to operate for 5 to 10 minutes.   ❏

8. *After run time of 5 to 10 minutes, record the system's low-side operating pressure.* _____ ❏

9. *Use a pressure–temperature chart and convert low-side psig to saturation temperature.* _____ ❏

EVALUATING THE CONDENSER'S OPERATING HIGH SIDE PRESSURE

10. *Record the system's operating high-side pressure.* _____ ❏
11. *Record the condenser's entering air temperature.* _____ ❏
12. *Is the system's operating high-side pressure in design range based on the manufacturer's design condensing temperature of 15 to 25 degrees above ambient temperature for high efficiency air-cooled condensers?* _____ ❏

EVALUATING THE CONDENSER'S SATURATION TEMPERATURE RANGE

13. *Use a pressure–temperature chart and convert high-side psig to saturation temperature.* _____ ❏
14. *Is the condenser's saturation temperature in design range based on the manufacturer's design condensing temperature of 15 to 25 degrees above ambient temperature for high-efficiency air-cooled condensers?* _____ ❏

EVALUATING THE CONDENSER'S SUB-COOLING RANGE

15. *Record the condenser's liquid line temperature.* _____ ❏
16. *Determine condenser sub-cooling:* ❏

Condenser saturation temperature _____
MINUS
Condenser liquid line temperature _____
EQUALS Condenser sub-cooling _____

17. *Record condenser sub-cooling.* _____ ❏
18. *Is the condenser's sub-cooling temperature within design range as that which is specified by the manufacturer for high-efficiency air-cooled condensers?* _____ ❏

EVALUATING THE CONDENSER'S TEMPERATURE DIFFERENCE

19. *Record the condenser entering air temperature.* _____ ❏
20. *Determine entering air and saturation temperature difference using the formula:* ❏

Condensing temperature _____
MINUS
Entering air temperature _____
EQUALS Temperature difference _____

21. *Record the temperature difference.* _____ ❏
22. *Is the condenser operating at less than a 15- to 25-degree (F) temperature difference?* _____ ❏

NOTE: *If the condenser is operating at less than a 25-degree (F) temperature difference, it is probably operating efficiently. If the temperature difference is more than 25 degrees (F), the condenser is having trouble rejecting heat and may need to be cleaned or evaluated for other conditions, which could prevent it from rejecting heat in the proper range.*

---

23. Turn the unit OFF. ❏
24. Have your instructor check your work. ❏
25. Use proper procedures and remove manifold gages and temperature tester. ❏
26. Return all tools and test equipment to their proper location. ❏

RESEARCH QUESTIONS (CONDENSER PERFORMANCE EVALUATION RESULTS)

1. Was the condenser's operating high-side pressure within design range? _____ (*If not, explain what you think could be the reason(s) as to why the high-side pressure is not within design range.*)

   _____

   _____

2. Was the condenser's saturation temperature within the design range? _____ (*If not, explain what you think could be the reason(s) as to why the condensers saturation temperature was not within design range.*)

   _____

   _____

3. Was the condenser's sub-cooling within the design range?_____ (*If not, explain what you think could be the reason(s) as to why the condensers sub-cool temperature was not within design range.*)

   _____

   _____

4. Was the condenser's temperature difference within the design range? _____ (*If not, explain what you think could be the reason(s) as to why the entering air temperature and saturation temperature difference was not within the design range.*)

   _____

   _____

5. How would you rate this condenser's performance?

   _____

   _____

Passed Competency _____   **Failed Competency** _____

Instructor Signature _____   **Grade** _____

Practical Competency 100

Water-Cooled Condenser Performance Evaluation (Figure 3–269)

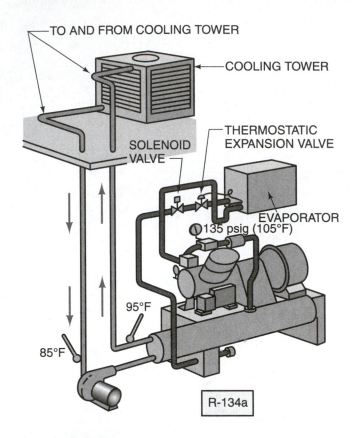

FIGURE 3–269 Water-cooled condenser performance under standard operating conditions.

SUGGESTED MATERIALS

Textbook
Refrigeration & Air Conditioning Technology, 5th Edition, Thomson Delmar Learning
Unit 22—Condensers
Unit 40—Typical Operating Conditions

Review Topics
Condensers; Water-Cooled Condensers; Documentation with the Unit

COMPETENCY OBJECTIVE

The student will be able to evaluate the performance of a water-cooled condenser.

OVERVIEW

The condenser of refrigeration sealed systems performs three functions in the refrigeration process. It rejects heat from the refrigerant vapor, condenses vapor to liquid, and sub-cools liquid refrigerant. (*Theory Lesson: Condensers, Figure 3–263.*)

How fast a condenser can perform these functions can be used to evaluate the condenser's performance and efficiency. During rejection of **latent heat** from the refrigerant vapor, the vapor reaches a point of saturation and starts to condense back to the liquid state. The temperature at which this process happens is determined by the pressure being exerted on the vapor.

The **saturation temperature** of a condenser can be determined by using the manufacturer's design condensing temperature range for the different type of condensers. In residential and light commercial air conditioning and heat pump applications, two types of condensers are used: air cooled and water cooled condensers.

Water-cooled condensers are designed to condense refrigerant vapor to liquid at 10 degrees above the leaving water temperature in summer applications (*Theory Lesson: Condensers, Figure 3–266*) and approximately 20 degrees above the leaving water temperature in winter applications. (*Theory Lesson: Condensers, Figure 3–267*.) The sub-cooling range for **water-cooled condensers** is **approximately 5 to 10 degrees** (F) below the saturation temperature.

Operating high-side pressure for water-cooled condensers can be determined by adding 10 degrees (F) for summer operation, or 20 degrees (F) for winter operation to the **condenser leaving water temperature**. This is the condenser's **saturation temperature**, which can be converted to high-side pressure by using a pressure–temperature chart for the type of refrigerant being used in the system.

EQUIPMENT REQUIRED

Manifold gage set
Temperature meter (*3 temperature probes*)
Insulation tape
Pressure–temperature chart
Safety goggles
Safety gloves
Refrigeration wrench (*if applicable*)

SAFETY PRACTICES

The student should be knowledgeable in the use of tools and testing equipment. Follow all EPA rules and regulations when working with refrigerants.

COMPETENCY PROCEDURES   Checklist

> NOTE: *This competency can be performed on any refrigeration equipment with a water-cooled condenser.*

1. Turn the system OFF. ❑
2. Make sure the manifold gage valves are closed. ❑
3. Attach the low-side manifold gage hose to the system's low-side service valve port. ❑
4. Attach the high-side manifold gage hose to the system's high-side service valve port. ❑
5. Use the temperature meter and locate a temperature probe ON or IN the entering condenser water line. ❑
6. Locate another temperature probe ON or IN the condenser's leaving water line. ❑
7. Attach a temperature probe to the condenser's liquid line. ❑
8. Turn the unit ON and allow unit to operate for 5 to 10 minutes. ❑
9. *Record the type of system refrigerant.* _____ ❑
10. *Is the system in summer or winter operation?* _____ ❑
11. *After a run time of 5 to 10 minutes, record the system's low-side operating pressure.* _____ ❑
12. *Use a pressure–temperature chart and convert low-side psig to saturation temperature.* _____ ❑

EVALUATING THE CONDENSER'S OPERATING HIGH-SIDE PRESSURE

13. *Record the system's operating high-side pressure.* _____ ❏
14. *Record the condenser's leaving-water temperature.* _____ ❏
15. *Is the system's operating high-side pressure in the design range based on the manufacturer's design condensing temperature for water-cooled condensers of 10 degrees for summer operation or 20 degrees for winter operation?* _____ ❏

EVALUATING THE CONDENSER'S SATURATION TEMPERATURE RANGE

16. *Use a pressure–temperature chart and convert high-side psig to saturation temperature.* _____ ❏
17. *Is the condenser's saturation temperature in the design range based on the manufacturer's design condensing temperature for water-cooled condensers of 10 degrees for summer operation or 20 degrees for winter operation?* _____ ❏

EVALUATING THE CONDENSER'S SUB-COOLING RANGE

18. *Record the condenser's liquid line temperature.* _____ ❏
19. *Determine condenser sub-cooling:* ❏

> *Condenser saturation temperature* _____
> **MINUS**
> *Condenser liquid line temperature* _____
> **EQUALS** *Condenser sub-cooling* _____

20. *Record the condenser sub-cooling.* _____ ❏
21. *Is the condenser's sub-cooling temperature within design range as that which is specified by the manufacturer for water-cooled condensers?* _____ ❏

EVALUATING THE CONDENSER'S TEMPERATURE DIFFERENCE

22. *Record the condenser's leaving-water temperature.* _____ ❏
23. *Determine the leaving-water and refrigerant saturation temperature difference using the formula:* ❏

> *Condensing saturation temperature* _____
> **MINUS**
> *Condenser leaving water temperature* _____
> **EQUALS** *Temperature difference* _____

24. *Record the temperature difference.* _____ ❏

NOTE: *The amount of temperature difference will be determined by seasonal operation (10-degree difference for summer operation and 20-degree difference for winter operation).*

---

25. *Is the condenser operating at less than a 10- or 20-degree (F) temperature difference based on seasonal operation?* _____ ❏

NOTE: *If the condenser is operating at less than a 10- or 20-degree (F) temperature difference based on seasonal operation, it is probably operating efficiently. If the temperature difference is more than then 10 or 20 degrees (F), based on seasonal operation, the condenser water is having trouble absorbing heat from the refrigerant and may need to be cleaned or evaluated for other conditions, which could prevent it from absorbing heat in the proper range.*

---

26. Turn the unit OFF. ❏
27. Have your instructor check your work. ❏
28. Use proper procedures and remove manifold gages and temperature tester. ❏
29. Return all tools and test equipment to their proper location. ❏

RESEARCH QUESTIONS (CONDENSER PERFORMANCE EVALUATION RESULTS)

1. Was the condenser's operating high-side pressure within the design range? _____ (*If not, explain what you think could be the reason(s) as to why the high-side pressure is not within the design range.*)

_____

_____

2. Was the condenser's saturation temperature within the design range? _____ (*If not, explain what you think could be the reason(s) as to why the condenser's saturation temperature was not within design range.*)

_____

_____

3. Was the condenser's sub-cooling within the design range?_____ (*If not, explain what you think could be the reason(s) as to why the condensers sub-cool temperature was not within the design range.*)

_____

_____

4. Was the condenser's temperature difference within the design range? _____ (*If not, explain what you think could be the reason(s) as to why the leaving water temperature and saturation temperature difference was not within the design range.*)

_____

_____

5. How would you rate this condenser's performance?

_____

_____

Passed Competency _____   **Failed Competency** _____

Instructor Signature _____   **Grade** _____

Practical Competency 101

Determining Correct Operating High-Side Pressure for Refrigeration Equipment with a Standard Air-Cooled Condenser (Figure 3–270)

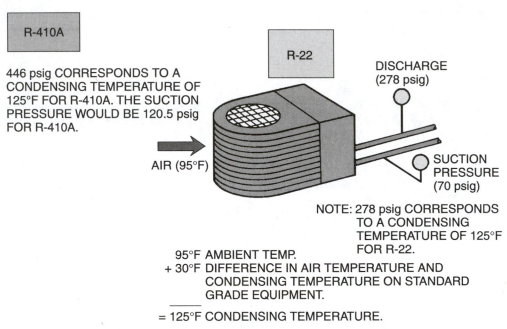

R-410A

446 psig CORRESPONDS TO A CONDENSING TEMPERATURE OF 125°F FOR R-410A. THE SUCTION PRESSURE WOULD BE 120.5 psig FOR R-410A.

AIR (95°F)

R-22

DISCHARGE (278 psig)

SUCTION PRESSURE (70 psig)

NOTE: 278 psig CORRESPONDS TO A CONDENSING TEMPERATURE OF 125°F FOR R-22.

95°F AMBIENT TEMP.
+ 30°F DIFFERENCE IN AIR TEMPERATURE AND CONDENSING TEMPERATURE ON STANDARD GRADE EQUIPMENT.
= 125°F CONDENSING TEMPERATURE.

FIGURE 3–270   Pressures and temperatures for standard air-cooled condenser.

SUGGESTED MATERIALS

Textbook
Refrigeration & Air Conditioning Technology, 5th Edition, Thomson Delmar Learning
Unit 22—Condensers
Unit 36—Refrigeration Applied to Air Conditioning
Unit 40—Typical Operating Conditions

Review Topics
Standard Air-Cooled Condensers; Typical Operating Conditions; Troubleshooting; Refrigeration Applied to Air Conditioning

COMPETENCY OBJECTIVE

The student will be able to determine the correct operating high-side for refrigeration equipment with a standard air-cooled condenser.

OVERVIEW

The condenser of refrigeration sealed systems performs three functions in the refrigeration process. It rejects heat from the refrigerant vapor, condenses vapor to liquid, and sub-cools liquid refrigerant. *(Theory Lesson: Condensers, Figure 3–263.)*

How fast a condenser can perform these functions can be used to evaluate the condenser's performance and efficiency. During rejection of **latent heat** from the refrigerant vapor, the vapor reaches a point of saturation and starts to condense back to the liquid state. The temperature at which this process happens is determined by the pressure being exerted on the vapor.

The **saturation temperature** of a condenser can be determined by using the manufacturer's design condensing temperature range for the different type of condensers. In residential and light commercial air conditioning and heat pump applications, two types of condensers are used: air cooled and water cooled condensers. There are two types of air-cooled condensers used, standard efficiency and high efficiency. Manufacturers have designed all **standard air-cooled condensers** to condense the refrigerant vapor to a liquid at a temperature of about 30 to 35 degrees above the outside ambient air temperature. *(Theory Lesson: Condensers, Figure 3–264.)*

New **high-efficiency air-cooled condensers** condense the refrigerant vapor back to a liquid at 15 to 25 degrees above the outside ambient air temperature based on the SEER rating of the equipment. The higher the SEER rating of the equipment, the lower the saturation temperature of the refrigerant. *(Theory Lesson: Condensers, Figure 3–265.)*

Water-cooled condensers are designed to condense refrigerant vapor to liquid at 10 degrees above the leaving water temperature in summer applications *(Theory Lesson: Condensers Figure 3–266)*, and approximately 20 degrees above the leaving water temperature in winter applications. *(Theory Lesson: Condensers Figure 3–267.)*

Knowing this information can assist technicians in evaluating a system's **high-side operating pressure and saturation temperature.** A system's **correct saturation temperature** can be determined by measuring the outside ambient air temperature and adding the manufacturer's designed condensing temperature to it, based on the type of condenser being used in the system. This tells the technician what the saturation temperature of the refrigerant should be for the condenser under current conditions. Using the saturation temperature can assist the technician in determining what the system's **high-side operating pressure** should be under current conditions. The technician would convert the saturation temperature to pressure for the type of system refrigerant being used. This would be what the system's operating head pressure should be under current conditions.

This information can then be used to evaluate actual system operating high-side pressure and saturation temperature. An actual system pressure and saturation temperature of **plus or minus 10%** would be acceptable. Any pressure and saturation temperature higher or lower than the allowable 10% range is an indication to the technician that the condenser is working too hard or below an acceptable range. Technicians should look for conditions that could cause the condenser to perform outside the determined acceptable range.

EQUIPMENT REQUIRED

Manifold gage set
Temperature meter
Pressure–temperature chart
Safety goggles
Safety gloves
Refrigeration wrench (*if applicable*)

SAFETY PRACTICES

The student should be knowledgeable in the use of tools and testing equipment. Follow all EPA rules and regulations when working with refrigerants.

COMPETENCY PROCEDURES

> **NOTE:** *This competency is to be performed on an air conditioning or refrigeration equipment with a standard air-cooled condenser.*

1. Turn the system ON and allow the system to operate for 10 minutes. ❑

DETERMINING CORRECT CONDENSER SATURATION TEMPERATURE
2. *Record the type of refrigerant used in system. _____* ❑
3. *Measure and record the air temperature entering the condenser. _____* ❑
4. *Determine what you think the maximum condenser temperature should be by using the following formula:* ❑

Condenser entering air temperature _____ + 30 degrees = saturation temperature _____
Condenser entering air temperature _____ + 35 degrees = saturation temperature _____

> **NOTE:** *Remember that standard air-cooled condensers are designed to condense the refrigerant vapor to a liquid at about 30 to 35 degrees above the outside ambient air temperature.*

5. *Record the approximate saturation temperature range from Step 4.* ❑
   Saturation temperature range _____ degrees to _____ degrees.

DETERMINING CORRECT OPERATING HIGH-SIDE PRESSURE RANGE
6. *Use the pressure–temperature chart and convert the outside air temperature (plus) 30 degrees to the corresponding refrigerant pressure. _____ psig* ❑
7. *Use the pressure–temperature chart and convert the outside air temperature (plus) 35 degrees to the corresponding refrigerant pressure. _____ psig* ❑
8. *Record the approximate operating head pressure ranges steps 6 and 7.* ❑
   Operating high-side pressure range _____ psig to _____ psig

> **NOTE:** *You have just determined what the approximate operating high-side pressure should be for the system you are checking.*

The following procedures will be used to see if you were correct and if the system's operating high-side pressure is correct.

EVALUATING ACTUAL SYSTEM CONDITIONS
9. Make sure the manifold gage valves are closed. ❑
10. Attach the high-side manifold gage hose to the system's high-side service valve port. ❑
11. Mid-seat the high-side valve (*if applicable*). ❑
12. *Record the system's actual operating high-side pressure. _____* ❑
13. *Is the system's operating high-side pressure within operating range as that determined in Step 8? _____* ❑
14. *Record actual condensing temperature by converting actual high-side pressure to temperature based on type of system refrigerant. _____* ❑
15. *Was the refrigerant condensing temperature in the correct range as determined in Step 5? _____* ❑

DETERMINING AIR TEMPERATURE TO CONDENSER SATURATION TEMPERATURE
16. *Record the actual condensing temperature as recorded in Step 14. _____* ❑
17. *Record the air temperature entering the condenser as recorded in Step 3. _____* ❑

18. *Determine the temperature difference by using the following formula:* ❑
   Condenser condensing temperature _____
   MINUS
   Condenser entering air temperature _____
   EQUALS *temperature difference* _____ *degrees* (F)

NOTE: *The difference from the saturation temperature and the entering air temperature should be in the range of 30 to 35 degrees. If the temperature difference is in the correct range, the operating head pressure is correct for the system under these conditions.*

   If the temperature difference is higher than 35 degrees, this is an indication that the systems high-side pressure and saturation temperature are out of range for the conditions listed. The condenser is working harder than it should be and technicians should look for conditions that could be creating a higher high-side pressure and saturation temperature—such things as a dirty condenser, higher than normal outside air temperature, restricted airflow, broken belt, burnout fan motor, etc. This situation should be corrected. If corrections are not made to bring high-side pressures into proper range, the system's efficiency will suffer, let alone cause possible damage to systems compressor.

---

19. *Were you correct in determining the correct operating high-side pressure for the system being checked?* _____ ❑
20. Have your instructor check your work. ❑
21. Disconnect equipment and return all tools and test equipment to their proper location. ❑

PERFORMANCE EVALUATION

1. Was the system's operating high-side pressure correct for the conditions? _____ *(If not, explain what you think could be the cause(s) as to the high-side pressure being incorrect.)*

   _____

   _____

2. Was the condenser saturation temperature correct for the conditions? *(If not, explain what you think could be the cause(s) as to the condensing temperature being incorrect.)*

   _____

   _____

3. Was the temperature difference between the condensing temperature and entering air temperature correct? _____ *(If not, explain what you think could be the cause(s) as to the temperature difference being incorrect.)*

   _____

   _____

4. Why does the low-side of the system affect the high side of the system?

   _____

   _____

5. Besides what was stated above, what else could cause the head pressure on system to be high?

   _____

   _____

Passed Competency _____ **Failed Competency** _____

Instructor Signature _____ **Grade** _____

Practical Competency 102

Determining the Correct High-Side Pressure of Air Conditioning or Heat Pump Equipment with a High-Efficiency Air-Cooled Condenser

SUGGESTED MATERIALS

Textbook
Refrigeration & Air Conditioning Technology, 5th Edition, Thomson Delmar Learning
Unit 22—Condensers
Unit 36—Refrigeration Applied to Air Conditioning
Unit 40—Typical Operating Conditions

Review Topics
High-Efficiency Air-Cooled Condensers; Typical Operating Conditions; Troubleshooting; Refrigeration Applied to Air Conditioning

COMPETENCY OBJECTIVE

The student will be able to determine the correct operating high-side air conditioning or heat pump equipment with a high-efficiency air-cooled condenser.

OVERVIEW

The condenser of refrigeration sealed systems performs three functions in the refrigeration process. It rejects heat from the refrigerant vapor, condenses vapor to liquid, and sub-cools liquid refrigerant. *(Theory Lesson: Condensers, Figure 3–263.)*

How fast a condenser can perform these functions can be used to evaluate the condenser's performance and efficiency. During rejection of **latent heat** from the refrigerant vapor, the vapor reaches a point of saturation and starts to condense back to the liquid state. The pressure being exerted on the vapor determines the temperature at which this process happens.

The **saturation temperature** of a condenser can be determined by using the manufacturer's design condensing temperature range for the different type of condensers. In residential and light commercial air conditioning and heat pump applications, two types of condensers are used: air cooled and water cooled condensers. There are two types of air-cooled condensers used, standard efficiency and high efficiency. Manufacturers have designed all **standard air-cooled condensers** to condense the refrigerant vapor to a liquid at a temperature of about 30 to 35 degrees above the outside ambient air temperature. *(Theory Lesson: Condensers, Figure 3–264.)*

New **high-efficiency air-cooled condensers** condense the refrigerant vapor back to a liquid at 15 to 25 degrees above the outside ambient air temperature based on the SEER rating of the equipment. The higher the SEER rating of the equipment, the lower the saturation temperature of the refrigerant. *(Theory Lesson: Condensers, Figure 3–265.)*

Water-cooled condensers are designed to condense refrigerant vapor to liquid at 10 degrees above the leaving water temperature in summer applications *(Theory Lesson: Condensers, Figure 3–266)* and approximately 20 degrees above the leaving water temperature in winter applications *(Theory Lesson: Condensers Figure 3–267).*

Knowing this information can assist technicians in evaluating a system's **high-side operating pressure** and **saturation temperature**. Determining a system's **correct saturation temperature** can be found by measuring the outside ambient air temperature and adding the manufacturer's designed condensing temperature to it, based on the type of condenser being used in the system. This tells the technician what the saturation temperature of the refrigerant should be for the condenser under current conditions. Using the saturation temperature can assist the technician in determining what the system's **high-side operating pressure** should be under current conditions. The technician would convert the saturation temperature to pressure for the type of system refrigerant being used. This would be what the system's operating head pressure should be under current conditions.

This information can then be used to evaluate actual system operating high-side pressure and saturation temperature. An actual system pressure and saturation temperature of **plus or minus 10%** would be acceptable. Any pressure and saturation temperature higher or lower than the allowable 10% range is an indication to the technician that the condenser is working too hard or below an acceptable range. Technicians should look for conditions that could cause the condenser to perform outside the determined acceptable range.

EQUIPMENT REQUIRED

Manifold gage set
Temperature meter
Pressure–temperature chart
Safety goggles
Safety gloves
Refrigeration wrench (*if applicable*)

SAFETY PRACTICES

The student should be knowledgeable in the use of tools and testing equipment. Follow all EPA rules and regulations when working with refrigerants.

COMPETENCY PROCEDURES   Checklist

> NOTE: *This competency is to be performed on an air conditioning or heat pump equipment with a high-efficiency air-cooled condenser.*

> NOTE: *Heat pumps should be operating in the cooling mode if used for competency.*

1. Turn the system ON and allow the system to operate for 10 minutes.   ❏

DETERMINING CORRECT CONDENSER SATURATION TEMPERATURE
2. *Record the type of refrigerant used in the system.* _____   ❏
3. *Measure and record the air temperature entering the condenser.* _____   ❏
4. Determine what you think the maximum condenser temperature should be by using the
   following formula: ❏
   Condenser entering air temperature _____ + 15 degrees = saturation temperature _____
   Condenser entering air temperature _____ + 25 degrees = saturation temperature _____

> NOTE: *Remember that high-efficiency air-cooled condensers are designed to condense the refrigerant vapor to a liquid at about 15 to 25 degrees above the outside ambient air temperature.*

5. Record the approximate saturation temperature range from Step 4. ❑
   Saturation temperature range
   _____ degrees to _____ degrees

DETERMINING CORRECT OPERATING HIGH-SIDE PRESSURE RANGE

6. Use the pressure–temperature chart and convert the outside air temperature (*plus*)
   15 degrees to the corresponding refrigerant pressure. _____ psig ❑
7. Use the pressure–temperature chart and convert the outside air temperature (*plus*)
   25 degrees to the corresponding refrigerant pressure. _____ psig ❑
8. Record the approximate operating head pressure ranges from Steps 6 and 7. ❑
   Operating high-side pressure range
   _____ psig to _____ psig

> **NOTE:** You have just determined what the approximate operating high-side pressure should be for the system you are checking.

The following procedures will be used to see if you were correct and if the system's operating high-side pressure is correct.

EVALUATING ACTUAL SYSTEMS CONDITIONS

9. Make sure the manifold gage valves are closed. ❑
10. Attach the high-side manifold gage hose to the system's high-side service valve port. ❑
11. Mid-seat the high-side valve (*if applicable*). ❑
12. Record the system's actual operating high-side pressure. _____ ❑
13. Is the system's operating high-side pressure within operating range as that determined in
    Step 8? _____ ❑
14. Record the actual condensing temperature by converting actual high-side pressure to
    temperature based on type of system refrigerant. _____ ❑
15. Was refrigerant condensing temperature in the correct range as determined in
    Step 5? _____ ❑

DETERMINING AIR TEMPERATURE TO CONDENSER SATURATION TEMPERATURE

16. Record the actual condensing temperature as recorded in Step 14. _____ ❑
17. Record the air temperature entering the condenser as recorded in Step 3. _____ ❑
18. Determine the temperature difference by using the following formula: ❑
    Condenser condensing temperature _____
    MINUS
    Condenser entering air temperature _____
    EQUALS Temperature difference _____ degrees (F)

> **NOTE:** The difference from the saturation temperature and the entering air temperature should be in the range of 15 to 25 degrees. If the temperature difference is in the correct range, the operating head pressure is correct for the system under these conditions.
>
> If the temperature difference is higher than 25 degrees, this is an indication that the system's high-side pressure and saturation temperature are out of range for the conditions listed. The condenser is working harder than it should be and technicians should look for conditions that could be creating a higher high-side pressure and saturation temperature—such things as a dirty condenser, higher than normal outside air temperature, restricted airflow, broken belt, burnout fan motor, etc.
>
> This situation should be corrected. If corrections are not made to bring high-side pressures into proper range, the systems efficiency will suffer, let alone cause possible damage to systems compressor.

19. *Were you correct in determining the correct operating high-side pressure for the system being checked?* _____ ❑
20. Have your instructor check your work. ❑
21. Disconnect equipment and return all tools and test equipment to their proper location. ❑

PERFORMANCE EVALUATION

1. Was the system's operating high-side pressure correct for the conditions? _____ *(If not, explain what you think could be the cause(s) as to the high-side pressure being incorrect.)*

   _____

   _____

2. Was the condenser saturation temperature correct for the conditions? _____ *(If not, explain what you think could be the cause(s) as to the condensing temperature being incorrect.)*

   _____

   _____

3. Was the temperature difference between the condensing temperature and entering air temperature correct? _____ *(If not, explain what you think could be the cause(s) as to the temperature difference being incorrect.)*

   _____

   _____

4. What makes a high-efficiency air-cooled condenser more efficient than a standard air-cooled condenser?

   _____

   _____

5. What does the Condenser SEER rating mean?

   _____

   _____

   Passed Competency _____ Failed Competency _____

   Instructor Signature _____ Grade _____

Practical Competency 103

Determining the Correct Operating High-Side Pressure for Equipment with a Water-Cooled Condenser

SUGGESTED MATERIALS

Textbook
Refrigeration & Air Conditioning Technology, 5th Edition, Thomson Delmar Learning
Unit 22—Condensers
Unit 40—Typical Operating Conditions

Review Topics
Condensers; Water-Cooled Condensers; Documentation with the Unit

COMPETENCY OBJECTIVE

The student will be able to determine the correct high-side operating pressure for refrigeration equipment with a water-cooled condenser.

OVERVIEW

The condenser of refrigeration sealed systems performs three functions in the refrigeration process. It rejects heat from the refrigerant vapor, condenses vapor to liquid, and sub-cools liquid refrigerant. (*Theory Lesson: Condensers, Figure 3–263.*)

How fast a condenser can perform these functions can be used to evaluate the condenser's performance and efficiency. During rejection of **latent heat** from the refrigerant vapor, the vapor reaches a point of saturation and starts to condense back to the liquid state. The pressure being exerted on the vapor determines the temperature at which this process happens.

The **saturation temperature** of a condenser can be determined by using the manufacturer's design condensing temperature range for the different type of condensers. In residential and light commercial air conditioning and heat pump applications, two types of condensers are used: air cooled and water-cooled condensers.

Water-cooled condensers are designed to condense refrigerant vapor to liquid at 10 degrees above the leaving water temperature in summer applications (*Theory Lesson: Condensers, Figure 3–266*), and approximately 20 degrees above the leaving water temperature in winter applications. (*Theory Lesson: Condensers Figure 3–267.*) The sub-cooling range for **water-cooled condensers** is **approximately 5 to 10 degrees** (F) below the saturation temperature.

Operating high-side pressure for water-cooled condensers can be determined by adding 10 degrees (F) for summer operation or 20 degrees (F) for winter operation to the **condenser leaving-water temperature**. This is the condensers **saturation temperature**, which can be converted to high-side pressure by using a pressure temperature chart for the type of refrigerant being used in the system.

EQUIPMENT REQUIRED

Manifold gage set
Temperature meter

Insulation tape
Pressure–temperature chart
Safety goggles
Safety gloves
Refrigeration wrench (*if applicable*)

SAFETY PRACTICES

The student should be knowledgeable in the use of tools and testing equipment. Follow all EPA rules and regulations when working with refrigerants.

COMPETENCY PROCEDURES

Checklist

> NOTE: *This competency can be performed on any refrigeration equipment with a water-cooled condenser.*

1. Turn the system OFF. ❑
2. Locate another temperature probe ON or IN the condensers leaving water line. ❑
3. Turn unit ON and allow unit to operate for 5 to 10 minutes. ❑
4. *Record the type of system refrigerant.* _____ ❑
5. *Is system in summer or winter operation?* _____ ❑

DETERMINING CORRECT CONDENSER SATURATION TEMPERATURE

6. *Record the leaving water temperature of the condenser.* _____ ❑
7. Determine what you think the maximum condenser temperature should be by using the following formula: ❑

Condenser leaving water temperature _____ *+ 10 degrees = saturation temperature* _____
Condenser leaving water temperature _____ *+ 20 degrees = saturation temperature* _____

> NOTE: *Remember that water-cooled condensers are designed to condense the refrigerant vapor to a liquid at about 10 degrees during summer and 20 degrees during winter operation above the condenser leaving water temperature.*

8. *Record the approximate saturation temperature range from Step 7.* ❑
   Saturation temperature range
   _____ *degrees to* _____ *degrees*

DETERMINING CORRECT OPERATING HIGH-SIDE PRESSURE RANGE

9. *Use the pressure–temperature chart and convert the leaving-water temperature (plus) 10 degrees to the corresponding refrigerant pressure.* _____ *psig* ❑
10. *Use the pressure–temperature chart and convert the leaving water temperature (plus) 20 degrees to the corresponding refrigerant pressure.* _____ *psig* ❑
11. *Record the approximate operating high-side pressure ranges from Steps 9 and 10.*
    Operating high-side pressure range
    _____ *psig to* _____ *psig* ❑

> NOTE: *You have just determined what the approximate operating high-side pressure should be for the system you are checking.*

The following procedures will be used to see if you were correct and if the system's operating high-side pressure is correct.

EVALUATING ACTUAL SYSTEM CONDITIONS

12. Make sure the manifold gage valves are closed. ❏
13. Attach the high-side manifold gage hose to the system's high-side service valve port. ❏
14. Mid-seat the high-side valve (*if applicable*). ❏
15. *Record the system's actual operating high-side pressure.* _____ ❏
16. *Is the system's operating high-side pressure within operating range as that determined in Step 11?* _____ ❏
17. *Record actual condensing temperature by converting actual high-side pressure to temperature based on type of system refrigerant.* _____ ❏
18. *Was refrigerant condensing temperature in the correct range as determined in Step 8?* _____ ❏

EVALUATING THE CONDENSER'S TEMPERATURE DIFFERENCE

19. *Record the condenser's leaving-water temperature.* _____ ❏
20. *Determine the leaving water and refrigerant saturation temperature difference using the formula:* ❏

Condensing saturation temperature _____

MINUS

Condenser leaving water temperature _____

EQUALS Temperature difference _____

21. *Record the temperature difference.* _____ ❏

NOTE: *The amount of temperature difference will be determined by seasonal operation (10-degree difference for summer operation and 20-degree difference for winter operation).*

---

22. *Is the condenser operating at less than a 10- or 20-degree (F) temperature difference based on seasonal operation?* _____ ❏

NOTE: *If the condenser is operating at less than a 10- or 20-degree (F) temperature difference based on seasonal operation, it is probably operating efficiently. If the temperature difference is more than than 10 or 20 degrees (F), based on seasonal operation, the condenser water is having trouble absorbing heat from the refrigerant and may need to be cleaned or evaluated for other conditions, which could prevent it from absorbing heat in the proper range.*

---

23. Turn unit OFF. ❏
24. Have your instructor check your work. ❏
25. Use proper procedures and remove manifold gages and temperature tester. ❏
26. Return all tools and test equipment to their proper location. ❏

PERFORMANCE EVALUATION

1. Was the system's operating high-side pressure correct for the conditions? _____ (*If not, explain what you think could be the cause(s) as to the high-side pressure being incorrect.*)

_____

_____

2. Was the condenser saturation temperature correct for the conditions? _____ (*If not, explain what you think could be the cause(s) as to the condensing temperature being incorrect.*)

_____

_____

3. Was the temperature difference between the condensing temperature and leaving water temperature correct? _____ (*If not, explain what you think could be the cause(s) as to the temperature difference being incorrect.*)

_____

_____

4. Would the water flow rate have to be increased or decreased to lower the high-side pressure of a system with a water-cooled condenser?

_____

_____

5. List two types of water-cooled condensers.

_____

_____

Passed Competency _____   **Failed Competency** _____

Instructor Signature _____   **Grade** _____

Practical Competency 104

Testing an Air Conditioner for Low Charge Before Installing Manifold Gages (Figure 3–271)

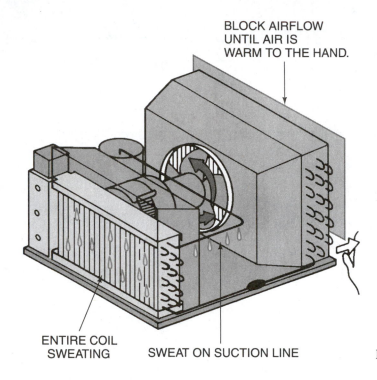

BLOCK AIRFLOW
UNTIL AIR IS
WARM TO THE HAND.

ENTIRE COIL
SWEATING

SWEAT ON SUCTION LINE

FIGURE 3–271   Testing for a low charge.

SUGGESTED MATERIALS

Textbook
Refrigeration & Air Conditioning Technology, 5th Edition, Thomson Delmar Learning
Unit 47—Room Air Conditioners

Review Topics
Checking Unit Charge; Maintaining and Servicing Room Units

COMPETENCY OBJECTIVE

The student will be able to evaluate a window air conditioner charge condition before determining to install manifold gages.

OVERVIEW

Window air conditioners have a fixed-bore metering device, normally a capillary tube, and have a critical charge for the system to perform efficiently. For this reason, if a low charge is suspected, technicians should perform the following test to confirm suspicions before adding line tap valves and gages to the system. This test can be performed in a shop or home area. While performing the test, the condenser airflow may have to be restricted to build system head pressure because the evaporator and condenser will be working in the same temperature area.

EQUIPMENT REQUIRED

Temperature tester
Clamp-on ammeter
Something to restrict condenser airflow if need be
Pieces of cardboard

SAFETY PRACTICES

The student should be knowledgeable in the use of tools and testing equipment. Follow all EPA rules and regulations when working with refrigerants.

COMPETENCY PROCEDURES   Checklist

1. Turn the unit OFF.   ❏
2. Remove the unit from its case.   ❏
3. Check to see that air flows through by the evaporator and condenser.   ❏

NOTE: *Cardboard may be positioned over places where panels force air to flow through the coils.*

---

4. Start the unit on high cool.   ❏
5. Let the unit operate for about 5 minutes.   ❏
6. After 5 minutes of run time, check the sweat on the suction line; the sweat line should be close to the compressor.   ❏
7. *Was the sweat line close to the compressor?* _____   ❏
8. Feel the evaporator from bottom to top, it should be cold from the bottom of the coil to the top of the coil.   ❏
9. *Was the evaporator coil cold from bottom to top?* _____   ❏
10. Place a temperature probe in the condenser's leaving air stream.   ❏
11. Block a portion of the condenser coil until the condenser leaving air temperature is approximately 110 degrees (F).   ❏
12. Let the system operate for another 5 minutes with a discharge air temperature of approximately 110 degrees (F).   ❏
13. If the unit is charged properly, the suction sweat line should move to the compressor.   ❏
14. Did the suction sweat line move to the compressor? _____   ❏

NOTE: *If the humidity is too low for the line to sweat, the suction line should be cold. The evaporator should be cold from bottom to top. If any part of the evaporator is cold (frost may form), the evaporator is starved, which could be due to a restriction or a low charged unit.*

---

15. Look at the unit nameplate and record the unit's full-load amperage or RLA amperage. _____   ❏
16. Clamp an ammeter around the compressor common wire.   ❏
17. Record the unit's actual full-load amperage. _____   ❏

NOTE: *If the compressor is not pumping to capacity and the charge is correct, the evaporator coil will not be cold anywhere.*
*   If the full-load or RLA amperage is very low, the compressor may be pumping on only one or two cylinders.*

---

18. *Was the unit's operating amperage in line with the unit's rated full-load amperage or RLA rating?* _____ ❏

CONCLUSION: *In this test procedure, if any of the following cannot be reached, the unit is probably low on charge or has a low-capacity compressor, and gages would have to be installed to check system charge.*

    A. Frost will not form on suction line all the way back to the compressor.
    B. Suction line does not feel cold to the touch.
    C. Evaporator is not cold from bottom to top.
    D. Compressor is pulling well below full-load amperage.

19. Have your instructor check your work. ❏
20. Return all test equipment to its proper location. ❏

RESEARCH QUESTIONS

1. Where does the sweat come from on the suction line?

_____

_____

2. A typical room air-conditioning unit has _____ (one or two) fan motors.

3. Why should gages be installed only after there is evidence that they are needed?

_____

_____

4. In a normally operating window unit, the suction pressure should be approximately 65 psig, which converts to _____ degrees (F).

5. What are two methods used to evaporate condensate from the evaporator?

_____

_____

Passed Competency _____   Failed Competency _____

Instructor Signature _____   Grade _____

Practical Competency 105

Using the Sweat Line Method to Adjust the Charge on a Window Air Conditioner (Figure 3–272)

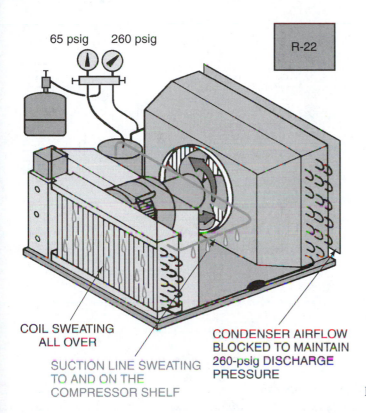

65 psig   260 psig

R-22

COIL SWEATING
ALL OVER

SUCTION LINE SWEATING
TO AND ON THE
COMPRESSOR SHELF

CONDENSER AIRFLOW
BLOCKED TO MAINTAIN
260-psig DISCHARGE
PRESSURE

FIGURE 3–272   Charging a unit by sweat line.

SUGGESTED MATERIALS

Textbook
Refrigeration & Air Conditioning Technology, 5th Edition, Thomson Delmar Learning
Unit 47—Room Air Conditioners

Review Topics
Charge a Unit by Sweat Line; Maintaining and Servicing Room Units

COMPETENCY OBJECTIVE

The student will be able to use the sweat line method for charging a window air conditioner.

OVERVIEW

Charging a window air conditioning unit by the sweat line method is not the most accurate method used for obtaining the exact refrigeration charge in a system but can be used to adjust a system charge rather than starting with a complete new charge. This method of charging will get the refrigerant charge as close

561

to the manufacturer's required charge as possible, although this method of charging does not take into account the condition of the actual air that the evaporator of the air conditioner is dealing with. Sweat line method of charging only approximates the load condition of the air being conditioned by the air conditioner. Depending on the condition of the air at the time this method of charging is used, the air conditioner could possibly be overcharged or undercharged.

EQUIPMENT REQUIRED

Manifold gage set
Permanent access valves (*if applicable*)
Refrigerant cylinder (*type used in unit*)
Clamp-on ammeter
Pieces of cardboard
Safety gloves
Safety goggles

SAFETY PRACTICES

The student should be knowledgeable in the use of tools and testing equipment. Follow all EPA rules and regulations when working with refrigerants.

COMPETENCY PROCEDURES

Checklist

> *NOTE: This competency can be completed on any packaged air conditioning unit.*

> *NOTE: If the system is already charged correctly, use a recovery cylinder and remove part of the charge so that the following procedures can be used to train the student on charging the system using the sweat line method.*

1. Turn OFF the power to the unit. ❑
2. Remove the air conditioning unit from the case (*if applicable*). ❑
3. Locate the high-side and low-side lines of the system (*if applicable*). ❑
4. Install permanent refrigerant access valve on the low-side line (*if applicable*). ❑
5. Install permanent refrigerant access valve on the high-side line (*if applicable*). ❑
6. Make sure the manifold gage valves are closed. ❑
7. Connect the high-side manifold gage hose to the refrigeration system's high-side service valve. ❑
8. Connect the low-side manifold gage hose to the refrigeration system's low-side service valve. ❑
9. Purge both the high- and low-side hose connections at the manifold gage bar by loosening the connection for a second and then closing. ❑
10. Set the refrigerant cylinder in the upright position and open the refrigerant cylinder valve. ❑
11. Purge the air from the center hose by loosening the center hose connection at the manifold gage bar for a second and then closing. ❑
12. Make sure that air is passing through the evaporator and the condenser coils; cardboard may be used for temporary panels. ❑
13. Start the unit on high cool. ❑

> *NOTE: Watch the unit's low-side pressure. Do not let the operating suction pressure go into a vacuum; add refrigerant through the suction gage if the pressure falls too low.*

14. *Record the low-side operating pressure.* _____ ❏
15. Adjust the airflow across the condenser until the operating high-side pressure is 260 psig for refrigerant R-22 systems. This is equal to a condensing temperature of 120 degrees (F). ❏

> **NOTE:** *Refrigerant may have to be added to obtain a high-side pressure of 260 psig.*

16. *Record the high-side operating pressure.* _____ ❏
17. Let the system operate for another 10 minutes. ❏
18. *After 10 minutes of operation, record the low-side operating pressure.* _____ ❏
19. *After 10 minutes of operation, record the high-side operating pressure.* _____ ❏
20. Notice the last area on the suction line where it is sweating. ❏
21. Is the suction line sweating back to the compressor? _____ (*If it is, the unit charge is OK; if it isn't, proceed.*) ❏

> **NOTE:** *If the suction line is not sweating back to the compressor, add refrigerant vapor through the suction line at intervals with the high-side (head pressure) at 260 psig until the suction line is sweating back to the compressor.*

> **NOTE:** *As refrigerant is being added to the system, the head pressure will rise; the condenser airflow will have to be adjusted over the condenser to maintain a head pressure of 260 psig.*

22. Open the low-side manifold gage and add refrigerant vapor to the suction line in intervals, added until the suction line is sweating back to the compressor. ❏

> **NOTE:** *Adjust condenser airflow to maintain a high-side pressure of 260 psig while adding refrigerant to system.*

23. Once the suction line is sweating back to the compressor, let the system operate for 5 minutes. ❏
24. *Record the low-side operating pressure.* _____ ❏

> **NOTE:** *For the system to be charged close to the manufacturer's charge, the low-side pressure should be about 65 psig for an R-22 system or an evaporator temperature of 40 degrees.*

25. *Was the suction pressure inline with 65 psig?* _____ ❏

> **NOTE:** *If suction pressure is not at or near 65 psig, check the suction line sweating again. It should be sweating back to the compressor if the high-side pressure is maintained at 260 psig. If sweat line does not stay at the compressor, continue to charge until the sweat line stays at the compressor.*

26. *Record the high-side operating pressure.* _____ ❏
27. *Once the suction line is sweating back to the compressor, check the unit nameplate and record the units full-load or RLA amperage.* _____ ❏
28. *Clamp the ammeter around the compressor common wire.* ❏
29. *Record the unit's actual operating amperage.* _____ ❏

> **NOTE:** *If the low-side pressure is about 65 psig, the compressor should be operating at or near the units rated full-load or RLA amperage.*

30. *Is the unit operating at or near the manufacturer's rated full-load or RLA amperage?* _____ ❑
31. Does the evaporator feel cold from bottom to top? _____ ❑
32. Have your instructor check your work. ❑
33. Make any adjustments as directed by your instructor. ❑
34. Disconnect equipment and return all tools and test equipment to their proper location. ❑
35. Install air conditioning back in case (*if applicable*). ❑

RESEARCH QUESTIONS

1. Where does the sweat on the suction line come from?

   _____

   _____

2. Why let the system run for 5 minutes after adding refrigerant to the unit?

   _____

   _____

3. If the airflow was not correct going across the evaporator coil during this charging procedure, what could happen to the charge on the system?

   _____

   _____

4. What creates the pressure on the low side of the system?

   _____

   _____

5. Does ice frozen on the suction line or side of the compressor indicate that the system has flood-back to the compressor?

   _____

   _____

Passed Competency _____ Failed Competency _____

Instructor Signature _____ Grade _____

Student Name _____ Grade _____ Date _____

Practical Competency 106

Using Air Temperature Measurements to Check the Charge on a Window Air Conditioner (Figure 3–273)

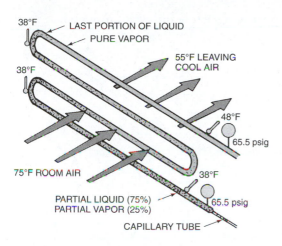

38°F — LAST PORTION OF LIQUID
— PURE VAPOR

55°F LEAVING COOL AIR

38°F

48°F

65.5 psig

75°F ROOM AIR

38°F

65.5 psig

PARTIAL LIQUID (75%)
PARTIAL VAPOR (25%)

CAPILLARY TUBE

FIGURE 3–273 Typical evaporator with pressure and temperatures.

SUGGESTED MATERIALS

Textbook

Refrigeration & Air Conditioning Technology, 5th Edition, Thomson Delmar Learning
Unit 47—Room Air Conditioners

Review Topics

The Refrigeration Cycle; Cooling; Maintaining and Servicing Room Units

COMPETENCY OBJECTIVE

The student will be able to use air temperature measurements to determine the charge of a window air conditioning unit, along with an ammeter to confirm the condition of the charge.

OVERVIEW

Air conditioning is the process of conditioning the air from the conditioned space by removing moisture from it and dropping the sensible heat temperature of the air. This is accomplished by passing the conditioned space air across the evaporator coil of the refrigeration system. The condition of the air after it has passed the evaporator coil can also be used to check the efficiency and charge of the air conditioning system.

EQUIPMENT REQUIRED

1 Temperature tester (two probes)
1 Clamp-on ammeter

SAFETY PRACTICES

The student should be knowledgeable in the use of tools and testing equipment. Follow all EPA rules and regulations when working with refrigerants.

> **NOTE:** *This competency can be completed on any packaged unit air conditioning unit.*

1. Remove the air conditioning unit from its case (*if applicable*). ❑
2. Make sure that there is correct airflow across each coil of the air conditioner. ❑

> **NOTE:** *This may require the placement of cardboard around areas where the air conditioner case would have covered.*

3. Place a temperature probe in the air stream of the air going across the evaporator coil. ❑
4. Place a temperature probe in the supply air stream coming off the evaporator. ❑
5. Turn ON the power to the unit and let it operate for 10 minutes. ❑
6. *How much of the evaporator coil is sweating? (percentage)* _____ ❑
7. *How much of the suction line is sweating?* _____ ❑

> **NOTE:** *If the suction line is not sweating back to the compressor, block off two thirds of the condenser and allow the unit to run for 10 minutes (**Figure 3–274**).*

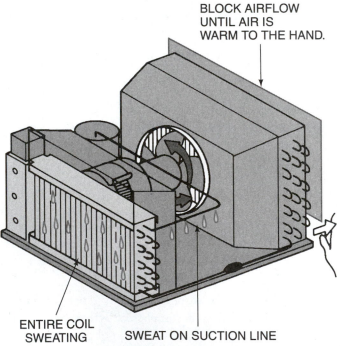

BLOCK AIRFLOW
UNTIL AIR IS
WARM TO THE HAND.

ENTIRE COIL
SWEATING

SWEAT ON SUCTION LINE

FIGURE 3–274 Restricting condenser airflow.

8. *Record the air temperature entering the evaporator coil.* _____ ❑
9. *Record the air temperature of the air leaving the evaporator coil.* _____ ❑
10. *Determine the temperature drop across the coil by using the following formula:* ❑
    Entering air temperature _____
    MINUS
    Leaving air temperature _____
    EQUALS *Temperature drop* _____
11. *Record the amount of temperature drop.* _____ ❑

12. Was the temperature drop within the 20-degree (F) range? _____ ❑
13. Check the unit's nameplate and record the rated full-load amperage or RLA. _____ ❑
14. Place the clamp-on ammeter around the common compressor motor lead. ❑
15. Record the unit's operating amperage. _____ ❑
16. Was the operating amperage higher or lower than the manufacturer's rated full-load or RLA amperage? _____ ❑

17. Turn OFF the power to the unit. ❑
18. Remove temperature probes and replace unit back into case (if applicable). ❑
19. Have your instructor check your work. ❑
20. Disconnect equipment and return all tools and test equipment to their proper location. ❑

RESEARCH QUESTIONS

1. What could cause the system's full-load amperage or RLA to be higher than normal?

   _____

   _____

2. What could cause the system's full-load amperage or RLA to be lower than normal?

   _____

   _____

3. What is the normal air temperature drop on an air conditioner that is operating in design load conditions?

   _____

   _____

4. How many degrees below the load are air-conditioning evaporator coils designed to operate?

   _____

   _____

5. What type of refrigeration is air conditioning?

   High Temperature—Medium Temperature—Low Temperature

   _____

   _____

Passed Competency _____   Failed Competency _____

Instructor Signature _____   Grade _____

4 HEAT PUMPS

Theory Lesson: Heat Pump Theory of Operation (Figure 4–1)

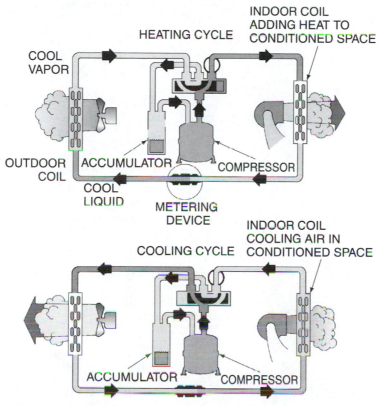

FIGURE 4–1   The heat pump refrigeration cycle shows direction of the refrigerant gas flow. (*Courtesy of Carrier Corporation*)

SUGGESTED MATERIALS

Textbook
Refrigeration & Air Conditioning Technology, 5th Edition, Thomson Delmar Learning
Unit 43—Air Source Heat Pumps

Review Topics
Reverse-Cycle Refrigeration

Key Terms
absolute zero • capillary tubes • cooling cycle • defrosting cycle • four-way valve • heat • heating cycle • indoor coil • outdoor coil • pilot valve • reversing valve

OVERVIEW

Refrigeration is the process of removing heat from an area where it is not wanted and rejected to an area where it is not objectionable. Air conditioning equipment allows for heat transfer in one direction only.

Heat is normally removed from the inside of a structure and then rejected to the outdoors. On the other hand, a heat pump has the ability to pump heat in two directions. **During the summer,** heat pumps absorb heat from the structure and discharge heat to the outdoors (**Figure 4–2**).

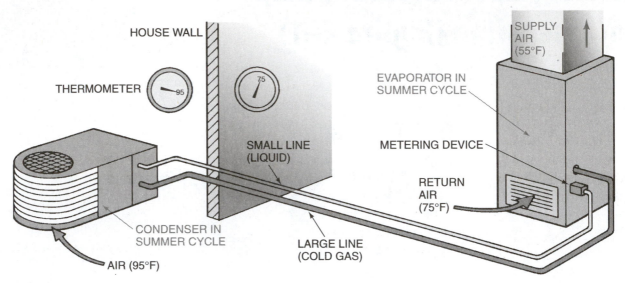

FIGURE 4–2   An air-to-air heat pump moving heat from the inside of a structure and rejecting the heat to the outdoors.

During the winter, heat pumps absorb heat from the outdoors and discharge heat to the inside of the structure (**Figure 4–3**).

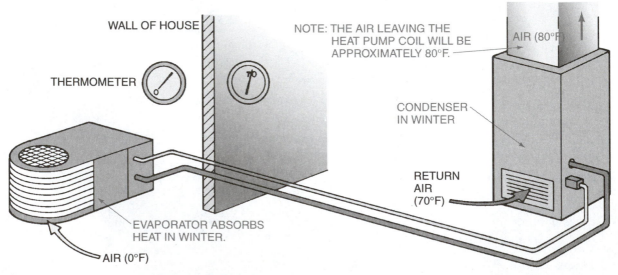

FIGURE 4–3   In the winter the heat pump absorbs heat from the outside air and rejects the heat to the inside of the structure.

You may wonder how it is possible for heat to be absorbed from an outside air temperature of 0 degrees. Remember that **heat** is molecules in motion and that there is heat in everything all the way down to a temperature of **–460 degrees (F).**

This temperature is referred to as **absolute zero.** Scientists believe that at this temperature molecular motion stops; thus there is no heat. Considering this fact, an outside air temperature of 0 degrees (F) contains a tremendous amount of heat, which can be absorbed by a heat pump and discharged into a structure or building. The absorption of heat from 0 degrees (F) is based on the **Second Law of the Thermodynamics of Heat,** which states that *heat will always travel from a higher temperature to a lower temperature.* Understanding this law makes it easy to understand how heat is absorbed from the outdoors in the wintertime. Look at **Figure 4–4.**

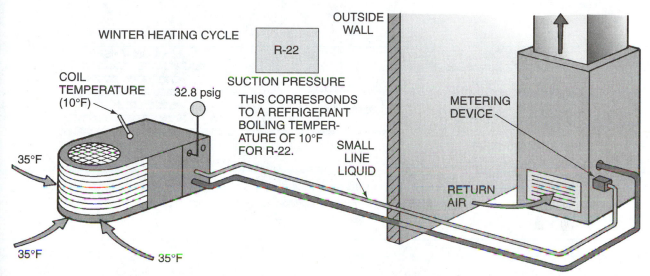

FIGURE 4–4   Outdoor temperature is 35 degrees (F) and outdoor coil is 10 degrees (F).

The outdoor temperature is 35 degrees, but the refrigerant in the outdoor coil is evaporating (boiling); the refrigerant in the coil is at 10 degrees (F). Based on the Second Law of the Thermodynamics of Heat, heat from the outside air temperature of 35 degrees (F) would move naturally toward the coil temperature of 10 degrees (F). To assist the absorption process the outdoor unit fan draws the 35-degree air across the 10-degree outdoor coil. Heat from the 35-degree air is absorbed as latent heat in the refrigerant evaporation process that is taking place inside the outdoor coil. Remember that **latent heat** is heat required to cause a substance to change state from one form of matter to another. In this case, heat from the outside air is being used to change the state of liquid refrigerant-to-refrigerant vapor by boiling (evaporating) within the coil of the outdoor unit. The low-temperature superheated refrigerant vapor is removed from the outdoor coil, compressed, and discharged to the indoor coil by the unit's compressor.

Now the Second Law of the Thermodynamics of Heat takes place at the indoor coil, but a reverse change of state takes place. Cooler air from the structure is drawn across the indoor coil by the unit's blower. The indoor coil has a high-temperature superheated refrigerant vapor passing through it; cooler air from the structure passes over the indoor coil, removing heat from the high-temperature superheated vapor, causing the refrigerant vapor to condense to high-temperature liquid. The latent heat removed from the refrigerant vapor is absorbed into the cooler structure air, raising its temperature, and then discharged into the structure.

Heat pumps contain the same four major sealed system components as an air conditioning system, except that the refrigerant coils of a heat pump are not identified as a condenser or evaporator. The refrigerant coils of a heat pump are identified as outdoor and indoor coils. The reason for this is based on the fact that each coil will perform the function of an evaporator or condenser depending on the heat pump mode of operation (**Figure 4–5**).

In addition to the four major sealed system components, heat pumps are also equipped with an accumulator (refer to *Figure 4–1*), and a four-way valve, also referred to as a reversing valve (**Figure 4–6**).

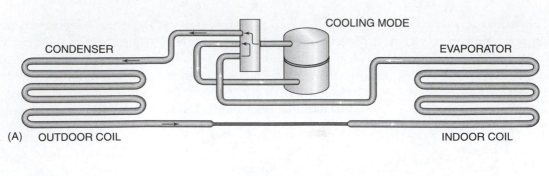

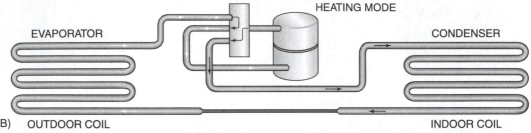

FIGURE 4–5   (A) Heat pump system operating in the cooling mode. The indoor coil performs as the evaporator and the outdoor coil performs as the condenser. (B) Heat pump system operating in the heating mode. The indoor coil performs as a condenser and the outdoor coil performs as an evaporator.

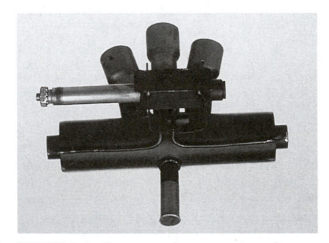

FIGURE 4–6   Four-way valve or reversing valve.

    The suction line accumulator is added to a heat pump to aid in preventing damage to the compressor due to possible liquid flood back during the heating cycle. The accumulator is located in the suction line between the compressor and the reversing valve. It allows any liquid refrigerant that may be present in the suction line to boil into a vapor before returning to the compressor (**Figure 4–7**).

    During winter operation ice or sweat may form on the accumulator because of the low saturation temperature of the refrigerant liquid and vapor in the outdoor coil.

    The **four-way or reversing valve** provides the heat pump the ability to absorb and reject heat in two different directions. The reversing valve has four refrigerant ports (**Figure 4–8**).

    Port 1 of the four-way valve is connected to the compressor's discharge port, or high side. High-temperature, high-pressure superheated vapor is discharged through this port. **Port 2** is connected to the compressors suction line. Low-temperature, low-pressure superheated vapor moves through this port. **Ports**

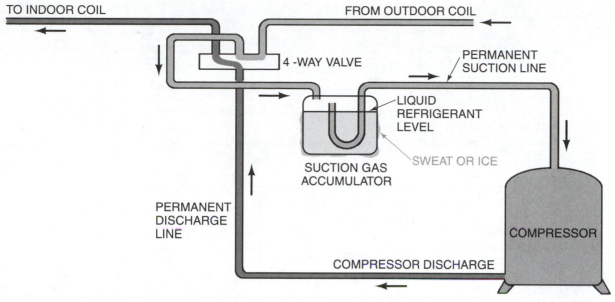

FIGURE 4-7 Suction accumulator prevents liquid flood-back to compressor.

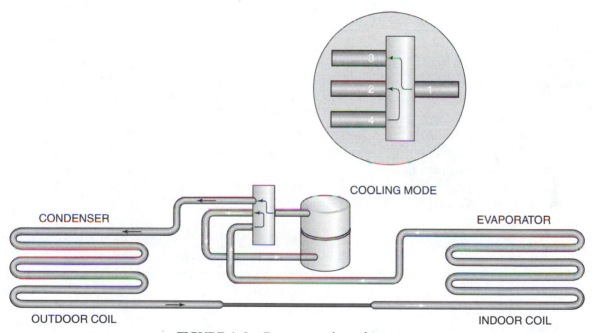

FIGURE 4-8 Four-way valve refrigerant port.

1 and 2 perform the same function regardless of the heat pump's operating mode. **Port 3** of the four-way valve is connected to the system's outdoor coil, and **Port 4** of the four-way valve is connected to the system's indoor coil. Depending on the mode of operation of the heat pump, high-temperature superheated vapor will be discharged through **Port 3** or **Port 4.**

During **summer operation**, high-temperature superheated vapor will be discharged to the outdoor coil at **Port 3** and low-temperature superheated vapor will be sucked from the indoor coil at **Port 4.**

During **winter operation**, high-temperature superheated vapor will be discharged to the outdoor coil at **Port 4** and low-temperature superheated vapor will be sucked from the indoor coil at **Port 3.** (**Figure 4-9**).

Figure 4-10 shows the state of the refrigerant at important locations in the heat pump during cooling and heating cycles.

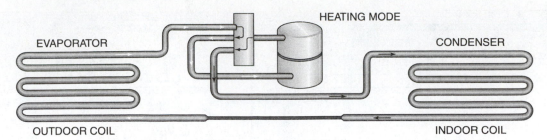

FIGURE 4–9   Refrigerant flow during winter operation.

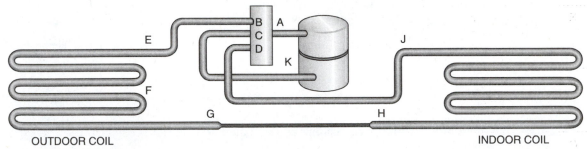

FIGURE 4–10   State of refrigerant at key locations during the cooling and heating cycles.

Figure 4–11 lists the state of the refrigerant at key points in heat pump system during the different heat pump cycles.

Once the thermostat of the system is set to a cycle of operation, the reversing valve is used to direct the discharge refrigerant vapor from the compressor to either the indoor coil or the outdoor coil (**Figure 4–12**).

On most heat pump thermostats, the **O terminal** is used to energize the four-way valve **holding coil** during the first stage of the cooling cycle (**Figure 4–13** and **Figure 4–14**).

Notice the location of the four-way valve holding coil. (Refer to **Figure 4–15**.) It is located on the four-way valve's **pilot valve** and is energized and de-energized by the system's thermostat through the **O termi-nal**. However, how does the four-way valve switch direction to send the high-temperature discharge gas to the correct coil?

Looking at a reversing valve, you will notice that there are four small capillary tubes connected to the pilot valve and also connected to the four-way valve's refrigerant ports (**Figure 4–15**).

This type of four-way valve is referred to as a **pilot-operated reversing valve** because it uses the difference between the system's high-side and low-side pressures to push the valve slide from one position to the other.

Notice that one of the capillary tubes from the pilot valve is attached to the center port of the revers-ing valve or compressor suction port. Another capillary tube is attached from the pilot valve to the com-pressor discharge port, and the other two capillary tubes are attached to the pilot valve and at opposite ends of the four-way valve body. The direction of the force created by the high-side pressure is determined by the pilot ports, which in turn are controlled by the four-way valve holding coil.

Figure 4–16 shows the heat pump four-way valve in the heating cycle.

Notice that the **holding coil is de-energized** during the heating cycle. Most manufacturers operate the reversing valve in this manner. Notice points **"A" and "B"** on the pilot valve. The capillary tube at point "A" has a low pressure and at point "B" the capillary tube has a high pressure. This forces the main valve to slide to the left, allowing the compressor's high-temperature discharge vapor to travel to the indoor coil for heating the structure. The cooler air from the structure removes heat from the high-temperature refrig-erant vapor as it passes over the indoor coil. This cause the high-temperature refrigerant vapor to condense to a high-temperature liquid which is then metered to the outdoor coil so that the low-temperature liquid refrigerant can be boiled by absorbing heat from the outside air.

	Cooling Mode			Heating Mode	
Letter Designation	Location in the System	State of the Refrigerant	Letter Designation	Location in the System	State of the Refrigerant
A	Compressor discharge port	High-temp, high-pressure superheated vapor	A	Compressor discharge port	High-temp high-pressure superheated vapor
B	Reversing-valve port leading to the outdoor coil	High-temp, high-pressure superheated vapor	D	Reversing-valve port leading to the indoor coil	High-temp high-pressure superheated vapor
E	Inlet of the outdoor coil	High-temp, high-pressure superheated vapor	J	Inlet of the indoor coil	High-temp high-pressure superheated vapor
F	Middle of outdoor coil	High-temp, high-pressure saturated liquid	H	Outlet of the indoor coil/inlet of the metering device	High-temp high-pressure subcooled liquid
G	Outlet of outdoor coil/ Inlet of the metering device	High-temp, high-pressure subcooled liquid	G	Inlet of the outdoor coil/ Outlet of the metering device	Low-temp, low-pressure saturated liquid
H	Outlet of the metering device	Low-temp, low-pressure saturated liquid	F	Middle of the outdoor coil	Low-temp, low-pressure saturated liquid
J	Outlet of indoor coil	Low-temp, low-pressure superheated vapor	E	Outlet of the outdoor coil	Low-temp, low-pressure superheated vapor
D	Inlet of the reversing valve	Low-temp, low-pressure superheated vapor	B	Inlet of the reversing valve	Low-temp low-pressure superheated vapor
C	Outlet of the reversing valve	Low-temp, low-pressure superheated vapor	C	Outlet of the reversing valve	Low-temp, low-pressure superheated vapor
K	Compression suction port	Low-temp, low-pressure superheated vapor	K	Compressor suction port	Low-temp, low-pressure superheated vapor

FIGURE 4–11   State of refrigerant at various points in the system.

Figure 4–17 shows the heat pump four-way valve in the cooling cycle.

Notice that the **holding coil is energized** during the cooling cycle. Most manufacturers operate the reversing valve in this manner. Notice points **"A" and "B"** on the pilot valve. The capillary tube at point "A" now has a high pressure and at point "B" the capillary tube has a low pressure. This forces the main valve to slide to the right, allowing the compressor's high-temperature discharge vapor to travel to the outdoor coil where heat from the structure is discharged to the outside air. The outside air is cooler than the high-temperature refrigerant vapor and removes heat from the vapor as it passes through the outdoor coil. This

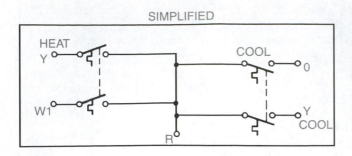

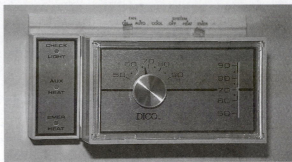

FIGURE 4–12 Unit thermostat controls the reversing valve during the heating and cooling mode function. (*Diagram Courtesy of Honeywell Corporation. Photo by Bill Johnson*)

FIGURE 4–13 Four-way valve holding coil.

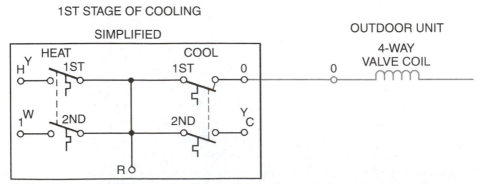

FIGURE 4–14 The first stage of the cooling cycle in which the four-way valve magnetic holding coil is energized.

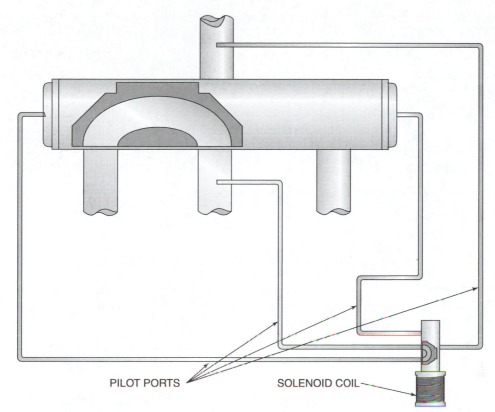

FIGURE 4–15   Four-way valve pilot valve and capillary tube connects.

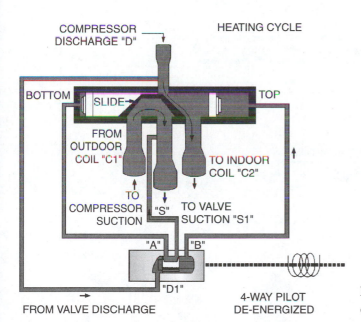

FIGURE 4–16   Pilot and four-way valve position during the heating cycle.

causes the high-temperature refrigerant vapor to condense to a high-temperature liquid, which is then metered to the indoor coil so that the low-temperature liquid refrigerant can be boiled by absorbing heat from the air inside the structure being cooled.

The line set of heat pumps are also identified differently than that of air conditioning equipment. In air conditioning, there is suction or low-sideline, sometimes referred to as a vapor line. There is also the high-side or discharge line, sometimes referred to as the liquid line. The low-sideline is always larger in diameter

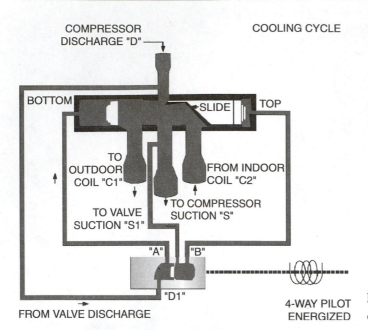

FIGURE 4–17 Pilot and four-way valve position during the cooling cycle.

than the high-sideline and has a low-temperature, low-pressure superheated refrigerant vapor moving through it. The high-side or discharge line is smaller in diameter and moves a high-temperature, high-pressure sub-cooled liquid through it.

In a heat pump, the larger diameter line is called the gas or vapor line and the smaller line is called the liquid line (**Figure 4–18**).

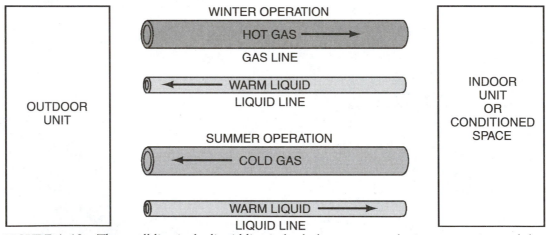

FIGURE 4–18 The small line is the liquid line in both the summer and winter operation and the larger line is the gas or vapor line in both summer and winter operation.

Almost all heat pump installations will have some source of auxiliary heat to assist in providing additional heat in cases where the heat pump cannot provide all the heat needed for the structure or to act as the emergency heating source if the heat pump becomes inoperative for some reason. For these reasons, most heat pump applications will also have a secondary or back-up heating source. Most heat pump installation use electrical heating elements or duct heaters (**Figure 4–19**) for this purpose. The secondary heating source is also used during the heat pump's defrosting cycle.

During winter operation, ice will form on the outdoor coil anytime the outside temperature falls below 45 degrees (F) (**Figure 4–20**).

FIGURE 4–19   Duct heater. (*Courtesy of Thermolec*)

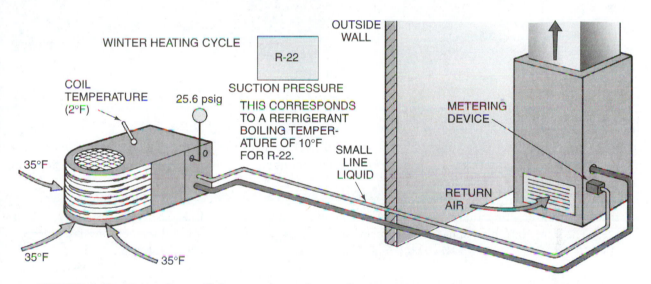

FIGURE 4–20   Ice or frost will form on the outdoor coil anytime the outside air temperature falls below 45 degrees.

Ice and frost buildup acts as an insulator reducing the refrigerant's ability to absorb heat from the outside air, thus reducing the heat pump's ability to provide enough heat to the structure. For this reason the ice or frost must be removed from the outdoor coil periodically through a manufacturer designed defrost cycle. During the defrosting cycle on all heat pumps, the unit is cycled into the cooling mode and the outdoor fan is stopped. This allows the high-side pressure to build, increasing the temperature of the refrigerant vapor as it passes through the outdoor coil, resulting in melted ice and frost from the coil.

The location of the outdoor coil and the height of the coil from ground are important. As the ice melts from the coil, the water needs to be able to drain away from the coil, but not into an area where it could be hazardous to the structure or individuals. When setting the outdoor coil, consideration must be given to the average snowfall level for the area where the heat pump is to be located. The outdoor coil should be raised above the average snowfall height for the area. This will allow drainage of water away from the outdoor so that the water doesn't ice the coil when it refreezes.

At the same time the unit is cycled into the defrost mode, the auxiliary heat is turned ON. This keeps warm air being provided to the conditioned space while the unit is operating in the cooling mode for defrosting purposes. The length of time for defrosting will vary from manufacturer to manufacturer, but most will not be in the defrost mode for more than 10 minutes as the maximum allowable time. Once the defrost cycle is complete, the heat pump is cycled back into the heating mode based on demand from the structure thermostat.

Heat pumps are designed to initiate the defrost cycle when the coil builds frost that will affect the heat pump's performance. Some heat pumps use **time and temperature** to start the defrost cycle. Many older heat pumps use a mechanical timer (**Figure 4–21**) and temperature sensing device (**Figure 4–22**).

FIGURE 4–21   Defrost timer. (*Photo by Bill Johnson*)

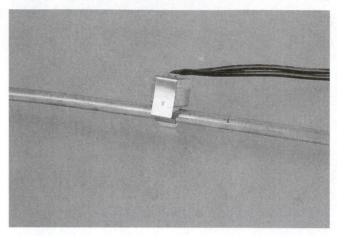

FIGURE 4–22   Temperature-sensing device. (*Photo by Bill Johnson*)

The defrost timer closes a set of contacts for 10 to 20 seconds for a trial defrost every 90 minutes. The timer contacts are in series with the temperature sensor contacts so that both must be closed at the same time before initiating the defrost cycle.

This means that two conditions must be made at the same time before defrost can be started. Typically, the timer runs any time the compressor runs, even in the cooling mode, and tries for defrost every 90 minutes. The temperature sensing device contacts will close when the outdoor coil temperature is 26 degrees (F) or lower, and open when the coil temperature reaches 50 degrees (F) (**Figure 4–23**).

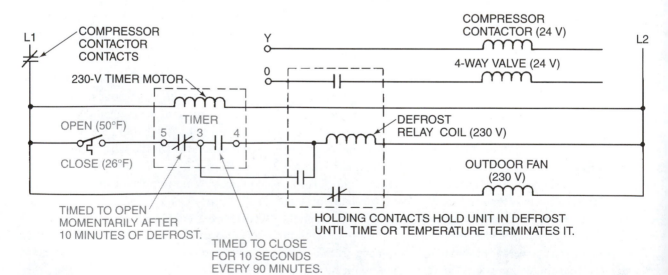

FIGURE 4–23   A wiring diagram of conditions to be met for defrost to start with a mechanical timer and temperature sensing device.

Some manufacturers will also incorporate an air-pressure switch that measures the air pressure drop across the outdoor coil. The air-pressure switch is wired in series with the timer contacts and the temperature sensing device contacts (**Figure 4–24**).

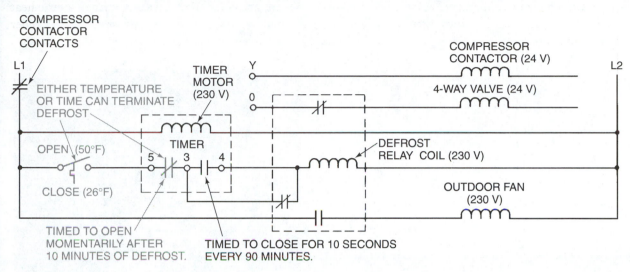

FIGURE 4–24   Air pressure switch wired in series with timer contacts and temperature sensor contacts. Used to prove the ice has formed on the coil before initiating the defrost cycle.

When the unit begins to build ice, the pressure drop occurs and the air switch contacts close. This combination of time, temperature, and pressure is very effective in initiating the defrost cycle when ice has actually formed on the coil.

After defrost has started, time, temperature, or pressure (*if used*) can terminate the defrost cycle. The temperature sensor can terminate once it senses approximately 50-degree coil temperature at its location. If the temperature contacts would not open due to cold outside temperatures, the timer will terminate defrost within a predetermined time period. Normally, 10 minutes is the maximum allowable time limit (**Figure 4–25**).

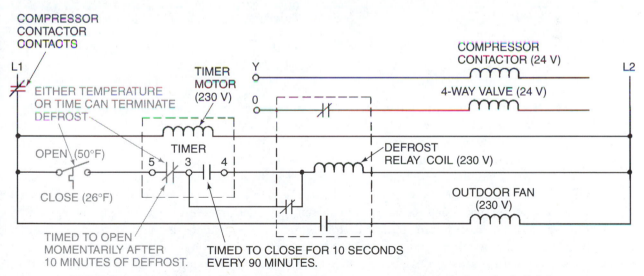

FIGURE 4–25   A wiring diagram for a time-or-temperature–terminated method of defrost.

Newer heat pump systems use solid-state technology to initiate and terminate the defrost cycle (**Figure 4–26**).

These controls are used by manufacturers of heat pumps to more closely control the defrost cycle with much more accuracy. The manufacturer may use *time and temperature* initiated and *time or temperature* terminated, like the mechanical timer and temperature sensor of the older systems. Some heat pumps use the *difference in the entering air temperature and the coil temperature* to arrive at a temperature split when defrost is desired. The coil temperature and the air temperature are normally sensed by thermistors (**Figure 4–27**) rather than bimetal temperature sensors. Thermistors sense changes in temperature and change their internal resistance according to theses temperature changes. In many cases, solid-state defrost systems employ two thermistors.

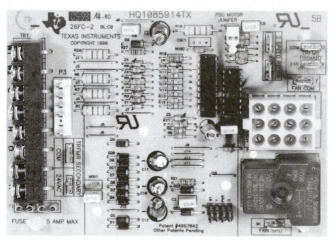

FIGURE 4–26   Solid-state heat pump defrost board. (*Photo by Bill Johnson*)

FIGURE 4–27   Thermistor attached to refrigerant coil.

One senses the temperature of the outdoor coil, and the other senses the temperature of the outdoor air. The differential between the resistance of the two thermistors is what initiates the defrost cycle. As frost accumulates on the outdoor coil, the temperature sensed by the thermistor attached to the coil will drop. This will cause the resistance of the thermistor to drop. A temperature differential of 15 to 20 degrees (F) will initiate the defrost cycle. Once the temperature differential is reached, the defrost board will energize the necessary contacts and holding coils to bring the system into defrost. With this type of system, defrost is often terminated by a high-side pressure switch that is connected to the solid-state defrost control board (**Figure 4–28**).

Some solid-state defrost boards **initiate** defrost by **time** and **terminate** the defrost cycle by **pressure**, **temperature**, or a combination of both. Solid-state timers are employed on the defrost board, and the time interval between defrost attempts can be altered in the field by repositioning a jumper wire on the defrost board (**Figure 4–29**).

Most *timed solid-state defrost boards* have interval times of 30, 60, or 90 minutes between defrost attempts. Most heat pump systems with time-initiated defrost use temperature to terminate the defrost cycle. There are units that use *pressure-initiated and terminated defrost systems*. These systems use a dual-pressure sensing switch connected to the circuit board that signals the system to enter defrost when the pressure differential rises to a predetermined point.

FIGURE 4–28   High-pressure control.

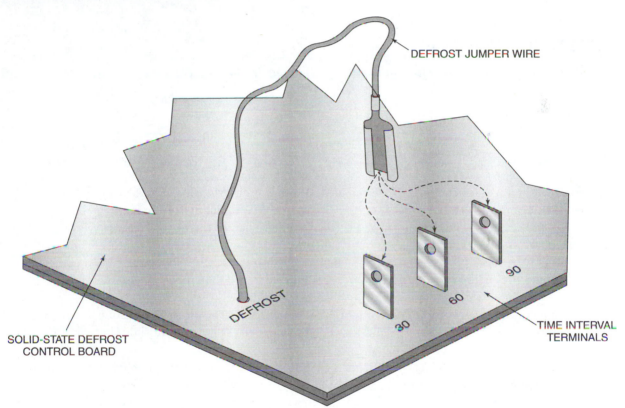

FIGURE 4–29   The time interval can be changed by repositioning the jumper wire on the circuit board.

NOTE: *Chapter 3 covers many service procedures on air conditioning equipment that would also be used when servicing heat pump equipment. Theory Lessons and Practical Competencies explicit to heats pumps are integrated in Chapter 4.*

Practical Competency 107

Checking a Four-Way (*Reversing Valve*) Valve Holding Coil

SUGGESTED MATERIALS

Textbook
Refrigeration & Air Conditioning Technology, 5th Edition, Thomson Delmar Learning
Unit 43—Air Source Heat Pumps

Review Topics
Troubleshooting the Four-Way Valve

COMPETENCY OBJECTIVE

The student will be able to use different methods of checking the holding coil of a four-way valve.

> NOTE: *The completion of this competency requires a complete operation heat pump.*

OVERVIEW

On most heat pump thermostats, the **O terminal** is used to energize the four-way valve **holding coil** during the first stage of the cooling cycle.

Notice the location of the four-way valve holding coil. It is located on the four-way valve's **pilot valve** and is energized and de-energized by the system's thermostat through the **O terminal**. However, how does the four-way valve switch direction to send the high-temperature discharge gas to the correct coil?

Looking at a reversing valve, you will notice that four small capillary tubes are connected to the pilot valve and also connected to the four-way valves refrigerant ports.

This type of four-way valve is referred to as a **pilot-operated reversing valve** because it uses the difference between the system's high-side and low-side pressures to push the valve slide from one position to the other.

Notice that one of the capillary tubes from the pilot valve is attached to the center port of the reversing valve or compressor suction port. Another capillary tube is attached from the pilot valve to the compressor discharge port, and the other two capillary tubes are attached to the pilot valve and at opposite ends of the four-way valve body. The direction of the force created by the high-side pressure is determined by the pilot ports, which in turn are controlled by the four-way valve holding coil.

Notice that the **holding coil is de-energized** during the heating cycle. Most manufacturers operate the reversing valve in this manner. Notice points **"A" and "B"** on the pilot valve. The capillary tube at point "A" has a low pressure and at point "B" the capillary tube has a high pressure. This forces the main valve to slide to the left, allowing the compressor's high-temperature discharge vapor to travel to the indoor coil for heating the structure. The cooler air from the structure removes heat from the high-temperature refrigerant vapor as it passes over the indoor coil. This cause the high-temperature refrigerant vapor to condense to a high-temperature liquid which is then metered to the outdoor coil so that the low-temperature liquid refrigerant can be boiled by absorbing heat from the outside air.

Notice that the **holding coil is energized** during the cooling cycle. Most manufacturers operate the reversing valve in this manner. Notice points **"A" and "B"** on the pilot valve. The capillary tube at point

"A" now has a high pressure and at point, "B" the capillary tube has a low pressure. This forces the main valve to slide to the right, allowing the compressor's high-temperature discharge vapor to travel to the outdoor coil where heat from the structure is discharged to the outside air. The outside air is cooler than the high-temperature refrigerant vapor and removes heat from the vapor as it passes through the outdoor coil. This causes the high-temperature refrigerant vapor to condense to a high-temperature liquid, which is then metered to the indoor coil so that the low-temperature liquid refrigerant can be boiled by absorbing heat from the air inside the structure being cooled.

There are a couple of different ways of checking the holding coil used on a four-way valve. When the valve is energized, technicians can use a screwdriver and hold it on top or near the valve; a magnetic pull should be felt when the valve is energized. The coil can also be checked by touch, when energized, the coil should feel warm to hot, which is an indication that the coil is actually energized and creating a magnetic field around the four-way valve's pilot valve. Technicians can also use a voltmeter and check for voltage at the holding coil electrical leads, or use an ohmmeter and check for resistance across the coil. Voltage at the coil means that the coil should be energized if the holding coil is good. A resistance value across the holding coil means that the coil is good and should create a magnetic field around the pilot vale when the correct voltage is applied. An infinity reading is an indication that the valve has an open coil and needs to be replaced. In most cases, the holding coil can be replaced without replacing the complete four-way valve.

EQUIPMENT REQUIRED

Split-system heat pump
1 Voltmeter or VOM
1 Ohmmeter or VOM
Screwdriver

SAFETY PRACTICES

The student should be knowledgeable in the use of tools and testing equipment and follow all safety rules when working with live voltages.

COMPETENCY PROCEDURES   Checklist

> NOTE: *The procedures in this competency are based on a split-system heat pump with the four-way valve holding coil energized in the cooling mode. Some manufacturers of heat pumps energize the four-way valve holding coil in the heating mode. Check the schematic of the equipment you are testing to see in what mode the reversing valve is being energized.*

1. Turn the power OFF to the outdoor coil.   ❑
2. Remove outdoor coil panels (*if applicable*) and gain access to the heat pump's four-way
   valve holding coil.   ❑
3. Remove the holding coil electrical plug or wires.   ❑
4. Make sure that the power is turned ON at the air handler.   ❑
5. Set the unit's thermostat to the HEATING cycle.   ❑
6. Use a VOM meter and set the meter to measure AC voltage in a range of 24 volts or greater. ❑
7. Place the meter leads across the holding coil connector or electrical wires.   ❑
8. *Record the voltage reading. _____*   ❑

> NOTE: *A measured voltage of 24 volts is an indication that the thermostat has closed and is supplying voltage to the holding coil. No recorded voltage means that there is a problem from the thermostat through the unit's defrost board.*

9. Set the VOM meter to measure resistance. ❏
10. Place the meter leads across the holding coil terminals. ❏
11. *Record the resistance value.* _____ ❏

> **NOTE:** *A measured resistance across the holding coil terminals means that the holding coil is good. An infinity reading is an indication of an OPEN COIL and would require that the holding coil be replaced.*

12. Attach the holding coil connector or electrical wires to the four-way valve coil. ❏
13. Place the screwdriver at or near the top of the holding coil. ❏
14. *Explain what effect you noticed when the screwdriver was placed near the top of the holding coil.* _____ ❏

> **NOTE:** *If voltage was applied to the holding coil and the coil is good, you should have noticed a slight magnetic pull on the screwdriver when it was placed on or near the top of the holding coil. NO magnetic pull is an indication of NO VOLTAGE or an OPEN COIL.*

15. Carefully touch the holding coil. ❏
16. *Explain what you felt when you touched the holding coil.* _____ ❏

> **NOTE:** *If the holding coil is good and voltage applied, the holding coil should have felt warm to the touch. You may have even felt some vibration. If the coil wasn't warm to the touch, the coil may be defective or voltage may not be applied.*

17. Turn the unit thermostat to the OFF position. ❏
18. Reassemble the outdoor coil and panels (*if applicable*). ❏
19. Turn the power back on to the outdoor coil. ❏
20. Place the thermostat to the HEATING cycle. ❏
21. Move the thermostat temperature selector switch high enough to call for heat. ❏
22. Let the system operate for 5 minutes. ❏
23. Feel the heat pump's gas/vapor line (*larger diameter tubing*). ❏
24. *Did the heat pumps gas/vapor line feel cool or hot?* _____ ❏

> **NOTE:** *In the heating cycle the gas/vapor line should have felt hot.*

25. Feel the heat pump's liquid line (*smaller diameter line*). ❏
26. *Did the liquid line feel cool or hot?* _____ ❏

> **NOTE:** *In the heating cycle the liquid line should have felt cooler than the gas/vapor line.*

27. Turn the thermostat OFF. ❏
28. Set the thermostat to COOL. ❏
29. Move the temperature selector switch to a temperature setting lower than ambient air temperature. (*Note: You may hear a woooshing sound.*) ❏
30. Let the unit operate for 5 minutes. ❏

31. Feel the gas/vapor line. ❏

32. Did the gas/vapor line feel cool or hot? _____ ❏

NOTE: *In the cooling cycle the gas/vapor line should feel cool to the touch.*

---

33. Feel the liquid line. ❏

34. *Did the liquid line feel cool or warm?* _____ ❏

NOTE: *In the cooling mode the liquid line should feel warm.*

---

35. Turn the thermostat to the OFF position. ❏

36. Have the instructor check your work. ❏

37. Return all tools and test equipment to their proper location. ❏

RESEARCH QUESTIONS

1. What large line connects the indoor unit to the outdoor unit?

   A. A discharge line
   B. A liquid line
   C. An equalizer line
   D. A gas line

2. Why is the indoor unit not called an evaporator?

   _____

   _____

3. When is the outdoor coil called an evaporator?

   A. Heating season
   B. Cooling season
   C. Defrost cycle
   D. Spring and fall seasons

4. The indoor coil is called a condenser during what season?

   A. Heating season
   B. Cooling season
   C. Defrost cycle
   D. Spring and fall seasons

5. Why is it important to have proper drainage for the outdoor unit?

   _____

   _____

Passed Competency _____   Failed Competency _____

Instructor Signature _____   Grade _____

Practical Competency 108

Mechanical Performance Check of a Four-Way Valve

SUGGESTED MATERIALS

Textbook
Refrigeration & Air Conditioning Technology, 5th Edition, Thomson Delmar Learning
Unit 43—Air Source Heat Pumps

Review Topics
Troubleshooting the Four-Way Valve; Troubleshooting Mechanical Problems

COMPETENCY OBJECTIVE

The student will be able to check the mechanical performance of a heat pump four-way valve.

> **NOTE:** *The completion of this competency requires a complete operation heat pump.*

OVERVIEW

A heat pump has the ability to pump heat in two directions. **During the summer,** heat pumps absorb heat from the structure and discharge heat to the outdoors.

During the winter, heat pumps absorb heat from the outdoors and discharge heat to the inside of the structure.

Looking at a reversing valve, you will notice that there are four small capillary tubes connected to the pilot valve and also connected to the four-way valves refrigerant ports.

This type of four-way valve is referred to as a **pilot-operated reversing valve** because it uses the difference between the system's high-side and low-side pressures to push the valve slide from one position to the other.

Notice that one of the capillary tubes from the pilot valve is attached to the center port of the reversing valve or compressor suction port. Another capillary tube is attached from the pilot valve to the compressor discharge port, and the other two capillary tubes are attached to the pilot valve and at opposite ends of the four-way valve body. The direction of the force created by the high-side pressure is determined by the pilot ports, which in turn are controlled by the four-way valve holding coil.

Notice that the **holding coil is de-energized** during the heating cycle. Most manufacturers operate the reversing valve in this manner. Notice points **"A" and "B"** on the pilot valve. The capillary tube at point "A" has a low pressure and at point, "B" the capillary tube has a high pressure. This forces the main valve to slide to the left, allowing the compressor's high-temperature discharge vapor to travel to the indoor coil for heating the structure. The cooler air from the structure removes heat from the high-temperature refrigerant vapor as it passes over the indoor coil. This causes the high-temperature refrigerant vapor to condense to a high-temperature liquid which is then metered to the outdoor coil so that the low-temperature liquid refrigerant can be boiled by absorbing heat from the outside air.

Notice that the **holding coil is energized** during the cooling cycle. Most manufacturers operate the reversing valve in this manner. Notice points **"A" and "B"** on the pilot valve. The capillary tube at point "A" now has a high pressure and at point "B" the capillary tube has a low pressure. This forces the main valve to slide to the right, allowing the compressor's high-temperature discharge vapor to travel to the outdoor coil where heat from the structure is discharged to the outside air. The outside air is cooler than the high-temperature refrigerant vapor and removes heat from the vapor as it passes through the outdoor coil. This causes the high-temperature refrigerant vapor to condense to a high-temperature liquid, which is then metered to the indoor coil so that the low-temperature liquid refrigerant can be boiled by absorbing heat from the air inside the structure being cooled.

EQUIPMENT REQUIRED

Split-system heat pump
Temperature meter with two or more temperature leads
Roll of insulation tape

SAFETY PRACTICES

The student should be knowledgeable in the use of tools and testing equipment. Follow all safety rules when working with live voltages.

COMPETENCY PROCEDURES
Checklist

> NOTE: Procedures in this competency are based on the four-way valve holding coil being energized in the cooling mode. Some manufacturers of heat pumps energize the holding coil in the heating mode. Check the equipment you are testing to see in what mode the holding coil is energized.

1. Turn unit thermostat to the OFF position. ❑
2. At the outdoor coil place a temperature probe on the gas/vapor line. ❑
3. At the outdoor coil place a temperature probe on the liquid line. ❑
4. Insulate the temperature probes. ❑
5. Place the thermostat to the HEATING cycle. ❑
6. Move thermostat temperature selector switch high enough to call for heat. ❑
7. Let the system operate for 5 minutes. ❑
8. Feel the heat pump's gas/vapor line (*larger diameter tubing*). ❑
9. Did the heat pump's gas/vapor line feel cool or hot? _____ ❑
10. Record the gas/vapor line temperature. _____ ❑

> NOTE: In the heating cycle, the gas/vapor line should have felt hot and the temperature should be higher than the liquid line temperature if the four-way valve is operating in the correct mode.

11. Feel the heat pump's liquid line (*smaller diameter line*). ❑
12. Did the liquid line feel cool or hot? _____ ❑
13. Record the liquid line temperature. _____ ❑

> NOTE: In the heating cycle the liquid line should have felt cooler and have a lower temperature than the gas/vapor line if the four-way valve is operating in the correct mode.

14. Turn heat pump thermostat to the OFF position. ❑
15. Set the thermostat to the COOL. ❑
16. Move the temperature selector switch to a temperature setting lower than ambient air temperature. (*Note: You may hear a woooshing sound.*) ❑
17. Let the unit operate for 5 minutes. ❑
18. Feel the gas/vapor line. ❑
19. *Did the gas/vapor line feel cool or hot?* _____ ❑
20. *Record the gas/vapor line temperature.* _____ ❑

> **NOTE:** *In the cooling cycle, the gas/vapor line should feel cool and its temperature should be lower than that of the liquid line if the four-way valve is operating in the correct mode.*

21. Feel the liquid line. ❑
22. *Did the liquid line feel cool or warm?* _____ ❑
23. *Record the temperature of the liquid line.* _____ ❑

> **NOTE:** *In the cooling mode the liquid line should feel warm or at about ambient temperature if the four-way valve is operating in the correct mode.*

> **NOTE:** *If the charge is correct and the four-way valve is working, the gas/vapor line and liquid line temperatures should have changed in temperature when the thermostat was changed from one mode to another. Properly charged systems with a correctly operating four-way valve would indicate a considerable temperature difference between the gas/vapor line and liquid line in either mode of operation. Little temperature difference between the two lines is an indication of a possible low charge or four-way valve, which may not be shifting properly. NO change in line temperature from one mode of operation to another is an indication the reversing valve is not shifting. Checks would have to be performed to determine the cause.*

24. Turn the thermostat to the OFF position. ❑
25. Have the instructor check your work. ❑
26. Remove temperature leads from the line set. ❑
27. Return all tools and test equipment to their proper location. ❑

RESEARCH QUESTIONS

1. Which metering device is most efficient in heat pumps?

    A. Thermostatic expansion valve
    B. Automatic expansion valve
    C. Capillary tube
    D. An orifice or restrictor

2. Where is the only permanent suction line on a heat pump?

    _____

    _____

3. Can a heat pump switch from heating to cooling and from cooling to heating automatically?

    _____

    _____

4. Name the three common sources of heat for heat pumps.

_____

_____

5. Why do frost and ice collect on the outdoor coil in the heating cycle?

_____

_____

Passed Competency _____ **Failed Competency** _____

Instructor Signature _____ **Grade** _____

Practical Competency 109

Checking a Four-Way Valve for Leaking from System High Side to System Low Side

SUGGESTED MATERIALS

Textbook
Refrigeration & Air Conditioning Technology, 5th Edition, Thomson Delmar Learning
Unit 43—Air Source Heat Pumps

Review Topics
Troubleshooting the Four-Way Valve; Troubleshooting Mechanical Problems

COMPETENCY OBJECTIVE

The student will be able to check a four-way valve suspected of leaking high pressure into the low side of the valve.

NOTE: The completion of this competency requires a complete operation heat pump.

OVERVIEW

Looking at a reversing valve, you will notice that there are four small capillary tubes connected to the pilot valve and also connected to the four-way valve's refrigerant ports.

This type of four-way valve is referred to as a **pilot-operated reversing valve** because it uses the difference between the system's high-side and low-side pressures to push the valve slide from one position to the other.

Notice that one of the capillary tubes from the pilot valve is attached to the center port of the reversing valve or compressor suction port. Another capillary tube is attached from the pilot valve to the compressor discharge port, and the other two capillary tubes are attached to the pilot valve and at opposite ends of the four-way valve body. The direction of the force created by the high-side pressure is determined by the pilot ports, which in turn are controlled by the four-way valve holding coil.

Notice that the **holding coil is de-energized** during the heating cycle. Most manufacturers operate the reversing valve in this manner. Notice points **"A" and "B"** on the pilot valve. The capillary tube at point "A" has a low pressure and at point "B" the capillary tube has a high pressure. This forces the main valve to slide to the left, allowing the compressor's high-temperature discharge vapor to travel to the indoor coil for heating the structure. The cooler air from the structure removes heat from the high-temperature refrigerant vapor as it passes over the indoor coil. This causes the high-temperature refrigerant vapor to condense to a high-temperature liquid which is then metered to the outdoor coil so that the low-temperature liquid refrigerant can be boiled by absorbing heat from the outside air.

Notice that the **holding coil is energized** during the cooling cycle. Most manufacturers operate the reversing valve in this manner. Notice points **"A" and "B"** on the pilot valve. The capillary tube at point "A" now has a high pressure and at point "B" the capillary tube has a low pressure. This forces the main valve to slide to the right, allowing the compressor's high-temperature discharge vapor to travel to the outdoor coil, where heat from the structure is discharged to the outside air. The outside air is cooler than the high-temperature refrigerant vapor and removes heat from the vapor as it passes through the outdoor coil. This causes the high-temperature refrigerant vapor to condense to a high-temperature liquid, which is then metered to the indoor coil so that the low-temperature liquid refrigerant can be boiled by absorbing heat from the air inside the structure being cooled.

A four-way valve that is leaking can have the same effects on system capacity in summer and winter operations as a low-capacity compressor. A leaking four-way valve is a valve in which hot gas from the high-pressure side of the system is leaking into the low-pressure side of the system through the four-way valve. When refrigerant gas is pumped around and around the system, work is accomplished in the compression process, but usable refrigerant is not available in the heating or cooling process of the structure. A suspected leaking four-way valve can be checked by using a good thermometer to check the temperature of the low-side line, the suction line from the evaporator (the indoor coil in summer or the outdoor coil in winter), and the permanent suction line between the four-way valve and the compressor. The temperature difference should not be more than 3 degrees (F). (**Figure 4–30**).

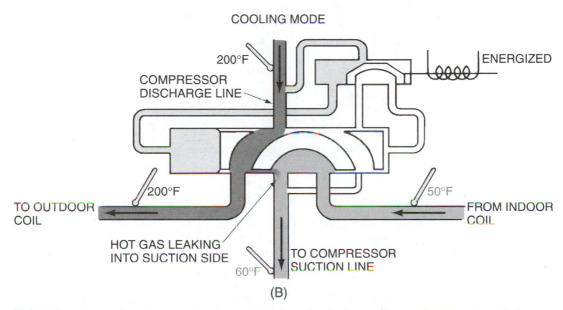

FIGURE 4–30   A line diagram showing a defective valve in the cooling mode. There is a 10-degree (F) difference between the compressor suction line and the gas line from the indoor coil.

EQUIPMENT REQUIRED

Split-system heat pump
Temperature meter with two or more temperature leads
Insulation tape

SAFETY PRACTICES

The student should be knowledgeable in the use of tools and testing equipment. Follow all safety rules when working with live voltages.

COMPETENCY PROCEDURES   Checklist

NOTE: *Procedures in this competency are based on the four-way valve holding coil being energized in the cooling mode. Some manufacturers of heat pumps energize the holding coil in the heating mode. Check the equipment you are testing to see in what mode the holding coil is energized.*

1. Turn the thermostat to the OFF position. ❑
2. Remove panels form the outdoor unit as needed to gain access to the four-way valve. ❑
3. Tightly attach a temperature probe on the permanent suction line (*center line*) at least 5" away from the four-way valve body (**Figure 4–31**). ❑

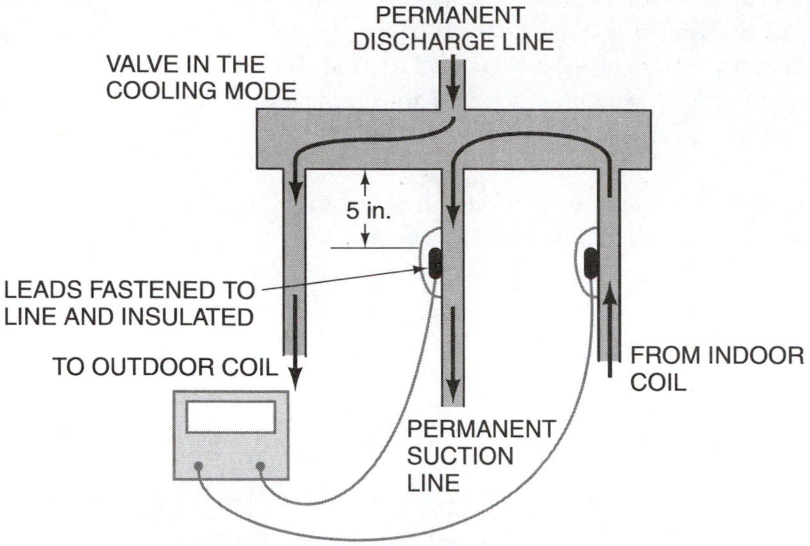

FIGURE 4–31   Attach and insulate temperature leads at least 5" away from the four-way valve body.

4. Insulate the temperature probe very well. ❑
5. Make sure that temperature probe lines are free of the outdoor fan. ❑
6. Set the thermostat to the **Cooling** mode. ❑
7. Set the temperature selector switch to a temperature lower than the surrounding ambient temperature. ❑
8. Let the unit operate in the **Cooling** mode for 10 minutes. ❑
9. *Record the permanent suction line temperature after 10 minutes of operation.* _____ ❑
10. *Record the gas/vapor line temperature from the indoor coil after 10 minutes of operation.* _____ ❑
11. *Determine the temperature difference:*
    Permanent suction line temperature _____
    MINUS
    Gas/vapor line temperature from indoor coil _____ ❑
12. *Record the temperature difference.* _____ ❑

NOTE: *A temperature difference between the two lines of 3 degrees (F) or more means that the four-way valve is leaking and should be replaced.*

---

13. Was the temperature difference 3 or more degrees (F)? _____ ❑
14. Turn the thermostat OFF. ❑
15. Leave the temperature probe attached to the permanent suction line. ❑
16. Remove the temperature probe from the four-way valve indoor coil line. ❑
17. Attach a temperature probe at least 5" away from the four-way valve body to the four-way valve line coming from the outdoor coil. ❑
18. Insulate the temperature probe well. ❑

19. Turn the thermostat to the HEATING mode. ❏
20. Set the thermostat temperature selector higher than the surrounding ambient air temperature. ❏
21. Let the unit operate in the HEATING mode for 10 minutes. ❏
22. Record the permanent suction line temperature after 10 minutes of operation. _____ ❏
23. Record the gas/vapor line temperature from the outdoor coil after 10 minutes of operation. _____ ❏
24. Determine the temperature difference:
    Permanent suction line temperature _____ ❏
    MINUS
    Gas/vapor line temperature from outdoor coil _____ ❏
25. Record the temperature difference. _____ ❏

> NOTE: *A temperature difference between the two lines of 3 degrees (F) or more means that the four-way valve is leaking and should be replaced.*

26. *Was the temperature difference 3 or more degrees (F)?* _____ ❏
27. According to the above test, was the four-way valve leaking? _____ ❏
28. Turn the thermostat OFF. ❏
29. Remove the temperature probe leads from the four-way valve tubes. ❏
30. Replace the outdoor coil panels. ❏
31. Have the instructor check your work. ❏
32. Return all tools and test equipment to their proper location. ❏

RESEARCH QUESTIONS

1. Which port of the four-way valve is the compressor discharge port?

   _____

   _____

2. What is the most common problem with four-way valves?

   _____

   _____

3. What is demand defrost?

   _____

   _____

4. During the heating mode, in which direction is the four-way valve sending the compressor discharge gas?

   _____

   _____

5. During the cooling mode, in which direction is the four-way valve sending the compressor discharge vapor?

   _____

   _____

Passed Competency _____   Failed Competency _____

Instructor Signature _____   Grade _____

Theory Lesson: Heat Pumps Using Scroll Compressors (Figure 4–32)

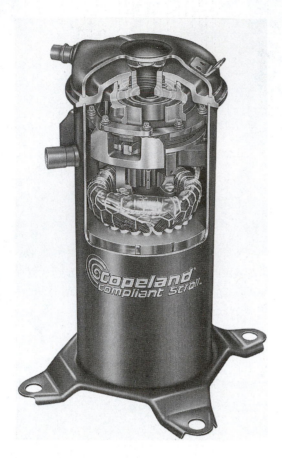

FIGURE 4–32 Scroll compressor.
(*Courtesy of Copeland Corporation*)

SUGGESTED MATERIALS

Textbook
Refrigeration & Air Conditioning Technology, 5th Edition, Thomson Delmar Learning
Unit 43—Air Source Heat Pumps

Review Topics
Scroll Compressors

Key Terms
absolute discharge pressure • absolute pressures • absolute suction pressure • compression ratio • compressor • heat of compression

OVERVIEW

All types of compressors serve two functions in the refrigeration system. One of its tasks is to remove the low-temperature, low-pressure refrigerant vapor from the evaporator, while reducing the evaporator pressure to a point where the desired evaporator temperature can be maintained. The second task of the compressor is to raise the pressure of the low-temperature, low-pressure suction vapor to a level where its saturation temperature is higher than the temperature of the condenser-cooling medium (**Figure 4–33**).

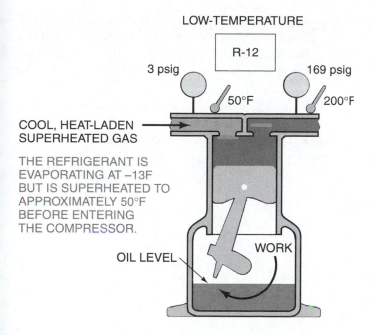

LOW-TEMPERATURE

R-12

3 psig   169 psig

50°F   200°F

COOL, HEAT-LADEN
SUPERHEATED GAS

THE REFRIGERANT IS
EVAPORATING AT −13F
BUT IS SUPERHEATED TO
APPROXIMATELY 50°F
BEFORE ENTERING
THE COMPRESSOR.

WORK

OIL LEVEL

FIGURE 4–33   Low-side superheated gas temperature and pressure and high-side superheated gas temperature and pressure.

Compression ratio is a technical expression of the pressure difference between the compressor's low-side pressure and the high-side pressure and is used to compare pumping conditions of a compressor.

The **compressor compression ratio** is determined by the dividing the compressor's *absolute discharge pressure* by the compressor's *absolute suction pressure*.

$$\text{Compression ratio} = \frac{\textit{Absolute discharge psia}}{\textit{Absolute suction psia}}$$

Absolute pressures are found by adding an atmospheric pressure of 14.7 psi to the system's operating low-side and high-side gage pressures (psig). Compression ratios of 3:1 are considered normal, although any situation that could cause an increase in a system's head pressure or a decrease in a system's suction pressure will cause higher compression ratios. When the compression ratio gets above the 12:1 range, the refrigerant gas temperature leaving the compressor can rise to a point at which the compressor's lubricating oil becomes overheated. Refrigerant oil that becomes overheated can turn into carbon and create acids within the system.

Because of the compressor compression ratio, *heat is pumped up the temperature scale.* During the compression of the low-temperature vapor from the evaporator, the vapor is compressed into a smaller area. This causes the temperature and pressure to rise. This is called **heat of compression.** The compressor takes the heat that is already in the vapor from the evaporator and pushes it into a small space.

By doing this, the temperature and pressure exerted on the vapor rises to a high-temperature and high-pressure vapor that is hotter than the area in which it is to be released. Once the compressor completes the heat of compression, it forces this high-temperature, high-pressure vapor into the condenser coil, which is normally located in an area where the heat from the evaporator is to be released.

Heat pumps have a special compressor. These compressors have to operate almost three times longer and over a broader range of temperature conditions than a cooling-only compressor used in air conditioning equipment. They also operate under severe low-ambient temperature conditions during start-up, cycle reversals, and defrost cycles.

In the air conditioning cycles, the compression ratio of the heat pump compressor hits a maximum of about 3:1. During the heating cycle at low-ambient temperature conditions, the compression ratio can reach 8:1 or higher. The motor, running gear, and compression parts are heavy-duty to withstand these operating conditions.

Scroll compressors are ideally suited for heat pump applications because of its pumping characteristics. The scroll compressor does not lose as much capacity as the reciprocating compressor at the higher head pressures of summer or the lower suction pressures of winter operation. This is because the scroll compressor does not have the top-of-stroke clearance volume loss that a reciprocating compressor has. The gas trapped in the clearance volume of a reciprocating compressor must re-expand before the cylinder starts to fill. This reduces the efficiency. The scroll compressor is about 15% more efficient than a reciprocating compressor. A scroll compressor's compression takes place between two specially designed spiral-shaped forms (**Figure 4–34**).

The spiral-shaped forms have multiple chambers, each of which is at a different stage of compression (**Figure 4–35**).

FIGURE 4–34   Two identical scrolls that form crescent-shaped pockets when nested together. (*Courtesy of Copeland Corporation*)

1. GAS ENTERS AN OUTER OPENING AS ONE SCROLL ORBITS THE OTHER.

2. THE OPEN PASSAGE IS SEALED AS GAS IS DRAWN INTO THE COMPRESSION CHAMBER.

3. AS ONE SCROLL CONTINUES ORBITING, THE GAS IS COMPRESSED INTO AN INCREASINGLY SMALLER "POCKET."

4. GAS IS CONTINUALLY COMPRESSED TO THE CENTER OF THE SCROLLS, WHERE IT IS DISCHARGED THROUGH PRECISELY MACHINED PORTS AND RETURNED TO THE SYSTEM.

5. DURING ACTUAL OPERATION, ALL PASSAGES ARE IN VARIOUS STAGES OF COMPRESSION AT ALL TIMES, RESULTING IN NEAR-CONTINUOUS INTAKE AND DISCHARGE.

FIGURE 4–35   Compression process in a scroll compressor. (*Courtesy of Copeland Corporation*)

Compression is created by the interaction of an orbiting spiral and a stationary spiral. Gas enters the outer openings as one of the spirals orbits. As the spiral orbits, gas is compressed into an increasingly smaller pocket, reaching discharge pressure at the center port. Actually, during operation, all six-gas passages are in various stages of compression at all times, resulting in nearly continuous suction and discharging of refrigerant vapor (**Figure 4–36**).

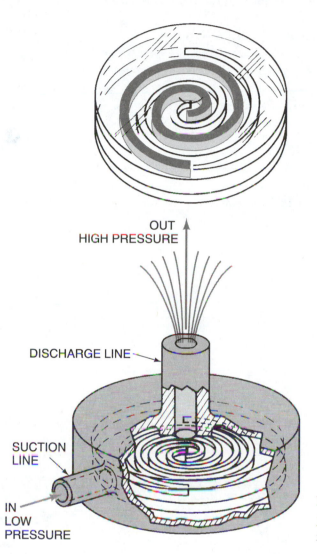

OUT
HIGH PRESSURE

DISCHARGE LINE

SUCTION
LINE

IN
LOW
PRESSURE

FIGURE 4–36   The orbiting action of the scroll plates provides continuous suction and discharging of refrigerant vapor.

Scroll compressors can handle liquid that may enter the scroll plates better than reciprocating compressors, which also increases the need for a crankcase heater, and in most cases a suction line accumulator. This is because one of the scroll plates is stationary and the other plate floats, which gives the plates some play. This play will allow the floating scroll plate to move if liquid should enter the compressor, thus preventing damage to the compressor. Although scroll compressors can handle liquid flood-back better than reciprocating compressors, liquid flood-back does reduce the system's capacity.

Scroll compressors are discharge gas cooled so the compressor shell would feel much hotter than suction gas cooled compressors. They also operate with less vibration and a lower noise level. Scrolls do not have suction or discharge valves; they are equipped with a low-mass, disc-type check valve at the discharge port that prevents the high-pressure refrigerant from traveling back through the compressor during the off cycle. When the compressor shuts off, the gas trapped between the check valve and the compressor scroll will back up through the scroll, causing a strange sound until the pressures between the check valve and scroll equalize. The sound resembles the rustling of a handful of marbles. The high-side pressure will then equalize through the metering device to the low side, like a reciprocating compressor system. Equalization between the check valve and the compressor scroll allows the compressor to start up without hard-start assistance because there is no pressure differential for the first starting revolutions of the compressor.

Considerations for air conditioning and heat pumps equipped with scroll compressors:

A. Do not operate a scroll compressor with the suction valve closed.
B. Avoid operating three-phase scroll compressors in reverse rotation. Reverse rotation is very recognizable because of the extreme noise scroll compressors make when operating in reverse rotation.
C. *Correct rotation of three-phase compressors can be checked with gage manifolds. Attach manifold gages to both high- and low-side valves, operate the compressor, and check to see if normal suction and discharge pressure are established quickly; if they are not, shut the unit OFF, reverse any two of the power leads, and recheck.*

NOTE: *Troubleshooting outdoor and indoor fan and blower motors, transformers, relays, and capacitors was covered in previous chapters. Practical Competencies contained in this chapter are specific to heat pumps.*

Practical Competency 110

Single-Phase Scroll Compressor Motor Winding and Overload Protector Evaluation

SUGGESTED MATERIALS

Textbook

Refrigeration & Air Conditioning Technology, 5th Edition, Thomson Delmar Learning
Unit 43—Air Source Heat Pumps

Review Topics

Heat Pumps Using Scroll Compressors

COMPETENCY OBJECTIVE

The student will be able to check to evaluate the motor windings and overload protector of a single-phase scroll compressor.

OVERVIEW

Heat pumps have a special compressor. These compressors have to operate almost three times longer and over a broader range of temperature conditions than a cooling-only compressor used in air conditioning equipment. They also operate under severe low-ambient temperature conditions during start-up, cycle reversals, and defrost cycles.

In the air conditioning cycles, the compression ratio of the heat pump compressor hits a maximum of about 3:1. During the heating cycle at low-ambient temperature conditions, the compression ratio can reach 8:1 or higher. The motor, running gear, and compression parts are heavy-duty to withstand these operating conditions.

Scroll compressors are ideally suited for heat pump applications because of their pumping characteristics. The scroll compressor does not lose as much capacity as the reciprocating compressor at the higher head pressures of summer or the lower suction pressures of winter operation. This is because the scroll compressor does not have the top-of-stroke clearance volume loss that a reciprocating compressor has. The gas trapped in the clearance volume of a reciprocating compressor must re-expand before the cylinder starts to fill. This reduces the efficiency. The scroll compressor is about 15% more efficient than a reciprocating compressor. A scroll compressor's compression takes place between two specially designed spiral-shaped forms.

The spiral-shaped forms have multiple chambers, each of which is at a different stage of compression.

Compression is created by the interaction of an orbiting spiral and a stationary spiral. Gas enters the outer openings as one of the spirals orbits. As the spiral orbits, gas is compressed into an increasingly smaller pocket, reaching discharge pressure at the center port. Actually, during operation, all six gas passages are in various stages of compression at all times, resulting in nearly continuous suction and discharging of refrigerant vapor.

Scroll compressors can handle liquid that may enter the scroll plates better than reciprocating compressors, which also increases the need for a crankcase heater, and in most cases a suction line accumulator. This is because one of the scroll plates is stationary and the other plate floats, which gives the plates some play. This play will allow the floating scroll plate to move if liquid should enter the compressor, thus preventing damage to the compressor. Although scroll compressors can handle liquid flood-back better than reciprocating compressors, liquid flood-back does reduce the system's capacity.

Scroll compressors are discharge gas cooled so the compressor shell would feel much hotter than suction gas cooled compressors. They also operate with less vibration and a lower noise level. Scrolls do not have suction or discharge valves; they are equipped with a low-mass, disc-type check valve at the discharge port that prevents the high-pressure refrigerant from traveling back through the compressor during the off cycle. When the compressor shuts off, the gas trapped between the check valve and the compressor scroll will back up through the scroll, causing a strange sound until the pressure between the check valve and scroll equalize. The sound resembles the rustling of a handful of marbles. The high-side pressure will then equalize through the metering device to the low side, like a reciprocating compressor system. Equalization between the check valve and the compressor scroll allows the compressor to start up without hard-start assistance because there is no pressure differential for the first starting revolutions of the compressor.

Considerations for air conditioning and heat pumps equipped with scroll compressors:

A. *Do not operate a scroll compressor with the suction valve closed.*
B. *Avoid operating three-phase scroll compressors in reverse rotation. Reverse rotation is very recognizable because of the extreme noise scroll compressors make when operating in reverse rotation.*
C. *Correct rotation of three-phase compressors can be checked with gage manifolds. Attach manifold gages to both high and low-side valves, operate compressor, and check to see if normal suction and discharge pressure are established quickly, if they are not, shut unit OFF, reverse any two of the power leads, and recheck.*

EQUIPMENT REQUIRED

Scroll compressor (*in or out of sealed system*)
1 Ohmmeter or VOM
Assorted hand tools

SAFETY PRACTICES

The student should be knowledgeable in the use of tools and testing equipment. Follow all safety rules when working with live voltages.

> NOTE: *For safety, motor winding value readings should be obtained through the motor's terminal wiring leads when possible. If not applicable, readings can be taken at the compressor motor terminals, but should be obtained with care.*

COMPETENCY PROCEDURES

Checklist

1. Make sure power has been removed from the system or scroll compressor. ❑
2. Gain access to the scroll compressor motor terminals or motor terminal wire leads. ❑
3. Discharge all motor circuit capacitors before removing scroll compressor terminal wires (*if applicable*). ❑
4. Remove scroll compressor motor wiring leads from associated components if necessary (*relays, capacitor, etc.*)(*if applicable*). ❑

> NOTE: *If removing motor wires from associated components in equipment, make notes for rewiring motor lead wires back into equipment.*

5. Draw a picture of the scroll compressor's terminal layout. ❑
6. Number the scroll compressor terminal or compressor wire leads from the terminal layout above as "1," "2," and "3." (Refer to *Figure 3–220*.) ❑

> **NOTE:** *If testing scroll compressor motor terminal wires, number wires and wire colors so that identification and resistance readings can be recorded and identified (for example, Yellow Wire 1, Black Wire 2, etc.).*

7. Set the ohmmeter the lowest ohms value scale ($R \times 1$, $R \times 200$, etc.). ❏
8. *Zero meter (if applicable).* ❏
9. *Place ohmmeter leads on scroll compressor terminal or wire leads 1 and 2 and check for a resistance value reading on meter.* ❏
10. *Record your ohms value reading at scroll compressor terminals or wire leads 1 and 2. _____* ❏
11. *Place ohmmeter leads on scroll compressor terminal or wire leads 1 and 3 and check for a resistance value reading on the meter.* ❏
12. *Record your ohms value reading at compressor terminals or wire leads 1 and 3. _____* ❏
13. *Place ohmmeter leads on scroll compressor terminals or wire leads 2 and 3 and check for a resistance value reading on meter.* ❏
14. *Record your ohms value reading at compressor terminals or wire leads 2 and 3. _____* ❏

GOOD SCROLL COMPRESSOR MOTOR WINDING RESISTANCE CHECK (SPLIT-PHASE)

*When checking **split-phase scroll compressor motor windings**, good motor windings would be found by a resistance check between the motor's winding terminals. If the motor windings are good, a resistance check across the motor terminals should indicate three separate and different resistance values, with one of those resistance values totaling the sum of the other two.*

> **NOTE:** *An infinity reading (no ohms value reading) is an indication of an open motor winding. A scroll compressor or any motor with an open motor winding would have to be replaced.*

15. *Were the scroll compressor motor windings OK? _____* ❏
16. Gain access to the overload protector lead wires and remove from connectors or electrical terminal connections. ❏
17. Place the ohmmeter leads across the overload protector leads. ❏
18. *Record the measured resistance valve. _____* ❏
19. *Is the overload protector switch OPEN or CLOSED according to resistance value in Step 18? _____* ❏
20. If the overload protector is good, replace the overload protector leads to the correct connectors or electrical terminal leads. ❏

> **NOTE:** *The overload protector of scroll compressors is located on top of the compressor motor housing. The overload protector is a normally closed switch and will open on a rise in temperature above its designed set point. When the overload protector opens, the compressor motor circuit is connected, shutting the compressor OFF. Once the motor cools to a temperature below the design set temperature, the overload resets so that the compressor can restart. If the overload is open on a performance check, feel the top of the compressor housing and see if it is "hot" to the touch. If it is, allow the compressor to cool down and recheck the overload to see if it resets.*

21. Have the instructor check your work. ❏
22. Return all equipment and supplies to their original location. ❏

RESEARCH QUESTIONS

1. What are the three most common compressors used in residential and light commercial air conditioning and refrigeration?

   _____

   _____

2. Which compressor uses a piston for compression of the refrigerant?

   _____

   _____

3. Which compressor is used in small equipment, such as air conditioners and household refrigerators?

   _____

   _____

4. Which compressor uses a stationary part that has a moving part that meshes and matches with the stationary part to complete the compression of the refrigerant vapor?

   _____

   _____

5. Which compressor can deal better with liquid flood-back?

   _____

   _____

Passed Competency _____   Failed Competency _____

Instructor Signature _____   Grade _____

Practical Competency 111

Identifying Common, Run, and Start Terminals of a Split-Phase Scroll Compressor

SUGGESTED MATERIALS

Textbook
Refrigeration & Air Conditioning Technology, 5th Edition, Thomson Delmar Learning
Unit 20—Troubleshooting Electric Motors
Unit 23—Compressors
Unit 41—Troubleshooting
Unit 43—Air Source Heat Pumps

Review Topics
Split-Phase Motors; Heat Pumps Using Scroll Compressors

COMPETENCY OBJECTIVE

The student will be able to identify the Common, Run, and Start terminals of a split-phase scroll compressor.

OVERVIEW

Heat pumps have a special compressor. These compressors have to operate almost three times longer and over a broader range of temperature conditions than a cooling-only compressor used in air conditioning equipment. They also operate under severe low-ambient temperature conditions during start-up, cycle reversals, and defrost cycles.

In the air conditioning cycles, the compression ratio of the heat pump compressor hits a maximum of about 3:1. During the heating cycle at low-ambient temperature conditions, the compression ratio can reach 8:1 or higher. The motor, running gear, and compression parts are heavy-duty to withstand these operating conditions.

Scroll compressors are ideally suited for heat pump applications because of their pumping characteristics. The scroll compressor does not lose as much capacity as the reciprocating compressor at the higher head pressures of summer or the lower suction pressures of winter operation. This is because the scroll compressor does not have the top-of-stroke clearance volume loss that a reciprocating compressor has. The gas trapped in the clearance volume of a reciprocating compressor must re-expand before the cylinder starts to fill. This reduces the efficiency. The scroll compressor is about 15% more efficient than a reciprocating compressor. A scroll compressor's compression takes place between two specially designed spiral-shaped forms.

The spiral-shaped forms have multiple chambers, each of which is at a different stage of compression. Compression is created by the interaction of an orbiting spiral and a stationary spiral. (Refer to *Figure 4–35.*) Gas enters the outer openings as one of the spirals orbits. As the spiral orbits, gas is compressed into an increasingly smaller pocket, reaching discharge pressure at the center port. Actually, during operation, all six gas passages are in various stages of compression at all times, resulting in nearly continuous suction and discharging of refrigerant vapor.

Scroll compressors can handle liquid that may enter the scroll plates better than reciprocating compressors, which also increases the need for a crankcase heater, and in most cases a suction line accumulator. This is because one of the scroll plates is stationary and the other plate floats, which gives the plates some play. This play will allow the floating scroll plate to move if liquid should enter the compressor, thus preventing damage to the compressor. Although scroll compressors can handle liquid flood-back better than reciprocating compressors, liquid flood-back does reduce the system's capacity.

Scroll compressors are discharge gas cooled so the compressor shell would feel much hotter than suction gas cooled compressors. They also operate with less vibration and a lower noise level. Scrolls do not have suction or discharge valves; they are equipped with a low-mass, disc-type check valve at the discharge port that prevents the high-pressure refrigerant from traveling back through the compressor during the off cycle. When the compressor shuts off, the gas trapped between the check valve and the compressor scroll will back up through the scroll, causing a strange sound until the pressure between the check valve and scroll equalize. The sound resembles the rustling of a handful of marbles. The high-side pressure will then equalize through the metering device to the low side, like a reciprocating compressor system. Equalization between the check valve and the compressor scroll allows the compressor to start up without hard-start assistance because there is no pressure differential for the first starting revolutions of the compressor.

Considerations for air conditioning and heat pumps equipped with scroll compressors:

 A. *Do not operate a scroll compressor with the suction valve closed.*
 B. *Avoid operating three-phase scroll compressors in reverse rotation. Reverse rotation is very recognizable because of the extreme noise scroll compressors make when operating in reverse rotation.*
 C. *Correct rotation of three-phase compressors can be checked with gage manifolds. Attach manifold gages to both high- and low-side valves, operate compressor, and check to see if normal suction and discharge pressure are established quickly; if they are not, shut unit OFF, reverse any two of the power leads, and recheck.*

GOOD SCROLL COMPRESSOR MOTOR WINDING RESISTANCE CHECK (SPLIT-PHASE)

*When checking **split-phase scroll compressor motor windings**, good motor windings would be found by a resistance check between the motor's winding terminals. If the motor windings are good, a resistance check across the motor terminals should indicate three separate and different resistance values, with one of those resistance values totaling the sum of the other two.*

If the scroll compressor motor windings are good, the measured resistance values can also be used to determine the common, run, and start terminals of the motor.

The **highest** resistance from Common is identified as the START terminal and the **lowest** resistance from the Common terminal is the RUN terminal.

EQUIPMENT REQUIRED

Scroll compressor (*in or out of system*)
Ohmmeter or VOM meter
Assorted hand tools (*if applicable*)

SAFETY PRACTICES

Make sure meter is set to the proper function and the electrical power is disconnected from the motor. **Discharge capacitors used in compressor motor circuit before removing motor terminal wires** (*if applicable*).

> *NOTE: For safety, scroll compressor motor winding value readings should be obtained through the motor's terminal wiring leads when possible. If not applicable, readings can be taken at the compressor motor terminals, but should be obtained with care.*

COMPETENCY PROCEDURES

Checklist

 1. Make sure power has been removed from the system or scroll compressor. ❏
 2. Gain access to the scroll compressor motor terminals or motor terminal wire leads. ❏

3. Discharge all motor circuit capacitors before removing compressor terminal wires (*if applicable*). ❑
4. Remove scroll compressor motor wiring leads from associated components (*relay, capacitor, etc.*) (*if applicable*). ❑

NOTE: *If removing scroll compressor motor wires from associated components in equipment, make notes for rewiring motor lead wires back into equipment.*

---

5. Draw a picture of the scroll compressor's terminal layout. ❑
6. Number the scroll compressor terminal or scroll compressor wire leads from the terminal layout above as "**1**," "**2**," and "**3**." ❑

NOTE: *If testing scroll compressor motor terminal wires, number wires and wire colors so that identification and resistance readings can be recorded and identified (for example, Yellow Wire 1, Black Wire 2, etc.).*

---

7. Set the ohmmeter the lowest ohms value scale ($R \times 1$, $R \times 200$, *etc.*). ❑
8. *Zero meter (if applicable).* ❑
9. *Place ohmmeter leads on scroll compressor terminal or wire* **leads 1 and 2** *and check for a resistance value reading on meter.* ❑
10. *Record your ohms value reading at scroll compressor terminals or wire* **leads 1 and 2.** _____ ❑
11. *Place ohmmeter leads on scroll compressor terminal or wire* **leads 1 and 3** *and check for a resistance value reading on meter.* ❑
12. *Record your ohms value reading at scroll compressor terminals or wire* **leads 1 and 3.** _____ ❑
13. *Place ohmmeter leads on scroll compressor terminals or wire* **leads 2 and 3** *and check for a resistance value reading on meter.* ❑
14. *Record your ohms value reading at scroll compressor terminals or wire* **leads 2 and 3.** _____ ❑

GOOD SCROLL COMPRESSOR MOTOR WINDING RESISTANCE CHECK (SPLIT-PHASE)

*When checking **split-phase scroll compressor motor windings**, good motor windings would be found by a resistance check between the motor's winding terminals. If the motor windings are good, a resistance check across the motor terminals should indicate three separate and different resistance values, with one of those resistance values totaling the sum of the other two.*

15. *Were the motor windings of the split-phase scroll compressor OK?* _____ ❑

IDENTIFYING THE COMMON—RUN—START MOTOR TERMINALS OF A SCROLL COMPRESSOR

16. *Between which two terminal numbers did you get the highest resistance value?*
    Terminals _____ *and* _____ ❑
17. The two terminals listed in Step 16 are the RUN and START terminals; the terminal left over is the COMMON terminal. ❑
18. *Record the COMMON terminal number.* _____ ❑
19. *From the COMMON terminal, which terminal has the highest resistance value?* _____ ❑
20. Terminal number listed in Step 19 is the START winding. ❑
21. *Record the START terminal.* _____ ❑
22. *From the COMMON terminal, which terminal has the lowest resistance value?* _____ ❑

23. The terminal with the lowest resistance from the common terminal is the RUN winding. ❑
24. *Record the RUN terminal.* _____ ❑

> **NOTE:** *Common terminal to the highest resistance is START. Common terminal to the lowest resistance is RUN.*

25. Have the instructor check your work. ❑
26. Return all equipment and supplies to their original location. ❑

RESEARCH QUESTIONS

1. The two types of motors used for hermetic compressors are _____ and _____.
2. Define BEMF.

   _____

   _____

3. The two types of relays used to start hermetic compressors are the _____ relay and the _____.
4. **True or False:** Scroll compressors are equipped with crankcase heaters.

   _____

   _____

5. **True or False:** Three-phase motors have low starting torque.

   _____

   _____

Passed Competency _____ **Failed Competency** _____

Instructor Signature _____ **Grade** _____

Practical Competency 112

Three-Phase Scroll Compressor Evaluation

SUGGESTED MATERIALS

Textbook

Refrigeration & Air Conditioning Technology, 5th Edition, Thomson Delmar Learning
Unit 18—Application of Motors
Unit 20—Troubleshooting Electric Motors
Unit 23—Compressors
Unit 41—Troubleshooting
Unit 43—Air Source Heat Pumps

Review Topics

Heat Pumps Using Scroll Compressors

COMPETENCY OBJECTIVE

The student will be able to check the motor windings of a three-phase scroll compressor.

OVERVIEW

Electric motors used in compressor applications are normally split-phase or three-phase motors. Split-phase compressor motors can also employ the benefits of the addition of a run, start, or both types of capacitors. Since three-phase motors do not have a common, run, or start winding, a resistance reading from motor terminal to motor terminal should indicate the same resistance value across all terminals (**Figure 4–37**).

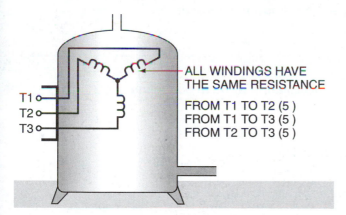

ALL WINDINGS HAVE
THE SAME RESISTANCE

FROM T1 TO T2 (5)
FROM T1 TO T3 (5)
FROM T2 TO T3 (5)

T1
T2
T3

FIGURE 4–37 A three-phase compressor with three leads for the three windings. The windings all have the same resistance.

Three-phase motors do not have run, start, and common windings as in a split-phase motor. These motors have the highest starting torque and best running efficiency without incorporating a run capacitor or start capacitor. There are three individual "hot legs" or motor windings that are 120 degrees out of phase with the next winding. Three-phase motors and equipment can be used only where three-phase voltage is supplied.

Three hot lines identified as L1, L2, and L3 feed the power to the motor. For a 240-volt, three-phase power supply, the voltage measured between any two of the hot legs is 240 volts.

The windings of three-phase motors can be connected in a couple of different ways. The **WYE configuration** three-phase motor has three motor terminals marked **T1**, **T2**, and **T3**. WYE connection motors are designed to operate at a specific voltage and are easily identifiable by the three designated motor terminal connections or motor leads with the motor windings displayed on a diagram in the form of a **Y**.

The directional rotation of these motors can be easily reversed by switching any two of the power lead connections or motor terminal connections.

GOOD SCROLL MOTOR WINDING RESISTANCE CHECK (THREE-PHASE MOTORS)

Three-phase motors do not have a common, run, or start winding. The motor windings are of the same resistance and depend on line voltage phasing to operate. A resistance check of these motors should indicate the same resistance across all motor terminals to reflect good motor windings.

EQUIPMENT REQUIRED

Three-phase compressor (*in or out of system*)
Ohmmeter or VOM
Assorted hand tools (*if applicable*)

SAFETY PRACTICES

Make sure the meter is set to the proper function and the electrical power is disconnected from the motor.

COMPETENCY PROCEDURES   Checklist

1. Make sure power has been removed from the system or scroll compressor. ❑
2. Gain access to the scroll compressors motor terminals or motor terminal wire leads. ❑
3. Remove compressor motor wiring leads from associated components if
   necessary (*relay, contactor, electronic board, etc.*) (*if applicable*). ❑

> **NOTE:** *If removing motor wires from associated components in equipment, make notes for rewiring motor lead wires back into equipment.*

4. Draw a picture of the compressor's terminal layout. ❑
5. Number the compressor terminal or compressor wire leads from the terminal
   layout above as "**1**," "**2**," and "**3**." ❑

> **NOTE:** *If testing compressor motor terminal wires, number wires and wire colors so that identification and resistance readings can be recorded and identified (for example, Yellow Wire 1, Black Wire 2, etc.).*

6. Set the ohmmeter the lowest ohms value scale ($R \times 1$, $R \times 200$, etc.). ❑
7. *Zero meter (if applicable).* ❑
8. *Place ohmmeter leads on compressor terminal or wire **leads 1 and 2** and check for a
   resistance value reading on meter.* ❑
9. *Record your ohms value reading at compressor terminals or **wire leads 1 and 2**. _____* ❑
10. *Place ohmmeter leads on compressor terminal or **wire leads 1 and 3** and check for a
    resistance value reading on meter.* ❑
11. *Record your ohms value reading at compressor terminals or **wire leads 1 and 3**. _____* ❑
12. *Place ohmmeter leads on compressor **terminal leads 2 and 3** and check for a resistance
    value reading on meter.* ❑
13. *Record your ohms value reading at compressor terminals or **wire leads 2 and 3**. _____* ❑

14. *Record resistance values:*

     Terminals 1 and 2 _____   ❑

     Terminals 1 and 3 _____   ❑

     Terminals 2 and 3 _____   ❑

GOOD SCROLL COMPRESSOR MOTOR WINDING RESISTANCE CHECK (THREE-PHASE MOTORS)

Three-phase motors do not have a common, run, or start winding. The motor windings are of the same resistance and depend on line voltage phasing to operate. A resistance check of these motors should indicate the same resistance across all motor terminals to reflect good motor windings.

> **NOTE:** *If an open motor winding were indicated on the motor winding test, it is important to remember that three-phase motors have internal overload protection on each winding. When checking motor windings, be sure to allow the motor to cool to ambient temperature before making a final decision on motor winding failure.*

15. *From resistance values taken, were the motor windings of the scroll compressor tested OK?* _____   ❑
16. Have the instructor check your work.   ❑
17. Return all equipment and supplies to their original location.   ❑

RESEARCH QUESTIONS

1. In the WYE or delta three-phase connection, which one is also referred to as a star-delta start connection?

_____

_____

2. List a couple of advantages of three-phase motors over single-phase motors.

_____

_____

3. Do three-phase motors need the assistance of capacitors and starting relays?

_____

_____

4. Between the WYE and the delta three-phase connection, which one operates where the line currents are 1.732 times higher than the phase current of each winding?

_____

_____

5. Between the WYE and delta three-phase connection, which one operates where the line voltages are equal to 1.732-times the phase voltage?

_____

_____

Passed Competency _____	**Failed Competency** _____
Instructor Signature _____	**Grade** _____

Theory Lesson: Sealed-System Pump Down

SUGGESTED MATERIALS

Textbook

Refrigeration & Air Conditioning Technology, 5th Edition, Thomson Delmar Learning
Unit 25—Special Refrigeration System Components
Unit 43—Air Source Heat Pumps

Review Topics

Refrigeration Line Services Valves

Key Terms

back seating • front seating • liquid line service valve • pump down • vapor service valve

OVERVIEW

Performing a pump down on split-system air conditioning or heat units requires that the condensing unit or outdoor coil be equipped with service valves that can be opened or closed manually.

Pumping down air conditioning or heat pump equipment is the process of using sealed-system service valves to isolate and trap the refrigerant of the system in a receiver, condenser, or outdoor coil of a heat pump. Performing a pump down requires the technician to attach manifold gages to the system's liquid and vapor lines, then use a refrigeration service or Allen wrench to front seat the system's liquid line valve clockwise the whole way in. The system's compressor is then started, allowing the compressor to suck refrigerant vapor from the indoor coil, line set, and metering device. The refrigerant is then compressed and discharged into the condenser, outdoor coil, or system receiver. Because the liquid line service valve is front-seated the whole way, refrigerant can no longer pass through the liquid line to the metering device and evaporator or indoor coil. This allows the refrigerant to back-up and collect in the condenser, outdoor coil, or system receiver. Since the vapor line service valve is still open, the compressor continues to suck refrigerant vapor from the evaporator, and vapor line until the system's vapor line is pulled to 0 psig or vacuum. Once 0 psig or a vacuum is indicated on the low-side gage manifold, the vapor line service valve is front seated clockwise the whole way, closing off the vapor. Once the low-sideline pressure is at 0 psig or vacuum, the system's compressor is shut off.

This permits sealed system work on the line set, evaporator, filter drier, and metering device, without having to recover the refrigerant from the system. Once repairs are made, the system can be pressure checked and evacuated through the vapor and liquid line service valves. On completion of a proper evacuation, the vapor and liquid line service valves can be back-seated the whole way by turning the valve counterclockwise. This allows the stored refrigerant to reenter the sealed system and the unit can be put back into operation.

Some sealed systems may have two liquid line service valves, one located at liquid line leaving the condenser and the other at the outlet port of the receiver. This type of arrangement allows the refrigerant to be stored in the liquid receiver, isolating the receiver from the compressor, condenser, metering device, line set, and evaporator. This allows repairs or replacement to take place on any component of the sealed system without having to recover the refrigerant.

In this type of system arrangement, the liquid line service valve at the receiver is front-seated the whole way before the system is started. The vapor and the other liquid line valve are left open while operating the system's compressor until a 0 psig reading is reached on the manifold low-side gage.

At this point, the vapor valve and the other liquid line valve are front-seated the whole way before the system's compressor is shut OFF. Once the system repairs or replacement is completed a vacuum can be pulled on the total system before the system service valves are back-seated, to allow refrigerant to reenter the system.

Practical Competency 113

Performing a Pump Down on a Split-System Heat Pump

SUGGESTED MATERIALS

Textbook
Refrigeration & Air Conditioning Technology, 5th Edition, Thomson Delmar Learning
Unit 43—Air Source Heat Pumps

Review Topics
Heat Pump Operation-Sealed System Components

COMPETENCY OBJECTIVE

The student will be able to pump down a split-system heat pump.

OVERVIEW

Performing a pump down on split-system air conditioning or heat units requires that the condensing unit or outdoor coil be equipped with service valves that can be opened or closed manually.

Pumping down air conditioning or heat pump equipment is the process of using sealed-system service valves to isolate and trap the refrigerant of the system in a receiver, condenser, or outdoor coil of a heat pump. Performing a pump down requires the technician to attach manifold gages to the system's liquid and vapor lines, then use a refrigeration service or Allen wrench to front seat the system's liquid line valve clockwise the whole way in. The system's compressor is then started, allowing the compressor to suck refrigerant vapor from the indoor coil, line set, and metering device. The refrigerant is then compressed and discharged into the condenser, outdoor coil, or system receiver. Because the liquid line service valve is front-seated the whole way, refrigerant can no longer pass through the liquid line to the metering device and evaporator or indoor coil. This allows the refrigerant to back-up and collect in the condenser, outdoor coil, or system receiver. Since the vapor line service valve is still open, the compressor continues to suck refrigerant vapor from the evaporator, and vapor line until the system's vapor line is pulled to 0 psig or vacuum. Once 0 psig or a vacuum is indicated on the low-side gage manifold, the vapor line service valve is front seated clockwise the whole way, closing off the vapor. Once the low-sideline pressure is at 0 psig or vacuum, the system's compressor is shut off.

EQUIPMENT REQUIRED

Manifold gage set
3/16" Allen wrench
5/16" Allen wrench
Adjustable wrench
Refrigeration service wrench (*if applicable*)
Safety glasses
Safety gloves

SAFETY PRACTICES

The student should be knowledgeable in the use of tools and testing equipment. Follow all EPA rules and regulations when working with refrigerants.

COMPETENCY PROCEDURES

1. Turn unit thermostat to the OFF position. ❏
2. Make sure manifold gage hose valves are closed. ❏
3. Connect the high-side manifold gauge hose to the unit's liquid line and vapor line service valves. ❏
4. Set the thermostat to the COOLING mode. ❏
5. Set thermostat temperature selector switch below ambient temperature. ❏
6. Let the unit operate for 5 minutes. ❏
7. *Record the operating high-side pressure.* _____ ❏
8. *Record the operating low-side pressure.* _____ ❏
9. Let the heat pump continue to operate. ❏
10. Remove liquid line and vapor line service valve protective caps. ❏
11. Use the refrigeration wrench or 3/16" Allen wrench and front-seat the liquid line service valve the whole way in by turn valve clockwise (**Figure 4–38**). ❏

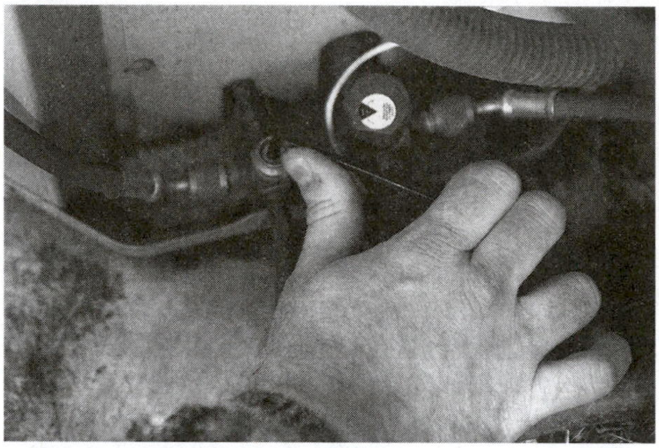

FIGURE 4–38   Front-seating the liquid line service valve.

12. Observe the manifold low-side pressure gage. ❏
13. Use the 5/16" Allen wrench or refrigeration service wrench and front-seat the vapor line service valve when the low-side manifold gage reaches 0 psig. ❏

> **NOTE:** *If applicable, the low-side pressure switch may have to be jumped out to complete the system pump down.*

14. Turn OFF the power to the unit. ❏
15. *Watch both manifold gauge pressures for about 5 minutes.* ❏

> **NOTE:** *If low-side manifold gage pressure rises higher than 0 psig, there is still refrigerant in the system. Open the vapor line service valve, start the unit, and continue the pump down process. Operate system until low-side pressure holds at 0 psig.*

> **NOTE:** *If pressures continue to rise once the unit is shut OFF, the service valves may be leaking, or valves within the compressor may be leaking.*

16. Once pump down is complete, shut power OFF to the unit for 5 minutes. ❏
17. After the system has been OFF for 5 minutes open the liquid line service valve by back-seating it the whole way out. (*Turn valve counterclockwise.*) ❏
18. Back-seat the vapor line service valve the whole way out. (*Turn valve counterclockwise.*) ❏

19. Let the system's pressures equalize for about 5 minutes. ❑
20. *Record the low-side manifold gage standing pressure.* _____ ❑
21. *Record the high-side manifold gage standing pressure.* _____ ❑
22. Turn ON the power to the unit. ❑
23. Let the system operate for 5 minutes. ❑
24. *Record the operating low-side manifold gage pressure.* _____ ❑
25. *Record the operating high-side manifold gage pressure.* _____ ❑
26. Set thermostat to the OFF position. ❑
27. Remove manifold gages from the sealed system. ❑
28. Replace service valve protective caps. ❑
29. Have the instructor check your work. ❑
30. Return all tools and test equipment to their proper location. ❑

RESEARCH QUESTIONS

1. When performing a pump down on a system, where does the refrigerant get stored?

   _____

   _____

2. A pump down performed on a sealed system would allow service work to be performed on what sealed-system components?

   _____

   _____

3. Once a pump down is performed on a sealed system, when would a vacuum need to be performed on the indoor coil and line set before the refrigerant could be returned to the complete sealed system?

   _____

   _____

4. Can a pump down be performed on all heat pump units?

   _____

   _____

5. What could prevent the low-side and high-side pressures from equalizing once the system has been off for a period of time?

   _____

   _____

Passed Competency _____   Failed Competency _____

Instructor Signature _____   Grade _____

Theory Lesson: Heat Pump Low-Voltage Thermostats

SUGGESTED MATERIALS

Textbook
Refrigeration & Air Conditioning Technology, 5th Edition, Thomson Delmar Learning
Unit 43—Air Source Heat Pumps

Review Topics
Controls for the Air-to-Air Heat Pump; Cooling Cycle Control; Space Heating Control; Heat Anticipator; Electronic Thermostats

Key Terms
combination thermostats • emergency heat • fan on-auto • heat-off-cool • RC and RH terminal thermostat modes • thermostat conductor wire • thermostat sub-base

OVERVIEW

All split-system packaged light commercial heat pumps are controlled by low-voltage thermostats. These thermostats route current through low-voltage control circuits that in turn, energize and de-energize high-voltage loads within a system (**Figure 4–39**).

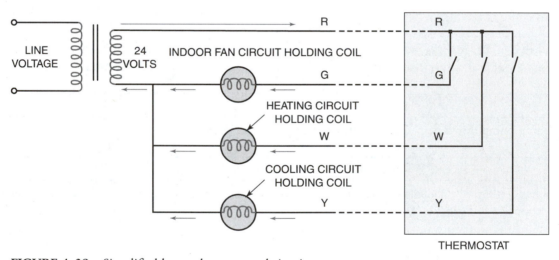

FIGURE 4–39   Simplified low-voltage control circuit.

Thermostats are passed through devices with no electrical loads with the exception of thermostats that may have emergency-heat lights, trouble lights, or electronic LED display. Thermostats are comprised of multiple switches that open and close according to temperature and the desired mode of operation, and are generally categorized by the type of HVAC equipment on which they are to be used. **Thermostat modes of operation are:**

Cooling only—*Used on systems not equipped with the addition of heating*
Heating only—*Used on systems not equipped with the addition of cooling*
Heating and cooling combination—*Used on systems equipped with both cooling and heating modes*

Combination thermostats may also have multistages for heating, cooling, or both.

Many different manufactured thermostats are used with different manufactured HVAC equipment, which makes wiring of combination thermostats confusing at times. For this reason it important that technicians understand universal thermostat terminal functions and identification.

Most thermostat manufacturers use the following thermostat terminals for the same functions and modes of operation:

R Terminal—Hot 24-volt power coming from the **control transformer**
RC Terminal—Hot 24-volt power feeding the **cooling circuit**
RH Terminal—Hot 24-volt power feeding the **heating circuit**

> *NOTE: Thermostats with **RC and RH terminals** can be used on HVAC equipment with dual low-voltage transformers, one for the cooling system and one for the heating system. For applications for both heating and cooling with one low-voltage thermostat, the RC and RH thermostat terminal would have to be "jumped" by passing a wire between the RC and RH terminals, so that low voltage is supplied for both the cooling and heating modes. Some combination thermostats come with the RC and RH terminals "jumped," and may have to be removed for single transformer installations.*

W Terminal—Thermostat **heating** circuit
Y Terminal—Thermostat **cooling** circuit
G Terminal—**Indoor fan (blower)** circuit
O Terminal—*Heat pump thermostats only*—Reversing (*four-way*) valve—**Cooling**
B Terminal—*Heat pump thermostats only*—Reversing (*four-way*) valve—**Heating**
E Terminal—*Heat pump thermostats only*—**Emergency heat**

These thermostat terminals and modes are universal in the industry for the most part. Combination thermostats may also have multiple modes of operation such as:

W1 Terminal—**First-stage heating**
W2 Terminal—**Second-stage heating**
W3 Terminal—**Third-stage heating**
Y1 Terminal—**First-stage cooling**
Y2 Terminal—**Second-stage cooling**
X or C Terminals—Are normally for the neutral 24-volt transformer leg

Although thermostat terminal letters are universal throughout the HVAC industry, it is always important to refer to the individual thermostat instructions and wiring diagram for actual terminal functions.

It is important to realize that **R thermostat terminals** feed the 24-volt power source to the thermostat; other thermostat terminals such as **G, W, and Y,** etc. carry the hot 24-volt power from the thermostat to the individual fan, heating, or cooling relay contactors, or sequencers.

Most low-voltage thermostats adhere to a thermostat colored conductor wire to thermostat terminal letter identification (**Figure 4–40**). This makes for easier thermostat terminal identification and system circuit evaluation.

Red conductor wire—wired to the thermostat R-terminal—*Hot 24 volts*
Green conductor wire—wired to the thermostat G-terminal—*Blower circuit*
Yellow conductor wire—wired to the thermostat Y-terminal—*Cooling circuit*
White conductor wire—wired to the thermostat W terminal—*Heating circuit*
Orange conductor wire—wired to heat pump thermostat O terminal–*Four-way valve*

Other thermostat conductor colors would be wired to additional thermostat terminals and would stay universal throughout the equipment control wiring. **Example:** *If the black conductor wire were wired at the*

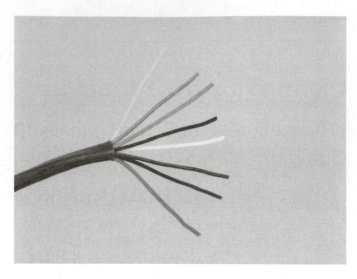

FIGURE 4–40   Thermostat conductor wire. (*Photo by Bill Johnson*)

***C-terminal* (*common 24 volt leg*)** *of the thermostat, it would also be connected to the **C-terminal** of the air handler and outdoor unit.*

It is important to note that technicians should learn thermostat modes of operation by terminal designation, rather than thermostat conductor wire color. Although wire color to thermostat terminal letters is flowed in most cases, there are times when it is not, so knowing the terminal function of the thermostat is important.

Some thermostats are supplied as complete units (**Figure 4–41**) and contain thermostat terminals and thermostat switch functions. Other thermostats come as two separate components, the thermostat, which controls temperature demand and temperature setting, and the thermostat sub-base (**Figure 4–42**).

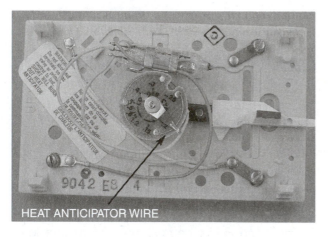

HEAT ANTICIPATOR WIRE

FIGURE 4–41   Single-unit thermostat. (*Photo by Bill Johnson*)

SOME OF THE FIELD WIRING CONNECTORS

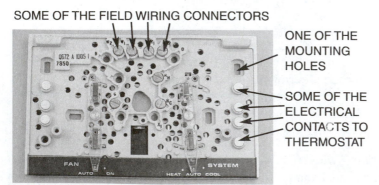

ONE OF THE MOUNTING HOLES

SOME OF THE ELECTRICAL CONTACTS TO THERMOSTAT

FIGURE 4–42   Thermostat sub-base. (*Photo by Bill Johnson*)

The thermostat sub-base normally mounts on the wall, and wiring is fastened inside the sub-base on terminals. These terminals are designed in such a way as to allow easy wire makeup. When the thermostat is screwed down onto the sub-base, electrical connections are made between the two. The sub-base normally contains the selector switches, such as **FAN ON-AUTO** and **HEAT-OFF-COOL** and **EMERGENCY HEAT** for heat pump thermostats.

Troubleshooting the thermostat is not as complicated as some would think. Before checking the thermostat it is important to confirm that 24 volts are being supplied to the thermostats **R, RC,** or **RH**. This can be done by a voltage check between the R and G terminal of the thermostat or thermostat sub-base, with the fan switch set in the AUTO position (**Figure 4–43**).

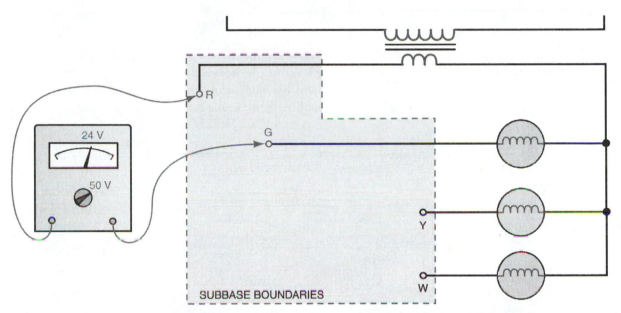

FIGURE 4–43 A voltage reading across thermostat R and G terminals with the fan switch set in the AUTO position.

A measured 24 volts between the **R** and **G** terminal is proof that the thermostat R terminal is being supplied with 24 volts from the system's transformer. No voltage between the R and G terminals is an indication of low voltage failure, either due to a defective step-down control transformer (**Figure 4–44**) or low-voltage wiring from the transformer to the thermostat.

FIGURE 4–44 Standard 24-volt–40 VA step-down control transformer.

A jumper wire placed from the thermostat R terminal and any function terminal such as **G, Y, W, E, O,** etc., should energize the system to operate in the selected mode (**Figure 4–45**).

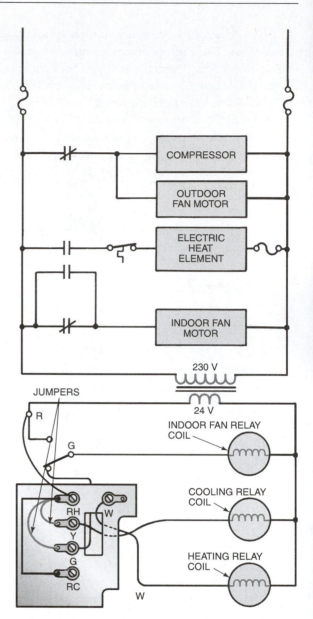

NOTE: THE FAN SELECTOR SWITCH IS NORMALLY A
PART OF THE SUBBASE.

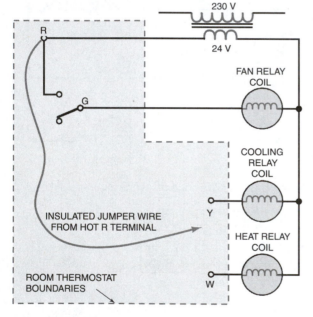

FIGURE 4–45 Jumper wire can be used to bypass the thermostat to test system component operation.

FIGURE 4–46 Jumper placed between the R and Y terminals should energize the cooling circuits. Jumper placed between the R and G terminals should energize the indoor fan circuit.

A jumper wired placed between **R** and any of the thermostat terminals should energize that particular system function (**Figure 4–46**). A jumper placed between the R and G terminals should energize the indoor fan circuit.

The space temperature for an air-to-air heat pump is not controlled in the same way as typical heating and cooling systems. Heat pumps have two complete heating systems and one cooling system. One of the heating systems is the **refrigerated heating,** and the other is the **auxiliary heat** from the supplemental heating system, which could be electric heating elements, oil, or gas furnace. Heat pump thermostats allow the auxiliary heating source to function with the refrigerated heating system or independently if the refrigerated heating system would fail for some reason. Any time the auxiliary heating source is operated on its own, it is referred to as **emergency heat.** Heat pump thermostats are the key to controlling heat pump systems. They are normally two-stage heating and two-stage cooling type thermostats. Understanding the internal workings of heat pumps thermostats can assist technicians in troubleshooting heat pump low-voltage control circuit problems and evaluating the proper mode of operation of a heat pump.

The following examples of a heat pump thermostat are discussed so that technicians can gain knowledge of how the thermostat functions and the internal working circuitry of the thermostat in different modes of operation. The heat pump thermostat is an **AUTOMATIC CHANGEOVER** thermostat. These are thermostats that will automatically change from cooling to heating cycle based on a set temperature selection (**Figure 4–47**).

Example: Cooling Cycle

Temperature sensing element is a bimetal that controls mercury bulb contacts. The auxiliary heat is electric. Fan switch is in AUTO (**Figure 4–48**).

On a rise in temperature, the **R** (*hot 24-volt*) terminal of the thermostat closes the contact with thermostat four-way valve holding coil terminal **0**.

When the space temperature rises about 1 degree (F), the second-stage contacts of the thermostat close from the **R** (*hot 24-volt*) terminal to the **Y** thermostat terminal, which energizes the compressor contactor coil (**Figure 4–49**).

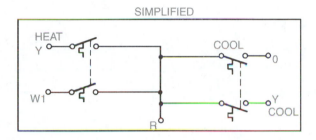

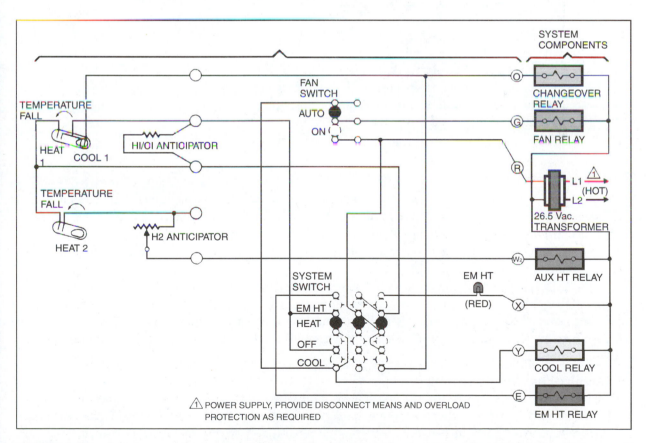

FIGURE 4–47   A heat pump thermostat with two-stage heating and two stage cooling. (*Courtesy of Honeywell Corporation*)

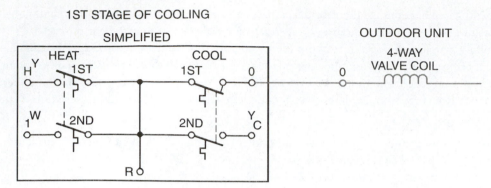

FIGURE 4–48 The first stage of the cooling cycle in which the four-way holding coil is energized between the thermostats **R** (*hot 24-volt*) and terminal **Y** (*four-way valve holding coil*)

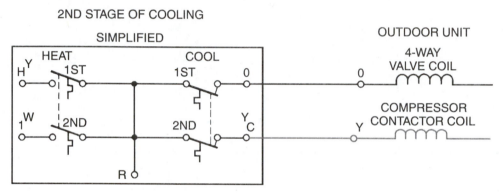

FIGURE 4–49 This diagram shows what happens with a rise in temperature. The first stage of the thermostat is already closed; then the second stage cooling contact closes from the thermostat **R** terminal and **Y** terminal, starting the compressor. The compressor was the last ON and will be the first OFF.

When the compressor contactor is energized, so is the outdoor fan through thermostat terminal **Y**. At the same time that thermostat terminal **Y** is closed through thermostat terminal **R**, the indoor blower is also energized through thermostat terminal **R** and terminal **G** (**Figure 4–50**).

Notice the closed thermostat contacts are **R** *to* **O**, **R** *to* **Y**, and **R** *to* **G**. This energizes the Changeover Relay, Indoor Fan Relay, and Cooling Relay.

When the space temperature is satisfied, the second-stage contacts of the thermostat terminals of **R** and **Y** open. This de-energizes the compressor contactor, stopping the compressor and outdoor fan motor. At the same time, thermostat terminals **R** and **G** open, de-energizing the indoor blower relay, stopping the indoor blower. The first stage thermostat contact terminals **R** and **O** remain **closed**, and the four-way valve holding coil energized.

Example: First-Stage Heating Cycle

When the space temperature falls, the **first-stage heating** contacts close. In the thermostat, terminal **R** (*hot 24-volt*) makes contact with thermostat terminal **Y**, *compressor contactor*, and terminal **G**, **indoor blower**. Notice that terminal contacts between thermostat terminal **R** and thermostat terminal **O** **open**, de-energizing the four-way valve holding coil, so that the compressor's hot discharge gas is directed to the indoor coil, providing heat to the structure through the refrigerated heating cycle only (**Figure 4–51**).

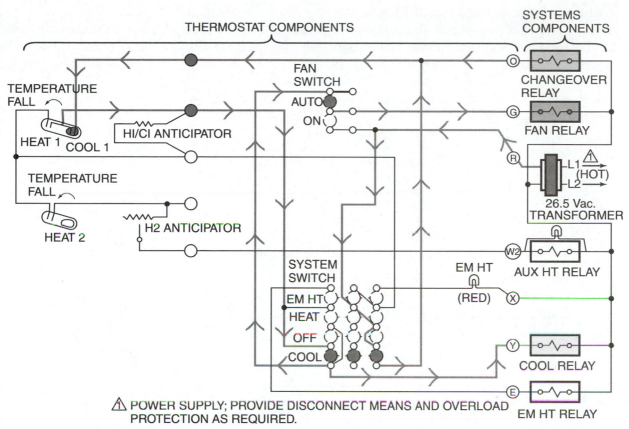

FIGURE 4–50   A generic HEAT–COOL heat pump thermostat in the cooling mode.

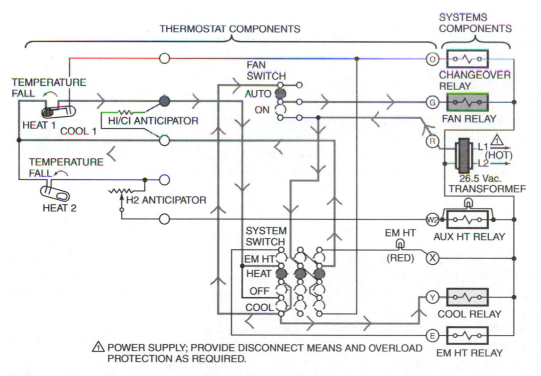

FIGURE 4–51   Thermostat low-voltage circuit with First-Stage Heating cycle energized.
Thermostat terminals **R** to **Y** (*energizing the compressor contactor*), **R** to **G** (*energizing the indoor blower*) are **closed**. Thermostat terminals **R** to **O** are **opened** (*de-energizing the four-way valve*).

Example: Second-Stage Heating Cycle

When the outdoor temperature is cold, below the **balance point** of the structure, the space temperature will fall because the refrigerated heating cycle cannot keep up with the heat loss of the structure; the **second-stage heating** contacts will close and bring on the auxiliary heat to assist the refrigerated heating system. In this example, the auxiliary heat is electric. Notice that besides the thermostat contacts between terminals **R** and **Y**, and **R** and **G**, terminals **R** and thermostat terminal **W2** closes, energizing the auxiliary heat relay (**Figure 4–52**).

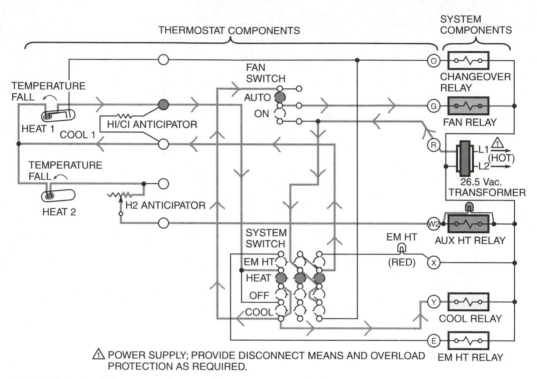

△ POWER SUPPLY; PROVIDE DISCONNECT MEANS AND OVERLOAD
PROTECTION AS REQUIRED.

FIGURE 4–52   Thermostat terminal contacts **R** to **W2** close to energize the auxiliary heat relay for the second stage of heating.

Balance point occurs when the heat pumps refrigerated heating cycle can pump in exactly as much heat as the structure is leaking out.

When the structure temperature rises to its balance point, the second stage heat will be de-energized by the thermostat opening contact terminals **R** and **W2**. If the structure temperature falls below the balance point again, the thermostat will once again close thermostat terminals **R** and **W2** (*second-stage heat*), auxiliary heat relay. The second-stage heat was the last component to be energized, so it will be first component de-energized.

Example: Emergency Heat Mode

Emergency heat is energized manually by the customer by sliding the thermostat selector switch over to the **emergency heat mode**. Most heat pump thermostats will also have an emergency heat LED light, indicating that the system is operating on auxiliary heat only. In most all cases, straight emergency heat is used when the refrigerated heating system fails for some reason. During the emergency heat mode, the thermostat terminal **R** closes contacts **R** to **X**, energizing the **emergency heat LED light**, and terminal **R** to **E**, energizing the **emergency heat relay**, which in turn brings ON the **electric heating banks** (**Figure 4–53**).

Notice that the thermostat terminals R and G are not closed to energize the indoor blower. With electric heat, a heat sequencer (**Figure 4–54**) or thermal fan control is used to energize the indoor blower during the emergency heating mode (**Figure 4–55**).

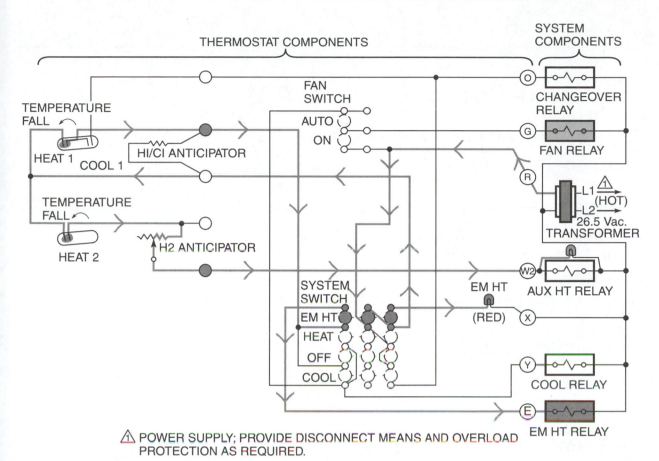

SYSTEM
COMPONENTS

THERMOSTAT COMPONENTS

△ POWER SUPPLY; PROVIDE DISCONNECT MEANS AND OVERLOAD
PROTECTION AS REQUIRED.

FIGURE 4–53 **Emergency heat** mode. Thermostat terminals **R** to **X** closed, energizing the emergency heat LED, and thermostat terminals **R** and **E** closed, energizing the emergency heat relay.

FIGURE 4–54 A multiple-type sequencer. (*Photo by Bill Johnson*)

FIGURE 4–55 Thermal disk-type fan speed control. (*Courtesy of Ferris State University. Photo by John Tomczyk*)

Practical Competency 114

Checking Internal Operation of a Heat Pump Thermostat (Voltage Checks)

SUGGESTED MATERIALS

Refrigeration & Air Conditioning Technology, 5th Edition, Thomson Delmar Learning
Unit 43—Air Source Heat Pumps

Review Topics
Controls for the Air-to-Air Heat Pumps; Cooling Cycle Control; Space Heating Control

COMPETENCY OBJECTIVE

The student will be able to check the performance of a heat pump thermostat by performing voltage checks at thermostat terminals during specific modes of operation.

OVERVIEW

All split-system packaged light commercial heat pumps are controlled by low-voltage thermostats. These thermostats route current through low-voltage control circuits that in turn energize and de-energize high-voltage loads within a system.

Thermostats are passed through devices with no electrical loads with the exception of thermostats that may have emergency-heat lights, trouble lights, or electronic LED display. Thermostats are comprised of multiple switches that open and close according to temperature and the desired mode of operation, and are generally categorized by the type of HVAC equipment on which they are to be used. **Thermostat modes of operation are:**

Cooling only—*Used on systems not equipped with the addition of heating*
Heating only—*Used on systems not equipped with the addition of cooling*
Heating and cooling combination—*Used on systems equipped with both cooling and heating modes*

Combination thermostats may also have multistages for heating, cooling, or both.

Many different manufactured thermostats are used with different manufactured HVAC equipment, which makes wiring of combination thermostats confusing at times. For this reason it important that technicians understand universal thermostat terminal functions and identification.

Most thermostat manufacturers use the following thermostat terminals for the same functions and modes of operation:

R Terminal—Hot 24-volt power coming from the **control transformer**
RC Terminal—Hot 24-volt power feeding the **cooling circuit**
RH Terminal—Hot 24-volt power feeding the **heating circuit**

W Terminal—Thermostat **heating** circuit

Y Terminal—Thermostat **cooling** circuit

G Terminal—**Indoor fan (blower)** circuit

O Terminal—*Heat pump thermostats only*—**Reversing (*four-way*) valve—Cooling**

B Terminal—*Heat pump thermostats only*—**Reversing (*four-way*) valve—Heating**

E Terminal—*Heat pump thermostats only*—**Emergency heat**

These thermostat terminals and modes are universal in the industry for the most part. Combination thermostats may also have multiple modes of operation such as:

W1 Terminal—**First-stage heating**

W2 Terminal—**Second-stage heating**

W3 Terminal—**Third-stage heating**

Y1 Terminal—**First-stage cooling**

Y2 Terminal—**Second-stage cooling**

X or C Terminals are normally for the neutral 24-volt transformer leg.

Although thermostat terminal letters are universal throughout the HVAC industry, it is always important to refer to the individual thermostat instructions and wiring diagram for actual terminal functions.

It is important to realize that **R thermostat terminals** feed the 24-volt power source to the thermostat; other thermostat terminals such as **G, W, and Y,** etc. carry the hot 24-volt power from the thermostat to the individual fan, heating, or cooling relay contactors, or sequencers.

Most low-voltage thermostats adhere to a thermostat colored conductor wire to thermostat terminal letter identification. This makes for easier thermostat terminal identification and system circuit evaluation.

Red conductor wire—wired to the thermostat **R** terminal—*Hot 24-volt*

Green conductor wire—wired to the thermostat **G** terminal—*Blower circuit*

Yellow conductor wire—wired to the thermostat **Y** terminal—*Cooling circuit*

White conductor wire —wired to the thermostat **W** terminal—*Heating circuit*

Orange conductor wire—wired to heat pump thermostat **O** terminal—*Four-way valve*

Other thermostat conductor colors would be wired to additional thermostat terminals and would stay universal throughout the equipment control wiring. **Example:** *If the black conductor wire were wired at the* **C terminal (Common 24-volt leg)** *of the thermostat, it would also be connected to the* **C terminal** *of the air handler and outdoor unit.*

It is important to note that technicians should learn thermostat modes of operation by terminal designation, rather than thermostat conductor wire color. Although wire color to thermostat terminal letters is flowed in most cases, there are times when it is not, so knowing the terminal function of the thermostat is important.

Some thermostats are supplied as complete units and contain thermostat terminals and thermostat switch functions. Other thermostats come as two separate components, the thermostat, which controls temperature demand and temperature setting, and the thermostat sub-base.

The thermostat sub-base normally mounts on the wall, and wiring is fastened inside the sub-base on terminals. These terminals are designed in such a way as to allow easy wire makeup. When the thermostat is screwed down onto the sub-base, electrical connections are made between the two. The sub-base normally contains the selector switches, such as **FAN ON-AUTO** and **HEAT-OFF-COOL** and **EMERGENCY HEAT** for heat pump thermostats.

Troubleshooting the thermostat is not as complicated as some would think. Before checking the thermostat it is important to confirm that 24 volts are being supplied to the thermostats **R, RC,** or **RH.** This can be done by a voltage check between the R and G terminal of the thermostat or thermostat sub-base, with the fan switch set in the AUTO position.

A measured 24 volts between the **R** and **G** terminal is proof that the thermostat R terminal is being supplied with 24 volts from the system's transformer. No voltage between the **R** and **G** terminal is an indication of low voltage failure, either due to a defective step-down control transformer or low-voltage wiring from the transformer to the thermostat.

A jumper wire placed from the thermostat **R** terminal and any function terminal such as **G, Y, W, E, O,** etc., should energize the system to operate in the selected mode. A jumper wired placed between **R** and any of the thermostat terminals should energize that particular system function.

The space temperature for an air-to-air heat pump is not controlled in the same way as typi-cal heating and cooling systems. Heat pumps have two complete heating systems and one cooling system. One of the heating systems is the **refrigerated heating,** and the other is the **auxiliary heat** from the supplemental heating system, which could be electric heating elements, oil, or gas furnace. Heat pump thermostats allow the auxiliary heating source to function with the refrigerated heating system or independently if the refrigerated heating system would fail for some reason. Any time the auxiliary heating source is operated on its own, it is referred to as **emergency heat.** Heat pump thermostats are the key to controlling heat pump systems. They are normally two-stage heating and two-stage cooling type thermostats. Understanding the internal workings of heat pump thermostats can assist technicians in troubleshooting heat pump low-voltage control circuit problems and evaluating the proper mode of operation of a heat pump.

EQUIPMENT REQUIRED

Voltmeter or VOM meter
Assorted hand tools (*if applicable*)

SAFETY PRACTICES

The student should be knowledgeable with the use of tools and testing equipment. Follow all EPA rules and regulations when working with refrigerants.

COMPETENCY PROCEDURES
Checklist

> NOTE: *This competency will require the technician to place the system in to different operating modes with voltage checks being taken and recorded at specific thermostat terminals. A recorded voltage of* **24 volts** *means that the set of thermostat contacts are* "OPEN"; **no voltage** *or* **0 volts** *is an indication that the thermostat contacts are* "CLOSED" *energizing the low-voltage circuit associated with the specific thermostat mode. Once readings are recorded for each specific thermostat mode, recorded voltage reading will be referenced with the "Contact & Voltage Reading Reference Chart"* (**Figure 4–56**).

COOLING MODE CHECK
1. Gain access to low-voltage thermostat terminals and conductor wire either at the thermostat, air handler, or furnace control board. ❑
2. Make sure power is supplied to both the indoor and outdoor units. ❑
3. Set thermostat to the **Cooling** mode. ❑
4. Set thermostat temperature selector switch below the structure's ambient air temperature. ❑
5. Set the VOM meter to measure A/C voltage in a range of at least 24 volts. ❑

> NOTE: *With thermostat set in the* **Cooling** *mode regardless if system is actually cooling or not, take and record the following voltage readings. Remember this test procedure is a method of testing and confirming that the thermostat* **is** *or* **is not** *operating properly.*

Contacts						
Mode	**R & G**	**R & W1**	**R & W2**	**R & E**	**R & C**	**R & Y1**
Cooling	0 volts	0 volts	24 volts	24 volts	24 volts	0 volts
First Stage Heating	0 volts	0 volts	24 volts	24 volts	24 volts	24 volts
Second Stage Heating	0 volts	0 volts	0 volts	24 volts	24 volts	24 volts
Defrost	0 volts	0 volts	24 volts	24 volts	24 volts	0 volts
Emergency Heat	0 volts	24 volts	0 volts	0 volts	24 volts	24 volts

FIGURE 4–56   Thermostat contacts and voltage reading reference chart.

6. *Place meter leads across the* **R** *and* **G** *terminals and record voltage reading.* _____ *Volts* ❏
7. *Place meter leads across the* **R** *and* **W1** *terminals and record voltage reading.* _____ *Volts* ❏
8. *Place meter leads across the* **R** *and* **W2** *terminals and record voltage reading.* _____ *Volts* ❏
9. *Place meter leads across the* **R** *and* **E** *terminals and record voltage reading.* _____ *Volts* ❏
10. *Place meter leads across the* **R** *and* **C** *terminals and record voltage reading.* _____ *Volts* ❏
11. *Place meter leads across the* **R** *and* **Y (Y1)** *terminals and record voltage reading.*
     _____ *Volts* ❏
12. *Compare the actual recorded voltage for* **Cooling** *mode with* **Figure 4–56.** ❏
     No voltage or 0 volts should have been indicated on the following contacts during the **Cooling** *mode (***closed thermostat contacts***):*

R and G–R and W–R and Y (Y1)

Voltage of 24 volts should have been indicated on the following contacts during the **Cooling** *mode (***open thermostat contacts***):*

R and W–R and E–R and C

NOTE: *Any recorded voltage across thermostat contacts that should have indicated NO volts or 0 volts is an indication that the thermostat is defective and should be replaced. No voltage or 0 volts across thermostat contact terminals which should have indicated a 24-volt reading is also an indication of a defective thermostat and requires replacement.*

---

13. *Were any of the thermostat contacts defective for the* **Cooling** *mode?* _____ ❏
14. *If thermostat contacts were defective, list which contacts were defective.*
     _____ _____ _____ ❏

FIRST-STAGE HEATING MODE CHECK
15. *Set thermostat to the* **First-Stage Heating** *mode.* ❏
16. *Set thermostat temperature selector switch 2 degrees (F) above the structures ambient air temperature.* ❏

17. *Place meter leads across the R and G terminals and record voltage reading.* _____ Volts ❑
18. *Place meter leads across the R and W1 terminals and record voltage reading.* _____ Volts ❑
19. *Place meter leads across the R and W2 terminals and record voltage reading.* _____ Volts ❑
20. *Place meter leads across the R and E terminals and record voltage reading.* _____ Volts ❑
21. *Place meter leads across the R and C terminals and record voltage reading.* _____ Volts ❑
22. *Place meter leads across the R and Y (Y1) terminals and record voltage reading.* _____ Volts ❑
23. *Compare recorded voltages for the **First-Stage Heating** mode with **Figure 4–56.*** ❑

    *No voltage or 0 volts should have been indicated on the following contacts during the **First-Stage Heating** mode (**closed thermostat contacts**):*

    R and G–R and W1

    *Voltage of 24 volts should have been indicated on the following contacts during the **First-Stage Heating** mode (**open thermostat contacts**):*

    R and W2–R and E–R and C–R and Y1

24. *Were any of the thermostat contacts defective for the **First-Stage Heating** mode?* _____ ❑
25. *If thermostat contacts were defective, list which contacts were defective.* ❑

    _____ _____ _____

SECOND-STAGE HEATING MODE CHECK

26. Set the thermostat to the **Second-Stage Heating** mode. ❑
27. Set the thermostat temperature selector switch at least 10 degrees (F) above the structures ambient air temperature. ❑

28. *Place meter leads across the R and G terminals and record voltage reading.* _____ Volts ❑
29. *Place meter leads across the R and W1 terminals and record voltage reading.* _____ Volts ❑
30. *Place meter leads across the R and W2 terminals and record voltage reading.* _____ Volts ❑
31. *Place meter leads across the R and E terminals and record voltage reading.* _____ Volts ❑
32. *Place meter leads across the R and C terminals and record voltage reading.* _____ Volts ❑
33. *Place meter leads across the R and Y (Y1) terminals and record voltage reading.* _____ Volts ❑
34. *Compare recorded voltages for the **Second-Stage Heating** mode with **Figure 4–56.*** ❑

    *No voltage or 0 volts should have been indicated on the following contacts during the **Second-Stage Heating** mode (**closed thermostat contacts**):*

    R and G–R and W1–R and W2

Voltage of 24 volts should have been indicated on the following contacts during the Second-Stage Heating mode (open thermostat contacts):

R and E–R and C–R and Y1

> **NOTE:** *Any recorded voltage across thermostat contacts that should have indicated NO volts or 0 volts is an indication that the thermostat is defective and should be replaced. No voltage or 0 volts across thermostat contact terminals that should have indicated a 24-volt reading is also an indication of a defective thermostat and requires replacement.*

35. Were any of the thermostat contacts defective for the **Second-Stage Heating** mode? _____ ❏
36. If thermostat contacts were defective, list which contacts were defective.
_____ _____ _____ ❏

DEFROST MODE CHECK

37. Activate the unit into a **Defrost mode.** (*Advance into defrost mode at the defrost board or defrost timer at the outdoor unit.*) ❏

> **NOTE:** *With the system in the **Defrost** mode, regardless if the system is actually defrosting or not, take and record the following voltage readings. Remember this test procedure is a method of testing and confirming that the thermostat **is** or **is not** operating properly.*

38. *Place meter leads across the **R and G** terminals and record voltage reading. _____ Volts* ❏
39. *Place meter leads across the **R and W1** terminals and record voltage reading. _____ Volts* ❏
40. *Place meter leads across the **R and W2** terminals and record voltage reading. _____ Volts* ❏
41. *Place meter leads across the **R and E** terminals and record voltage reading. _____ Volts* ❏
42. *Place meter leads across the **R and C** terminals and record voltage reading. _____ Volts* ❏
43. *Place meter leads across the **R and Y (Y1)** terminals and record voltage reading. _____ Volts* ❏
44. *Compare recorded voltages for the **Defrost** mode with **Figure 4–56.*** ❏
*No voltage or 0 volts should have been indicated on the following contacts during the **Defrost** mode (**closed thermostat contacts**):*

R and G–R and W1–R and Y (Y1)

*Voltage of 24 volts should have been indicated on the following contacts during the **Defrost** mode (**open thermostat contacts**):*

R and W2–R and E–R and C

> **NOTE:** *Any recorded voltage across thermostat contacts that should have indicated NO volts or 0 volts is an indication that the thermostat is defective and should be replaced. No voltage or 0 volts across thermostat contact terminals that should have indicated a 24 volt reading is also an indication of a defective thermostat and requires replacement.*

45. Were any of the thermostat contacts defective for the **Second-Stage Heating mode?** _____ ❏
46. If thermostat contacts were defective, list which contacts were defective.
_____ _____ _____ ❏

EMERGENCY HEATING MODE CHECK

47. Set thermostat to the **Emergency Heat** mode. ❏

48. Set the thermostat temperature selector switch at least 10 degrees (F) above the structure's ambient air temperature. ❏

> **NOTE:** *With thermostat set in the* **Emergency Heat** *mode, regardless if the system is actually heating or not, take and record the following voltage readings. Remember this test procedure is a method of testing and confirming that the thermostat* **is** *or* **is not** *operating properly.*

49. *Place meter leads across the R and G terminals and record voltage reading. _____ Volts* ❏
50. *Place meter leads across the R and W1 terminals and record voltage reading. _____ Volts* ❏
51. *Place meter leads across the R and W2 terminals and record voltage reading. _____ Volts* ❏
52. *Place meter leads across the R and E terminals and record voltage reading. _____ Volts* ❏
53. *Place meter leads across the R and C terminals and record voltage reading. _____ Volts* ❏
54. *Place meter leads across the R and Y (Y1) terminals and record voltage reading. _____ Volts* ❏
55. *Compare recorded voltages for the* **Emergency Heat mode** *with* **Figure 4–56.** ❏

    No voltage or 0 volts should have been indicated on the following contacts during the **Emergency Heat** *mode (***closed thermostat contacts***):*

    ## R and G–R and W2–R and E

    Voltage of 24 volts should have been indicated on the following contacts during the **Emergency heat** *mode (***open thermostat contacts***)*

    ## R and W1–R and C–R and Y1

> **NOTE:** *Any recorded voltage across thermostat contacts that should have indicated NO volts or 0 volts is an indication that the thermostat is defective and should be replaced. No voltage or 0 volts across thermostat contact terminal should have indicated a 24-volt reading is also a indication of a defective thermostat and requires replacement.*

56. *Were any of the thermostat contacts defective for the* **Emergency Heat mode?** _____ ❏
57. *If thermostat contacts were defective, list which contacts were defective.*
    _____ _____ _____ ❏
58. Turn the thermostat to the OFF position. ❏
59. Have the instructor check your work. ❏
60. Replace all equipment panels (*if applicable*). ❏
61. Return tools and test equipment to their proper location. ❏

RESEARCH QUESTIONS

1. During the heating cycle of a heat pump system, the hot gas leaving the compressor will flow to _____ after leaving the four-way valve.

    A. Indoor coil
    B. Outdoor coil
    C. Metering device
    D. None of the above is correct.

2. A heat pump system that initiates its defrost cycle by temperature alone is most likely equipped with which of the following?

    A. Thermistor
    B. Pressure-sensing switch
    C. Defrost timer
    D. Time delay relay

3. Ideally, there will be low-pressure, low-temperature, superheated vapor at:
    A. The outlet of the outdoor coil in the heating mode
    B. The outlet of the indoor coil in the cooling mode
    C. The inlet of the compressor in the heating mode
    D. All are correct
4. What port of the four-way valve is the true suction port?

_____

_____

5. At the highest COP, what would be the approximate air temperature leaving the indoor coil?

_____

_____

Passed Competency _____   **Failed Competency** _____

Instructor Signature _____   Grade _____

Practical Competency 115

Checking Internal Operation of a Heat Pump Thermostat (Continuity Check) (Figure 4–57)

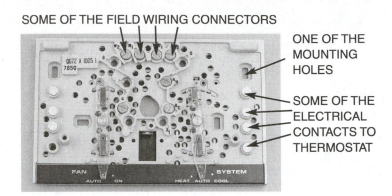

SOME OF THE FIELD WIRING CONNECTORS

ONE OF THE MOUNTING HOLES

SOME OF THE ELECTRICAL CONTACTS TO THERMOSTAT

FIGURE 4–57   Heat pump thermostat sub-base.

SUGGESTED MATERIALS

Refrigeration & Air Conditioning Technology, 5th Edition, Thomson Delmar Learning
Unit 43—Air Source Heat Pumps

Review Topics
Controls for the Air-to-Air Heat Pumps; Cooling Cycle Control; Space Heating Control

COMPETENCY OBJECTIVE

The student will be able to check the performance of a heat pump thermostat by performing voltage checks at thermostat terminals during specific modes of operation.

OVERVIEW

All split-system packaged light commercial heat pumps are controlled by low-voltage thermostats. These thermostats route current through low-voltage control circuits that in turn energize and de-energize high-voltage loads within a system.

Thermostats are passed through devices with no electrical loads with the exception of thermostats that may have emergency-heat lights, trouble lights, or electronic LED display. Thermostats are comprised of multiple switches that open and close according to temperature and the desired mode of operation, and are generally categorized by the type of HVAC equipment on which they are to be used. **Thermostat modes of operation are:**

Cooling only—*Used on systems not equipped with the addition of heating*
Heating only—*Used on systems not equipped with the addition of cooling*
Heating and cooling combination—*Used on systems equipped with both cooling and heating modes*

Combination thermostats may also have multistages for heating, cooling, or both.

Many different manufactured thermostats used with different manufactured HVAC equipment, which makes wiring of combination thermostats confusing at times. For this reason it important that technicians understand universal thermostat terminal functions, and identification.

Most thermostat manufacturers use the following thermostat terminals for the same functions and modes of operation:

R Terminal—Hot 24-volt power coming from the **control transformer**
RC Terminal—Hot 24-volt power feeding the **cooling circuit**
RH Terminal—Hot 24-volt power feeding the **heating circuit**

---

*NOTE: Thermostats with **RC and RH terminals** can be used on HVAC equipment with dual low voltage transformers, one for the cooling system and one for the heating system. For applications for both heating and cooling with one low-voltage thermostat, the RC and RH thermostat terminal would have to be "jumped" by passing a wire between the RC and RH terminals, so that low voltage is supplied for both the cooling and heating modes. Some combination thermostats come with the RC and RH terminals "jumped," and may have to be removed for single transformer installations.*

W Terminal—Thermostat **heating** circuit
Y Terminal—Thermostat **cooling** circuit
G Terminal—**Indoor fan (blower) circuit**
O Terminal—*Heat pump thermostats only*—**Reversing (four-way) valve**—**Cooling**
B Terminal—*Heat pump thermostats only*—**Reversing (four-way) valve**—**Heating**
E Terminal—*Heat pump thermostats only*—**Emergency heat**

These thermostat terminals and modes are universal in the industry for the most part. Combination thermostats may also have multiple modes of operation such as:

W1 Terminal—**First-stage heating**
W2 Terminal—**Second-stage heating**
W3 Terminal—**Third-stage heating**
Y1 Terminal—**First-stage cooling**
Y2 Terminal—**Second-stage cooling**
X or C Terminals are normally for the neutral 24-volt transformer leg.

Although thermostat terminal letters are universal throughout the HVAC industry, it is always important to refer to the individual thermostat instructions and wiring diagram for actual terminal functions.

It is important to realize that **R thermostat terminals** feed the 24-volt power source to the thermostat; other thermostat terminals such as **G, W, and Y**, etc. carry the hot 24-volt power from the thermostat to the individual fan, heating, or cooling relay contactors, or sequencers.

Most low-voltage thermostats adhere to a thermostat colored conductor wire to thermostat terminal letter identification. This makes for easier thermostat terminal identification and system circuit evaluation.

Red conductor wire—wired to the thermostat R terminal—*Hot 24-volt*
Green conductor wire—wired to the thermostat G terminal—*Blower circuit*
Yellow conductor wire—wired to the thermostat Y terminal—*Cooling circuit*
White conductor wire—wired to the thermostat W terminal—*Heating circuit*
Orange conductor wire—wired to heat pump thermostat O terminal—*Four-way valve*

Other thermostat conductor colors would be wired to additional thermostat terminals and would stay universal throughout the equipment control wiring. **Example:** *If the black conductor wire were wired at the*

C-terminal (**Common 24-volt leg**) of the thermostat, it would also be connected to the **C-terminal** of the air handler and outdoor unit.

It is important to note that technicians should learn thermostat modes of operation by terminal designation, rather than thermostat conductor wire color. Although wire color to thermostat terminal letters is flowed in most cases, there are times when it is not, so knowing the terminal function of the thermostat is important.

Some thermostats are supplied as complete units and contain thermostat terminals and thermostat switch functions. Other thermostats come as two separate components, the thermostat, which controls temperature demand and temperature setting, and the thermostat sub-base.

The thermostat sub-base normally mounts on the wall, and wiring is fastened inside the sub-base on terminals. These terminals are designed in such a way as to allow easy wire makeup. When the thermostat is screwed down onto the sub-base, electrical connections are made between the two. The sub-base normally contains the selector switches, such as **FAN ON-AUTO** and **HEAT-OFF-COOL** and **EMERGENCY HEAT** for heat pump thermostats.

Troubleshooting the thermostat is not as complicated as some would think. Before checking the thermostat it is important to confirm that 24 volts are being supplied to the thermostats **R, RC,** or **RH**. This can be done by a voltage check between the R and G terminal of the thermostat or thermostat sub-base, with the fan switch set in the AUTO position.

A measured 24 volts between the **R** and **G** terminal is proof that the thermostat R terminal is being supplied with 24 volts from the system's transformer. No voltage between the **R** and **G** terminal is an indication of low voltage failure, either due to a defective step-down control transformer or low-voltage wiring from the transformer to the thermostat.

A jumper wire placed from the thermostat **R** terminal and any function terminal such as **G, Y, W, E, O,** etc., should energize the system to operate in the selected mode. A jumper wired placed between **R** and any of the thermostat terminals should energize that particular system function.

The space temperature for an air-to-air heat pump is not controlled in the same way as typical heating and cooling systems. Heat pumps have two complete heating systems and one cooling system. One of the heating systems is the **refrigerated heating**, and the other is the **auxiliary heat** from the supplemental heating system, which could be electric heating elements, oil, or gas furnace. Heat pump thermostats allow the auxiliary heating source to function with the refrigerated heating system or independently if the refrigerated heating system would fail for some reason. Any time the auxiliary heating source is operated on its own, it is referred to as **emergency heat**. Heat pump thermostats are the key to controlling heat pump systems. They are normally two-stage heating and two-stage cooling type thermostats. Understanding the internal workings of heat pumps thermostats can assist technicians in troubleshooting heat pump low-voltage control circuit problems and evaluating the proper mode of operation of a heat pump.

EQUIPMENT REQUIRED

Voltmeter or VOM meter
Assorted hand tools (*if applicable*)

SAFETY PRACTICES

The student should be knowledgeable with the use of tools and testing equipment. Follow all EPA rules and regulations when working with refrigerants.

COMPETENCY PROCEDURES

Checklist

> NOTE: *This competency will require that the technician remove thermostat conductor wires from the thermostat sub-base and place the thermostat in different operating modes with continuity checks being taken and recorded at specific thermostat terminals. When taking continuity readings on thermostat terminals, a measured resistance (**Low-Resistance reading**) means that the thermostat contacts are "**Closed.**" An infinity reading (**High-Resistance reading**) with a digital meter or no needle movement with an analog meter means that the set of thermostat contacts are "**Open.**"*

COOLING MODE CHECK

1. Make sure power to the unit is turned OFF. ❑
2. Remove thermostat conductor wires from the heat pump thermostat terminals. ❑
3. Set thermostat to the **Cooling mode.** ❑
4. Set thermostat temperature selector switch below the structures ambient air temperature. ❑
5. Set the VOM meter to measure resistance. ❑

NOTE: With thermostat set in the **Cooling mode,** *even though the thermostat conductor wires have been removed. Measure and record the continuity readings at the specified thermostat terminals. Remember this test procedure is a method of testing and confirming that the thermostat* **is** *or* **is not** *operating properly.*

---

6. *Place meter leads across the* **R** *and* **G** *terminals and record the continuity reading.* _____ ❑
7. *Place meter leads across the* **R** *and* **W1** *terminals and record the continuity reading.* _____ ❑
8. *Place meter leads across the* **R** *and* **W2** *terminals and record the continuity reading.* _____ ❑
9. *Place meter leads across the* **R** *and* **E** *terminals and record the continuity reading.* _____ ❑
10. *Place meter leads across the* **R** *and* **C** *terminals and record the continuity reading.* _____ ❑
11. *Place meter leads across the* **R** *and* **Y (Y1)** *terminals and record the continuity reading.* _____ ❑
12. *Compare recorded continuity readings for* **Cooling mode** *with* **Figure 4–58.** ❑

	Contacts					
Mode	**R & G**	**R & W1**	**R & W2**	**R & E**	**R & C**	**R & Y1**
Cooling	Low resistance	Low resistance	High resistance	High resistance	High resistance	Low resistance
First Stage Heating	Low resistance	Low resistance	High resistance	High resistance	High resistance	High resistance
Second Stage Heating	Low resistance	Low resistance	Low resistance	High resistance	High resistance	High resistance
Defrost	Low resistance	Low resistance	High resistance	High resistance	High resistance	High resistance
Emergency Heat	Low resistance	Low resistance	Low resistance	Low resistance	High resistance	High resistance

FIGURE 4–58 Thermostat Contacts and Resistance Reading Reference Chart.

*A measurable resistance or needle movement (***low resistance***) with an analog meter should have been indicated on the following contacts during the Cooling mode indicating closed thermostat contacts:*

R and G–R and W1–R and Y (Y1)

*Infinity or no needle movement (***high resistance***) with an analog meter should have been indicated on the following contacts during the Cooling mode indicating open thermostat contacts:*

R and W2–R and E–R and C

NOTE: Any **low-resistance reading** across thermostat contacts that should have indicated **high resistance** is an indication of a defective thermostat that should be replaced. **High-resistance readings** across thermostat contact terminals that should have indicated a **low-resistance reading** is also an indication of a defective thermostat and requires replacement.

13. *Were any of the thermostat contacts defective for the* **Cooling mode**? _____ ❏
14. *If thermostat contacts were defective, list which contacts were defective.*

_____ _____ _____ ❏

FIRST-STAGE HEATING MODE CHECK

15. Set thermostat to the **First-Stage Heating** mode. ❏
16. Set thermostat temperature selector switch 2 degrees (F) above the structures ambient air temperature. ❏

NOTE: With thermostat set in the **First-Stage Heating** mode even though the thermostat conductor wires have been removed, measure and record the continuity readings at the specified thermostat terminals. Remember this test procedure is a method of testing and confirming that the thermostat **is** or **is not** operating properly.

17. *Place meter leads across the* **R** *and* **G** *terminals and record the continuity reading.* _____ ❏
18. *Place meter leads across the* **R** *and* **W1** *terminals and record the continuity reading.* _____ ❏
19. *Place meter leads across the* **R** *and* **W2** *terminals and record the continuity reading.* _____ ❏
20. *Place meter leads across the* **R** *and* **E** *terminals and record the continuity reading.* _____ ❏
21. *Place meter leads across the* **R** *and* **C** *terminals and record the continuity reading.* _____ ❏
22. *Place meter leads across the* **R** *and* **Y (Y1)** *terminals and record the continuity reading.* _____ ❏
23. *Compare recorded continuity readings for the* **First-Stage Heating** *with* **Figure 4–58.**
    *A measurable resistance or needle movement (***low-resistance***) with an analog meter should have been indicated on the following contacts during the* **First-Stage Heating mode** *indicating* **closed thermostat contacts:**

R and G–R and W1

*Infinity or no needle movement (***high-resistance***) with an analog meter should have been indicated on the following contacts during* **First-Stage Heating Mode** *indicating* **open thermostat contacts:**

R and W2–R and E–R and C–R and Y1

NOTE: Any **low-resistance reading** across thermostat contacts that should have indicated **high-resistance** is an indication of a defective thermostat that should be replaced. **High-resistance readings** across thermostat contact terminals that should have indicated a **low-resistance reading** is also an indication of a defective thermostat and requires replacement.

24. *Were any of the thermostat contacts defective for the* **First-Stage Heating mode**? _____ ❏
25. *If thermostat contacts were defective, list which contacts were defective.*

_____ _____ _____ ❏

SECOND-STAGE HEATING MODE CHECK

26. Set thermostat to the **Second-Stage Heating** mode. ❏
27. Set thermostat temperature selector switch at least 10 degrees (F) above the structures ambient air temperature. ❏

28. *Place meter leads across the* **R** *and* **G** *terminals and record the continuity reading.* _____ ❑
29. *Place meter leads across the* **R** *and* **W1** *terminals and record the continuity reading.* _____ ❑
30. *Place meter leads across the* **R** *and* **W2** *terminals and record the continuity reading.* _____ ❑

> **NOTE:** *With thermostat set in the* **Second-Stage Heating mode,** *even though the thermostat conductor wires have been removed, measure and record the continuity readings at the specified thermostat terminals. Remember this test procedure is a method of testing and confirming that the thermostat* **is** *or* **is not** *operating properly.*

31. *Place meter leads across the* **R** *and* **E** *terminals and record the continuity reading.* _____ ❑
32. *Place meter leads across the* **R** *and* **C** *terminals and record the continuity reading.* _____ ❑
33. *Place meter leads across the* **R** *and* **Y (Y1)** *terminals and record the continuity reading.* _____ ❑
34. *Compare recorded continuity readings for the* **Second-Stage Heating mode** *with*
    Figure 4–58. ❑
    A measurable resistance or needle movement (low-resistance) with an analog meter should have been indicated on the following contacts during the Second-Stage Heating mode indicating closed thermostat contacts:

    ### R and G–R and W1–R and W2

    *Infinity or no needle movement (**high-resistance**) with an analog meter should have been indicated on the following contacts during the **Second-Stage Heating** mode indicating **open thermostat contacts:***

    ### R and E–R and C–R and Y

> **NOTE:** *Any* **low-resistance reading** *across thermostat contacts that should have indicated* **high resistance,** *is an indication of a defective thermostat that should be replaced.* **High-resistance readings** *across thermostat contact terminals, which should have indicated a* **low-resistance reading** *is also an indication of a defective thermostat and requires replacement.*

35. *Were any of the thermostat contacts defective for the* **Second-Stage Heating mode?** _____ ❑
36. *If thermostat contacts were defective, list which contacts were defective.* _____
    _____   _____ ❑

DEFROST MODE CHECK
37. Activate the unit into a **Defrost mode.** (*Advance into defrost mode at the defrost board or defrost timer at the outdoor unit.*) ❑
38. *Place meter leads across the* **R** *and* **G** *terminals and record the continuity reading.* _____ ❑

> **NOTE:** *With thermostat set in the* **Defrost mode,** *even though the thermostat conductor wires have been removed. Measure and record the continuity readings at the specified thermostat terminals. Remember this test procedure is a method of testing and confirming that the thermostat* **is** *or* **is not** *operating properly.*

39. *Place meter leads across the* **R** *and* **W1** *terminals and record the continuity reading.* _____ ❑
40. *Place meter leads across the* **R** *and* **W2** *terminals and record the continuity reading.* _____ ❑
41. *Place meter leads across the* **R** *and* **E** *terminals and record the continuity reading.* _____ ❑
42. *Place meter leads across the* **R** *and* **C** *terminals and record the continuity reading.* _____ ❑
43. *Place meter leads across the* **R** *and* **Y (Y1)** *terminals and record the continuity reading.* _____ ❑

44. *Compare recorded continuity readings for the **Defrost mode** with **Figure 4–58**.* ❑
*A measurable resistance or needle movement (**low resistance**) with an analog meter should have been indicated on the following contacts during the **Defrost mode** indicating **closed thermostat contacts**:*

R and G–R and W1–R and Y (Y1)

*Infinity or no needle movement (**high resistance**) with an analog meter should have been indicated on the following contacts during the **Defrost mode** indicating **open thermostat contacts**:*

R and W2–R and E–R and C

> **NOTE:** *Any **low-resistance reading** across thermostat contacts that have indicated **high resistance** is an indication of a defective thermostat that should be replaced. **High-resistance readings** across thermostat contact terminals that should have indicated a **low-resistance reading** is also an indication of a defective thermostat and requires replacement.*

45. *Were any of the thermostat contacts defective for the **Second-Stage Heating mode**?* _____ ❑
46. *If thermostat contacts were defective, list which contacts were defective.* ❑
_____ _____ _____

EMERGENCY HEATING MODE CHECK

47. *Set thermostat to the **Emergency Heat** mode.* ❑
48. *Set thermostat temperature selector switch at least 10 degrees (F) above the structure's ambient air temperature.* ❑

> **NOTE:** *With thermostat set in the **Emergency Heat mode**, even though the thermostat conductor wires have been removed. Measure and record the continuity readings at the specified thermostat terminals. Remember this test procedure is a method of testing and confirming that the thermostat **is** or **is not** operating properly.*

49. *Place meter leads across the **R and G** terminals and record the continuity reading.* _____ ❑
50. *Place meter leads across the **R and W1** terminals and record the continuity reading.* _____ ❑
51. *Place meter leads across the **R and W2** terminals and record the continuity reading.* _____ ❑
52. *Place meter leads across the **R and E** terminals and record the continuity reading.* _____ ❑
53. *Place meter leads across the **R and C** terminals and record the continuity reading.* _____ ❑
54. *Place meter leads across the **R and Y (Y1)** terminals and record the continuity reading.* _____ ❑
55. *Compare recorded continuity readings for the **Emergency Heat mode** with **Figure 4–58**.* ❑
*A measurable resistance or needle movement (**low resistance**) with an analog meter should have been indicated on the following contacts during the **Emergency Heat mode** indicating **closed thermostat contacts**:*

R and G–R and W2–R and E

*Infinity or no needle movement (**high resistance**) with an analog meter should have been indicated on the following contacts during the **Emergency Heat mode** indicating **open thermostat contacts**.*

R and W1–R and C–R and Y1

> **NOTE:** *Any **low-resistance reading** across thermostat contacts that should have indicated **high resistance** is an indication of a defective thermostat that should be replaced. **High-resistance readings** across thermostat contact terminals that should have indicated a **low-resistance reading** is also an indication of a defective thermostat and requires replacement.*

56. *Were any of the thermostat contacts defective for the **Emergency Heat mode**?* _____ ❏

57. *If thermostat contacts were defective, list which contacts were defective.* ❏
    _____ _____ _____
58. Turn thermostat to the OFF position. ❏
59. Have the instructor check your work. ❏
60. Replace all equipment panels (*if applicable*). ❏
61. Return tools and test equipment to their proper location. ❏

RESEARCH QUESTIONS

1. What is the most likely cause if a heat-pump system provides heat to the occupied space when the thermostat is set to operate in the cooling mode?
   A. A defective solenoid coil on the four-way reversing valve
   B. A defective compressor
   C. A defective indoor fan motor
   D. A defective outdoor fan motor

2. Excessive ice formation on the outdoor coil of a heat-pump system can be caused by which of the following?
   A. A defective holdback thermostat
   B. A defective indoor fan motor
   C. A defective defrost control
   D. Both A and C

3. If the indoor fan motor is operating, which of the following must be true?
   A. The system is operating in the cooling mode.
   B. The fan switch is in the ON position.
   C. The control transformer is functioning.
   D. The supplementary electric-strip heaters are energized.

4. If the supplementary electric-strip heaters are energized, which of the following must be true?
   A. The indoor fan is energized.
   B. The system is operating in defrost.
   C. The system is calling for second-stage heat.
   D. All of the above must be true.

5. Excessive ice buildup on the outdoor coil may be the result of:
   A. A system that will not switch over to the heating mode.
   B. A system that has defective supplementary electric-strip heaters.
   C. A system that has a defective or improperly adjusted holdback thermostat.
   D. All of the above are possible causes of excessive ice buildup on the outdoor unit.

Passed Competency _____   Failed Competency _____

Instructor Signature _____   Grade _____

Practical Competency 116

Troubleshooting Mechanical Defrost Timers

> NOTE: *This Practical Competency is based on a scenario. Since most defrost systems operate on concepts similar to this, an understanding of this scenario will help make troubleshooting of this defrost timer and other types of defrost timers and defrost systems much easier.*

SUGGESTED MATERIALS

Refrigeration & Air Conditioning Technology, 5th Edition, Thomson Delmar Learning
Unit 43—Air Source Heat Pumps

Review Topics
The Defrost Cycle; Initiating the Defrost Cycle; Terminating the Defrost Cycle; Electronic Control of Defrost

COMPETENCY OBJECTIVE

The student will be able to check the operation of a mechanical defrost timer.

OVERVIEW

Heat pumps are designed to initiate the defrost cycle when the coil builds frost that will affect the heat pump's performance. Some heat pumps use **time and temperature** to start the defrost cycle. Many older heat pumps use a mechanical timer and temperature sensing device.

The defrost timer closes a set of contacts for 10 to 20 seconds for a trial defrost every 90 minutes. The timer contacts are in series with the temperature sensor contacts so that both must be closed at the same before initiating the defrost cycle.

This means that two conditions must be made at the same time before defrost can be started. Typically, the timer runs any time the compressor runs, even in the cooling mode and tries for defrost every 90 minutes. The temperature sensing device contacts will close when the outdoor coil temperature is 26 degrees (F) or lower, and open when the coil temperature reaches 50 degrees (F).

Some manufacturers will also incorporate an air-pressure switch that measures the air pressure drop across the outdoor coil. The air pressure switch is wired in series with the timer contacts and the temperature sensing device contacts.

When the unit begins to build ice, the pressure drop occurs and the air switch contacts close. This combination of time, temperature, and pressure is very effective in initiating the defrost cycle when ice has actually formed on the coil.

After defrost has started, time, temperature, or pressure (*if used*) can terminate the defrost cycle. Temperature sensor can terminate once it senses approximately 50-degree coil temperature at its location. If the temperature contacts would not open due to cold outside temperatures, the timer will terminate defrost within a predetermined time period. Normally, 10 minutes is the maximum allowable time limit (**Figure 4–59**).

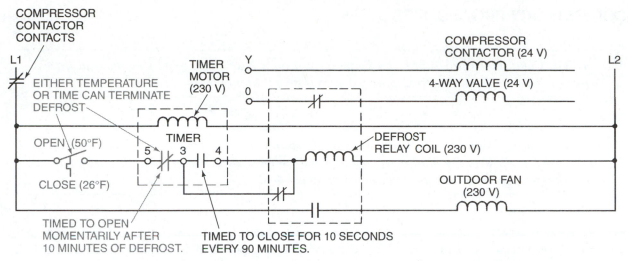

FIGURE 4–59 A wiring diagram for a time-or-temperature–terminated method of defrost.

The defrost timer is made up of several different internal components including the motor, as well as normally open and normally closed contacts within the control. Evaluating the motor can be done visually or with an ohmmeter. Some timer motors are designed to have the defrost timer motor energized continuously, while others are designed to operate only when the compressor is in operation. Most mechanical defrost timers have a timer dial or knob that allows for manual advancement into defrost in the field. Defrost timer contacts and electrical terminals numbers may be different from manufacturer to manufacturer. Technicians should review the schematic for the specific defrost timer being checked.

SYMPTOMS OF FAILURE OF SYSTEM DEFROST

The symptoms of a defective defrost timer or defrost system include excessive ice buildup on the outdoor coil and complaint of reduced heating by the customer. **If this were a real situation,** make sure that proper voltage is being supplied to the system. Then advance the mechanical defrost timer by turning the defrost timer knob until the outdoor fan stops. This should happen if the outdoor coil temperature is below 26 degrees (F) and the temperature sensor is closed. If the sensor and defrost timer contacts are closed, the outdoor fan should stop, and a "whooshing sound" should be heard, indicating that the four-way valve holding coil was energized, placing the heat pump into the cooling cycle so that ice can be removed from the outdoor coil. **If the unit manually was placed into defrost, the problem is definitely in the mechanical timer.** It could be that the timer motor is or is not receiving voltage, the timer motor could have an open coil, the timer cam could be broken or jammed, or the contacts of the mechanical timer could be defective.

If the system did not go into defrost after advancing the timer, the problem could be an open temperature sensor, defective four-way valve holding coil, no voltage at the four-way valve holding coil, or open conductor wire somewhere in the defrosting circuits.

EQUIPMENT REQUIRED

Heat pump with mechanical defrost timer or mechanical defrost timer
Voltmeter or VOM
Assorted hand tools

SAFETY PRACTICES

The student should be knowledgeable with the use of tools and testing equipment. Follow all EPA rules and regulations when working with refrigerants.

> *NOTE: Competency procedures can be performed on a heat pump with a mechanical defrost timer or on a mechanical defrost timer removed from a heat pump.*

> *NOTE: Practical Competency procedures will be used on an example of a time- and temperature-initiated defrost. Referenced defrost timer terminal numbers are associated with the defrost timer numbers used in the wiring diagram shown in* **Figure 4–59.**

Defrost timer contacts and electrical terminals numbers may be different from manufacturer to manufacturer. **Technicians should review the schematic for the specific defrost timer being checked.** If a schematic is available for the particular defrost timer this Competency is being performed on, substitute those terminal numbers for the ones listed in the Competency Procedures.

Notice that in **Figure 4–59**, the defrost timer terminal numbers are 5 - 3 - 4. Contact numbers **3 and 5 are normally closed** contacts and are designed to open after 10 minutes of defrost time. Terminals **3 and 4 are normally open** contacts that are designed to **close for 10 seconds after every 90 minutes** of timer operation. Notice also that even though **contacts 3 and 4 will close every 90 minutes**, defrost will not be initiated unless the defrost thermostat is closed at the same time. Once in defrost, either time or temperature will terminate the defrost cycle.

1. If testing a mechanical defrost timer in an actual heat pump, turn power to the outdoor unit OFF. ❏
2. Gain access to the mechanical defrost timer. ❏
3. Note the color of wires and terminal location on the defrost timer. ❏
4. Remove the wire from the timer. ❏
5. Locate the power wires to the mechanical timer motor. ❏
6. Set the VOM meter to measure resistance. ❏
7. Place meter leads across the defrost timer motor terminals. ❏
8. *Record defrost timer motor resistance value. _____* ❏

> *NOTE: A high-resistance valve (infinity reading) means that the timer motor has an open coil and would need to be replaced individually if possible, or the complete timer assembly would have to be replaced.*

9. Place VOM meter leads at defrost timer terminals 3 and 5. ❏
10. *Record the resistance value of terminals of 3 and 5. _____* ❏
11. *According to resistance value recorded in Step 10, are contact terminals 3 and 5 OPEN or CLOSED? _____* ❏
12. Place VOM meter leads at defrost timer terminals 3 and 4. ❏
13. *Record the resistance value of terminals 3 and 4. _____* ❏
14. *According to the resistance value recorded in Step 13, are contact terminals 3 and 4 OPEN or CLOSED? _____* ❏
15. Locate the defrost timer knob. ❏
16. Rotate the timer knob clockwise until a "click" is heard. ❏
17. Place the VOM meter leads across defrost terminals 3 and 5 again. ❏
18. *Record the resistance value of terminals 3 and 5. _____ .* ❏
19. *According to the resistance value recorded in Step 18, are contact terminals 3 and 5 OPEN or CLOSED? _____* ❏

20. Place the VOM meter leads across defrost terminals 3 and 4. ❑
21. *Record the resistance value of terminals 3 and 4.* _____ ❑
22. *According to resistance value recorded in Step 21, are contacts terminals 3 and 5*
    OPEN or CLOSED? _____ ❑

NOTE: *If timer cam and timer contacts are good, they should have changed position from OPENED to CLOSED or CLOSED to OPEN once the defrost timer knob was advanced to the point that a "click" was heard. If the terminal contacts did not change position on advancement of the defrost timer knob, the timer contacts or defrost timer cam are defective and would require that the defrost timer be replaced.*

NOTE: *If the timer motor has a measured (low resistance) and the timer contacts switch from OPEN to CLOSED when the timer knob was advanced, the defrost timer is good and should work within the heat pump system.*

---

23. If the defrost timer being checked was in an actual heat pump, replace defrost
    timer terminal wires to their proper terminal. ❑
24. Replace any panels removed. ❑
25. Have the instructor check your work. ❑
26. Return tools and test equipment to their proper location. ❑

RESEARCH QUESTIONS

1. Why is the outdoor fan de-energized when the heat cycles into defrost?

   _____

   _____

2. Which of the following is correct with respect to readings taken across a set of closed contacts?

   A. A voltage reading of 0 volts should be obtained across the contacts.
   B. A low-resistance reading should be obtained across the contacts.
   C. Continuity should register between the contacts.
   D. All of the above are correct.

3. What temperature does the defrost temperature control close?

   _____

   _____

4. Determining that a defrost timer motor is operating can be accomplished by:

   A. Determining that voltage is being supplied to the motor.
   B. Visually inspecting the motor.
   C. Establishing that continuity exists through the motor.
   D. Both A and B are correct.

5. Why does the efficiency of heat pump decrease if the outdoor coil ices over too much?

   _____

   _____

Passed Competency _____   Failed Competency _____

Instructor Signature _____   Grade _____

Practical Competency 117

Troubleshooting Solid-State Defrost Circuit Board and Temperature Sensor

SUGGESTED MATERIALS

Refrigeration & Air Conditioning Technology, 5th Edition, Thomson Delmar Learning
Unit 43—*Air Source Heat Pumps*

Review Topics
The Defrost Cycle; Initiating the Defrost Cycle; Terminating the Defrost Cycle;
Electronic Control of Defrost

COMPETENCY OBJECTIVE

The student will be able to check the operation of a solid-state defrost circuit board.

OVERVIEW

Newer heat pump systems use solid-state technology to initiate and terminate the defrost cycle (**Figure 4–60**).

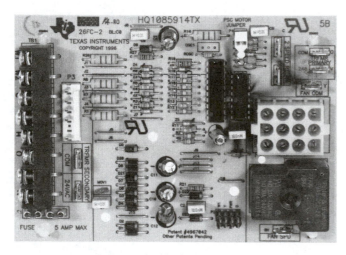

FIGURE 4–60   Solid-state heat pump defrost board. (*Photo by Bill Johnson*)

These controls are used by manufacturers of heat pumps to more closely control the defrost cycle with much more accuracy. The manufacturer may use *time and temperature* initiated and *time or temperature* terminated, like the mechanical timer and temperature sensor of the older systems. Some heat pumps use the *difference in the entering air temperature and the coil temperature* to arrive at a temperature split when defrost is desired. The coil temperature and the air temperature are normally sensed by thermistors rather than bimetal temperature sensors. Thermistors sense changes in temperature and change their internal resistance according to theses temperature changes. In many cases, solid-state defrost systems employ two thermistors. One senses the temperature of the outdoor coil, and the other senses the temperature of the outdoor air. The differential between the resistance of the two thermistors is what initiates the defrost cycle. As frost accumulates on the outdoor coil, the temperature sensed by the thermistor attached to the coil will drop. This will cause the resistance of the thermistor to drop. A temperature differential of 15 to 20 degrees (F) will initiate the defrost cycle. Once the temperature differential is reached, the defrost board will energize the necessary contacts and holding coils to bring the system into defrost. With this type of system, defrost is often terminated by a high-side pressure switch that is connected to the solid-state defrost control board.

Some solid state defrost boards **initiate** defrost by **time** and **terminate** the defrost cycle by **pressure, temperature,** or a combination of both. Solid-state timers are employed on the defrost board and the time interval between defrost attempts can be altered in the field by repositioning a jumper wire on the defrost board.

Most *timed solid-state defrost boards* have interval times of 30, 60, or 90 minutes between defrost attempts. Most heat pump systems with time initiated defrost use temperature to terminate the defrost cycle. There are units that use *pressure-initiated and terminated defrost systems*. These systems use a dual pressure-sensing switch connected to the circuit board and signals the system to enter defrost when the pressure differential rises to a predetermined point.

NOTE: Before a defrost board is suspected of being defective, technicians should first make sure that the heat pump operates properly in the heating and cooling modes. If heat pumps cooling and heating, modes of operation are confirmed. The system's defrost cycle and controls should be checked.

EQUIPMENT REQUIRED

VOM meter
Jumper wire (*if applicable*)
Assorted hand tools

SAFETY PRACTICES

The student should be knowledgeable with the use of tools and testing equipment. Follow all EPA rules and regulations when working with refrigerants.

COMPETENCY PROCEDURES   Checklist

NOTE: Each manufacturer utilizes different means by which defrost boards are checked and technicians should refer to the manufacturer's guidelines for specific troubleshooting procedures if available for the equipment this Competency is being performed on. An understanding of the procedures in this Competency will help make troubleshooting of other types of solid-state defrost systems much easier.

NOTE: In actual situations before checking the system defrost board, the outdoor coil thermostat or thermistor should be checked.

CHECKING THE OUTDOOR COIL TEMPERATURE SENSOR (IN OR OUT OF A HEAT PUMP)
1. Turn heat pump thermostat to the OFF position (*if applicable*).   ❏
2. Turn power to the outdoor unit OFF (*if applicable*).   ❏
3. Gain access to the heat pump defrost board (*if applicable*). Locate the outdoor coil
   temperature sensor (*if applicable*).   ❏
4. Locate outdoor coil temperature sensor wire leads on the units defrost board
   (*if applicable*). Remove outdoor coil temperature sensor leads from the defrost board
   (*if applicable*).   ❏

NOTE: *Outdoor coil temperature sensors should indicate a low resistance value, meaning the sensor contacts are CLOSED when sensing a temperature of approximately 26 degrees (F) or colder. An infinite (high resistance) reading, meaning the sensor contacts are OPEN, should be indicated when sensing temperatures of approximately 50 degrees (F) or higher. If sensor contacts are OPEN when measured temperatures indicate the sensor contacts should be CLOSED, or vice versa, the sensor is defective and should be replaced.*

5. Set VOM to measure resistance (*continuity*). ❑
6. Place VOM meter leads across the temperature sensor wire leads. ❑
7. *Record the measured resistance.* _____ ❑

NOTE: *Remember the temperature sensor is **a normally open switch** and CLOSES when the outdoor coil temperature reaches approximately **26 degrees** (F). A reading of **infinite resistance** is an indication that the thermostat's contacts are OPEN and not calling for defrost. If sufficient frost appears on the outdoor coil, and the sensor is mounted properly, an infinite reading is an indication that the sensor is probably defective.*

A **measured resistance** (**low resistance**) means that the temperature sensor contacts are CLOSED and should OPEN if the sensor is warmed to temperatures above **50 degrees** (F).

If the **coil temperature is not cold enough** or the temperature sensor is being checked in temperatures above 50 degrees (F), confirm that the sensor will close by placing the sensor in a freezer (if applicable), or use some other means to drop the sensor temperature below 26 degrees (F); then recheck for continuity across the temperature sensor leads. Under these conditions, if an infinite resistance is still measured, the temperature sensor is defective and needs to be replaced.

CONFIRMING TEMPERATURE SENSOR IS DEFECTIVE (REAL SERVICE FIELD SITUATION)

8. Place a jumper wire across the terminals on the defrost board where the outdoor temperature sensor leads were connected. ❑
9. Turn power on to the outdoor unit. ❑
10. Set thermostat to the **Heating** cycle. ❑
11. Set temperature selector switch above structure temperature ❑

NOTE: *If the heat pump goes into the defrost cycle, the outdoor temperature sensor is definitely defective and should be replaced. Let the system continue to defrost until outdoor coil is clear of frost and ice. Turn outdoor unit OFF and replace the outdoor temperature sensor.*

12. *With the temperature sensor terminals leads jumped on the defrost board, did unit cycle into defrost once restarted?* _____ ❑

NOTE: *If the unit DID NOT GO INTO DEFROST CYCLE—proceed to Step 15.*

13. Turn power to the outdoor unit OFF. ❑
14. Remove jumper wire across temperature sensor terminals on the defrost board. ❑
15. Replace the outdoor temperature sensor wire leads to the unit defrost board. ❑

16. Locate the defrost boards "**Test Terminals.**" ❑
17. Place a jumper wire across the "**Test Terminals**" on the defrost board. ❑

18. Turn power back ON to the outdoor coil. ❑
19. Wait 60 to 90 seconds for system to switch over to defrost. ❑
20. Remove jumper wire from defrost "Test Terminals" if unit cycles into defrost. ❑
21. *Did heat pump cycle into defrost with the defrost board "Test" terminals jumped?* _____. ❑
22. If the system **DID** advance to the defrost cycle, allow the system to operate through the defrost cycle and return to the **Heating** cycle (*if applicable*). ❑
23. If the system **DID NOT** advance to a defrost cycle, use proper procedures and replace the unit defrost board (*if applicable*). ❑
24. Once repair work is complete and checked, set unit thermostat to the OFF position. (*if applicable*). ❑
25. Have the instructor check your work. ❑
26. Replace unit panels (*if applicable*). ❑
27. Return all tools and test equipment to their proper location. ❑

RESEARCH QUESTIONS

1. Where must a suction-line drier be placed in a heat pump after a compressor burnout?

_____

_____

2. What controls the heat pump to determine whether it is in the heating cycle or the cooling cycle?

_____

_____

3. Can a heat pump switch from heating to cooling and from cooling to heating automatically?

_____

_____

4. Where is the only permanent suction line on a heat pump?

_____

_____

5. What type of drier may be used in the liquid line of a heat pump?

_____

_____

Passed Competency _____   Failed Competency _____

Instructor Signature _____   Grade _____

Practical Competency 118

Checking Heat Pump Auxiliary Heat—Electric Heating Elements (Limit Switch, Fuse Link, Heating Element, Sequencer)

SUGGESTED MATERIALS

Textbook

Refrigeration & Air Conditioning Technology, 5th Edition, Thomson Delmar Learning
Unit 43—Air Source Heat Pumps
Unit 30—Electric Heat

Review Topics

The Defrost Cycle; Initiating the Defrost Cycle; Terminating the Defrost Cycle; Electronic Control of Defrost; Auxiliary Heat

COMPETENCY OBJECTIVE

The student will be able to check the operation of an electric heating bank and the auxiliary heat circuit.

OVERVIEW

Almost all heat pump installations will have some source of auxiliary heat to assist in providing additional heat in cases where the heat pump cannot provide all the heat needed for the structure or to act as the emergency heating source if the heat pump would become inoperative for some reason. For these reasons, most heat pump applications will also have a secondary or back-up heating source. Most heat pump installation use electrical heating elements or duct heaters for this purpose. The secondary heating source is also used during the heat pump's defrosting cycle.

During the defrosting cycle on all heat pumps, the unit is cycled into the cooling mode and the outdoor fan is stopped. At the same time the unit is cycled into the defrost mode, the auxiliary heat is turned ON. This keeps warm air being provided to the conditioned space while the unit is operating in the cooling mode for defrosting purposes. The length of time for defrosting will vary from manufacturer to manufacturer, but most will not be in the defrost mode for more than 10 minutes as the maximum allowable time. Once the defrost cycle is complete, the heat pump is cycled back into the heating mode based on demand from the structure thermostat.

Depending on the heat pump installation, there may be one set of strip heaters, or multiple sets of strip heaters. The decision for this is based on two factors. Some contractors will install enough strip heat to totally heat the structure in the event of complete heat pump failure, while others will install enough strip heat to help the heat pump during low temperatures. This is a decision, which should be discussed with the customer during installation.

Strip heaters are energized through the **W1** terminal, or what is referred to as the **Second Stage Heating Terminal** of the thermostat.

When the electric heater operates along with the heat pump, it is called **auxiliary heat**. On a call for heat from the structure, the first-stage heat is energized through the heat pump thermostat's **Y** terminal.

The compressor contactor is energized; bring on the compressor and outdoor unit fan. Since the four-way valve is de-energized through the thermostat **O terminal**, the system is sending the compressor discharge refrigerant vapor to the indoor coil, providing heat to the structure. If the space temperature continues to fall, approximately 1.5 to 2 degrees (F), the thermostat will close the **W1 terminals** and energize the second-stage heating contacts, energizing the auxiliary electric heat strips to assistance the compressor. Once the structure temperature rises to within 1 degree (F) of the thermostat set point, the thermostat **W1 terminal** contacts OPEN, de-energizing the auxiliary heat.

Some manufacturers use **outside thermostats** to energize the auxiliary heat instead of the heat pump thermostat.

Outdoor thermostats allow the auxiliary heat on only when needed, using the structure balance point as a guideline. Depending on the sophistication of the heat pump installation, there may be several stages of electric strip heat applied to the system. Remember that "**balance point**" occurs when the heat pump can pump in exactly as much heat as the structure is leaking out. When the structure is leaking more heat than the heat pump can pump in, the auxiliary heat is energized to assist the heat pump. Systems with outdoor thermostats do not depend on the heat pump thermostat to energize the auxiliary heat; the electric heaters are energized based on outdoor air temperatures. Outdoor thermostats are low-voltage, and are mounted with the sensing bulbs outside to monitor the outdoor temperature. The temperature settings of the outdoor thermostats are normally set for the first strip of heaters to be energized when the outdoor temperature reaches approximately 20 degrees, with other thermostats set to energize other strips of heaters at a differential of 2- to 3-degree (F) intervals. Systems that use outdoor thermostats also will have an **emergency heat relay** which allows the customer to shut off the heat pump in cases of failure, and use all the strip heaters for providing heat to the structure (**Figure 4–61**).

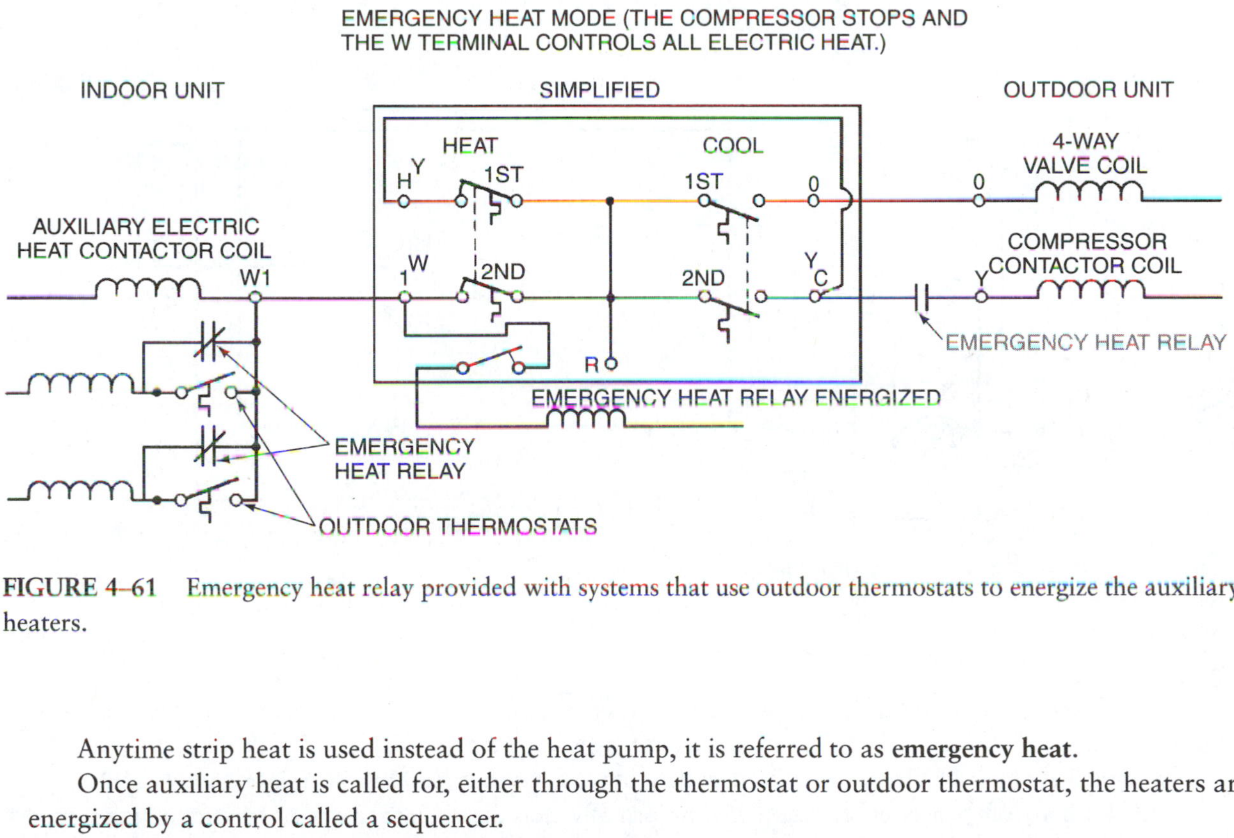

FIGURE 4–61 Emergency heat relay provided with systems that use outdoor thermostats to energize the auxiliary heaters.

Anytime strip heat is used instead of the heat pump, it is referred to as **emergency heat**.

Once auxiliary heat is called for, either through the thermostat or outdoor thermostat, the heaters are energized by a control called a sequencer.

A sequencer uses low voltage to energize a bimetal strip with a low-voltage wire wrapped around it.

The bimetal strip has one or multiple sets of contacts attached to it. As 24 volts pass through the low-voltage wire, the bimetal is heated causing the bimetal to warp, closing the sequencer contacts.

Since sequencers are **temperature activated** rather than magnetically activated like relay contactors, there is a delay with the sequencer contacts CLOSING and OPENING, due to the time required to heat the bimetal strip heater (**Figure 4–62**).

Once the bimetal strip is heated enough, it warps and energizes the sequencer contacts, energizing the heating elements (**Figure 4–63**).

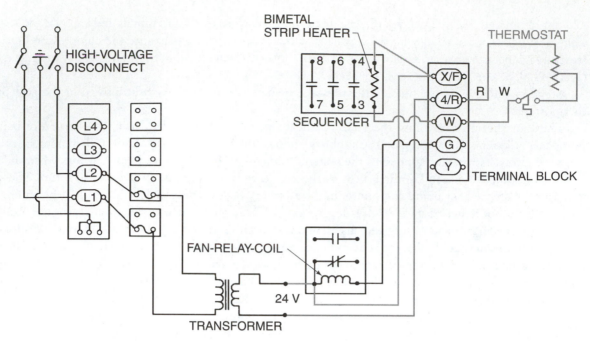

FIGURE 4–62 Bimetal strip heater must heat up before sequencer contacts are closed.

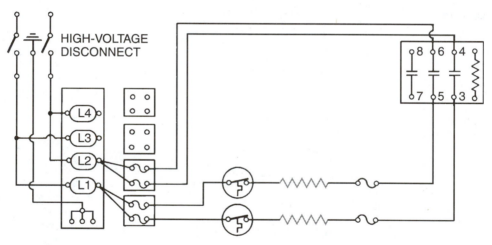

FIGURE 4–63 Two heating element circuit.

Sequencers are also used to delay the fan from coming ON and going OFF in other types of heating equipment.

Electric heat element circuit has a limit switch, heating element, and fusible link (**Figure 4–64**).

The **limit switch** is a safety device that is a normally closed switch that will open due to high temperature and de-energize the heating element. Once the limit switch cools, it will reset and re-energize the heating element. If unsafe temperature occurs again the limit switch will once again warp open and de-energize the heating element. The cycling of the limit switch could continue until service is requested due to inefficient structure heating.

The **fuse link** is also a safety device that will open when extreme unsafe temperatures are reached. The fuse link is a normally closed link that when opened will not reset automatically. The loss of efficient heating to the structure will be noticed, and service will be requested. The cause of the problem would need to be resolved before a new fuse link is installed.

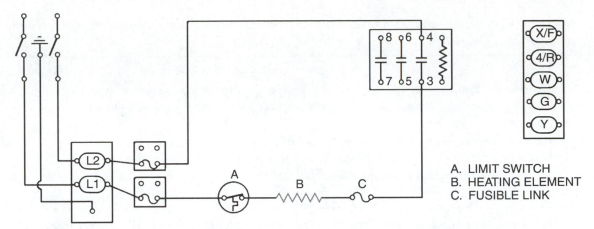

FIGURE 4–64   A single heating element circuit with limit switch, heating element, fusible link, and sequencer.

A. LIMIT SWITCH
B. HEATING ELEMENT
C. FUSIBLE LINK

EQUIPMENT REQUIRED

VOM meter
Clamp-on ammeter
Assorted hand tools
Heat pump with strip heat or strip heat assembly
Sequencer (*if applicable*)
Limit switch (*if applicable*)
Fuse link (*if applicable*)
Step-down transformer—24-volt secondary heat gun (*if applicable*)
Jumper cord
Two jumper wires

SAFETY PRACTICES

The student should be knowledgeable with the use of tools and testing equipment. Follow all EPA rules and regulations when working with refrigerants.

COMPETENCY PROCEDURES

Checklist

> NOTE: *Competency procedure can be performed on a heat pump with strip heat or on a strip heater bank.*

CLAMP-ON AMMETER TEST

1. Make sure unit thermostat is in the OFF position (*if applicable*). ☐
2. Turn power ON to the heat pump (*if applicable*). ☐
3. If using a strip heater bank for this test, wire heater bank so proper line voltage is applied. **(DO NOT APPLY VOLTAGE TO THE STRIP HEATER BANK AT THIS STEP.)** ☐
4. Gain access to the unit's strip heaters (*if applicable*). ☐
5. *If using a heat pump unit with strip heat for this competency, record the number of strip heaters equipped with the heat pump (if applicable).* _____ ☐
6. *Record the KW of the strip heater being checked.* _____ ☐
7. Set unit thermostat to the Emergency Heat setting (*if applicable*). ☐
8. Place the clamp-on ammeter around the one of the legs of power of the first strip heater. ☐

> **NOTE:** *If strip heater bank is being used in this competency, place the clamp-on ammeter to one of the heater's power legs.*

9. Set thermostat temperature selector switch 2 degrees over the structure temperature. ❏
10. If strip heater bank is being used for this competency, apply line voltage and observe the clamp-on ammeter. ❏
11. *Once an amperage reading is indicated, record heater amp draw.* _____ ❏
12. *If heater being tested is in an actual heat pump, was there a delay before the clamp-on ammeter indicated an amperage reading?* _____ ❏

> **NOTE:** *An observed amperage reading while testing the strip heater is an indication the heater circuit is good. This would include the limit switch, heating element, and fuse link are all-OK. If the heater strip being checked is in an actual heat pump, the measured amp draw would also indicate that the sequencer was also OK.*

*NO **measure amperage** is an indication of the following: either no line voltage, no control voltage (24 volts) from the E terminal of the thermostat, open limit switch, heating element, fuse link, or sequencer bimetal strip heater. Each would have to be checked to confirm the failure of the circuit.*

13. *Using the VOM meter, measure and record the applied voltage.* _____ ❏

CONFIRMING THE BTU OUTPUT OF THE STRIP HEATER

> **NOTE:** *1 Electrical Watt 5 3.413 Btu's*

14. Determine the Btu output of the strip heater by the following formula:
    Volts × Amps = Watts × 3.413 = Btu's ❏
15. *Fill in the information recorded in Step 11 and Step 13.*
    Volts _____ × _____ **Amps** = _____ **Total Watts** ❏
16. *Determine Btu's by filling in the following information*
    Total Watts _____ × **3.413** = _____ **Total BTU Output** ❏
17. Determine strip heater KW by filling in the following information.
    Total Watts _____ **(DIVIDED BY) 1000** = _____ **Total KW** ❏
18. *Does the total KW match the Recorded KW in Step 6?* _____ ❏
19. *If this an actual heat pump that has additional strip heaters, use an additional piece of paper and repeat above procedures for all strip heater banks. Record all information and provide to instructor on completion of this competency.* ❏

> **NOTE:** *If no additional strip heaters are to be checked, proceed to Step 19.*

CONTINUITY CHECKS OF STRIP HEATER AND CIRCUIT

20. Turn power OFF to heat pump (*if applicable*). ❏
21. Gain access to the unit strip heaters. ❏
22. Locate the strip heater sequencer(s). ❏
23. Remove the low-voltage wires to the sequencer bimetal heater. ❏
24. Set VOM to measure resistance. ❏

25. Place meter leads across the sequencer bimetal heater terminals. ❏
26. *Record resistance value.* _____ ❏

NOTE: *A measured resistance (low resistance) is an indication that the sequencer bimetal heater is GOOD and should work when low voltage is applied. An infinite reading (high resistance) is an indication that the sequencer bimetal heater is OPEN, and would require replacement.*

27. *Observe and record the number of sequencer terminal contacts.*
    Terminals _____ **and** _____
    Terminals _____ **and** _____
    Terminals _____ **and** _____ ❏
28. Place the meter leads across the first set of sequencer contacts. ❏
29. *Record the measured resistance.* _____ ❏
30. Place the meter leads across the second set of sequencer contacts (*if applicable*). ❏
31. *Record the measured resistance of the second set of sequencer contacts (if applicable).* _____ ❏
32. Place the meter leads across the third set of sequencer contacts (*if applicable*). ❏
33. *Record the measured resistance of the third set of sequencer contacts (if applicable).* _____ ❏

NOTE: *Sequencer contacts are **normally open** and should indicate an infinite resistance value if OPEN. A measured resistance (low resistance) is an indication that the contacts are CLOSED. This is an indication that the contacts are welded together. In this case the sequencer would require replacement.*

CHECKING TO SEE IF SEQUENCER CONTACTS CLOSE
34. If Competency is being performed on an actual heat pump with strip heat, remove heating element leads from the sequencer. ❏

NOTE: *Be sure to mark down on a piece of paper the heating element wires and associated sequencer terminal numbers.*

35. Replace sequencer low-voltage bimetal heater wires. ❏
36. If sequencer is being checked outside of a heat pump, use the low-voltage step-down transformer and wire the transformer so that 24 volts will be supplied to the sequencer bimetal heater terminals. ❏
37. Turn power on to heat pump and set thermostat to the Emergency Heat setting, and raise the temperature switch high enough to call for heat. ❏
38. If sequencer is being checked out of a heat pump, apply voltage to the step-down transformer so that 24 volts will be supplied to the sequencer bimetal heater terminals. ❏
39. Allow the sequencer bimetal heater time to heat and close the sequencer contact(s). ❏
40. Place the meter leads across the first set of sequencer contacts. ❏
41. *Record the measured resistance.* _____ ❏
42. Place the meter leads across the second set of sequencer contacts (*if applicable*). ❏
43. *Record the measured resistance of the second set of sequencer contacts (if applicable).* _____ ❏
44. Place the meter leads across the third set of sequencer contacts (*if applicable*). ❏
45. *Record the measured resistance of the third set of sequencer contacts (if applicable).* _____ ❏

NOTE: *With the sequencer bimetal heater energized, the sequencer contacts should have CLOSED, which should be indicated by a measurable resistance value, indicating that the sequencer contacts are OK. An infinite resistance value reading indicates that the contacts are OPEN when they should have CLOSED. This is an indication that the contacts are not working properly, and would require replacement of the sequencer. Again, this is after making sure that proper voltage is being supplied to the sequencer bimetal heater and that the bimetal heater of the sequencer is good.*

46. Turn power OFF to the heat pump or remove power from the step-up transformer if sequencer is being checked outside of a heat pump. ❏
47. Replace bimetal strip heater low-voltage wires (*if applicable*). ❏
48. Replace electric heater lead wires on to the proper sequencer terminals (*if applicable*). ❏

CHECKING THE LIMIT SWITCH
49. Remove the electrical wires attached to the limit switch (*if applicable*). ❏
50. Place the meter leads across the limit switch terminal. ❏
51. *Record the measured resistance.* _____ ❏

NOTE: *The limit switch is a safety switch and is a **normally closed switch**. When checking resistance across the limit switch terminals, a measurable resistance should be indicated. This is an indication that the switch contacts are CLOSED. An infinite resistance reading means the limit switch contacts are OPEN, and would not allow current flow through the heater circuit. If this is the case, and the limit switch is at ambient temperature, it is defective and would require replacement.*

52. *Remove limit switch from strip heater (if applicable).* ❏
53. Use the heat gun and heat the limit switch until you hear a "click." ❏
54. Turn the heat gun OFF. ❏
55. Place the meter leads across the limit switch terminals again. ❏
56. *Record the measured resistance.* _____ ❏

NOTE: *In this test the limit switch contacts should have OPENED indicating an infinite resistance value. If a measured resistance was indicated, this means that the limit switch contacts are still CLOSED, when they should have OPENED due to high temperature. In this case, the limit switch would require replacement.*

57. *After the limit switch cooled, did you hear another "click"?* _____ ❏

NOTE: *The limit switch resets automatically when it cools; if it does not the limit switch is defective and should be replaced.*

CHECKING THE FUSE LINK
58. *Isolate one lead of the fuse link from the electric element circuit (if applicable).* ❏
59. Place the meter leads across the fuse link terminals. ❏
60. *Record the measured resistance.* _____ ❏

> **NOTE:** *The fuse link is a **normally closed** link that protects against extreme high temperature. If the link is good, a measurable resistance (low resistance) should be indicated during a continuity check. An infinite reading means that the fuse link is OPEN and would require replacement. It does not reset automatically like the limit switch.*

61. Reconnect the fuse link to the heater circuit (*if applicable*). ❏

CHECKING THE HEATING ELEMENT
62. Isolate the heating element from the heater limit switch, fuse link, and sequencer (*if applicable*). ❏
63. Place the meter leads across the heating element terminal leads. ❏
64. *Record the measured resistance.* _____ ❏

> **NOTE:** *A measured resistance (low resistance) is an indication that the heating element is good. An infinite reading indicates that the heating element has a break in its circuit and would require replacement.*

65. Replace all electrical connections to put strip heaters back into operation (*if applicable*). ❏
66. Have the instructor check your work. ❏
67. Replace all panels (*if applicable*). ❏
68. Return all tools, test equipment, and supplies to their proper location. ❏

RESEARCH QUESTIONS

1. The type of material often used for the resistance wire in electric heat is _____
2. Three common controls used in central electric heat applications are thermostats, contactors (or relays), and:

   A. Capacitors
   B. Sequencers
   C. Cool anticipators

3. The sequencer uses _____ (high- or low-) voltage control power to start and stop electric heaters.
4. What is the formula for finding watts?

   _____

   _____

5. How many Btu's are generated by one electrical watt?

   _____

   _____

Passed Competency _____   Failed Competency _____

Instructor Signature _____   Grade _____

Practical Competency 119

Using Manufacturer's Charging Chart to Check or Charge a Heat Pump (Figure 4–65)

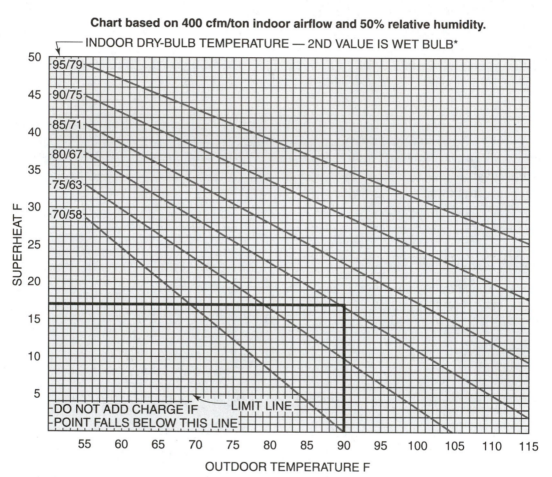

FIGURE 4–65 A charging curve used for charging a split-system air-to-air heat pump incorporating a capillary tube or fixed-bore metering device.

SUGGESTED MATERIALS

Textbook

Refrigeration & Air Conditioning Technology, 5th Edition, Thomson Delmar Learning
Unit 43—Air Source Heat Pumps
Unit 10—System Charging
Unit 9—Refrigerant and Oil Chemistry and Management—Recovery, Recycling, Reclaiming, and Retrofitting

Review Topics

Liquid Refrigerant Charging; Weighing Refrigerant; Using Charging Devices; Using Charging Charts; Using Charging Curves; Azeotropic Refrigerants; Vapor Refrigerant Charging

COMPETENCY OBJECTIVE

The student will be able to charge or adjust the charge of a heat pump system by using a manufacturer's charging chart.

OVERVIEW

Adding refrigerant to refrigeration sealed system is referred to as charging. Air conditioning, heat pumps, and refrigeration equipment must have the correct amount of refrigerant charge for the equipment to operate and perform as efficiently as it was designed to. With today's high-efficiency equipment, ensuring that a system is charged with the correct amount of refrigerant is one of the most important procedures a technicians is required to perform. This requires that technicians have a thorough understanding of the system operation and operating conditions which indicate whether a system is properly *charged*, *overcharged*, or *undercharged*. Refrigerant can be added to a sealed system in the vapor or liquid state, which can be accomplished by *weighing the refrigerant charge* into the sealed system, measuring the charge into the sealed system (*Chapter 3, Theory Lesson: Sealed System Charging, Figure 3–116*), using system operating charts (*Theory Lesson: Sealed System Charging, Figure 3–117*), using manufacturer's superheat charging tables (*Theory Lesson: Sealed System Charging, Figure 3–118*), suction line temperature tables (*Theory Lesson: Sealed System Charging, Figure 3–119*), or manufacturer's sub-cooling charging tables for systems with a TXV metering device (*Theory Lesson: Sealed System Charging, Figure 3–120*).

Manufacturers will often supply charts or curves to assistant in correctly checking or charging air-conditioning and heat pumps. The charts are referred to as **charging charts** or **charging curves**. These charts are based on superheat or sub-cool readings under certain load conditions in which the air conditioning or heat pump may be operating. The type of metering device is also taken into consideration by the manufacturer when supplying these charts with the equipment. Most air conditioning and heat pump equipment is designed with capillary tube, fixed orifice, or piston type metering devices. For this type of metering device, the charts or curves will be based on system **superheat values**. For systems with thermostatic expansion devices (TXV), charts and curves will be based on system **sub-cooling values**. The use of the charts requires technicians to evaluate the load conditions the air conditioning or heat pump is operating in. Evaluating the load refers to the amount of sensible and latent heat in the air being conditioned. The procedures to use manufacturer's charging charts or charging curves are universal, although the safest and most accurate way to charge any unit is to follow the manufacturer's charging instructions that come attached to the unit or supplied in the unit Installation Manual.

EQUIPMENT REQUIRED

1 Manufacturer's charging chart or charging chart (*Figure 4–65*)
Air conditioning or heat pump unit
1 Sling psychrometer
1 Temperature tester
1 Manifold gage set
1 Refrigerant cylinder, which matches system refrigerant

SAFETY PRACTICES

Wear goggles or safety glasses and gloves when working with refrigerant in sealed systems. Beware of high pressures and follow all electrical safety rules. Follow all EPA rules and regulations while working with refrigerants.

> **NOTE:** *Competency can be performed on a heat pump operating in the cooling mode. The charging chart shown in Figure 4–65 is based on 400 CFMs (cubic feet per minute) of air per ton or air flow with 50% relative humidity across the evaporator coil.*

COMPETENCY PROCEDURES

Checklist

1. Turn the system ON to the **Cooling** mode. ❑
2. *Measure and record the indoor return air dry-bulb temperature. _____* ❑

> **NOTE:** *Use the indoor air wet-bulb temperature if the percentage of relative humidity in the air is above 70% or below 20%.*

3. *Measure and record the out door dry-bulb temperature at the outdoor unit. _____* ❑
4. Make sure manifold gage valves are closed. ❑
5. Purge any refrigerant from refrigerant hoses ❑
6. Attach manifold low-side gage hose to the sealed system's vapor line valve. ❑
7. Attach manifold high-side gage hose to the sealed system's liquid line valve. ❑
8. Attach and insulate a temperature probe lead at or near the vapor line service valve of the outdoor unit. ❑
9. Let operate for 15 minutes. ❑
10. *Record the type of system refrigerant. _____* ❑
11. *Record the temperature of the vapor line. _____* ❑
12. *Record the system's low-side operating pressure (psig). _____* ❑
13. *Use a pressure–temperature chart and convert and record the system low-side pressure to refrigerant saturation temperature. _____* ❑
14. Determine system superheat by subtracting the system's low-side saturation temperature from the system's vapor line temperature. ❑

Vapor line temperature

MINUS

Low-side saturation temperature

EQUALS *system superheat*

15. *Record the amount of superheat. _____* ❑
16. *Refer to system charging chart or charging chart shown in Figure 4–65.* ❑
17. Locate the indoor dry-bulb temperature lines on the charging chart. ❑
18. Locate the outdoor dry-bulb temperature lines on the charging chart. ❑
19. Locate where the indoor dry-bulb temperature and outdoor dry-bulb temperature intersect on the charging chart. ❑
20. Follow the intersecting line over to the superheat values. ❑
21. *Record the required superheat according to the charging chart. _____* ❑
22. *Does the recorded superheat value recorded in Step 15 match the charging chart required superheat value recorded in Step 21? _____* ❑
23. *Is the recorded superheat value in Step 15 HIGHER or LOWER than the required superheat listed in Step 21? _____* ❑

> **NOTE:** *If the superheat of the system is more than 5 degrees (F) ABOVE what the charging chart requires, add refrigerant through the system's vapor line service valve until the system's superheat is within 5 degrees (F) of the required superheat according to the charging chart.*

If the superheat of the system is more than 5 degrees (F) BELOW what the charging charts call for, use a recovery unit and recover some refrigerant from the system until the superheat is within 5 degrees of what is required according to the charging chart.

> **NOTE:** *Make adjustments to system refrigerant charge slowly and let the system operate for an additional 15 minutes before recalculating system superheat.*

> **NOTE:** *Any adjustments to system charge will require recalculation of the system's superheat.*

ONCE SUPERHEAT IS WITHIN RANGE RECALCULATE SYSTEM SUPERHEAT

24. *Record the temperature of the vapor line.* _____ ❑
25. *Record the system's low-side operating pressure (psig).* _____ ❑
26. *Using a pressure–temperature chart, convert and record the system low-side pressure to refrigerant saturation temperature.* _____ ❑
27. *Determine system superheat by subtracting the system's low-side saturation temperature from the system vapor line temperature.* ❑

Vapor line temperature _____

MINUS

Low-side saturation temperature _____

EQUALS system superheat _____

28. *Record the amount of superheat.* _____ ❑
29. *Once the system charge is within required range, have your instructor check your work.* ❑
30. With the instructor's approval, turn the unit OFF. ❑
31. Remove the temperature probe from system vapor line. ❑
32. Remove low- and high-side manifold hoses from the system service valves. ❑
33. Replace the system service valve caps. ❑
34. Return equipment to its proper location. ❑

RESEARCH QUESTIONS

1. How is liquid refrigerant added to a heat pump when the system is out of refrigerant?

   _____

   _____

2. How is the refrigerant cylinder pressure kept above the system pressure when charging with vapor from a cylinder?

   _____

   _____

3. Why does the refrigerant pressure decrease in a refrigerant cylinder while charging with vapor?

   _____

   _____

4. What other methods besides weighing and measuring are used for charging heat pump systems?

   _____

   _____

5. What is the main difference between zeotropic and azeotropic refrigerant blends?

   _____

   _____

Passed Competency _____   **Failed Competency** _____

Instructor Signature _____   **Grade** _____

Practical Competency 120

Using the Sub-Cool Method of Charging R-22 Heat Pumps Equipped with a Thermostatic Expansion Device (TXV)

SUGGESTED MATERIALS

Textbook

Refrigeration & Air Conditioning Technology, 5th Edition, Thomson Delmar Learning
Unit 10—*System Charging*
Unit 21—*Evaporators and Refrigeration Systems*
Unit 24—*Expansion Devices*
Unit 36—*Refrigeration Applied to Air Conditioning*
Unit 40—*Typical Operating Conditions*
Unit 43—*Air Source Heat Pumps*

Review Topics

TXV Metering Devices; Air-Cooled Condensers; Condenser Sub-cooling; Standard Air-Cooled Condensers; High-Efficiency Air-Cooled Condensers

COMPETENCY OBJECTIVE

The student will be able to properly charge or check the charge a heat pump equipped with a TXV metering device by using the sub-cool method.

OVERVIEW

Some air conditioning and heat pump equipment is equipped with a **thermostatic expansion valve** (TXV). These types of metering devices are designed to maintain a constant amount of superheat on the evaporator (*indoor coil*) under any load conditions, including changing outdoor ambient temperatures. TXV are provided mainly on high-efficiency equipment. When used, a refrigerant charging method known as the **sub-cooling method** must be used to charge or to check the charge on air conditioning or heat pump equipment so equipped. This method of charging or checking a systems charge involves measuring the amount of liquid line sub-cooling existing the condenser (*outdoor coil*). Because of the different SEER ratings on the market today, the actual sub-cooling range for particular equipment varies. Manufacturers have made an effort to provide the required sub-cool amount for air cool condensers (*outdoor coil*) on the system data plate or technical manual. In the absence of such information, use a sub-cooling value of 10 to 15 degrees (F).

EQUIPMENT REQUIRED

Manufacturer's sub-cooling chart or *Figure 4–66*
1 Temperature tester (with temperature leads)
1 Manifold gage set
1 Refrigerant cylinder, which matches system refrigerant
Insulation tape
Assorted hand tools

SAFETY PRACTICES

Wear goggles or safety glasses and gloves when working with refrigerant in sealed systems. Beware of high pressures and follow all electrical safety rules. Follow all EPA rules and regulations while working with refrigerants.

COMPETENCY PROCEDURES

Checklist

1. Turn unit ON to the **Cooling** mode. ❏
2. Allow the unit to operate for 15 minutes. ❏
3. Make sure manifold gage valves are closed. ❏
4. Remove the system's liquid line service valve cap. ❏
5. Attach manifold high-side hose to the system's liquid line service valve port. ❏
6. Attach a temperature probe to the outdoor unit's liquid line. ❏
7. Insulate the temperature probe. ❏
8. *After 15 minutes of operation, record the system's high-side pressure.* _____ ❏
9. Use a gage manifold or a pressure temperature chart and convert high-side pressure to saturation temperature. ❏
10. *Record the high-side saturation temperature.* _____ ❏
11. *Record the temperature of the outdoor unit's liquid line.* _____ ❏
12. Determine outdoor coil sub-cool amount by the following formula: ❏

   High-side saturation temperature _____

   MINUS

   Outdoor unit's liquid line temperature _____ ❏

   EQUALS *outdoor coil sub-cooling* _____ ❏
13. *Record the amount of the outdoor unit's sub-cooling.* _____ ❏

NOTE: *Required sub-cooling used for this competency is based on 15 degrees. If required sub-cooling is known for the equipment competency is being performed on, use the manufacturer's required rather than 15 degrees. Refer to Figure 4–66.*

---

14. Find the approximate high-side pressure recorded in **Step 8** on the sub-cooling chart in **Figure 4–66.** ❏
15. Follow the high-side pressure recorded in Step 8 across the sub-cool chart to the 15-degree (F) required sub-cooling column. ❏
16. *Record the required liquid line temperature according to the sub-cooling chart.* _____ ❏

NOTE: *According to the sub-cooling chart, the liquid line temperature recorded in Step 17 is the required liquid line temperature for this unit if the system is charged correctly.*

---

17. Record the actual liquid line temperature recorded in *Step 11.* _____ ❏
18. *Was the actual recorded liquid line temperature within 3 to 4 degrees of the temperature required according to the sub-cooling chart?* _____ ❏

NOTE: *Allow a 3- to 4-degree (F) temperature tolerance between required and actual liquid line temperatures.*

*If the measured liquid line temperature recorded in Step 11 is **higher** than the desired liquid line temperature recorded in Step 17, **add refrigerant** to the unit.*

*If the measured liquid line temperature recorded in Step 11 is **lower** than the desired liquid line temperature recorded in Step 17, **recover refrigerant** from the unit.*

Pressure (psig) at Service Fitting	Required Subcooling Temperature (F)					
	0	5	10	15	20	25
134	76	71	66	61	56	51
141	79	74	69	64	59	54
148	82	77	72	67	62	57
156	85	80	75	70	65	60
163	88	83	78	73	68	63
171	91	86	81	76	71	66
179	94	89	84	79	74	69
187	97	92	87	82	77	72
196	100	95	90	85	80	75
205	103	98	93	88	83	78
214	106	101	96	91	86	81
223	109	104	99	94	89	84
233	112	107	102	97	92	87
243	115	110	105	100	95	90
253	118	113	108	103	98	93
264	121	116	111	106	101	96
274	124	119	114	109	104	99
285	127	122	117	112	107	102
297	130	125	120	115	110	105
309	133	128	123	118	113	108
321	136	131	126	121	116	111
331	139	134	129	124	119	114
346	142	137	132	127	122	117
359	145	140	135	130	125	120

FIGURE 4–66 Manufacturer's charging chart. *(Courtesy of Carrier Corporation)*

MAKING CHARGE ADJUSTMENTS

> **NOTE:** *When adding or removing refrigerant from the system, it should be done in small increments. Allow at least 15 minutes of system operation between refrigerant adjustments before retaking system pressures and temperatures.*

19. Adjust system charge if required. ❏
20. Once the system is within tolerance according to the sub-cooling chart, have the instructor check your work. ❏
21. Turn the system OFF. ❏
22. Remove all equipment and return materials to their proper location. ❏

RESEARCH QUESTIONS

1. What is the difference between standard and high-efficiency air-cooled condensers?

_____

_____

2. Is the designed sub-cooling range for high efficiency air-cooled condensers (*higher or lower*) than that which is established for standard air-cooled condensers?

_____

_____

3. The SEER rating of air conditioning and heat pump equipment is based on what factors?

_____

_____

4. What factors could cause outdoor coil sub-cooling to be higher than the designed range?

_____

_____

5. What factors could cause outdoor coil sub-cooling to be lower than the designed range?

_____

_____

Passed Competency _____   **Failed Competency** _____

Instructor Signature _____   Grade _____

Practical Competency 121

Charging an R-22 Heat Pump by Using an R-22 Superheat/ Sub-cooling Calculator—Superheat Method

SUGGESTED MATERIALS

Textbook
Refrigeration & Air Conditioning Technology, 5th Edition, Thomson Delmar Learning
Unit 10—System Charging
Unit 40—Typical Operating Conditions
Unit 43—Air Source Heat Pumps

Review Topics
Vapor Refrigerant Charging; Liquid Refrigerant Charging

COMPETENCY OBJECTIVE

The student will be able to properly charge a heat pump system using an R-22 charging calculator and the superheat method.

OVERVIEW

It is important for HVAC technicians to understand that systems that are undercharged or overcharged will remove heat from a conditioned area, although they will not perform efficiently doing it!

It has been reported by national air conditioning associations that more than 70% of air conditioning and heat pump equipment in use in the United States is either *overcharged* or *undercharged*. This is a staggering number when efficiency performance of air conditioning and heat pump equipment relies on operating with the correct refrigerant charge. Additional factors that affect the efficiency of such equipment are making sure that the system is moving the correct airflow in CFMs (*cubic feet per minute*) for both the evaporator and condenser. For forced air systems, making sure that the ductwork is sized correctly and operating with the correct static pressure is important. Making sure that the evaporator and condenser coils are clean and that the equipment is sized properly are all important to the efficiency performance of air conditioning and heat pump equipment. Yet while having all these additional factors correct, having the correct refrigerant charge is paramount in comparison, especially with the new higher rated SEER equipment. *Remember, it is the refrigerant within the mechanical refrigeration system that actually absorbs heat and rejects heat from the conditioned space*. Having too much or too little refrigerant within the system affects the equipment's ability and efficiency performance in absorbing and rejecting heat within an area where comfort is trying to be sustained, let alone shortening the life of the equipment.

Charging and checking the charge on air conditioning and heat pump equipment has become one of the most important aspects of the HVAC industry, and a certified technician's duties.

Using and understanding how to use a **R-22 superheat/sub-cooling calculator** takes the *"guess work"* out of checking or charging air conditioning and heat pump equipment along with checking for proper system air flow. The calculator is designed for use on TXV and non-TXV equipment. Technicians should use the **Required Superheat (Cooling, non-TXV) side for systems with fixed-bore metering devices,** and the

Sub-cooling (Cooling, TXV) for systems with a TXV metering device. The calculator can also be used to check system's airflow by using the **Proper Airflow Range (Cooling)** portion of the calculator.

EQUIPMENT REQUIRED

R-22 Superheat/sub-cooling charging calculator
1 Sling psychrometer
1 Temperature tester (with temperature leads)
1 Manifold gage set
1 Refrigerant cylinder, which matches system refrigerant
Insulation tape
Assorted hand tools

SAFETY PRACTICES

Wear goggles or safety glasses and gloves when working with refrigerant in sealed systems. Beware of high pressures and follow all electrical safety rules. Follow all EPA rules and regulations while working with refrigerants.

> **NOTE:** *The system being charged should have been leak checked and evacuated prior to performing this competency.*

COMPETENCY PROCEDURES   **Checklist**

1. The sealed system should be OFF. ❑
2. Use the sling psychrometer and measure the indoor entering air wet-bulb temperature. ❑
3. *Record the indoor entering air wet-bulb temperature.* _____ ❑
4. Use the required superheat calculator (*Cooling, non-TXV*) and slide the slide rule of the charging calculator down to the recorded indoor entering air wet-bulb temperature at **Number 1.** ❑
5. Use the temperature tester and measure the outdoor unit's entering air dry-bulb temperature. ❑
6. *Record the outdoor unit's entering air dry-bulb temperature.* _____ ❑
7. Find the outdoor unit's (*condenser*) entering air dry-bulb temperature at **Number 2** of the charging calculator. ❑
8. Directly below the condenser entering air dry-bulb temperature is the required superheat temperature at **Number 3** of the charging calculator. ❑
9. *Record the required superheat temperature.* _____ ❑
10. Attach and insulate a temperature probe near the system's vapor line service valve. ❑
11. Add refrigerant through the system's low-side valve in the vapor state until the system's low-side pressure is brought out of a vacuum. ❑
12. Turn the sealed system ON to the **Cooling** mode. ❑
13. Continue to add refrigerant in the vapor state until the system's low-side pressure is operating at least **64 psig.** ❑
14. Once the system's operating low-side pressure is at least *64 psig*, stop charging and allow the system to operate for 10 minutes. ❑
15. At **Number 5** of the charging calculator, find the required superheat temperature as recorded in **Step 9.** ❑
16. Directly below the required superheat at **Number 5** is the required vapor line temperature at **Number 6.** ❑
17. *Record the required vapor line temperature as indicated by the charging calculator at **Number 6.*** _____ ❑
18. *Check the system's vapor line temperature and record actual vapor line temperature.* _____ ❑

> **NOTE:** *Under the Instructions of the Charging Calculator at Number 6 it states the following:*
> *If the measured vapor line temperature does not agree with the required vapor line temperature,* **ADD** *refrigerant to* **LOWER** **temperature** *or* **REMOVE** *refrigerant to* **RAISE** **temperature.** *(Allow tolerance of + or −5 degrees [F].)*

19. Add or remove refrigerant in small amounts to the system until the system's suction (vapor) line temperature falls within (+ *or −5 degrees*) of the required vapor line temperature as indicated by the charging calculator. ❏

> **NOTE:** *Any change in refrigerant charge to the system will change the system's low-side operating pressure. The vapor pressure at Number 4 will have to be readjusted as the system's pressures change due to the addition or removal of refrigerant to the system. The same required superheat temperature would continue to be used to determine required vapor line temperature unless there has been a considerable time lapse or temperature change from the original temperature readings taken at Numbers 1 and 2.*

20. Once system charge is correct, have the instructor check your work. ❏
21. On completion of system charge, remove all equipment and return to its proper location. ❏

RESEARCH QUESTIONS

1. Compressor oil is used for lubrication; what is used to cool the compressor motor windings?

   _____

   _____

2. What is a starved evaporator?

   _____

   _____

3. What causes refrigerant inside an evaporator to evaporate?

   _____

   _____

4. What is a flooded evaporator?

   _____

   _____

5. How many fins are there per inch for evaporators used in air conditioning and heat pump equipment?

   _____

   _____

Passed Competency _____ Failed Competency _____

Instructor Signature _____ Grade _____

Practical Competency 122

Charging an R-22 Heat Pump Equipped with a TXV Metering Device by Using an R-22 Superheat/Sub-cooling Calculator—Sub-cool Method

SUGGESTED MATERIALS

Textbook

Refrigeration & Air Conditioning Technology, 5th Edition, Thomson Delmar Learning
Unit 10—*System Charging*
Unit 40—*Typical Operating Conditions*
Unit 43—*Air Source Heat Pumps*

Review Topics

Vapor Refrigerant Charging; Liquid Refrigerant Charging

COMPETENCY OBJECTIVE

The student will be able to properly charge a heat pump equipped with a TXV metering device by using a charging calculator and the sub-cool method.

OVERVIEW

It is important for HVAC technicians to understand that systems that are undercharged or overcharged will remove heat from a conditioned area, although they will not perform efficiently doing it!

It has been reported by national air conditioning associations that more than 70% of air conditioning and heat pump equipment in use in the United States is either *overcharged* or *undercharged*. This is a staggering number when efficiency performance of air conditioning and heat pump equipment relies on operating with the correct refrigerant charge. Additional factors that affect the efficiency of such equipment are making sure that the system is moving the correct airflow in CFMs (*cubic feet per minute*) for both the evaporator (*indoor coil*) and condenser (*outdoor coil*). For forced air systems, making sure that the ductwork is sized correctly and operating with the correct static pressure is important. Making sure that the evaporator (*indoor coil*) and condenser coils (*outdoor coil*) are clean and that the equipment is sized properly are all important to the efficient performance of air conditioning and heat pump equipment. Yet while having all these additional factors correct, having the correct refrigerant charge is paramount in comparison, especially with the new higher rated SEER equipment. *Remember, it is the refrigerant within the mechanical refrigeration system that actually absorbs heat and rejects heat from the conditioned space.* Having too much or too little refrigerant within the system affects the equipment's ability and efficiency performance in absorbing and rejecting heat within an area where comfort is trying to be sustained, let alone shortening the life of the equipment.

Charging and checking the charge on air conditioning and heat pump equipment has become one of the most important aspects of the HVAC industry, and a certified technician's duties.

Using and understanding how to use an **R-22 superheat/sub-cooling calculator** takes the *"guess work"* out of checking or charging air conditioning and heat pump equipment along with checking for proper system air flow. The calculator is designed for use on TXV and non-TXV equipment. Technicians should use the **Required Superheat (Cooling, non-TXV) side for systems with fixed-bore metering devices,** and the **Sub-cooling (Cooling, TXV),** for systems with a TXV metering device. The calculator can also be used to check the system's airflow by using the **Proper Airflow Range (Cooling)** portion of the calculator.

EQUIPMENT REQUIRED

R-22 Superheat/sub-cooling charging calculator
1 Temperature tester (with temperature leads)
1 Manifold gage set
1 Refrigerant cylinder, which matches system refrigerant
Insulation tape
Assorted hand tools

SAFETY PRACTICES

Wear goggles or safety glasses and gloves when working with refrigerant in sealed systems. Beware of high pressures and follow all electrical safety rules. Follow all EPA rules and regulations while working with refrigerants.

> NOTE: *The system being charged should have been leak checked and evacuated prior to performing this competency.*

COMPETENCY PROCEDURES   Checklist

1. The sealed system should be OFF.   ❏
2. Check the system's nameplate or service literature and determine the sub-
   cooling range for the air-cooled outdoor unit.   ❏

> NOTE: *If the system sub-cooling range is unavailable use 10 or 15 degrees of sub-cooling.*

3. At **Number 1** slide the calculator's slide rule down until the arrow lines with desired
   sub-cool temperature.   ❏
4. Attach and insulate a temperature probe to the system liquid line just ahead of the
   system's liquid line service valve.   ❏
5. Add refrigerant to the system in either the vapor or liquid state and bring the system's
   low-side pressure out of a vacuum.   ❏
6. Turn the system ON to the **Cooling mode.**   ❏
7. Continue to charge system until a low-side operating pressure of 62 psig is established.   ❏
8. Let the system operate for 10 minutes.   ❏
9. *Check and record the system's operating pressure.* _____   ❏
10. Find the system's high-side pressure at **Number 2** of the charging calculator.   ❏
11. Determine the liquid line temperature by looking directly underneath the System's
    operating high-side pressure at **Number 2.**   ❏
12. *Record the required liquid line temperature according to the charging calculator.* _____   ❏

> NOTE: *Under the Charging Calculator Instructions at Number 4 it states the following:*
> *If the measured liquid line temperature does not agree with the required liquid line temperature,*
> **ADD** *refrigerant to LOWER temperature or REMOVE refrigerant to RAISE temperature. (Allow a tolerance of + or −3 degrees [F].)*

> **NOTE:** *Any change in refrigerant charge to the system will change the system's high-side operating pressure. The liquid pressure (psig) at Number 2 will have to be readjusted as the system's pressures change due to the addition or removal of refrigerant to the system. The same required sub-cool temperature would continue to be used.*

13. *Add additional refrigerant as needed to reach the required sub-cool range.* ❏
14. Once system charge is correct, have the instructor check your work. ❏
15. On completion of system charge, remove all equipment and return to its proper location. ❏

RESEARCH QUESTIONS

1. What is the definition of sub-cooling?

   _____

2. What would happen to a system's operating head pressure if the condenser were operating in low ambient temperature conditions?

   _____

3. What are a couple of methods of maintaining a system's operating head pressure during low ambient temperature conditions?

   _____

4. What are the three functions of an air-cooled condenser?

   _____

5. Piston type metering devices for split-system air conditioning or heat pump system gets sized to which component of the sealed system?

   _____

   _____

Passed Competency _____ Failed Competency _____

Instructor Signature _____ Grade _____

Practical Competency 123

Checking for Proper Airflow Range of a Split-System Heat Pump Using an Airflow Range Charging Calculator

SUGGESTED MATERIALS

Textbook
Refrigeration & Air Conditioning Technology, 5th Edition, Thomson Delmar Learning
Unit 10—System Charging
Unit 40—Typical Operating Conditions

Review Topics
Vapor Refrigerant Charging; Liquid Refrigerant Charging

COMPETENCY OBJECTIVE

The student will be able check a split-system heat pump for proper airflow by using a charging calculator.

OVERVIEW

One of the additional factors that affect the efficiency of air conditioning and heat pump equipment is making sure that the system is moving the correct airflow in CFMs *(cubic feet per minute)* for both the evaporator and condenser. For most installations 400 to 450 CFMs of air are required per ton of cooling.

EQUIPMENT REQUIRED

R-22 Superheat/sub-cooling charging calculator
1 Sling psychrometer
1 Temperature tester (with temperature leads)
1 Manifold gage set
1 Refrigerant cylinder, which matches system refrigerant
Insulation tape
Assorted hand tools

SAFETY PRACTICES

Wear goggles or safety glasses and gloves when working with refrigerant in sealed systems. Beware of high pressures and follow all electrical safety rules. Follow all EPA rules and regulations while working with refrigerants.

COMPETENCY PROCEDURES   Checklist

1. Turn the system ON to the **Cooling** mode.   ❑
2. Allow the system to operate for 10 minutes.   ❑
3. Use the PROPER AIRFLOW RANGE (Cooling) of the charging calculator.   ❑
4. Use a temperature tester and measure the indoor entering air dry-bulb temperature. ❑
5. *Record the indoor entering air dry-bulb temperature.* _____   ❑
6. Slide the calculator slide rule to the proper indoor entering air dry-bulb temperature at **Number 1** of the Proper Airflow Range.   ❑

7. Use the sling psychrometer and measure the indoor entering air wet-bulb temperature. ❑
8. *Record the indoor entering air wet-bulb temperature.* _____ ❑
9. Find the indoor entering air wet-bulb temperature at **Number 2** of the charging calculator. ❑
10. *Read the **proper evaporator leaving air dry-bulb temperature** directly below the measured indoor entering air wet-bulb temperature recorded at Step 8.* ❑
11. *Record the PROPER evaporator leaving air dry-bulb temperature as established by the charging calculator.* _____ ❑
12. *Measure the actual leaving indoor coil air dry-bulb temperature.* ❑
13. *Record the actual leaving indoor coil air dry-bulb temperature.* _____ ❑
14. Compare the actual leaving evaporator air dry-bulb temperature to the PROPER evaporator coil leaving air dry-bulb temperature established by the charging calculator. ❑

> ***NOTE:*** *Under the Charging Calculator Instructions at Number 4 it states the following:*
> *If the measured leaving air-dry-bulb temperature is 3 degrees (F) or more **LOWER** than the proper leaving air temperature, **INCREASE** evaporator fan speed. If the measured leaving air dry-bulb temperature is 3 degrees (F) or more **HIGHER** than the proper leaving air temperature, **DECREASE** evaporator fan speed.*

15. If airflow is not correct, make adjustments as needed to bring system airflow into proper range. ❑
16. Have the instructor check your work. ❑
17. Turn the system OFF. ❑
18. Return all supplies and equipment to proper location. ❑

RESEARCH QUESTIONS

1. What effect would restricted airflow have on a system's evaporator?

   _____

   _____

2. What effect would a blocked air filter have on a system's evaporator?

   _____

   _____

3. What effect would restricted airflow have on an air-cooled condenser?

   _____

   _____

4. Heat pump air handlers that have a multispeed blower should have the blower operating at what speed during the cooling cycle?

   _____

   _____

5. Heat pump air handlers that have a multispeed blower should have the blower operating at what speed for the heating cycle?

   _____

   _____

Passed Competency _____   Failed Competency _____

Instructor Signature _____   Grade _____

Student Name _____ Grade _____ Date _____

Practical Competency 124

Charging an R-410A Heat Pump by Using an R-410A Superheat/Sub-cooling Calculator—Superheat Method

SUGGESTED MATERIALS

Textbook
Refrigeration & Air Conditioning Technology, 5th Edition, Thomson Delmar Learning
Unit 10—System Charging
Unit 40—Typical Operating Conditions
Unit 43—Air Source Heat Pumps

Review Topics
Vapor Refrigerant Charging; Liquid Refrigerant Charging

COMPETENCY OBJECTIVE

The student will be able to properly charge an R-410 heat pump system using an R-410A Charging Calculator using the Superheat Method.

OVERVIEW

It is important for HVAC technicians to understand that systems that are undercharged or overcharged will remove heat from a conditioned area, although they will not perform efficiently doing it!

It has been reported by national air conditioning associations that more than 70% of air conditioning and heat pump equipment in use in the United States is either *overcharged* or *undercharged*. This is a staggering number when efficiency performance of air conditioning and heat pump equipment relies on operating with the correct refrigerant charge. Many of the new refrigerants are referred to as *near-azeotropic* or zeotropic refrigerants and do not behave like azeotropic refrigerants. These refrigerants are blends, but referred to as *refrigerant mixtures* because refrigerants used within the blend can still separate into individuals refrigerants, water and oil, which can be mixed, but if allowed to sit for any period of time the water and oil will separate. This is where a difference in the pressure–temperature relationship arises versus the pressure–temperature relationship of azeotropic refrigerants.

Near-azeotropic blends experience a *temperature glide (Chapter 3, Figure 3–149, Theory Lesson: Superheat and Sub-cooling Calculations for Systems with Near-Azeotropic [Zeotropic] Refrigerant Blends).*

This temperature glide occurs when the blend has many temperatures as it evaporates and condenses at a given pressure. This means that refrigerants used in the mixture will change phase from liquid to vapor or vapor back to a liquid faster than other refrigerants used to make the mixture. Therefore, it possible with these refrigerant mixtures for one refrigerant in the mixture to be boiled away to 100% vapor and get *superheated* while other refrigerants in the mixture are still boiling from liquid to vapor. *(Chapter 3, Figure 3–150, Theory Lesson: Superheat and Sub-cooling Calculations for Systems with Near-Azeotropic [Zeotropic] Refrigerant Blends).*

Likewise it is possible for one or more of the refrigerants to be condensed from a vapor to 100% liquid and start to *sub-cool*, while other refrigerants within the mixture are still being condensed from vapor back to a liquid. Near-azeotropic refrigerants may also experience what is referred to as *fractionation*. This occurs when one or more of the refrigerants in the blend will condense or evaporate at different rates than other refrigerants making up the mixture. Fractionation only happens when the leak occurs as a vapor leak within the sealed system; it will not occur when the system is leaking pure liquid. This means that during a vapor leak within a system charged with near-azeotropic refrigerant, refrigerants within the mixture will leak out at different rates, which can have an effect on sealed system efficiency if system charge is *"topped off"* during a leak situation, rather than being completely removed from the system and a new charge weighed into the system. Fractionation also occurs if refrigerant is removed from a near-azeotropic refrigerant cylinder. For this reason, near-azeotropic or zeotropic blends should be *charged as a liquid* into a sealed system. New pressure–temperature charts have been designed and make it easier for technicians to check, charge, and perform superheat and sub-cool calculations for equipment charged with near-azeotropic refrigerant blends. (*Chapter 3, Figure 3–151, Theory Lesson: Superheat and Sub-cooling Calculations for Systems with Near-Azeotropic [Zeotropic] Refrigerant Blends*

For system superheat calculations, technicians are instructed to use the **dew point** values of the pressure temperature chart. *Dew point is the temperature at which saturated vapor first starts to condense.* For sub-cool calculations, technicians are instructed to use the **bubble point** values of the pressure temperature chart. *Bubble point is the temperature where the saturated liquid starts to boil off its first bubble of vapor.*

Using and understanding how to use an **R-410A superheat/sub-cooling calculator** takes the *"guess work"* out of checking or charging air conditioning and heat pump equipment along with checking for proper system air flow. The calculator is designed for use on TXV and non-TXV equipment. Technicians should use the **Required Superheat (Cooling, non-TXV) side for systems with fixed-bore metering devices,** and the **Sub-cooling (Cooling, TXV) for systems with a TXV metering device.** The calculator can also be used to check a system's airflow by using the **Proper Airflow Range (Cooling)** portion of the calculator.

EQUIPMENT REQUIRED

R-410A Superheat/sub-cooling charging calculator
1 Sling psychrometer
1 Temperature tester (with temperature leads)
1 Manifold gage set
1 Refrigerant cylinder, which matches system refrigerant
Insulation tape
Assorted hand tools

SAFETY PRACTICES

Wear goggles or safety glasses and gloves when working with refrigerant in sealed systems. Beware of high pressures and follow all electrical safety rules. Follow all EPA rules and regulations while working with refrigerants.

> **NOTE:** *System being charged should have been leak checked and evacuated prior to performing this competency.*

COMPETENCY PROCEDURES

Checklist

1. The sealed system should be OFF. ☐
2. Use the sling psychrometer and measure the indoor entering air wet-bulb temperature. ☐
3. *Record the indoor entering air wet-bulb temperature.* _____ ☐
4. Use the required superheat calculator (*Cooling, non-TXV*) and slide the slide rule of the charging calculator down to the recorded indoor entering air wet-bulb temperature at **Number 1.** ☐
5. Use the temperature tester and measure the (*outdoor unit*) condenser entering air dry-bulb temperature. ☐

6. *Record the condenser entering air dry-bulb temperature.* _____ ❏
7. Find the (*outdoor unit*) condenser entering air dry-bulb temperature at **Number 2** of the charging calculator. ❏
8. Directly below the (*outdoor unit*) condenser entering air dry-bulb temperature is the required superheat temperature at **Number 3** of the charging calculator. ❏
9. *Record the required superheat temperature.* _____ ❏
10. Attach and insulate a temperature probe near the system's vapor line service valve. ❏
11. Add refrigerant through the system's vapor line service valve in the vapor state until the system's low-side pressure is brought out of a vacuum. ❏
12. Turn the heat pump ON to operate in the **Cooling** mode. ❏
13. Continue to add refrigerant in the vapor state until the system's low-side pressure is operating at least **64 psig.** ❏
14. Once the system's operating low-side pressure is at least *64 psig,* stop charging and allow the system to operate for 10 minutes. ❏
15. At **Number 5** of the charging calculator, find the required superheat temperature as recorded in **Step 9.** ❏
16. Directly below the required superheat at **Number 5** is the required vapor line temperature at **Number 6.** ❏
17. *Record the required vapor line temperature as indicated by the charging calculator at **Number 6.*** _____ ❏
18. *Check the system's vapor line temperature and record actual vapor line temperature.* _____ ❏

NOTE: *Underneath the Charging Calculator Instructions at Number 6 it states the following:*
If the measured vapor line temperature does not agree with the required vapor line temperature, **ADD** *refrigerant to* **LOWER temperature** *or* **REMOVE** *refrigerant to* **RAISE temperature.** *(Allow a tolerance of + or –5 degrees [F].)*

---

19. Add or remove refrigerant in small amounts to the system until the system's vapor line temperature falls within (+ *or –5 degrees*) of the required vapor line temperature as indicated by the charging calculator. ❏

NOTE: *Any change in refrigerant charge to the system will change the system's low-side operating pressure. The vapor pressure at Number 4 will have to be readjusted as systems pressures change due to the addition or removal of refrigerant to the system. The same required superheat temperature would continue to be used to determine required vapor line temperature unless there has been a considerable time lapse or temperature change from the original temperature readings taken at Numbers 1 and 2.*

---

20. Once the system charge is correct, have the instructor check your work. ❏
21. On completion of system charge, remove all equipment and return to their proper location. ❏

RESEARCH QUESTIONS

1. What is the design superheat range for a heat pump system with a fixed bore metering device, operating under standard conditions?

_____

_____

2. Heat pumps with a TXV metering device will also be equipped with what sealed system component?

_____

_____

3. What is the main difference between a zeotropic and azeotropic refrigerant blend?

_____

_____

4. When should a service technician use the sub-cooling method of charging a heat pump system?

_____

_____

5. What causes fractionation to happen in certain blends of refrigerant?

_____

_____

Passed Competency _____   Failed Competency _____

Instructor Signature _____   Grade _____

Practical Competency 125

Charging an R-410A Heat Pump Equipped with a TXV by Using an R-410A Superheat/Sub-cooling Calculator — Sub-cool Method

SUGGESTED MATERIALS

Textbook
Refrigeration & Air Conditioning Technology, 5th Edition, Thomson Delmar Learning
Unit 10—System Charging
Unit 40—Typical Operating Conditions
Unit 43—Air Source Heat Pumps

Review Topics
Vapor Refrigerant Charging; Liquid Refrigerant Charging

COMPETENCY OBJECTIVE

The student will be able to properly charge a heat pump unit equipped with a TXV metering device by using a R-410A charging calculator and the sub-cool method.

OVERVIEW

It is important for HVAC technicians to understand that systems that are undercharged or overcharged will remove heat from a conditioned area, although they will not perform efficiently doing it!

It has been reported by national air conditioning associations that more than 70% of air conditioning and heat pump equipment in use in the United States is either *overcharged* or *undercharged*. This is a staggering number when efficiency performance of air conditioning and heat pump equipment relies on operating with the correct refrigerant charge. Many of the new refrigerants are referred to as *near-azeotropic* or zeotropic refrigerants and do not behave like azeotropic refrigerants. These refrigerants are blends, but referred to as *refrigerant mixtures* because refrigerants used within the blend can still separate into individuals refrigerants, like a mixture of water and oil, which can be mixed, but if allowed to sit for any period of time the water and oil will separate. This is where a difference in pressure–temperature relationship arises versus the pressure–temperature relationship of azeotropic refrigerants.

Near-azeotropic blends experience a *temperature glide (Chapter 3, Figure 3–149, Theory Lesson: Superheat and Sub-cooling Calculations for Systems with Near-Azeotropic [Zeotropic] Refrigerant Blends).*

This temperature glide occurs when the blend has many temperatures as it evaporates and condenses at a given pressure. This means that refrigerants used in the mixture will change phase from liquid to vapor or vapor back to a liquid faster than other refrigerants used to make the mixture. Therefore, it possible with these refrigerant mixtures for one refrigerant in the mixture to be boiled away to 100% vapor and get *superheated* while other refrigerants in the mixture are still boiling from liquid to vapor *(Chapter 3, Figure 3–150, Theory Lesson: Superheat and Sub-cooling Calculations for Systems with Near-Azeotropic (Zeotropic) Refrigerant Blends).*

Likewise it is possible for one or more of the refrigerants to be condensed from a vapor to 100% liquid and start to *sub-cool,* while other refrigerants within the mixture are still being condensed from vapor back to a liquid. Near-azeotropic refrigerants may also experience what is referred to as *fractionation.* This occurs when one or more of the refrigerants in the blend will condense or evaporate at different rates than other refrigerants making up the mixture. Fractionation only happens when the leak occurs as a vapor leak within the sealed system; it will not occur when the system is leaking pure liquid. This means that during a vapor leak within a system charged with near-azeotropic refrigerant, refrigerants within the mixture will leak out at different rates, which can have an effect on sealed system efficiency if system charge is *"topped off"* during a leak situation, rather than being completely removed from the system and a new charge weighed into the system. Fractionation also occurs if refrigerant is removed from a near-azeotropic refrigerant cylinder. For this reason, near-azeotropic or zeotropic blends should be *charged as a liquid* into a sealed system. New pressure–temperature charts have been designed and make it easier for technicians to check, charge, and perform superheat and sub-cool calculations for equipment charged with near-azeotropic refrigerant blends. (*Chapter 3, Figure 3–151 Theory Lesson: Superheat and Sub-cooling Calculations for Systems with Near-Azeotropic [Zeotropic] Refrigerant Blends*).

For system superheat calculations, technicians are instructed to use the **dew point** values of the pressure–temperature chart. *Dew point is the temperature at which saturated vapor first starts to condense.* For sub-cool calculations, technicians are instructed to use the **bubble point** values of the pressure temperature chart. *Bubble point is the temperature where the saturated liquid starts to boil off its first bubble of vapor.*

Using and understanding how to use an **R-410A superheat/sub-cooling calculator** takes the *"guess work"* out of checking or charging air conditioning and heat pump equipment along with checking for proper system air flow. The calculator is designed for use on TXV and non-TXV equipment. Technicians should use the **Required Superheat (Cooling, non-TXV)** side for systems with fixed-bore metering devices, and the **Sub-cooling (Cooling, TXV)** for systems with a TXV metering device. The calculator can also be used to check the system's airflow by using the **Proper Airflow Range (Cooling)** portion of the calculator.

EQUIPMENT REQUIRED

R-410A Superheat/sub-cooling charging calculator
1 Temperature tester (with temperature leads)
1 Manifold gage set
1 Refrigerant cylinder, which matches system refrigerant
Insulation tape
Assorted hand tools

SAFETY PRACTICES

Wear goggles or safety glasses and gloves when working with refrigerant in sealed systems. Beware of high pressures and follow all electrical safety rules. Follow all EPA rules and regulations while working with refrigerants.

NOTE: *If the system's sub-cooling range is unavailable use 10 or 15 degrees of sub-cooling.*

COMPETENCY PROCEDURES

Checklist

1. The sealed system should be OFF. ❏
2. Check the system's nameplate or service literature and determine the sub-cooling range for the outdoor air cooled coil. ❏

NOTE: *The system being charged should have been leak checked and evacuated prior to performing this competency.*

---

3. At **Number 1,** slide the calculators slide rule down until the arrow lines with desired sub-cool temperature. ❏

4. Attach and insulate a temperature probe to the system liquid line just ahead of the system's liquid line service valve. ❏

5. Add refrigerant to the system in either the vapor or liquid state and bring the system's low-side pressure out of a vacuum. ❏

6. Turn system ON to operate in the **Cooling mode.** ❏

7. Continue to charge system until a low-side operating pressure of 62 psig is established. ❏

8. Let the system operate for 10 minutes. ❏

9. *Check and record the system's operating high-side pressure.* _____ ❏

10. Find the system's high-side pressure at **Number 2** of the charging calculator. ❏

11. Determine the liquid line temperature by looking directly underneath the system's operating high-side pressure at **Number 2.** ❏

12. *Record the required liquid line temperature according to the charging calculator.* _____ ❏

NOTE: *Under the Charging Calculator Instructions at Number 4 it states the following:*
If the measured liquid line temperature does not agree with the required liquid line temperature, **ADD** *refrigerant to* **LOWER** *temperature or* **REMOVE** *refrigerant to* **RAISE** *temperature. (Allow a tolerance of + or −3 degrees [F].)*

NOTE: *Any change in refrigerant charge to the system will change the system's high-side operating pressure. The liquid pressure (psig) at Number 2 will have to be readjusted as the system's pressures change due to the addition or removal of refrigerant to the system. The same required sub-cool temperature would continue to be used to determine.*

---

13. Continue to add refrigerant if needed. ❏

14. Once system charge is correct, have the instructor check your work. ❏

15. On completion of system charge, remove all equipment and return to its proper location. ❏

RESEARCH QUESTIONS

1. How is condenser high efficiency obtained?

_____

_____

2. The evaporator design temperature may in some cases operate at a slightly _____ (*higher or lower*) pressure and temperature on high-efficiency equipment because the evaporator is larger.

3. The three major power-consuming devices on a heat pump system that may have to be analyzed are the _____, _____, and the _____.

4. Heat pump systems normally move about _____ CFM of air per ton.

5. The compressor amperage of a 3-ton system operating on 230 volts is approximately _____ A.

Passed Competency _____   **Failed Competency** _____

Instructor Signature _____   **Grade** _____

Practical Competency 126

Checking for Proper Airflow Range of a Split-System Heat Pump Using an R-410A Proper Airflow Range Charging Calculator

SUGGESTED MATERIALS

Textbook
Refrigeration & Air Conditioning Technology, 5th Edition, Thomson Delmar Learning
Unit 10—System Charging
Unit 40—Typical Operating Conditions
Unit 43—Air Source Heat Pumps

Review Topics
Vapor Refrigerant Charging; Liquid Refrigerant Charging

COMPETENCY OBJECTIVE

The student will be able check a split-system heat pump for proper airflow by using an R-410A charging calculator.

OVERVIEW

One of the additional factors that affect the efficiency of air conditioning and heat pump equipment is making sure that the system is moving the correct airflow in CFMs (*cubic feet per minute*) for both the evaporator and condenser. For most installations 400 to 450 CFMs of air are required per ton of cooling.

EQUIPMENT REQUIRED

R-410A Superheat/sub-cooling charging calculator
1 Sling psychrometer
1 Temperature tester (with temperature leads)
1 Manifold gage set
1 Refrigerant cylinder, which matches system refrigerant
Insulation tape
Assorted hand tools

SAFETY PRACTICES

Wear goggles or safety glasses and gloves when working with refrigerant in sealed systems. Beware of high pressures and follow all electrical safety rules. Follow all EPA rules and regulations while working with refrigerants.

COMPETENCY PROCEDURES

Checklist

1. Turn the system ON to operate in the **Cooling** mode. ☐
2. Allow the system to operate for 10 minutes. ☐
3. Use the PROPER AIRFLOW RANGE (Cooling) of the charging calculator. ☐

681

4. Use a temperature tester and measure the indoor entering air dry-bulb temperature. ❑

5. *Record the indoor coil entering air dry-bulb temperature.* _____ ❑

6. Slide the calculator slide rule to the proper indoor entering air temperature at Number 1 of the Proper Airflow Range. ❑

7. Use the sling psychrometer and measure the indoor entering air wet-bulb temperature. ❑

8. *Record the indoor entering air wet-bulb temperature.* _____ ❑

9. Find the indoor entering air wet-bulb temperature at **Number 2** of the charging calculator. ❑

10. *Read the **Proper evaporator** (**indoor coil**) **leaving air dry-bulb temperature** directly below the measured indoor entering air wet-bulb temperature recorded at Step 8.* ❑

11. *Record the PROPER evaporator (indoor coil) leaving air dry-bulb temperature as established by the charging calculator.* _____ ❑

12. *Measure the actual leaving indoor coil air dry-bulb temperature.* ❑

13. *Record the actual leaving indoor coil air dry-bulb temperature.* _____ ❑

14. Compare the actual leaving indoor coil air dry-bulb temperature to the PROPER evaporator (*indoor coil*) leaving air dry-bulb temperature established by the charging calculator. ❑

NOTE: Under the Charging Calcultor Instructions at Number 4 it states the following:
    *If the measured leaving air-dry-bulb temperature is 3 Degrees (F) or more **LOWER** than the proper leaving air temperature, **INCREASE** evaporator fan speed. If the measured leaving air-dry-bulb temperature is 3 degrees (F) or more **HIGHER** than the proper leaving air temperature, **DECREASE** indoor coil fan speed.*

---

15. If airflow is not correct, make adjustments as needed to bring system airflow into proper range. ❑

16. Have the instructor check your work. ❑

17. Turn the system OFF. ❑

18. Return all supplies and equipment to their proper location. ❑

RESEARCH QUESTIONS

1. Why do heat pumps require a defrost system?

_____

_____

2. The typical temperature difference between the entering air and boiling refrigerant in the indoor coil is _____ (F) under standard conditions.

3. **True or False:** When troubleshooting a small system, the first step is to attach the high- and low-side manifold pressure gages.

_____

_____

4. The suction gas may have a high superheat if the unit has an _____ charge.

5. What two types of metering devices are used on heat pumps? _____ and _____

Passed Competency _____   Failed Competency _____

Instructor Signature _____   Grade _____

Practical Competency 127

Field Charging or Checking the Charge of a Heat Pump System with a Fixed-Bore Metering Device Operating in Low Ambient Temperatures

SUGGESTED MATERIALS

Textbook

Refrigeration & Air Conditioning Technology, 5th Edition, Thomson Delmar Learning
Unit 41—Troubleshooting
Unit 43—Air Source Heat Pumps

Review Topics

Charging Procedures in the Field; Fixed-Bore Metering Devices; Capillary Tube and Orifice Type

COMPETENCY OBJECTIVE

The student will be able to charge or check the charge of heat pump during low ambient temperature by using field service procedures.

OVERVIEW

When charging or checking the charge of any air conditioning or heat pump system it is important that technicians follow manufacturer's guidelines when available. There are times when the technician needs to check or make system charge adjustments and the manufacturer's guidelines are not available. Besides using a charging calculator, charging charts, or charging curves, technicians are left with making personal judgments as to a system's charge condition. Technicians can accurately determine or adjust a system's charge without the aid of such tools, but to do so requires an understanding of refrigeration fundamentals, the manufacturer's design of such equipment, and how the equipment should perform under certain conditions.

Field charging procedures are best established by knowing how the equipment is to perform under *Standard Operating Conditions.* When a system is charged correctly and operating within Standard Conditions, a prescribed amount of refrigerant should be in the condenser, the evaporator, and the system's liquid line. **It is important that technicians understand the following facts about air conditioning and heat pump equipment:**

1. *The amount of refrigerant in the evaporator (indoor coil) can be measured by the superheat method.*
2. *The amount of refrigerant in the condenser (outdoor coil) can be measured by the sub-cooling method.*
3. *The amount of refrigerant in the liquid line may be determined by measuring the length and calculating the refrigerant charge.*
4. *If the evaporator (indoor coil) is performing correctly, the liquid line has the correct charge.*

When system charge adjustments are required, technicians need to establish charging procedures in the field for all types of equipment. The following are a couple of methods used for different type of equipment.

683

PROCEDURES FOR SYSTEMS WITH FIXED-BORE METERING DEVICES: CAPILLARY TUBE AND PISTON (ORIFICE) TYPE

Fixed-bore metering devices allow refrigerant flow based on the difference in the inlet (high-side) and outlet (low-side) pressures. Ideally, the best situation would be to check for a system's correct charge under Standard Operating Conditions, 75-degree (F) return air with 50% humidity, and an outside temperature of 95 degrees. Unfortunately, rarely do these exact conditions exist. When a system is operating outside of these conditions, different system pressures superheat readings, and sub-cooling temperatures will occur. The condition that most affects system readings is the outside ambient air temperature. In lower than normal outside air temperatures, the condenser (*outdoor coil*) will become more efficient in rejecting heat from the refrigerant vapor, causing condensing of vapor to liquid to happen sooner, raising the condenser's (*outdoor coil*) sub-cooling temperature, and in turn lowering the system's high-side operating pressure.

Since fixed-bore metering devices rely on a pressure difference between the system's high-side and low-side pressures to feed the correct amount of refrigerant into the evaporator (*indoor coil*), lower outside temperature can create the effect of partially starving the evaporator (*indoor coil*) for refrigerant. In other words, because the condenser (*outdoor coil*) is operating more efficiently, there is more liquid refrigerant in the condenser (*outdoor coil*) than that which is being fed into the evaporator (*indoor coil*) by the metering device, in turn creating a situation where the evaporator (*indoor coil*) becomes starved.

The lower than normal high-side pressure under this condition is what makes checking a system's charge hard to evaluate, although typical operating conditions **can be simulated by reducing the airflow across the condenser (*outdoor coil*), which will cause the system's high-side pressure to rise (Figure 4–67).**

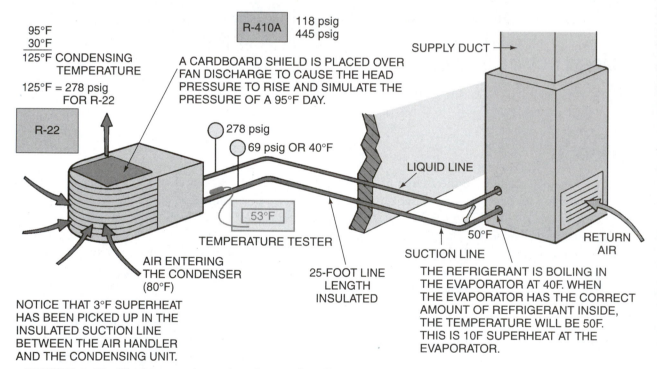

FIGURE 4–67 Blocking condenser (*outdoor coil*) airflow to simulate typical outside air temperature of 95 degrees (F), increasing system head pressure to correlate with established conditions, so the superheat method can be used to check the charge of a system equipped with a fixed-bore metering device.

On a 95-degree (F) day the highest condenser (*outdoor coil*) head pressure for **R-22 systems is 278 psig**, which equals to a condensing temperature of 125 degrees for standard air-cooled condensers (*outdoor coil*). Remember that standard air-cooled condensers (*outdoor coil*) condense the refrigerant vapor to liquid at approximately 30 degrees (F) above the outside air temperature. (*95-degree day plus 30 degrees equals 125 degrees saturation temperature*). *R-410A systems would operate with a high-side pressure of approximately **445 psig** under standard conditions. For **high-efficiency equipment**, a head pressure of 250 psig would be sufficient for an R-22 system, and 390 psig for an R-410A system.*

The higher high-side pressure pushes the refrigerant through the metering device at the correct rate, rather than allowing refrigerant liquid buildup in the condenser as with lower ambient conditions.

> NOTE: *Once the condenser's (outdoor coil) airflow is restricted, allow time for the system's high-side pressure to build. If the correct high-side pressure is not in range as that which should be under standard conditions, refrigerant will have to added or removed from the system to bring the high-side pressure into range.* Once the correct pressure is established, the following additional procedures can be used to bring the system's charge into range.

Using this procedure allows the technician to perform a superheat check on the system's evaporator (*indoor coil*), although a superheat check at the actual evaporator (*indoor coil*) is not always easy with split systems, so a superheat check at the condensing unit (*outdoor coil*) vapor line can be made. When using the system's vapor line, the length of the line set would need to be taken into consideration. Remember that under standard conditions, 75 degrees (F) inside air with 50% humidity, and an outside air temperature of 95 degree (F), the evaporator superheat should be around 10 degrees. Since the superheat is actually measured at the system's vapor line, the length of the system's line set needs to be taken into consideration.

With the proper high-side pressure and a properly insulated vapor line, **a line set length of 10 to 30 feet should indicate a superheat value of 10 to 15 degrees (F) if the system's charge is close to being correct. Line sets of 30 to 50 feet should indicate a superheat value of 15 to 18 degrees.**

EQUIPMENT REQUIRED

Heat pump system
1 Piece of cardboard/wood shield large enough to restrict outdoor coil airflow
1 Temperature tester (with temperature leads)
1 Manifold gage set
1 Refrigerant cylinder, which matches system refrigerant
Insulation tape
Assorted hand tools

SAFETY PRACTICES

Wear goggles or safety glasses and gloves when working with refrigerant in sealed systems. Beware of high pressures and follow all electrical safety rules. Follow all EPA rules and regulations while working with refrigerants.

> NOTE: *This competency is best performed on a system operating with an outside air temperature below 95 degrees (F).*

COMPETENCY PROCEDURES

Checklist

1. Turn the system OFF. ❏
2. Remove vapor and liquid line service valve caps. ❏
3. Make sure that manifold gage valves are closed. ❏
4. Attach the manifold high-side hose to the system's liquid line valve service port. ❏
5. Attach the manifold low-side hose to the system's vapor valve service port. ❏
6. Attach a temperature lead to the system's vapor line, near the system's vapor line service valve. ❏
7. Insulate the temperature lead. ❏
8. *Estimate and record the systems line set length.* _____ ❏

9. *According to the Competency Overview and considering the approximate length of systems line set, what should be the approximate amount of superheat if the systems charge is correct?* _____ ❏
10. *Record the type of refrigerant used in the system.* _____ ❏
11. Turn the system ON to raise structure temperature to at least 75 degrees. ❏

> **NOTE:** *To raise the structure temperature, place heat pump in either the heating or emergency heat mode. The unit can also be placed in the Emergency Heat mode, and jumper the thermostat R–Y terminals to operate the heat pump in the cooling mode at the same time. This will keep structure temperature up while the heat pump is operating in the Cooling mode.*

12. Once structure temperature has risen to at least 75 degrees (F), switch the unit over to the **Cooling** mode and allow system to operate for 5 to 10 minutes. ❏
13. *Record the system's operating low-side pressure.* _____ ❏
14. *Use a pressure–temperature chart and record the system's low-side saturation temperature.* _____ ❏
15. *Record the vapor line temperature.* _____ ❏
16. Determine the amount of the system's superheat:

    Vapor line temperature _____ ❏
    MINUS
    Low-side saturation temperature _____ ❏
    EQUALS *Superheat amount* _____ ❏
17. *Record the system's superheat.* _____ ❏
18. *Does the superheat amount fall in range with that which is stated in the Overview based on a 95-degree (F) outside temperature and length of the system's line set?* _____ ❏
19. *Record the system's operating high-side pressure.* _____ ❏
20. *Does the system's high-side pressure fall in line with an outside temperature of 95-degree (F) day as described in the Competency Overview?* _____ ❏

> **NOTE:** *If the system's superheat and high-side pressures **are within range** as described in the Competency Overview, the system's charge is correct. **Do not proceed**; have the instructor check your work.*
> *If the system's superheat and high-side pressure **are not within range** as described in the Competency Overview, **proceed to the next steps** of the competency.*

21. Place the cardboard/wood shield over the top of the outdoor coil to restrict outdoor coil airflow (**Figure 4–68**). ❏
22. Allow the system to operate for 10–15 minutes with restricted outdoor coil airflow. ❏
23. *Record the system's high-side pressure.* _____ ❏
24. *Does the system's high-side pressure fall in line for the type of refrigerant as described in the Competency Overview, based on a 95-degree day?* _____ ❏

> **NOTE:** *With outdoor coil air restricted, if the high-side pressure is higher or lower than that which is expected on a 95-degree (F) day according to the Competency Overview, **ADD Refrigerant** to **RAISE** high-side pressure, or **REMOVE** Refrigerant to **LOWER** high-side pressure.*

25. Adjust the system's charge as necessary to bring the system's operating high-side pressure into line based on a 95-degree (F) day. ❏

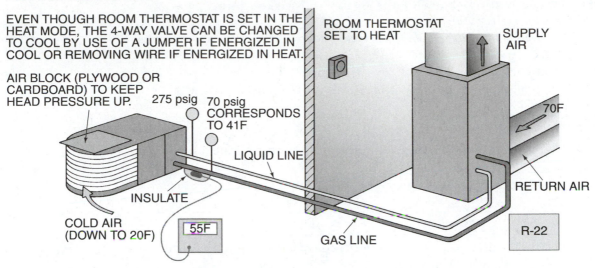

OUTDOOR UNIT   INDOOR UNIT

TO KEEP THE CONDITIONED SPACE FROM BECOMING
TOO COOL, THIS PROCEDURE WILL CAUSE THE
REFRIGERANT UNIT TO COOL AND THE STRIP
HEAT TO PRODUCE HEAT.

EVEN THOUGH ROOM THERMOSTAT IS SET IN THE
HEAT MODE, THE 4-WAY VALVE CAN BE CHANGED
TO COOL BY USE OF A JUMPER IF ENERGIZED IN
COOL OR REMOVING WIRE IF ENERGIZED IN HEAT.

ROOM THERMOSTAT
SET TO HEAT

SUPPLY
AIR

AIR BLOCK (PLYWOOD OR
CARDBOARD) TO KEEP
HEAD PRESSURE UP.   275 psig 70 psig
CORRESPONDS
TO 41F

70F

LIQUID LINE

INSULATE

RETURN AIR

COLD AIR
(DOWN TO 20F)   55F

GAS LINE

R-22

FIGURE 4–68   Build outdoor head pressure by restricting airflow to the outdoor coil.

26. *Record the system's operating high-side pressure once charge adjustment is complete.* _____ ❑

27. *Record the system's operating low-side pressure.* _____ ❑

28. *Use a pressure–temperature chart and record the system's low-side saturation temperature.* _____ ❑

29. *Record the suction line temperature.* _____ ❑

30. Determine the amount of the system's superheat:
    Vapor line temperature _____ ❑
    MINUS
    Low-side saturation temperature _____ ❑
    EQUALS Superheat amount _____ ❑

31. *Record the system's superheat.* _____ ❑

32. *Does the superheat amount fall in range with that which is stated in the Overview based on a 95-degree (F) outside temperature and length of the system's line set?* _____ ❑

33. Make an additional adjustment to system charge if needed. ❑

34. Once the system charge is in line based on a 95-degree day, have the instructor check your work. ❑

35. Remove the outdoor coil airflow restrictor. ❑

36. Let the system operate for 10 minutes. ❑

37. *Record the system's operating low-side pressure.* _____ ❑

38. *Use a pressure–temperature chart and record the system's low-side saturation temperature.* _____ ❑

39. *Record the vapor line temperature.* _____ ❑

40. Determine the amount of the system's superheat:
    Vapor line temperature _____ ❑
    MINUS
    Low-side saturation temperature _____ ❑
    EQUALS *superheat amount* _____ ❑

41. *Record the system's superheat.* _____ ❑

42. *Does the superheat amount fall in range with that which is stated in the Overview based on a 95-degree (F) outside temperature and length of the system's line set?* _____ ❏
43. *Record the system's operating high-side pressure.* _____ ❏
44. Have the instructor check your work. ❏
45. Turn the system OFF. ❏
46. Remove manifold gage hoses from high- and low-side system service valves. ❏
47. Replace service valve caps. ❏
48. Remove the temperature lead probe from the vapor line. ❏
49. Return all supplies and equipment to their proper location. ❏

RESEARCH QUESTIONS

1. The three materials of which condensers are normally made are _____, _____, and _____.

2. When is the most heat removed from the refrigerant in the condensing process?

   _____

   _____

3. After heat is absorbed into a condenser medium in a water- cooled condenser, the heat can be deposited in one of two places. They are _____ and _____.

4. When a standard efficiency air-cooled condenser is used, the condensing refrigerant will normally be _____ (F) higher in temperature than the entering air temperature.

5. Four methods for controlling head pressure in an air-cooled condenser are _____, _____, _____, and _____.

Passed Competency _____   Failed Competency _____

Instructor Signature _____   Grade _____

Practical Competency 128

Field Charging or Checking the Charge of a Heat Pump System with a TXV Metering Device Operating in Low Ambient Temperatures

SUGGESTED MATERIALS

Textbook
Refrigeration & Air Conditioning Technology, 5th Edition, Thomson Delmar Learning
Unit 41—Troubleshooting
Unit 43—Air Source Heat Pumps
Unit 22—Condensers

Review Topics
Charging Procedures in the Field; Fixed-Bore Metering Devices; Capillary Tube and Orifice Type

COMPETENCY OBJECTIVE

The student will be able to charge or check the charge of an air conditioner or heat pump during low ambient temperature by using field service procedures.

OVERVIEW

When charging or checking the charge of any air conditioning or heat pump system it is important that technicians follow manufacturer's guidelines when available. There are times when the technician needs to check or make system charge adjustments and the manufacturer's guidelines are not available. Besides using a charging calculator, charging charts, or charging curves, technicians are left with making personal judgments as to a system's charge condition. Technicians can accurately determine or adjust a system's charge without the aid of such tools, but to do so requires an understanding of refrigeration fundamentals, the manufacturer's design of such equipment, and how the equipment should perform under certain conditions.

Field charging procedures are best established by knowing how the equipment is to perform under *Standard Operating Conditions*. When a system is charged correctly and operating within Standard Conditions, a prescribed amount of refrigerant should be in the condenser, the evaporator, and the system's liquid line. **It is important that technicians understand the following facts about air conditioning and heat pump equipment:**

1. *The amount of refrigerant in the indoor coil can be measured by the superheat method.*
2. *The amount of refrigerant in the outdoor coil can be measured by the sub-cooling method.*
3. *The amount of refrigerant in the liquid line may be determined by measuring the length and calculating the refrigerant charge.*
4. *If the indoor coil is performing correctly, the liquid line has the correct charge.*

When system charge adjustments are required, technicians need to establish charging procedures in the field for all types of equipment. The following are a couple of methods used for different type of equipment.

PROCEDURES FOR FIELD CHARGING A TXV SYSTEM

A TXV metering device is designed to maintain a constant superheat on an evaporator of 8 to 12 degrees under all load conditions. Depending on the load on the evaporator (*indoor coil*), the TXV may not need to feed the system's total charge to the evaporator, so its needs space to store the additional refrigerant when not in use. Liquid receivers are equipped with TXV systems for this very reason, and work in conjunction with the system's condenser to store liquid refrigerant in low ambient temperature conditions. This process always ensures enough liquid refrigerant is available to the TXV for feeding the evaporator (*indoor coil*). Because of this, a TXV system will not be affected as much as a system with a fixed-bore metering device during low ambient temperatures. In fact a TXV system could be undercharged or overcharged, yet have enough refrigerant to maintain designed superheat on the evaporator (*indoor coil*). For this reason, checking or adjusting the charge of a system with a TXV metering device cannot be performed by using the superheat method.

The charge of air conditioning and heat systems equipped with a TXV metering device can be checked in much the same way as a system with a fixed-bore metering device during low ambient temperature conditions. The only difference is that once pressures are established to represent a 95-degree (F) day, the sub-cool method of checking or charging the system would be used.

Once again, typical operating conditions **would need to be simulated by reducing the airflow across the outdoor coil, which will cause the system's high-side pressure to rise (Figure 4–69).**

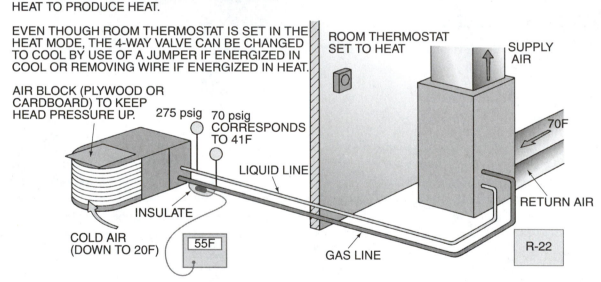

FIGURE 4–69 Blocking outdoor coil airflow to simulate typical outside air temperature of 95 degrees (F), increasing system head pressure to correlate with established conditions so that the sub-cool method can be used to check the charge of a TXV metering device system.

On a 95-degree (F) day, the highest condenser **head pressure for R-22 systems is 278 psig,** which equals to a condensing temperature of 125 degrees for standard air-cooled condensers. Remember that standard air-cooled condensers condense the refrigerant vapor to liquid at approximately 30 degrees (F) above the outside air temperature. (*95-degree day plus 30 degrees equals 125 degrees saturation temperature*). *R-410A systems would operate with a high-side pressure of approximately **445 psig** under standard conditions. For **high-efficiency equipment**, a head pressure of **250 psig** would be sufficient for an R-22 system, and **390 psig** for a R-410A system.*

Once the correct high-side pressure for the type of refrigerant is simulated, checking the condensers subcooling can be used to evaluate the system's charge. A **typical sub-cooling** circuit will sub-cool the liquid refrigerant **from 10 to 20 degrees** (F) than the condenser's (*outdoor coil*) condensing temperature. Under the simulated conditions, a temperature lead should be attached to the condenser's (*outdoor coil*) **liquid line and a temperature of 105 to 115 degrees (F)**, equivalent to 10 to 20 degrees (F) of sub-cooling should be established. (Refer to *Figure 4–69*.)

> *NOTE: Establishing the proper sub-cooling range under these simulated conditions means that the systems charge is quite close to accurate.*

> *NOTE: (OVERCHARGED SYSTEM) A sub-cooling temperature higher than 20 degrees (F) is an indication of an overcharged system. This will require that some of the refrigerant from the system be recovered until the sub-cooling temperature is brought within the design range.*

> *NOTE: (UNDERCHARGED SYSTEM) Sub-cooling temperatures lower than the 10 to 20 degrees (F) range are an indication that the system is undercharged. Refrigerant would have to be added to the system until the sub-cooling temperature is brought within the design range.*

> *NOTE: Refrigerant charge adjustments should be made guardedly. Once an adjustment is made, allow system to stabilize before making additional adjustments.*

EQUIPMENT REQUIRED

Air conditioner or heat pump system
1 Piece of cardboard/wood shield large enough to restrict outdoor coil airflow
1 Temperature tester (with temperature leads)
1 Manifold gage set
1 Refrigerant cylinder, which matches system refrigerant
Insulation tape
Assorted hand tools

SAFETY PRACTICES

Wear goggles or safety glasses and gloves when working with refrigerant in sealed systems. Beware of high pressures and follow all electrical safety rules. Follow all EPA rules and regulations while working with refrigerants.

COMPETENCY PROCEDURES

Checklist

1. Turn the system OFF. ❑
2. Remove the liquid line service valve cap. ❑
3. Make sure that manifold gage valves are closed. ❑
4. Attach manifold high-side hose to the system's liquid line valve service port. ❑
5. Attach a temperature lead to the system's liquid line, near the system's liquid line service valve. ❑
6. Insulate the temperature lead. ❑
7. *Record the type of refrigerant used in the system.* _____ ❑
8. Turn the system ON to raise structure temperature to at least 75 degrees. ❑

9. Once structure temperature has risen to at least 75 degrees (F), switch the unit over to the **Cooling** mode and allow system to operate for 5 to 10 minutes. ❑❑
10. *Record the system's operating high-side pressure.* _____ ❑
11. *Use a pressure–temperature chart and record the system's high-side saturation temperature.* _____ ❑
12. *Record the liquid line temperature.* _____ ❑
13. Determine the system's sub-cool amount:

High-side saturation temperature _____ ❑

MINUS

Liquid line temperature _____ ❑

EQUALS sub-cool amount _____ ❑

14. *Record the system's sub-cool.* _____ ❑
15. *Does the sub-cool amount fall in range with that which is stated in the Overview based on a 95-degree (F) outside temperature?* _____ ❑
16. *Record the system's operating high-side pressure.* _____ ❑
17. *Does the system's high-side pressure fall in line with an outside temperature of a 95-degree (F) day as described in the Competency Overview?* _____ ❑

18. Place the cardboard/wood shield over the top of the outdoor coil to restrict Airflow. (Refer to *Figure 4–69*.) ❑
19. Allow the system to operate for 10–15 minutes with restricted outdoor coil airflow. ❑
20. *Record the system's high-side pressure.* _____ ❑
21. *Does the system's high-side pressure fall in line for the type of refrigerant as described in the Competency Overview, based on a 95-degree day?* _____ ❑

22. Adjust the system's charge as necessary to bring the system's operating high-side pressure into line based on a 95-degree (F) day. ❑
23. *Record the system's operating high-side pressure once charge adjustment is complete.* _____ ❑
24. *Use a pressure–temperature chart and record the system's high-side saturation temperature.* _____ ❑
25. *Record the liquid line temperature.* _____ ❑
26. Determine the amount of system's sub-cool:

High-side saturation temperature _____ ❑

MINUS

Liquid line temperature _____ ❑

EQUALS sub-cool amount _____ ❑

27. *Record the system's sub-cool amount.* _____ ❑
28. *Does the sub-cool amount fall in range with that which is stated in the Overview based on a 95-degree (F) outside temperature and length of the system's line set?* _____ ❑
29. Make an additional adjustment to system charge if needed. ❑
30. Once system charge is in line based on a 95-degree day, have the instructor check your work. ❑
31. Remove the outdoor coil airflow restrictor. ❑
32. Let the system operate for 10 minutes. ❑
33. *Record the system's operating high-side pressure.* _____ ❑
34. *Use a pressure–temperature chart and record the system's high-side saturation temperature.* _____ ❑
35. *Record the liquid lines temperature.* _____ ❑
36. *Determine the amount of system's sub-cool:*

        High-side saturation temperature _____ ❑
        MINUS
        Liquid line temperature _____ ❑
        EQUALS *sub-cool amount* _____ ❑

37. *Record the system's sub-cool.* _____ ❑
38. *Does the sub-cool amount fall in range with that which is stated in the Overview based on a 95-degree (F) outside temperature and length of the system's line set?* _____ ❑
39. Have the instructor check your work. ❑
40. Turn the system OFF. ❑
41. Remove manifold gage hose from liquid line service valve. ❑
42. Replace the service valve cap. ❑
43. Remove the temperature lead probe from the liquid line. ❑
44. Return all supplies and equipment to their proper location. ❑

RESEARCH QUESTIONS

1. The prevailing winds can affect which of the air-cooled condensers?

   _____

   _____

2. **True or False:** Air-cooled condensers are much more efficient than water-cooled condensers.

   _____

   _____

3. The type of condenser that has the lowest operating head pressure is the _____ cooled condenser.
4. Name two ways in which the heat from an air-cooled condenser can be used for heat reclaim.

   _____

   _____

5. What is the approximate condensing temperature for high-efficiency condensers?

   _____

   _____

Practical Competency 129

Determining Correct Operating High-Side Pressure for Heat Pump Equipment with a Standard Air-Cooled Condenser (Figure 4–70)

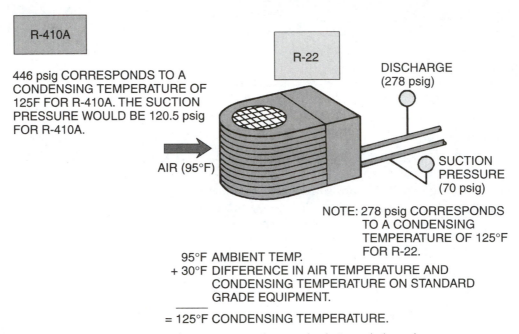

R-410A

446 psig CORRESPONDS TO A CONDENSING TEMPERATURE OF 125F FOR R-410A. THE SUCTION PRESSURE WOULD BE 120.5 psig FOR R-410A.

R-22

DISCHARGE (278 psig)

AIR (95°F)

SUCTION PRESSURE (70 psig)

NOTE: 278 psig CORRESPONDS TO A CONDENSING TEMPERATURE OF 125°F FOR R-22.

95°F AMBIENT TEMP.
+ 30°F DIFFERENCE IN AIR TEMPERATURE AND CONDENSING TEMPERATURE ON STANDARD GRADE EQUIPMENT.
_____
= 125°F CONDENSING TEMPERATURE.

FIGURE 4–70   Pressures and temperatures for standard air-cooled condenser.

SUGGESTED MATERIALS

Textbook
Refrigeration & Air Conditioning Technology, 5th Edition, Thomson Delmar Learning
Unit 22—Condensers
Unit 36—Refrigeration Applied to Air Conditioning
Unit 40—Typical Operating Conditions
Unit 43—Air Source Heat Pumps

Review Topics
Standard Air-Cooled Condensers; Typical Operating Conditions; Troubleshooting; Refrigeration Applied to Air Conditioning

COMPETENCY OBJECTIVE

The student will be able to determine the correct operating high-side for heat pump equipment with a standard air-cooled condenser.

OVERVIEW

The condenser (*outdoor coil*) of refrigeration sealed systems performs three functions in the refrigeration process. It rejects heat form the refrigerant vapor, condenses vapor to liquid, and sub-cools liquid refrigerant (*Chapter 3, Theory Lesson: Condensers, Figure 3–263*).

How fast a condenser (*outdoor coil*) can perform these functions can be used to evaluate the condenser's performance and efficiency. During rejection of **latent heat** from the refrigerant vapor, the vapor reaches a point of saturation and starts to condense back to the liquid state. The temperature at which this process happens is determined by the pressure being exerted on the vapor.

The **saturation temperature** of a condenser (*outdoor coil*) can be determined by using the manufacturer's design condensing temperature range for the different type of condensers. In residential and light commercial air conditioning and heat pump applications, two types of condensers (*outdoor coil*) are used, air-cooled and water-cooled condensers. There are two types of air-cooled condensers (*outdoor coil*) used, standard efficiency and high efficiency. Manufacturers have designed all **standard air-cooled condensers** to condense the refrigerant vapor to a liquid at a temperature of about 30 to 35 degrees above the outside ambient air temperature (*Chapter 3, Theory Lesson: Condensers, Figure 3–264*).

New **high-efficiency air-cooled condensers** (*outdoor coil*) condense the refrigerant vapor back to a liquid at 15 to 25 degrees above the outside ambient air temperature based on the SEER rating of the equipment. The higher the SEER rating of the equipment, the lower the saturation temperature of the refrigerant (*Chapter 3, Theory Lesson: Condensers, Figure 3–265*).

Water-cooled condensers are designed to condense refrigerant vapor to liquid at 10 degrees above the leaving water temperature in summer applications (*Chapter 3, Theory Lesson: Condensers, Figure 3–266*), and approximately 20 degrees above the leaving water temperature in winter applications (*Chapter 3, Theory Lesson: Condensers, Figure 3–267*).

Knowing this information can assist technicians in evaluating a system's **high-side operating pressure** and **saturation temperature**. A system's **correct saturation temperature** can be determined by measuring the outside ambient air temperature and adding the manufacturer's designed condensing temperature to it, based on the type of condenser (*outdoor coil*) being used in the system. This tells the technician what the saturation temperature of the refrigerant should be for the condenser under current conditions. Using the saturation temperature can assist the technician in determining what the system's **high-side operating pressure** should be under current conditions. The technician would convert the saturation temperature to pressure for the type of system refrigerant being used. This would be what the system's operating head pressure should be under current conditions.

This information can then be used to evaluate actual system operating high-side pressure and saturation temperature. An actual system pressure and saturation temperature of **plus or minus 10%** would be acceptable. Any pressure and saturation temperature higher or lower than the allowable 10% range is an indication to the technician that the condenser is working too hard or below an acceptable range. Technicians should look for conditions that could cause the condenser to perform outside the determined acceptable range.

EQUIPMENT REQUIRED

Manifold gage set
Temperature meter
Pressure–temperature chart
Safety goggles
Safety gloves
Refrigeration wrench (*if applicable*)

SAFETY PRACTICES

The student should be knowledgeable in the use of tools and testing equipment. Follow all EPA rules and regulations when working with refrigerants.

> **NOTE:** *This competency is to be performed on heat pump equipment with a standard air-cooled condenser.*

1. Turn the system ON to the **Cooling** mode and allow system to operate for 10 minutes. ❑

DETERMINING CORRECT CONDENSER SATURATION TEMPERATURE

2. *Record the type of refrigerant used in system.* _____ ❑
3. *Measure and record the dry-bulb air temperature entering the outdoor coil.* _____ ❑
4. Determine what you think the outdoor coil saturation temperature should be by using the following formula:
   Outdoor coil entering dry-bulb air temperature _____ + 30 degrees = saturation
   temperature _____ ❑
   Outdoor coil entering dry-bulb air temperature _____ + 35 degrees = saturation
   temperature _____ ❑

> **NOTE:** *Remember that standard air-cooled condensers (outdoor coil) are designed to condense the refrigerant vapor to a liquid at about 30 to 35 degrees above the outside ambient air temperature.*

5. *Record the approximate saturation temperature range from Step 4. Saturation temperature range* _____ *degrees to* _____ *degrees.* ❑

DETERMINING CORRECT OPERATING HIGH-SIDE PRESSURE RANGE

6. *Use the pressure–temperature chart and convert the outside dry-bulb air temperature (plus) 30 degrees to the corresponding refrigerant pressure.* _____ *psig* ❑
7. *Use the pressure–temperature chart and convert the outside dry-bulb air temperature (plus) 35 degrees to the corresponding refrigerant pressure.* _____ *psig* ❑
8. *Record the approximate operating head pressure ranges from Steps 6 and 7. Operating high-side pressure range* _____ *psig to* _____ *psig* ❑

> **NOTE:** *You have just determined what the approximate operating high-side pressure should be for the system you are checking.*

The following procedures will be used to see if you were correct and if the system's operating high-side pressure is correct.

EVALUATING ACTUAL SYSTEMS CONDITIONS

9. Make sure the manifold gage valves are closed. ❑
10. Attach the high-side manifold gage hose to the system's liquid line service valve port. ❑
11. *Record the system's actual operating high-side pressure.* _____ ❑
12. *Is the system's operating high-side pressure within operating range as that determined in Step 8?* _____ ❑
13. *Record actual condensing temperature by converting actual high-side pressure to temperature based on type of system refrigerant.* _____ ❑
14. *Was refrigerant condensing temperature in the correct range as determined in Step 5?* _____ ❑

DETERMINING AIR TEMPERATURE TO CONDENSER SATURATION TEMPERATURE

15. Record the actual condensing temperature as recorded in Step 14. _____ ❑
16. Record the air temperature entering the condenser as recorded in Step 3. _____ ❑
17. Determine the temperature difference by using the following formula:

    Outdoor coil condensing temperature _____ ❑

    MINUS

    Outdoor coil entering air dry-bulb temperature _____ ❑
    EQUALS temperature difference _____ degrees (F) ❑

NOTE: The difference from the saturation temperature and the entering air temperature should be in the range of 30 to 35 degrees. If the temperature difference is in the correct range, the operating head pressure is correct for the system under these conditions.

  If the temperature difference is higher than 35 degrees, this is an indication that the system's high-side pressure and saturation temperature are out of range for the conditions listed. The outdoor coil is working harder than it should be and technicians should look for conditions that could be creating a higher high-side pressure and saturation temperature—such things as dirty outdoor coils, higher than normal outside air temperature, restricted airflow, broken belt, burnout fan motor, etc.

  This situation should be corrected; if corrections are not made to bring high-side pressures into proper range, the systems efficiency will suffer, let alone cause possible damage to the systems compressor.

---

18. Were you accurate in determining the correct operating high-side pressure for the system being checked? _____ ❑
19. Have the instructor check your work. ❑
20. Disconnect equipment and return all tools and test equipment to their proper location. ❑

PERFORMANCE EVALUATION

1. Was the system's operating high-side pressure correct for the conditions? _____ (If not, explain what you think could be the cause(s) as to the high-side pressure being incorrect.)

   _____

   _____

2. Was the condenser saturation temperature correct for the conditions? _____ (If not, explain what you think could be the cause(s) as to the condensing temperature being incorrect.)

   _____

   _____

3. Was the temperature difference between the condensing temperature and entering air temperature correct? _____ (If not, explain what you think could be the cause(s) as to the temperature difference being incorrect.)

   _____

   _____

4. Why does the low-side of the system affect the high side of the system?

   _____

   _____

5. Besides what was stated above, what else could cause the head pressure to be high on the system?

_____

_____

Practical Competency 130

Determining the Correct High-Side Pressure of Heat Pump Equipment with a High-Efficiency Air-Cooled Condenser (Figure 4–71)

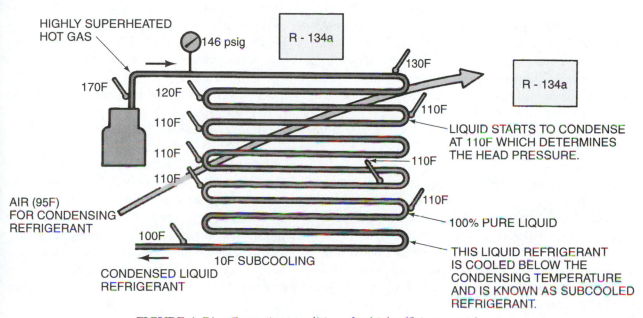

FIGURE 4–71   Operating conditions for high-efficiency condenser.

SUGGESTED MATERIALS

Textbook
Refrigeration & Air Conditioning Technology, 5th Edition, Thomson Delmar Learning
Unit 22—*Condensers*
Unit 36—*Refrigeration Applied to Air Conditioning*
Unit 40—*Typical Operating Conditions*
Unit 43—*Air Source Heat Pumps*

Review Topics
High-Efficiency Air-Cooled Condensers; Typical Operating Conditions; Troubleshooting; Refrigeration Applied to Air Conditioning

COMPETENCY OBJECTIVE

The student will be able to determine the correct operating high-side air conditioning or heat pump equipment with a high-efficiency air-cooled condenser.

OVERVIEW

The condenser (*outdoor coil*) of refrigeration sealed systems performs three functions in the refrigeration process. It rejects heat form the refrigerant vapor, condenses vapor to liquid, and sub-cools liquid refrigerant (Chapter 3, *Theory Lesson: Condensers, Figure 3–263*).

How fast a condenser (*outdoor coil*) can perform these functions can be used to evaluate the condenser's (*outdoor coil*) performance and efficiency. During rejection of **latent heat** from the refrigerant vapor, the vapor reaches a point of saturation and starts to condense back to the liquid state. The pressure being exerted on the vapor determines the temperature at which this process happens.

The **saturation temperature** of a condenser (*outdoor coil*) can be determined by using the manufacturer's design condensing temperature range for the different type of condensers (*outdoor coil*). In residential and light commercial air conditioning and heat pump applications, two types of condensers are used, air-cooled and water-cooled condensers. There are two types of air-cooled condensers (*outdoor coils*) used, standard efficiency and high efficiency. Manufacturers have designed all **standard air-cooled condensers** (*outdoor coil*) to condense the refrigerant vapor to a liquid at a temperature of about 30 to 35 degrees above the outside ambient air temperature (*Chapter 3, Theory Lesson: Condensers, Figure 3–264*).

New **high-efficiency air-cooled condensers** (*outdoor coil*) condense the refrigerant vapor back to a liquid at 15 to 25 degrees above the outside ambient air temperature based on the SEER rating of the equipment. The higher the SEER rating of the equipment, the lower the saturation temperature of the refrigerant (*Chapter 3, Theory Lesson: Condensers, Figure 3–265*).

Water-cooled condensers are designed to condense refrigerant vapor to liquid at 10 degrees above the leaving water temperature in summer applications (*Chapter 3, Theory Lesson: Condensers, Figure 3–266*), and approximately 20 degrees above the leaving water temperature in winter applications (Chapter 3, *Theory Lesson: Condensers, Figure 3–267*).

Knowing this information can assist technicians in evaluating a system's **high-side operating pressure** and **saturation temperature**. A system's **correct saturation temperature** can be determined by measuring the outside ambient air temperature and adding the manufacturer's designed condensing temperature to it, based on the type of condenser (*outdoor coil*) being used in the system. This tells the technician what the saturation temperature of the refrigerant should be for the condenser (*outdoor coil*) under current conditions. Using the saturation temperature can assist the technician in determining what the system's **high-side operating pressure** should be under current conditions. The technician would convert the saturation temperature to pressure for the type of system refrigerant being used. This would be the system's operating head pressure should be under current conditions.

This information can then be used to evaluate actual system operating high-side pressure and saturation temperature. An actual system pressure and saturation temperature of **plus or minus 10%** would be acceptable. Any pressure and saturation temperature higher or lower than the allowable 10% range is an indication to the technician that the condenser (*outdoor coil*) is working too hard or below an acceptable range. Technicians should look for conditions that could cause the condenser (*outdoor coil*) to perform outside the determined acceptable range.

EQUIPMENT REQUIRED

Manifold gage set
Temperature meter
Pressure–temperature chart
Safety goggles
Safety gloves
Refrigeration wrench *(if applicable)*

SAFETY PRACTICES

The student should be knowledgeable in the use of tools and testing equipment. Follow all EPA rules and regulations when working with refrigerants.

COMPETENCY PROCEDURES

> **NOTE:** *This competency is to be performed on an air conditioning or heat pump equipment with a high-efficiency air-cooled condenser.*

1. Turn system ON to operate in the **Cooling** mode and allow the system to operate for 10 minutes. ❏

DETERMINING CORRECT CONDENSER SATURATION TEMPERATURE

2. *Record the type of refrigerant used in system.* _____ ❏
3. *Measure and record the dry-bulb air temperature entering the outdoor coil.* _____ ❏
4. Determine what you think the maximum outdoor coil saturation temperature be by using the following formula:
   Outdoor coil entering dry-bulb air temperature _____ *+ 15 degrees = saturation temperature* _____ ❏
   Outdoor coil entering dry-bulb air temperature _____ *+ 25 degrees = saturation temperature* _____ ❏

> **NOTE:** *Remember that high-efficiency air-cooled condensers (outdoor coil) are designed to condense the refrigerant vapor to a liquid at about 15 to 25 degrees above the outside ambient air temperature.*

5. *Record the approximate saturation temperature range from Step 4.*
   Saturation temperature range _____ *degrees to* _____ *degrees.* ❏

DETERMINING CORRECT OPERATING HIGH-SIDE PRESSURE RANGE

6. *Use the pressure–temperature chart and convert the outside air temperature (plus) 15 degrees to the corresponding refrigerant pressure.* _____ *psig* ❏
7. *Use the P/T chart and convert the outside air temperature (plus) 25 degrees to the corresponding refrigerant pressure.* _____ *psig* ❏
8. Record the approximate operating head pressure ranges from Steps 6 and 7.
   Operating high-side pressure range _____ *psig to* _____ *psig* ❏

> **NOTE:** *You have just determined what the approximate operating high-side pressure should be for the system you are checking.*

The following procedures will be used to see if you were correct and if the systems operating high-side pressure is correct.

EVALUATING ACTUAL SYSTEM CONDITIONS

9. Make sure the manifold gage valves are closed. ❏
10. Attach the high-side manifold gage hose to the system's liquid line service valve port. ❏
11. *Record the system's actual operating high-side pressure.* _____ ❏
12. *Is the system's operating high-side pressure within operating range as that determined in Step 8?* _____ ❏
13. *Record actual condensing temperature by converting actual high-side pressure to temperature based on type of system refrigerant.* _____ ❏
14. *Was refrigerant condensing temperature in the correct range as determined in Step 5?* _____ ❏

DETERMINING AIR TEMPERATURE TO CONDENSER SATURATION TEMPERATURE

15. *Record the actual condensing temperature as recorded in Step 14.* _____ ❑

16. *Record the air temperature entering the condenser as recorded in Step 3.* _____ ❑

17. Determine the temperature difference by using the following formula:

Outdoor coil condensing temperature _____ ❑

MINUS

Outdoor coil entering air temperature _____ ❑

EQUALS temperature difference _____ *degrees (F)* ❑

> **NOTE:** *The difference from the saturation temperature and the entering air temperature should be in the range of 15 to 25 degrees. If the temperature difference is in the correct range, the operating head pressure is correct for the system under these conditions.*
>
> *If the temperature difference is higher than 25 degrees, this is an indication that the system's high-side pressure and saturation temperature are out of range for the conditions listed. The condenser is working harder than it should be and technicians should look for conditions that could be creating a higher high-side pressure and saturation temperature—such things as dirty coils, higher than normal outside air temperature, restricted airflow, broken belt, burnout fan motor, etc.*
>
> *This situation should be corrected; if corrections are not made to bring high-side pressures into the proper range, the systems efficiency will suffer, let alone cause possible damage to systems compressor.*

---

18. Were you accurate in determining the correct operating high-side pressure for the system being checked? _____ ❑

19. Have the instructor check your work. ❑

20. Disconnect the equipment and return all tools and test equipment to their proper location. ❑

PERFORMANCE EVALUATION

1. Was the system's operating high-side pressure correct for the conditions? _____ *(If not, explain what you think could be the cause(s) as to the high-side pressure being incorrect.)*

   _____

   _____

2. Was the outdoor coil saturation temperature correct for the conditions? _____ *(If not, explain what you think could be the cause(s) as to the condensing temperature being incorrect.)*

   _____

   _____

3. Was the temperature difference between the condensing temperature and entering air temperature correct? _____ *(If not, explain what you think could be the cause(s) as to the temperature difference being incorrect.)*

   _____

   _____

4. What makes a high-efficiency air-cooled condenser more efficient than a standard air-cooled condenser?

   _____

   _____

5. What does the condenser SEER rating mean?

   _____

   _____

Passed Competency _____   **Failed Competency** _____

Instructor Signature _____   **Grade** _____

OIL HEAT

Theory Lesson: Oil Heat (Figure 5–1)

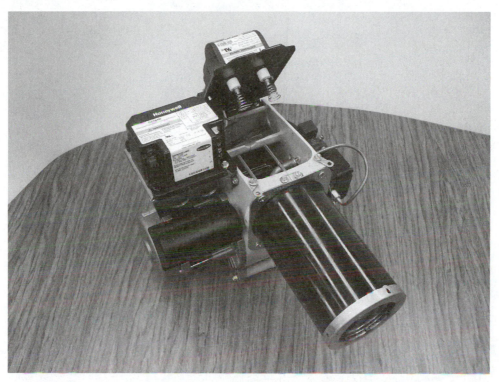

FIGURE 5–1   High-pressure oil burner. (*Photo by Cecil Johnson*)

SUGGESTED MATERIALS

Textbook

Refrigeration & Air Conditioning Technology, 5th Edition, Thomson Delmar Learning
Unit 32—Oil Heat

Review Topics

Oil Heat; Fuel Oil; Combustion; Oil Burner

Key Terms

arc • atomization • atomizing • blast tube • cad cell • choke • electrodes • fire point • flash point • fossil fuel • head • high-pressure burner • hot forced air system • hydrocarbon • hydronic heating • ignition point • nozzle • nozzle orifice • oil pump • primary control • stack relay retention static pressure disc • step-up transformer tangential slots

OVERVIEW

Oil as a fuel source can be used to provide heat to an occupied space either by heating the structure air and using a blower to circulate the heated air through designed ductwork, known as a *hot forced air system*

(**Figure 5–2**), or by heating water or making steam in a boiler (**Figure 5–3**) and circulating the hot water or steam (**Figure 5–4**) through terminal heating units (**Figure 5–5**).

Using hot water or steam to carry heat through pipes in the areas to be heated is referred to as **hydronic heating.**

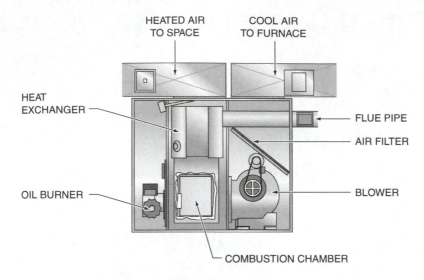

FIGURE 5–2   Oil forced hot air heating system.

FIGURE 5–3   Oil fired boiler. (*Courtesy of Well-Mclain*)

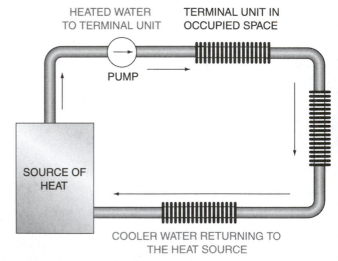

FIGURE 5–4   Series loop piping circuit.

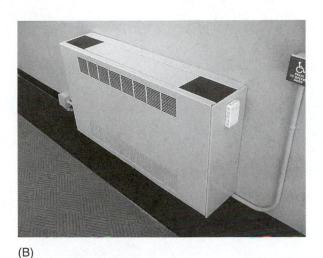

(A)   (B)

FIGURE 5–5 Terminal hydronic heating units. (A) Water in the radiator radiates heat out into the room. (B) The coil units are similar to the baseboard units but have a fan that blows the hot air into the room. (*A, Photo by Bill Johnson, B, Courtesy of Ferris State University. Photo by John Tomczyk*)

Fuel oil is a **fossil fuel**, which is a product of decayed prehistoric plants and animals. There are six grades of fuel oil that are numbered to identify the grade, with No. 2 being the most common used in residential and commercial heating systems. No. 2 fuel oil has an average heat content of 140,000 British thermal units (Btu's) per gallon. Its normal color resembles that of champagne, with a red dye added to distinguish it from diesel fuel. No. 2 fuel oil is made from hydrogen (15%) and carbon (85%), which classifies it as a **hydrocarbon**. The **flash point**, or temperature at which fuel oil will momentarily flash and immediately extinguish, is a minimum of 100 degrees (F), with a typical flash point range from 130 to 214 degrees (F). The **ignition point**, or **fire point**, is a minimum of 637 degrees (F). This is the lowest temperature at which the fuel will rapidly burn in the presence of air. The combustion process for any fuel to burn requires a supply of oxygen, heat, and fuel. If any of the three components are missing, combustion cannot take place. During combustion, oxygen reacts with the fuel oil to produce heat, carbon dioxide, and water vapor (**Figure 5–6**).

Incomplete combustion results in carbon monoxide, soot, and other undesirable substances, along with heat, carbon dioxide, and water vapor (**Figure 5–7**).

The better the fuel and air are mixed together, the better the combustion process. Air is the biggest factor to consider when there is incomplete combustion. Too much air can cause lower system efficiency, but insufficient air results in the formation of carbon monoxide.

Oil in the liquid state is hard to ignite, but can be ignited easily in the vapor state. The most common method used to convert oil to a vapor is called atomizing. **Atomization** is the process of breaking oil into small particles and then mixing these particles with the air required for combustion (**Figure 5–8**). This is the purpose of the oil burner.

Two types of oil burners are used in fuel oil systems: low-pressure and high-pressure burners. Low-pressure burners mix some air with low-pressure oil as it is blown into the combustion chamber (**Figure 5–9**).

Some low-pressure burners are still in use today, although the most common type of burner in use is the high-pressure burner (**Figure 5–10**).

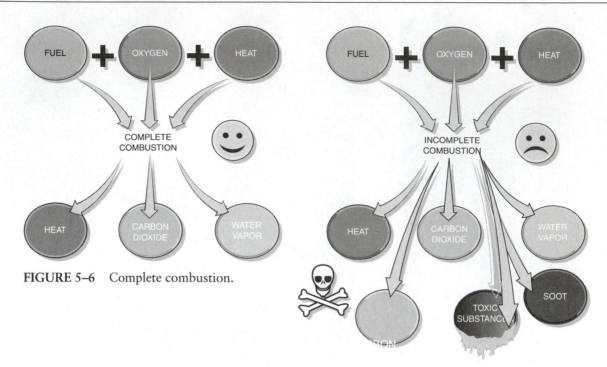

FIGURE 5–6   Complete combustion.

FIGURE 5–7   Incomplete combustion.

FIGURE 5–8   Atomized fuel oil droplets and air leaving the burner nozzle. (*Photo by Cecil Johnson*)

High-pressure burners create atomization of the fuel oil under a pressure of 100 psig; newer high-pressure burners atomize the oil at around 140 psig. The common components of an oil burner are (**Figure 5–11**):

- Oil burner motor
- Blower wheel
- Air tube

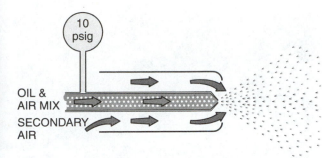

FIGURE 5–9   A low-pressure burner.

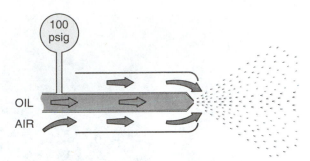

FIGURE 5–10   A high-pressure burner.

FIGURE 5–11   Exploded view of an oil burner assembly. (*Photo by Cecil Johnson*)

- Static pressure disc
- Flame retention head
- Oil pump
- Oil nozzle
- Ignition transformer
- Electrodes
- Primary control
- Cad cell

The power for the oil burner assembly is an electric motor (**Figure 5–12**) that drives both a squirrel cage fan and an oil pump (**Figure 5–13**). A blower in the burner assembly pushes the combustion air through the burner assembly (**Figure 5–14**).

The fan forces combustion air through the air tube (**Figure 5–15**), static pressure disc (**Figure 5–16**), and choke or flame retention head (**Figure 5–17**), where it is mixed with the fuel oil vapor (**Figure 5–18**).

The static pressure disc is located within the burner air tube and is designed to increase both the velocity and turbulence of the combustion air within the tube. The flame retention head is inserted at the end of the air tube and is designed to take the already turbulent air and further increase the air velocity (**Figure 5–19**).

FIGURE 5–12   Oil burner motors. (*Photo by Cecil Johnson*)

FIGURE 5–13   Burner motor drives the blower and oil pump. (*Photo by Cecil Johnson*)

FIGURE 5–14   The blower wheel. (*Photo by Cecil Johnson*)

FIGURE 5–15   Air tube. (*Photo by Cecil Johnson*)

FIGURE 5–16   Static pressure disc. (*Photo by Cecil Johnson*)

FIGURE 5–17   Increased velocity of oil–air mixture through flame retention head. (*Courtesy of Ducane Corporation*)

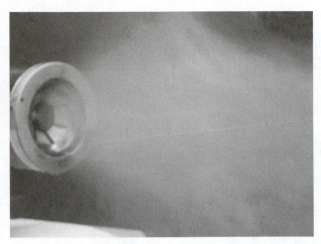

FIGURE 5–18   Mixing of atomized oil and air. (*Photo by Cecil Johnson*)

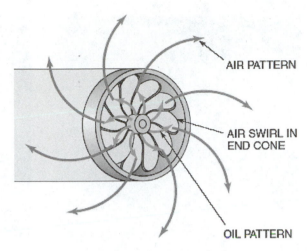

FIGURE 5–19   The end cone swirls the air into the swirling oil for the best oil–air mixture.

The retention head also creates a swirling of the high-velocity atomized oil and air mixture, making it ready for ignition (**Figure 5–20**).

An adjustable air inlet collar is provided on the burner assembly to allow for the proper amount of air mixing with the atomized fuel (**Figure 5–21**).

FIGURE 5–20   Increased velocity of oil–air mixture creates swirling of oil–air mixture through retention head. (*Photo by Cecil Johnson*)

FIGURE 5–21   Adjustable air inlet collar. (*Photo by Cecil Johnson*)

The motor drives the oil pump, which draws the oil from the oil tank (**Figure 5–22**), where it is pressurized and forced through the burner nozzle (**Figure 5–23**), at a velocity that ensures complete atomizing of the fuel oil (**Figure 5–24**).

The nozzle of the fuel oil burner prepares the oil for mixing with the combustion air. It does this by breaking the oil up into small droplets that can be easily mixed with the combustion air. When the oil enters the nozzle, it first passes through the nozzle strainer, which filters the oil. Then it goes through tangential slots, which creates a swirling and whirling action of the oil. This oil is then forced through the nozzle orifice, which increases the oil velocity and determines the spray pattern of the oil vapor. This causes the oil to break up into a fine spray or mist. This oil mist is mixed with the high velocity air, leaving the air tube and flame retention head where an arc across the electrodes ignites the fuel.

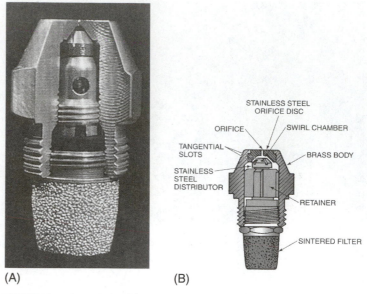

(A)   (B)

FIGURE 5–23   An oil burner nozzle. (*Courtesy of Delavan Corporation*)

FIGURE 5–22   Oil pump. (*Courtesy of R.W. Beckett Corporation*)

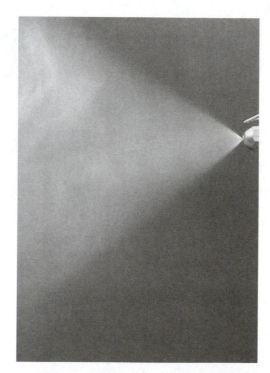

FIGURE 5–24   Atomized oil from nozzle. (*Photo by Cecil Johnson*)

There are three basic spray patterns: hollow, semi-hollow, and solid—with angles from 30 to 90 degrees spray pattern (**Figure 5–25**). The fuel air mixture created by the fan, and fuel nozzle is ignited by an electrical spark arcing between the two electrodes of the burner assembly (**Figure 5–26**) ignites the atomized oil (**Figure 5–27**). A 120-volt step-up transformer, which steps the transformers secondary voltage up to 10,000 volts, creates the spark across the electrodes (**Figure 5–28**).

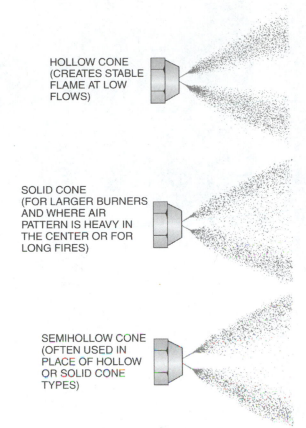

HOLLOW CONE
(CREATES STABLE
FLAME AT LOW
FLOWS)

SOLID CONE
(FOR LARGER BURNERS
AND WHERE AIR
PATTERN IS HEAVY IN
THE CENTER OR FOR
LONG FIRES)

SEMIHOLLOW CONE
(OFTEN USED IN
PLACE OF HOLLOW
OR SOLID CONE
TYPES)

FIGURE 5–25   Nozzle spray patterns.
(*Courtesy of Delavan Corporation*)

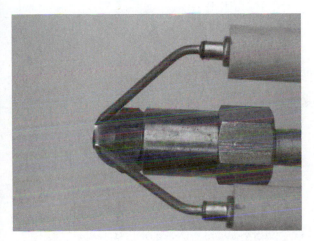

FIGURE 5–26   Spark across electrodes. (*Photo by Cecil Johnson*)

FIGURE 5–27   Ignition of atomized oil. (*Photo by Cecil Johnson*)

The secondary terminals of the transformer make contact with the electrode elements. When the transformer is energized an arc is generated across the tips of the electrodes (**Figure 5–29**). The arc created across the electrode tips is about 20 milliamps and creates enough heat to cause the atomized fuel oil to ignite (**Figure 5–30**).

The electrodes are not inserted in the oil–air spray. The arc across the electrode tips is blown into the oil vapor and air mixture. The ceramic insulators of the electrodes are designed to insulate the electrodes and serve the purpose of positioning the electrodes in the static pressure disc assembly (**Figure 5–31**).

FIGURE 5–28   Ignition transformer's step-up 120 V/10,000 volts. (*Photo by Cecil Johnson*)

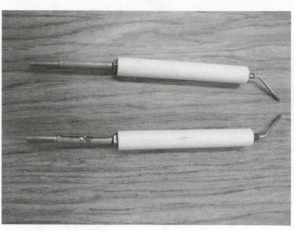

FIGURE 5–29   Electrodes. (*Photo by Cecil Johnson*)

FIGURE 5–30   Electrode arc about 20 milliamps. (*Photo by Cecil Johnson*)

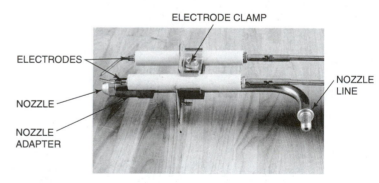

FIGURE 5–31   Electrode and nozzle assembly.

To ensure that the correct arc is generated, the electrodes should be positioned in the static disc correctly. The gap between the electrodes should be approximately 1/8" to 3/16". The electrodes should extend out about 1/8" to 3/8" past the end of the burner nozzle and should be 3/8" to 1/2" high above the center of the nozzle orifice (**Figure 5–32**).

The oil burner **primary control** provides a means for operating the burner and safety functions used to turn the burner ON and OFF in response to the room temperature demand through the system thermostat, ignition failure, or high temperature conditions. In cases of any safety failure, the primary control safety switch will trip, preventing ignition of the oil burner until the safety switch is reset (**Figure 5–33**).

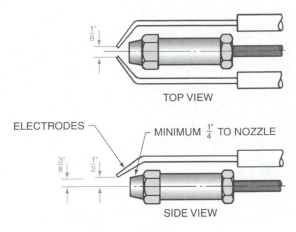

TOP VIEW

ELECTRODES

MINIMUM $\frac{1}{4}$" TO NOZZLE

SIDE VIEW

HEIGHT ADJUSTMENT

$\frac{1}{2}$" RESIDENTIAL INSTALLATION

$\frac{3}{8}$" COMMERCIAL INSTALLATION

POSITION OF ELECTRODES IN FRONT OF NOZZLE
IS DETERMINED BY SPRAY ANGLE OF NOZZLE.

FIGURE 5–32   Electrode adjustments. Electrodes cannot be closer than 1/4" to any metal part.

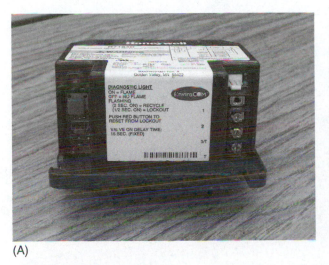

(A)

(B)

FIGURE 5–33   Cad cell oil primary controls. (*A, Photo by Cecil Johnson. B, Photo by Bill Johnson*)

NOTE: *Any time the primary control safety switch trips and shuts the oil burner down, efforts should be made to determine what caused the primary control safety switch to trip before resetting the oil burner primary control.*

Modern furnaces incorporate a Cad Cell Oil Primary Control. The cad cell or electric eye is designed to prove to the primary control that ignition of the atomized oil has taken place in the furnace combustion chamber (**Figure 5–34**).

The **cad cell** is located beneath the ignition system and positioned in a manner that it can sight flame through the oil burner air tube. The cad cells is a photocell that has a high electrical resistance when it is in the dark and a low resistance in visible light. The resistance of the cad cell is about 100,000 ohms when it is in the dark. When light is reflected on the cad cell its resistance drops to about 1,600 ohms.

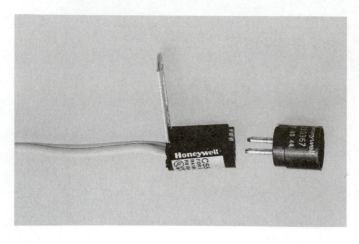

FIGURE 5–34   Cad cell. (*Photo by Bill Johnson*)

Older oil furnaces used a **stack switch** or **stack relay** to turn ON and OFF the oil burner on a call for heat or safety failure (**Figure 5–35**).

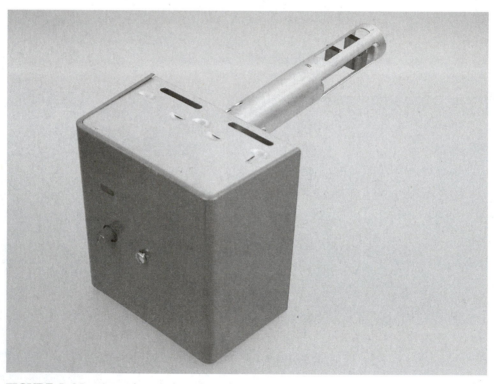

FIGURE 5–35   A stack switch. (*Photo by Bill Johnson*)

The stack relay is mounted in the furnace stack pipe. It is designed to sense the furnace flue gas temperature to determine whether ignition of the atomized oil had taken place in the combustion chamber.

Theory Lesson: Oil Furnace Sequence of Operation

SUGGESTED MATERIALS

Textbook
Refrigeration & Air Conditioning Technology, 5th Edition, Thomson Delmar Learning
Unit 32—Oil Heat

Review Topics
Oil Furnace Wiring Diagram

Key Terms
blower • burner motor • fan–limit switch • fuel oil pump • ignition transformer • primary control • safety switch • thermostat

OVERVIEW

On a call for heat, the thermostat closes a circuit between the R and W terminals on the thermostat (**Figure 5–36**).

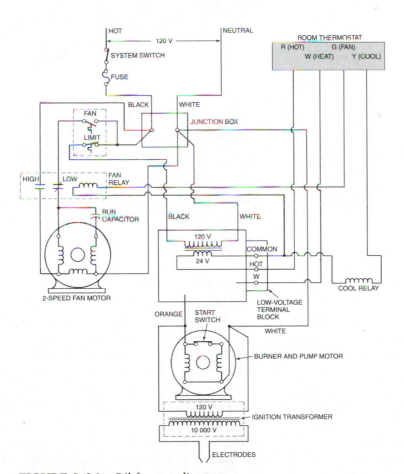

FIGURE 5–36   Oil furnace diagram.

This circuit sends a 24-volt power source to the oil burner primary control (**Figure 5–37**).

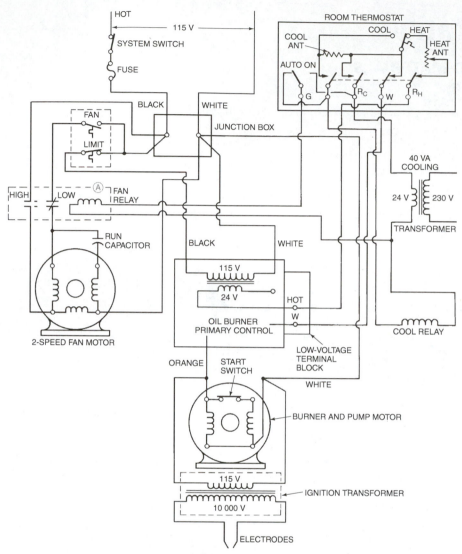

FIGURE 5–37   Low-voltage circuit energized from thermostat contacts R and W.

The primary control checks to make sure that all systems on the furnace and burner are ready to start. The primary control then turns on the burner motor, which drives the blower, which blows excess air into the combustion chamber. The burner motor also drives the fuel oil pump, which delivers oil under pressure to the fuel nozzle in the burner's blast tube (**Figure 5–38**).

A spark jumping across the ignition electrodes ignites the mixtures of oil and air. A step-up transformer known as the ignition transformer produces the high voltage necessary to create the spark by stepping up the supplied voltage of 120 volts to 10,000 volts or higher. A flame detector known as the cad cell eye sees the flame in the combustion chamber and sends a signal back to the primary control that ignition has taken place and the primary control can continue to allow the burner to operate until the need for heat has been satisfied and the thermostat shuts the burner down. The cad cell is a photocell that has a high electrical resistance when it is in the dark and a low resistance in visible light. The resistance of the cad cell is about 100,000 ohms when it is in the dark. When light is reflected on the cad cell its resistance drops to about 1,600 ohms.

If for some reason the cad cell eye does not see flame in a certain amount of time, a safety switch in the primary control opens automatically, breaks the ignition circuit, and shuts down the burner (**Figure 5–39**).

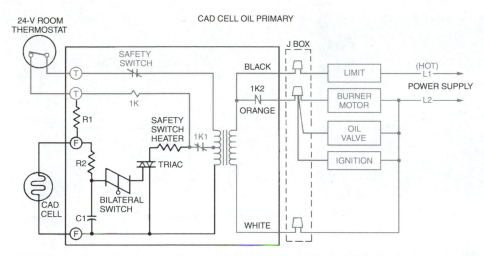

FIGURE 5–38   Primary control circuit energized through primary control T–T terminals.

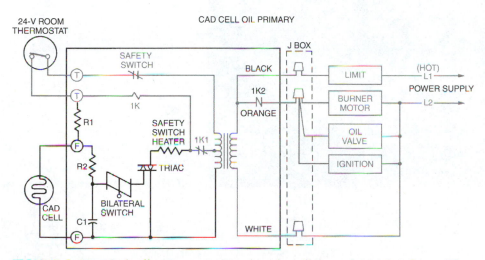

FIGURE 5–39   Cad cell primary circuit when the cell does not "sight" flame. The safety switch heater will heat and open the safety switch.

The blower of the furnace does not come on right away when the thermostat calls for heat. The burner will operate for a short time without the blower on to allow the furnace heat exchanger to heat up. This prevents cold air from being circulated into the conditioned space. A fan–limit switch is placed above the heat exchanger and senses the heat coming off of the heat exchanger (**Figure 5–40**).

Once the fan–limit switch senses a predetermined temperature it will close the electrical circuit to the furnace blower. The blower draws the cold air from the conditioned space, moves it across the furnace heat exchanger, heating the air, and then returns it to the conditioned space. The fan–limit switch also services as a safety switch in the event that the furnace overheats for any reason (**Figure 5–41**).

The safety switch of the high-limit switch would break the power supply to the oil burner primary control, which will shut down the furnace operation.

Some newer oil furnaces do not use a fan-limit switch as shown in Figure 5–39, but use a bimetal device (**Figure 5–42**).

These bimetal switches can be used to activate the furnace blower (**Figure 5–43**).

On the other hand, they can be used as a fan-limit switch; this controls the furnace blower operation and acts as a high-limit switch, shutting the burner down due to high-temperature conditions (**Figure 5–44**).

FIGURE 5–40   Fan–limit switch.

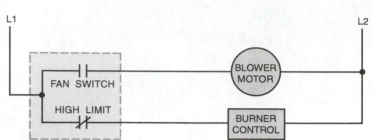

FIGURE 5–41   Schematic for fan–limit switch.

FIGURE 5–42   Bimetal limit switches.

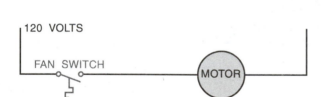

FIGURE 5–43   Bimetal fan switch controls the operation of the fan motor.

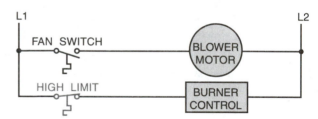

FIGURE 5–44   Fan switch and limit switch connected in series with the heating source.

When the conditioned space temperature is satisfied, the thermostat will open the electrical circuit between the R and W terminals of the thermostat. This de-energizes the oil burner primary control, shutting OFF burner operation. The furnace blower will continue to operate until the heat exchanger has been cooled to a preset temperature. Once the heat exchanger is cooled enough, the fan–limit switch will open the circuit to the furnace blower, shutting the blower OFF.

Practical Competency 131

Oil Burner Motor Evaluation

SUGGESTED MATERIALS

Textbook
Refrigeration & Air Conditioning Technology, 5th Edition, Thomson Delmar Learning
Unit 32—Oil Heat

Review Topics
Gun-Type Oil Burners; Burner Motor; Burner Fan or Blower

COMPETENCY OBJECTIVE

The student will be able to check the motor performance of an oil burner.

OVERVIEW

The power for the oil burner assembly is an electric motor that drives both a squirrel cage fan and an oil pump.

The burner motor drives the blower, which forces combustion air through the burner air tube. A flexible coupler is used to connect the burner motor shaft and oil pump shaft (**Figure 5–45**).

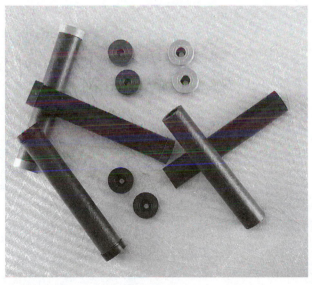

FIGURE 5–45   Flexible coupler.

The shaft connection allows the burner motor to drive the oil pump which draws oil from the oil tank, pressurizes the oil, and forces it through the burner nozzle at a velocity that ensures complete atomizing of the fuel oil.

Burner motors are split-phase motors that incorporate a RUN and START winding.

These motors have a high starting torque and when used as split-phase or capacitor start motors, require that the motor starting winding be removed from the motor circuit once the motor reaches three quarters of its rated speed. In most oil burner split-phase motors, a centrifugal switch is used to accomplish this procedure.

Burner motors also incorporate a motor overload reset button, which will trip and open the motor circuit due to high motor current draw or motor over heating (**Figure 5–46**).

FIGURE 5–46 Burner motor overload reset button. (*Photo by Cecil Johnson*)

A burner motor that will not start may have a "tripped overload" and require resetting. If this is the case and the motor runs after the reset switch has been reset, technicians should evaluate the condition of the motor and make a determination as to what caused the motor overload to trip in the first place. Avoiding this evaluation may result in an additional service call or permanent damage to the motor.

The motor windings, centrifugal switch, and overload protector can be evaluated by performing an ohms check on individual components.

EQUIPMENT REQUIRED

Oil burner motor
Standard screwdriver
VOM or ohmmeter
Clamp-on ammeter
Electrical cord jumper
Two wire nuts
Motor oil lubricant

SAFETY PRACTICES

Be sure to follow all electrical safety rules when working with live circuits. Make sure all electrical test equipment is set to proper function and ranges for testing.

COMPETENCY PROCEDURES

RESISTANCE PERFORMANCE CHECK

1. Turn OFF the power supply to furnace (*if applicable*). ❏
2. Loosen primary control holding screws. ❏
3. Locate primary control so burner motor electrical leads are accessible. ❏
4. Remove wire nuts holding burner motor electrical leads. ❏

> NOTE: *Make a note of where power leads are reconnected.*

5. Remove burner motor mounting screws. ❏
6. Remove motor from oil pump coupler and burner housing. ❏
7. Check to see the motor shaft turns freely by rotating the motor shaft with your hand. ❏
8. Add droplets of motor oil lubricant to the motor bearing ports (*if applicable*). ❏
9. Set VOM or ohmmeter to measure motor winding resistance. ❏
10. Place the ohmmeter or VOM meter leads across the burner motor electrical wires. ❏
11. *Record the measured resistance.* _____ ❏

> NOTE: *A measured resistance value across the burner motor leads is an indication that the motor windings are good. This also indicates that the motor overload circuit and centrifugal switch are closed, allowing the motor to operate if proper power is applied.*

> NOTE: *An infinite resistance reading is an indication that the motor windings, motor overload, or centrifugal switch is open. Make sure the motor overload reset button is "reset," by pushing it in if need be. If the overload is not tripped, the problem is probably internal to the motor. This would require that the motor be taken apart so that evaluation of the centrifugal switch, motor overload switch, and motor windings can take place.*

CLAMP-ON AMMETER MOTOR CHECK

12. If the motor windings check OK, use the wire nuts and connect the jumper cord leads to the burner motor power leads. ❏
13. Set the ammeter to measure at least 30 amps or higher. ❏
14. Clamp the ammeter around one of the power legs of the jumper cord. ❏
15. Hold the motor housing and make sure that the blower wheel is clear of hitting anything. ❏
16. Observe the clamp-on ammeter and plug the jumper cord into the proper power supply. ❏
17. *Record the motor's start-up amperage.* _____ ❏
18. *While the motor is operating, observe and record the motor's running amperage.* _____ ❏
19. Unplug the jumper cord and allow the motor to completely stop operating. ❏
20. Was the amperage recorded in Step 17 higher than the amperage recorded in Step 18? _____ ❏
21. *If the recorded amperage in Step 17 was higher than the amperage recorded in Step 18, how many times higher was the amperage in Step 17 compared to the recorded amperage Step 18?* _____ ❏
22. *Was the recorded amperage in Step 18 lower than the amperage recorded in Step 17?* _____ ❏

NOTE: *In the amperage performances check you should have noticed the following if the motor windings and centrifugal switch are working OK. On start-up you should have observed a high amperage reading for a split second and noticed that the amperage dropped lower amperage readings. This is an indication that the start and run windings are being used for "start-up" and the centrifugal switch removed the start winding from the motor circuit once the motor reached three quarters of its rated speed. The start-up amperage should have been approximately three to five times higher than the motor's operating amperage. If there was not a change in the amperage once the motor started, the centrifugal switch is defective and would require replacement or complete replacement of the motor.*

23. If the motor checks out, replace the motor back into the burner (*if applicable*). ❑

NOTE: *During motor replacement, make sure to align the motor and pump coupler (if applicable).*

24. Have the instructor check your work. ❑
25. Return all tools and test equipment to their proper location. ❑

RESEARCH QUESTIONS

1. What control of the oil burner is responsible for turning on the burner motor?

   _____

   _____

2. What is the function of the burner motor?

   _____

   _____

3. What control of the burner assembly shuts the burner motor off?

   _____

   _____

4. What function of the primary control shuts the burner motor off if a flame is not detected in a certain amount of time?

   _____

   _____

5. List five mechanical problems that could prevent the burner motor from operating.

   _____

   _____

Passed Competency _____   Failed Competency _____

Instructor Signature _____   Grade _____

Theory Lesson: Fuel Oil Pumps

SUGGESTED MATERIALS

Textbook
Refrigeration & Air Conditioning Technology, 5th Edition, Thomson Delmar Learning
Unit 32—Oil Heat

Review Topics
Fuel Oil Pumps

Key Terms
automatic pressure valve • B bypass plug • fuel pump • horizontal run • positive cut-off • rotary pumps • single-pipe system • single-stage pump • standing pressure test • two-stage pump • vertical lift

OVERVIEW

The fuel pump on an oil burner provides three functions:

    A. *Conditions the oil from the tank to the nozzle.*
    B. *Controls the rate at which the oil is fed to the nozzle.*
    C. *Controls the cut-off of the oil to the nozzle when the burner shuts off.*

Most oil pumps used in oil burners are rotary pumps that use gears, cams, or both to create suction and discharge oil pressure. The oil pump's suction pressure is used to lift the oil from the oil supply tank to the pump so that the oil can be pressurized and forced to the burner nozzle for atomization.

Oil pumps that are located above the oil tank will develop a suction pressure that is used to lift the oil from the oil supply tank to the pump so that the oil can be pressurized and forced to the burner nozzle for atomization.

> NOTE: *The rule of thumb is that an oil pump requires 1 inch Hg of vacuum to lift oil 1 foot vertically; 1 inch Hg vacuum to pull oil 10 feet horizontally; and 1 inch Hg for filters, valves, and other accessories.*

EXAMPLE: *A pump that is 6 feet above the oil tank and 30 feet away from the tank with an oil filter installed would create a 10-inch Hg vacuum.*

In cases where the **oil pump is located below the oil tank**, a vacuum test should indicate a gage reading of 0 inch Hg. If the vacuum gage indicates a vacuum, one of the following may be the problem:

    A. *The fuel line may be kinked.*
    B. *The line filter may be clogged or blocked.*
    C. *The tank shutoff valve may be partially closed.*

The pump is also equipped with an internal automatic pressure valve that is used to control the amount of pressure exerted on the fuel oil to the nozzle and also bleed off any excess pressure. This valve also has a positive cut-off so that if the pressure on the pump drops for any reason, it would shut off the oil supply to the burner nozzle. This prevents oil from leaking out of the nozzle and burning with a lack of air.

    There are two types of oil pumps used on oil furnaces: *single-stage* and *two-stage* (**Figure 5–47**).

    The single-stage pump is used when the fuel oil tank is stored above the furnace burner and when the oil lift requirement is no more than 10 inch Hg of vacuum on a two-pipe system and 6 inch Hg of vacuum on a single-pipe system. The single-stage pump is used to increase oil pressure for use in the burner and is normally supplied with a single-pipe system from the oil tank. For the single-pipe installation, the *B bypass*

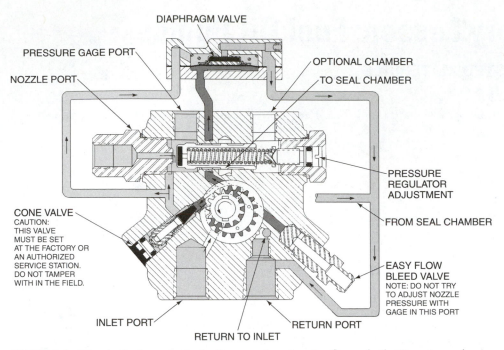

FIGURE 5–47 A single-stage fuel oil pump showing the flow of oil. (*Courtesy of Suntec Industries Inc.*)

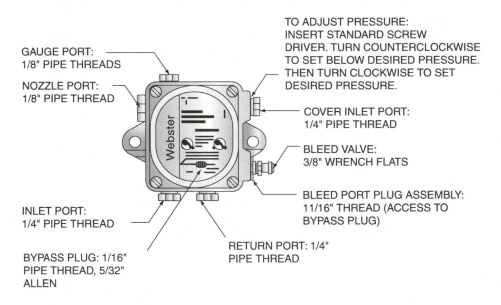

FIGURE 5–48 Bypass plug needs to be removed on a single pipe from oil tank installation.

plug must be removed from the pump so that the surplus oil from the oil pressure valve can be relieved to the low-pressure side of the burner (**Figure 5–48**).

Furnaces equipped with single-stage pumps can be set up with a two-pipe oil supply system. In this case, the B bypass plug would have to be installed so that excess oil not used during atomization can be returned to the oil tank.

The two-stage pump is used where the tank of fuel oil is located below the oil burner or where there is a high lift from the oil tank to the burner with a vacuum requirement of 15 inches Hg (**Figure 5–49**).

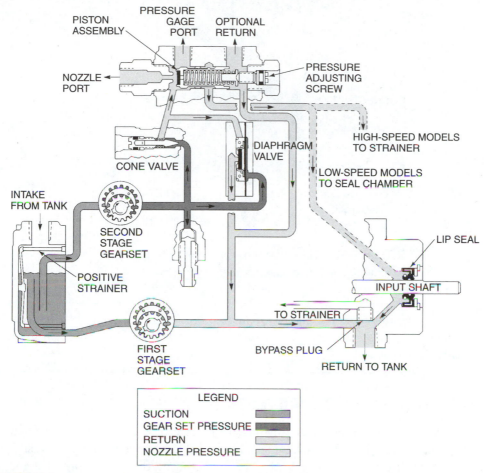

FIGURE 5–49   A two-stage fuel oil pump. (*Courtesy of Suntec Industries Inc.*)

The two-stage pumps have two internal gears. One of the gears is used to draw the fuel oil from the oil tank into the pump. The second gear of the pump is used to supply the nozzle pressure to the burner. Two-stage pumps should be set up with a two-pipe oil supply system from the oil tank. One of the pipes is used to supply from the oil tank to the pump and the other is to return the unused oil back to the oil supply tank. This requires that the *B bypass plug be installed* into the two-stage pump. The two-pipe system has an advantage over a single-pipe supply oil system in that the two-pipe systems are self-venting, self-priming, and self-purging. If air is drawn into a single-pipe system, the air gets trapped in the fuel pump and can cause oil foaming in the pump, causing intermittent burner shut-off.

Installing gages on the pump and checking the vacuum pressure and discharge pressure can check oil pump performance (**Figure 5–50**).

OIL PUMP VACUUM TEST

A vacuum pump check on a furnace where the oil tank is above the burner should indicate a 0-inch Hg vacuum. If the vacuum gage indicates a vacuum below 0 inch Hg, the following could be the cause: The fuel line could be kinked, the oil filter could be clogged, or the oil tank shut-off valve could be closed or partially closed. If the oil tank is located below the burner, a vacuum gage should indicate a vacuum reading. The rule is that there should be a 1-inch Hg vacuum for every foot of vertical lift and a 1-inch Hg vacuum for every 10 feet of horizontal run of the oil lines, but should not exceed a 17-inch Hg vacuum.

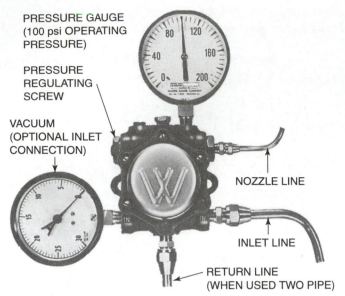

FIGURE 5–50 Gage locations for pump testing and servicing. (*Courtesy of websterfuelpumps.com*)

NOTE: Some new burners and pumps operate with a nozzle pressure of approximately 140 psi. So, check burner installation or service manual for correct operating pressure before making oil pump pressure adjustments.

OIL PUMP DISCHARGE PRESSURE TEST

Oil pump discharge pressure can be checked by installing a pressure gage on the pump and observing the pump's supply pressure. The pressure regulator should be adjusted to supply 100-psi nozzle pressure. During oil pump operation, oil pressure to the nozzle should be the correct pressure and steady. On a pressure test of the pump, a pulsating gage pressure reading may be created by one of the following:

A. *Partially clogged supply filter element*
B. *Partially clogged unit filter or screen*
C. *Air leak in fuel oil supply line*
D. *Air leak in fuel oil pump cover*

STANDING PRESSURE TEST

With the burner in operation, the oil pump should be able to supply a constant operating pressure to the burner nozzle. When the burner is shut off, the oil pump pressure should drop about 15 psi and hold. This is called the **standing pressure.** If the gage pressure drops more than 15 psi below the operating pressure, the following could be the problem:

A. *Possible air leak in fuel unit (pump)*
B. *Possible air leak in fuel oil supply system*
C. *Clogged nozzle strainer*

Repair attempts on defective oil pumps should be avoided; an oil pump that is defective should be replaced.

Practical Competency 132

Checking and Adjust the Oil Pump's Discharge Pressure

SUGGESTED MATERIALS

Textbook
Refrigeration & Air Conditioning Technology, 5th Edition, Thomson Delmar Learning
Unit 32—Oil Heat

Review Topics
Oil Pumps

COMPETENCY OBJECTIVE

The student will be able to check and adjust an oil pump's discharge pressure.

OVERVIEW

The fuel pump on an oil burner provides three functions:

A. *Conditions the oil from the tank to the nozzle.*
B. *Controls the rate at which the oil is fed to the nozzle.*
C. *Controls the cut-off of the oil to the nozzle when the burner shuts OFF.*

Most oil pumps used in oil burners are rotary pumps that use gears, cams, or both to create suction and discharge pressures. Oil pumps are also equipped with an internal automatic pressure valve that is used to control the amount of pressure exerted on the fuel oil to the nozzle and bleed off any excess pressure. This valve also has a positive cut-off so that if the pressure on the pump drops for any reason, it would shut off the oil supply to the burner nozzle. This prevents oil from leaking out of the nozzle and burning with a lack of air.

OIL PUMP DISCHARGE PRESSURE TEST

Oil pump discharge pressure can be checked by installing a pressure gage on the pump and observing the pump's supply pressure. The pressure regulator should be adjusted to supply 100-psi nozzle pressure. Some new burners and pumps operate with a nozzle pressure of approximately 140 psi, so check burner installation or service manual for correct operating pressure before making oil pump pressure adjustments.

During oil pump operation, oil pressure to the nozzle should be the correct pressure and steady. On a pressure test of the pump, a pulsating gage pressure reading may be created by one of the following:

A. *Partially clogged supply filter element*
B. *Partially clogged unit filter or screen*
C. *Air leak in fuel oil supply line*
D. *Air leak in fuel oil pump cover*

STANDING PRESSURE TEST

With the burner in operation, the oil pump should be able to supply a constant operating pressure to the burner nozzle. When the burner is shut off, the oil pump pressure should drop about 15 psi and hold. This is called the **standing pressure test.** If the gage pressure drops more than 15 psi below the operating pressure, the following could be the problem:

A. *Possible air leak in fuel unit (pump)*
B. *Possible air leak in fuel oil supply system*
C. *Clogged nozzle strainer*

Repair attempts on defective oil pumps should be avoided; an oil pump that is defective should be replaced.

EQUIPMENT REQUIRED

Pressure gage (0–300 psig)
Adjustable wrench
Standard screwdriver
Oil drain pan
Clean rag

SAFETY PRACTICES

Care should be taken not to spill fuel oil on hot surfaces. Be aware of working with lines and pumps under high pressure.

COMPETENCY PROCEDURES

Checklist

> NOTE: *Pump suction pressure should be correct to obtain the correct oil discharge pressures for single-stage and two-stage pumps.*

1. Turn power OFF to unit. ❏
2. Place the oil pan underneath the bleeder valve for a couple of seconds and release the pump pressure, then close the bleeder valve (**Figure 5–51**). ❏

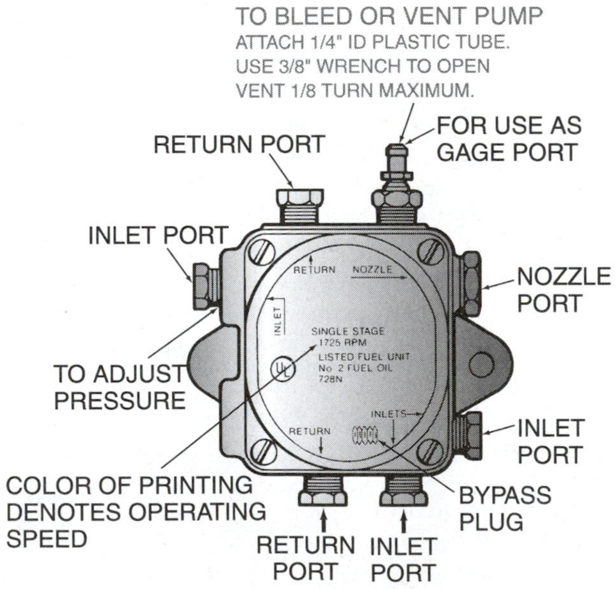

TO BLEED OR VENT PUMP
ATTACH 1/4" ID PLASTIC TUBE.
USE 3/8" WRENCH TO OPEN
VENT 1/8 TURN MAXIMUM.

FOR USE AS GAGE PORT

RETURN PORT

INLET PORT

NOZZLE PORT

RETURN   NOZZLE

INLET

SINGLE STAGE
1725 RPM
LISTED FUEL UNIT
No. 2 FUEL OIL
728N

TO ADJUST PRESSURE

RETURN

INLETS

INLET PORT

COLOR OF PRINTING DENOTES OPERATING SPEED

RETURN PORT   INLET PORT

BYPASS PLUG

FIGURE 5–51 Diagram of a fuel oil pump. Take note of bleeder port. (*Courtesy Webster Electric Company*)

3. Find the gage port plug on the fuel pump and remove it. ❏
4. Install the pressure gage in the gage port on the fuel pump (**Figure 5–52**). ❏

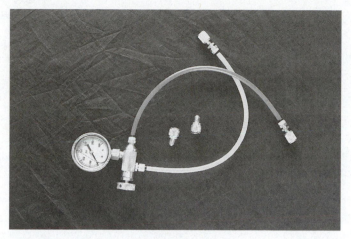

FIGURE 5–52   Fuel pump pressure gage.

5. Make sure the oil filter is cleaned and that the oil line(s) are clear of restrictions.   ❏
6. Turn power ON to furnace.   ❏
7. Set the thermostat to heating mode and temperature selector high enough to energize furnace burner.   ❏
8. *Record the pump's discharge gage reading.*   ❏

> **NOTE:** *The high-pressure pumps are designed to deliver a high pressure of 100 psig or higher for proper nozzle pressure. Some new burners are equipped with oil pumps that require a higher operating pressure for proper nozzle operation, normally 140 psig. Check with the manufacturer's required pump discharge pressure for the oil pump being checked.*

9. *Is the pump's discharge pressure in design range?* _____   ❏
10. *If the pump's discharge pressure is not within design range, does the pressure need to be increased or decreased?* _____   ❏
11. If the pressure needs to be raised or lowered, insert a screwdriver into the pressure adjustment screw.   ❏

> **NOTE:** *To increase pressure, turn the pressure adjustment screw clockwise. To decrease pump pressure, turn pressure adjustment screw counterclockwise to obtain required operating pump pressure.*

12. Once pressure is within design range, turn power OFF to unit.   ❏
13. Place the oil pan underneath the bleeder valve for a couple of seconds and release the pump pressure, then close the bleeder valve. (Refer to *Figure 5–51*.)   ❏
14. Remove the pressure gage.   ❏
15. Replace the gage port plug on the oil pump.   ❏
16. Have the instructor check your work.   ❏
17. Disconnect equipment and return all tools and test equipment to their proper location.   ❏

RESEARCH QUESTIONS

1. The valve piston of the pressure adjustment screw assist to prevent what from happening with the burner nozzle?

_____

_____

2. If the piston valve allows the oil to continue to spray out after the burner shuts down, what problems can this create in the combustion chamber?

_____

_____

3. What part of the burner operates the oil pump?

_____

_____

4. What should be done with an oil pump that has a piston valve that will not shut the oil off to the nozzle on burner shutdown?

_____

_____

5. What is the output voltage range of ignition transformers used with oil burners?

_____

_____

Passed Competency _____   **Failed Competency** _____

Instructor Signature _____   **Grade** _____

Practical Competency 133

Oil Pump Standing Pressure Test (Figure 5–53)

FIGURE 5–53   A pressure gage in place to check valve differential. (*Photo by Cecil Johnson*)

SUGGESTED MATERIALS

Textbook
Refrigeration & Air Conditioning Technology, 5th Edition, Thomson Delmar Learning
Unit 32—Oil Heat

Review Topics
Fuel Oil Pump

COMPETENCY OBJECTIVE

The student will be able to evaluate the valve differential of the oil pump by performing a standing pressure test.

OVERVIEW

The fuel pump on an oil burner provides three functions:

    A. *Conditions the oil from the tank to the nozzle.*
    B. *Controls the rate at which the oil is fed to the nozzle.*
    C. *Controls the cut-off of the oil to the nozzle when the burner shuts OFF.*

Most oil pumps used in oil burners are rotary pumps that use gears, cams, or both to create suction and discharge pressures. Oil pumps are also equipped with an internal automatic pressure valve that is used to control the amount of pressure exerted on the fuel oil to the nozzle and bleed off any excess pressure. This valve also has a positive cut-off so that if the pressure on the pump drops for any reason, it would shut off the oil supply to the burner nozzle. This prevents oil from leaking out of the nozzle and burning with a lack of air.

With the burner in operation, the oil pump should be able to supply a constant operating pressure to the burner nozzle. When the burner is shut off, the oil pump pressure should drop about 15 psi and hold. This is called the **standing pressure.** If the gage pressure drops more than 15 psi below the operating pressure, the following could be the problem:

A. *Possible air leak in fuel unit (pump)*
B. *Possible air leak in fuel oil supply system*
C. *Clogged nozzle strainer*

EQUIPMENT REQUIRED

Pressure Gage (0–300 psig)
Adjustable wrench
Standard screwdriver
Oil drain pan
Clean rag

SAFETY PRACTICES

Care should be taken not to spill fuel oil on hot surfaces. Beware of working with lines and pumps under high pressure.

COMPETENCY PROCEDURES   Checklist

1. Turn power OFF to furnace.   ❑
2. Place the oil pan underneath the bleeder valve for a couple of seconds and release the pump pressure, then close the bleeder valve.   ❑
3. Remove the nozzle port line at the oil pump.   ❑
4. Install the pressure gage at the nozzle line port. (Refer to *Figure 5–53.*)   ❑
5. Make sure the oil filter is cleaned and that the oil line(s) are clear of restrictions.   ❑
6. Turn power ON to furnace.   ❑
7. Set thermostat to heating mode and temperature selector high enough to energize furnace burner.   ❑
8. *Record the pump's discharge gage reading.* _____   ❑
9. *Turn power OFF to furnace and observe pressure gage.*   ❑
10. *When the pressure stops dropping, record gage pressure.* _____   ❑
11. *Record the pressure differential between the pump's operating pressure and shut-off pressure.* _____   ❑

> **NOTE:** When the burner is shut off, the oil pump pressure should drop about 15 psi and hold.

12. *Was the pump's differential pressure within 15 psi?* _____   ❑

> **NOTE:** If the gage pressure drops more than 15 psi below the operating pressure, oil will leak into the combustion chamber after burner shutdown. This can cause "after-burn," or allow fuel oil vapors to collect in the combustion chamber. If enough time has not elapsed for the vapors to escape the chamber before the next ignition process, an explosion or "backfire" can take place in the combustion process.

13. Was the pump's oil pressure differential greater than 15 psi? _____ ❑

NOTE: *If the pressure differential is greater than 15 psi, the following could be the problem:*
  A. *Possible air leak in fuel unit (pump)*
  B. *Possible air leak in fuel oil supply system*
  C. *Clogged nozzle strainer*

---

14. If the differential were greater than 15 psi, check the possible problems as listed, and recheck the pump's standing pressure. ❑

NOTE: *If the gage drops to 0 inch psi, the fuel oil cut-off inside the pump is defective, and the pump should be replaced if none of the corrective actions listed above correct the problem.*

---

15. Place the oil pan underneath the bleeder valve for a couple of seconds and release the pump pressure, then close the bleeder valve. ❑
16. Remove the pressure gage. ❑
17. Replace the nozzle line to the oil pump. ❑
18. Have the instructor check your work. ❑
19. Disconnect equipment and return all tools and test equipment to their proper location. ❑

RESEARCH QUESTIONS

1. When should a single-pipe system be converted to a two-pipe oil supply system?

   _____

   _____

2. When should the bypass plug be installed into the oil pump?

   _____

   _____

3. What happens to the excess oil in the fuel pump on a two-pipe oil supply system?

   _____

   _____

4. Is repairing a leaking fuel pump on an oil furnace recommended?

   _____

   _____

5. A one-pipe system may be used when the oil tank is higher or lower than the oil pump?

   _____

   _____

Passed Competency _____   Failed Competency _____

Instructor Signature _____   Grade _____

Practical Competency 134

Oil Pump Vacuum Test (Figure 5–54)

FIGURE 5–54   Vacuum gage attached to measure oil pump's suction capabilities. (*Photo by Cecil Johnson*)

SUGGESTED MATERIALS

Textbook
Refrigeration & Air Conditioning Technology, 5th Edition, Thomson Delmar Learning
Unit 32—Oil Heat

Review Topics
Fuel Oil Pumps

COMPETENCY OBJECTIVE

The student will be able to check the suction performance of a single-stage and two-stage fuel oil pump.

OVERVIEW

The fuel pump on an oil burner provides three functions:

A. *Conditions the oil from the tank to the nozzle.*

B. *Controls the rate at which the oil is fed to the nozzle.*

C. *Controls the cut-off of the oil to the nozzle when the burner shuts OFF.*

Most oil pumps used in oil burners are rotary pumps that use gears, cams, or both to create suction and discharge pressures. **Oil pumps that are located above the oil tank** will develop a suction pressure that is used to lift the oil from the oil supply tank to the pump so that the oil can be pressurized and forced to the burner nozzle for atomization.

NOTE: *The rule of thumb is that an oil pump requires 1 inch Hg of vacuum to lift oil 1 foot vertically; 1 inch Hg vacuum to pull oil 10 feet horizontally; and 1 inch Hg for filters, valves, and other accessories.*

EXAMPLE: A pump that is 6 feet above the oil tank and 30 feet away from the tank with an oil filter installed would create a 10-inch Hg vacuum.

In cases where the **oil pump is located below the oil tank,** a vacuum test should indicate a gage reading of 0 inch Hg. If the vacuum gage indicates a vacuum, one of the following may be the problem:

A. *The fuel line may be kinked.*

B. *The line filter may be clogged or blocked.*

C. *The tank shutoff valve may be partially closed.*

Installing gages on the pump and checking the vacuum pressure and discharge pressure can check oil pump performance (**Figure 5–55**).

FIGURE 5–55   Vacuum gage.

OIL PUMP VACUUM TEST

A vacuum pump check on a furnace where the **oil tank is above the burner** should indicate a 0-inch Hg vacuum. If the vacuum gage indicates a vacuum below 0 inch Hg, the following could be the cause: The fuel line could be kinked, the oil filter could be clogged, or the oil tank shut-off valve could be closed or partially closed.

For burners where the **oil tank is below the oil pump,** there should be a 1-inch Hg vacuum for every foot of vertical lift and a 1-inch Hg vacuum for every 10 feet of horizontal run of the oil lines, but should not exceed a 17-inch Hg vacuum. Repair attempts on defective oil pumps should be avoided; an oil pump that is defective should be replaced.

EQUIPMENT REQUIRED

Vacuum gage (0–30 inches Hg or compound gage with 0–30 inches Hg)
Adjustable wrench
Oil drain pan

Clean rag
Tape rule

SAFETY PRACTICES

Care should be taken not to spill fuel oil on hot surfaces. Beware of working with lines and pumps under high pressure.

COMPETENCY PROCEDURES

Checklist

NOTE: *Pump discharge operating pressure should be correct to obtain the correct suction levels for single-stage and two-stage pumps. Make sure the oil filter is cleaned and that the oil line(s) are clear of restrictions.*

NOTE: *In a single-stage pump with a single- or two-pipe system, the vacuum on the pump should not exceed 10 inches Hg of vacuum. In a two-stage pump with a two-pipe system, the vacuum on the pump should not exceed 15 inches Hg of vacuum.*

---

1. Shut power OFF to furnace. ❏
2. *Is this a single- or a two-pipe oil supply system?* _____ ❏
3. *What is the maximum allowable lift for the pump based on supply line installation? (Refer to notes above.)* _____ ❏

DETERMINE THE AMOUNT OF PUMP VACUUM REQUIRED (IF APPLICABLE)
4. *Measure and record the amount of vertical lift of the oil supply line.* _____ ❏
5. Determine the inches of vacuum required for the vertical lift by the following:
   The pump requires 1 inch Hg of vacuum to lift oil 1 foot vertically. ❏
6. *Record the number of inches Hg of vacuum required for the oil supply lines vertical lift.* _____ ❏
7. *Measure and record the amount of horizontal run of the oil supply line.* _____ ❏
8. Determine the inches Hg of vacuum required for the horizontal run by the following:
   The oil pump requires 1 inch Hg of vacuum to pull oil 10 feet horizontally. ❏
9. *Record the number of inches Hg of vacuum required for the oil supply line's horizontal run.* _____ ❏
10. *Does the oil supply line have an oil filter?* _____ ❏
11. *Does the oil supply line have a shut-off valve?* _____ ❏
12. Determine additional inches Hg of vacuum required for accessories based on the following:
   Add 1 inch Hg of vacuum for oil filter, shut-off valve, or other accessories. ❏
13. *Add and record the additional inch Hg of vacuum required for oil supply line accessories.* _____ ❏
14. *Determine total inch of vacuum requirement* ❏
   Vertical lift vacuum requirement _____ ❏
   Horizontal run vacuum requirement _____ ❏
   Accessories' vacuum requirement _____ ❏
15. *Record total oil pump's suction in vacuum requirement for system.* _____ ❏

NOTE: *If the burner assembly is located below the oil tank, the pump may show little or NO vacuum during the test.*

MEASURE THE PUMP'S ACTUAL SUCTION PRESSURE

16. On single-pipe supply oil systems, place the oil pan underneath the bleeder valve and release the pump pressure for a couple of seconds, then close the bleeder valve. ❏

NOTE: *Pumps with a two-pipe oil supply system do not require releasing of the oil pressure because the excessive oil is returned to the oil tank.*

---

17. Look at the back of the pump body and find the location of a suction port plug that is not being used, and remove the plug. ❏
18. Install the vacuum gage at the suction port where the oil pump plug was removed. ❏
19. Start the furnace and observe the vacuum gage. ❏
20. *Record the actual vacuum pump gage reading.* _____ ❏
21. Shut the unit off. ❏
22. *Was the suction pressure of the oil pump adequate for the oil supply system?* _____ ❏
23. Did the oil pump's suction pressure exceed the minimum suction requirement as listed in Step 3 above? _____ ❏

NOTE: *If the oil pump's suction pressure did not meet the required amount of suction pressure as determined in Step 13, or if the oil pump's suction pressure exceeded the maximum suction pressure as stated in Step 3, check for leaks at supply line fittings, clogged oil filter, and proper oil pump discharge pressure. If suction pressure cannot be brought in to line with that which is required for the installation, the pump may be defective and require replacement.*

---

24. Place the oil pan underneath the bleeder valve for a couple of seconds and release the pump pressure, then close the bleeder valve. ❏
25. Remove the vacuum pump gage. ❏
26. Replace the suction port plug on the oil pump. ❏
27. Have the instructor check your work. ❏
28. Disconnect equipment and return all tools and test equipment to their proper location. ❏

RESEARCH QUESTIONS

1. What could cause a single-stage pump to pull more than a 10-inch Hg vacuum while in operation?

   _____

   _____

2. When should a single-stage pump be replaced with a two-stage pump?

   _____

   _____

3. When changing a single-pipe oil system to a two-pipe oil system, what must be installed in the pump to ensure correct operation of the two-pipe system?

   _____

   _____

4. What could cause a two-stage pump to pull more than a 15-inch Hg vacuum?

   _____

   _____

5. What could be done in a two-stage pump with a two-pipe oil system to lower the pump's vacuum level below a 15-inch vacuum?

_____

_____

Passed Competency _____   Failed Competency _____

Instructor Signature _____   Grade _____

Practical Competency 135

Bleeding a One-Pipe System

SUGGESTED MATERIALS

Textbook
Refrigeration & Air Conditioning Technology, 5th Edition, Thomson Delmar Learning
Unit 32—Oil Heat

Review Topics
Fuel Oil Pumps

COMPETENCY OBJECTIVE

The student will be able to bleed air from a single-pipe fuel system.

OVERVIEW

Since single-pipe oil supply systems are not self-venting, air trapped within the fuel pump can create noisy operation, improper ignition, or inadequate nozzle pressure. When starting an oil burner with a one-pipe system for the first time or whenever the fuel-line filter or pump is serviced, or the oil tank has been allowed to run empty, air will collect within the system piping and will need to be bled from the piping and fuel oil pump. Oil pumps have a bleed or vent port for this very reason (**Figure 5–56**).

To bleed the system, a 1/4-inch clear flexible tubing should be placed on the vent or bleed port, and the other end of the tubing placed in a container to catch the oil during the bleeding process. Using an

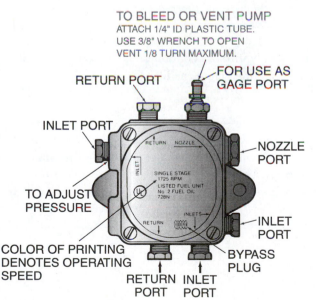

FIGURE 5–56   A diagram of a fuel oil pump. Take note of the bleed or vent port. (*Courtesy Webster Electric Company*)

adjustable wrench or box-end wrench, turn the bleeder or vent valve counterclockwise a quarter to half of a turn. Start the burner and allow the fuel oil to flow into the oil container. If air is in the system, the flowing oil through the tube will be foamy. Continue the oil flow until a steady stream of clear oil is flowing into the container, then shut off the bleeder or vent valve. The unit should perform properly.

EQUIPMENT REQUIRED

Adjustable wrench or proper sized box-end wrench
Clear flexible tubing
Oil drain pan
Clean rag
Jumper wire (*if applicable*)

SAFETY PRACTICES

Care should be taken not to spill fuel oil on hot surfaces. Beware of working with lines and pumps under high pressure.

COMPETENCY PROCEDURES

Checklist

1. Turn power OFF to the furnace. ❏
2. Remove furnace cover and gain access to the burner assembly. ❏
3. Locate the fuel oil pump. ❏
4. Locate the fuel oil pump's bleed or vent valve. ❏
5. Place the flexible hose over the bleed or vent valve. ❏
6. Place the other end of the flexible hose into the oil container ❏
7. Use the adjustable wrench or box-end wrench and turn the bleeder or vent port counterclockwise 1/4-in to 1/2-in turn. ❏
8. Turn power ON to furnace. ❏
9. Set thermostat to the heating cycle. ❏
10. Set thermostat selector switch high enough to turn on the burner. ❏

> **NOTE:** *Once power is applied to the furnace, the burner can be turned on by placing a jumper wire across the oil burner's primary control T–T terminals, or R and W terminals either on the primary control or electronic control board. This bypasses the furnace thermostat.*

11. Allow the burner to run until the oil runs clear and steady into the container. ❏
12. Once the oil is running clear and steady into the oil container, close the bleeder or vent port on the oil pump. ❏

> **NOTE:** *During the bleeding process the ignition may not take place, but will take place once the bleeder valve is closed. Some oil vapor may collect in the combustion chamber during the bleeding process so caution should be taken to avoid any furnace dampers or view-port panels when the bleeder valve port is closed. The furnace may backfire and blow flame out the view-port panel or furnace damper.*

13. Once the system is bled of air, turn power to the furnace OFF. ❏
14. Remove the flexible tubing from the oil pump bleeder port. ❏
15. Use a rag to wipe up any spilt oil. ❏

16. *Did the system have air in the oil lines?* _____ ❏
17. If so, explain what you noticed about the oil as it was being bled from the fuel
    pump. _____ ❏
18. *If the fuel oil system had oil in it, was the burner noisy during operation?* _____ ❏
19. *Did the burner operation quiet down once the air was bled from the fuel oil
    system?* _____ ❏
20. Have the instructor check your work. ❏
21. Return all tools and equipment to their proper location. ❏

RESEARCH QUESTIONS

1. The most common fuel oil used for residential and light commercial is:

    A. No. 1
    B. No. 2
    C. No. 3
    D. No. 4

2. The motor in the oil burner:

    A. Turns only the fan
    B. Turns only the fuel pump
    C. Turns both the fan and fuel pump
    D. Turns neither the fan or the fuel pump

3. The nozzle assembly is comprised of:

    A. Fuel pump and nozzle line
    B. Nozzle line, nozzle adapter, nozzle, and electrodes
    C. Nozzle line and electrodes
    D. Fuel pump and electrodes

4. The nozzle is responsible for:

    A. Metering the correct flow of fuel into the combustion area
    B. Creating the correct spray pattern and angle of the fuel oil
    C. Atomizing the fuel as it enters the combustion area
    D. All the above are correct

5. Describe the purpose of the heat exchanger.

    _____

    _____

Passed Competency _____   Failed Competency _____

Instructor Signature _____   Grade _____

Practical Competency 136

Ignition Transformer (Screwdriver Method)

SUGGESTED MATERIALS

Textbook
Refrigeration & Air Conditioning Technology, 5th Edition, Thomson Delmar Learning
Unit 32—Oil Heat

Review Topics
Ignition Transformer

COMPETENCY OBJECTIVE

The student will be able to check the performance of the ignition transformer of an oil burner.

OVERVIEW

The fuel–air mixture created by the fan and fuel pump is ignited by a spark arcing between the two electrodes of the burner assembly.

The spark is created by using a step-up transformer with a primary voltage of 120 volts, with a secondary voltage of 10,000 volts. The secondary terminals of the transformer make contact with the burner electrodes.

On the call for heat, the burner primary control energizes the step-up transformer which applies 10,000 volts to the electrodes. An electrical arc of about 20 milliamps is created across the electrode tips. The electrical arc generates enough heat to ignite the atomized oil. Newer and higher efficiency oil burners have an ignition transformer with a step-up voltage of 14,000 volts.

A typical voltmeter cannot be used to check the voltage output of high-voltage transformers; high-voltage meters are designed for this purpose (**Figure 5–57**).

Technicians can check transformers in the field by the use of a well-insulated screwdriver. Although this procedure does not indicate actual transformer voltage output, it can be used to determine if the output voltage is high enough to generate an arc that is hot enough across the electrodes to ignite the fuel oil (**Figure 5–58**).

With the transformer energized, a screwdriver is placed across the secondary terminals until an electrical arc is generated on the screwdriver shaft. A transformer generating a voltage high enough to create an arc hot enough to ignite the fuel oil should allow the electrical arc to be lifted and held about 1/2 to 3/4 inches. If the arc cannot be lifted to this height and held, the transformer is weak and should be replaced.

> *NOTE: Care should be taken to avoid touching the screwdriver shaft while performing this test.*

EQUIPMENT REQUIRED

Ignition transformer
120-volt jumper cord (*if applicable*)
Standard screwdriver
2 Jumper wires
Wire nuts

FIGURE 5–57 A high-voltage transformer tester. (*Photo by Bill Johnson*)

FIGURE 5–58 Checking a high-voltage transformer using a well-insulated screwdriver. (*Photo by Cecil Johnson*).

SAFETY PRACTICES

Care should be taken when working with high-voltage equipment. Make sure the voltmeter has a range to handle at least 10,000 volts. Make sure screwdriver handles are well insulated.

COMPETENCY PROCEDURES

> NOTE: *If the transformer being tested is in an actual furnace, you will need to place a jumper wire across the F–F terminals of the primary control once the burner has started, or the safety switch of the primary control will de-energize power to the transformer within 15 to 45 seconds after the burner has started. If this is the case, allow the combustion chamber to clear of oil vapor before resetting the primary control.*

SET-UP: TRANSFORMER IN ACTUAL FURNACE

1. Turn power OFF to furnace. ☐
2. Turn oil supply line OFF (*if applicable*). ☐
3. Remove transformer holding screws. ☐
4. Lay back transformer. ☐
5. Use the jumper wire and jump the T–T terminals of the primary control. ☐

SET-UP: BENCH TESTING TRANSFORMER

6. Use the wire nuts and connect the jumper cord to the transformer's primary wire leads. ☐
7. Turn power ON to furnace or apply power to jumper cord. ☐
8. Hold the screwdriver by insulated handle. ☐
9. Carefully place the shaft of the screwdriver across the transformer secondary terminals. ☐
10. Raise the screwdriver shaft until an arc is created between the screwdriver shaft and the transformer's secondary terminals. ☐
11. Once the arc is established, try to the lift the arc 1/2" to 3/4" away from the transformer's secondary terminals. ☐
12. Once the test is complete, remove the screwdriver from the transformer's secondary terminals. ☐
13. Turn power OFF to transformer. ☐

> **NOTE:** *If arc was lifted 1/2 to 3/4 of an inch or more, the transformer is generating enough voltage to create a spark hot enough that the oil should ignite. If the arc cannot be lifted to at least 1/2", the transformer is weak and may not be able to generate an arc that is hot enough to ignite the atomized oil. In this case, the transformer should be replaced.*

14. During the test, could the arc be lifted 1/2" to 3/4" away from the transformer's secondary terminals? _____ ❏

15. From the test results, would the transformer be considered good or defective? _____ ❏

16. Have the instructor check your work. ❏

17. Disconnect equipment and return all tools and test equipment to their proper location. ❏

RESEARCH QUESTIONS

1. A weak transformer can cause what to happen during ignition?

_____

_____

2. Puff-back in an oil burner happens when?

_____

_____

3. How many amps are created by the step-up transformer's secondary voltage?

_____

_____

4. What should be done with an ignition transformer that has a low-output voltage?

_____

_____

5. A step-up transformer has more turns of wire in _____ coil and less turns of wire in the _____ coil.

Passed Competency _____   **Failed Competency** _____

Instructor Signature _____   **Grade** _____

Practical Competency 137

Checking and Setting the Electrodes of an Oil Burner

SUGGESTED MATERIALS

Textbook
Refrigeration & Air Conditioning Technology, 5th Edition, Thomson Delmar Learning
Unit 32—Oil Heating

Review Topics
Electrodes

COMPETENCY OBJECTIVE

The student will be able to check and set the electrodes of an oil burner to proper alignment.

OVERVIEW

Electrodes are responsible for generating the spark for ignition of the atomized oil on a call for heat from the furnace thermostat. There are two types of ignition in use: *constant or continuous,* and *intermittent.* Modern furnaces incorporate the *constant ignition system,* in which the high-voltage *spark across the electrodes is being supplied during the complete burning cycle.* Older furnaces incorporated the *intermittent ignition systems,* which supply the spark across the electrodes for a short amount of time during the start-up and de-energize.

Electrodes should not be inserted into the actual oil–air spray. They should be positioned so that the arc across the electrode tips is blown into the atomized oil and air mixture.

The ceramic insulators of the electrodes are designed to insulate the electrodes and serve the purpose of positioning the electrodes in the static pressure disc assembly.

To ensure that the correct arc is generated, the electrodes should be positioned in the static disc correctly. The gap between the electrodes should be approximately 1/8" to 3/16". The electrodes should extend out about 1/8" to 3/8" past the end of the burner nozzle and should be 3/8" to 1/2" high above the center of the nozzle orifice (**Figure 5–59**).

When checking electrodes ensure that the three spark gap settings are correctly set, the distance of the tips forward from the nozzle center is correct and back of the oil spray. Insulators should be cleaned with a cloth moistened with a solvent that removes oil and soot. If electrodes remain discolored after being cleaned, they are filled with carbon throughout their porous surface and should be replaced.

> **NOTE:** *The electrode settings described in this competency are typical electrode settings. Refer to manufacturers' actual required settings for the burner assembly used in completing this competency. If the required manufacturer settings are different, use manufacturers' recommended settings.*

EQUIPMENT REQUIRED

Oil burner assembly (*if applicable*)
Electrode assembly (*if applicable*)
Standard screwdriver

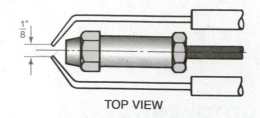

TOP VIEW

ELECTRODES —— —— MINIMUM $\frac{1}{4}''$ TO NOZZLE

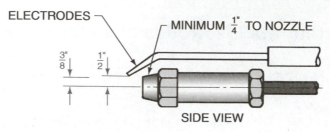

SIDE VIEW

HEIGHT ADJUSTMENT

$\frac{1}{2}''$ RESIDENTIAL INSTALLATION

$\frac{3}{8}''$ COMMERCIAL INSTALLATION

POSITION OF ELECTRODES IN FRONT OF NOZZLE
IS DETERMINED BY SPRAY ANGLE OF NOZZLE.

FIGURE 5–59 Electrode settings.

Adjustable wrench
T-gage or ruler
Piece of soft sanding cloth
Oil drain pan
Clean rag
Cleaning solvent
Jumper wire

SAFETY PRACTICES

Be sure to follow all electrical safety rules when working with live circuits. Make sure electrical test equipment is set to proper function and ranges for testing.

COMPETENCY PROCEDURES

Checklist

REMOVING THE ELECTRODE AND NOZZLE ASSEMBLY FROM THE BURNER (IF APPLICABLE)

1. Turn power OFF to furnace. ❑
2. Shut oil supply line valve OFF. ❑
3. Place the oil drain pan underneath the burner pump assembly. ❑
4. Remove screws holding the ignition transformer and flip it back. ❑
5. Remove the oil connection line from the electrode, nozzle, and static disc assembly by using the adjustable wrench to loosen the flare nut (**Figure 5–60**). ❑
6. Remove the electrode, nozzle, and static disc holding nut. ❑
7. Remove the transformer holding clips. ❑
8. Lay the transformer back. ❑
9. Pull the nozzle and electrode assembly from the air tube of the burner. Hold your thumb over the hole to keep oil from dripping. ❑
10. Drain the oil from the nozzle assembly into the drain pan. ❑
11. Loosen the screw(s) that hold the electrodes to the nozzle assembly and remove the electrodes. ❑

FIGURE 5–60 Removing oil line connection from electrode, nozzle, and static disc assembly. (*Photo by Cecil Johnson*)

12. Using the cleaning solvent, wipe the electrodes' insulators off with a rag. ❑
13. Inspect the electrode insulators for cracks. ❑
14. Check the color of the insulators once they are cleaned. If they remain discolored, they need to be replaced. ❑

> **NOTE:** *Replace electrodes as a pair if one or both has cracked porcelain.*

15. Use a fine sand cloth and clean the electrode tips. ❑
16. Place the electrodes back into the nozzle support bracket of the static disc. (**Do not tighten completely.**) ❑

> **NOTE:** *Refer to the manufacturer's desired setting for the electrodes if available; if not available use settings in the following procedures.*

17. Use the T-gage or ruler to set the electrodes. ❑
18. Move the electrodes forward so the tips of the electrodes extend out over the nozzle 1/8" to 3/8". (Refer to *Figure 5–59*.) ❑
19. Set the electrode gap setting by moving the electrode tips together to from a 1/8" gap area centered over the nozzle port. (Refer to *Figure 5–59*.) ❑
20. Set the electrode tips so that they are 3/8" to 1/2" above the center of the nozzle. (Refer to *Figure 5–59*.) ❑

> **NOTE:** *Electrode height adjustments:*
> *1/2" for residential installation*
> *3/8" for commercial installation*

21. Once electrodes are aligned and set, tighten the electrode-supporting bracket. ❑
22. Have the instructor check your work (*if applicable*). ❑
23. Slide the nozzle and electrode assembly back into the air tube of the burner and adjust. ❑
24. Screw oil line flare nut back on to the nozzle assembly. (*Tighten enough to avoid leaks.*) ❑
25. Flip the ignition transformer back into position and tighten the mounting clips. ❑
26. Open the oil line shut-off valve. ❑
27. Check for leaks at all line connections. ❑

28. Make sure control thermostat is turned OFF. ❏
29. Take the jumper wire and jump the T–T terminals on the burner primary control. ❏
30. Turn the oil furnace's power supply back ON. ❏

> **NOTE:** *If the burner does not fire before the primary control safety switch trips, wait at least 5 minutes before resetting the safety switch. This will allow any oil vapor to clear the combustion chamber. If the safety switch trips again, wait another 5 minutes, bleed the oil pump, and try restarting the burner.*

31. Once the burner fires, use the inspection mirror and inspect the flame and adjust if needed. ❏

> **NOTE:** *A clean fire should have bright (almost a glowing white) tips as you look straight across the fire.*

32. Let the burner operate until the fan–limit or thermal switch turns on the furnace blower. ❏
33. Pull the jumper wire from the T–T terminals of the primary control. ❏
34. Let the furnace operate until the fan–limit or thermal switch shuts the blower OFF. ❏
35. *What was the color of the electrodes once they were cleaned?* _____ ❏
36. *Were there any cracks or chips in the electrode insulators?* _____ ❏
37. *Were the electrode insulators discolored at all?* _____ ❏
38. *Were the tips of the electrodes melted at all from the heat of the arc?* _____ ❏
39. *Did the burner fire off on the first restart?* _____ ❏
40. *What was the color of the burner flame?* _____ ❏
41. Have the instructor check your work. ❏
42. Disconnect equipment and return all tools and test equipment to their proper location. ❏

RESEARCH QUESTIONS

1. What is the approximate amperage of the spark across the electrode?

   _____

   _____

2. How is the spark of the electrodes used to ignite the fuel oil?

   _____

   _____

3. What effect can a cracked insulator have on an electrode?

   _____

   _____

4. The purpose of the cad cell in an oil burner control circuit is to:

   A. Make sure that the burner ignites by sensing the flame.
   B. Turn the burner OFF when the room temperature is satisfied.
   C. Start the burner when the thermostat calls for heat.
   D. Prevent the furnace from overheating.

5. What are the three different nozzle spray patterns?

   _____   _____   _____

Passed Competency _____   Failed Competency _____

Instructor Signature _____   Grade _____

Theory Lesson: Operation of Cad Cell Oil Primary Controls

SUGGESTED MATERIALS

Textbook
Refrigeration & Air Conditioning Technology, 5th Edition, Thomson Delmar Learning
Unit 32—Oil Heat

Review Topics
Cad Cell Primary Controls

Key Terms
cad cell • primary control • safety switch heater

OVERVIEW

The oil burner **primary control** provides a means for operating the burner and safety functions used to turn the burner *ON* and *OFF* in response to the room temperature demand through the system thermostat, ignition failure, or high-temperature conditions. In cases of any safety failure, the primary control safety switch will trip, preventing ignition of the oil burner until the safety switch is reset.

> **NOTE:** *Any time the primary control safety switch trips and shuts the oil burner down, efforts should be made to determine what caused the primary control safety switch to trip before resetting the oil burner primary control.*

Modern furnace incorporates a cad cell oil primary control. The cad cell or electric eye is designed to prove to the primary control that ignition of the atomized oil has taken place in the furnace combustion chamber.

The **cad cell** is located beneath the ignition system and positioned in a manner that it can sight flame through the oil burner air tube. The cad cell is a photocell that has a high electrical resistance when it is in the dark and a low resistance in visible light. The resistance of the cad cell is about 100,000 ohms when it is in the dark. When light is reflected on the cad cell its resistance drops to about 1,600 ohms.

On a call for heat, the thermostat closes a circuit between the R and W terminals on the thermostat. This circuit sends a 24-volt power source to the oil burner primary control.

The primary control checks to make sure that all systems on the furnace and burner are ready to start. The primary control then turns on the burner motor which drives the blower, which blows excess air into the combustion chamber. The burner motor also drives the fuel oil pump, which delivers oil under pressure to the fuel nozzle in the burner's blast tube.

A spark jumping across the ignition electrodes ignites the mixtures of oil and air. A step-up transformer known as the ignition transformer produces the high voltage necessary to create the spark by stepping up the supplied voltage of 120 volts to 10,000 volts or higher. A flame detector known as the cad cell eye sees the flame in the combustion chamber and sends a signal back to the primary control that ignition has taken place and the primary control can continue to allow the burner to operate until the need for heat has been satisfied and the thermostat shuts the burner down.

If for some reason the cad cell eye does not see flame in a certain amount of time, a safety switch in the primary control opens automatically, breaks the ignition circuit, and shuts down the burner (**Figure 5–61**).

The blower of the furnace does not come on right away when the thermostat calls for heat. The burner will operate for a short time without the blower on to allow the furnace heat exchanger to heat up. This

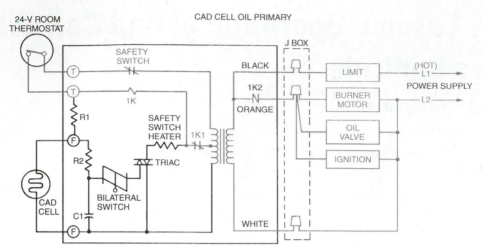

FIGURE 5–61 Cad cell primary circuit when the cell does not "sight" flame. The safety switch heater will heat and open the safety switch.

prevents cold air from being circulated into the conditioned space. A fan–limit switch is placed above the heat exchanger and senses the heat coming off the heat exchanger. (Refer to *Theory Lesson: Oil Furnace Sequence of Operation, Figure 5–40*.)

Once the fan–limit switch senses a predetermined temperature it will close the electrical circuit to the furnace blower. The blower draws the cold air from the conditioned space, moves it across the furnace heat exchanger, heating the air, and then returns it to the conditioned space. The fan–limit switch also services as a safety switch in the event that the furnace overheats for any reason. (Refer to *Theory Lesson: Oil Furnace Sequence of Operation, Figure 5–41*.)

The safety switch of the high-limit switch would break the power supply to the oil burner primary control, which will shut down the furnace operation.

Some newer oil furnaces do not use a fan-limit switch as shown in *Figure 5–39 (Theory Lesson: Oil Furnace Sequence of Operation)*, but use a bimetal device. (Refer to *Theory Lesson: Oil Furnace Sequence of Operation, Figure 5–42*.) These bimetal switches can be used to activate the furnace blower. On the other hand, they can be used as a fan-limit switch; this controls the furnace blower operation and acts as a high-limit switch, shutting the burner down due to high-temperature conditions.

When the conditioned space temperature is satisfied, the thermostat will open the electrical circuit between the R and W terminals of the thermostat. This de-energizes the oil burner primary control, shutting off burner operation. The furnace blower will continue to operate until the heat exchanger has been cooled to a preset temperature. Once the heat exchanger is cooled enough, the fan–limit switch will open the circuit to the furnace blower, shutting the blower off.

Troubleshooting the primary control can be broken down into two areas:

1. *The oil burner will not start.*
2. *The oil burner will start and then lock out of safety in troubleshooting the primary control. First try to start the burner and observe at what point the primary control fails. Then start your troubleshooting from that point.*

Student Name _____ Grade _____ Date _____

Practical Competency 138

Wiring a Primary Control into an Oil Burner Assembly

SUGGESTED MATERIALS

Textbook
Refrigeration & Air Conditioning Technology, 5th Edition, Thomson Delmar Learning
Unit 32—Oil Heat

Review Topics
Cad Cell Oil Burner Control

COMPETENCY OBJECTIVE

The student will be able to wire a cad cell primary control into an oil burner assembly.

> *NOTE: To complete this competency, the instructor can unwire a primary control from an oil burner assembly and have the student follow the procedures to rewire the control back into the burner assembly.*

OVERVIEW

The primary control of the burner is used to turn the burner on and off in response to the room thermostat and to shut down the burner in the event of a flame failure or high-temperature limit switch trip out. The primary control has a reset button that must be reset any time the primary control shuts the unit down. On a call for heat, the thermostat closes a circuit between the R and W terminals on the thermostat. This circuit sends a 24-volt power source to the oil burner primary control.

The primary control checks to make sure that all systems on the furnace and burner are ready to start. The primary control then turns on the burner motor which drives the blower, which blows excess air into the combustion chamber. The burner motor also drives the fuel oil pump, which delivers oil under pressure to the fuel nozzle in the burner's blast tube. Older cad cell primary controls had three colored wires attached—one white, one black, and one orange. Newer primary controls have spade terminals for connecting line voltage, transformer, burner motor, and oil valve. With these primary controls, follow instructions for making electrical connects as indicated on the back of the control.

> *NOTE: Any time the primary control trips and shuts the burner down, effort should be made to determine what caused the primary control to trip before resetting the reset switch.*

EQUIPMENT REQUIRED

Burner assembly
Primary control (*if replacing a control on a burner assembly*)
Wire nuts

Jumper wire
Standard screwdriver
Phillips screwdriver

SAFETY PRACTICES

Be sure to follow all electrical safety rules when working with live circuits. Make sure electrical test equipment is set to proper function and ranges.

COMPETENCY PROCEDURES   Checklist

1. Make sure power to the unit is OFF. ❏
2. The primary control should already be removed from the burner assembly. ❏
3. Use a wire nut and connect the hot leg (**L-1**) of the power supply to the black wire of the primary control. ❏
4. Connect the common leg (**L-2**) of the power supply to one wire lead of each or the following components: ❏
   A. One wire lead of the burner motor ❏
   B. One wire lead of the transformer ❏
   C. One wire lead of the oil valve (*if used*) ❏
5. Use a wire nut and attach all the wires connected to the power leg (**L-2**), to the white wire of the primary control. ❏
6. Use a wire nut and connect the orange wire of the primary control the opposite power wires of each of the following burner components: ❏
   A. The opposite leg of the burner motor ❏
   B. The opposite leg of the ignition transformer ❏
   C. The opposite leg of the oil valve (if used) ❏
7. Make sure all wire nuts are tight. ❏
8. Wrap wire nuts with electrical tape. ❏
9. Mount primary control back on to the burner. (*Make sure cad cell socket leads are accessible.*) ❏
10. Mount the cad cell wire leads to the **F–F** terminal of the primary control. ❏
11. Make sure oil supply valve is turned ON. ❏
12. Turn the control thermostat ON or use the jumper wire and jump the **T–T** terminals of the primary control. ❏
13. Make sure the safety-reset button on the primary control is reset. ❏
14. Turn ON the power supply to the unit (burner should start.) ❏
15. Let the unit operate for a minute. ❏
16. Let the unit continue to operate and turn the oil supply valve off. ❏
17. Explain what happened to the burner operation after 15 to 45 seconds of operation. ❏

NOTE: *If the primary control was wired into the burner circuit correctly, the burner should have come ON and operated OK for the 1 minute of run time. When the oil supply valve turned OFF, the burner should have continued to run for about 15 to 45 seconds and then shut OFF. The primary control's reset button should have tripped at this point.*

NOTE: *If the following sequence of operation did not happen, something is wrong with your primary control wiring. Recheck procedures 3 through 10.*

18. Turn the oil supply valve back ON. ❏
19. After 5 minutes, push the reset button of the primary control. (*The burner should start again.*) ❏
20. Turn the control thermostat OFF or shut the power down to the unit. ❏
21. Leave equipment as instructed by the instructor. ❏
22. Have the instructor check your work. ❏
23. Disconnect equipment and return all tools and test equipment to their proper location. ❏

RESEARCH QUESTIONS

1. The low-voltage transformer of the primary control is supplied line voltage by which two wires of the primary control?

_____

_____

2. Which two terminals of the primary control are used to operate the safety switch function of the primary control?

_____

_____

3. Which wire of the primary control supplies the line voltage to the ignition transformer, burner motor, and oil valve if used?

_____

_____

4. An ignition system that keeps the ignition transformer on during the total burn cycle is called what type of ignition system?

_____

_____

5. An ignition system that shuts the ignition transformer OFF once the fuel oil is ignited is called?

_____

_____

Passed Competency _____ Failed Competency _____

Instructor Signature _____ Grade _____

Practical Competency 139

Cad Cell Primary Control Check Out

SUGGESTED MATERIALS

Textbook
Refrigeration & Air Conditioning Technology, 5th Edition, Thomson Delmar Learning
Unit 32—Oil Heat

Review Topics
Cad Cell Primary Controls

COMPETENCY OBJECTIVE

The student will be able to evaluate the performance of a cad cell primary control.

OVERVIEW

The purpose of the primary control is to turn the oil burner ON and OFF in response to the conditioned space thermostat and to shut down the burner in the event of a flame failure or high-temperature limit switch trip-out. The primary control normally uses a flame detector called a cad cell eye to prove ignition of the atomized oil. The cad cell is a photocell that has a high electrical resistance when it is in the dark and a low resistance in visible light. The resistance of the cad cell is about 100,000 ohms when it is in the dark. When light is reflected on the cad cell its resistance drops to about 1,600 ohms.

On a call for heat, the thermostat closes a circuit between the **R** and **W** terminals on the thermostat. This circuit sends a 24-volt power source to the oil burner primary controls relay.

The primary control relay turns on the burner motor which drives the blower and fuel oil pump, which delivers oil under pressure to the fuel nozzle in the burner's blast tube.

The ignition transformer is also energized by the primary control relay to produce the high voltage necessary to create the spark at the element ends of the electrodes. At the same time all this is going on, the primary control has also energized the safety switch heater of the primary control. This switch consists of a heater and a bimetal element set to warp out at a preset period of time when heated. This time factor is normally 15 to 30 or 45 seconds. This is the time factor given to allow the cad cell eye to see the ignition of the atomized oil.

If no flame is detected by the cad cell in this time period, current continues to pass through the safety switch heater. This allows the safety switch heater to heat the bimetal element to a point that the bimetal warps open the primary control safety switch and shuts the burner down on safety lock out.

This switch trips the primary control reset button open. The reset button must be pushed to reset the primary control safety switch circuit so that the burner can be restarted again.

When the burner is operating and the cad cell detects flame in the allowable time, the safety switch heater is bypassed to prevent it from continuing to heat the bimetal switch element. This is accomplished in the newer primary controls by having the safety switch heater wired in series with a triac and an electrical circuit made up of a bilateral switch and a resister-capacitor network. When the cad cell sees flame, its resistance drops and shuts off the triac, forcing the current to bypass the safety switch heater.

In the newer primary controls, if the cad cell does not detect flame in the allowable time factor, the triac stays on and current continues to flow through the safety switch heater, causing heat to cause the bimetal element to warp and trigger the primary safety switch lockout. In the older model primary controls the safety switch heater was bypassed by a sensitive electromechanical relay that opened or closed in response to the signals from the cad cell flame detector.

Troubleshooting the primary control can be broken down into two areas:

1. *The oil burner will not start.*
2. *The oil burner will start and then lock out of safety. In troubleshooting the primary control, first try to start the burner and observe at what point the primary control fails. Then start your troubleshooting from that point.*

EQUIPMENT REQUIRED

2 Jumper wires
Phillip screwdriver
Standard screwdriver

SAFETY PRACTICES

Be sure to follow all electrical safety rules when working with live circuits. Make sure electrical test equipment is set to proper function and ranges.

COMPETENCY PROCEDURES

Checklist

> **NOTE:** *These procedures are used to test the function of the primary control to see if it is at fault for no burner operation or intermittent burner operation.*

1. The furnace should be completely operational. ❑
2. Shut power OFF to furnace. ❑

> **NOTE:** *Assume that the furnace had the proper power to operate, but the burner would not come on when a call for heat was set by the thermostat, and the following test was performed.*

3. Remove the low-voltage control wiring from the **T–T or W–B** terminals of the primary control. ❑
4. Use one of the electrical jumper wires and jumper the **T–T or W–B** terminals on the primary control. ❑
5. Turn power ON to furnace. ❑
6. *Did the burner start?* _____ ❑

IF THE BURNER STARTED DURING THIS TEST

7. *If the burner started and continued to operate, what would this indicate about the operation of the primary control?* _____
_____ ❑

8. *What component of the system is likely to be at fault for no burner operation on a call for heat from the furnace thermostat?* _____ ❑

BURNER DID NOT START DURING THIS TEST

9. *If the burner did not start during this test, what would this indicate about the primary control?* _____ ❑

10. *Would the furnace thermostat be at fault for "no operation" in this test?*
_____ ❑

11. Shut power OFF to the furnace. ❏
12. Remove the jumper wire from the **T–T** or **W–B** terminals of the primary control. ❏
13. Rewire the thermostat wires back on to the correct **T–T** or **W–B** terminals. ❏
14. Remove the cad cell lead wires from the **F–F** terminals. ❏
15. Use a jumper wire and jumper the **F–F** terminals of the primary. ❏
16. Set the control thermostat to heat mode and temperature at a setting to call for heat. ❏
17. Turn power ON to the furnace. ❏
18. *Did the burner come ON during this test?* _____ ❏
19. *Did the burner continue to operate during this test?* _____ ❏

BURNER CAME ON AND CONTINUED TO OPERATE
20. *What would this indicate about the primary control?* _____ ❏
21. *What would this indicate about the cad cell?* _____ ❏
22. *What would this indicate about the furnace thermostat?* _____ ❏

BURNER CAME ON AND SHUT DOWN AFTER A FEW SECONDS OF OPERATION
23. *What would this indicate about the primary control?* _____ ❏
24. *What would this indicate about the cad cell?* _____ ❏
25. *What would this indicate about the furnace thermostat?* _____ ❏
26. Shut power OFF to furnace. ❏
27. Leave the jumper wire on the **F–F** terminals. ❏
28. Remove the thermostat wires from the **T–T** or **W–B** terminals and jump them with the other electrical jumper wire. ❏
29. Turn power ON to the furnace. ❏
30. *Did the burner start?* _____ ❏

BURNER CAME ON AND CONTINUED TO OPERATE
31. *What would this indicate about the primary control?* _____ ❏
32. *What would this indicate about the cad cell?* _____ ❏
33. *What would this indicate about furnace thermostat?* _____ ❏

BURNER CAME ON AND LOCKED OUT ON THE PRIMARY SAFETY SWITCH
34. *What would this indicate about the primary control?* _____ ❏
35. *What would this indicate about the cad cell?* _____ ❏
36. *What would this indicate about the furnace thermostat?* _____ ❏

BURNER DID NOT COME ON
37. Remove jumper wires and rewire cad cell and thermostat wires to proper primary control terminals. ❏

SUMMARY OF TEST PROCEDURES
In the **first test** with the **T–T** terminals jumped, if the burner started, the problem is in the thermostat or thermostat low-voltage wiring. If the burner did not start, the problem could be in the cad cell or primary control.

In the **second test** with the **F–F** terminals jumped, if the unit burner started, the problem is in the cad call or cad cell circuit. If the burner did not start, the problem could be in the primary control.

In the **third test,** with the **T–T** and **F–F** terminals jumped, if the burner started, the problem is in the primary control. If the burner did not start, check for proper voltage to primary control, and correct wiring to primary control and furnace.

38. Have the instructor check your work. ❏
39. Disconnect equipment and return all tools and test equipment to their proper location. ❏

RESEARCH QUESTIONS

1. If the cad cell is defective, should the technician jump the F–F terminals to keep the system in operation for the customer?

   _____

   _____

2. What should a technician do before firing a burner where the firebox is saturated with oil?

   _____

   _____

3. What is the danger of firing a burner off when the fire chamber is saturated with oil?

   _____

   _____

4. Are all primary controls capable of supplying the control voltage for an add-on air conditioning system?

   _____

   _____

5. The thermostat heat anticipator should be set to the amp draw of which control of the furnace?

   _____

   _____

Passed Competency _____   Failed Competency _____

Instructor Signature _____   Grade _____

Practical Competency 140

Checking Low-Voltage Output of Cad Cell Primary Control

SUGGESTED MATERIALS

Textbook
Refrigeration & Air Conditioning Technology, 5th Edition, Thomson Delmar Learning
Unit 32—Oil Heat

Review Topics
Cad Cell Primary Controls

COMPETENCY OBJECTIVE

The student will be able to measure and check the low-voltage output of a cad cell primary control.

OVERVIEW

On oil furnaces, the low-voltage control power source comes from the burner primary control. The primary control has a built-in step-down transformer that steps down the line voltage of 125 volts to 24 volts. The 24 volts from the primary control is used to supply control voltage to the furnace control thermostat. On most primary controls, the **T–T** terminals or **R–C** terminals are the 24-volt secondary of the internal step-down transformer. All low-voltage checks can be made at the terminals.

EQUIPMENT REQUIRED

Operation oil burner primary control
Voltmeter or VOM meter

SAFETY PRACTICES

Be sure to follow all electrical safety rules when working with live circuits. Make sure electrical test equipment is set to proper function and ranges.

COMPETENCY PROCEDURES   Checklist

1. Make sure that the proper power is applied to the furnace.   ❑
2. Make sure that the reset button on the primary control is reset.   ❑
3. Remove the thermostat wires at the primary control **T–T** terminals.   ❑
4. Set the voltmeter on the AC volt function and a scale to read at least 24 volts. ❑
5. Turn power ON to furnace.   ❑
6. Place one of the meter leads on one of the (**T**) terminals of the primary control. ❑
7. Place the other meter lead to the opposite (**T**) terminal of the primary control. ❑
8. *Record the voltage output reading.* _____   ❑
9. Turn the power OFF to the furnace.   ❑

> **NOTE:** *A 24-volt reading should be indicated across these terminals if the primary control is wired correctly and the step-down transformer of the primary control is good. No voltage is an indication of improper supply voltage, improper wiring of primary control, or defective primary control step-down transformer.*

10. Have the instructor check your work. ❏
11. Disconnect equipment and return all tools and test equipment to their proper location. ❏

RESEARCH QUESTIONS

1. Where does the low voltage of the primary control come?

   _____

   _____

2. Why is a step-down transformer needed in a primary control?

   _____

   _____

3. On the low-voltage panel board of the primary control, which terminal identification is used for the heating mode of the thermostat?

   _____

   _____

4. On the low-voltage panel board of the primary control, which terminal identification is used for the hot wire of the step-down transformer?

   _____

   _____

5. Some primary controls have a **G** terminal on the low-voltage board; this terminal is used as a hot wire for what furnace operation?

   _____

   _____

Passed Competency _____ **Failed Competency** _____

Instructor Signature _____ **Grade** _____

Theory Lesson: Fan and Limit Safety Switch (Figure 5–62)

FIGURE 5–62   Fan–limit switch. (*Photo by Bill Johnson*)

SUGGESTED MATERIALS

Textbook
Refrigeration & Air Conditioning Technology, 5th Edition, Thomson Delmar Learning
Unit 31—Gas Heat
Unit 32—Oil Heat

Review Topics
Fan Switch; Safety Limit Switch; Combination Fan–Limit Switch

Key Terms
bimetal helix coil • fan–limit switch • heat exchanger • primary control • safety switch • temperature dial tabs

OVERVIEW

The fan–limit switch is located on the furnace above the furnace heat exchanger. It is designed to turn the furnace fan ON and OFF once; it senses a predetermined temperature above the heat exchanger. It also serves as a safety limit switch for a high-limit situation on the furnace. The fan–limit switch contains two sets of contacts. One set of contacts is normally open and operates the ON and OFF cycling of the furnace blower. The other set of contacts is normally closed and operates the safety limit function of the fan–limit (**Figure 5–63**).

The fan–limit uses a bimetal helix coil, which is attached to a temperature dial that rotates as the helix coil is heated (**Figure 5–64**).

The bimetal metal element of the switch is located above the furnace heat exchanger and expands as it senses heat coming from the exchanger (**Figure 5–65**).

Heat from the exchanger causes the temperature dial of the limit switch to rotate, which opens and closes the fan switch at a preset temperature. It also shuts the fan OFF at a preset temperature and/or opens the high-limit switch, shutting down the burner in an oil furnace or gas valve in a gas furnace.

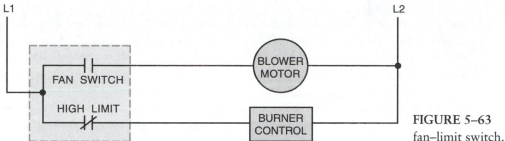

FIGURE 5–63   Schematic for fan–limit switch.

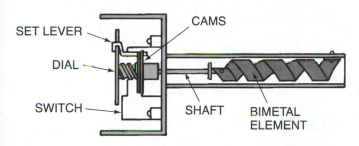

FIGURE 5–64   Helical bimetallic fan and limit controller.

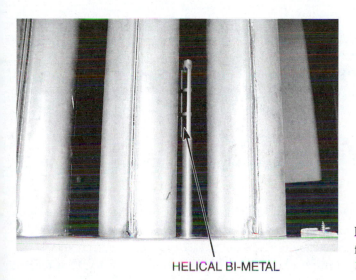

HELICAL BI-METAL

FIGURE 5–65   Helical bimetal located above furnace heat exchanger. (*Courtesy of Ferris State University. Photo by John Tomczyk*)

When the fan is operating, it will keep the temperature of the heat exchanger down so that it does not overheat enough for the limit switch to open. This is accomplished by moving the cool air from the conditioned space across the heat exchanger while the furnace blower is operating.

Once the thermostat calls for heat, the primary control starts the burner. The heat from the combustion chamber rises and heats the heat exchanger. The bimetal helix of the fan–limit switch starts to expand and rotate turning the cam of the fan–limit switch. Once the helix is hot enough, the temperature dial will have rotated far enough to reach the preset temperature to close the fan switch and turn ON the blower of the furnace. Once the blower comes ON, cool air from the structure is circulated across the heat exchanger, heated, and delivered back into the conditioned space.

The air from the conditioned space moving across the heat exchanger helps to keep the heat exchanger from heating to a high-limit temperature. Once the thermostat in the conditioned space is satisfied, it opens the circuit to the primary control of the furnace burner, which shuts OFF the oil burner. The furnace blower continues to operate, moving conditioned space air across the heat exchanger. This assists in cooling the heat exchanger and provides additional heat to the structured space to satisfy the heat anticipator of the furnace.

As the cooler air moves across the heat exchanger, the bimetal helix of the fan–limit starts to contract, turning the temperature dial in the opposite direction. The temperature dial will continue to rotate as the bimetal cools to a point at which the temperature dial will open the fan contacts and shut the fan OFF at the preset temperature.

If the furnace fan does not operate or stops working during the heating cycle, the heat exchanger will continue to get hot. The higher temperature of heat coming off of the heat exchanger will cause the bimetal helix to rotate the cam of the fan–limit to the point that the normally closed limit safety switch contacts will open, breaking the electrical circuit to the oil burner. When this happens, the primary control safety switch opens on lockout so that the burner cannot be restarted until the problem for high limit is corrected.

Setting the fan ON and OFF temperature is done by adjusting the tabs on the temperature dial of the fan–limit switch (**Figure 5–66**).

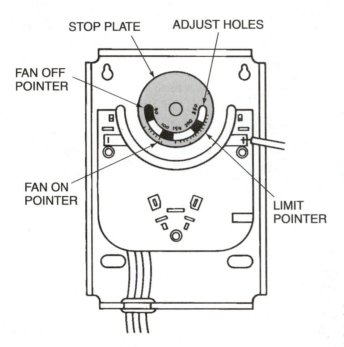

FIGURE 5–66   Fan ON–OFF and limit safety switch pointers.

This is also where the high-limit temperature is set. When setting the temperature, care should be taken to avoid turning the temperature dial. The pointer temperature indicators can be moved by holding the dial stationary and moving the temperature indicators to the desired temperature setting.

The first tab is to be set at the fan OFF temperature. The fan will stop after the burner has shut OFF at this temperature. In most cases, the fan OFF temperature setting is 90–100 degrees (F). The second tab is the setting for the fan ON temperature. This is the air temperature reached from the heating of the heat exchanger during the beginning of the call for heat from the thermostat. In most cases the fan ON temperature is set at 125–150 degrees (F). The third tab is the setting for the safety limit switch temperature. The limit switch will open and shut the burner down in case of a high-limit situation at this temperature. Different state codes have different safety limit temperature cutout requirements; normally the safety limit temperature is set between 200–250 degrees (F) (**Figure 5–67**).

> NOTE: *Refer to Figure 5–67: Some combination fan limit controls can be easily retrofitted for different voltages in the field. An example would be in a gas furnace installation; the high limit safety would be in circuit with the gas valve, which operates on 24 volts, and the fan–limit would be in circuit with the furnace blower, which operates on 120 volts. In circuits where **high voltage is being used for both high limit and fan–limit**, the combination control **tab must remain in place**. In dual-voltage applications such as in the example stated, the combination control **tab must be removed**.*

JUMPER
TAB

FIGURE 5–67   A combination fan and limit control settings. (*Courtesy of Ferris State University, Photo by John Tomczyk*)

As stated, these are recommended temperature settings. Some manufacturers may require different settings for their equipment. Always check the manufacturer's literature for recommended temperature settings on the fan–limit.

Troubleshooting the fan limit switch is a simple operation. Remember that the fan switch contacts are normally open and the limit switch contacts are normally closed. The use of an ohmmeter can indicate if these switch contacts are opening and closing as required.

With advances in technology, newer furnaces are being equipped with bimetal limit switched for both furnace blower operation and safety limit (**Figure 5–68**).

FIGURE 5–68   Bimetal limit switches.

These bimetal switches can be used to activate the furnace blower (**Figure 5–69**). On the other hand, they can be used as a fan–limit switch; this controls the furnace blower operation and acts as a high-limit switch, shutting the burner down due to high-temperature conditions (**Figure 5–70**). The bimetal switches are located near or over the furnace heat exchanger (**Figure 5–71**).

When the conditioned space temperature is satisfied, the thermostat will open the electrical circuit between the **R** and **W** terminals of the thermostat. This de-energizes the oil burner primary control, shutting

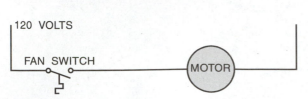

FIGURE 5–69 Bimetal fan switch controls the operation of the fan motor.

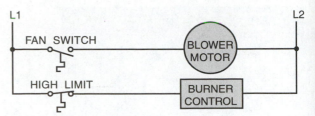

FIGURE 5–70 Fan switch and limit switch connected in series with the heating source.

(A)

(B)

FIGURE 5–71 (A) A snap-action, bimetallic, fan or limit switch. (B) Limit switches are located close to the heat exchanger. (*Courtesy Ferris State University. Photos by John Tomczyk*)

OFF burner operation. The furnace blower will continue to operate until the heat exchanger has been cooled to a preset temperature. Once the heat exchanger is cooled enough, the fan–limit switch will open the circuit to the furnace blower, shutting the blower OFF.

Practical Competency 141

Checking a Fan Limit Switch on a Fossil Fuel Furnace

SUGGESTED MATERIALS

Textbook

Refrigeration & Air Conditioning Technology, 5th Edition, Thomson Delmar Learning
Unit 31—Gas Heat
Unit 32—Oil Heat

Review Topics

Fan Switch; Safety Limit Switch; Combination Fan–Limit Switch

COMPETENCY OBJECTIVE

The student will learn the proper procedures for checking a fan–limit switch on a fossil fuel furnace.

OVERVIEW

The fan–limit switch is located on the furnace above the furnace heat exchanger. It is designed to turn the furnace fan ON and OFF once; it senses a predetermined temperature above the heat exchanger. It also serves as a safety limit switch for a high-limit situation on the furnace. The fan limit switch contains two sets of contacts. One set of contacts is normally open and operates the ON and OFF cycling of the furnace blower. The other set of contacts is normally closed and operates the safety limit function of the fan–limit.

The fan–limit uses a bimetal helix coil, which is attached to a temperature dial that rotates as the helix coil is heated.

The bimetal metal element of the switch is located above the furnace heat exchanger and expands as it senses heat coming from the exchanger.

Heat from the exchanger causes the temperature dial of the limit switch to rotate which opens and closes the fan switch at a preset temperature. It also shuts the fan OFF at a preset temperature and/or opens the high-limit switch, shutting down the burner in an oil furnace or gas valve in a gas furnace.

When the fan is operating, it will keep the temperature of the heat exchanger down so that it does not overheat enough for the limit switch to open. This is accomplished by moving the cool air from the conditioned space across the heat exchanger while the furnace blower is operating.

If the furnace blower does not operate or stops working during the heating cycle, the heat exchanger will continue to get hot. The higher temperature of heat coming off of the heat exchanger will cause the bimetal helix to rotate the cam of the fan–limit to the point that the normally closed limit safety switch contacts will open, breaking the electrical circuit to the oil burner. When this happens, the primary control safety switch opens on lockout so that the burner cannot be restarted until the problem for high limit is corrected.

EQUIPMENT REQUIRED

Combination fan–safety limit switch
Ohmmeter or VOM
Voltmeter or VOM
Heat gun (*heats to at least 200 degrees*)

SAFETY PRACTICES

Be sure to follow all electrical safety rules when working with live circuits. Make sure electrical test equipment is set to proper function and ranges.

COMPETENCY PROCEDURES

Checklist

1. Turn power OFF to furnace. ❏
2. Allow the furnace to cool (*if applicable*). ❏
3. Locate the fan–limit switch. ❏
4. Remove the cover to the fan–limit switch. ❏
5. Remove one of the wires from the safety limit side of the switch (*normally the red wire*). ❏
6. Remove one of the wires from the fan–limit side of the switch (*normally the black wire*). ❏
7. Set the ohmmeter to the lowest ohms scale or buzzer setting. ❏

CHECKING THE SAFETY LIMIT SWITCH

8. Place the ohmmeter leads across the safety limit switch terminals (*right side of the switch*). ❏
9. *Record the ohms value reading across the safety limit switch.* _____ ❏

NOTE: *The limit switch is a NC (normally closed) switch. A good limit switch would have shown (0) or some ohms value reading. An infinity reading would indicate the safety limit switch is OPEN. This could be due to the fact that the bimetallic element is hot enough to open or that the bimetallic element is dirty and sticking open, or that the switch is defective. Before replacing the switch, allow it to cool down if applicable and carefully clean the bimetallic element. Add a little lubricating oil at the back of the control where the bimetallic element enters the switch and attaches to the switch cam or dial; then recheck the resistance across the safety limit side of the switch. If the switch is still open, it is defective and requires replacement of the control.*

---

10. Remove the combination switch from the furnace (*if applicable*). ❏
11. *Look at pointer number 3 of the switch and record the set safety limit temperature of the switch.* _____ ❏
12. If not already set, firmly push in the combination switch dial and move pointer number 3 to a temperature of 175 degrees (F). ❏
13. Place the ohmmeter leads across the safety limit switch terminals again. ❏
14. Turn the heat gun on high and heat the bimetal element of the combination switch. ❏
15. Continue to apply heat until the temperature dial rates to indicate a temperature of 175 degrees (F) and the tripping open of the switch is heard. ❏
16. *Record your ohms value reading once limit switch trips open.* _____ ❏

NOTE: *In this test, a good limit switch would have shown infinity ohms reading once the bimetal element reached the set limit temperature. If the limit switch did not open once the bimetal element reached the set limit temperature, the safety limit switch is defective or not working properly. This could be due to the fact that the bimetallic element is hot enough to open or that the bimetallic element is dirty and sticking open, or that the switch is defective. Before replacing the switch, allow it to cool down if applicable and carefully clean the bimetallic element. Add a little lubricating oil at the back of the control where the bimetallic element enters the switch and attaches to the switch cam or dial; then recheck the resistance across the safety limit side of the switch. If the safety limit switch still does not open when heated again, it is defective and requires replacement of the control.*

---

17. Leave the ohmmeter leads across the safety limit switch terminals and allow the bimetal element to cool. ❑

18. Once the safety limit switch cools below the 175-degree (F) tab setting, a click should be heard and the ohmmeter should have shown a (0) ohms value reading. ❑

> **NOTE:** *Once the safety limit switch cools to a temperature below the 175-degree (F) setting, the switch should close again. If it does not, before condemning the switch, carefully clean the bimetallic element. Add a little lubricating oil at the back of the control where the bimetallic element enters the switch and attaches to the switch cam or dial; then recheck the resistance across the safety limit side of the switch. If the safety limit switch still does not close when allowed to cool, it is defective and requires replacement of the control.*

19. Reset the safety limit switch to the temperature recorded in procedure 11 by pushing firmly in on the dial cam and moving pointer tab number 3 to the correct temperature. ❑

CHECKING THE FAN SWITCH

20. *Record the preset temperature of tab pointer number 1.* _____ ❑
21. *Record the preset temperature of tab pointer number 2.* _____ ❑

> **NOTE:** *Pointer tab number 1 is the furnace blower OFF temperature and pointer tab number 2 is the furnace blower ON temperature.*

22. Place the ohmmeter leads across the fan switch terminals. ❑

> **NOTE:** *If the fan–limit switch has a manual and auto fan button, make sure the switch is pulled out for auto fan operation for this test. Combination switches with this feature allow the customer to operate the furnace blower for circulating air only, without having the furnace operating in the heating or air conditioning cycle.*

23. *Record your ohms value reading.* _____ ❑

> **NOTE:** *The fan switch is a NO (normally open) switch. A good fan switch would have shown an infinity ohms reading in this test. If you recorded any ohms value reading on the fan switch during this test, the switch contacts are closed when they should be open. This means that the furnace blower would operate all the time. If this is a combination switch, which has a Fan-On—Auto Fan button, make sure the fan button is pulled out. If it is and the fan switch still indicates a measured resistance, the switch is closed when it should be open, indicating a defective switch. This would require replacement of the control.*

24. If the fan switch is OK, keep the ohmmeter leads on the fan–limit switch terminals. ❑
25. Use the heat gun and apply heat to the bimetal element. ❑
26. Observe the bimetal element cam dial temperature scale and number 2 tab pointer setting. ❑
27. *When a "click" is heard, take notice to the temperature on the cam dial and record the approximate temperature at which the "click" was heard.* _____ ❑
28. *Record the ohms value reading once the fan–limit switch closes.* _____ ❑

NOTE: *A good fan–limit switch would have shown an infinity ohms value reading when the bimetal element was cool and a measured ohms value when the bimetal element was heated above the set fan–limit temperature of pointer tab number 2. If the limit switch did not close once the bimetal element was heated above the fan–limit set temperature of tab pointer number 2, the fan switch is defective or out of calibration. Before condemning the switch, carefully clean the bimetallic element. Add a little lubricating oil at the back of the control where the bimetallic element enters the switch and attaches to the switch cam or dial, then perform the test again. If the fan switch still does not close when heated, it is defective and requires replacement of the control.*

29. When the bimetal element was heated, did the fan switch close? _____ ❏
30. *According to the dial temperature scale, at what approximate temperature did the fan switch close? _____* ❏
31. *Was this within 10 degrees of the set temperature of pointer tab number 2? _____* ❏

NOTE: *If the temperature calibration of the control is out of range of 10 to 20 degrees or more, the control should be replaced.*

32. Leave the ohmmeter leads on the fan switch terminals and allow the bimetal element to cool to a point at which the fan switch contacts open. ❏
33. Observe the bimetal element cam dial temperature scale and number 1 tab pointer setting. ❏
34. *When a "click" is heard, take notice to the temperature on the cam dial and record the approximate temperature at which the "click" was heard. _____* ❏
35. *Record the ohms value reading once the fan–limit switch opens. _____* ❏

NOTE: *Once the fan–limit switch opens, an infinity ohm reading should be indicated on the meter. If a measured resistance is indicated, the fan switch contacts did not open which would allow the fan to operate all the time. Before condemning the switch, carefully clean the bimetallic element. Add a little lubricating oil at the back of the control where the bimetallic element enters the switch and attaches to the switch cam or dial; then perform the test again. If the fan switch still does not open when cooled, it is defective and requires replacement of the control.*

36. When the bimetal element was heated, did the fan switch close? _____ ❏
37. *According to the dial temperature scale, at what approximate temperature did the fan switch open? _____* ❏
38. *Was this within 10 degrees of the set temperature of pointer tab number 1? _____* ❏

NOTE: *If the temperature calibration of the control is out of range of 10 to 20 degrees, or more, the control should be replaced.*

39. Replace the fan–limit switch back into the furnace *(if applicable)*. ❏
40. Re-hook the leads wires to the limit switch and fan switch *(if applicable)*. ❏
41. Have the instructor check your work. ❏
42. Disconnect equipment and return all tools and test equipment to their proper location. ❏

RESEARCH QUESTIONS

1. What are the three functions of the fan–limit switch?

   _____

   _____

2. What could cause the fan–limit to shut the unit down on the safety switch function?

   _____

   _____

3. Why is the fan delayed from coming on right away on a call for heat from the thermostat?

   _____

   _____

4. If the furnace had an A/C system added on to the furnace, does the fan–limit switch bring the blower on when there is a call for cooling by the thermostat?

   _____

   _____

5. The jumper tab in the center of the fan–limit switch would be removed when?

   _____

   _____

Passed Competency _____   Failed Competency _____

Instructor Signature _____   Grade _____

Practical Competency 142

Setting and Checking a Fan Limit Temperature Switch

SUGGESTED MATERIALS

Textbook
Refrigeration & Air Conditioning Technology, 5th Edition, Thomson Delmar Learning
Unit 31—Gas Heat
Unit 32—Oil Heat

Review Topics
Fan–limit; Safety Limit; Combination Fan–Safety Limit Switch

COMPETENCY OBJECTIVE

The student will be able to set and check temperature limits on a combination fan ON/OFF and safety limit switch.

OVERVIEW

Setting the fan ON and OFF temperature is done by adjusting the tabs on the temperature dial of the fan–limit switch (**Figure 5–72**).

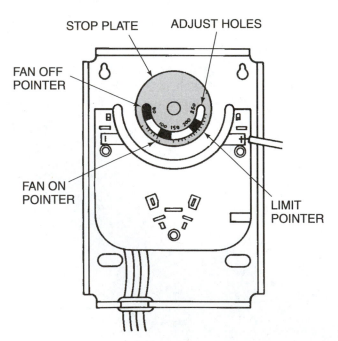

FIGURE 5–72   Fan ON–OFF and limit safety switch pointers.

This is also where the high-limit temperature is set. When setting the temperature, care should be taken to avoid turning the temperature dial. The pointer temperature indicators can be moved by holding the dial stationary and moving the temperature indicators to the desired temperature setting.

The first tab is to be set at the fan OFF temperature. The fan will stop after the burner has shut OFF at this temperature. In most cases, the fan OFF temperature setting is 90–100 degrees (F). The second tab is the setting for the fan ON temperature. This is the air temperature reached from the heating of the heat exchanger during the beginning of the call for heat from the thermostat. In most cases the fan ON temperature is set at 125–150 degrees (F). The third tab is the setting for the safety limit switch temperature. The limit switch will open and shut the burner down in case of a high-limit situation at this temperature. Different state codes have different safety limit temperature cutout requirements; normally the safety limit temperature is set between 200 and 250 degrees (F).

As stated, these are recommended temperature settings. Some manufacturers may require different settings for their equipment. Always check the manufacturer's literature for recommended temperature settings on the fan–limit.

EQUIPMENT REQUIRED

Temperature meter
Standard screwdriver
Phillips screwdriver
Wire nut
Roll of electrical tape

SAFETY PRACTICES

Be sure to follow all electrical safety rules when working with live circuits. Make sure electrical test equipment is set to proper function and ranges.

COMPETENCY PROCEDURES   Checklist

> NOTE: *To complete this competency, the fan–limit switch should be tested in equipment that is operational. The instructor may want to move all the fan–limit settings to give the student the opportunity to reset fan and safety limit temperature settings.*

1. Turn power OFF to furnace.   ❏
2. Remove the combination fan–safety switch cover.   ❏

> NOTE: *The temperature settings in this competency are only recommended settings. Different states and equipment manufacturers may require different settings to be used. Follow the state or manufacturer's temperature settings if different from those of this competency.*

> NOTE: *When setting the tab pointers on the temperature dial, firmly hold the dial in a stationary position while making settings. Never rotate the dial by hand.*

3. Firmly hold the temperature dial and slide the first pointer (*Number 1*) to align with a
   temperature of 90 degrees (F).   ❏
4. Firmly hold the temperature dial and slide the second pointer (*Number 2*) to align with a
   temperature of 125 degrees (F).   ❏
5. Firmly hold the temperature dial and slide the third pointer (*Number 3*) to align with a
   temperature of 225 degrees (F).   ❏
6. Locate a temperature meter lead above the heat exchanger of the furnace.   ❏
7. Turn power ON to furnace.   ❏
8. Set the thermostat and temperature control to call for heat.   ❏

9. *Record the return air temperature going across the heat exchanger when the burner starts up.* _____ ❑

10. *Record the supply air temperature going over the heat exchanger once the furnace blower comes ON.* _____ ❑

11. Set the thermostat temperature setting back until the burner stops. ❑

12. *Record the air temperature going over the heat exchanger once the blower of the furnace stops.* _____ ❑

13. *Did the temperature that you recorded when the furnace blower came on match the temperature setting of the tab pointer number 2?* _____ ❑

> **NOTE:** *Your temperature reading and the set temperature may be off a couple of degrees depending on the location of the temperature probe. If the recorded temperature reading and set temperature of the pointer are off by more than 10 degrees, the fan–limit may be defective and need to be replaced. Clean fan–limit and check again.*

14. *Did the temperature recorded at the time blower shut OFF match the temperature set at pointer tab 1 on the limit switch?* _____ ❑

> **NOTE:** *Your temperature reading and the set temperature may be off a couple of degrees depending on the location of the temperature probe. If the recorded temperature reading and set temperature of the pointer are off by more than 10 degrees, the fan–limit may be defective and need to be replaced. Clean the fan–limit and check again.*

15. Turn power OFF to furnace. ❑

16. Remove one of the power lead wires of the fan switch terminals (*normally a black wire*) on the fan–limit switch. (*Fan switch wires are located on the left of the combination switch.*) ❑

17. Put a wire nut or use electrical tape to tape the end of this wire. ❑

18. Turn power back ON to the furnace. ❑

19. Move the thermostat temperature setting to call for heat. ❑

20. *Watch the temperature meter and record the temperature once the burner of the furnace shuts down.* _____ ❑

21. *Did the blower come on in this test?* _____ ❑

22. Did the record temperature at which the burner shut OFF match the temperature setting of the tab pointer number 3? _____ ❑

> **NOTE:** *In this test, the blower should not have come ON and the burner of the furnace should have shut OFF once the air temperature crossing over the heat exchanger reached the temperature of the third tab pointer setting.*

> **NOTE:** *Your temperature reading and the set temperature may be off a couple of degrees depending on the location of the temperature probe. If the recorded temperature reading and set temperature of the pointer are off by more than 10 degrees, the fan–limit may be defective and need to be replaced. Clean the fan–limit and check again.*

23. Turn power OFF to furnace. ❑

24. Remove the tape or wire nut from the fan wire that was removed from the fan–limit switch terminal and reinstall the wire. ❑

25. Replace the fan–limit switch cover.

26. Have the instructor check your work.

27. Disconnect equipment and return all tools and test equipment to their proper location.

RESEARCH QUESTIONS

1. Which pointer setting on the fan–limit switch turns the fan ON once the air temperature passing over the heat exchanger has gotten hot enough to start circulating air through the conditioned space?

   _____

   _____

2. Which pointer setting turns the fan OFF once the air temperature passing over the heat exchanger has cooled enough?

   _____

   _____

3. Which pointer settings act as a safety temperature setting?

   _____

   _____

4. Besides the blower not operating, what other factors could cause the heat exchanger to overheat and trip the safety limit switch?

   _____

   _____

5. Can the temperature range of bimetal limit switches be adjusted?

   _____

   _____

Passed Competency _____ **Failed Competency** _____

Instructor Signature _____ **Grade** _____

Theory Lesson: Stacked-Mounted Safety Control (Figure 5–73)

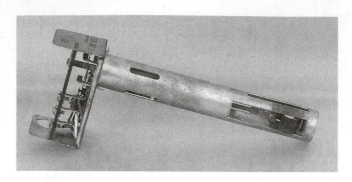

FIGURE 5–73   Stack switch.
(*Photo by Bill Johnson*)

SUGGESTED MATERIALS

Textbook
Refrigeration & Air Conditioning Technology, 5th Edition, Thomson Delmar Learning
Unit 32—Oil Heat

Review Topics
Stack Switch Safety Controls

Key Terms
bimetal element • cold contacts • drive shaft • hot contacts • safety lock out • safety switch heater • stack-mounted safety control • thermal detection

OVERVIEW

Stack switch safety controls perform the same function as the burner-mounted primary controls, with the only difference being how they detect and prove the presence of the burner flame. Current cad cell primary controls use a cad cell eye to detect flame and are much faster at proving burner flame than the stack switch safety control. Stack switch safety controls use thermal detection to prove that ignition of the atomized oil has taken place. This type of safety control is inserted in the furnace stack pipe near the breech of the furnace. This permits the stack safety control to sense the temperature of flue gases leaving the combustion chamber.

When the thermostat first calls for heat, the stack switch safety control checks to make sure all start-up conditions are met and then signals the burner motor to start. Ignition occurs and flame is established. The bimetal element in the stack switch safety control expands in response to the rise of the stack temperature. Turning the bimetal element pushes the detector drive shaft, causing the contacts of the stack switch safety control to open and close (**Figure 5–74**).

When the stack of the furnace is cold, the **cold contacts** of the stack switch safety control **must be closed** at the beginning of the heat cycle and the **hot contacts** of the stack switch safety control **must be open** (**Figure 5–75**).

As the stack temperature rises, the **hot contacts close** before the **cold contacts open**. This proves that the flame is properly established in the combustion chamber, allowing the burner to continue to operate until the thermostat is satisfied (**Figure 5–76**).

This happens when the flue gas temperature reaches approximately 600 degrees (F). At the same time, the **hot contacts close** and the **cold contacts open**. The furnace blower is also turned on when the hot contacts close.

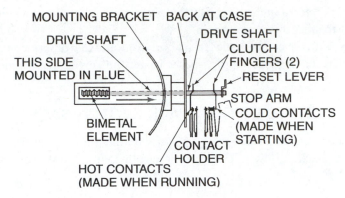

FIGURE 5–74   An illustration of a stack switch.

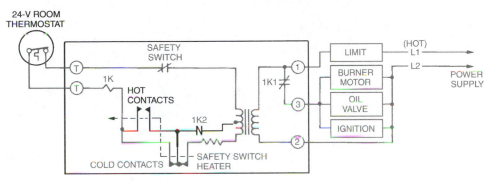

FIGURE 5–75   Stack switch circuit when the burner first starts. Note current flow through the safety switch heater.

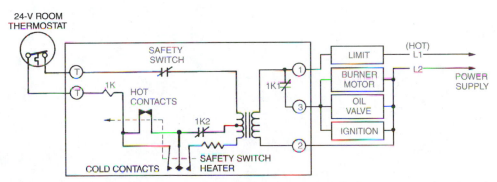

FIGURE 5–76   Typical stack switch circuitry with flame ON, hot contacts closed, and cold contacts open. The safety switch heater is not in the circuit.

When the thermostat opens the circuit to the burner, and as the stack cools, the bimetal element reverses, allowing the stack switch safety control drive shaft to reset the hot and cold contacts. If the fuel oil fails to ignite for any reason, the thermal element detector cannot expand without heat, so the cold contacts remain closed and the hot contacts remain open. This permits current to flow through the safety switch heater.

Once the heater heats the bimetal element to a set temperature it warps open and shuts down the system on safety lock out. This normally happens after about 70 seconds of burner operation. In this situation, the hot and cold contacts must be reset manually.

If the flame fails during burner operation, the stack temperature would cool to a temperature at which the hot contacts would open, shutting down the burner. After 2 or 3 minutes the cold contacts close and the safety switch heater begins to heat. If the thermostat is still calling for heat, the furnace will attempt to re-ignite the flame; if not detected in the established in 70 seconds, the safety switch locks out on safety and must be manually reset to start the burner again.

Stack switch safety controls can be replaced by burner-mounted primary controls. Technicians should check with the manufacturer or parts supply houses to get the correct replacement burner-mounted primary control.

RESETTING THE CONTACTS OF THE STACK SWITCH SAFETY CONTROL

Reset the hot and cold contacts on a stack relay by gently pulling the reset lever arm out about 3/4" slowly and release the lever arm back in slowly. This closes the cold contacts and opens the hot contacts.

Practical Competency 143

Checking the Operational Components of a Stack Switch Safety Control

SUGGESTED MATERIALS

Textbook
Refrigeration & Air Conditioning Technology, 5th Edition, Thomson Delmar Learning
Unit 32—Oil Heat

Review Topics
Stack Switch Safety Control

COMPETENCY OBJECTIVE

The student will be able to check the operation of a stack switch safety control components.

OVERVIEW

When the thermostat first calls for heat, the stack switch safety control checks to make sure all start-up conditions are met and then signals the burner motor to start. Ignition occurs and flame is established. The bimetal element in the stack switch safety control expands in response to the rise of the stack temperature. Turning the bimetal element pushes the detector drive shaft, causing the contacts of the stack switch safety control to open and close. When the stack of the furnace is cold, the cold contacts of the stack switch safety control must be closed at the beginning of the heat cycle and the hot contacts of the stack switch safety control must be open (*Theory Lesson: Stacked-Mounted Safety Control*).

As the stack temperature rises, the hot contacts close before the cold contacts open. This proves that the flame is properly established in the combustion chamber, allowing the burner to continue to operate until the thermostat is satisfied. This happens at around 600 degrees (F) flue gas temperature. At the same time the hot contacts close and the cold contacts open, the blower fan of the furnace is also turned on by the hot contacts.

When the thermostat opens the circuit to the burner, and as the stack cools, the bimetal element reverses, allowing the safety switch control drive shaft to reset the hot and cold contacts. If the fuel oil fails to ignite for any reason, the thermal element detector cannot expand without heat so the cold contacts remain closed and the hot contacts remain open. This permits current to flow through the safety switch heater.

Once the heater heats the bimetal element to a set temperature it warps open and shuts down the system on safety lock out. This normally happens after about 70 seconds of burner operation. In this situation, the hot and cold contacts must be reset manually.

If the flame fails during burner operation, the stack temperature would cool to a temperature at which the hot contacts would open, shutting down the burner. After 2 or 3 minutes the cold contacts close and the safety switch heater begins to heat. If the thermostat is still calling for heat, the furnace will attempt to re-ignite the flame, but if it is not established within 70 seconds, the safety switch locks out on safety and must be manually reset to start the burner again.

Stack switch safety controls can be replaced by burner-mounted primary controls. Technicians should check with the manufacturer or parts supply houses to get the correct replacement burner-mounted primary control.

EQUIPMENT REQUIRED

Stack switch safety control
Ohmmeter or VOM
Phillips screwdriver
Standard screwdriver
Piece of sanding cloth

SAFETY PRACTICES

Be sure to follow all electrical safety rules when working with live circuits. Make sure electrical test equipment is set to proper function and ranges.

COMPETENCY PROCEDURES

Checklist

> *NOTE: For this competency, the stack switch safety control may or may not be in an operating furnace system.*

1. Turn off the power to the furnace (*if applicable*). ❏
2. Remove the stack switch cover. ❏
3. Make sure that the hot and cold contacts are put in step. ❏

> *NOTE: Set the hot and cold contacts on a stack relay by gently pulling the reset lever arm out about 3/4" slowly and release the lever arm back in slowly. This closes the cold contacts and opens the hot contacts.*

4. Set the ohmmeter to a low ohms scale or buzzer setting. ❏
5. Touch the meter leads across the cold contact support blades. ❏
6. *Did you get an ohms value reading?* _____ ❏

> *NOTE: A good set of closed cold contacts would show an ohms value reading in the above test.*

7. Touch the meter leads across the hot contact support blades. ❏
8. *Did you get an ohms value reading?* _____ ❏

> *NOTE: A good set of opened hot contacts would show infinity ohms value reading in the above test.*

9. Remove the field wiring from terminals 1 and 2 of the stack relay. ❏
10. Place the ohmmeter leads across the stack relay terminals 1 and 2. ❏
11. *Record your ohms value reading.* _____ ❏

> *NOTE: In the above test, you tested the primary coil of the step-down transformer of the stack relay. A recorded ohms reading in this test would tell you that the primary side of the transformer was good. An infinity ohms value reading would tell you that the primary coil of the transformer was defective. This would require replacement of the stack relay.*

12. Place the ohmmeter leads across terminals 1 and 3. ❏
13. *Record your ohms value reading.* _____ ❏

NOTE: *In this test you were checking the 1K-1 set of normally opened contacts. This is the set of contacts used to bring on the burner, transformer, and oil valve if used once the thermostat called for heat. In this test with no power applied to the stack relay, you should have gotten an infinity reading. If an ohms value was indicated, the contacts are welded together. You could try to free the contacts or replace the stack relay.*

14. Remove the thermostat wires from the stack relay terminals marked **T–T** or **R–W–B**. ❑
15. Place the ohmmeter leads across the **T–T** or **R–W** of the stack relay. ❑
16. *Record your ohms value reading.* _____ ❑

NOTE: *A recorded ohms value reading in this test would tell you that the 1K relay coil, the cold contacts, the safety switch heater, the safety switch contact, and the secondary of the step-down transformer were all OK. No recorded ohms value reading in this test would tell you that one of the above components of the stack relay was defective. Try resetting the contacts before condemning the stack relay.*

17. Replace all wiring to the stack relay primary control. ❑
18. Replace the stack relay cover. ❑
19. Have the instructor check your work. ❑
20. Disconnect equipment and return all tools and test equipment to their proper location. ❑

RESEARCH QUESTIONS

1. What is the purpose of the stack switch safety control?

_____

_____

2. What system has been designed to replace the stack switch safety control?

_____

_____

3. What advantage does a cad cell primary control system have over the stack switch safety control?

_____

_____

4. Which set of contacts on the stack switch safety control was used to get the burner or gas valve going on start-up?

_____

_____

5. Which set of contacts were used on the stack switch safety control to keep the burner going?

_____

_____

Passed Competency _____ Failed Competency _____

Instructor Signature _____ Grade _____

Practical Competency 144

Measuring and Setting the Heat Anticipator (Figure 5–77)

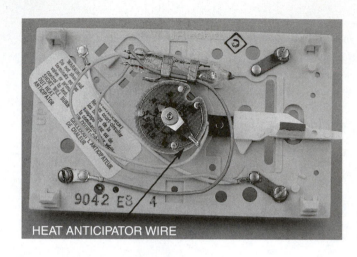

HEAT ANTICIPATOR WIRE

FIGURE 5–77 Heat anticipator wire. (*Photo by Bill Johnson*)

SUGGESTED MATERIALS

Textbook
Refrigeration & Air Conditioning Technology, 5th Edition, Thomson Delmar Learning
Unit 14—Automatic Control Components and Applications
Unit 32—Oil Heat

Review Topics
Thermostats; Heat Anticipator

COMPETENCY OBJECTIVE

The student will able to measure current draw of heating system and correctly set the heat anticipator.

OVERVIEW

The **mercury bulb thermostat** is the most common low-voltage thermostat used for heating and cooling applications (**Figure 5–78**), although solid-state heating and cooling thermostats are taking the lead as replacements for mercury bulb type thermostats. In mercury bulb type thermostats the **mercury bulb** is used to make and break the electrical circuits contacts. The mercury bulb is fastened to a wound spiral bimetal coil. The **bimetal coil** will move in response to temperature change. This causes the mercury in the bulb to move back and forth over the electrical probes inserted in the bulb. As the mercury settles over the particular probes in the bulb, a 24-volt electrical circuit is completed across the probes. This action completes the circuit for a particular function of the heating or cooling equipment.

The mercury bulb thermostats are designed with a **heat anticipator** that is designed to shut the heating equipment off a couple of degrees, normally 2 degrees (F), before the thermostat is satisfied at the desired temperature. This is to allow the additional heat in the heat exchanger to be used to bring the temperature of the conditioned space up to the thermostat set point. If this were not incorporated in the thermostat, there would be an overshoot of the conditioned space temperature.

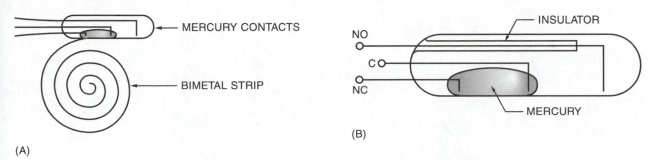

FIGURE 5–78 Bimetal spiral and mercury bulb.

The **heat anticipator** is a small coil of wire that is wired in series with the heating contacts of the thermostat (**Figure 5–79**). This allows a small amount of current to pass through the heat anticipator coil, which heats the thermostat bimetal, a couple degrees higher than the space temperature. The current flow through the heat anticipator coil must be set at the proper setting for the heating equipment being used (**Figure 5–80**).

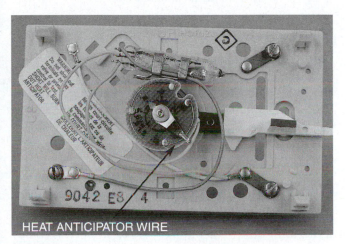

FIGURE 5–79 Heat anticipator wire. (*Photo by Bill Johnson*)

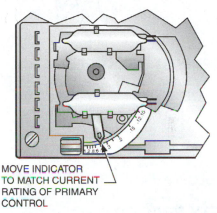

FIGURE 5–80 Adjustable heat anticipator. The indicator must match the current rating of the heating control circuit. (*Courtesy of Honeywell, Inc.*)

The **cooling anticipator** is designed to prevent the conditioned space temperature from rising higher than the desired temperature setting before turning on the air conditioning equipment. The cooling anticipator coil or resistor is wired parallel with the cooling thermostat contacts.

When the thermostat is in the OFF position, a small amount of current passes through the resistor or anticipator coil, heating the bimetal coil a couple of degrees higher than the room temperature. This creates a false temperature reading by the thermostat and brings the air conditioning equipment on a couple of degrees lower than the thermostat setting.

Cooling anticipators are preset within the thermostat and do not require any field adjustment, although the heat anticipator of the thermostat is not preset and requires field adjustment by technicians during installation of thermostat replacement of the same type. Because the heating anticipator of mercury bulb thermostats is adjustable, technicians must determine what the current flow through the heating control circuit is for the particular heating unit being controlled by the thermostat. Sometimes this information can be found on the primary control of an oil burner, or gas valve of a gas furnace. If the this information is not available, one of the most accurate ways to measure current flow through the heating anticipator is with a clamp-on ammeter, anticipator meter or modern digital meter, between the **R** and **W** terminals of the thermostat (**Figure 5–81**).

A lower anticipator setting will cause the burner to short cycle; with higher anticipator settings the burner will run longer.

Solid-state thermostat are self-setting and do not require any anticipator adjustment due to the built-in microprocessor and software that will determine the cycle rate.

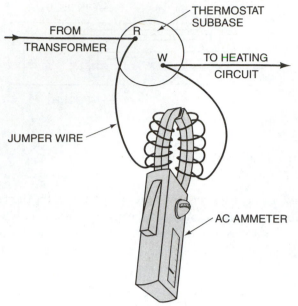

FROM TRANSFORMER

THERMOSTAT SUBBASE

R

W

TO HEATING CIRCUIT

JUMPER WIRE

AC AMMETER

FIGURE 5–81 An AC ammeter measuring current through the heating circuit. Often, ten turns of wire wrapped around an ammeter will give a more accurate reading. Do not forget to divide the actual reading by 10. Modern digital ammeters can read amperages without the wire turns.

EQUIPMENT REQUIRED

Clamp-on ammeter (*if applicable*)
Anticipator meter (*if applicable*)
Digital VOM (*if applicable*)
18" Piece of electrical wire (*if applicable*)
Standard screwdriver
Temperature tester

SAFETY PRACTICES

Be sure to follow all electrical safety rules when working with live circuits. Make sure electrical test equipment is set to proper function and ranges.

COMPETENCY PROCEDURES

Checklist

1. Turn OFF power to furnace. ☐
2. Gain access to the thermostat terminals ☐
3. If using a clamp-on ammeter use the electrical wire and coil it around one of the jaws of the meter 10 times. ☐
4. Attach one end of the coil of wire from around the ammeter jaw to the **R** terminal of the thermostat. ☐
5. Attach the opposite end of the coil of wire around the ammeter jaw to the **W** terminal of the thermostat. ☐

> **NOTE:** *If using an anticipator meter or digital meter, set to proper function and scale range. Attach meter leads to the **R** and **W** terminals of the thermostat.*

6. Observe the meter when ignition takes place. ☐
7. *Record your ammeter reading.* _____ ☐
8. *Divide your ammeter reading from procedure 7 by 10 and record the actual amperage draw of heating circuit.* _____ ☐

> **NOTE:** *This is the heat anticipator setting number.*

9. Find the heating anticipator on the thermostat. ❏
10. Set the thermostat heat anticipator by moving the pointer to the correct amperage value. ❏
11. Remove meter leads from **R** and **W** thermostat terminals. ❏
12. Replace the thermostat panel. ❏
13. Set the thermostat to the OFF position. ❏
14. Turn power ON to furnace. ❏
15. Set thermostat to call for heat. ❏
16. *Measure and record space temperature.* _____ ❏
17. Set temperature selector 3 degrees (F) above conditioned space temperature. ❏
18. Let furnace operate until thermostat shuts furnace OFF. ❏
19. *Measure and record space temperature.* _____ ❏
20. *Did the thermostat allow the furnace to bring space temperature up to within a 1 degree of thermostat set temperature?* _____ ❏

NOTE: *If the thermostat did not turn furnace OFF to within 1 degree (plus or minus) of thermostat set temperature, the heat anticipator may require further adjustment.*

NOTE: *A lower anticipator setting will cause the burner to short cycle; with higher anticipator settings the burner will run longer.*

---

21. Shut the power OFF. ❏
22. Have the instructor check your work. ❏
23. Disconnect equipment and return all tools and test equipment to their proper location. ❏

RESEARCH QUESTIONS

1. What is the purpose of the heat anticipator on a thermostat?

_____

_____

2. Is the heat anticipator wired in parallel or series with the heating contacts of the thermostat?

_____

_____

3. What would happen to the heating cycle of the thermostat if the heating element of the heat anticipator burnt out?

_____

_____

4. Does the heat anticipator give a "false heat" reading during the ON cycle or the OFF cycle of the furnace?

_____

_____

5. About how many degrees does the heat anticipator "fake" the thermostat out?

_____

_____

Passed Competency _____   Failed Competency _____

Instructor Signature _____   Grade _____

Theory Lesson: Smoke in Fossil Fuel Combustion

SUGGESTED MATERIALS

Textbook
Refrigeration & Air Conditioning Technology, 5th Edition, Thomson Delmar Learning
Unit 32—Oil Heat

Review Topics
Combustion

Key Terms
air leaks • CO_2 (carbon dioxide) • heat exchanger • insulator •smoke • soot • unburned carbon

OVERVIEW

Excessive smoke is created from insufficient excess air used in the combustion process of fossil fuels, and can cause smoke deposits, called soot, to collect on a furnace heat exchanger. This soot acts as an insulator on the heat exchanger of the furnace and affects the rate of heat transfer from the heat exchanger to the air medium being used to heat the conditioned space.

A 1/8" thick coating of soot on the heat exchanger wall has the same insulating ability as a 1" thick sheet of fiberglass. This insulating factor results in the waste of fuel. A 1/16 of an inch of soot can cause a 4.5% increase in fuel consumption. The more soot on the walls of a heat exchanger, the higher the fuel consumption and heat loss (**Figure 5–82**).

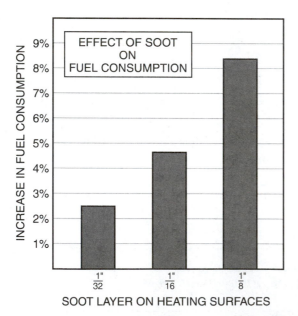

FIGURE 5–82   The effect of soot on fuel consumption. (*Courtesy of Bacharach, Inc., Pittsburgh, Pennsylvania*)

During normal combustion there will always be a certain amount of smoke because some of the oil droplets do not contact enough oxygen to complete the reaction that forms CO_2 (carbon dioxide). The smoke consists of small particles of unburned carbon, which can stick to the heat exchanger surface, acting as insulation and can eventually clog up the flue passages of the heat exchanger if not removed through cleaning periodically. The other particles of unburned carbon are emitted through the stack of the furnace and add to the pollution of the air.

Controlling the amount of smoke generated during the combustion process relies on sufficient excess air to provide good mixing of combustion air and fuel oil (**Figure 5–83**).

Without the proper amount of excess air, incomplete combustion occurs and smoke is generated (**Figure 5–84**).

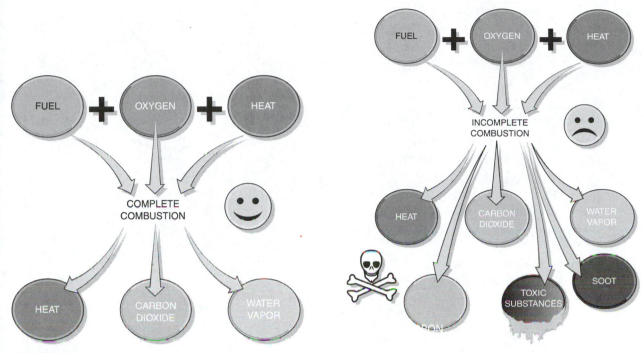

FIGURE 5–83   Complete combustion.

FIGURE 5–84   Incomplete combustion.

Lowering the amount of smoke normally requires the addition of more excess air into the combustion process, but unfortunately this can create other problems in the combustion process. As the excess air is increased, the transfer of heat from the furnace heat exchanger to the medium being used to heat the conditioned space is reduced. A delicate balance must be achieved between smoke generation and reduced heat transfer due to reduced combustion gas temperature caused by unnecessary excess air. Note that smoke and efficiency increase as the excess air is decreased. The highest efficiency occurs when there is a proper balance between the smoke and the excess air.

Air leaks that are in the combustion gases before they pass through the furnace heat exchanger act like excess air. This air will dilute the combustion gases, cooling them and increasing their volume so that they pass through the heat exchanger too quickly. This does not allow the air being conditioned to have sufficient contact time with the heat exchanger. Having air leaks is a worse situation than having excess air in the combustion chamber because air leaks will not reduce the smoke formed in the combustion area. During service and repair, every effort should be made to seal any air leaks that would allow additional air to enter the combustion area of the burner.

Excessive smoke in combustion and can be measured by performing a smoke test of the furnace combustion gases. This is accomplished by using a smoke tester (**Figure 5–85**).

A smoke test is accomplished by drawing a prescribed number of cubic inches of smoke-laden flue products through a specific area of filter paper. The residue on the filter paper is then compared with a scale furnished with the smoke tester. The degree of sooting can be read off the scale. A good smoke reading would be in the range Number 0 to Number 1. Smoke readings higher than this should be addressed by the technician, and attempts made to lower the number of smoke reading into this range or as close as possible.

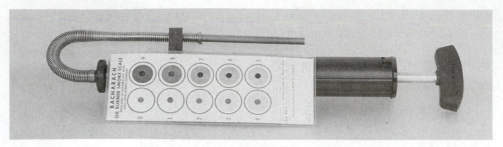

FIGURE 5–85   A smoke tester. (*Photo by Bill Johnson*)

The following can cause excessive smoke:

- Improper fan collar setting (*burner air adjustment*)
- Improper draft adjustment (*draft regulator may be required or need adjustment*)
- Poor oil supply (*pressure*)
- Oil pump not functioning properly
- Defective or incorrect type of nozzle
- Excessive air leaks in furnace (*air diluting flame*)
- Improper fuel-to-air ratio
- Defective firebox
- Improper burner air-handling parts

Practical Competency 145

Performing a Smoke Test on Fossil Fuel Equipment (Figure 5–86)

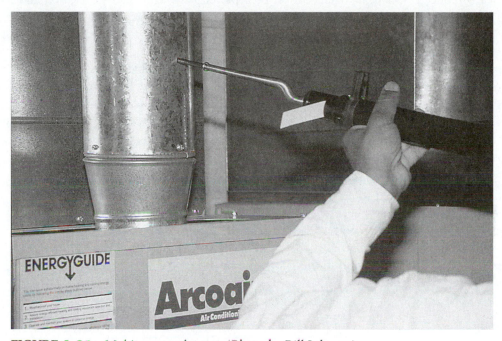

FIGURE 5–86   Making a smoke test. (*Photo by Bill Johnson*)

SUGGESTED MATERIALS

Textbook
Refrigeration & Air Conditioning Technology, 5th Edition, Thomson Delmar Learning
Unit 32—Oil Heat

Review Topics
Smoke Test

COMPETENCY OBJECTIVE

The student will be able to perform a smoke test on a fossil fuel furnace and evaluate the condition of excess air being used during combustion.

OVERVIEW

A 1/8" thick coating of soot on the heat exchanger wall has the same insulating ability as a 1" thick sheet of fiberglass. This insulating factor results in the waste of fuel. A 1/16 of an inch of soot can cause a 4.5% increase in fuel consumption. The more soot on the walls of a heat exchanger, the higher the fuel consumption and heat loss.

During normal combustion there will always be a certain amount of smoke because some of the oil droplets do not contact enough oxygen to complete the reaction that forms CO_2 (carbon dioxide). The

787

smoke consists of small particles of unburned carbon, which can stick to the heat exchanger surface, acting as insulation and can eventually clog up the flue passages of the heat exchanger if not removed through cleaning periodically. The other particles of unburned carbon are emitted through the stack of the furnace and add to the pollution of the air.

Controlling the amount of smoke generated during the combustion process relies on sufficient excess air to provide good mixing of combustion air and fuel oil. Without the proper amount of excess air, incomplete combustion occurs and smoke is generated.

Lowering the amount of smoke normally requires the addition of more excess air into the combustion process, but unfortunately this can create other problems in the combustion process. As the excess air is increased, the transfer of heat from the furnace heat exchanger to the medium being used to heat the conditioned space is reduced. A delicate balance must be achieved between smoke generation and reduced heat transfer due to reduced combustion gas temperature caused by unnecessary excess air. Note that smoke and efficiency increase as the excess air is decreased. The highest efficiency occurs when there is a proper balance between the smoke and the excess air.

A smoke test is accomplished by drawing a prescribed number of cubic inches of smoke-laden flue products through a specific area of filter paper. The residue on the filter paper is then compared with a scale furnished with the smoke tester. The degree of sooting can be read off the scale. A good smoke reading would be in the range Number 0 to Number 1. Smoke readings higher than this should be addressed by the technician, and attempts made to lower the number of smoke reading into this range or as close as possible.

The following can cause excessive smoke:

- Improper fan collar setting (*burner air adjustment*)
- Improper draft adjustment (*draft regulator may be required or need adjustment*)
- Poor oil supply (*pressure*)
- Oil pump not functioning properly
- Defective or incorrect type of nozzle
- Excessive air leaks in furnace (*air diluting flame*)
- Improper fuel-to-air ratio
- Defective firebox
- Improper burner air-handling parts

EQUIPMENT REQUIRED

Smoke tester
Smoke test paper
Smoke scale chart
Electric drill
3/8" Drill bit

SAFETY PRACTICES

Care should be taken to avoid touching the stack pipe of the furnace and the spout of the smoke tester. These areas will be hot.

COMPETENCY PROCEDURES

Checklist

1. Drill a 3/8" hole in the furnace stack pipe between the furnace breech or elbow and draft regulator. ❑
2. Start the furnace and let it operate for 15 minutes. ❑
3. Insert a piece of smoke test paper into the front of the smoke pump by loosening the nut at the front of pump and sliding a piece of smoke paper into the slot and retighten the nut. ❑
4. After furnace has operated for 15 minutes, insert the sampler tube into the 3/8" hole in the furnace stack pipe. (Refer to *Figure 5–86*.) ❑

5. Slowly withdraw the pump tester handle fully and hold for 3 seconds and then slowly push the smoke handle back in. ❏
6. Repeat the previous procedure ten (10) times. ❏
7. Remove the sampler tube from the stack pipe. ❏
   CAUTION: *Be careful not to touch the sampler tube; it will be hot.*
8. Remove the smoke test paper from the smoke pump. ❏
9. Hold the test paper behind the smoke scale so that the spot on the paper fills the center hole in the spot on the smoke scale. ❏
10. Match the smoke spot from the test paper to the closest scale on the smoke scale. ❏
11. *Record the matching smoke scale number of test paper.* _____ ❏
12. *Was smoke reading within acceptable range?* _____ ❏
13. *If the number of smoke was not in acceptable range, make necessary adjustments and recheck smoke.* ❏
14. Once smoke is in acceptable range, shut furnace OFF. ❏
15. Have the instructor check your work. ❏
16. Disconnect equipment, return all tools, and test equipment to their proper location. ❏

RESEARCH QUESTIONS

1. What is considered a good smoke test reading?

   _____

   _____

2. During the burning of oil, too much excessive air during the burning process will cause what to happen to the furnace performance?

   _____

   _____

3. Smoke is created in the burning process due to the lack of what?

   _____

   _____

4. An excessive smoke reading, is a good indication of what?

   _____

   _____

5. Describe the purpose of the heat exchanger.

   _____

   _____

Passed Competency _____   Failed Competency _____

Instructor Signature _____   Grade _____

Practical Competency 146

Overfire Draft Test (Figure 5–87)

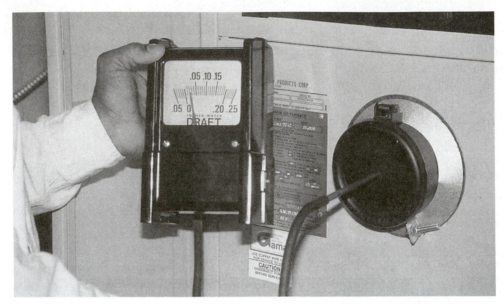

FIGURE 5–87   Checking overfire draft. (*Photo by Bill Johnson*)

SUGGESTED MATERIALS

Textbook
Refrigeration & Air Conditioning Technology, 5th Edition, Thomson Delmar Learning

Review Topics
Combustion; Draft; Overfire Draft

COMPETENCY OBJECTIVE

The student will be able to check the overfire draft of a furnace.

OVERVIEW

Draft is used to describe the slight vacuum, or suction, that exists inside most heating units. The measurement of the amount of vacuum is called draft intensity and is measured in inches of water by using a draft gage (**Figure 5–88**).

Draft volume is used to measure the volume gas in cubic feet that a chimney can handle in a given time. There are three types of draft: **natural draft, currential draft**, and **induced draft. Natural draft** works based on one of the Laws of the Thermodynamics of Heat, which states: "Heat will travel from a higher temperature to a lower temperature." This is how natural draft is created. Heated gas expands and the volume of the gas weighs less than an equal amount of the same gas at a cool temperature. Heated gas weighs less than the room air or outdoor air and therefore rises.

FIGURE 5–88　Draft gage. (*Courtesy of Bacharach, Inc., Pittsburgh, Pennsylvania*)

How fast these gases rise up the chimney depends on the temperature difference of the fuel gases and the outside air temperature. The speed of a volume of gas rising in the chimney creates a suction or vacuum that is referred to as the **draft**. Installing a blower into the furnace can increase combustion and increase the amount of draft. This is referred to as **induced draft**, direct venting, or power venting (**Figure 5–89**).

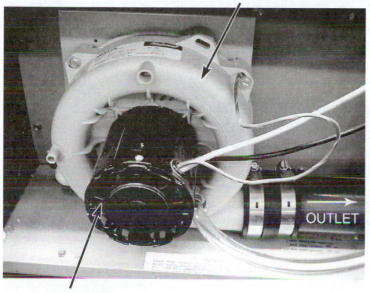

COMBUSTION
BLOWER HOUSING

OUTLET

COMBUSTION
BLOWER MOTOR

FIGURE 5–89　A combustion and direct vent motor assembly. (*Courtesy of Ferris State University. Photo by John Tomczyk*)

With a direct vent system, pressure is created by the oil burner fan motor to vent the products of combustion (**Figure 5–90**). The air introduced to the oil burner is supplied directly from outside and the products of combustion are immediately vented outside.

Power venting uses a blower motor that pulls the products of combustion from the appliance and vents them outside the structure (**Figure 5–91**).

Currential draft works according to the same Law of Heat that affects natural draft, except that the draft occurs when high winds or air currents moving across the top of a chimney creates suction in the stack and draws the flue gases up the chimney. The amount of draft a chimney can make is based on the following three factors:

A. *The height of the chimney*
B. *The weight per unit volume of the hot combustion products*
C. *The weight per unit volume of the air outside the building or home*

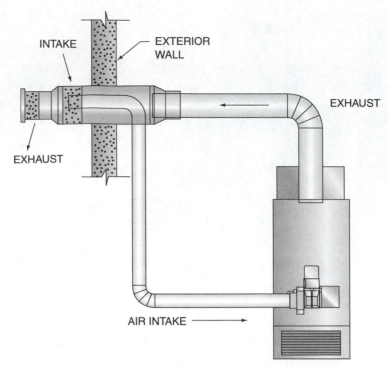

FIGURE 5–90   Pipe-in-pipe direct vent.

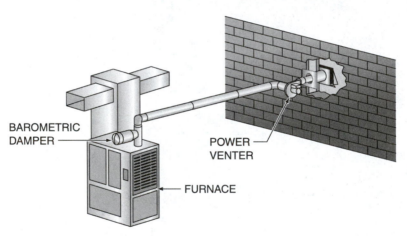

FIGURE 5–91   Power venting.

The higher the chimney, the greater the draft; the hotter the combustion gases, the greater the draft; and the cooler the outside air compared to the temperature of the combustion gases, the greater the draft. The amount of draft will change due to temperature changes of the outside air and the flue gas temperatures. After the furnace has operated for a while, the amount of draft will increase due to a temperature rise in the flue gases and the inner lining of the chimney.

Incorrect draft can create the following problems:

A. *Too little draft can reduce the combustion air delivery of the burner and can result in an increase in the amount of smoke that is created.*
B. *Excessively high draft increases the air delivery of the burner and can increase air leakage into the combustion chamber, reducing the amount of CO_2 (carbon dioxide), causing a rise in the stack temperature, and reducing the furnace operating efficiency.*
C. *High draft when the burner is off increases the stand-by heat loss up the chimney.*

During cold start-up, there will not be much draft for the furnace burner to obtain air to be used in the combustion process. The best way to ensure that the burner has enough air for combustion during cold start-up is to set the burner for a smoke-free combustion with a low overfire draft of .01 to .02 inches of water.

A constant draft is required at all times for proper burner operation and efficient combustion. With the advent of induced draft and forced draft furnaces, the proper draft is assured in most cases. With natural draft systems, draft will change based on flue gas temperatures and outside temperatures. To assist in providing the proper draft with these systems, a draft regulator or barometric damper can be installed in the flue pipe of the furnace (**Figure 5–92**).

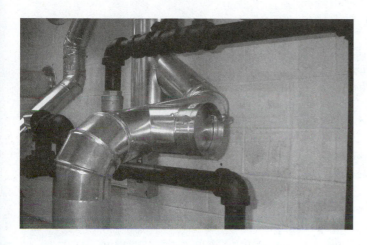

FIGURE 5–92   Draft regulator or barometric damper location. (*Photo by Eugene Silberstein*)

The function of the draft regulator is to maintain a stable or fixed draft through the furnace within the draft limits available for the chimney. Draft regulators are sized by the size of the outlet breech of the furnace.

Draft is measured by using a draft gage and should be measured at two locations of the furnace. The **first draft measurement** and most important reading and setting should be the overfire draft, which should be adjusted to measure .01 or .02 inches of water column. This draft level will be high enough to prevent leakage of combustion products into the home or building.

> *NOTE: If using a draft regulator on the furnace system and the overfire draft is higher than .02" of water column, adjust the regulator weight to allow the regulator door to open more. If the regulator door is adjusted wide open and the overfire draft is still higher than .02" of water column, install a second draft regulator and adjust to obtain an overfire draft of .01" or .02" of water column. If the overfire draft is less than .01" of water column, the draft regulator should be adjusted to close the regulator door. The door should not be closed any farther than required to obtain the correct overfire draft.*

The *second draft measurement* should be made at the breech connection of the furnace or stack draft. The draft reading at the breech of the furnace should be slightly higher than the overfire draft. A clean heat exchanger will cause the breech draft or stack draft to be in the range of .03" to .06" water column when the overfire draft is set at .01" to .02" water column.

EQUIPMENT REQUIRED

Draft gage
Electric drill
3/8" Drill bit

SAFETY PRACTICES

Care should be taken to avoid touching the draft spout of the draft gage once it is removed from the view port window of the furnace. The draft spout will be extremely hot.

COMPETENCY PROCEDURES

1. If it does not already exist, drill a 3/8" hole in the furnace view port window door. (Refer to *Figure 5–87*.) ❑

> NOTE: *Some furnaces are equipped with a hole already in the furnace view port window.*

2. Start the furnace and let it operate for 15 minutes. ❑
3. Set the draft gage on a level surface near the furnace and adjust the draft gage pointer to 0 inch reading on the draft scale. ❑
4. Insert the draft sampling tube into the 3/8" hole in the view port window to check the overfire draft of the furnace. (Refer to *Figure 5–87*.) ❑
5. *Record the overfire draft reading from the draft gage.* _____ ❑
6. Was the draft reading within .01" to .02" of water column. _____ ❑
7. If the draft reading was not in acceptable range, make air adjustments as need be to bring overfire draft into proper range. ❑
8. Remove the draft-sampling spout from the view port window. ❑
   CAUTION: The draft spout will be extremely hot.
9. Shut the furnace OFF. ❑
10. Plug the 3/8" hole in the view port window with a bolt or metal cap. ❑
11. Have the instructor check your work. ❑
12. Disconnect equipment and return all tools and test equipment to their proper location. ❑

RESEARCH QUESTIONS

1. Why is having the correct draft important in a furnace?

   _____

   _____

2. The draft determines what in a furnace?

   _____

   _____

3. Excessive overfire draft can increase what?

   _____

   _____

4. Insufficient draft will cause what to happen in the combustion chamber?

   _____

   _____

5. What other furnace draft reading affects the overfire draft?

   _____

   _____

Passed Competency _____ Failed Competency _____

Instructor Signature _____ Grade _____

Practical Competency 147

Performing a Stack Draft or Breech Test on Fossil Fuel Equipment

SUGGESTED MATERIALS

Textbook
Refrigeration & Air Conditioning Technology, 5th Edition, Thomson Delmar Learning
Unit 32—Oil Heat

Review Topics
Draft; Stack Draft–Breech Draft

COMPETENCY OBJECTIVE

The student will be able to measure the stack draft or breech draft of a furnace.

OVERVIEW

Draft is used to describe the slight vacuum, or suction, that exists inside most heating units. The measurement of the amount of vacuum is called draft intensity and is measured in inches of water by using a draft gage.

Draft volume is used to measure the volume gas in cubic feet that a chimney can handle in a given time. There are three types of draft: **natural draft, currential draft,** and **induced draft. Natural draft** works based on one of the Laws of the Thermodynamics of Heat, which states: "Heat will travel from a higher temperature to a lower temperature." This is how natural draft is created. Heated gas expands and the volume of the gas weighs less than an equal amount of the same gas at a cool temperature. Heated gas weighs less than the room air or outdoor air and therefore rises.

How fast these gases rise up the chimney depends on the temperature difference of the fuel gases and the outside air temperature. The speed of a volume of gas rising in the chimney creates a suction or vacuum that is referred to as the **draft.** Installing a blower into the furnace can increase combustion and increase the amount of draft. This is referred to as **induced draft,** direct venting, or power venting.

With a direct vent system, pressure is created by the oil burner fan motor to vent the products of combustion. The air introduced to the oil burner is supplied directly from outside and the products of combustion are immediately vented outside.

Power venting uses a blower motor that pulls the products of combustion from the appliance and vents them outside the structure.

The blower is located close to the flue pipe termination point, near the shell of the structure.

Currential draft works according to the same Law of Heat that affects natural draft, except that the draft occurs when high winds or air currents moving across the top of a chimney creates suction in the stack and draws the flue gases up the chimney. The amount of draft a chimney can make is based on the following three factors:

A. *The height of the chimney.*
B. *The weight per unit volume of the hot combustion products.*
C. *The weight per unit volume of the air outside the building or home.*

The higher the chimney, the greater the draft; the hotter the combustion gases, the greater the draft; and the cooler the outside air compared to the temperature of the combustion gases, the greater the draft. The amount of draft will change due to temperature changes of the outside air and the flue gas temperatures. After the furnace has operated for a while, the amount of draft will increase due to a temperature rise in the flue gases and the inner lining of the chimney.

Incorrect draft can create the following problems:

A. *Too little draft can reduce the combustion air delivery of the burner and can result in an increase in the amount of smoke that is created.*
B. *Excessively high draft increases the air delivery of the burner and can increase air leakage into the combustion chamber, reducing the amount of CO_2 (carbon dioxide), causing a rise in the stack temperature, and reducing the furnace operating efficiency.*
C. *High draft when the burner is off increases the stand-by heat loss up the chimney.*

During cold start-up, there will not be much draft for the furnace burner to obtain air to be used in the combustion process. The best way to assure that the burner has enough air for combustion during cold start-up is to set the burner for a smoke-free combustion with a low over fire draft of .01 to .02 inches of water.

A constant draft is required at all times for proper burner operation and efficient combustion. With the advent of induced draft and forced draft furnaces, the proper draft is ensured in most cases. With natural draft systems, draft will change based on flue gas temperatures and outside temperatures. To assist in providing the proper draft with these systems, a draft regulator or barometric damper can be installed in the flue pipe of the furnace.

The function of the draft regulator is to maintain a stable or fixed draft through the furnace within the draft limits available for the chimney. Draft regulators are sized by the size of the outlet breech of the furnace.

Draft is measured by using a draft gage and should be measured at two locations of the furnace. The **first draft measurement** and most important reading and setting should be the overfire draft, which should be adjusted to measure .01 or .02 inches of water column. This draft level will be high enough to prevent leakage of combustion products into the home or building.

> **NOTE:** *If using a draft regulator on the furnace system and the overfire draft is higher than .02" of water column, adjust the regulator weight to allow the regulator door to open more. If the regulator door is adjusted wide open and the overfire draft is still higher than .02" of water column, install a second draft regulator and adjust to obtain an overfire draft of .01" or .02" of water column. If the overfire draft is less than .01" of water column, the draft regulator should be adjusted to close the regulator door. The door should not be closed any farther than required to obtain the correct overfire draft.*

The *second draft measurement* should be made at the breech connection of the furnace or stack draft. The draft reading at the breech of the furnace should be slightly higher than the overfire draft. A clean heat exchanger will cause the breech draft or stack draft to be in the range of .03" to .06" water column when the overfire draft is set at .01" to .02" water column.

EQUIPMENT REQUIRED

Draft gage
Electric drill
3/8" Drill bit
2" Piece of aluminum duct tape

SAFETY PRACTICES

Care should be taken to avoid touching the draft spout of the draft gage once it is removed from the stack pipe of the furnace. The draft spout will be extremely hot.

COMPETENCY PROCEDURES   **Checklist**

1. Drill a 3/8" hole in the furnace stack pipe between the furnace breech or elbow and draft regulator. ❑

> *NOTE: There may already be a 3/4" test hole in the furnace stack pipe due to previous tests.*

2. Start the furnace and let it operate for 15 minutes. ❑
3. Set the draft gage on a level surface near the furnace and adjust the draft gage pointer to 0 inch reading on the draft scale. ❑
4. Insert the draft sampling tube into the 3/8" hole in the furnace stack pipe. ❑
5. *Record the stack draft reading from the draft gage. _____* ❑
6. *Was the stack draft within .03" to .06" of water column? _____* ❑
7. *If stack draft was not within desired range, make air adjustments and check again.* ❑
8. Once stack draft is within proper range, remove the draft-sampling spout from the stack pipe. ❑
   CAUTION: The draft spout will be extremely hot.
9. Shut the furnace OFF. ❑
10. Plug the 3/8" hole in the stack pipe with aluminum duct tape. ❑
11. Have the instructor check your work. ❑
12. Disconnect equipment and return all tools and test equipment to their proper location. ❑

RESEARCH QUESTIONS

1. The stack draft should always be higher than what other draft of the furnace?

   _____

   _____

2. A good breech draft should be in what range?

   _____

   _____

3. In the oil heating industry, what is a draft?

   _____

   _____

4. Draft intensity is measured in what scale?

   _____

   _____

5. What is a currential draft?

   _____

   _____

Passed Competency _____   Failed Competency _____

Instructor Signature _____   Grade _____

Practical Competency 148

Determining Net-Stack Temperature of Fossil Fuel Equipment (Figure 5–93)

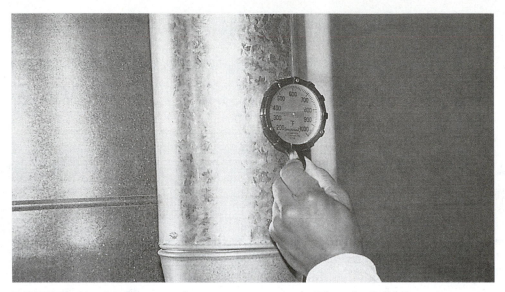

FIGURE 5–93   Making a stack gas temperature test. (*Photo by Bill Johnson*)

SUGGESTED MATERIALS

Textbook
Refrigeration & Air Conditioning Technology, 5th Edition, Thomson Delmar Learning
Unit 32—Oil Heat

Review Topics
Stack Temperature; Net-Stack Temperature

COMPETENCY OBJECTIVE

The student will able to check and determine the net stack temperature of a furnace.

OVERVIEW

The fuel and air normally enter the burner at the room temperature in which the furnace equipment is located. The temperature of the gases in the flue will vary from furnace to furnace and can be measured with a stack thermometer. The difference between the flue gas temperature and the furnace surrounding room temperature is called the net-stack temperature. The net-stack temperature is important because an abnormally high temperature is an indication that the furnace may not be operating efficiently as possible. Net-stack temperature is determined by subtracting the surrounding air temperature where the furnace is located from the actual furnace temperature. Stack temperature of the flue gas is also required in determining the efficiency of fossil fuel equipment.

The following can cause higher than normal stack temperature:

- Excessive draft through the combustion chamber
- Dirty or soot-covered heat exchanger
- Lack of baffling
- Incorrect or defective combustion chamber
- Overfiring (check nozzle size and pressure)

EQUIPMENT REQUIRED

Stack thermometer
Temperature meter
Electric drill
3/8" Drill bit
2" Piece of aluminum duct tape

SAFETY PRACTICES

Care should be taken to avoid touching the thermometer once it is removed from the stack pipe of the furnace. The thermometer will be extremely hot.

COMPETENCY PROCEDURES   Checklist

1. Drill a 3/8" hole in the furnace stack pipe between the furnace breech or elbow and
   draft regulator.   ❏

NOTE: *There may already be a test hole in the furnace stack pipe from previous testing.*

---

2. Start the furnace and let the burner operate for 15 minutes.   ❏
3. Insert the stack thermometer into the stack of the furnace and leave it there until the stack
   temperature reaches a steady state. (Refer to *Figure 5–93.*)   ❏

NOTE: *It is extremely important that the flue gas temperature is measured in a stabilized condition. This usually requires the furnace burner to operate for at least 15 minutes. The best way to determine if the flue gas temperature is stabilized is to watch the stack thermometer and when the temperature rises less than 5 degrees (F) during a 1-minute period, a stabilized condition exists.*

---

4. *Once the stack temperature reaches a steady state, record the stack temperature.* _____   ❏
5. *Use the temperature recorder and measure and record the air temperature around
   the furnace.* _____   ❏
6. Subtract the recorded temperature from around the furnace from the stack
   temperature.

   Stack temperature ____   ❏
   MINUS
   Air temperature around furnace ____   ❏
   EQUALS net-stack temperature _____   ❏

NOTE: *To determine if the furnace net-stack temperature is OK, you should compare the stack temperature reading with the manufacturer's specifications for the furnace you are working with.*

---

7. Remove the stack thermometer from the stack pipe.   ❏
   CAUTION: The stack thermometer will be hot.

8. Shut the furnace OFF. ❑
9. Plug the 3/8" hole with aluminum duct tape. ❑
10. Have the instructor check your work. ❑
11. Disconnect equipment and return all tools and test equipment to their proper location. ❑

RESEARCH QUESTIONS

1. What can the net stack temperature tell us about the furnace?

   _____

   _____

2. List a couple of things that could cause a higher than normal stack temperature.

   _____

   _____

3. List a couple of by-products of oil combustion.

   _____

   _____

4. Steady-state efficiency is a reference to what?

   _____

   _____

5. Describe three functions of the oil burner nozzle.

   _____

   _____

Passed Competency _____ Failed Competency _____

Instructor Signature _____ Grade _____

Practical Competency 149

Performing a Carbon Dioxide (CO_2) Test on Fossil Fuel Equipment (Figure 5–94)

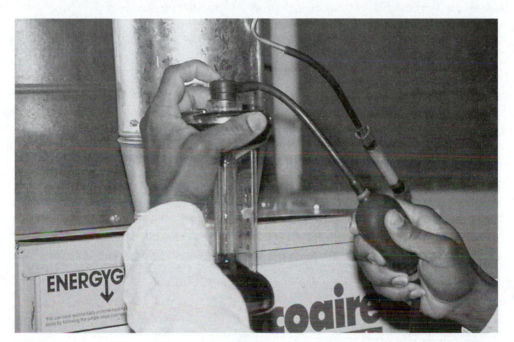

FIGURE 5–94 Using a CO_2 Fyrite analyzer to check CO_2 percentage. (*Photo by Bill Johnson*)

SUGGESTED MATERIALS

Textbook
Refrigeration & Air Conditioning Technology, 5th Edition, Thomson Delmar Learning
Unit 32—Oil Heat

Review Topics
Carbon Dioxide Test

COMPETENCY OBJECTIVE

The student will be able to check the CO_2 percentage of a fossil fuel furnace.

OVERVIEW

For combustion of the fuel oil, a proper mixture of fuel, oxygen, and a source of ignition are required (**Figure 5–95**).

The efficiency of the combustion process is determined by comparing the amount of useful heat produced with the total heat produced. For example, there are approximately 140,000 Btu's of heat available in a gallon of number 2 fuel oil, yet not all the heat produced provides heat to the structure; some of it is

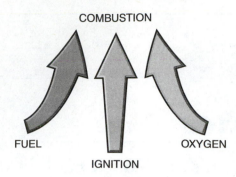

COMBUSTION

FUEL

IGNITION

OXYGEN

FIGURE 5–95 Combustion requires the proper mixture of fuel, oxygen, and some form of ignition.

lost up the chimney. A furnace that is operating at 85% efficiency is providing 85% of the heat produced to the structure, with the other 15% of heat produced being lost up the chimney.

Complete combustion is achieved when carbon that is released during the combustion process bonds with the oxygen from the air used for combustion and forms **carbon dioxide** (CO_2). **Carbon dioxide** *is non-toxic* and is safely vented to the atmosphere, along with other byproducts such as water vapor, nitrogen, and air, which are neither toxic nor poisonous.

The lack of the proper amount of oxygen during the combustion process affects the formation of the amount or percentage of CO_2, and carbon monoxide (CO) is created. **Carbon monoxide** *is a toxic and poisonous gas.* Carbon monoxide is a result of incomplete combustion.

The higher the CO_2 percentage, the less excess air being used in the combustion process. Therefore, it is important to keep the amount of excess air used for combustion of fossil fuels to a minimum to ensure a higher CO_2 percentage. Under most normal conditions, a carbon dioxide reading greater than 10% should be obtained. There are circumstances where a 10% or higher CO_2 percentage cannot be obtained. If this is the case, an 8% CO_2 would be acceptable as long as the stack furnace stack temperature was at 400 degrees or less. Low CO_2 percentages may be caused by one or more of the following:

- High draft or draft regulator not working properly
- Excess combustion air
- Air leakage into the combustion chamber
- Poor oil atomization
- Worn, clogged, or incorrect nozzle oil-pressure regulator set incorrectly

Too much excess air can cause the following:

    A. *Lower flame temperature*
    B. *Lower combustion gas temperature*
    C. *Higher flue stack gas temperature*
    D. *Poorer heat exchange to the air being conditioned*

EQUIPMENT REQUIRED

CO_2 Fyrite® analyzer (red fluid)
Electric drill
3/8" Drill bit
2" Piece of aluminum duct tape

SAFETY PRACTICES

Care should be taken to avoid touching the metal sampling tube once it is removed from the furnace stack pipe. The metal sampling tube will be extremely hot.

CAUTION: The Fyrite fluids used in the CO_2 and O_2 analyzers are corrosive and contain poisonous elements, which must not be taken internally. In the event of a spill or accidental body contact with Fyrite fluid, flood with water and then wash area with vinegar. If Fyrite fluid enters the eye(s), flood eye(s) with water, then wash with 5% boric acid solution and call a physician. If an internal contact is made with Fyrite fluid, give vinegar or juice of lemon, grapefruit, or orange. Follow with olive oil and contact a physician.

COMPETENCY PROCEDURES

1. Turn furnace OFF (*if applicable*). ❑
2. Drill a 3/8" hole in the furnace stack pipe between the furnace breech or elbow and draft regulator. ❑

NOTE: *There may already be a test hole in the furnace stack pipe.*

---

3. Start the furnace and let the burner operate for 15 minutes. ❑
4. Make sure the CO_2 analyzer is at room temperature. If not, let it warm up. ❑
5. Wet the wool-packing filter in the saturator tube of the Fyrite sampling pump. ❑

NOTE: *Avoid saturating the wool-packing filter—should be damp.*

---

6. Hold Fyrite upright and away from face. ❑
7. Depress plunger valve at the top of the Fyrite cylinder momentarily to purge the Fyrite of air and previous gas samples. ❑
8. Invert the Fyrite cylinder by holding it at a slight angle to drain the CO_2 fluid into the top of the cylinder reservoir. ❑
9. Turn the Fyrite cylinder in an upright position and hold at a 45-degree angle momentarily to allow the fluid to drain into the bottom of the reservoir. ❑
10. Again, depress plunger valve at the top of the Fyrite cylinder momentarily to purge the Fyrite of air and previous gas samples. ❑
11. Hold the Fyrite upright and loosen the scale locknut. ❑
12. Slide the scale plate until the top of the fluid line lines up with zero (0) on the scale plate and then tighten the scale nut. ❑
13. Insert the Fyrite sampling tube into the stack pipe 3/8" sampling hole. ❑
14. Hold the Fyrite cylinder in an upright position and place the sampling assembly rubber connector tip over the Fyrite plunger valve. ❑
15. Push the rubber connector tip down on the Fyrite plunger valve and hold firm. ❑
16. Slowly squeeze and release the aspirator bulb **18 times.** ❑
17. During the 18th aspirator depression (with the aspirator bulb deflated), release the rubber connector tip from the Fyrite cylinder plunger valve. ❑
18. Absorb sample of gas into the Fyrite fluid by inverting the Fyrite upside down at a 45-degree angle and allow the Fyrite fluid to flow to the top reservoir. ❑
19. Once Fyrite fluid has drained to the top reservoir, turn Fyrite upright at a 45-degree angle and allow the Fyrite fluid to drain to the bottom reservoir. ❑
20. Repeat this process three times. ❑
21. After the third time, permit the Fyrite fluid in the cylinder to stabilize a few seconds, then immediately read the percentage of CO_2 on the scale at the point corresponding to the top of the fluid column. ❑
22. *Record percentage of CO_2 of test you performed.* _____ ❑

NOTE: *A delay in reading of 5 to 10 seconds may decrease the accuracy of reading slightly. Longer delays in obtaining reading may cause substantial error. If this is the case, resample the flue gas by following previous procedures.*

---

23. Have the instructor check your work. ❑
24. Remove the rubber sampling hose assembly from the stack pipe sampling hose. ❑
    CAUTION: Sampler spout will be extremely hot. Avoid contact.

Disconnect equipment and return all tools and test equipment to their proper location. ❑

RESEARCH QUESTIONS

1. What is CO_2 the formula for?

_____

_____

2. The percentage of CO_2 relates to the amount of what during the combustion process?

_____

_____

3. A low CO_2 reading indicates what?

_____

_____

4. A high CO_2 reading indicates what?

_____

_____

5. What are the byproducts of perfect combustion?

_____

_____

Passed Competency _____   Failed Competency _____

Instructor Signature _____   Grade _____

Practical Competency 150

Complete Combustion Test on Fossil Fuel Equipment (Figure 5–96)

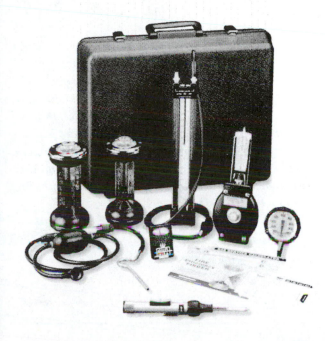

FIGURE 5–96   Combustion test kit. (*Courtesy of Bacharach, Inc.*)

SUGGESTED MATERIALS

Textbook
Refrigeration & Air Conditioning Technology, 5th Edition, Thomson Delmar Learning
Unit 32—Oil Heat

Review Topics
Overfire Draft; Stack Draft; Stack Temperature; Net-Stack Temperature; Smoke Test; CO_2 Test

COMPETENCY OBJECTIVE

The student will able to perform a complete combustion test on an oil furnace.

OVERVIEW

To check the performance and efficiency of an oil furnace requires a technician to analyze the furnace's performance and combustion efficiency. This requires the technician to check the furnace overfire draft, stack draft, stack temperature, number of smoke, and CO_2 percentage. The results of these tests can be used to analyze and make adjustments as required to bring the furnace performance into the most efficient operation allowable.

EQUIPMENT REQUIRED

Combustion kit (*if applicable*)
CO_2 Fyrite® analyzer (red fluid)

gage

ke analyzer

.noke paper

Smoke scale

Oil efficiency calculator

Stack thermometer

Temperature tester

Electric drill

3/8" Drill bit

2" Piece of aluminum duct tape

SAFETY PRACTICES

Care should be taken to avoid touching metal sampling tubes once they are removed from the furnace stack pipe and view port window of heating equipment. The metal sampling tube will be extremely hot.

CAUTION: The Fyrite fluids used in the CO_2 and O_2 analyzers are corrosive and contain poisonous elements, which must not be taken internally. In the event of a spill or accidental body contact with CO_2 Fyrite fluid, flood with water and then wash area with vinegar. If CO_2 Fyrite fluid enters the eye(s), flood eye(s) with water, then wash with 5% boric acid solution, and call a physician. If internal contact is made with Fyrite fluid, give vinegar or juice of lemon, grapefruit, or orange. Follow with olive oil and contact a physician. In the event of a spill or accidental body contact with O_2 Fyrite fluid, flood with water and then treat affected area with baking soda. If internal contact is made with O_2 Fyrite fluid, induce vomiting; administer milk or white of eggs. Combat collapse and contact a physician.

COMPETENCY PROCEDURES   Checklist

1. Turn furnace OFF.   ❏
2. Drill a 3/8" hole in the furnace stack pipe between the furnace breech or elbow and draft regulator.   ❏

NOTE: *There may already be a test hole in the furnace stack pipe.*

---

3. Start the furnace and let the burner operate for 15 minutes.   ❏
4. Look through the view port window at the flame and make adjustments to bring the flame into a good burn.   ❏
5. Drill a 3/8' hole in furnace view port window (*if applicable*).   ❏
6. Take an overfire draft measurement by sticking the draft tube through a 3/8" hole in the view port window.   ❏
7. *Record overfire draft reading.* _____   ❏
8. *What overfire draft reading is considered acceptable?* _____   ❏
9. *Was overfire draft reading within acceptable range?* _____   ❏

NOTE: *If overfire draft is not in acceptable range, make adjustments to bring it into acceptable range.*

---

10. Measure the stack draft.   ❏
11. *Record your stack draft reading.* _____   ❏
12. *What stack draft reading is considered acceptable?* _____   ❏
13. *Was the stack draft reading within acceptable range?* _____   ❏

NOTE: *If the stack draft is not in an acceptable range, make adjustments to bring it into an acceptable range.*

---

14. Obtain a smoke reading. ❏
15. Compare the smoke sample with the smoke scale. ❏
16. *Record number of smoke readings.* _____ ❏
17. *What smoke number is considered acceptable?* _____ ❏
18. *Was smoke number within acceptable range?* _____ ❏

NOTE: *If the smoke number is not in acceptable range, make adjustments to bring it into an acceptable range.*

---

19. Measure the furnace stack temperature. ❏
20. *Record the furnace stack temperature.* _____ ❏
21. Measure the surrounding air temperature near the furnace. ❏
22. *Record the surrounding air temperature.* _____ ❏
23. Determine the net-stack temperature. ❏
24. *Record the net-stack temperature.* _____ ❏
25. Make sure the CO_2 analyzer is at room temperature. If not, let it warm up. ❏
26. Obtain the CO_2 percentage by taking gas sample from stack pipe flue gages. ❏
27. *Record the percentage of CO_2 obtained from gage sample.* _____ ❏
28. What CO_2 percentage is considered acceptable? _____ ❏
29. Was CO_2 percentage is within an acceptable range? _____ ❏

NOTE: *If the CO_2 percentage was not in an acceptable range, make adjustments to bring it into an acceptable range or as close as possible.*

---

30. Determine the heating equipment efficiency by using the stack or net stack temperature as required by the efficiency calculator and the CO_2 percentage. ❏
31. *Record the furnace efficiency as determined by the efficiency calculator.* _____ ❏
32. *What percentage of the heat produced during combustion is used to heat the structure?* _____ ❏
33. *What percentage of the heat produced during combustion is being lost up the chimney?* _____ ❏
34. *Do you feel the furnace is working as efficiently as it can under current conditions?* _____ ❏
35. *If YES, explain why.* ❏
36. *If NO, explain why.* ❏
37. Have the instructor check your work. ❏
38. Disconnect equipment and return all tools and test equipment to their proper location. ❏

RESEARCH QUESTIONS

1. What is considered a good overfire draft reading?

   _____

   _____

2. What is considered a good stack draft reading?

   _____

   _____

3. What is considered a good smoke value reading?

   _____

   _____

What is considered a good percentage of CO_2 for oil equipment?

_____

_____

5. What is considered a good percentage of CO_2 for natural gas heating equipment?

_____

_____

Passed Competency _____ **Failed Competency** _____

Instructor Signature _____ **Grade** _____

6 GAS HEAT

Theory Lesson: Gas Heat (Figure 6–1)

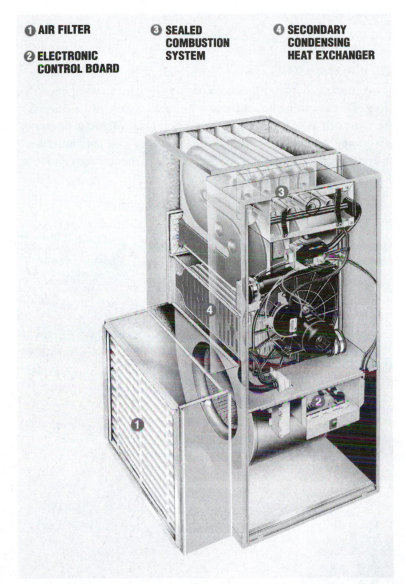

1 AIR FILTER

2 ELECTRONIC CONTROL BOARD

3 SEALED COMBUSTION SYSTEM

4 SECONDARY CONDENSING HEAT EXCHANGER

FIGURE 6–1 A high-efficiency furnace with burners on top of the heat exchanger. (*Courtesy of Bryant Heating and Cooling Systems*)

SUGGESTED MATERIALS

Textbooks

Refrigeration & Air Conditioning Technology, 5th Edition, Thomson Delmar Learning
Unit 31—Gas Heat

Introduction to Gas-Fired Forced-Hot Air Furnaces

Key Terms

automatic combination gas valve • bimetal limit switch • cold junction • direct-spark ignition (DSI) • dual-rod rectification • fan–limit switch • flame rectification • flame sensor rod • hard lockout • heat exchanger • hot junction • hot surface ignition (HSI) • integrated furnace controllers (IFC) • intermittent pilot (IP) • interpurge • liquefied petroleum • liquid propane • lockout • natural gas • postpurge • prepurge • primary air • secondary air • single-rod rectification • smart valve • soft lockout • spud/orifice • standing pilot • thermocouple • thermopiles • twinning • venting categories• venturi

OVERVIEW

Gas furnaces use two different types of fuel sources: **natural gas** or **liquefied petroleum,** which is also referred to as LP gas. On the average one cubic foot of **natural gas** can be converted to approximately **1,050 Btu's of heat energy per cubic foot.** LP gas can be liquid propane, liquid butane, or a liquefied mixture of both, with liquid propane being the most common for use in heating equipment. **Liquid propane** gives off approximately **2,500 Btu's of heat energy per cubic foot.** Most gas appliances can be converted to operate on either fuel source with minimal manufactured required retrofitting. The heat energy from either fuel source can be used to heat water in hydronic heating equipment (**Figure 6–2**), or condition air in forced air furnaces (**Figure 6–3**).

FIGURE 6–2 Gas-fired boiler. (*Courtesy of Well-Mclain*)

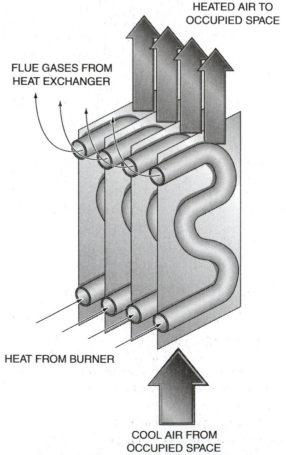

HEATED AIR TO OCCUPIED SPACE

FLUE GASES FROM HEAT EXCHANGER

HEAT FROM BURNER

COOL AIR FROM OCCUPIED SPACE

FIGURE 6–3 The air from the conditioned space and the products of combustion are separated from each other by a heat exchanger.

Gas as a fuel source is used mostly in forced air heating equipment, which is manufactured in different styles:

1. **Upflow (Figure 6–4)**

Upflow furnaces take cool air from the conditioned space from the rear, bottom, or sides near the bottom. It discharges the heated air out the top of the furnace.

2. **Downflow (Figure 6–5)**

Downflow furnaces are often referred to as counterflow furnaces. The air intake for this type of furnace is at the top and the conditioned air is discharged at the bottom.

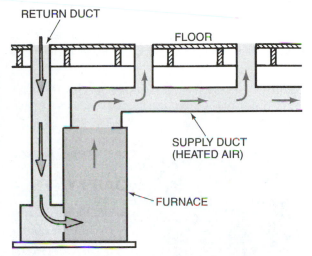

FIGURE 6–4   The airflow for an upflow gas furnace.

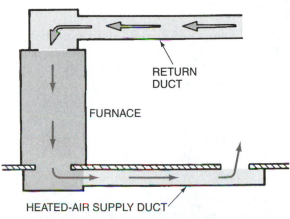

FIGURE 6–5   The airflow for a downflow or counterflow gas furnace.

3. **Low-Boy (Figure 6–6)**

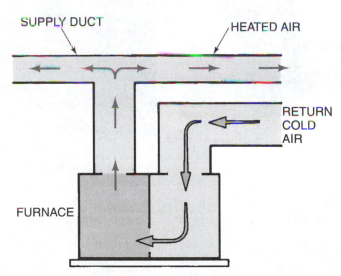

FIGURE 6–6   A lowboy gas furnace.

Low-boy furnaces are approximately 4 feet high and used primarily in basement installations where the ductwork for the supply and return air are located under the first floor. Both the air intake and discharge are at the top of the furnace.

4. Horizontal (Figure 6–7)

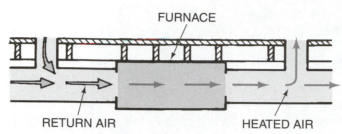

FIGURE 6–7   A horizontal gas furnace.

Horizontal furnaces are positioned on their side and normally installed in crawl spaces such as attics, or suspended from the floor joists in a basement. The air intake is at one end and the discharge air is at the other end.

Years ago most gas-fired equipment used a standing pilot to ignite the fuel source (**Figure 6–8**), although today's advances in electronics and solid-state components have allowed for modern advances in gas furnace design, efficiency, and ignition system technology (**Figure 6–9**).

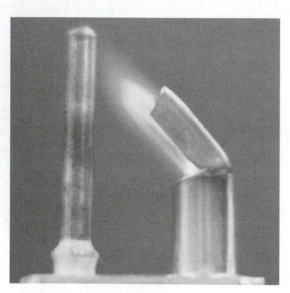

FIGURE 6–8   Standing pilot.

FIGURE 6–9   Hot surface igniter. (*Courtesy of Ferris State University. Photo by John Tomczyk*)

Gas furnace components are the same from manufacturer to manufacturer, with the differences being in the type of ignition system, safety controls, and efficiency percentages. Basic gas furnace components include the following:

- Furnace heat exchanger
- Fuel control and delivery system
- Fuel ignition system
- Means by which heat is transferred to conditioned space air
- Furnace combustion monitoring
- Furnace venting
- Fan motor and blower
- Fan and limit switches
- Safety control system

*Fuel combustion takes place inside of the furnace **heat exchanger** (**Figure 6–10** and **Figure 6–11**), **with the byproducts of** combustion gasses heating the heat exchanger as the gasses pass through the heat exchanger and out the furnace flue for venting to the atmosphere. The heated surface area of the exchanger is in turn used to transfer heat to the conditioned space air as it passes over the exchanger.*

FIGURE 6–10   Furnace heat exchanger. (*Courtesy of Ferris State University. Photo by John Tomczyk*)

FIGURE 6–11   Ignition of fuel inside furnace heat exchanger. (*Courtesy of Ferris State University. Photo by John Tomczyk*)

Fuel control and delivery system to the furnace are controlled by the **gas valve** (**Figure 6–12**). Several types of gas valves are in use, although most residential and light commercial furnaces use an automatic combination gas valve that controls the flow of gas to both the pilot and main gas line that feeds gas through the manifold and then through the burner orifices (**Figure 6–13**).

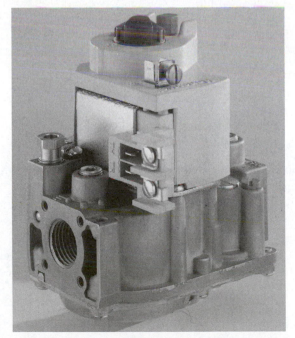

FIGURE 6–12   Automatic combination gas valve. (*Courtesy of Honeywell, Inc., Residential Division*)

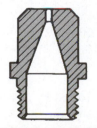

CROSS SECTION

FIGURE 6–13   A spud showing the orifice through which the gas enters the burner.

The pressure at which the gas is supplied to the manifold and **spud** or **orifice** is controlled by the gas valve and is different depending on the fuel source. The desired gas manifold pressure for natural gas furnaces is 3 to 3.5 in. W.C. (*inches of water column*) and 11 in. W.C. for liquefied petroleum. Gas valve pressure can be checked and adjusted by using a water manometer (**Figure 6–14**), a digital manometer (**Figure 6–15**), or a gas gage (**Figure 6–16**).

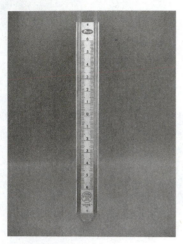

FIGURE 6–14 A water manometer. (*Photo by Bill Johnson*)

FIGURE 6–15 A digital manometer being used to measure gas pressure in inches of water column. (*Courtesy of Ferris State University. Photo by John Tomczyk*)

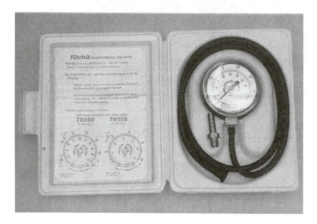

FIGURE 6–16 A gage calibrated in inches of water column. (*Photo by Bill Johnson*)

Adjustments to manifold pressure can be accomplished by tightening or loosening an adjustment spring on the gas valve, which is used to increase or decrease gas pressure to the furnace.

Burners used in residential applications are atmospheric burners, which means that air and gas mixture required for combustion is under atmospheric pressure. The gas is metered to the burner through the orifice.

The velocity of the gas pulls in the **primary air** around the orifice.

The burner tube diameter is reduced where the gas is passing through it to induce the air for mixing with the gas. This reduction area is called the **venturi**. The gas–air mixture moves through the venturi into the mixing tube, where the gas is forced by its own velocity through the burner ports or slots where it is ignited as it leaves the burner ports (**Figure 6–17**).

When the gas is ignited at the burner ports, **secondary air** is drawn in around the burner ports to support combustion. A good flame should be well-defined blue in color with slightly orange tips (**Figure 6–18**).

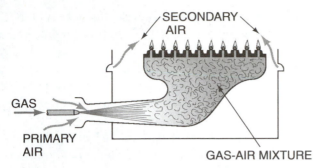

FIGURE 6–17   Ignition of the gas takes place on the top of the burner.

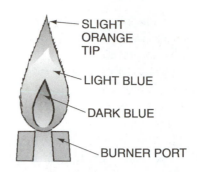

PROPER FLAME COLOR

FIGURE 6–18   Proper flame color.

The orange streaks in the flame are dust particles and should not be confused with yellow streaks. Yellow streaks, flame, or a tip is an indication of a lack of sufficient primary air that will emit poisonous carbon monoxide gas (**Figure 6–19**).

Too much primary air will cause the flames to lift off the burners (**Figure 6–20**).

 **FIGURE 6–19**   Yellow flame indicates insufficient primary air.

 **FIGURE 6–20**   Lifting flame from burner.

Different ignition systems are used to ignite the gas–air mixture, which can be divided into four categories:

- Intermittent pilot (IP)
- Direct-spark ignition (DSI)
- Hot surface ignition (HSI)
- Standing pilot

Intermittent Pilot Spark-to-Pilot Type of Ignition Systems

Intermittent pilot or spark-to-pilot type of ignition systems (**Figure 6–21**) use a spark generated from an electronic module to ignite the pilot, which is used to ignite the main gas burners. The pilot in this type of

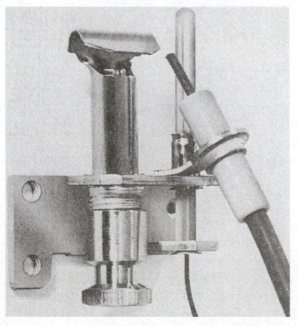

FIGURE 6–21   A spark-to-pilot ignition system. (*Courtesy of Invensys Climate Controls America*)

system is lit only when the system is calling for heat. During the furnace off-cycle, the pilot is OFF. There are two types of electronic modules used in this type of ignition system: 100% shut-off module and non-100% shut-off module.

The **non-100% shutoff** intermittent pilot control system (**Figure 6–22**) is *used only on natural gas heating equipment*. With this type of system, if the gas pilot does not ignite, the pilot valve will remain open and the electronic spark will continue to try to ignite the pilot valve. The main gas valve will not open until the pilot is lit and proved.

The **100% shutoff** intermittent pilot control system (**Figure 6–23**) is *used with LP gas and some natural gas systems*. This control system if there is no pilot ignition, the gas valve will close and may go into safety lockout after approximately 90 seconds, and the spark will stop. Once the ignition system is in safety lockout, it must be reset manually by either turning power OFF to the furnace or switching the thermostat from a call for heat to the OFF position and then back to a call for heat.

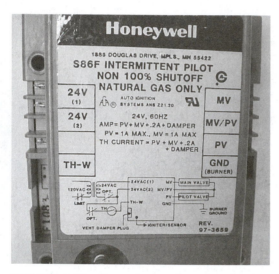

FIGURE 6–22 Intermittent pilot control with non-100% shutoff. (*Courtesy of Ferris State University. Photo by John Tomczyk*)

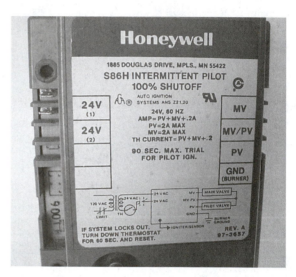

FIGURE 6–23 Intermittent pilot control with 100% shutoff.

On a call for heat from the thermostat, a 24-volt circuit is closed between the thermostat **R** and **W** terminals. 24 Volts are sent to the electronic control module terminals, which sends power to the pilot igniter and to the pilot valve coil.

The coil opens the pilot valve, and the electronic spark ignites the pilot. The spark is intermittent and arcs approximately 100 times per minute with the module generating approximately 10,000 volts with low amperage. The 10,000 volts generates a spark hot enough to ignite the gas–air mixture. A direct path to ground must be provided because the arc actually arcs to ground. A ground strap near the pilot assembly or the pilot hood often acts as a ground (**Figure 6–24**).

Once the pilot flame is lit, ignition of the main gas must be proved. This is done through a process called **flame rectification**. This is a process in which the pilot flame changes the normal alternating current to direct current, by the use of a single- (**Figure 6–25**) or dual-rod sensing system (**Figure 6–26**).

The single-rod rectification systems uses a single rod to accomplish both ignition and sensing. With this type of system, only one large wire will be connected to the pilot assembly and ignition module. (Refer to *Figure 6–25*.)

The **dual-rod rectification system** uses one rod as the igniter and the other as the flame sensor. This system will have two wires running from the pilot assembly to the control module. (Refer to *Figure 6–26*.) The flame sensing rod wires goes to the **sensing terminal** of the control module. The flame sensor works with

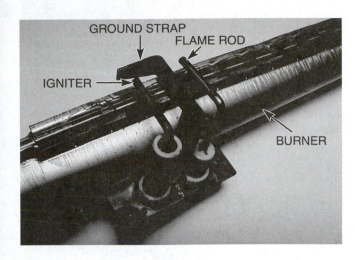

FIGURE 6–24   A direct-spark ignition assembly near the gas burner. (*Courtesy of Ferris State University. Photo by John Tomczyk*)

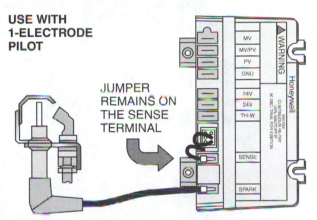

FIGURE 6–25   A single-rod or local sensing system connected to furnace electronic module. (*Courtesy of Honeywell, Inc.*)

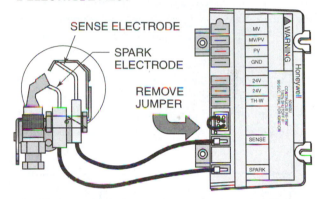

FIGURE 6–26   A dual-rod or remote sensing flame rectification system showing two separate wires going to the electronic furnace module. (*Courtesy of Honeywell, Inc.*)

the grounded burner hood and the ground strap to achieve flame rectification. The igniter wire goes to the **spark terminal** of the control module.

In the flame rectification concept, the pilot flame that results from burning gas contains combustion gases with charged particulate matter. When flame is present between the two electrodes, electrical current is able to flow freely between them (**Figure 6–27**).

Current flow between the two electrodes completes the rectification circuit in the control board, generating DC voltage that powers the main gas valve (**Figure 6–28**).

The electronic components in the system will energize and open the main gas valve only with a direct current (DC) measured in microamps from the flame-sensing rod. *The generated DC single normally ranges between 1 and 25 microamps* and can be measured with a microammeter. If there is no pilot, the main gas valve will not open.

Direct-Spark Ignition (DSI) Systems

The direct-spark ignition system does not incorporate a pilot to ignite the fuel from the main gas valve. DSI systems use an assembly that incorporates the igniter–sensor assembly along with a DSI module, which is located close to the burner (Refer to *Figure 6–27*.) With this type of ignition system, the spark from the igniter–sensor assembly is used to ignite the main gas. The design of the system is that on a call for heat the

**FLAME RECTIFICATION
PILOT & PROBE**

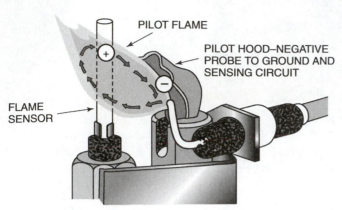

RECTIFICATION CIRCUIT

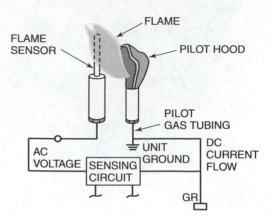

FIGURE 6–28   Rectification circuit.

FIGURE 6–27   An intermittent pilot dual-rod system showing the pilot hood and the flame sensor rod as the two electrodes for the flame rectification. Dual-rod systems are often referred to as remote sensing systems. (*Courtesy of Johnson Controls*)

R and **W** terminals of the thermostat close, sending a 24-volt signal to the igniter module. The igniter module in turn supplies voltage to the igniter assembly and opens the main gas valve. Ignition must take place very quickly and at the same time send a signal to the igniter module verifying that the furnace has fired, or the furnace will shut down and go into a *safety lockout mode* (**Figure 6–29**).

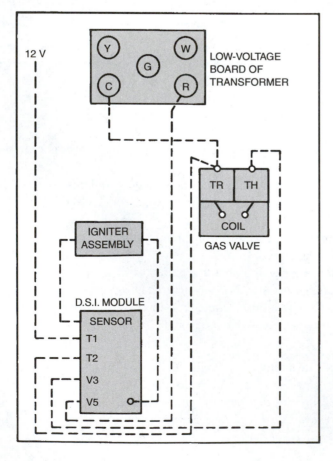

FIGURE 6–29   Control circuit and components of a DSI system. (*Courtesy of International Comfort Products, LLC*)

The **flame sensor rod** is used to verify that the furnace has fired and has 4 to 11 seconds to send a microamp signal through flame rectification to the DSI module which allows the furnace to continue to operate. If a flame is not verified within the allowable time period, the furnace goes into lockout and can be reset only by turning OFF power to the system control and waiting 1 minute before turning power back ON.

Most ignition failures with the DSI system are associated with the igniter–sensor assembly, with the main causes due to improperly adjusted spark gap, igniter positioning, and bad grounding (**Figure 6–30**).

FLAME SENSOR AND SPARK IGNITER LOCATIONS

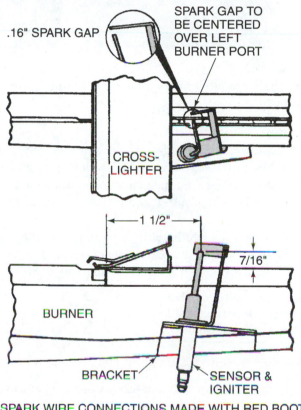

.16" SPARK GAP

SPARK GAP TO BE CENTERED OVER LEFT BURNER PORT

CROSS-LIGHTER

1 1/2"

7/16"

BURNER

BRACKET   SENSOR & IGNITER

SPARK WIRE CONNECTIONS MADE WITH RED BOOT TO MODULE AND BLACK BOOT TO IGNITER

FIGURE 6–30   Spark gap and igniter position for a DSI system. (*Courtesy Heil- Quaker Corporation*)

NOTE: *Most manufacturers provide specific troubleshooting instructions for DSI systems along with wiring diagrams and operating instructions. Technicians should review this information if available during service procedures.*

Hot Surface Ignition (HSI) (Figure 6–31)

The HSI ignition system does not use an open flame to ignite the main fuel; instead an igniter made of silicon carbide is used to generate heat that is higher than the ignition temperature of gas when the correct voltage is applied, normally 120 volts. The igniter is placed in the gas stream and is allowed to get very hot before the gas is allowed to flow and impinge on the glowing surface of the igniter. If the igniter is hot enough, immediate ignition should occur when the main gas valve is opened. Most furnaces that incorporate the HSI system energize the igniter for a set period before the main gas valve is energized. Once ignition is proven by the flame sensor, the HSI is de-energized by the control board and the furnace continues to operate.

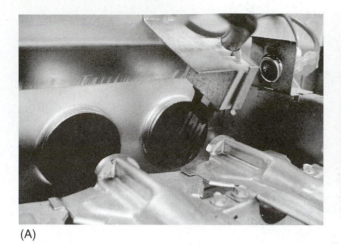

(A)

(B)

FIGURE 6–31   (A) A hot surface igniter de-energized. (B) The hot surface igniter system energized before main gas is allowed to flow.

If ignition is not proven in the designed time cycle, the HSI will be cycled OFF and back ON again in an attempt to provide ignition of the main gas. Most HSI furnaces will allow HSI cycling for a number of times and then go into safety lockout if ignition does not take place. Furnaces that incorporate the HSI system operate on the same sequencing procedures, although technicians should review the manufacturer's instructions for the furnace being serviced.

Care should be taken when handling or working around the HSI because they are very brittle and can be broken very easily. Newer hot surface igniters are made of a different material that does not break as easily as older ones. The following should be taken into consideration when repeated failure of an HIS exists:

- Higher than normal applied voltage. In excess of 125 volts can shorten the igniter's life.
- Accumulation of drywall dust, fiberglass insulation, or sealant residue.
- Delayed ignition can stress the igniter due to the small explosion.
- Overfiring condition.
- Furnace short cycling as with a dirty filter may cause the furnace to cycle on high limit.

Some furnaces use the HSI to ignite the pilot or for direct ignition of the main furnace burner (**Figure 6–32**).

Systems that use an HSI to ignite the furnace pilot generally operate on a 24-volt power supply, and if used to ignite the main fuel, will have 4 to 11 seconds to do so before the system will cycle into safety lockout.

FIGURE 6–32   The 24-volt surface igniter energized. (*Courtesy of Ferris State University. Photo by John Tomczyk*)

Ignition Module Wiring and Terminal Naming

On the faceplate of electronic ignition modules are terminal markings and abbreviations that can assist the service technician in wiring and troubleshooting a furnace's electrical and mechanical system. These terminal markings and abbreviations are standard to most modules and technicians should become familiar with their interpretation. (Refer to *Figures 6–25 and 6–26.*)

- **MV**   Main valve (gas valve)
- **PV**   Pilot valve
- **MV/PV**   Common terminal for main and pilot valves
- **GND**   Burner ground
- **24V**   24-volt source out of the transformer
- **24V(GND)**   Common or ground out of the 24-volt transformer
- **TH-W**   Thermostat lead to module
- **Igniter/Sensor** High-voltage igniter and flame rectification rod or sensor for local sensing (single-rod system)
- **Sense**   Flame rectification rod or sensor for remote sensing (dual-rod systems)
- **Spark**   High-voltage igniter

Standing Pilot Systems

The standing pilot system is not widely used in modern gas equipment although there is equipment still in use that incorporates this type of ignition system. In this type of system a pilot light is lit all the time regardless if the main gas is energized or not. In older systems the pilot must be lit manually; newer pilot systems will incorporate a glow coil or spark to re-ignite the pilot. In most standing pilot systems **thermocouples** are used to prove the pilot is lit before the main gas valve is opened, although other methods such as **thermopiles, bimetallic strips,** and **liquid-filled remote bulbs** are also in use. Since thermocouples are the most widely used, they will be discussed in more detail.

Thermocouples act as the safety device to prove that the pilot is lit before the main gas valve is energized. The thermocouple sits directly in the pilot flame (**Figure 6–33**). Thermocouples consist of two dissimilar metals welded together at one end; this is called the **hot junction** (**Figure 6–34**).

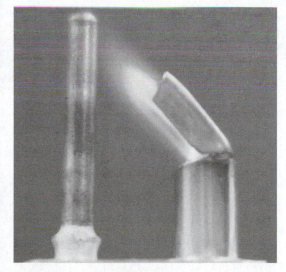

FIGURE 6–33   Electric current is generated when the thermocouple is heated. If the thermocouple does not sense any heat, the gas valve cannot open. (*Photo by Bill Johnson*)

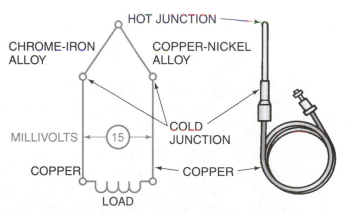

FIGURE 6–34   Thermocouple hot junction with two dissimilar metals. (*Courtesy of Robertshaw Controls Company*)

When the hot junction is heated, a small voltage is generated across the two wires or metals at the other end, referred to as the **cold junction** (**Figure 6–35**).

The millivolts DC generated by the thermocouple will be approximately 30 millivolts DC. This is enough voltage to hold the power unit of the gas valve open so that when the gas valve main gas valve solenoid is energized, gas can flow through the power unit to the burner for ignition (**Figure 6–36**).

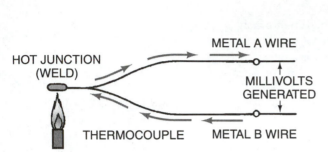

FIGURE 6–35 When heated, electrons flow in one direction in one type of metal (A) and in the opposite direction in the other type (B). (*Courtesy of Robertshaw Controls Company*)

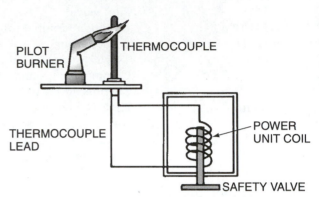

FIGURE 6–36 Millivolt signal from the thermocouple creates a magnetic field in the pilot unit that holds the valve open.

As long as the thermocouple generates the proper millivolts DC, gas can flow through the main valve. If the pilot flame goes out, the thermocouple will cool in about 30 to 120 seconds, de-energizing the power unit to the gas valve and stopping gas flow from the gas valve. Thermopiles are a group of thermocouples wired in series to increase the voltage supplied to the gas valve power unit, and work the same way as one thermocouple (**Figure 6–37**).

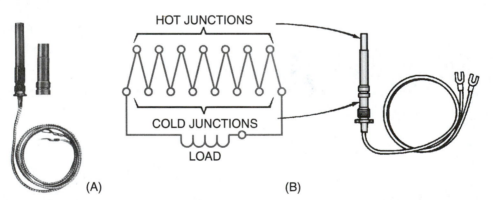

FIGURE 6–37 (A) Thermopile. (*Courtesy of Honeywell, Inc.*) (B) A thermopile consists of a series of thermocouples in one housing.

Checking thermocouples is a matter of measuring the DC millivolt output by heating the hot junction of the thermocouple for at least 5 minutes and placing a voltmeter lead at the thermocouple cold junction and the other meter lead at the thermocouple end (**Figure 6–38**).

Any DC voltage less than 20 millivolts would require replacement of the thermocouple.

Electronic Ignition Modules and Integrated Furnace Controller
With advances in technology, other types of electronic ignition modules and integrated furnace controllers are being used besides those discussed previously. These modules are designed to control the ignition

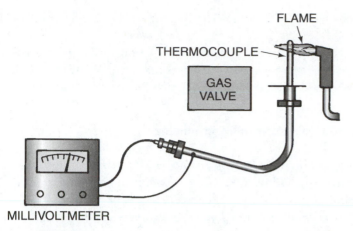

FIGURE 6–38   Checking the millivolts DC of a thermocouple.

(A)

(B)

FIGURE 6–39   (A) Four types of electronic ignition modules. (B) A integrated furnace controller (IFC). (*Courtesy Ferris State University. Photos by John Tomczyk*)

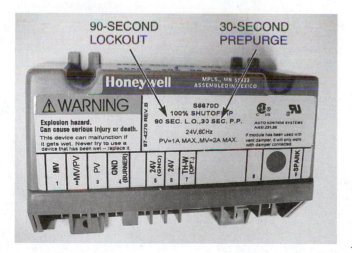

FIGURE 6–40   A 100% shutoff module with a 90-second lockout and 30-second prepurge. (*Courtesy of Ferris State University. Photo by John Tomczyk*)

(**Figure 6–39**) and sequence of operation of most modern furnaces. Some manufacturers of integrated controllers will indicate the control's sequence of operation on the control faceplate (**Figure 6–40**).

Integrated furnace controllers (IFC) provide all of the ignition, safety, and sequence of operation of the furnace. The controller carefully monitors all thermostat, temperature, resistance, and current inputs to ensure stable and safe operation of the combustion blower, gas valve, ignition, flame rectification system, and warm air blower.

Without a basic understanding of these types of controls, their sequence of operation, or control terminology, technicians will find it difficult to evaluate and diagnosis furnaces that incorporate these types of controls. For this reason the following information related to these additional controls and terminology will be discussed:

1. **Electronic ignition modules and integrated furnace controller terminology**
2. **100% Shutoff system** (Refer to *Figure 6–23*)
   A system is 100% shutoff when there is a failure in the flame proving device system and both the pilot valve and main gas valve are shut down (closed). This system was used mainly on older LP equipment.
3. **Non-100% shutoff system** (Refer to *Figure 6–22*)
   A system is non-100% shutoff when there is a failure in the flame proving device system and the main gas valve shuts down but the pilot valve continues to bleed gas as the system continually tries to relight the pilot to prevent nuisance shutdowns.
4. **Continuous retry with 100% shutoff (Figure 6–41)**

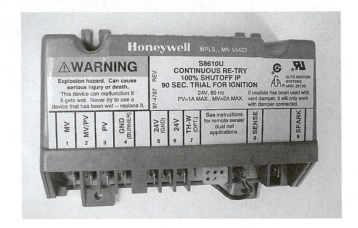

FIGURE 6–41   A continuous retry 100% shutoff electronic ignition module. (*Courtesy Ferris State University. Photo by John Tomczyk*)

   This control has a 100% shutoff with a built-in 90-second trial ignition cycle. The control will continuously try to relight the pilot after the shutoff period. During the shutoff period, no gas will bleed from the pilot.
5. **Lockout**
   Lockout is a time frame built into a module that will allow a certain length of time to light or relight the pilot and main burner. If the allowable time is exceeded, the module will go into a soft or hard lockout.
6. **Soft lockout**
   A time period built into the module that allows for a certain time to light or relight the pilot or main burner. If the time period is exceeded, the module will go into a semi-shutdown for a certain time period, but will eventually keep trying to relight the system. Depending on the design of the control the soft-lockouts time period can range from 5 minutes to 1 hour. Eventually the control may go into a hard lockout once several soft lockouts have been attempted and no flame has been established.
7. **Hard lockout**
   A time period built into the module that will allow for a certain time period to light or relight the pilot or main burner. If the time is exceeded, the module will go into a hard lockout and shut the ignition system down. The system can be taken out of lockout only by interrupting the power source to the module and then turning the power source back ON. A hard lockout normally requires the customer to arrange a service call through a local contractor.

8. **Prepurge (Figure 6–42).**

   Used on intermittent pilot ignition systems (IP). The module is designed with a 90-second lockout (LO) with a 30-second prepurge (PP). On a call for heat, the combustion blower motor will be energized first and run for 30 seconds to clear or prepurge the heat exchanger of any unwanted flue gases, household fumes, or dust that may have accumulated during the last heating cycle (Figure 6–43).

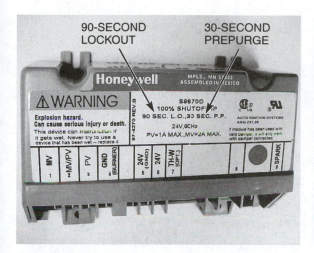

FIGURE 6–42   A 100% shutoff module with a 90-second lockout and a 30-second prepurge. (*Courtesy of Ferris State University. Photo by John Tomczyk*)

FIGURE 6–43   Combustion blower motor used to prepurge the furnace heat exchanger of any flue gases, household fumes, or dust prior to ignition of the main burner. (*Courtesy of Ferris State University. Photo by John Tomczyk*)

9. **Interpurge**

   Allows the combustion blower to operate for a certain time period between ignition tries. Generally used when the furnace does not light or does not sense a flame in the allowable time period.

10. **Postpurge**

    Allows the combustion blower motor to operate for a certain time period after the end of each heating cycle. This procedure along with an incorporated prepurge (PP) cycle is double insurance that the heating cycle will start with only air in the furnace heat exchanger, allowing for a quicker and safe ignition of the main gas. Both the prepurge and postpurge control strategies can be used alone or with one another.

    Integrated furnace controllers (IFC) often come with many DIP switches, which allow the controller to be programmed for different timing functions and control scheme changes (**Figure 6–44**).

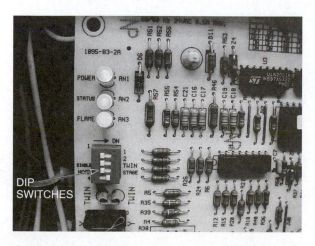

FIGURE 6–44   An integrated furnace controller showing DIP switches and status lights. (*Courtesy of Ferris State University. Photo by John Tomczyk*)

Referring to *Figure 6–44* and **Figure 6–45**, DIP switches 1 and 2 are setting the warm air blower's off-delay timing in seconds. DIP switch 3 is for furnace twinning applications. **Twinning** involves the operation of two furnaces side by side, connected by a common ducting system. High-voltage power is supplied by a common source. Low-voltage power is supplied off a transformer controlled by one common thermostat for both furnaces. DIP switch 4 is not used in this application (**Figure 6–45**).

IFC status lights monitor incoming power, furnace setup modes, diagnostic status for troubleshooting, and the flame rectification signal. The status check can be self-diagnostic flashing light with a code to help with systematically troubleshooting the furnace (**Figure 6–46**).

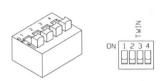

BLOWER OFF TIMES (SECONDS)

LOW FIRE	HIGH FIRE	COOLING	SWITCH 1	SWITCH 2
90	60	30	OFF	ON
120	90	45	OFF	OFF
160	130	60	ON	OFF
180	150	90	ON	ON

FIGURE 6–45 DIP switches for setting the warm air blower off-delay timing. (*Courtesy of Rheem Manufacturing Company*)

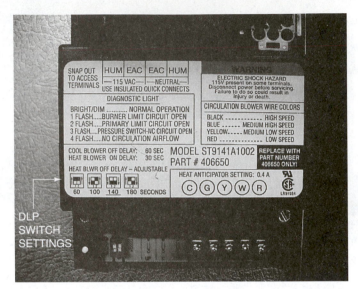

FIGURE 6–46 An integrated furnace controller (IFC) with self-diagnostic lights and flashing fault codes.

11. **Smart valve gas control system**

*Modern furnaces are also incorporating the smart valve (**Figure 6–47**), which incorporates the gas valve and electronic control module in one enclosure. These valves combine the features of an intermittent pilot (IP) and hot surface ignition (HSI) system into one. The smart valve's logic (**Figure 6–48**) lights the pilot with a 24-volt hot surface igniter system. The pilot in turn lights the main burners (**Figure 6–49**).*

FIGURE 6–47 A smart valve gas control system. (*Courtesy of Ferris State University. Photo by John Tomczyk*)

FIGURE 6–48 Internals of a smart valve showing electronic module and gas valve. (*Courtesy of Ferris State University. Photo by John Tomczyk*)

FIGURE 6–49 A 24-volt hot surface igniter that lights a pilot flame. (*Courtesy of Ferris State University. Photo by John Tomczyk*)

These valves are continuous trial for pilot for both natural and LP gas equipment. Once the flame is proved by the flame rod, the 24-volt HSI is de-energized and the electronic fan timer within the module is energized, which starts timing the fan ON delay and the fan OFF delay.

Troubleshooting the smart valve system is made very easy because of the simplified wiring and modular electrical connections. A voltmeter can be used at the valve and electrical plug connections (**Figure 6–50**).

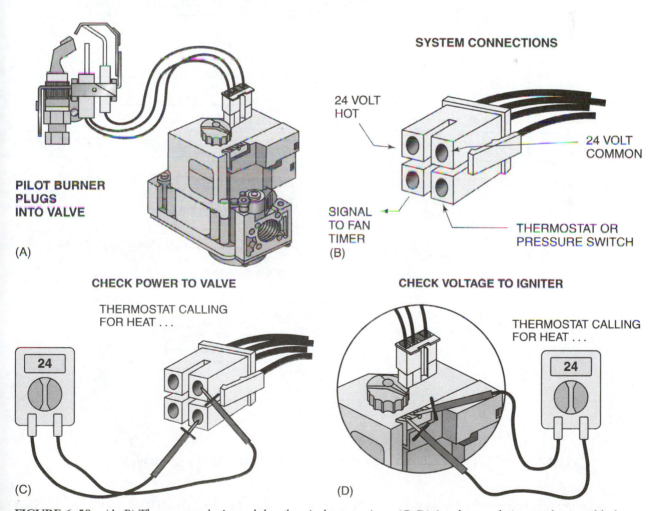

FIGURE 6–50 (A, B) The smart valve's modular electrical connections. (C, D) A voltmeter being used to troubleshoot the smart valve system. (*Courtesy of Honeywell, Inc.*)

The electronics of the smart valve provide an ignition sequence of operation, which is similar to most hot surface ignition, intermittent pilot ignition, and direct-spark ignition systems. Learning the ignition sequence of operation of modern furnaces can aid the technician in identifying and isolating the location as to where a problem lies when troubleshooting ignition failure (**Figure 6–51**).

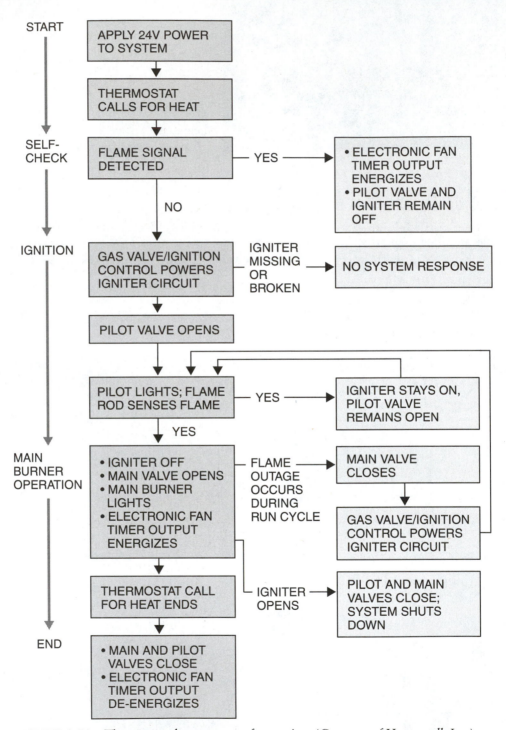

FIGURE 6–51 The smart valve sequence of operation. (*Courtesy of Honeywell, Inc.*)

Some manufacturers will also supply a troubleshooting chart for the valve (**Figure 6–52**).

The blowers of the all-fossil-fuel-forced air furnaces do not come on right away when the thermostat calls for heat. The burner of fossil fuel equipment will operate for a short time without the blower on to

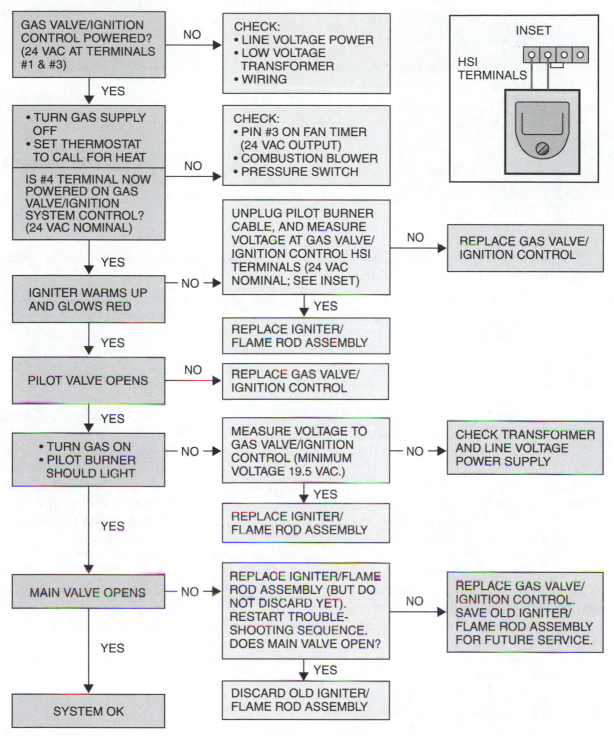

FIGURE 6–52   A smart valve-troubleshooting chart. (*Courtesy of International Comfort Products*)

allow the furnace heat exchanger to heat up. This prevents cold air from being circulated into the conditioned space. On some fossil-fuel-forced air equipment a fan–limit switch is placed above the heat exchanger, which senses the heat coming off the heat exchanger (**Figure 6–53**).

Once the **fan–limit switch** senses a preset temperature it will close the electrical circuit to the furnace blower. The blower draws the cold air from the conditioned space, moves it across the furnace heat exchanger, which heats the air, and then returns it to the conditioned space. The fan–limit switch also serves as a safety switch in the event that the furnace overheats for any reason (**Figure 6–54**).

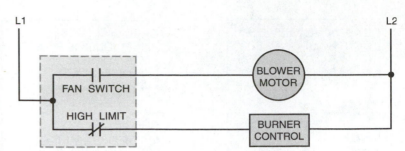

FIGURE 6–54 Schematic for fan–limit switch.

FIGURE 6–53 Fan–limit switch.

The safety switch of the fan-limit breaks the power supply to the gas valve or ignition control system, which in turn shuts down the furnace burner operation.

> **NOTE:** *Additional information on the fan limit safety switch can be reviewed in Chapter 5, Theory Lesson: Fan and Limit Safety Switch. Checking a fan limit switch on a fossil fuel furnace can be performed by reviewing Chapter 5, Practical Competency 141. Setting and checking a fan–limit temperature switch can be performed by reviewing Chapter 5, Practical Competency 142.*

Some modern gas furnaces do not use a fan–limit switch as shown in *Figure 6–53,* but use a bimetal device. These **bimetal switches** can be used to activate the furnace blower.

On the other hand, they can be used as a fan-limit safety switch; this controls the furnace blower operation and acts as a high-limit switch, shutting the burner down due to high-temperature conditions.

When the conditioned space temperature is satisfied, the thermostat will open the electrical circuit between the **R** and **W** terminals of the thermostat. This de-energizes the oil burner primary control, shutting OFF burner operation. The furnace blower will continue to operate until the heat exchanger has been cooled to a preset temperature. Once the heat exchanger is cooled enough, the fan–limit switch will open the circuit to the furnace blower, shutting the blower OFF.

When installing a gas furnace always refer to the installation information supplied with the equipment. This information will describe procedures and requirements for electrical connections to the furnace, gas piping size and requirements, along with proper venting requirements. When installing the furnace gas piping, technicians should be familiar with local codes because they may vary from national codes. All codes require that pipe dope or similar joint compound is required on all connections. All codes require a shut-off valve, drip trap, and union be placed on the gas feed line and within a certain distance of the furnace gas valve (**Figure 6–55**).

Venting of fossil fuel equipment should follow local codes and manufacturer's installation requirements. Some rules to consider when venting are:

- Vent pipe should be at least the same size as the furnace flue pipe fitting on the furnace.
- Horizontal flue pipe runs should be as short as possible.
- The minimum rise should be 1/4 inch per foot of horizontal run (**Figure 6–56**).

Use as few fittings as possible.

- Flue piping should be supported every 4 to 5 feet.
- Vent pipe should be flush with the inside surface of the chimney (**Figure 6–57**).
- Chimney should be completely clear and free of obstructions.
- If venting directly to atmosphere, each furnace vent should have an approved vent cap (**Figure 6–58**).

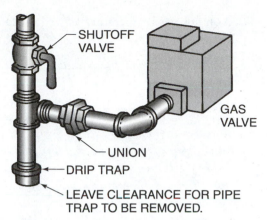

FIGURE 6–55   A shutoff valve, drip trap, and union should be installed ahead of the gas valve.

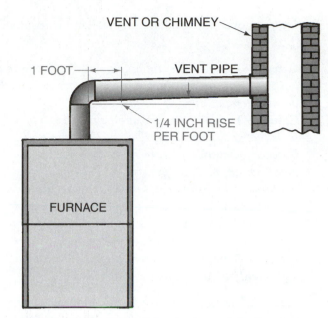

FIGURE 6–56   The minimum slope of the flue pipe should be 1/4 inch per 1 foot of horizontal run.

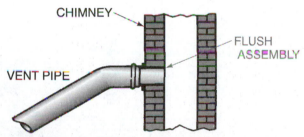

FIGURE 6–57   The vent pipe should not extend beyond the inside of the chimney lining.

FIGURE 6–58   Gas vent caps.

Vent caps prevent rain and debris from entering the chimney and furnace vent system. They also prevent wind from blowing the gases back into the residence or building through the draft hood.

Vent piping requirements for high-efficiency condensing furnaces are a little different compared to guidelines to follow for conventional furnaces. Most vent pipe for condensing furnaces will either be polyvinyl chloride (PVC) or chlorinated polyvinyl chloride (CPVC) because they are noncorrosive. Along with the guidelines listed above to follow when venting furnaces, the following additional guidelines should be followed for condensing furnaces.

- There should be a minimum of 6 inches of vertical run off the top of the furnace before any bends are made.
- All plastic pipe joints should be cleaned with purple primer prior to cementing.
- The venting pipe should be supported at least every 3 feet to extraction.
- Approved vent caps should be used at the structure penetration.
- Vent piping passing through unheated space must be insulated.

Venting categories are established for fossil fuel equipment based on flue gas temperature as well as static vent pressure.

Category I appliances *are appliances that operate with a negative static vent pressure and typically have flue gas temperatures above 275 degrees (F), which is roughly 150 degrees (F) higher than the condensing temperature of the flue gases.*

Category II appliances *are appliances that operate with a negative static vent pressure and typically have flue gas temperatures below 275 degrees (F). These appliances operate with natural draft.*

Category III appliances *are appliances that operate with a positive static vent pressure and flue gas temperatures above 275 degrees (F). These appliances rely on the heat of the flue gases to aid in the venting process and often reuse the heat in the flue, resulting in lower flue gas temperatures and condensing of the flue gases. Aluminum piping is often used for venting, given the possibility of condensation.*

Category IV appliances *are appliances that operate with a positive vent pressure and operate with flue gas temperatures well below 275 degrees (F), very often below the dew point temperature of the flue gases. Condensing furnaces as category IV appliances and use plastic PVC or CPVC materials for venting, mainly PVC because it is noncorrosive.*

Typically, B-vent venting materials are used for venting flue gases from Category I appliances and **NEVER** *used on Category III or IV appliances* (**Figure 6–59**).

Type B vent pipe is usually of a double-wall construction with air space between the inner and outer walls (**Figure 6–60**).

SECTION OF TYPE B GAS VENT

FIGURE 6–59   Section of type B gas vent.

FIGURE 6–60   Double-wall, insulated pipe.

The inner wall may be made of aluminum and the outer wall may be constructed of steel or aluminum. The air gap between the inner and outer walls acts as an insulator that helps keep the flue gases warm and helps reduce the surface temperature of the vent itself.

Category IV appliances are typically vented with PVC or chlorinated CPVC piping material, which does not corrode when subjected to condensing flue gases.

New products are currently on the market for venting category II, III, or IV appliances, which make venting a much easier task (**Figure 6–61**).

FIGURE 6–61   Single-wall venting material.
(*Courtesy of Martin Wawrla, Protech Systems, Inc.*)

These piping sections are constructed of stainless steel and have mechanical locking bands and gaskets. The locking bands are part of the pipe itself, once connected, an air and watertight seal is created. These newer products are also available for double wall vent pipe requirements (**Figure 6–62**).

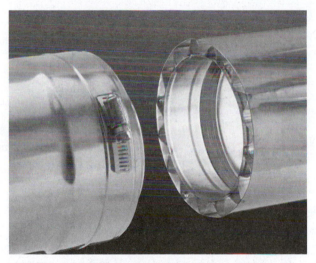

FIGURE 6–62   Double-wall venting materials.
(*Courtesy of Martin Wawla, Protech Systems, Inc.*)

NOTE: *Fossil fuel furnaces, which use natural draft for venting, may also require the addition of a draft diverter in the furnace venting system (**Figure 6–63**).*

Diverters are added to prevent a downdraft through the furnace venting system due to high winds pushing down the chimney and blown across the gas rack and pilot assembly. Diverters re-route any downdraft air out of the furnace.

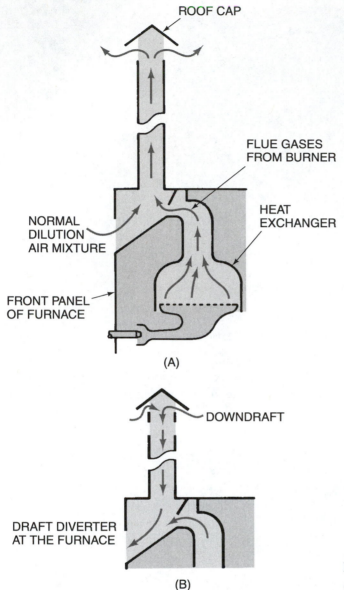

ROOF CAP

FLUE GASES
FROM BURNER

HEAT
EXCHANGER

NORMAL
DILUTION
AIR MIXTURE

FRONT PANEL
OF FURNACE

(A)

DOWNDRAFT

DRAFT DIVERTER
AT THE FURNACE

(B)

FIGURE 6–63 Draft diverters prevent a down-draft from extinguishing the pilot or burner flames.

NOTE: When venting appliances, be sure to refer to review the furnace installation instructions provided by the furnace manufacturer. Proper venting tables for fossil fuel equipment are available from GAMA (Gas Appliance Manufacturers Association). The organization provides installation information as well as tables to calculate and properly size venting systems.

Practical Competency 151

Threading Steel Pipe (Figure 6–64)

FIGURE 6–64   Threading steel pipe.
(*Photo by Bill Johnson*)

SUGGESTED MATERIALS

Textbook
Refrigeration & Air Conditioning Technology, 5th Edition, Thomson Delmar Learning
Unit 7—Tubing and Piping

Review Topics
Steel and Wrought Iron Pipe; Joining Steel Pipe; Installing Steel Pipe

COMPETENCY OBJECTIVE

The student will learn the proper procedures for using the proper tools for cutting, burring, threading, and joining steel pipe.

OVERVIEW

Steel pipe is normally used for water lines, drains, and natural gas supply lines for gas operating equipment. Steel piping is also used in accessories for refrigeration equipment such as refrigeration air-cooled condensers, water-cooled condensers, water-cooling towers, and condensate drains, but rarely used to act as refrigerant lines to carry refrigerant for sealed systems.

Steel pipe can be obtained in black pipe or galvanized pipe. The black pipe is uncoated pipe. Galvanized pipe is zinc coated and does not rust like black pipe, and is used mainly in water and drain lines. The grade of black pipe is based on the thickness and strength of the pipe, with three different grades available: **standard, extra strong,** and **double extra strong,** with standard pipe being used for most plumbing applications (**Figure 6–65**).

Steel pipe is measured by the inside diameter (ID). This is not an accurate measurement and is referred to as the **nominal size** because the wall thickness of the pipe can vary during the forming of the pipe during the manufacturing process. For pipe sizes 12 inches or less in diameter the nominal size is approximately the size of the ID of the pipe. Sizes larger than 12 inches in diameter the OD (*outside diameter*) is considered the nominal size. Fittings for black pipe, type L, K, and DWV copper tubing are measured by the nominal size or inside diameter.

Black pipe can be threaded due to the wall thickness of the pipe. Black pipe threads have been standardized with each thread v-shaped with an angle of 60 degrees. The diameter of the thread has a taper of 3/4" per foot or 1/16" per inch (**Figure 6–66**).

The tapered V-thread makes a conical spiral, which causes the threads to bind together and make a tight seal. To avoid leaks, pipe threads should not be nicked or broken. A minimum of seven perfect threads is required for a good, leakproof joint even if a couple of additional threads are imperfect. Some pipe sizes and number of threads per inch are shown in **Figure 6–67**.

STEEL PIPE
NOMINAL SIZE 2 in.

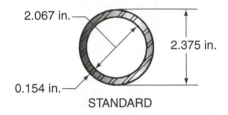

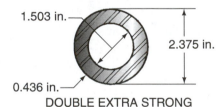

FIGURE 6–65   A cross section of standard, extra strong, and double extra strong steel pipe.

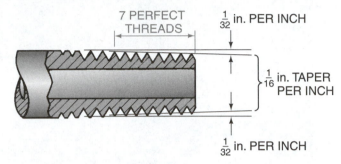

FIGURE 6–66   The cross section of a pipe thread.

PIPE SIZE (INCHES)	THREADS PER INCH
$\frac{1}{8}$	27
$\frac{1}{4}$, $\frac{3}{8}$	18
$\frac{1}{2}$, $\frac{3}{4}$	14
1 to 2	$11\frac{1}{2}$
$2\frac{1}{2}$ to 12	8

FIGURE 6–67   Threads per inch for some pipe sizes.

When tightening the threads to a fitting, about 1/2" of the threads will be used before the fitting begins to tighten on the taper of the joining pipe and fitting. When joining threaded pipe to a fitting, pipe compound or Teflon tape should be placed on the threads of both joining threads. This helps to create a strong leak-proof joint. Do not apply the pipe compound or Teflon tape closer than two threads from the end of the pipe to avoid getting compound of tape into the piping system (**Figure 6–68**).

USE MODERATE AMOUNT OF DOPE

LEAVE TWO END THREADS BARE

FIGURE 6–68   Applying pipe dope or Teflon tape.

Here are some "rules of thumb" that should be followed when working with steel pipe. When cutting steel pipe you should allow about 1/2" of extra pipe length at each end of the pipe for threading pipe and fittings together (**Figure 6–69**).

When the pipe is cut, it should always be deburred so that there are no restrictions on the inside of the pipe (**Figure 6–70**).

FIGURE 6–69   Holding the pipe and turning the fitting with pipe wrenches. (*Photo by Bill Johnson*)

FIGURE 6–70   Using a reamer to remove a burr. (*Photo by Bill Johnson*)

When threading the pipe, it is important to use the proper pipe diameter die along with thread-cutting oil.

Special fittings are made for connecting threaded black pipe, which are made of cast iron that has been annealed. Annealing is a process of heating the metal to a bright cherry red and then allowing it to cool, which makes the metal softer. These fittings are referred to as malleable fittings because the annealing process allows pipe fittings to withstand bending, pounding, and internal pressure more than ordinary cast iron.

Pipe fittings are threaded and sized according to the nominal size (ID). Fittings can be threaded with male or female threads. **Female threads** are those where a joining pipe would be screwed into the fitting. Female threads are **inside of the fitting**. **Male threads** are on the **outside of the fitting** and would be screwed into the inside of a joining female thread fitting.

Fittings can be purchased as elbows, couplings, tees, reducing tees, reducing bushings, caps, plugs, unions, and nipples at different angles and sizes (**Figure 6–71**).

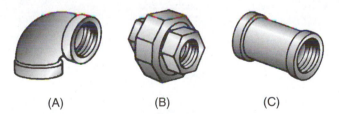

(A)   (B)   (C)

FIGURE 6–71   Steel pipe fittings. (A) A 90-degree elbow. (B) A union. (C) A coupling.

EQUIPMENT REQUIRED

Pipe cutter (**Figure 6–72**).
Pipe reamer (Refer to *Figure 6–70*)
Tri-stand vise (*chain or yoke type*) (**Figure 6–73**)
1/2" Pipe die and ratchet (**Figure 6–74**)

FIGURE 6–72   Three- and four-wheel cutters. (*Courtesy of Ridge Tool Company*)

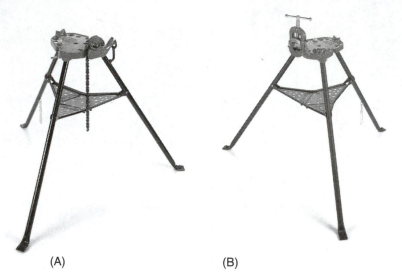

(A)   (B)

FIGURE 6–73   (A) A tri-stand with chain vise. (B) A tri-stand with a yoke vise. (*Courtesy of Ridge Tool Company*)

FIGURE 6–74   Pipe die and ratchet. (*Photo by Bill Johnson*)

FIGURE 6–75   Pipe wrench. (*Courtesy of Ridge Tool Company*)

2-12" Pipe wrenches (**Figure 6–75**)
Cutting oil
1-6' Piece of black pipe
4-1/2" × 1/2" 90-degree elbows
1-1/2" × 1/2" Union
12' Tape rule
Square
Pipe joint compound or Teflon tape
Rag

SAFETY PRACTICES

The student should be knowledgeable in the use of tools. Beware of sharp metals.

COMPETENCY PROCEDURES

1. Place the 6' piece of black pipe in the vise. ❑
2. Measure 12" from the end of the pipe and place a mark. ❑
3. Move the pipe in the vise so that there is enough sticking out the end to use the pipe cutter to cut the marked 12" piece. ❑
4. Use the reaming tool and deburr the end of the pipe before cutting (**Figure 6–76**). ❑

FIGURE 6–76   Using a reamer to remove a burr. (*Photo by Bill Johnson*)

5. Apply pressure against the reamer and turn the reamer clockwise. ❑

NOTE: *Ream the pipe only until the burr is removed.*

---

6. Place the pipe cutter blade over the mark to be cut. ❑
7. Turn the T-handle of the cutter until a moderate amount of pressure is applied to the pipe. (*Make sure cutter wheel is placed at the mark or be cut.*) ❑
8. Make a revolution with the pipe cutter around the pipe. ❑
9. Turn the T-handle one-quarter turn for each revolution around the pipe. ❑
10. Continue until pipe is completely cut. ❑
11. Cut three more pieces of pipe the same size using the same procedures. ❑
12. Once the pieces are cut, deburr the pipes as in procedures 4 and 5. ❑
13. Place one of the 12" pieces of pipe in the vise. ❑
14. Allow the pipe to extend out far enough to get the pipe die over the end. ❑
15. Place the 1/2" pipe die in the ratchet handle. ❑
16. Place the die over the end of the pipe and make sure it lines up square with the pipe end (**Figure 6–77**). ❑
17. Apply cutting oil on the pipe next to where the die is going to do the cutting. ❑
18. Set the ratchet of the die so that the die will advance clockwise (*forward*) on the pipe. ❑
19. Turn the die once or twice clockwise and then reverse (*counterclockwise*) one-quarter turn. ❑
20. Rotate the die once or twice clockwise again. ❑
21. Reverse one-quarter turn counterclockwise again. ❑
22. Apply cutting oil as needed and continue the same procedures until the threaded pipe reaches the end of the die. ❑
23. Turn the die counterclockwise to remove the die form the pipe threads of the pipe. ❑
24. Use the same procedures and thread both ends of all (4) four pieces of black pipe. ❑
25. Use a rag and wipe the threaded area of each pipe. ❑

FIGURE 6–77 Aligning pipe die on steel pipe.

26. Apply pipe compound or Teflon tape to one of the ends of one of the 12" pieces of pipe. (*Remember to leave two end threads bare.*) Refer to *Figure 6–7*. ❑
27. Clamp the doped piece of pipe in the vise with the doped end sticking out of the vise. ❑
28. Take one of the 1/2" × 90-degree elbows and screw it onto the pipe. ❑
29. Use the pipe wrench to tighten the elbow onto the pipefitting. ❑
30. Turn the pipe around and apply pipe compound of Teflon tape to other end, then connect another 1/2" × 90 degree elbow. ❑
31. Follow the same procedures and connect two 1/2" × 90-degree elbows to another piece of 12" pipe. ❑
32. Take one of the pieces of pipe 12" pipe and measure 6" from one of the ends and place a mark. ❑
33. Place this piece in the vise and cut the pipe at the 6" mark. ❑
34. Follow threading procedures as stated above and thread each end of the two pieces of 6" pipe. ❑
35. Leave one of the pieces of 6" pipe in the vise and apply pipe compound or Teflon tape to the end. ❑
36. Separate the 1/2" union. ❑
37. Slide the union locking nut over the pipe. ❑
38. Attach the compression end of the union to the pipe. ❑
39. Remove the pipe and union half from the vise. ❑
40. Place the other 6" piece of pipe in the vise. ❑
41. Apply pipe compound or Teflon tape to the threaded end of the pipe extruding from the vise. ❑
42. Attach the other half of union to the pipe (*threaded end*). ❑
43. Remove a 6" piece from the vise and dope the other ends of the 6" pieces of pipe. ❑
44. Place one of the pieces with the 90 degree elbows in the vise with the extruding 90 degree elbow straight up. ❑
45. Take the last piece of 12" pipe and dope the ends of the pipe with pipe compound or Teflon tape. ❑

46. Screw and tighten one end of the 12" piece of pipe into the 90 degree elbow extruding from the vise. ❏

47. Take the other piece of pipe with the 90-degree elbows and screw the one of on the elbows and pipe on the other end of the 12" pipe already screwed into the other elbow. ❏

48. Remove the assembly form the vise. ❏

49. Turn the assembly around and place and clamp the assembly back into the vise. ❏

> **NOTE:** *The assembly should be aligned in the vise so that both elbows are aligned and centered over each other.*

50. Place and tighten one of the 6" piece of pipe with half of the union attached into the opened end of one of the 90-degree elbows. ❏

51. Place and tighten the other 6" piece of pipe containing the other half of the union into the other open end of the 90-degree elbow. ❏

> **NOTE:** *You may have to move the pipe assembly some to attach the last piece of 6" pipe with the other half of the union.*

52. Align the half-union and tighten the two 6" pieces together to complete the assembly. ❏

53. Make adjustments as needed to get the pipe assembly square. ❏

54. Wipe the pipe assembly. ❏

55. Have the instructor check your work. ❏

56. Clean and return all tools and equipment to their proper location. ❏

RESEARCH QUESTIONS

1. What are the two methods of joining steel pipe?

_____

_____

2. What are the two types of American Standard Pipe threads?

_____

_____

3. In the HVAC industry, which pipe threads are used?

_____

_____

4. What is the angle degree of the V-shaped tapered threads?

_____

_____

5. Thread diameters are referred to by the steel pipe's what?

_____

_____

Passed Competency _____   Failed Competency _____

Instructor Signature _____   Grade _____

Practical Competency 152

Standing Pilot Ignition Procedures (Figure 6–78)

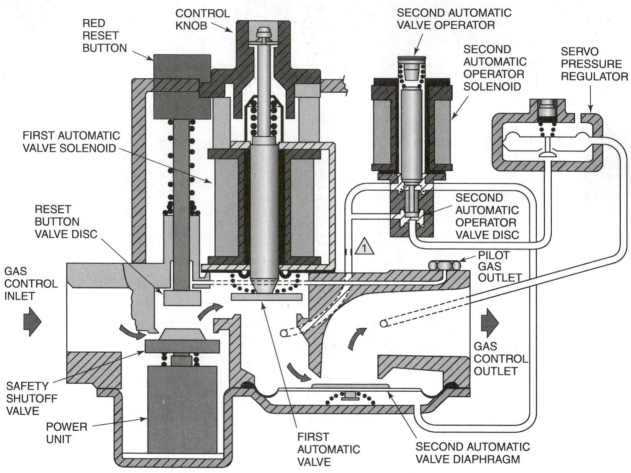

NOTE: SECOND AUTOMATIC VALVE OPERATOR AND SERVO PRESSURE REGULATOR
SHOWN OUTSIDE GAS CONTROL FOR EASE IN TRACING GAS FLOW.

⚠1 SLOW-OPENING GAS CONTROL HAS A GAS FLOW RESTRICTOR IN THIS PASSAGE.

FIGURE 6–78   A standing pilot combination gas valve with servo pressure regulation. (*Courtesy of Honeywell, Inc., Residential Division*)

SUGGESTED MATERIALS

Textbook
Refrigeration & Air Conditioning Technology, 5th Edition, Thomson Delmar Learning
Unit 31—Gas Heat

Review Topics
Safety Devices at the Standing Pilot; Thermocouples and Thermopiles

COMPETENCY OBJECTIVE

The student will be able to light a standing pilot gas system.

OVERVIEW

To light the pilot of a standing pilot system, the gas valve knob must be turned to pilot and depressed which opens the safety valve disc, allowing gas to flow through the pilot tube.

At this point, the pilot flame can manually be lit by using a match or sticker. Once the pilot flame is lit, the pilot button or knob will have to be depressed for an additional 60 to 90 seconds. This gives the pilot flame time to heat the hot junction of the thermocouple to generate the voltage and current necessary to create an electromagnet to hold the pilot valve open.

Once the pilot flame stays lit without depressing the gas valve knob or pilot button, the pilot button or knob can be released and gas valve knob turned to the ON position. If the pilot flame goes out for any reason, the electrical current needed to hold the pilot valve open would be lost, and the gas valve would shut down on safety.

The pilot flame should cover 3/8" to 1" of the top of the thermocouple. Too little of a pilot flame will decrease the thermocouple's electrical output. Too much of a pilot flame will oxidize the hot junction end of the thermocouple and decrease the thermocouple life.

EQUIPMENT REQUIRED

Fuel source to ignite the pilot

SAFETY PRACTICES

Care should be taken when working with hot surfaces and open flames around gas appliances. If for some reason the furnace overheats or the gas supply fails to shut off, turn the main gas valve OFF before shutting the electrical power to the furnace OFF. Make sure that the main gas valve has been OFF for at least 5 minutes before lighting the standing pilot.

COMPETENCY PROCEDURES   Checklist

1. Remove the gas furnace access panel. ☐
2. Turn ON the electrical supply to the gas furnace (*if applicable*). ☐
3. Turn ON the gas supply to the furnace (*if applicable*). ☐
4. Make sure that the manual gas control has been OFF for 5 minutes. ☐
5. Set the thermostat to the heating mode. ☐
6. Set the temperature control on the thermostat to the lowest setting. ☐
7. Turn the gas control valve to the pilot position or set position. ☐
8. Push down the red pilot button or knob. ☐
9. Use a match or striker to light the pilot burner. ☐
10. Continue to hold the pilot button down for 60 to 90 seconds. ☐
11. Release the pilot button. ☐
12. *Did the pilot light stay lit?* _____ ☐

NOTE: *If the pilot did not stay lit, repeat the above procedures.*

---

13. Turn the gas control valve knob or lever to the ON position. ☐
14. Replace the gas furnace access panel. ☐
15. Turn the room thermostat temperature control above the conditioned space temperature. ☐
16. The main burners should ignite. ☐
17. *Did the main burner ignite?* _____ ☐

18. Turn the room thermostat below the conditioned space temperature. ❑
19. *Did the main burners go out?* _____ ❑
20. *Did the pilot light stay lit once the main burners went out?* _____ ❑
21. Remove the access panel to the gas furnace. ❑
22. Turn the gas valve control knob to the OFF position. ❑
23. *Did the pilot go OFF?* _____ ❑
24. Replace the gas furnace access panel. ❑
25. Have the instructor check your work. ❑
26. Return all tools and test equipment to their proper location (*if applicable*). ❑

RESEARCH QUESTIONS

1. Explain in detail the sequence of operation of the standing pilot gas valve system.

   _____

   _____

2. To light the pilot, the gas valve knob should be turned to what setting?

   _____

   _____

3. How long should the knob be pressed while lighting the pilot?

   _____

   _____

4. Why must the gas valve knob be depressed for a certain amount of time before releasing the knob?

   _____

   _____

5. The flame of the pilot should be directed at what junction of the thermocouple?

   _____

   _____

Passed Competency _____   Failed Competency _____

Instructor Signature _____   Grade _____

Practical Competency 153

Pilot Flame Adjustment—Standing Pilot Ignition System

SUGGESTED MATERIALS

Textbooks
Refrigeration & Air Conditioning Technology, 5th Edition, Thomson Delmar Learning
Unit 31—Gas Heat

Review Topics
Safety Devices at the Standing Pilot; Thermocouples and Thermocouples

COMPETENCY OBJECTIVE

The student will be able to make proper adjustments to the pilot flame on a standing pilot ignition system.

OVERVIEW

The pilot flame should cover 3/8" to 1" of the top of the thermocouple. Too little of a pilot flame will decrease the thermocouple's electrical output. Too much of a pilot flame will oxidize the hot junction end of the thermocouple and decrease the thermocouple life.

EQUIPMENT REQUIRED

Fuel source to ignite pilot gas
Standard screwdriver

SAFETY PRACTICES

Care should be taken when working with hot surfaces and open flame around gas appliances. If for some reason the furnace overheats or the gas supply fails to shut off, turn the main gas valve to the furnace off before shutting off the electrical power to the furnace.

COMPETENCY PROCEDURES   Checklist

1. Remove the gas furnace access panel.   ❑
2. Turn the electrical supply ON to the gas furnace (*if applicable*).   ❑
3. Turn the gas supply ON to the furnace (*if applicable*).   ❑
4. Make sure that the manual gas control has been OFF for 5 minutes.   ❑
5. Set thermostat to the heating mode.   ❑
6. Set the temperature control on the thermostat to the lowest setting.   ❑
7. Turn the gas control valve to the pilot position or set position.   ❑
8. Push down the red pilot button.   ❑
9. Use a match of other fuel source to light the pilot.   ❑
10. Continue to hold the pilot button down for 60 to 90 seconds.   ❑
11. Release the pilot button.   ❑
12. *Did the pilot light stay lit?* _____   ❑

> *NOTE: If the pilot did not stay lit, repeat the above procedures until the burner lights.*

13. Locate the pilot adjustment screw on the gas valve. ❑
14. Remove the cap from the pilot adjustment screw. ❑
15. Turn the pilot adjustment screw clockwise until the pilot almost goes out. ❑
16. Now make a good adjustment for the pilot flame. ❑
17. Turn the pilot adjustment screw counterclockwise to make the adjustment of the pilot flame. ❑

> **NOTE:** *A good pilot flame should be adjusted to provide a soft blue flame that surrounds the tip of the thermocouple approximately 1/2" to 1". The pilot burner hood must be 1/16" to 1/8" above the flame runner.*

18. Turn the gas control valve knob or lever to the ON position. ❑
19. Replace the gas furnace access panel. ❑
20. Turn the room thermostat temperature control above the conditioned space temperature. ❑
21. The main burners should have ignited. ❑
22. *Did the main burner ignite?* _____ ❑
23. Turn the room thermostat below the conditioned space temperature. ❑
24. *Did the main burners go out?* _____ ❑
25. *Did the pilot light stay lit once the main burners went out?* _____ ❑
26. Remove the access panel to the gas furnace. ❑
27. Turn the gas valve control knob to the OFF position. ❑
28. *Did the pilot go OFF?* _____ ❑
29. Replace the gas furnace access panel. ❑
30. Have the instructor check your work. ❑
31. Return all tools and test equipment to their proper location. ❑

RESEARCH QUESTIONS

1. The thermocouple generates electrical current for what component in the gas valve?

   _____

   _____

2. How high should the pilot flame cover the top of the thermocouple?

   _____

   _____

3. If the pilot flame goes out during the main burner operation, what would happen?

   _____

   _____

4. How much current should a good thermocouple generate?

   _____

   _____

5. Too much pilot flame on the thermocouple can cause what to happen to the thermocouple?

   _____

   _____

Passed Competency _____ Failed Competency _____

Instructor Signature _____ Grade _____

Practical Competency 154

Checking a Thermocouple

SUGGESTED MATERIALS

Textbook
Refrigeration & Air Conditioning Technology, 5th Edition, Thomson Delmar Learning
Unit 31—Gas Heat

Review Topics
Safety Devices at the Standing Pilot; Thermocouples and Thermopiles

COMPETENCY OBJECTIVE

The student will be able to check and evaluate a thermocouple.

NOTE: *Competency can be completed with a thermocouple from an operating gas furnace or from an available thermocouple. If using just a thermocouple, skip procedures for removal and installation of the thermocouple.*

OVERVIEW

Thermocouples act as the safety device to prove that the pilot is lit before the main gas valve is energized. The thermocouple sits directly in the pilot flame.

Thermocouples consist of two dissimilar metals welded together at one end and is called the **hot junction**.

When the hot junction is heated, a small voltage is generated across the two wires or metals at the other end, referred to as the **cold junction**.

The millivolts DC generated by the thermocouple will be approximately 30 millivolts DC. This is enough voltage to hold the power unit of the gas valve open so that when the gas valve main gas valve solenoid is energized, gas can flow through the power unit to the burner for ignition.

As long as the thermocouple generates the proper millivolts DC, gas can flow through the main valve. If the pilot flame goes out, the thermocouple will cool in about 30 to 120 seconds, de-energizing the power unit to the gas valve and stopping gas flow from the gas valve. Thermopiles are a group of thermocouples wired in series to increase the voltage supplied to the gas valve power unit, and work the same way as one thermocouple.

Checking thermocouples is a matter of measuring the DC millivolt output by heating the hot junction of the thermocouple for at least 5 minutes and placing a voltmeter lead at the thermocouple cold junction and the other meter lead at the thermocouple end (**Figure 6–79**).

Any DC voltage less than 20 millivolts would require replacement of the thermocouple.

EQUIPMENT REQUIRED

Millivoltmeter or voltmeter that has a range to measure millivolts
Adjustable wrench
Propane or acetylene torch
Flint striker

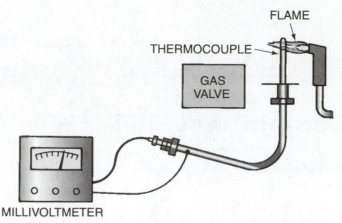

FIGURE 6–79   Checking the millivolts DC of a thermocouple.

SAFETY PRACTICES

Care should be taken when working with hot surfaces and open flames around gas appliances. If for some reason the furnace would overheat or the gas supply fail to shut off, turn the main gas valve off to the furnace before shutting off the electrical power to the furnace.

COMPETENCY PROCEDURES   Checklist

1. Remove the gas furnace access panel.   ❑
2. Turn the electrical supply to the gas furnace OFF (*if applicable*).   ❑
3. Turn the gas supply to the furnace OFF (*if applicable*).   ❑
4. Turn the main gas valve knob to the OFF position (*if applicable*).   ❑
5. Use the adjustable wrench and loosen the thermocouple nut on the gas valve.   ❑
6. Remove the thermocouple from the gas valve (*if applicable*).   ❑

   CAUTION: The thermocouple may be hot.

7. Set the voltmeter on the millivolt scale.   ❑
8. Locate the thermocouple and position it where heat can be applied by a torch. ❑
9. Light the torch.   ❑
10. Apply the heat from the torch to the hot junction of the thermocouple.   ❑

NOTE: *Do not overheat the thermocouple.*

---

11. Place one of the millivoltmeter leads at the hot junction of the thermocouple.   ❑
12. Place the opposite millivoltmeter lead to the cold junction of the thermocouple. ❑
13. Heat the thermocouple until the millivoltmeter reads 30 millivolts.   ❑

NOTE: *Do not allow the thermocouple to reach 90 millivolts or higher.*

NOTE: *A good thermocouple should be able to produce at least 30 millivolts when heated. A millivolt reading of 20 millivolts or less is an indication that the thermocouple should be replaced.*

---

14. Turn the torch OFF and allow the thermocouple to cool.   ❑
15. Once the thermocouple is cool to the touch, place it into the gas valve assembly (*if applicable*). ❑

> **NOTE:** *Do not over tighten the thermocouple nut on the gas valve.*

> **NOTE:** *The following procedures are to be followed if thermocouple tested was used from an operational gas furnace.*

16. Turn on the electrical power to the gas furnace. ❑
17. Turn the gas supply valve on. ❑
18. Replace the furnace access panel. ❑
19. Set the thermostat to the heating mode. ❑
20. Set the temperature control on the thermostat to the lowest setting. ❑
21. Turn the gas control valve to the pilot position or set position. ❑
22. Push down the red pilot button. ❑
23. Use a match or striker to light the pilot burner. ❑
24. Continue to hold the pilot button down for 60 to 90 seconds. ❑
25. Release the pilot button. ❑
26. *Did the pilot light stay lit?* _____ ❑

> **NOTE:** *If the pilot did not stay lit, repeat the above procedures until the burner lights.*

27. Turn the gas control valve knob or lever to the ON position. ❑
28. Replace the gas furnace access panel. ❑
29. Turn the room thermostat temperature control above the conditioned space temperature. ❑
30. The main burner should have ignited. ❑
31. *Did the main burner ignite?* _____ ❑
32. Turn the room thermostat below the conditioned space temperature. ❑
33. *Did the main burner go out?* _____ ❑
34. *Did the pilot light stay lit once the main burner went out?* _____ ❑
35. Remove the access panel to the gas furnace. ❑
36. Turn the gas valve control knob to the OFF position. ❑
37. *Did the pilot light go off?* _____ ❑
38. Replace the gas furnace access panel. ❑
39. Have the instructor check your work. ❑
40. Disconnect equipment and return all tools and test equipment to their proper location. ❑

RESEARCH QUESTIONS

1. The thermocouple is used to operate which valve of the main gas valve?

_____

_____

2. What is a thermopile?

_____

_____

3. What controls the gas flow needed for the pilot valve to stay on all the time?

_____

_____

4. The millivoltage of the thermocouple is used to do what?

_____

_____

5. Thermocouples and thermopiles
    A. Generate a low voltage when heated.
    B. Send a "signal" to the gas valve when the pilot is lit.
    C. Will close the gas valve if the pilot goes out.
    D. All of the above are correct.

Passed Competency _____   Failed Competency _____

Instructor Signature _____   Grade _____

Practical Competency 155

Checking for Correct Electrical Polarity (Figure 6–80)

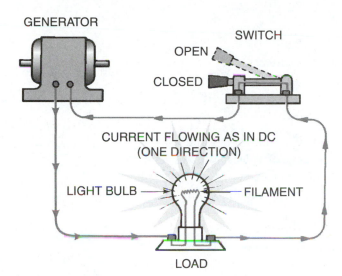

FIGURE 6–80   An electrical circuit.

SUGGESTED MATERIALS

Textbook
Refrigeration & Air Conditioning Technology, 5th Edition, Thomson Delmar Learning
Unit 31—Gas Heat

Review Topics
Gas Furnaces; Hot Surface Ignition Operation; Electricity

COMPETENCY OBJECTIVE

The student will be able to check for the correct electrical polarity.

NOTE: *The correct electrical polarity is necessary for correct furnace operation.*

OVERVIEW

All materials, solids, liquids, and gases contain two basic particles of electrical charges. The electron particle has a negative electrical charge, which is referred to as negative polarity. The proton has a positive electrical charge, which is referred to as positive polarity. The negative and positive polarities of these two particles create two opposite characteristics just like the north and south poles of magnets. The negative and positive polarities in electrical charges react to one another the same as the north and south poles of a magnet.

It is the arrangement of electrons and protons as the basic particles of electricity that determine the electrical characteristics of all substances. When an element has the same number of electrons and protons, its polarity is neutral. When we want to use electrical forces associated with the negative and positive charges in all matter, work must be done to separate the electrons and protons. Changing the balance of the negative and positive polarities in matter will produce evidence of electricity.

Electricity is the flow of electrons in a good conductor. Electrons are negatively charged particles and will always flow toward positively charged particles, or protons. It is important that correct electrical polarity is maintained in electrical equipment or electrical circuits. If the electrical polarity were reversed, electron flow may not take place or electrical controls and circuits could be damaged.

EQUIPMENT REQUIRED

Voltmeter or VOM

SAFETY PRACTICES

Care should be taken when making electrical measurements.

COMPETENCY PROCEDURES

Checklist

1. Turn ON power to the furnace. ❑
2. Gain access to the furnaces main line voltage terminals (**L-1** and **L-2**). ❑
3. Set voltmeter to measure AC voltage at a range of at least 125 volts. ❑
4. Place one of the voltmeter leads to the **L-1** terminal. ❑
5. Place the opposite voltmeter lead to any metal ground on the furnace. ❑
6. *Record the voltage reading.* _____ ❑
7. Place one of the voltmeter leads to the **L-2** terminal. ❑
8. Place the opposite voltmeter lead to any metal ground on the furnace. ❑
9. *Record the voltage reading.* _____ ❑

> **NOTE:** *The correct polarity should show a voltage reading of 115 to 125 volts from terminal **L-1** to a metal furnace ground. The voltage reading from the **L-2** terminal to the metal ground on the furnace should show a zero (0), or a very small amount of voltage.*

10. *Was the polarity of the unit you checked OK?* _____ ❑
11. Have the instructor check your work. ❑
12. Disconnect equipment and return all tools and test equipment to their proper location. ❑

> **NOTE:** *To correct the problem with reverse polarity, either reverse the hot and neutral wires to the furnace or have a licensed electrician check the property wiring.*

RESEARCH QUESTIONS

1. Reverse polarity refers to what?

   _____

   _____

2. How could the polarity of equipment become reversed?

   _____

   _____

3. What problem could reverse polarity cause to furnace components?

_____

_____

4. What could be done to correct a reverse polarity problem on electrical equipment?

_____

_____

5. What are the three types of opposition to current flow that impedance represents?

_____

_____

Practical Competency 156

Checking the Igniter of an HSI Ignition System (Figure 6–81)

FIGURE 6–81   Hot surface igniter.

SUGGESTED MATERIALS

Textbook
Refrigeration & Air Conditioning Technology, 5th Edition, Thomson Delmar Learning
Unit 31—Gas Heat

Review Topics
Hot Surface Ignition; Hot Surface Igniter

COMPETENCY OBJECTIVE

The student will be able to check the hot surface igniter.

OVERVIEW

The hot surface ignition systems (HSI) use an electrical resistive heating element to ignite the main burner. These heating elements are made of silicon carbide, which is a material that has a high resistance to current flow and is very stable at high temperature. When 120 volts AC energizes the heating element, the element's surface temperature gets hotter than the ignition temperature of the gas. The igniter element is located and positioned in such a way that it is in the gas stream of the carryover port. The electronic circuit for the hot surface ignition system is controlled through the equipment's control board.

EQUIPMENT REQUIRED

Ohmmeter or VOM

SAFETY PRACTICES

Care should be taken when working with hot surfaces and open flame around gas appliances. If for some reason the furnace overheats or the gas supply fails to shut off, turn the main gas valve to the furnace off before shutting off the electrical power to the furnace.

COMPETENCY PROCEDURES

Checklist

> NOTE: *Competency Procedures can be performed on an igniter in or out of the furnace.*

1. Turn power to the furnace OFF (*if applicable*). ❑
2. Remove the furnace compartment panel (*if applicable*). ❑
3. Locate and remove the HSI two-prong electrical connector or remove the igniter power leads from the igniter leads (*if applicable*). ❑

> NOTE: *Handle the igniter with care.*

4. Set the ohmmeter to measure high resistance. ❑

> NOTE: *The igniter should be at room temperature, around 70 to 75 degrees (F).*

5. Place the ohmmeter leads on the HSI electrical pins in the connector of wire leads. ❑
6. *Record the ohms value reading.* _____ ❑

> NOTE: *A good igniter will show an ohms value reading in the range of 30 to 300 ohms at room temperature.*

7. Plug the HSI connector back into the electrical connector or reconnect the wire leads to the igniter power leads (*if applicable*). ❑
8. Replace the furnace compartment panel (*if applicable*). ❑
9. Turn power to the furnace ON (*if applicable*). ❑
10. Set the furnace thermostat to the heating mode (*if applicable*). ❑
11. Raise thermostat temperature setting to call for heat (*if applicable*). ❑
12. Watch the furnace cycle until the main burner is lit (*if applicable*). ❑
13. *Did the igniter come on to ignite the pilot, main burner or both (if applicable)?* _____ ❑
14. *Turn thermostat to the OFF position (if applicable).* ❑
15. *Turn power to the furnace OFF.* ❑
16. Have the instructor check your work. ❑
17. Disconnect equipment and return all tools and test equipment to their proper location. ❑

RESEARCH QUESTIONS

1. What materials make up the HSI?

_____

_____

2. How hot does the HSI have to be to ignite the gas?

_____

_____

3. What happens to a hot surface ignition system if the igniter fails to ignite the gas?

_____

_____

4. Why should the HSI be handled with care?

_____

_____

5. Does the igniter stay heated during the burner operation?

_____

_____

Passed Competency _____   Failed Competency _____

Instructor Signature _____   Grade _____

Practical Competency 157

Checking a Flame Sensing Rod (Figure 6–82)

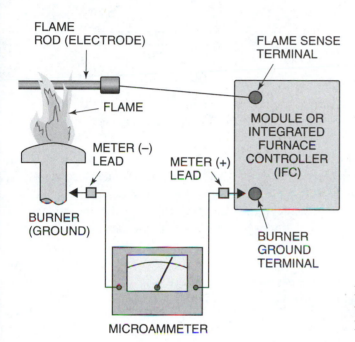

FLAME
ROD (ELECTRODE)

FLAME SENSE
TERMINAL

FLAME

METER (–)
LEAD

METER (+)
LEAD

MODULE OR
INTEGRATED
FURNACE
CONTROLLER
(IFC)

BURNER
(GROUND)

BURNER
GROUND
TERMINAL

MICROAMMETER

FIGURE 6–82   A microammeter placed in series with the flame rod or system ground will measure DC microamps of flame rectification.

SUGGESTED MATERIALS

Textbook
Refrigeration & Air Conditioning Technology, 5th Edition, Thomson Delmar Learning
Unit 31—Gas Heat

Review Topics
Hot Surface Ignition; Hot Surface Igniter

COMPETENCY OBJECTIVE

The student will be able to check for proper operation of a flame-sensing rod.

OVERVIEW

The flame sensor uses the principle of flame rectification to prove that the pilot, main burner, or both have been lit. This is a process where heat from the pilot flame, main flame, or both changes the normal alternating current being supplied to the flame sensor from the furnace electronic module is changed to direct current (DC). Electronic components used in modern ignition systems will energize and open the main gas valve only with direct current. Combustible gases from the pilot flame or main burner are able to conduct electricity because the gases are ionized, and contain positively and negatively charged particles. In a DSI ignition system, the flame is located between two electrodes of different sizes.

857

In the DSI system the electrodes are made of different sizes, which allows current to flow easier in one direction than the other (**Figure 6–83**).

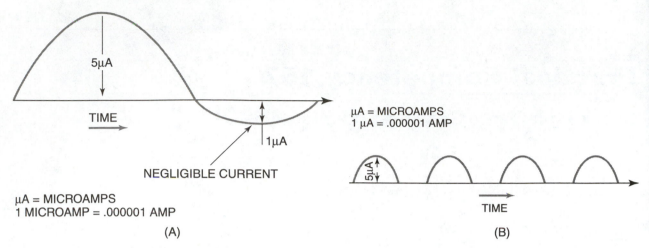

FIGURE 6–83   (A) the different electrode sizes cause current to flow better in one direction than the other. (B) A pulsating direct current (DC) signal that can be measured in microamps with a microammeter in series with one of the electrodes.

The flame actually acts as a switch. If there is a flame, the switch conducts electricity and closes a switch. An open switch is created when there is no flame because electricity cannot flow.

The DC current generated by the flame sensor is usually in the range of 1 to 25 microamps, and can be measured with a microammeter. The microammeter can be connected in series with the flame sensor and control module or in series with the furnace burner and burner ground terminal on the IFC. (Refer to *Figure 6–82.*)

The amount of microamps generated by the flame sensor will depend on the quality, size, and stability of the flame and on the electronic module and electrode design. Always consult with the furnace manufacturer information for specific flame rectification measurements.

EQUIPMENT REQUIRED

Microammeter or VOM that can measure microamps
Jumper wire with alligator clips at each end (*if applicable*)

SAFETY PRACTICES

Care should be taken when working with hot surfaces and open flame around gas appliances. If for some reason the furnace overheats or the gas supply fails to shut off, turn the main gas valve to the furnace off before shutting off the electrical power to the furnace.

COMPETENCY PROCEDURES   Checklist

1. Turn OFF power to the furnace.   ❑
2. Make sure the gas supply line valve is open.   ❑
3. Remove the furnace compartment panel.   ❑
4. Locate flame sensor lead wire at the control module or IFC.   ❑
5. Disconnect the flame sensor lead wire from the control module or IFC. ❑
6. Set microammeter to read at least 30 microamps.   ❑
7. Connect one lead of the microammeter to the flame sensor wire.   ❑
8. Place the second microammeter lead to the terminal on the control module or IFC. ❑

NOTE: *Using a jumper wire with alligator clips can make connection from the meter lead to the module easier.*

9. Set the thermostat to the heating mode. ❏
10. Turn the thermostat selector switch above the room temperature. ❏

> **NOTE:** *The thermostat can be bypassed by jumping the **R** and **W** terminals of the thermostat or at the furnace control board.*

11. Turn power ON to furnace. ❏
12. Once the main burner lights, note the microamp reading on the meter. ❏
13. *Record the microamp reading.* _____ ❏

> **NOTE:** *A good flame sensing micro amp reading would be in the micro amp range of 2.0 to 4.0 DC microamps, typically. The microamp reading at times can be as high as 25 microamps.*

14. *Was the microamp reading on the flame sensor you checked within these ranges?* _____ ❏
15. Turn the thermostat OFF or remove jumper wire between **R** and **W** terminals. ❏
16. Let furnace operate unit blower shuts OFF (*if applicable*). ❏
17. Turn power to the furnace OFF. ❏
18. Remove the microamp meter leads. ❏
19. Replace the sensor lead wire to the control module or IFC. ❏
20. Replace furnace compartment panel. ❏
21. Have the instructor check your work. ❏
22. Disconnect equipment and return all tools and test equipment to their proper location. ❏

RESEARCH QUESTIONS

1. What are used as the two electrodes in a spark-to-pilot (intermittent pilot) ignition system?

   _____

   _____

2. In a DSI system, what two things are used as the electrodes?

   _____

   _____

3. In a dual-rod system, what are each of the rods used for?

   _____

   _____

4. What is a one-rod pilot?

   _____

   _____

5. What is flame rectification?

   _____

   _____

   _____

Passed Competency _____   Failed Competency _____

Instructor Signature _____   Grade _____

Practical Competency 158

Checking Draft or Combustion Blower Pressure Switch (Figure 6–84)

FIGURE 6–84 A two-stage gas furnace showing the two-stage gas valve, the two-stage combustion blower motor, and dual-pressure controls. (*Courtesy of Ferris State University. Photo by ohn Tomczyk*)

SUGGESTED MATERIALS

Textbook
Refrigeration & Air Conditioning Technology, 5th Edition, Thomson Delmar Learning
Unit 31—Gas Heat

Review Topics
High-Efficiency Gas Furnaces

COMPETENCY OBJECTIVE

The student will be able to check for proper operation of a draft and combustion blower pressure switch.

OVERVIEW

Combustion blower motors use a pressure switch to prove that the mechanical draft has been established before allowing the main burner to light (**Figure 6–85**).

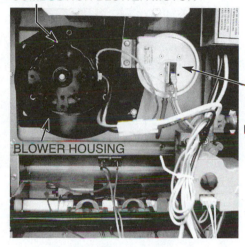

COMBUSTION BLOWER MOTOR

PRESSURE SWITCH & DIAPHRAGM FOR PROVING DRAFT

BLOWER HOUSING

FIGURE 6–85   A combustion blower motor assembly with a pressure switch and diaphragm used for safety venting a furnace. (*Courtesy of Ferris State University. Photo by John Tomczyk*)

The pressure switches are connected to the combustion blower housing or heat exchanger by rubber hoses and have a diaphragm that is sensitive to pressure differential. Some furnaces have a single-port pressure switch that senses the negative pressure inside the heat exchanger caused by an induced draft combustion blower. Other furnaces use a dual-tap or differential pressure switch that senses the difference in pressure between the combustion blower and the heat exchanger (**Figure 6–86** and **Figure 6–87**).

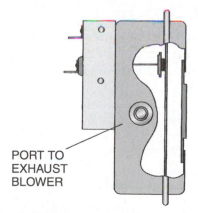

PORT TO EXHAUST BLOWER

FIGURE 6–86   A single-port pressure switch. (*Courtesy of International Comfort Products Corporation*)

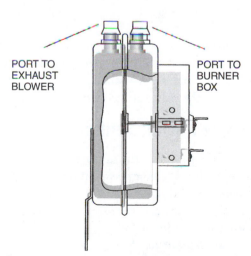

PORT TO EXHAUST BLOWER

PORT TO BURNER BOX

FIGURE 6–87   A dual-port or differential pressure switch. (*Courtesy of International Comfort Products Corporation*)

The induced draft blower pressure switch is a NO (normally opened) switch that is used to sense a negative pressure in the inducer blower housing. Once the induced draft blower is operating, it causes a negative pressure in the inducer housing, which closes the pressure switch contacts. The induced draft blower must generate a negative pressure of at least 0.1" of negative pressure to close its switch contacts.

Once the contacts are closed, electrical power is allowed to flow to the integrated furnace control board to start the furnace ignition sequence. If there is a problem in the air intake system, or vent system, or with the induced draft blower, the negative pressure switch will not close its contacts. This prevents the system from going into the ignition sequence and the furnace will shut down on lock out.

Condensing furnaces are normally equipped with a drain pressure switch, which is also used on upflow models. These pressure switches are connected to the drain in the secondary heat exchanger collector box. The drain pressure switch has a set of NC (normally closed) contacts. The induced draft blower causes a negative pressure in the heat exchanger. If there is a blockage in the drain line, trap, or secondary heat exchanger, the switch contacts of the drain pressure switch open. This prevents the unit from starting because it is wired in series with the gas valve. The furnace will try to reignite and will then shut down on lockout.

Two-stage gas furnaces use a dual pressure switches (refer to *Figure 6–103*), to prove draft. One pressure switch is a low-pressure switch and the other is a high-pressure switch to prove a low- and high-speed on the combustion blower motor.

Different furnace manufacturers use different pressure settings for their pressure switches. The settings are usually measured in inches of water column and are sometimes indicated on the pressure switch housing. The pressures can be measured using a digital or liquid-filled manometer.

NOTE: *Never replace a pressure switch with another pressure switch without knowing its operating pressures. Always consult with the furnace manufacturer or use a factory-authorized substitute pressure switch.*

EQUIPMENT REQUIRED

High-efficiency furnaces
Manometer or draft gage
Ohmmeter
1/4" × 1/4" × 1/4" Plastic tee
1/4" Piece of rubber or plastic hose
Furnace schematic

SAFETY PRACTICES

The student should follow all safety procedures when working with electricity. Make sure that meters are set at the proper function and proper value scale. Care should be taken when working with hot surfaces and open flames around gas appliances. If for some reason the furnace overheats or the gas supply fails to shut off, turn off the main gas valve to the furnace before shutting off the electrical power to the furnace.

COMPETENCY PROCEDURES   Checklist

1. Shut OFF the power to the furnace.   ❑
2. Remove the furnace compartment door.   ❑
3. *The furnace is equipped with how many pressure switches?* _____   ❑
4. *Review the furnace schematic and record what each pressure switch is being used for.*
   Pressure switch 1 (used for) _____   ❑
   Pressure switch 2 (used for) _____ *(if applicable)*   ❑
6. *Observe pressure switch number 1 and see if the working pressure is identified on the*
   pressure switch and record. _____   ❑

7. *Observe pressure switch number 2 and see if the working pressure is identified on the pressure switch and record (if applicable).* _____ ❏

8. *Is pressure switch number 1 an NO or NC switch?* _____ ❏

9. *Is pressure switch number 2 an NO or NC switch?* _____ ❏

10. *Remove the hose from pressure switch number 1 and make an arrangement using the plastic tee and additional hose so as to be able to incorporate the draft gage or manometer into the connection to be able to read pressure.* ❏

11. *Remove the wire leads from pressure switch number 1.* ❏

12. *Set the ohmmeter of VOM to measure resistance.* ❏

13. *Place the meter leads across the pressure switch terminals.* ❏

14. *Record the resistance value.* _____ ❏

15. *According to the resistance reading in Step 13, is the pressure switch open or closed?* _____ ❏

16. *Turn power ON to the furnace.* ❏

17. *Set thermostat to call for heat.* ❏

18. *Once furnace starts, place meter leads across the pressure switch terminals and observe the ohmmeter to see if pressure switch changes position.* ❏

19. *Did the pressure switch change position?* _____ ❏

20. *Observe the draft gage or manometer and record the pressure on the pressure switch.* _____ ❏

21. *Does the measured pressure match the recorded pressure of pressure switch number 1 (if applicable)?* _____ ❏

22. Turn thermostat to the OFF position. ❏

23. Turn power OFF to furnace. ❏

24. *Did pressure switch number one perform properly?* _____ ❏

25. Remove materials used for checking pressure switch number 1. Replace wires and hose to pressure switch number 1 to original set-up. ❏

IF APPLICABLE: *Set up and check pressure switch number 2 and record information. If system has only one pressure switch, proceed to Step 26.*

26. *Place the meter leads across the pressure switch terminals.* ❏

27. *Record the resistance value.* _____ ❏

28. *According to the resistance reading in Step 25, is the pressure switch open or closed?* _____ ❏

29. *Turn power ON to the furnace.* ❏

30. *Set thermostat to call for heat.* ❏

31. *Once furnace starts, place meter leads across pressure switch terminals and observe the ohmmeter to see if pressure switch changes position.* ❏

32. *Did the pressure switch change position?* _____ ❏

33. *Observe the draft gage or manometer and record the pressure on the pressure switch.* _____ ❏

34. *Does the measured pressure match the recorded pressure of pressure switch number two (if applicable)?* _____ ❏

35. Turn thermostat to the OFF position. ❏

36. Turn power OFF to the furnace. ❏

37. *Did pressure switch number two perform properly?* _____ ❏

38. Remove materials used for checking pressure switch number 2. Replace wires and hose to pressure switch number 1 to the original set-up. ❏

39. Have the instructor check your work. ❏

40. Disconnect equipment and return all tools and test equipment to their proper location. ❏

RESEARCH QUESTIONS

1. What is the function of the draft pressure switch?

_____

_____

2. What causes the switch of the draft pressure switch to close?

_____

_____

3. List a couple of reasons that could cause the draft pressure switch not to close.

_____

_____

4. Are draft pressure switches NO or NC switches?

_____

_____

5. What is the purpose of a drain pressure switch?

_____

_____

Passed Competency _____   Failed Competency _____

Instructor Signature _____   Grade _____

Practical Competency 159

Checking a Furnace High-Limit Switch

SUGGESTED MATERIALS

Textbook
Refrigeration & Air Conditioning Technology, 5th Edition, Thomson Delmar Learning
Unit 31—Gas Heat

Review Topics
High-Limit Switch

COMPETENCY OBJECTIVE

The student will learn the proper procedures for checking the operation and function of a high-limit control used on a gas furnace (see *Figure 6–15.*)

> NOTE: *The limit switch can be checked in or out of an operational gas furnace.*

OVERVIEW

These **bimetal switches** can be used to activate the furnace blower.

On the other hand, they can be used as a fan–limit safety switch; this controls the furnace blower operation and acts as a high-limit switch, shutting the burner down due to high-temperature conditions.

When the conditioned space temperature is satisfied, the thermostat will open the electrical circuit between the **R** and **W** terminals of the thermostat. This de-energizes the oil burner primary control, shutting OFF burner operation. The furnace blower will continue to operate until the heat exchanger has been cooled to a preset temperature. Once the heat exchanger is cooled enough, the fan–limit switch will open the circuit to the furnace blower, shutting the blower OFF.

EQUIPMENT REQUIRED

1 Ohmmeter or VOM
1 Voltmeter or VOM

SAFETY PRACTICES

The student should follow all safety procedures when working with electricity. Make sure that meters are set at the proper function and proper value scale. Care should be taken when working with hot surfaces and open flame around gas appliances. If for some reason the furnace overheats or the gas supply fails to shut off, turn the main gas valve to the furnace off before shutting off the electrical power to the furnace.

COMPETENCY PROCEDURES

1. Shut the power OFF to the furnace. ❏
2. Remove the furnace compartment door. ❏
3. Disconnect one of the power lead wires from the high-limit switch. ❏
4. Set the ohmmeter or VOM to the lowest ohms scale. ❏
5. Place the meter leads across the high-limit switch terminal leads. ❏
6. *Record the ohms value reading.* _____ ❏

> **NOTE:** *A high-limit switch should show a zero (0) ohms value reading when below its opening temperature. If an infinity reading was recorded, the switch is open and defective which requires replacement.*

---

7. Place the power lead back on the high-limit switch terminal. ❏
8. Turn the power to the furnace ON. ❏
9. Set the thermostat to heating mode. ❏
10. Set the temperature control of the thermostat above conditioned space temperature. ❏
11. Set the voltmeter to the AC voltage scale and at a voltage range to measure 24 volts or higher. ❏
12. Place the voltmeter leads across the high-limit switch terminals. ❏
13. *Record the voltage reading.* _____ ❏

> **NOTE:** *When using a voltmeter to check the high-limit switch, a voltage reading would indicate an open switch causing the main burner to shut down. A zero (0) volt reading would indicate the high-limit switch is closed, allowing the burner to operate.*

---

14. *Locate one of the furnace blower motor leads and carefully remove the motor lead from furnace control board or line voltage.* ❏
15. *Place the voltmeter leads across the high-limit switch terminals.* ❏
16. *Allow the burner to operate until a voltage reading is indicated.* ❏
17. *What happened to the furnace burner when voltage was indicated on the voltmeter?*

    _____ ❏

> **NOTE:** *A measured voltage means that the high-limit switch opened and should have de-energized the gas valve.*

---

18. *Did the gas valve de-energize when the high-limit switch opened?* _____ ❏
19. Once a voltage reading is indicated across the high-limit switch, carefully reconnect the blower motor lead. (Blower should come ON.) ❏
20. *Place the voltmeter leads across the high-limit switch again and observe voltmeter until voltage is no longer indicated.* ❏
21. *What happened to the gas valve and burner when voltage was no longer indicated on the voltmeter?* _____ ❏

> **NOTE:** *When voltage was no longer indicated on the voltmeter, the high-limit switch cooled down and closed. In a short time frame the gas valve should have opened and the burner should have been re-ignited.*

22. *Did the gas valve open and the burner ignited once the high-limit switch closed?* _____ ❑
23. Turn the thermostat off. ❑
24. Shut OFF power to the furnace once the blower stops operation. ❑
25. Replace the furnace compartment panel. ❑
26. Have the instructor check your work. ❑
27. Disconnect equipment and return all tools and test equipment to their proper location. ❑

RESEARCH QUESTIONS

1. What is the purpose of the high-limit switch on a gas furnace?

   _____

   _____

2. When the high-limit switch opens, what component of the furnace shuts down?

   _____

   _____

3. Is the high-limit switch an NO or NC switch?

   _____

   _____

4. Is the limit switch an auto reset or manual reset switch?

   _____

   _____

5. List a couple of things that could cause the furnace to overheat, causing the high-limit switch to open.

   _____

   _____

Passed Competency _____	**Failed Competency** _____
Instructor Signature _____	**Grade** _____

Practical Competency 160

Checking the Draft Induced or Combustion Blower Motor

SUGGESTED MATERIALS

Textbook
Refrigeration & Air Conditioning Technology, 5th Edition, Thomson Delmar Learning
Unit 31—Gas Heat

Review Topics
Condensing Gas Furnaces; Ignition Systems

COMPETENCY OBJECTIVE

The student will be able to check a draft or combustion blower motor.

OVERVIEW

When the thermostat calls for heat and closes the heating contacts, it sends a signal to the microprocessor that heat is required. The microprocessor energizes the induced draft or combustion blower fan. The blower will operate to purge the combustion chamber of any gas or combustion products. The draft pressure switch signals the microprocessor when adequate airflow is established through the combustion chamber. If a flame is established it will operate throughout the burner operation to remove flue gases and continue to provide combustion air.

These motors are shaded pole motors and have only one motor winding. A resistance reading across the motor lead terminals can indicate a good or open motor winding.

EQUIPMENT REQUIRED

Ohmmeter

SAFETY PRACTICES

The student should follow all safety procedures when working with electricity. Make sure that meters are set at the proper function and proper value scale. Care should be taken when working with hot surfaces and open flames around gas appliances. If for some reason the furnace overheats or the gas supply fails to shut off, turn the main gas valve to the furnace off before shutting off the electrical power to the furnace.

COMPETENCY PROCEDURES

Checklist

1. Shut OFF power to the furnace.   ❑
2. Remove the furnace compartment panel.   ❑
3. Disconnect the induced draft or combustion blower motor electrical connector.   ❑
4. Set the ohmmeter to measure resistance.   ❑
5. Place the ohmmeter leads across the motor lead wires of terminal connector.   ❑
6. *Record the ohms value reading.* _____   ❑

> **NOTE:** *A good induced blower motor will show a recorded ohms value reading. The actual ohms value reading of the motor will vary depending on the size of motor used by the different manufacturers. On any motor, an infinity reading means that the motor is defective and needs to be replaced. In addition, a good motor winding does not mean that the motor should operate. The motor requires proper voltage and a free rotating rotor.*

7. Reconnect the motor connection. ❏
8. Replace the furnace compartment panel. ❏
9. *Were the motor windings checked properly?* _____ ❏ ❏
10. Have the instructor check your work. ❏
11. Disconnect equipment and return all tools and test equipment to their proper location. ❏

RESEARCH QUESTIONS

1. What is the purpose of the induced draft fan motor?

   _____

   _____

2. How long does the induced draft motor operate during the heating cycle?

   _____

   _____

3. What is used to sense the pressure of the induced draft fan motor?

   _____

   _____

4. What is the purpose of a combustion blower motor?

   _____

   _____

5. If the combustion motor fails to operate, what will happen to the furnace?

   _____

   _____

Passed Competency _____ Failed Competency _____

Instructor Signature _____ Grade _____

Practical Competency 161

Checking a Gas Valve

SUGGESTED MATERIALS

Textbook
Refrigeration & Air Conditioning Technology, 5th Edition, Thomson Delmar Learning
Unit 31—Gas Heat

Review Topics
Automatic Combination Gas Valve; Standing Pilot Automatic Gas Valves; Intermittent Pilot Automatic Gas Valve; Direct Burner Automatic Gas Valves

COMPETENCY OBJECTIVE

The student will be able to check a gas valve.

> *NOTE: These gas valve-testing procedures can be used for gas furnaces with standing pilot, hot surface ignition systems, or electronic ignition system. Testing procedures for furnaces that incorporate the smart valve will be covered in another Competency. When checking the valve, all other components involved in the ignition and combustion process should be operational.*

> *NOTE: This competency requires an operational gas furnace with a standing pilot or hot-surface ignition, or electronic spark ignition system.*

OVERVIEW

Fuel control and delivery system to the furnace is controlled by the **gas valve**.

There are several types of gas valves in use, although most residential and light commercial furnaces use an automatic combination gas valve that controls the flow of gas to both the pilot and main gas line that feeds gas through the manifold and then through the burner orifices.

The pressure at which the gas is supplied to the manifold and **spud** or **orifice** is controlled by the gas valve and is different depending on the fuel source. The desired gas manifold pressure for natural gas furnaces is 3 to 3.5 in. W.C. (*inches of water column*), and 11 in. W.C. for liquefied petroleum. Gas valve pressure can be checked and adjusted by using a water manometer, a digital manometer (**Figure 6–88**), or a gas gage.

Adjustments to manifold pressure can be accomplished by tightening or loosening an adjustment spring on the gas valve that is used to increase or decrease gas pressure to the furnace.

EQUIPMENT REQUIRED

VOM meter

FIGURE 6–88 A digital manometer being used to measure gas pressure in inches of water column. (*Courtesy of Ferris State University. Photo by John Tomczyk*)

SAFETY PRACTICES

The student should follow all safety procedures when working with electricity. Make sure that meters are set at the proper function and proper value scale. Care should be taken when working with hot surfaces and open flames around gas appliances. If for some reason the furnace overheats or the gas supply fails to shut off, turn the main gas valve to the furnace off before shutting off the electrical power to the furnace.

COMPETENCY PROCEDURES

Checklist

1. Turn ON the gas supply for the gas furnace. ❏
2. Turn ON the electrical power to the furnace. ❏
3. Remove the furnace compartment panel. ❏
4. Set the thermostat to the heating mode. ❏
5. Set the thermostat temperature switch to a setting higher than the ambient temperature of the conditioned space. ❏
6. Because of the different types of ignition systems used on gas furnaces, allow the furnace about 45 seconds to energize the gas valve before checking the gas valve solenoid coil operational voltage. ❏
7. Set the voltmeter to the AC volt function and at a scale to measure at least 24 volts AC. ❏
8. Locate the 24-volt gas valve electrical leads on the gas valve. ❏
9. Place the voltmeter leads across the gas valve electrical leads. ❏
10. *Record the voltage reading.* _____ ❏
11. *Did the gas valve open?* _____ ❏
12. *Did the main burner ignite?* _____ ❏
13. Turn the thermostat OFF. ❏

NOTE: *A good valve should show a 24-volt reading and the main burner should ignite; this means the gas valve opened and is operating. If there is a 24-volt reading and the main burner is not ignited, the valve is closed and has to be replaced.*

NOTE: *If there is not a 24-volt reading at the gas valve terminals, do not replace the gas valve. The problem is elsewhere in the furnace's ignition system. The furnace total ignition system has to be completely checked.*

14. Turn power to furnace OFF. ❏
15. Remove power lead wires to the gas valve solenoid coil. ❏
16. Set the VOM meter or ohmmeter to measure resistance. ❏
17. Place meter leads across the gas valve solenoid coil terminals. ❏
18. *Record the measured resistance.* _____ ❏

> **NOTE:** *A measured resistance means the gas valve coil is good. An infinity reading indicates the gas valve solenoid coil is open and requires replacement.*

---

19. *Was there a measurable resistance across the gas valve coil?* _____ ❏
20. Replace gas valve terminal wires. ❏
21. Replace the furnace compartment panel. ❏
22. Have the instructor check your work. ❏
23. Disconnect equipment and return all tools and test equipment to their proper location. ❏

RESEARCH QUESTIONS

1. What is the purpose of the gas valve in a furnace?

   _____

   _____

2. When does the main gas valve open and allow gas to flow to the main burner?

   _____

   _____

3. What is a combination gas valve?

   _____

   _____

4. What is a redundant gas valve?

   _____

   _____

5. What is a smart valve?

   _____

   _____

Passed Competency _____ Failed Competency _____

Instructor Signature _____ Grade _____

Practical Competency 162

Checking a Smart Valve Control System

SUGGESTED MATERIALS

Textbook
Refrigeration & Air Conditioning Technology, 5th Edition, Thomson Delmar Learning
Unit 31—Gas Heat

Review Topics
Smart valve

COMPETENCY OBJECTIVE

The student will able to check a smart valve control system of a gas furnace.

OVERVIEW

Modern furnaces also include the smart valve, which incorporates the gas valve and electronic control module in one enclosure. These valves combine the features of an intermittent pilot (IP) and hot surface ignition (HSI) system into one. The smart valve's logic lights the pilot with a 24-volt hot surface igniter system. The pilot in turn lights the main burners.

These valves are continuous trial for pilot for both natural and LP gas equipment. Once the flame is proved by the flame rod, the 24-volt HSI is de-energized and the electronic fan timer within the module is energized, which starts timing the fan ON delay and the fan OFF delay.

Troubleshooting the smart valve system is made very easy owing to the simplified wiring and modular electrical connections. A voltmeter can be used at the valve and electrical plug connections (**Figure 6–89**).

The electronics of the smart valve provide an ignition sequence of operation that is similar to most hot surface ignition, intermittent pilot ignition, and direct-spark ignition systems. Learning the ignition sequence of operation of modern furnaces can aid the technician in identifying and isolating the location as to where a problem lies when troubleshooting ignition failure.

NOTE: Because there are different models and makes of gas valves, electronic modules, and IFCs, always consult with the gas valve and electronic module or IFC manufacturer for specific information on systematic troubleshooting and sequence of operation to avoid personal injury or damage to system components.

EQUIPMENT REQUIRED

VOM meter

SAFETY PRACTICES

The student should follow all safety procedures when working with electricity. Make sure that meters are set at the proper function and proper value scale. Care should be taken when working with hot surfaces

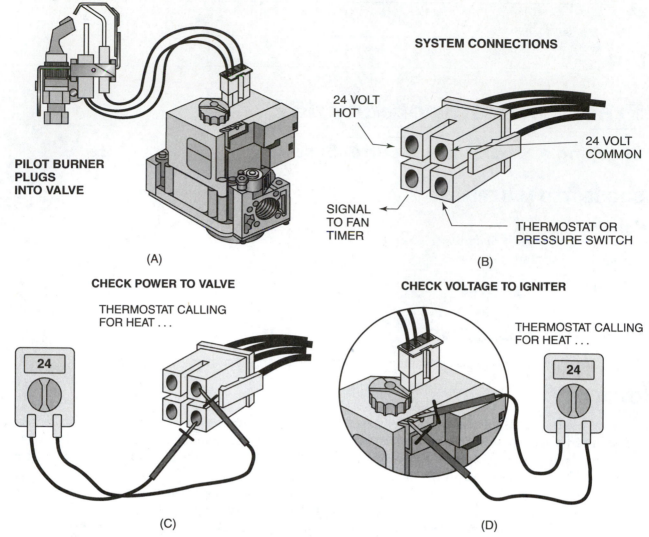

SYSTEM CONNECTIONS

PILOT BURNER
PLUGS
INTO VALVE

24 VOLT
HOT

24 VOLT
COMMON

SIGNAL
TO FAN
TIMER

THERMOSTAT OR
PRESSURE SWITCH

(A)

(B)

CHECK POWER TO VALVE

THERMOSTAT CALLING
FOR HEAT . . .

24

(C)

CHECK VOLTAGE TO IGNITER

THERMOSTAT CALLING
FOR HEAT . . .

24

(D)

FIGURE 6–89   (A, B) The smart valve's modular electrical connections. (C, D) A voltmeter being used to troubleshoot the smart valve system. (*Courtesy of Honeywell, Inc.*

and open flames around gas appliances. If for some reason the furnace overheats or the gas supply fails to shut off, turn the main gas valve to the furnace off before shutting off the electrical power to the furnace.

COMPETENCY PROCEDURES

Checklist

> NOTE: *Review furnace wiring schematic and identify smart valve 24-volt terminals, signal to fan timer terminal, and thermostat or pressure switch terminals (if applicable). At the pilot burner plug, identify which terminals of the connector are for the HSI and flame sensor rod.*

CHECKING POWER TO THE VALVE FROM THE THERMOSTAT
(Refer to Figure 6–89 B and C)
1. Turn power OFF to furnace (*if applicable*).   ❑
2. Remove the furnace compartment panel.   ❑
3. Remove the electrical connector from the gas valve.   ❑
4. Identify the 24-volt common terminal of the gas valve connector.   ❑
5. Identify the thermostat terminal of the gas valve connector.   ❑
6. Set voltmeter to measure 24 volts.   ❑

7. Place meter leads in the 24-volt common terminal and thermostat terminal connection of the gas valve connector. ❏
8. Turn ON the electrical power to the furnace. ❏
9. Set the thermostat to the heating mode (*if applicable*). ❏
10. Set the thermostat temperature switch to a setting higher than the ambient temperature of the conditioned space (*if applicable*). ❏
11. Allow the system to cycle through the ignition cycle (*if applicable*). ❏
12. *Record the measured voltage.* _____ ❏

> **NOTE:** *A 24-volt reading indicates the system control board is supplying voltage to the gas valve. If the valve fails to operate with a 24-volt power source supplied from the thermostat, the valve is defective and would need to be replaced. A 0-volt reading is an indication that the system control board is not providing the gas valve with the proper voltage. Further checks would have to be performed on the system control board or ignition system.*

13. *Was there a measured voltage across the thermostat and 24-volt common terminals?* _____ ❏
14. Turn power OFF to the furnace. ❏
15. Replace the gas valve connector. ❏

CHECKING VOLTAGE TO THE IGNITER (*Refer to Figure 6–89 D*)
16. Remove the pilot burner assembly connector from the gas valve. ❏
17. Identify the gas valve igniter terminals on the gas valve. ❏
18. Turn power ON to the furnace. ❏
19. Set the meter to measure at least 24 volts AC. ❏
20. Place the meter leads on the gas valve igniter terminals. ❏
21. Allow the system to sequence (*if applicable*). ❏
22. *Record the voltage supplied to the igniter terminals from the gas valve.* _____ ❏

> **NOTE:** *A measured reading of 24 volts is an indication that the thermostat and smart valve are working and the igniter should operate unless defective. No voltage is an indication that 24 volts is not being supplied to the igniter by the smart valve. There could be many reasons for this. First make sure that 24 volts are being supplied to the valve through the thermostat. Proceed to check all safety ignition devices such as the high-limit switch, rollout switches, pressures switches, etc. If all are confirmed to be OK and 24 volts are still not being supplied through the valve to the igniter terminals, the valve is probably defective and would require replacement.*

23. *Was there a 24-volt power source indicated at the gas valve igniter terminals?* _____ ❏

CHECKING THE IGNITER OF A SMART VALVE (*Refer to Figure 6–89*)
24. Turn power OFF to the furnace. ❏
25. *Set the VOM meter to measure resistance.* ❏
26. *With the pilot burner plug removed from the gas valve, place the meter leads at the plug terminals for the igniter.* ❏
27. *Record the measured resistance.* _____ ❏

> **NOTE:** *A measured resistance means the igniter is OK. Infinity would mean that the igniter is defective and would require replacement.*

28. Replace the igniter connector at gas valve. ❏

CHECKING SIGNAL TO THE FAN TIMER *(Refer to Figure 6–89)*

29. Remove the power supply plug from the gas valve. ❏
30. Identify the fan timer terminal. ❏
31. Identify the 24-volt connector terminal. ❏
32. Set the VOM meter to measure at least 24 volts. ❏
33. Place meter leads in the fan signal terminal and 24-volt common terminal. ❏
34. Turn power ON to the furnace. ❏
35. Set the thermostat to call for heat. ❏
36. *Record the measured voltage at the fan timer terminal.* _____ ❏

> **NOTE:** *A measured voltage indicates that voltage is supplied to start the fan timer. No voltage is an indication of possible control board failure. Further checks should be performed to isolate the problem.*

37. *Was 24 volts indicated at the fan timer terminal?* _____ ❏
38. Turn power OFF to the furnace. ❏
39. Replace the smart valve plug connector. ❏
40. Replace the furnace compartment panels. ❏
41. Have the instructor check your work. ❏
42. Return all tools and test equipment to their proper location. ❏

RESEARCH QUESTIONS

1. The smart valve combines which two ignition systems into one control?

   _____

   _____

2. On a smart valve system, when is the fan timer energized?

   _____

   _____

3. On a smart valve system, when is the igniter de-energized?

   _____

   _____

4. What voltage does the igniter of a smart valve system operate?

   _____

   _____

5. What material is the igniter of a smart valve system made of?

   _____

   _____

Passed Competency _____   Failed Competency _____

Instructor Signature _____   Grade _____

Practical Competency 163

Checking and Adjusting Manifold Gas Pressure

SUGGESTED MATERIALS

Textbook
Refrigeration & Air Conditioning Technology, 5th Edition, Thomson Delmar Learning
Unit 31—Gas Heat

Review Topics
Gas Valves

COMPETENCY OBJECTIVE

The student will be able to check and adjust gas manifold pressure into proper range.

> *NOTE: Properly checking the supply gas pressure to an individual gas appliance requires that all other gas appliances that are on the same gas supply line be operated at the same time while checking the gas supply pressure to the individual appliance.*

OVERVIEW

The pressure at which the gas is supplied to the manifold and **spud** or **orifice** is controlled by the gas valve and is different depending on the fuel source. The desired gas manifold pressure for natural gas furnaces is 3 to 3.5 in. W.C. (*inches of water column*) and 11 in. W.C. for liquefied petroleum. Gas valve pressure can be checked and adjusted by using a water manometer, a digital manometer, or a gas gage.

Adjustments to manifold pressure can be accomplished by tightening or loosening an adjustment spring on the gas valve that is used to increase or decrease gas pressure to the furnace.

To increase the pressure on the manifold, turn the pressure regulator adjustment screw clockwise. To decrease the manifold pressure, turn the pressure regulator adjustment screw counterclockwise.

EQUIPMENT REQUIRED

U-tube manometer or gas pressure test kit

SAFETY PRACTICES

The student should follow all safety procedures when working with electricity. Make sure that meters are set at the proper function and proper value scale. Care should be taken when working with hot surfaces and open flames around gas appliances. If for some reason the furnace overheats or the gas supply fails to shut off, turn the main gas valve to the furnace off before shutting off the electrical power to the furnace.

Note to the instructor:
The instructor may want to adjust the manifold gas pressure so it is not correct before student completes competency.

COMPETENCY PROCEDURES   Checklist

1. Turn OFF the gas supply to the gas furnace.   ❑
2. Turn OFF the electrical power to the furnace.   ❑

3. Remove the furnace compartment panel. ❑

4. Locate and remove the 1/2" pipe thread pressure tap plug on the outlet side of the gas valve. ❑

5. Install the male fitting from the manometer or gas pressure test kit. ❑

6. Connect the U-tube or gas test pressure gauge to male fitting. ❑

7. Turn ON the gas supply to the furnace. ❑

8. Turn ON the electrical power to the furnace. ❑

9. Set the thermostat to call for heat. ❑

10. *Does the furnace you are checking use LP or natural gas?* _____ ❑

11. *Once the burner ignites, record the operating gas manifold pressure.* _____ ❑

12. *Is the operating gas pressure in proper range for the type of gas being used by the furnace?*

    _____ ❑

13. *What should be the correct operating pressure for the equipment being tested?* _____ ❑

TO ADJUST MANIFOLD GAS PRESSURE

14. Remove the gas pressure regulator adjustment cap. ❑

> **NOTE:** *Make adjustments to obtain the correct gas manifold pressure for the system being serviced. To increase the manifold pressure, turn the manifold adjustment screw clockwise. To decrease manifold pressure, turn the manifold adjustment screw counterclockwise.*

15. Make adjustments slowly and continue to monitor the manometer or pressure gage reading. ❑

16. Once operating pressure is established, turn the gas valve to the OFF position. ❑

17. Turn thermostat to the OFF position. ❑

18. Remove the manometer of gas pressure gage from the male fitting. ❑

19. Remove the male pipe fitting from the gas valve. ❑

20. Replace the gas valve plug. ❑

21. Replace the gas valve pressure adjustment cap. ❑

22. Replace the furnace compartment panel. ❑

23. Have the instructor check your work. ❑

24. Return all tools and test equipment to their proper location. ❑

RESEARCH QUESTIONS

1. What is the recommended manifold gas pressure for a natural gas furnace? _____

2. What is the recommended manifold gas pressure for a LP gas furnace? _____

3. To increase the manifold pressure the pressure regulator adjustment screw should be turned in which direction? _____

4. To decrease the manifold pressure the pressure regulator adjustment screw should be turned in which direction? _____

5. A pressure reading of 10 in. W.C. is approximately equal to:

   A. 277 psig
   B. 27.7 psig
   C. 3 psig
   D. 0.36 psig

Passed Competency _____ Failed Competency _____

Instructor Signature _____ Grade _____

Practical Competency 164

Checking Proper Airflow Using the Temperature Rise Method

SUGGESTED MATERIALS

Textbook

Refrigeration & Air Conditioning Technology, 5th Edition, Thomson Delmar Learning
Unit 31—Gas Heat

Review Topics

Gas-Fired Forced-Hot Air Furnaces

COMPETENCY OBJECTIVE

The student will be able to check and make adjustments to furnace airflow by using the temperature rise method.

OVERVIEW

Proper airflow over the heat exchanger of the furnace is very important. One of the most common failures of the furnace heat exchanger is the lack of proper airflow. Most manufacturers will supply an airflow table on the blower door of the furnace, or in the installation instruction manual for the furnace. The temperature rise check can be used to determine if the airflow is correct.

If after performing a temperature rise check, the measured temperature rise is above the desired temperature range, the airflow is too low and more air must be moved across the heat exchanger. This can be accomplished by speeding up the furnace blower, removing any restrictions in the duct system, or by adding more supply or more return air duct.

If the measured temperature rise is below the desired temperature range, the air is moving too fast across the heat exchanger and will have to be slowed down. Slowing the furnace blower down either by belt adjustment or changing the blower speed tap could accomplish this. When using the air flow chart to set the furnace with the proper airflow, ideally try to set the temperature rise in the middle of the range.

EQUIPMENT REQUIRED

1 Temperature meter (2 sensing probes)

SAFETY PRACTICES

The student should follow all safety procedures when working with electricity. Make sure that meters are set at the proper function and proper value scale. Care should be taken when working with hot surfaces and open flames around gas appliances. If for some reason the furnace overheats or the gas supply fails to shut off, turn the main gas valve to the furnace off before shutting off the electrical power to the furnace.

COMPETENCY PROCEDURES

Checklist

1. Turn ON the power to the furnace. ❏
2. Turn the thermostat to the heating mode and set temperature selector to call for heat. ❏
3. Insert a temperature probe in the supply duct as close to the furnace as possible, but out of the area of the furnace heat exchanger. ❏

4. Insert another temperature probe into the return duct as close to the furnace as possible. ❑
5. *Watch the temperature meter and when the supply air duct temperature stops rising, record the stabilized temperature.* _____ ❑
6. *Record the return air temperature.* _____ ❑
7. Subtract the recorded return air temperature from the recorded supply air temperature. ❑

NOTE: *The difference of the two recorded temperatures equals the furnace temperature rise.*

---

8. *Record the temperature rise temperature.* _____ ❑
9. Compare the measured temperature rise to the manufacturer's approved temperature rise. ❑

NOTE: *The furnace approved temperature rise range is found on the furnace nameplate.*

---

10. *Was the airflow proper for the furnace you are checking according to the temperature rise?*
_____ ❑
11. Make adjustments if needed to bring the airflow of the furnace into the proper airflow range. ❑
12. Turn the thermostat OFF. ❑
13. Remove the temperature probes and meter. ❑
14. Have the instructor check your work. ❑
15. Disconnect equipment and return all tools and test equipment to their proper location. ❑

RESEARCH QUESTIONS

1. If the temperature rise of a furnace is higher than it should be, what is wrong with the airflow in the furnace?

_____

_____

2. If the temperature rise of the furnace is lower than it should be, what is wrong with the airflow in the furnace?

_____

_____

3. If the temperature rise is too high, what should be done with the furnace blower speed?

_____

_____

4. If the temperature rise is too low, what should be done with the furnace blower speed?

_____

_____

5. List a couple of other factors that could affect the furnace temperature rise besides the furnace blower.

_____

_____

Passed Competency _____ **Failed Competency** _____

Instructor Signature _____ **Grade** _____